Food Engineering Series

Springer's *Food Engineering Series* is essential to the Food Engineering profession, providing exceptional texts in areas that are necessary for the understanding and development of this constantly evolving discipline. The titles are primarily reference-oriented, targeted to a wide audience including food, mechanical, chemical, and electrical engineers, as well as food scientists and technologists working in the food industry, academia, regulatory industry, or in the design of food manufacturing plants or specialized equipment.

Anet Režek Jambrak
Editor

Nonthermal Processing in Agri-Food-Bio Sciences

Sustainability and Future Goals

Springer

Editor
Anet Režek Jambrak
Faculty of Food Technology and Biotechnology
University of Zagreb
Zagreb, Croatia

ISSN 1571-0297
Food Engineering Series
ISBN 978-3-030-92417-1 ISBN 978-3-030-92415-7 (eBook)
https://doi.org/10.1007/978-3-030-92415-7

This Springer imprint is published by the registered company Springer Nature Switzerland AG
The registered company address is: Gewerbestrasse 11, 6330 Cham, Switzerland

Contents

Part I
Introduction

Chapter 1
Sustainability in Food Science and Food Industry: Where Are We Now? – Viewpoints of the EFFoST Working Group on Sustainable Food Systems

Felix Schottroff, Henry Jaeger, Sergiy Smetana, Arthur Robin, Kelly Fourtouni, Anet Režek Jambrak, and Hugo de Vries

1 State of the Art: Sustainable Food Processing

A sustainable food system is a system that delivers food security and nutrition for all in such a way that the economic, social, and environmental bases to generate food security and nutrition for future generations are not compromised. There are three pillars of sustainability: economic, social, and environmental sustainability.

This chapter was prepared by the European Federation of Food Science and Technology (EFFoST) Working Group "Sustainable Food Systems". Further information can be found at: https://www.effost.org/members/effost+working+groups/wg+sustainable+food/default.aspx

---

F. Schottroff (✉)
Department of Food Science and Technology, Institute of Food Technology, University of Natural Resources and Life Sciences (BOKU), Vienna, Austria

BOKU Core Facility Food & Bio Processing, University of Natural Resources and Life Sciences (BOKU), Vienna, Austria
e-mail: felix.schottroff@boku.ac.at

H. Jaeger
Department of Food Science and Technology, Institute of Food Technology, University of Natural Resources and Life Sciences (BOKU), Vienna, Austria

S. Smetana
DIL Deutsches Institut für Lebensmitteltechnik e. V., Quakenbrück, Germany

A. Robin
Bioengineering Environmental Laboratory, Porter School of Environmental Sciences, Tel Aviv University, Tel Aviv, Israel

K. Fourtouni
E.J. Papadopoulos S.A., Athens, Greece

A. Režek Jambrak (ed.), *Nonthermal Processing in Agri-Food-Bio Sciences*, Food Engineering Series, https://doi.org/10.1007/978-3-030-92415-7_1

Food production is one of the most ancient and essential human activities. From early hunters and gatherers, we have now built a clustered global food production network on which we rely to feed the current – and the future – population. However, the technological and industrial developments that have led us to be able to sustain a population of several billion individuals are also placing a great burden on ecosystems through various adverse environmental impacts. Although the food production sector is not the only human activity responsible for such a burden, it plays a major role in the contribution to global greenhouse gas emissions (GHG; 26%), water resource depletion (>90% scarcity weighted water use), loss of biodiversity (livestock represents 94% of the global mammal biomass, excluding humans), land use (40–50%), eutrophication (78%), and terrestrial acidification (32%) (Poore and Nemecek 2018; Ritchie and Roser 2020). The upstream processes of food production (agriculture, animal farming, fishing, etc.) are by far the main sources of detrimental environmental effects in the food sector (Poore and Nemecek 2018; Ritchie and Roser 2020). Indeed, the supply chain accounts for only 18% of the total GHG emissions of the food sector, including food processing (4% of the total GHG emissions of the food sector, Fig. 1.1) (Ritchie and Roser 2020). However, since the majority of consumed foods are processed terrestrial or marine resources, the attention for the sustainability performance of manufacturing, packaging, distribution, and consumption is primordial for the entire food systems; this also includes e.g. considerations about the use of packaging materials, added value and employment, and the social value of food consumption.

In order to achieve the GHG emission reductions required to limit the global warming to +2 °C at the end of the century and to reach most of the United Nations 2030 Sustainable Development Goals, the food sector must undergo profound transformations (Masson-Delmotte et al. 2018; Poore and Nemecek 2018; Ritchie and Roser 2020; UNEP 2019; United Nations 2015). In addition to the reduction of their environmental footprints, the food production stakeholders need to ensure their economic resilience (the food sector is the largest employer in the world), food safety, the sufficient supply of nutrients without excess, and increasing health benefits, while maintaining cultural heritages and social practices linked to the consumption and production of food (Chaudhary et al. 2018; United Nations 2015). Such a complex and ambitious goal can only be achieved with a change of paradigm at each food production stage while encompassing the complexity and inter-dependency of food systems. Even if food processing is responsible for only a fraction of the global food production footprints, it plays a key role in the supply chain, such as increasing shelf-life and ensuring food safety, preserving health benefits, and reducing food waste generation (Poore and Nemecek 2018; Raak et al. 2017). As food wasted

A. Režek Jambrak
Laboratory for Sustainable Development, Faculty of Food Technology and Biotechnology, University of Zagreb, Zagreb, Croatia

H. de Vries
INRAE – French National Research Institute for Agriculture, Food and Environment, UMR IATE (University of Montpellier, INRAE, Institut Agro), Montpellier, France

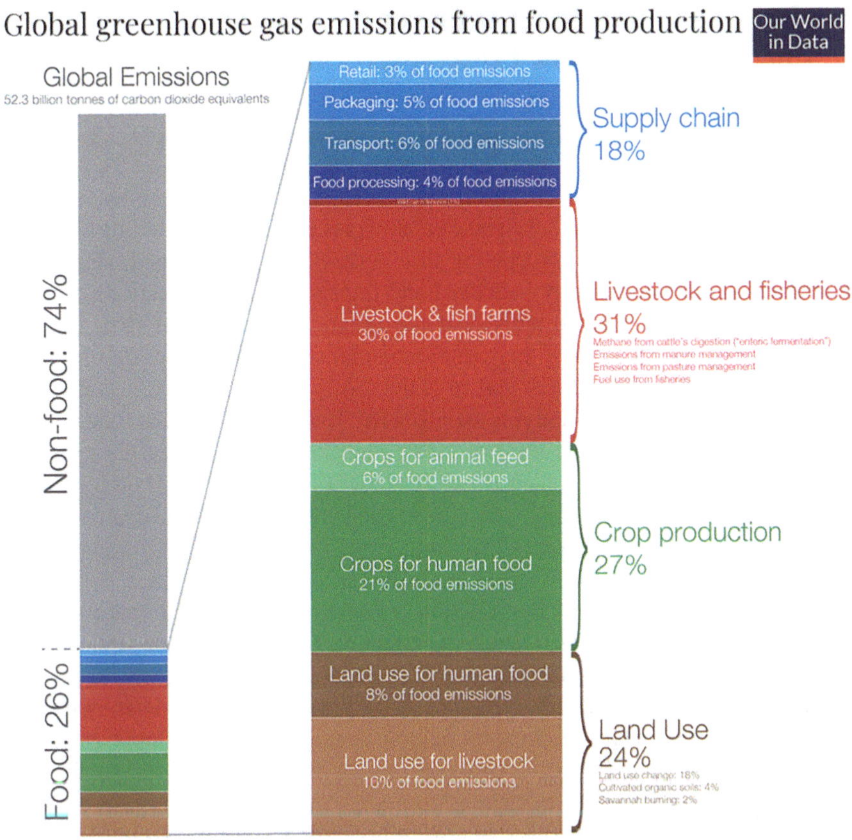

Fig. 1.1 Greenhouse gas emissions from food production (image reproduced from Ritchie and Roser (2020) (CC-BY), data source: Poore and Nemecek (2018))

throughout the food production chain caused about a quarter of the total food sector GHG emissions (6% of the global world GHG emissions), the weight of the holistic role of food processing (notably through food preservation) far exceeds its own GHG footprint of 4% (Raak et al. 2017; Ritchie and Roser 2020). Besides, improvement of the environmental dimension of sustainability of food processing includes the reduction of food waste downstream, the reduction of energy and water consumption, the reduction of the production of solid wastes and wastewater, and providing technological opportunities for up-recycling of food waste, all of which may be achieved, relatively to each case, by using emerging technologies including non-thermal processing technologies (Pereira and Vicente 2010; Picart-Palmade et al. 2019). In this context, the present chapter focuses on the environmental aspect of sustainability, although social and economic aspects are also addressed.

2 How Do the Individual Actors in the Food Value Chain Influence Sustainable Processing?

The food value chain consists of a variety of different stakeholders, all of them exerting an environmental impact. Raw materials for food production are provided by the agricultural, as well as the marine and freshwater sectors (fisheries, mariculture, seaweed harvesting, etc.). Logistics companies ensure that materials are transported along the chain. Food processing aims at the transformation and stabilization of products. Foods are usually packaged and transported to retail and distribution units, where they are sold to individual consumers. All individual steps bear the chance for an improvement of their environmental status (see Chap. 6), ultimately contributing to increasing sustainability of the food system (de Oliveira et al. 2018; Diamond et al. 2014; Knorr and Augustin 2021b). Besides the fact that food production accounts for around one third of GHG emissions, one third of the produced food is lost along the value chain (FAO 2011; REFRESH 2020). Therefore, apart from the reduction of the carbon footprint, food losses upstream of unavoidable cases during raw material processing should also be minimized (Obersteiner and Pilz 2020).

Although food raw material production is known to cause the greatest share of GHG emissions and food loss (Obersteiner and Pilz 2020), the interdependencies of the different actors of the value chain should be considered and set screws to enhance sustainability of the food system should be identified for each actor. Passing responsibilities to political actors would be too simple in such complex systems, characterized by the unpredictability of events and uncertainties in outcomes. In contrast, each actor may individually contribute to a shift towards more sustainable food systems. To improve sustainability of **agriculture and livestock production**, digitalization and a shift towards novel and greener technologies are often suggested. However, it remains unclear how farmers will be able to cover the potentially arising costs for such a transition, given the fact that distribution of income is distinctly imbalanced for differently sized farms, and farmers experience a pronounced volatility of their income (European Parliament 2015). Ultimately, such changes may be triggered by financial support from political actors, governments, or non-governmental organizations or may reflect in distinctly increasing prices for raw materials, which at the end would have to be compensated by the consumer. However, **consumers**' willingness to pay may not be given ab initio, but may rather have to be triggered, e.g. by the appeal of a product – driven by food processors; or marketing and information campaigns – driven by food processors, retailers and political actors. Another trigger for a transition towards more sustainable food systems initiated by the consumer are changing diets. This may be subject to certain trends, e.g. caused by social media influencers or marketing and information campaigns or a paradigm shift in the perception of certain behaviors. However, although this would be a powerful driver for innovation, consumer choices are a complex and individual phenomenon, which are also influenced by sociological and psychological aspects, economic considerations, convenience, confirmed habits, personal preferences, and the gap between attitude and actual behavior, among others. Moreover, this notion also relies on the fact that consumers are well educated in terms of their personal contribution towards sustainability, as well as the transfer of information towards the consumers (Elhoushy 2020; SAPEA 2020; Vermeir and Verbeke 2006). Further, distribution of sustainable alternatives to conventionally produced foods towards the consumer has to be ensured through **wholesale and retail businesses**. Especially food retail businesses are known to possess distinctive market power, acting in a highly competitive environment. As such, they are able to drive down the prices for the offered goods, which may positively affect the actions of customers (potentially increasing demand), but may have negative implications for food processors, and raw material producers, due to economic drawbacks (EC 2014; Obersteiner and Pilz 2020; SAPEA 2020). Despite its relative power, consumers may influence retail by a changing demand and food processors may convince markets to take new and sustainably produced products into their portfolios. Moreover, by communication and marketing of sustainably produced foods, but also the presentation of the goods and the selection of their assortment, supermarkets can contribute to actively influence buying behavior of consumers (SAPEA 2020).

Due to the fact that processes are often optimized and operated at high throughput levels, trying to operate in a resource efficient way, minimizing energy and water consumption as well as the generation of by-products and waste, the impact of **food processing** itself on GHG emission and food waste generation is limited. However, processors can influence almost all players in the value chain. Thus, food processors should not only react to the demand of consumers for sustainable diets, but also actively trigger such a change of consumer behavior, e.g. by developing appealing sustainable food products and a corresponding marketing strategy (incl. the adaption of sustainability labels for those foods). Moreover, existing products may be adapted, where possible, towards the use of more sustainably sourced raw materials, e.g. locally and seasonally produced food constituents and avoidance of animal-derived components. Also, the implementation of novel and more efficient technologies, e.g. non-thermal technologies, may contribute to a reduction of energy consumption and food losses, as well as favoring circular approaches (Knorr and Augustin 2021a; Knorr et al. 2020; Obersteiner and Pilz 2020; SAPEA 2020). Subsequent to production, products typically undergo **food packaging** operations. Here, the strategies in terms of increasing sustainability are not as straightforward, as packaging may also help to distinctly reduce food loss throughout the supply chain. A reduction or adaptation of packaging may make sense for certain products, especially such foods usually wrapped in multiple layers. However, where unavoidable, the development of novel, more sustainable packaging concepts, e.g. biodegradable plastics, may be favorable (Licciardello 2017; Obersteiner and Pilz 2020). Similar observations and conclusions may be drawn for the **food logistics** sector, as food storage, packaging and transport together only account for ~1% of GHG emissions (Quantis 2020). Therefore, those operations are hardly avoidable and necessary for the supply chain to function properly. In many cases it makes sense to reduce transportation routes of food products for the sake of sustainability. The responsibility for the implementation lies with the food processors and raw material producers but also with the logistic companies, e.g. by using digital tools for supply chain optimization. In addition, logistics companies may ensure that the transport of foods is implemented as resource efficient as possible (Bloemhof et al. 2015; Soysal et al. 2012). The last player, able to exert influence on all actors of the food value chain are **policy makers**. However, as previously mentioned, politics may not solve the transition towards sustainable food systems on their own. Instead, politics should build a framework that enables the development of the individual actors along the food value chain into the direction of enhanced sustainability. For this purpose, actions may be undertaken by implementation of labels and standards in different categories, such as voluntary vs. compulsory or directed at the customer vs. B2B. Moreover, corresponding laws, suitable information campaigns and educational efforts will contribute to raise awareness for the issue. Lastly, providing appropriate funding and subsidies will help players along the value chain to obtain these goals (EC 2021a; Knorr and Augustin 2021a; Leach et al. 2020; SAPEA 2020; Weber et al. 2020).

A more in-depth view on the topic of sustainability along the food value chain may be found in SAPEA (2020), Knorr and Augustin (2021b), Halberg and Westhoek (2019), and Weber et al. (2020).

3 What Do Companies Do to Investigate How Sustainable They Are?

Sustainable food production, i.e. the processes or means companies use to minimize their environmental impact, is a crucial matter which needs to be addressed by the food industry. For this purpose, food companies assess how sustainable they currently are, by measuring e.g. their use of energy, natural resources (water, land), materials (ingredients, packaging) and GHG emissions during the production and transportation of their products. They are benchmarked against their peers and competitors using internal and external tools as specified in the next section and in Fig. 1.2.

3.1 Labels

In order to communicate sustainable food production towards the consumer, and to consequently raise awareness and influence the demand for sustainably sourced foods, labels seem to be an appropriate measure.

Eco Labels The increasing concerns from the consumer side drove the emergence of organic food, as well as eco and fair-trade labelling. Though the awareness of the above, companies can attract more consumers purchasing their products, since they

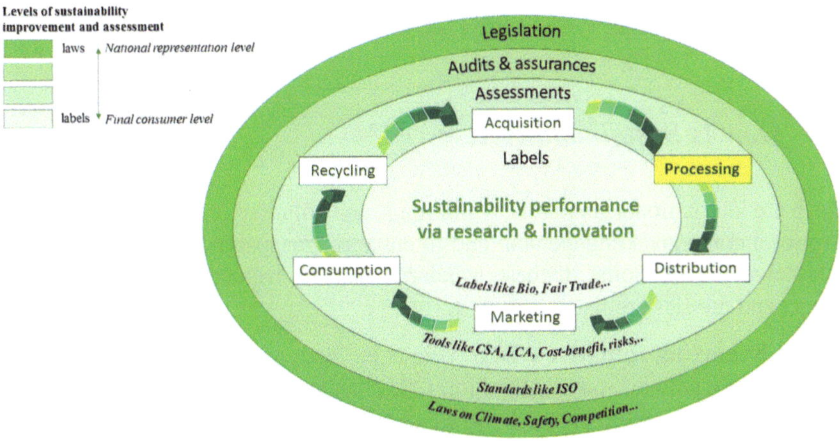

Fig. 1.2 From consumers to parliamentary chambers, from labels to laws: how sustainability performance in the food sector can be communicated, assessed, assured, and legislated, with a focus on research and innovation in the processing stage of food systems

have the adequate certifications contributing to a more sustainable food industry and system (http://www.ecolabelindex.com/).

Some of the following are voluntary certifications that can be assured by third parties and verify their compliance with specific conditions:

Rain Forest Alliance seal Promotes collective action for people and nature (Rainforest Alliance 2020). What it really means is that the product that has been certified with the Rain Forest Alliance was produced with systems and methods that do not compromise the viability of the three pillars of sustainable development; economy, environment, society. The seal focuses on four themes: Livelihoods, Human Rights, Climate Forests (https://www.rainforest-alliance.org/business/tag/2020-certification-program/).

The organic logo The European Union "Organic logo", gives an identity to European Union organic products, that have been certified and contain at least 95% organic ingredients (https://ec.europa.eu/info/food-farming-fisheries/farming/organic-farming/organic-logo_en).

Animal Welfare Approved Is a certified standard for animal welfare where animals must be able to live in their natural inhabitant and be in a state of physical and psychological well-being (http://www.ecolabelindex.com/ecolabel/animal-welfare).

Fair Trade Certified Is a standard promoting and supporting the lives and every day work of farmers and agricultural workers, through fair trade. Products that are promoted under the “Fair Trade seal” are coffee, cocoa, banana, flowers, tea and sugar (https://www.fairtradecertified.org/).

3.2 How Is Sustainability Assessed?

There are international standards, regulations, and voluntary codes for assessing the level of a company’s engagement and to enhance transparency, related to sustainable development. Some of these standards used to evaluate sustainable actions, impact and metrics are listed hereafter.

Global Reporting Initiative (GRI) GRI is an international not-for-profit organization, with a network-based structure. GRI Standards create a common language for organizations – large or small, private or public – to report on their sustainability impacts in a consistent and credible way (https://www.globalreporting.org/).

ISO 26000 An international standard, launched in 2010, providing guidance to those who recognize that respect for society and environment is a critical success factor. The application of ISO 26000 is increasingly viewed as a way of assessing an organization’s commitment to sustainability and its overall performance (https://www.iso.org/iso-26000-social-responsibility.html).

ISO 14044 Is used to specify the necessary requirements and to provide guidelines for life cycle assessment (LCA). It provides guidelines on the definition and scope of the LCA, the life cycle inventory phase (LCI), the life cycle impact assessment phase (LCIA), the life cycle interpretation phase, and the final phase of LCA reporting and review. It also covers limitations of the LCA and the relationship between the LCA phases (ISO 14044 2006).

Many existing tools can be used in the process of Sustainability Assessment. Some of those tools and methodologies include:

- Risk assessment
- Life cycle assessment
- Cost and benefit analysis
- Ecosystem services valuation
- Scenario analysis, including sensitivity analysis

Task Force on Climate-Related Financial Disclosures (TCFD) TCFD is an organization with the goal of developing a set of voluntary climate-related financial risk disclosures which can be adopted by companies so that those companies can inform investors and other members of the public about the risks they face related to climate change (https://www.fsb-tcfd.org/).

Carbon Disclosure Program A nonprofit project aiming at the disclosure of environmental information to assist companies address and minimize their environmental impact and footprint (https://www.cdp.net/en).

Life Cycle Assessment (LCA) Is a methodology where a company can assess its environmental impact at all the stages of the lifecycle of a product. This includes all the emissions and natural resources used by a company, concluding on the environmental, human health and resource depletion issues that may arise. LCA is a mean of identifying any environmental issue that relates to the production, assisting companies to take decisions based on robust results and set targets on mitigating those issues. Through the impact assessment, a company can identify the environmental and human health impact, set goals and targets, as well as an action plan, on how it can overcome and minimize these impacts. Performance indicators and corelated metrics can help in quantifying the goals and set more realistic targets, in order to achieve them and transit towards Sustainable Development faster, in comparison with other companies that lacking quantified targets (Muralikrishna and Manickam 2017).

Sustainability Reporting A mandatory process in the European Union (according to Directive 2014/95/EU), also known as Non-Financial Reporting Directive (NFRD). Since 2018 companies are required to include non-financial statements in their annual reports (EC 2020b).

According to the Directive, companies should disclose information on the following (EC 2020b):

- Environmental protection
- Social responsibility and treatment of employees
- Respect for human rights
- Anti-corruption and bribery
- Diversity on company boards (in terms of age, gender, educational and professional background, etc.)

3.3 External Audits and Assurance

Companies can also seek for assurance on their Sustainability Report and in particular also their ISO certification schemes, by Auditing Providers. Voluntary codes, such as GRI have issued specific guidance manuals used by the auditors, to verify and assure the disclosed data.

AA1000AS Assurance Standards Is a methodology used to evaluate sustainability-related assurance activities, and it assesses the nature and extent to which an organization adheres to the AccountAbility Principles (https://www.accountability.org/standards/aa1000-assurance-standard/).

Finally, a universally accepted standard is not yet in place, therefore companies follow one, or a combination of the above-mentioned standards and disclosures, with the auditors and assurance providers facing the challenge of credibility and consistency (Coyne 2006).

4 Legislation on Sustainable Food Processing

Sustainability as defined in Sect. 1 relies, in States of Law, on a set of legislations dealing with one or several of its aspects (SAPEA 2020).

At an international scale, the International Organization for Standardization (ISO) issued 14044:2006, Environmental management; Life cycle assessment (LCA) and Requirements and guidelines (see above). These standards contribute to the Sustainable Development Goals #12 and #13. The European Parliament and of the Council established the framework for achieving climate neutrality and amending Regulation (EU) 2018/1999 (European Climate Law), COM (2020) 80 final, 2020/0036 (COD). Also, the Parliament welcomes the plan for a sustainable food system strategy and highlighted the need to use natural resources more efficiently while supporting the agricultural sector (Waage et al. 2015). They reiterated calls to reduce pesticides dependency, and the use of fertilizers and antibiotics in agriculture, while striving for 25% of organic production. They also wanted higher animal welfare standards and an EU-wide food waste reduction target of 50%. Even though, the European Union currently does not possess an overarching framework for governing the European food system in a holistic manner (Candel 2016).

The waste reduction target directly concerns the processing part of food systems striving for sustainable outcomes (EC 2016). However, also the increase of organic production will have consequences on the environmentally friendly way of processing (Barabanova and Moeskops 2019). These consequences are not yet translated into appropriate legislation. Also scientific evidence underlying legislation and harmonized procedures, avoiding disperse private standards, should be reviewed (van der Meulen 2017).

Current scientific evidence reveals that food processing should be focused on energy reduction and low GHG emissions. In this regard, non-thermal technologies can be considered as green technologies (de Vries and Matser 2011). However, further calculations and justification are needed to consider them as sustainable. There are electrotechnologies (cold plasma processing, pulsed electric fields—PEFs, pulsed light—PL, e-beam processing); pressure-based technologies like high pressure processing—HPP (cold preservation—Pascalization), high pressure homogenization and mechanical technologies (high power ultrasound—US, hydrodynamic cavitation) (Režek Jambrak et al. 2021). There are some studies that conducted life-cycle assessment and environmental models developed by environmental scientists that are applied on food. There are also models developed by food scientists in order to analyze food-environmental interactions (Djekic et al. 2018).

If current legislation of food safety as described in the General Food Law (Regulation (EC) No 178/2002) – and overviewed by e.g. the European Food Safety Agency (EFSA) – is extended towards sustainability of food systems, the role of non-thermal processing is interesting to discuss in more detail. Nonthermal processing can be used in pre-treatments and treatment of food in order to assure on one hand food safety and on the other hand reduction of energy consumption and CO_2 emission. This can bring value to new green business models that can contribute to removal of CO_2 (non-increasing contribution) from the atmosphere, and contribute to the climate neutrality objective (carbon market).

Food packaging also plays a key role both in the safety as well as in the sustainability of food systems (EU Regulation (EC) No 1935/2004), which leads to an amended Regulation (EU) 2019/1381 on the transparency and sustainability of the EU risk assessment in the food chain (EC 2021b). The current rules need to be revised (e.g. the one on food contact materials) in legislation to improve food safety and public health (in particular in reducing the use of hazardous chemicals), support the use of innovative and sustainable packaging solutions using environmentally-friendly, re-usable and recyclable materials, and contribute to food waste reduction. Also the communication via labels towards consumers needs then a revision. The International Organization for Standardization (ISO) defines green labels as symbols printed on products or their packaging to advertise environmental quality or characteristics. From producers to consumers, following their green intentions by buying certified products there can be a new, appropriate, labeling system. The regulators should take action to expand and promote sustainable production methods and circular business models in food processing and retail, including specifically for Small and Medium-sized Enterprises (SMEs). Requiring in the food processing industry can promote sustainability by considering the 4 R's: reducing the amount used, re-thinking whether it is actually a genuine requirement, re-using existing or ensuring products can be recycled. Legislative initiatives to enhance cooperation of primary producers to support their position in the food chain and non-legislative initiatives to improve transparency; however, this may ask for a revision of the European Competition Law.

Ideally, the future directive is towards "zero waste" agenda, promote a low-carbon and eco-friendly economy and support economic growth in rural areas (Queenan et al. 2017).

5 How Can Sustainability Be Calculated and Evaluated?

Sustainability is a complex and wide concept, which requires specific approaches for the analysis. Progress measurement towards sustainable state of functioning requires a quantitative resource-based approach (Fig. 1.3). Quantification and analysis of sustainability from such a perspective would need to deal with natural (environmental), human and economic capital accounting.

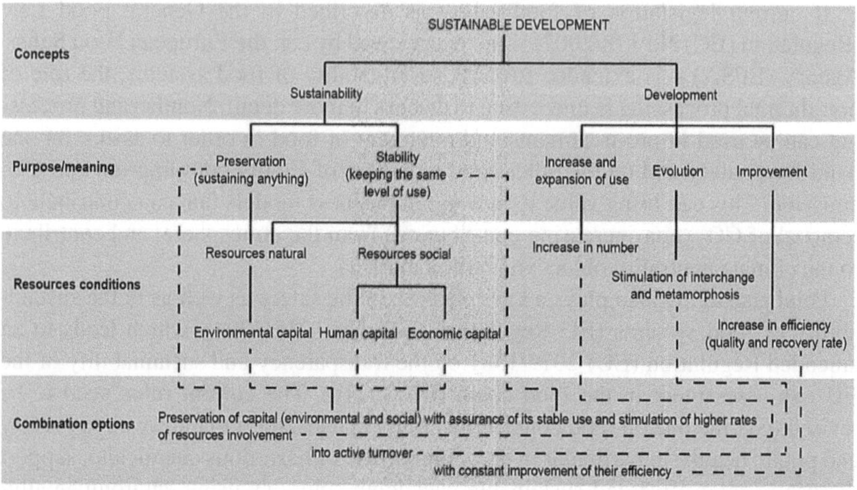

Fig. 1.3 Conceptualization of "sustainable development" through resource use approach (Smetana 2017)

There are four main groups of studies, oriented towards sustainability assessment, which have quite various approaches. The first group relies on integrated indicators (Environmental Sustainability Index, Happy Planet Index, Environment Vulnerability Index, Index of Sustainable Economic Welfare, Sustainable Governance Indicators, Sustainable Society Index, etc.), which are providing a very integrative information hardly usable for the detailed food technologies analysis (Fig. 1.4).

The second group includes a multi-criteria quantitative assessment of multiple indicators (Alrøe et al. 2016; El Gibari et al. 2019). Multiple criteria analyses are widely used for practical solutions of decision-making. At the same time such analyses require separate calculations of multiple indicators which have the disadvantages of separate benchmarks, which cannot be combined into a single integrated unit, unless multi-objective optimization is applied (Gazan et al. 2018).

Another group of indicators relates to input-output table analysis (IOTA), known as a precise method for the identification of links for national and even global economies (Leontief 1951). Combination of monetized (economic) and physical IOTA at the regional level results in analyses of waste generation and distribution (Meyer et al. 2020; Salemdeeb et al. 2016), ecosystem services determination (Fridman and Kissinger 2018; Smetana et al. 2016), and the determination of interdependencies between economy and physical resources tables (Liu et al. 2021; Wachs and Singh 2018). IOTA is related to the LCA method (see Sect. 3) and is in the basis of regional and national environmental impact accounting systems like EXIOBASE (Merciai and Schmidt 2016; Merciai and Schmidt 2018).

The fourth group of sustainability assessment systems is related to the previous, as it uses quite similar calculation methods. Life Cycle Assessment approaches are well developed and extensively applied on multiple levels of food systems analysis

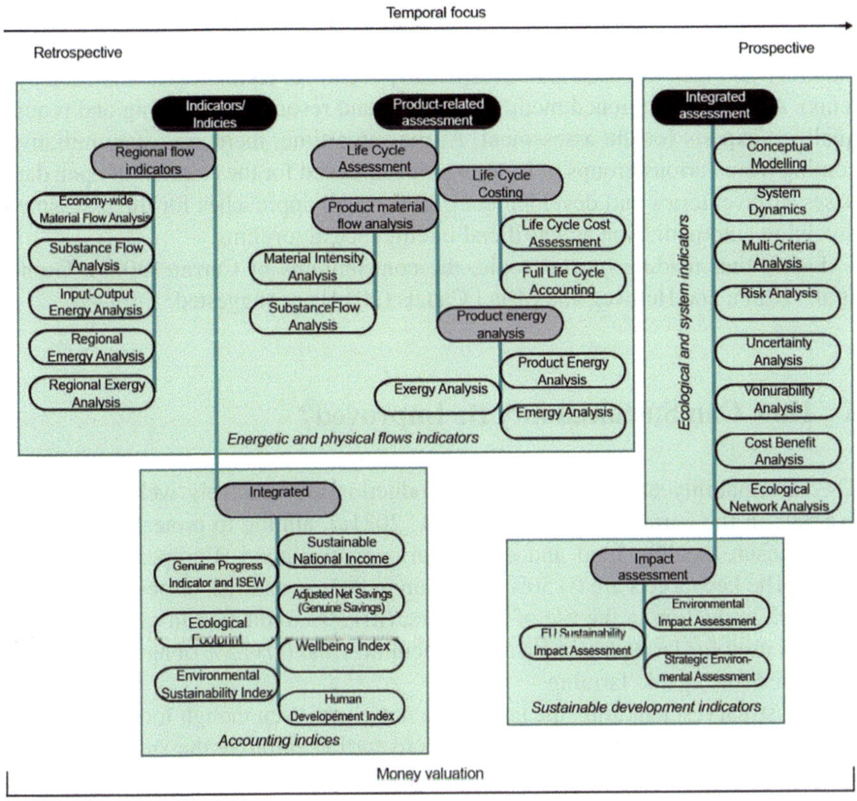

Fig. 1.4 Main groups of sustainability indicators (Smetana et al. 2015) (reprinted with permission from John Wiley and Sons)

(Finkbeiner et al. 2010; Patterson et al. 2017; Sala et al. 2015). Literature sources indicate the possibility of sustainability estimations for food technologies or products via Life Cycle Sustainability Assessment (LCSA) which includes LCA, Life Cycle Costing Analysis (LCC) and newly developing Social LCA (SLCA) (Costa et al. 2019; Finkbeiner et al. 2010). LCA is a tool that can be used to evaluate the environmental load of a food product, technical process, or food production-consumption system. LCA is a standardized methodology (ISO 14040 2006; ISO 14044 2006) and if the assessment follows them precisely it provides reliable results. LCC is a cost management method that considers the development of a product from the product idea to withdrawal from the market (product life cycle), i.e. from the "cradle to grave". Numerous guidelines exist to support the successful implementation of LCC (DIN EN 2005; Norris 2001; Verein Deutscher Ingenieure 2005; Verein Deutscher Ingenieure 2010). For SLCA UNEP developed Guidelines for SLCA of products (Benoît-Norris et al. 2011; UNEP 2020), which can be effectively applied for the indicator-based social factors calculations in food production chains.

Despite all the developments the scientists face the problems of databases absence, wide impact categories selection, qualitative and quantitative data incompatibility and lack of case studies (especially for novel technologies and food systems). All above mentioned methods are time and resource consuming and require qualified experts for the assessment. At the same time, there are a few initiatives coming from various groups of LCA experts, intended for the creation of open databases of inventories and development of integrated approaches for the assessment involving automatization and artificial intelligence algorithms.

For further reading on this topic, the contributions of Curran (2016), Guinée et al. (2011), and Hellweg and Milà i Canals (2014) are suggested.

6 How Can Sustainability Be Improved?

The sustainability targets for primary production are relatively well defined, for example in the Farm to Fork Strategy (EC 2021a), aiming to protect the environment, ensure healthy food and ensure farmers livelihood (European Parliament 2020). The key targets are (i) 50% reduction in the use and risk of pesticides, (ii) at least 20% reduction in the use of fertilizers, (iii) 50% reduction in sales of antimicrobials used for farmed animals and aquaculture, and (iv) 25% of agricultural land to be used for organic farming.

In post-harvest handling, the targets are not given, even though food science and technologies may substantially contribute to further improve the sustainability of food systems. This should be integrally considered from environmental, economic, and social dimensions, since food science connects eco-friendly production of resources (environmental dimension), with market value of food (economic dimension) and well-being in different cultural contexts (social dimension). In particular, it concerns the appreciation of food quality, the environmental-friendly way of processing, and the creation of added value. Food quality is hereby expressed in terms of a tasteful commodity, a common good, a right, a cultural heritage, etc. (social dimension; SAPEA (2020); Jackson et al. (2021)). Eco-friendly processing targets at most efficiently optimizing energy; water and resource usage (environmental; Lillford and Hermansson (2020); Knorr and Augustin (2021a, b)). Added value is created by new (circular) business and jobs, among others (economic; Donner; De Vries and Donner (2021)). They all should go hand in hand via a scenario-driven strategic approach for improving sustainability in the food domain (de Vries et al. 2018).

Examples are (i) combining healthy and sustainable diets, (ii) transforming food systems for a safe and fair planet, (iii) striving for 100% resource efficiency and circularity, (iv) innovating and empowering communities to make approaches inclusive (the four Food 2030 priorities; EC (2020a)). If one integrally takes into account the three dimensions for the first focus area, food science may then contribute to sourcing forgotten crops for developing new protein foods contributing to local new business and traditional culinary dishes. In the second focus area, zero-water-loss processing of biodiverse sources for multiple healthy dishes served by a multitude of actors could be envisaged. In the third domain, new industrial-ecological-community initiatives for 100% valorization of resources locally could be considered as a priority. Finally, in the fourth domain, consumer-driven innovations in 3D-food design is an option, both physically experimented within shared kitchens and virtually accessible thanks to informatics service suppliers to guarantee zero waste.

Thanks to the creativity of food scientists, many more examples will emerge at the crossroads of the planet, people, and profit. These are to be documented, visualized, and shared around the globe. They serve as input for new life long training programs in the wider food arena. Lessons learned are to be translated into policy options guiding the transitions towards more sustainable foods and broader bio-economy systems (De Vries et al. 2021; Directorate-General for Research and Innovation (European Commission) 2021).

7 Future Needs

Due to the interconnected nature of the topic as well as the "natural" growth of our current systems, a variety of different issues may be addressed now, as well as in the future, covering various fields, including research and development, economic and legal considerations, consumer behavior and the interests of the agricultural and food industries. It is strongly recommended to consider the various fields, tools, labels, and (policy and legal) measures, from a conceptual point of view. Even though there seems to be consensus about the need for holistic views, in particular a consistent framework is needed. Such a framework should allow revealing whether food systems are evolving sustainably or unsustainably, and in particular what are the consequences of interventions in food science and technology for overall sustainability of food systems. The interventions that may serve as leverage points for substantially changing the course of current food systems evolutions towards more sustainable outcomes are then addressed preferentially.

Only through the collaboration of all stakeholders – preferentially in well-defined and inter-connected cases around the globe – can the transformation towards more sustainable food systems be reached. We believe such collaborative works encompass a change in the education of current and future food stakeholders, such as food engineers, food technologists, food scientists, food chemists, among others. Not only can education in sustainable development help to achieve sustainability goals in practice, but it will also ensure that all food engineers have the required knowledge, values, and learning capacity to explore new paradigms and face future hurdles (Askfood 2021; Halbe et al. 2015; Serhan and Yannou-Lebris 2021; Sterling 2010).

Acknowledgements The authors acknowledge the input of Christina Zachakou (Global Sustain) towards Chap. 3 and the support of the European Federation of Food Science and Technology (EFFoST).

References

Alrøe HF, Moller H, Læssøe J, Noe E (2016) Opportunities and challenges for multicriteria assessment of food system sustainability. Ecol Soc 21(1). Retrieved from https://www.ecologyandsociety.org/vol21/iss1/art38/. https://doi.org/10.5751/ES-08394-210138

Askfood (2021) Alliance for skills and knowledge to widen food sector-related open innovation, optimization and development. Retrieved from https://www.askfood.eu/

Barabanova Y, Moeskops B (2019) Strategic research & innovation agenda for organics and agroecology leading the transition to sustainable food and farming in Europe. Retrieved from https://tporganics.eu/wp-content/uploads/2020/01/ifoam-sria-full-version-final.pdf

Benoît-Norris C, Vickery-Niederman G, Valdivia S, Franze J, Traverso M, Ciroth A, Mazijn B (2011) Introducing the UNEP/SETAC methodological sheets for subcategories of social LCA. Int J Life Cycle Assess 16(7):682–690. Retrieved from https://doi.org/10.1007/s11367-011-0301-y

Bloemhof JM, van der Vorst JGAJ, Bastl M, Allaoui H (2015) Sustainability assessment of food chain logistics. Int J Log Res Appl 18(2):101–117. Retrieved from https://doi.org/10.1080/13675567.2015.1015508

Candel JJL (2016) Putting food on the table: the European Union governance of the wicked problem of food security. Wageningen University, The Netherlands. Retrieved from https://edepot.wur.nl/375781

Chaudhary A, Gustafson D, Mathys A (2018) Multi-indicator sustainability assessment of global food systems. Nat Commun 9:848. https://doi.org/10.1038/s41467-018-03308-7

Costa D, Quinteiro P, Dias AC (2019) A systematic review of life cycle sustainability assessment: current state, methodological challenges, and implementation issues. Sci Total Environ 686:774–787. Retrieved from https://www.sciencedirect.com/science/article/pii/S0048969719324891. https://doi.org/10.1016/j.scitotenv.2019.05.435

Coyne KL (2006) Sustainability auditing. Environ Qual Manag 16(2):25–41. Retrieved from https://onlinelibrary.wiley.com/doi/abs/10.1002/tqem.20119

Curran MA (2016) Life cycle assessment. In: Kirk-Othmer encyclopedia of chemical technology. Wiley, New York, pp 1–28

de Oliveira UR, Espindola LS, da Silva IR, da Silva IN, Rocha HM (2018) A systematic literature review on green supply chain management: research implications and future perspectives. J Clean Prod 187:537–561. Retrieved from https://www.sciencedirect.com/science/article/pii/S0959652618307406. https://doi.org/10.1016/j.jclepro.2018.03.083

De Vries H, Donner M (2021) How to innovate business models for a circular bio-economy?. https://doi.org/10.1002/bse.2725. hal-03111854

de Vries H, Matser A (2011) NovelQ: novel processing methods for the production and distribution of high-quality and safe foods. Retrieved from https://library.wur.nl/WebQuery/wurpubs/fulltext/202496

de Vries H, Axelos MAV, Sarni-Manchado P, O'Donohue M (2018) Meeting new challenges in food science technology: the development of complex systems approach for food and biobased research. Innovative Food Sci Emerg Technol 46:1–6. Retrieved from https://www.sciencedirect.com/science/article/pii/S1466856418304223. https://doi.org/10.1016/j.ifset.2018.04.004

De Vries H, Donner M, Axelos M (2021) A new conceptual 'cylinder' framework for sustainable bioeconomy systems and their actors. J Agric Environ Ethics 34:1–26

Diamond A, Tropp D, Barham J, Muldoon MF, Kiraly S, Cantrell P (2014) Food value chains: creating shared value to enhance marketing success. United States Department of Agriculture

DIN EN (2005) 60300-3-3 Zuverlässigkeitsmanagement–Teil 3-3: Anwendungsleitfaden–Lebenszykluskosten

Directorate-General for Research and Innovation (European Commission) (2021) Innovation for accelerating FOOD system transformation – operationalising FOOD 2030 through living labs. Publications Office of the European Union. Retrieved from https://data.europa.eu/doi/10.2777/122836

Djekic I, Sanjuán N, Clemente G, Jambrak AR, Djukić-Vuković A, Brodnjak UV et al (2018) Review on environmental models in the food chain - current status and future perspectives. J Clean Prod 176:1012–1025. Retrieved from https://www.sciencedirect.com/science/article/pii/S0959652617329153. https://doi.org/10.1016/j.jclepro.2017.11.241

EC (2014) European Commission (EC): the economic impact of modern retail on choice and innovation in the EU food sector. Retrieved from https://ec.europa.eu/competition/publications/KD0214955ENN.pdf

EC (2016) European Commission (EC): EU platform on food losses and food waste Terms of Reference (ToR). Retrieved from https://ec.europa.eu/food/sites/food/files/safety/docs/fw_eu-actions_flw-platform_tor.pdf

EC (2020a) European Commission (EC): European Research & Innovation for Food & Nutrition Security. Retrieved from https://ec.europa.eu/info/sites/info/files/research_and_innovation/research_by_area/images/ec_rtd_food-2030-infographic.jpg

EC (2020b) European Commission (EC): non-financial reporting. Retrieved from https://ec.europa.eu/info/business-economy-euro/company-reporting-and-auditing/company-reporting/non-financial-reporting_en

EC (2021a) Farm to Fork Strategy for a fair, healthy and environmentally-friendly food system. Retrieved from https://ec.europa.eu/food/farm2fork_en

EC (2021b) Legislation. Retrieved from https://ec.europa.eu/food/safety/chemical_safety/food_contact_materials/legislation_en

El Gibari S, Gómez T, Ruiz F (2019) Building composite indicators using multicriteria methods: a review. J Bus Econ 89(1):1–24. Retrieved from https://doi.org/10.1007/s11573-018-0902-z

Elhoushy S (2020) Consumers' sustainable food choices: antecedents and motivational imbalance. Int J Hosp Manag 89:102554. Retrieved from https://www.sciencedirect.com/science/article/pii/S0278431920301067. https://doi.org/10.1016/j.ijhm.2020.102554

European Parliament (2015) Comparison of farmers' incomes in the EU member States

European Parliament (2020) Creating a sustainable food system: the EU's strategy. Retrieved from https://www.europarl.europa.eu/news/en/headlines/society/20200519STO79425/creating-a-sustainable-food-system-the-eu-s-strategy

FAO (2011) Global food losses and food waste – extent, causes and prevention. Food and Agriculture Organization of the United Nations, Rome

Finkbeiner M, Schau EM, Lehmann A, Traverso M (2010) Towards life cycle sustainability assessment. Sustainability 2(10). https://doi.org/10.3390/su2103309

Fridman D, Kissinger M (2018) An integrated biophysical and ecosystem approach as a base for ecosystem services analysis across regions. Ecosyst Serv 31:242–254. Retrieved from https://www.sciencedirect.com/science/article/pii/S2212041617305570. https://doi.org/10.1016/j.ecoser.2018.01.005

Gazan R, Brouzes CMC, Vieux F, Maillot M, Lluch A, Darmon N (2018) Mathematical optimization to explore tomorrow's sustainable diets: a narrative review. Adv Nutr 9(5):602–616. Retrieved from https://doi.org/10.1093/advances/nmy049

Guinée JB, Heijungs R, Huppes G, Zamagni A, Masoni P, Buonamici R et al (2011) Life cycle assessment: past, present, and future. Environ Sci Technol 45(1):90–96. Retrieved from https://doi.org/10.1021/es101316v

Halbe J, Adamowski J, Pahl-Wostl C (2015) The role of paradigms in engineering practice and education for sustainable development. J Clean Prod 106:272–282. https://doi.org/10.1016/j.jclepro.2015.01.093

Halberg N, Westhoek H (2019) SCAR SWG food systems policy brief: the added value of a food systems approach in research and innovation. Retrieved from https://scar-europe.org/index.php/food-deliverables

Hellweg S, Milà i Canals L (2014) Emerging approaches, challenges and opportunities in life cycle assessment. Science 344(6188):1109. Retrieved from http://science.sciencemag.org/content/344/6188/1109.abstract. https://doi.org/10.1126/science.1248361

ISO 14040 (2006) Environmental management — Life cycle assessment — Principles and framework. ISO Standard

ISO 14044 (2006) Environmental management — Life cycle assessment — Requirements and guidelines. ISO Standard

Jackson P, Rivera Ferre MG, Candel J, Davies A, Derani C, de Vries H et al (2021) Food as a commodity, human right or common good. Nature Food 2(3):132–134. Retrieved from https://doi.org/10.1038/s43016-021-00245-5

Knorr D, Augustin MA (2021a) Food processing needs, advantages and misconceptions. Trends Food Sci Technol 108:103–110. Retrieved from https://www.sciencedirect.com/science/article/pii/S0924224420307068. https://doi.org/10.1016/j.tifs.2020.11.026

Knorr D, Augustin MA (2021b) From value chains to food webs: the quest for lasting food systems. Trends Food Sci Technol 110:812–821. Retrieved from https://www.sciencedirect.com/science/article/pii/S092422442100145X. https://doi.org/10.1016/j.tifs.2021.02.037

Knorr D, Augustin MA, Tiwari B (2020) Advancing the role of food processing for improved integration in sustainable food chains. Front Nutr 7(34). Retrieved from https://www.frontiersin.org/article/10.3389/fnut.2020.00034

Leach M, Nisbett N, Cabral L, Harris J, Hossain N, Thompson J (2020) Food politics and development. World Dev 134:105024. Retrieved from https://www.sciencedirect.com/science/article/pii/S0305750X20301509. https://doi.org/10.1016/j.worlddev.2020.105024

Leontief WW (1951) Input-Output Economics. Sci Am 185(4):15–21. Retrieved from http://www.jstor.org/stable/24945285

Licciardello F (2017) Packaging, blessing in disguise. Review on its diverse contribution to food sustainability. Trends Food Sci Technol 65:32–39. Retrieved from https://www.sciencedirect.com/science/article/pii/S0924224417301644. https://doi.org/10.1016/j.tifs.2017.05.003

Lillford P, Hermansson A-M (2020) Global missions and the critical needs of food science and technology. Trends Food Sci Technol. Retrieved from https://www.sciencedirect.com/science/article/pii/S0924224420304386. https://doi.org/10.1016/j.tifs.2020.04.009

Liu Z, Huang Q, He C, Wang C, Wang Y, Li K (2021) Water-energy nexus within urban agglomeration: an assessment framework combining the multiregional input-output model, virtual water, and embodied energy. Resour Conserv Recycl 164:105113. Retrieved from https://www.sciencedirect.com/science/article/pii/S0921344920304304. https://doi.org/10.1016/j.resconrec.2020.105113

Masson-Delmotte V, Zhai P, Pörtner H-O, Roberts D, Skea J, Shukla PR, Pirani A, Moufouma-Okia W, Péan C, Pidcock R, Connors S, Matthews JBR, Chen Y, Zhou X, Gomis MI, Lonnoy E, Maycock T, Tignor M, Waterfield T (eds.) (2018) PCC, 2018: Global Warming of 1.5 °C. An IPCC Special Report on the impacts of global warming of 1.5 °C above pre-industrial levels and related global greenhouse gas emission pathways, in the context of strengthening the global response to the threat of climate change, sustainable development, and efforts to eradicate poverty. https://www.ipcc.ch/site/assets/uploads/sites/2/2019/06/SR15_Full_Report_Low_Res.pdf

Merciai S, Schmidt J (2016) Physical/hybrid supply and use tables. Retrieved from http://fp7desire.eu/documents/category/3-public-deliverables

Merciai S, Schmidt J (2018) Methodology for the construction of global multi-regional hybrid supply and use tables for the EXIOBASE v3 database. J Ind Ecol 22(3):516–531. Retrieved from https://onlinelibrary.wiley.com/doi/abs/10.1111/jiec.12713

Meyer DE, Li M, Ingwersen WW (2020) Analyzing economy-scale solid waste generation using the United States environmentally-extended input-output model. Resour Conserv Recycl 157:104795. Retrieved from https://www.sciencedirect.com/science/article/pii/S0921344920301166. https://doi.org/10.1016/j.resconrec.2020.104795

Muralikrishna IV, Manickam V (2017) Life Cycle Assessment. In: Muralikrishna IV, Manickam V (eds) Environmental management: science and engineering for industry. Butterworth-Heinemann, Oxford, UK, pp 57–76

Norris GA (2001) Integrating life cycle cost analysis and LCA. Int J Life Cycle Assess 6(2):118–120. Retrieved from https://doi.org/10.1007/BF02977849

Obersteiner G, Pilz H (2020) Food packaging sustainability: a guide for packaging manufacturers, food processors, retailers, political institutions & NGOs, based on the results of the project "STOP waste – SAVE food". Retrieved from https://www.ecoplus.at/media/20683/guideline_stopwaste_en.pdf

Patterson M, McDonald G, Hardy D (2017) Is there more in common than we think? Convergence of ecological footprinting, emergy analysis, life cycle assessment and other methods of environmental accounting. Ecol Model 362:19–36. Retrieved from https://www.sciencedirect.com/science/article/pii/S0304380017300613. https://doi.org/10.1016/j.ecolmodel.2017.07.022

Pereira RN, Vicente AA (2010) Environmental impact of novel thermal and non-thermal technologies in food processing. Food Res Int 43:1936–1943. https://doi.org/10.1016/j.foodres.2009.09.013

Picart-Palmade L, Cunault C, Chevalier-Lucia D, Belleville M-P, Marchesseau S (2019) Potentialities and limits of some non-thermal technologies to improve sustainability of food processing. Front Nutr 5:130

Poore J, Nemecek T (2018) Reducing food's environmental impacts through producers and consumers. Science 360:987–992. https://doi.org/10.1126/science.aaq0216

Quantis (2020) DIG IN – a landscape of business actions to cultivate a sustainable + resilient food system. Retrieved from https://quantis-intl.com/report/dig-in-food-report/

Queenan K, Garnier J, Rosenbaum N, Buttigieg S, de Meneghi D, Holmberg M, Zinsstag J, Rüegg SR, Häsler B, Kock R (2017) Roadmap to a one health agenda 2030. CAB Rev Perspect Agric Vet Sci Nutr Nat Resour 12(014):1–12

Raak N, Symmank C, Zahn S, Aschemann-Witzel J, Rohm H (2017) Processing- and product-related causes for food waste and implications for the food supply chain. Waste Manag 61:461–472. https://doi.org/10.1016/j.wasman.2016.12.027

Rainforest Alliance (2020) What does "rainforest alliance certified" mean? Retrieved from https://www.rainforest-alliance.org/faqs/what-does-rainforest-alliance-certified-mean

REFRESH (2020) REFRESH FinalResults: The REFRESH Project: Resource Efficient Food and dRink for the Entire Supply cHain. Retrieved from https://eu-refresh.org/sites/default/files/REFRESH%20Final%20Results%20Brochure_200225_0.pdf

Režek Jambrak A, Nutrizio M, Djekić I, Pleslić S, Chemat F (2021) Internet of nonthermal food processing technologies (IoNTP): food industry 4.0 and sustainability. Appl Sci 11(2):686. Retrieved from https://www.mdpi.com/2076-3417/11/2/686

Ritchie H, Roser M (2020). Environmental impacts of food production. Our world in data. Retrieved from https://ourworldindata.org/environmental-impacts-of-food#

Sala S, Ciuffo B, Nijkamp P (2015) A systemic framework for sustainability assessment. Ecol Econ 119:314–325. Retrieved from https://www.sciencedirect.com/science/article/pii/S0921800915003821. https://doi.org/10.1016/j.ecolecon.2015.09.015

Salemdeeb R, Al-Tabbaa A, Reynolds C (2016) The UK waste input-output table: linking waste generation to the UK economy. Waste Manag Res 34(10):1089–1094. https://doi.org/10.1177/0734242x16658545

SAPEA (2020) Science Advice for Policy by European Academies (SAPEA): a sustainable food system for the European Union. SAPEA, Berlin

Serhan H, Yannou-Lebris G (2021) The engineering of food with sustainable development goals:policies, curriculums, business models, and practices. Int J Sustain Eng 14:12–25. https://doi.org/10.1080/19397038.2020.1722765

Smetana SM (2017) Bioeconomy and regional sustainability: potential of life cycle assessment for food production (PhD Thesis). Vechta, Germany

Smetana S, Tamásy C, Mathys A, Heinz V (2015) Sustainability and regions: sustainability assessment in regional perspective. Reg Sci Policy Pract 7(4):163–186. Retrieved from https://rsaiconnect.onlinelibrary.wiley.com/doi/abs/10.1111/rsp3.12068

Smetana S, Tamásy C, Mathys A, Heinz V (2016) Measuring relative sustainability of regions using regional sustainability assessment methodology. Geogr Anal 48(4):391–410. Retrieved from https://onlinelibrary.wiley.com/doi/abs/10.1111/gean.12102

Soysal M, Bloemhof-Ruwaard JM, Meuwissen MPM, van der Vorst JG (2012) A review on quantitative models for sustainable food logistics management. Int J Food Syst Dyn 3(2):136–155

Sterling S (2010) Learning for resilience, or the resilient learner? Towards a necessary reconciliation in a paradigm of sustainable education. Environ Educ Res 16:511–528. https://doi.org/10.1080/13504622.2010.505427

UNEP (2019) Emissions gap report 2019, Nairobi

UNEP (2020) United Nations Environment Programme (UNEP) guidelines for social life cycle assessment of products and organizations. Retrieved from https://www.lifecycleinitiative.org/wp-content/uploads/2021/01/Guidelines-for-Social-Life-Cycle-Assessment-of-Products-and-Organizations-2020-22.1.21sml.pdf

United Nations (2015) Transforming our world: the 2030 agenda for sustainable development, New York

van der Meulen B (2017) Food legal diversity – harmonising across cultural divides? Retrieved from https://www.globalharmonization.net/sites/default/files/gfx/wg/2017-12_New-Food-Bernd_vd_Meulen.pdf

Verein Deutscher Ingenieure (2005) Purchase, operating and maintenance of production equipment using Life Cycle Costing (LCC). VDI-Gesellschaft Produktion und Logistik, Duesseldorf

Verein Deutscher Ingenieure (2010) VDI-Richtlinie: VDI 2067 Blatt 1 Wirtschaftlichkeit gebäudetechnischer Anlagen – Grundlagen und Kostenberechnung. VDI-Gesellschaft Produktion und Logistik, Duesseldorf

Vermeir I, Verbeke W (2006) Sustainable food consumption: exploring the consumer "attitude – behavioral intention" gap. J Agric Environ Ethics 19(2):169–194. Retrieved from https://doi.org/10.1007/s10806-005-5485-3

Waage J, Yap C, Bell S, Levy C, Mace G, Pegram T, Unterhalter E, Dasandi N, Hudson D, Kock R, Mayhew SH, Marx C, Poole N (2015) Governing sustainable development goals: interactions, infrastructures, and institutions. In: Waage J, Yap C (eds) Thinking beyond sectors for sustainable development. Ubiquity Press, London, pp 79–88

Wachs L, Singh S (2018) A modular bottom-up approach for constructing physical input–output tables (PIOTs) based on process engineering models. J Econ Struct 7(1):26. Retrieved from https://doi.org/10.1186/s40008-018-0123-1

Weber H, Poeggel K, Eakin H, Fischer D, Lang DJ, Wehrden HV, Wiek A (2020) What are the ingredients for food systems change towards sustainability?—insights from the literature. Environ Res Lett 15(11):113001. Retrieved from https://doi.org/10.1088/1748-9326/ab99fd

Chapter 2
Innovative Processing: From Raw Material, Post Harvesting, Processing, and Applications

Alexandrina Sirbu

1 Innovative Processing, a Driver for the Sustainable Food Supply Chain

In the context of sustainable development, the balance between environmental protection, economic growth, and social cohesion is ensured by constantly embracing alternative solutions for carrying out economic activities. In the market economy, sustainability in the food chain, in terms of production and trade of the food commodities as well as management of environmental factors (referring to energy, water, wastes, greenhouse gases, etc.) for the growth of the economic activities often involve the innovation of products, processes, or the development of new technologies. Innovation has become an utmost important driver of sustainability that helps companies face the increased market challenges nowadays and convert new ideas into business opportunities for the food industry within the supply chain.

In its essential meaning, sustainability is a concept that is applied through a continuous process; that is why many times has been quoted that sustainability is more a journey than a destination. Consequently, lasting development has become more and more connected to the innovation process and environmental issues entailed finding innovative solutions to be integrated into the supply chains and develop *eco-innovation* practices (e.g., eco-design, life cycle assessment, cleaner production, energy recovering, packages recycling, waste reducing, water treatment, etc.). In this regard, *sustainable innovation* has targeted both economic and sustainability performances of individual companies that show to advantage various opportunities on existent or new markets (Kneipp et al. 2019), being either reactive or proactive to technology advancements and new business systems. Tebaldi et al. (2018) reviewed the Sustainable Supply Chain Innovation (SSCI), in which sustainable

A. Sirbu (✉)
Constantin Brancoveanu University of Pitesti, FMMAE, Ramnicu Valcea, Romania

A. Režek Jambrak (ed.), *Nonthermal Processing in Agri-Food-Bio Sciences*, Food Engineering Series, https://doi.org/10.1007/978-3-030-92415-7_2

innovation is approached beyond a company level but it addresses the whole supply chain by integrating all kinds of innovations from incremental to radical ones and from the product and technological innovations to any resource allocation and organizations, promoting innovation beneficial for all stakeholders associated with the supply chain. Learned lessons from the twentieth century have shown new knowledge should be produced considering scientific advancements and economic added value, and additionally the moral and social responsibilities. Even full of promises or incentives, sometimes have arisen concerns about innovation in science and technology regarding unforeseen hazards, loss of control on research outcomes, unpredictable consequences of emerging technologies, or other ethical effects at the societal level in a long term. Stilgoe et al. (2013) commented on the governance of innovation in science and emerging technology and supported *responsible innovation*, which has four dimensions (i.e., anticipation, reflexivity, inclusion, and responsiveness).

The food industry involves a large number of horizontal and vertical relationships between stakeholders (business people, governmental representatives, technologists, researchers, etc.), the very dynamic nature of these relationships playing an important role in the field of innovation incorporated into the food supply chain (Beske et al. 2014). In recent decades, end-consumers have gained a central role in the food market, currently being increasingly promoted in the management and implementation of consumer-oriented innovation. So, the food industry is a very good example of how dynamic capabilities such as product and process development among the transparency, knowledge acquisition and assessment, ability development, etc., is used by companies on the food supply chain to gain a market competitive advantage and face more challenges of customers' demands. The literature reviewed in areas of innovation and sustainability within the supply chain reveals a consistent interest for agricultural and food sectors (over 17% of total industries), the technological innovation being identified in the top 4 of various types of innovation (Rajeev et al. 2017; Tebaldi et al. 2018). Most studies have been performed on sustainable farming, agro-chemicals, alternative technologies to improve the performance of the food industry, and so forth. Concerning food processing, as an alternative to thermal conventional technologies for food preservation, many novel thermal and nonthermal emerging technologies have already proved their efficiency, e.g., in terms of food safety, food quality, the shelf life of agro-food, and energy decrease. In the aim to promote sustainable food systems, Gimenez-Escalante and Rahimifard (2019) proposed an innovative strategy, i.e., Distributed and localised food manufacturing (DLM), which consists of those food technologies' selection as the best alternative (in terms of processing, yields at the local scale, exploitation of operational capabilities, business flexibility, and response to market demands) to be implemented in a manufacturing company based on its individual requirements. As Khouryieh (2020) found, food processors from the U.S. considered that equipment manufacturers and government research are the main drivers for innovation in emerging food technologies. But, from those emerging food technologies, further on discusses "hurdle" and nonthermal technologies, which are applied in the processing of agro-food products "from farm to fork".

Globally, the use of technology remains one of the greatest challenges to the food supply chain. It is important to notice that radical innovation, in time, has become an incremental one, and emerging or young technology from yesterdays may come to a mature technology tomorrow. Van den Oord and van Witteloostuijn (2018) studied the case of biotechnologies' evolution from the 1970s to 2000s (including co-evolution of technologies and organizations) and they defined a multi-level model of emerging technology from seed to its growth. As Hernández-Hernández et al. (2019) quoted, some emerging food processing technologies have been studied more than others in the last decades, a few are well known but not of all with commercial applications. However, an in-depth analysis of the stage of nonthermal technology evolution for agro-food processing (incl. Technology Readiness Level TRL) is out of scope within this chapter, when it refers to the term of emerging or novel technologies onwards.

Since the twentieth-century, consumers have expressed their demands related to foodstuffs that should be at the same time, if it is possible, convenient, safe, healthy, with an extended shelf life, and having a fresh-like quality as well. Innovative solutions promoted in the 1990s in food technology envisaged minimal processing technologies that could be applied in various stages of the food supply chain (Ohlsson 1994). Minimal processing consists of a combination of technologies, methods, or processing procedures by using the "hurdle" concept to ensure adequate stability, sensory, safety, or shelf life of food commodities. That means a processing system or fine-tuning preservation by using different barriers such as controlled temperature, water activity, pH, redox agents, preservatives, etc., aimed at having a mild or minimal effect on food quality changes.

Designing the minimal processing technologies evolved from combinations' optimization of traditional preservation techniques to combinations of traditional methods with emerging preservations, or novel combinations of emerging technologies, all of them applied without afflicting safety nor altering quality attributes of food (Alzamora et al. 2016; Khan et al. 2017). A suitable selection of stressor factors and combinations of intervention technologies (e.g., emerging ones) to perform minimal processing continues to be innovative; this is related to its foreseen purpose, available resources, type of food, and balance between advantages and limitations of the chosen solution.

2 Nonthermal Emerging Technologies Along the Food Supply Chain

Nonthermal technologies allow food preservation or processing at milder temperatures, below the temperatures used during thermal pasteurization, or without generating very high temperatures whilst the products are exposed to the specific, intense treatment during a short period of time. This kind of technology is considered nonthermal processing (NTP) because numerous applications studied on microbial degradations proved that biological inactivation occurs at a sub-lethal temperature of the microorganisms or room temperature (Pereira and Vicente 2010; Hassoun et al. 2020). Also, as regard food quality, NTPs proposed food processing aimed at minimizing the changes in flavor, color, texture, odor, or residuals in comparison with other treatments, but preliminary results were not always so outstanding. Hence, several processing techniques and fine-tuning of the experimental setup continue to be under development, and scaling up may neither be technically nor economically feasible yet. It seems that the oldest nonthermal technologies applied to the biological matrices are Pulsed Electric Fields (PEF) from the 1950s (Zhang et al. 2019) and High Hydrostatic Pressure (HHP). HHP is also known as ultra-high pressure (UHP) or high pressure processing (HPP). The applicability of HHP was firstly proved for milk preservation in 1899, scientifically reapproached (on other food matrices) in the mid-1980s, but it has started to be industrially used for fruit juices, jams, sauces, desserts, etc., in the 1990s in Japan; whilst the first commercial PEF technology was implemented for fruit juice treatment in 2005 in the US (Yaldagard et al. 2008; Barba et al. 2016). According to Khan et al. (2017), one of the most researched novel technology is food irradiation, the first patent of ionizing radiation for pathogen spoilage dating from 1905; and then assessed against pork trichinae in 1921. Also, the bactericidal effect of pressurized carbon dioxide was discovered by Valley and Rettger in 1927 (Yu et al. 2020).

Nonthermal emerging technologies can be grouped based on different criteria such as engineering aspects, novelty, scope, and food applications. Engineering aspects refer to the scientific principles and technical solutions applied in engineering to perform that technology. The advantages and limitations of each nonthermal food technology are explained by its mechanisms of action, with consequences on potential applications in the agro-food sector (Barbosa-Cánovas et al. 2005; Barba et al. 2016; Khan et al. 2017; Režek Jambrak et al. 2018; Picart-Palmade et al. 2019). As mentioned above, nonthermal technologies under discussion are groundbreaking technologies, and their novelty is straightly related to TRLs or their life stage since they were developed and used. Based on processing principles and mechanism of action, technological purposes envisage their applications on the food supply chain related to their scope and operations to be done. Nonthermal food processing applications are considered from the agricultural raw materials towards end-consumption on food categories or areas.

Depending on their novelty, namely, if relevant technologies are industrial available for the manufacture of food products enabling food processes cost-efficient (according to DLM concept), Gimenez-Escalante and Rahimifard (2019) distinguished the novel nonthermal food processings in two categories, namely: applicable (as HHP/HPP, irradiation, ultrasound, PEF, and membrane separation) and emerging technologies (e.g., Ultraviolet (UV) and pulsed light, pulsed X-Rays, cold plasma, oscillating magnetic fields (OMF), electric arc discharges, microfluidics, nanotechnologies, and membrane emulsification). For example, HHP is the most widely applied novel nonthermal food technology at the industrial level, followed by PEF (Hernández-Hernández et al. 2019; Khouryieh 2020). According to Khouryieh (2020), OMF is one of the most common nonthermal technologies under development. In the next decade is foreseen that cold plasma to come on second place within the top of the most significant emerging technologies available, whilst ionizing radiation will decrease on the tenth position (Hernández-Hernández et al. 2019).

The main purpose of developing these nonthermal technologies focused on inhibition of microbial growth and food preservation in ways to minimally alter the sensorial properties of the food matrix and ensure food stability during storage or processing with an adequate shelf life increment (Keyser et al. 2008; Yaldagard et al. 2008; Niemira 2012; Barba et al. 2016; Alzamora et al. 2016; Khan et al. 2017; Brodowska et al. 2018; Hernández-Hernández et al. 2019; Zhang et al. 2019). In this respect, the effects of these nonthermal technologies have been studied either on food substrate or microbiota, or both. Additionally, much research on emerging nonthermal processings encompassed inactivation of enzymes, modification of physicochemical and colloidal properties (e.g., hydrophilic/hydrophobic changes, solute diffusion), denaturation of macromolecular compounds (proteins, starch), partial modifications of nutritive substances, bioactive compounds, and sensory attributes. Consequently, depending on the scope of novel NTP and their potential technological effects, these technologies in the food industry can be grouped as:

- Technologies for food preservation, shelf life extending and food safety ensuring
- Green technologies for food processing
- Technologies type "clean-room" for food safety assurance
- Technologies for the valorization of food by-products, and other bioresources.

In the classification above, a distinctive approach is proposed between food preservation, which is mainly focused on food products (as part of Hazard Analysis Critical Control Points (HACCP) method), and all necessary conditions, as part of prerequisite programs within a food safety management system (FSMS) to ensure safe production and distribution of food commodities. So, the first group of NTP technologies refers to minimal damages of food products and their preservation (e.g., food decontamination, cleaning, washing, pre-treatment, pasteurization, sterilization, etc.), whilst the third category comprised the adequate technologies for keeping a clean and safe environment in plants, warehouses, and any manufacturing facilities (e.g., sanitation, disinfection, controlled environment, etc.). Emerging nonthermal technologies try to solve food safety issues by several means, namely: reducing the microbial population on the surface of food, inactivation of foodborne pathogens and spoilage microorganisms, nonthermal neutralization of certain biological hazards, surfaces decontamination and other food-contact materials, sanitation, and disinfection, hygienically removal of wastes, etc. Zhang et al. (2019) reviewed different mechanisms of some nonthermal technologies (e.g., PEF and ultrasound) against food spoilage and explained them by destroying genetic materials or changing the structure of the cell membrane of the microorganisms.

As discussed, new agricultural and food strategies, practices, and technologies integrated along a sustainable food supply chain have claimed innovation for green technology development. As an alternative to conventional or traditional technologies, the green nonthermal technologies allow to the same extent the food manufacturing and lower environmental impact that means food transformation with energy efficiency, reducing carbon footprint, saving water, reducing waste generation along the food chain, etc. Many emerging NTP food technologies under discussion proved that are green or at least potentially green (López-Pedrouso et al. 2019; Picart-Palmade et al. 2019). But food processing is not limited to technologies and operations aiming at food preservation or thermal effect on the food products. In literature, there are different other criteria for defining food technologies based on their role (e.g., conditioning, preparation, cooking) or type of processes or unit operations (e.g., mass and energy transfer). So, this class of green technologies for food processing comprises those NTP with the primary technological purpose other than preservation or decontamination, namely: extraction, drying, cooking, boiling, tenderization, maturation (aging), separation, gelatinization, mixing, emulsification, homogenization, thawing, blanching, forming, fermentation, distillation, etc. Green nonthermal processing address the main challenges of market demands as regards the new formulation of food, fresh-like food products, quality improvement, better sensory characteristics, chemical or additive-free, nutrition keeping or increasing, integrity, and in a certain manner, environmental-friendly. For sure, some of these green technologies may influence also food stability during storage, shelf life extending, or preservative character but no necessarily as a first or unique goal of their implementation into a food processing plant.

Food by-products are secondary, sometimes unintended products that resulted during the food manufacture, and their removal, recovery, and attempts to harness through processing became of interest for a high valorization and diversifying of food materials, food manufacture cost-efficiency, and reducing waste, at least. Nowadays, various high-value compounds can be capitalized within the food supply chain, as bioresources, by recovering not only from food by-products but also food raw materials, waste, or food wastewater. Also, the separation of valuable bioactive compounds from food opened the gates for new food branches as nutraceuticals, food supplements, etc. In conclusion, concerning this specific technological scope, the novel nonthermal technologies implemented for the valorization of food by-products, and other products (e.g., edible films) should represent a distinct category, apart from food production itself.

Based on a literature review, a few applications of nonthermal emerging technologies are introduced in Table 2.1. As mentioned above, there is a difference between opportunities of usage and commercialization of emerging NTP; that is why Table 2.1 does not necessarily consider the TRL of each technology in terms of technological effects nor food types.

Main applications, which are foreseen on the food chain for emerging nonthermal technologies refer to at least two or many of the following: enhance the microbiological, nutritional, and sensorial quality of food, keep the fresh-like quality properties almost or as of food's native state, develop new food products, give food with a prolonged shelf life, partially solve the issues arisen in other types of traditional or novel thermal processing, promote sustainable innovation and address to new niche markets, increase yield, reduce environmental impact, etc. Table 2.1 presents some applications of novel NTPs such as treatment of post-harvest agricultural products, food preservation, food processing, sanitation or disinfection, food

Table 2.1 Emerging nonthermal processing food technologies applied within the food supply chain

Food supply chain (FSC)	Nonthermal technology (NTP)	Scope/technological purpose	Food type application	References
Post harvesting (raw materials)	Magnetic field (MF)/oscillatory magnetic field (OMF)	Maintain quality/integrity	Cantaloupe, melon	Miñano et al. (2020)
	Electrohydrodynamic (EHD) processing	Drying	Fruits and vegetables	Singh et al. (2012)
	Ionizing radiation (IOR)	Minimizes the post-harvest loss, food decontamination postharvest	Fruits and vegetables (e.g., potatoes, onions, garlic, etc.)	Sunil et al. (2018), Priyadarshini et al. (2019)
	High hydrostatic pressure (HHP) or high pressure processing (HPP)	Microbial decontamination	Oyster	Barba et al. (2016)
	Ozonation/ozone technology	Decontamination postharvest to extend the shelf life	Fruits and vegetables	Brodowska et al. (2018)
	Ultraviolet (UV) light	Decontamination postharvest	(Whole and cut) fruit surfaces	Alzamora et al. (2016)
	Photosensitization	Microbial control of food	Fresh fruit and vegetables	Luksiene and Brovko (2013)

(continued)

Table 2.1 (continued)

Food supply chain (FSC)	Nonthermal technology (NTP)	Scope/technological purpose	Food type application	References
Food processing (incl. Food preservation)	Pulsed electric field (PEF)	Nonthermal pasteurization (preservation)	Fruit juices	Alzamora et al. (2016), Priyadarshini et al. (2019), Zhang et al. (2019)
			Fruit (apple) juices and smoothies	Barba et al. (2016)
		Microbial decontamination (preservation)	Milk	Alzamora et al. (2016), Barba et al. (2016), Zhang et al. (2019), Shabbir et al. (2021)
		Liquid food preservation (in general)	Dairy fluid food products (milk, yogurt), liquid eggs, soups, and other pumpable food	Zhang et al. (2019), Barba et al. (2016), Priyadarshini et al. (2019)
		Conditioning (peeling pretreatment)	Tomatoes	Arnal et al. (2018)
		Tenderization, and aging	Fish and (frozen) meat	Gómez et al. (2019)
	Cold plasma (CPL)/ atmospheric cold plasma (CAP)	Preservation (sterilization, food decontamination)	Cherry tomatoes, strawberries, almonds	Zhang et al. (2019)
			Chicken muscle, brown rice, red pepper powder	Sonawane et al. (2020)
			Various food (e.g., fresh-cut apples, melon, etc.)	Pankaj et al. (2018), Priyadarshini et al. (2019), Hernández-Hernández et al. (2019)

Food supply chain (FSC)	Nonthermal technology (NTP)	Scope/technological purpose	Food type application	References
	MF/static magnetic field (SMF)	Processing	Honey	Sakdatorn et al. (2018), Miñano et al. (2020)
		Freezing	Various liquid and solid food	Yıkmış (2016)
	EHD processing	Drying	Fruit	Defraeye and Martynenko (2018)
	IOR (e.g., electron beam EB, irradiation gamma ray)	Dry food surfaces' decontamination	Dry food (herbs, spices)	Hertwig et al. (2018)
		Preservation (pasteurization, sterilization, foodstuffs decontamination)	Fresh fruits and vegetables, legumes, tubers, cereals, fish and seafood, meat, poultry, cheese, spices, herbs, dry vegetable seasonings, nuts	Farkas (2006), Khan et al. (2017), Priyadarshini et al. (2019), Hernández-Hernández et al. (2019), Zhang et al. (2019)
			Ready to-eat meat products or meat products	Zhang et al. (2019), Sunil et al. (2018)

(continued)

Table 2.1 (continued)

Food supply chain (FSC)	Nonthermal technology (NTP)	Scope/technological purpose	Food type application	References
	High hydrostatic pressure (HHP, HPP)	Preservation (nonthermal pasteurization, microbial decontamination)	Various food products, especially from or based on fruits: jellies, jams, juices and purees, fruit yogurts, dairy based fruit smoothie, sauces	Barbosa-Cánovas et al. (2005), Alzamora et al. (2016)
			Juices, meats (e.g., dry cured hams), vegetable and seafood products, pasta and salsa sauces, fillings (e.g., for Rodilla sandwich)	Tonello (2010)
			Milk	Shabbir et al. (2021)
		Food processing (based on protein denaturation, solute diffusion in salting or sugaring operations, assisted freezing-thawing processes, modification of functional properties of macromolecules, etc.)	Smoked salmon, cheese manufacturing (curd formation)	Zhang et al. (2019)
			Various food – e.g., fruits and legume juice (orange, mandarin, grape, apple, carrot, broccoli), smoothies, Avocado purée, prepared vegetable dishes, desserts, jams, sauces, jellies	Barba et al. (2016), Chao et al. (2013), Li and Zhu (2018)
	Ultra high pressure homogenization (UHPH)	Food processing (for stabilization of disperse system, obtaining of stable emulsion, etc.)	Milk, juice	Picart-Palmade et al. (2019)

Food supply chain (FSC)	Nonthermal technology (NTP)	Scope/technological purpose	Food type application	References
	Carbon dioxide processes (e.g., dense phase carbon dioxide DPCD, high-pressure carbon dioxide HPCD, supercritical carbon dioxide SCF)	(cold)-pasteurization	Milk, fruit juices (e.g., carrot, red grapefruit, orange), and vegetables juices	Sunil et al. (2018), Picart-Palmade et al. (2019)
			Liquid food (e.g. apple juice), beverages	Balaban and Duong (2014), López-Pedrouso et al. (2019), Yu et al. (2020)
			Fresh-cut carrots	Spilimbergo et al. (2013)
	Ultrasound (US) (incl. power/ high intensity US)	Food preservation (e.g., inactivation of microorganisms/decontamination, sterilisation)	Various foodstuffs (e.g., raw vegetables and liquid whole eggs)	Zhang et al. (2019)
		Various processings, such as food emulsification, degassing, defoaming, homogenization, filtration, US-assisted freezing, sonocrystallization	Different food (e.g., milk and dairy products, chocolate, drinks and beverages)	Musielak et al. (2016), Zhang et al. (2019), López-Pedrouso et al. (2019), Chávez-Martínez et al. (2020), Shabbir et al. (2021)
		Vinification (i.e., grape crushing and grape must preparation)	Red wine	Bautista-Ortín et al. (2017)
		Tenderisation, accelerates maturation	Meat	Alarcon-Rojo et al. (2015), Barba et al. (2016)
		Cutting	Food (e.g., vegetables, meat, nuts, berries)	López-Pedrouso et al. (2019), Chávez-Martínez et al. (2020)

(continued)

Table 2.1 (continued)

Food supply chain (FSC)	Nonthermal technology (NTP)	Scope/technological purpose	Food type application	References
	Non-thermal hydrodynamic cavitation	Preservation/processing	Orange juice	Katariya et al. (2020)
	Ozonation/ozone technology	Preservation (e.g., against microbial spoilage/ decontamination)	Fresh and dried fruits (whole or cut), juices	Alzamora et al. (2016)
			Fruits, vegetables, spices, meat, seafood products, poultry, dairy products, beverages	Brodowska et al. (2018)
	Nanotechnologies	Processing (with controlled release of active nanoencapsulated compounds, physicochemical properties changes at nanoscale matter) – e.g., dairy processing	Various food product applications	Singh et al. (2017), Nazir and Azaz Ahmad Azad (2019)
	UV light	Preservation (e.g., liquid pasteurization, food surface decontamination, and disinfection against microorganisms)	Apple cider and clear juices, whole and cut fruit surfaces	Alzamora et al. (2016)
			Fresh fruit juices (e.g., apples and citrus juices, strawberry nectar)	Keyser et al. (2008), Priyadarshini et al. (2019)
	Pulsed light (PL)	Microbial load reduction (preservation)	Whole and cut fruits and in clear juices	Alzamora et al. (2016)
			Cheese, cooked meat products, cut mushrooms	Zhang et al. (2019)
	Membrane technologies (e.g., Membrane filtration MF, ultrafiltration UF, nanofiltration NF, and reverse osmosis RO)	Preservation (decontamination, purification)	Milk and dairy products	Shabbir et al. (2021)
			Fruit juices	Echavarría et al. (2011)
		Processing (e.g., fractionation and separation, clarification, membrane distillation, concentration, etc.)	Milk and dairy products	Hausmann et al. (2014), Shabbir et al. (2021)
			Fruit juices	Echavarría et al. (2011)

Food supply chain (FSC)	Nonthermal technology (NTP)	Scope/technological purpose	Food type application	References
Whole FSC-Food safety assurance (e.g., prerequisite programs PRPs and operational ones OPRPs)	Plasma-activated water (PAW)	Sanitizers/disinfectant characteristics –various surfaces		Sonawane et al. (2020)
	CPL/CAP	Microbial decontamination/sterilization of food package materials (incl. post-packaging decontamination)		Pankaj et al. (2018), Priyadarshini et al. (2019), Zhang et al. (2019), Sonawane et al. (2020)
		Disinfection		Hernández-Hernández et al. (2019)
		Wastewater treatment		Pankaj et al. (2018)
	IOR	Pest disinfection – Phytosanitary crop protection	Tropical fruits	Farkas (2006)
		Disinfestation, microbial inactivation in packaging materials		Hernández-Hernández et al. (2019)
	Ultrasound (US)	Waste treatment		Zhang et al. (2019)
	Ozonation/ozone technology	Water disinfection and purification		Brodowska et al. (2018)
	Nanotechnologies	Food packaging	Edible and biodegradable films for active packaging	Singh et al. (2017), Nazir and Azaz Ahmad Azad (2019)
	UV	Wastewater treatment and water disinfection	Water	Priyadarshini et al. (2019)
	PL	Sanitation (equipment surface, knives, food packaging materials)		Zhang et al. (2019)
	Photosensitization	Decontamination of food-packaging and edible films for food-packaging		Luksiene and Brovko (2013)

(continued)

Table 2.1 (continued)

Food supply chain (FSC)	Nonthermal technology (NTP)	Scope/technological purpose	Food type application	References
By-product valorization and other bioresources	PEF	Extraction of food by-products/recovery of bioactive compounds (e.g., plant oil, polyphenols, pigments)	Plant material	Zhang et al. (2019), Barba et al. (2016)
	High-voltage electrical discharges (HVED)	Recovery of polyphenols	Plant material	Yıkmış (2016)
	HHP (HPP)	Extraction of some cellular compounds		Barba et al. (2016)
	UHPH ultra high pressure homogenization	Extraction of natural compounds from suspended particles	Plant material	Picart-Palmade et al. (2019)
	CO_2 processes	Extraction of compounds (terpenes, carotenoids, polyphenols, fatty acids, tocopherols, flavonoids, etc.)	Plant material	Picart-Palmade et al. (2019), López-Pedrouso et al. (2019)
		Encapsulation of natural compounds (e.g., antioxidants)		López-Pedrouso et al. (2019)
	Ultrasound (US)	Extraction of active ingredients	Various foodstuffs (e.g., vegetal and algae matters)	Barba et al. (2016), Zhang et al. (2019)
	Nanotechnologies	Nanoencapsulation of various compounds (e.g., essential oils, carotenoids)	Nanoencapsulated food additives	Singh et al. (2017), Nazir and Azaz Ahmad Azad (2019)
	Membrane technologies (MF, UF, NF, RO)	Separation (e.g., fractionation, and extraction), recovery and processing of food by-products	Milk and dairy products, and others	Hausmann et al. (2014), Picart-Palmade et al. (2019), Shabbir et al. (2021)

packaging, and shelf life extension, valorization of by-products, extraction of various compounds, etc. As is introduced in Table 2.1, research of novel nonthermal techniques under development or already applicable in agriculture, food industry, and trade encompassed a high diversity of food products on categories, namely:

- cereals, grains, pulses, and farinaceous foodstuffs,
- fruits, vegetables and legumes, fruit and vegetable juices and other derivated products,
- spices and herbs,
- milk and dairy products,
- eggs,
- meat, poultry, and meat products,
- fish and seafood,
- beverage, beer, wine
- other compounds (nutrients, etc.).

Different authors pointed out that novel nonthermal technologies have restrictive applications as regards the food types to be processed, as a consequence of operating principles and processing conditions (Barba et al. 2016; Picart-Palmade et al. 2019; López-Pedrouso et al. 2019). For example, Barbosa-Cánovas et al. (2005) commented that some nonthermal technologies, such as HHP, OMF, and light pulses, can be used for processing various liquid and solid food but other NTPs allow processing only for solid or fluid food. A few examples of food-type applications using some nonthermal processes are depicted in Table 2.1, too. Emerging nonthermal technologies remain an attractive solution for food manufacturers that may select optimal processing alternatives if they understand NTP potentiality for food processing, know certainly product requirements, and have access to necessary resources.

As regards the overall engineering or technical aspects, it is difficult to find in the literature a comprehensive classification of all novel nonthermal technologies. Considering the source of energy transfer and type of action on microorganisms, Režek Jambrak et al. (2018) grouped nonthermal emerging technologies in: electric and magnetic fields (pulsed electric fields, oscillating magnetic fields, cold plasma, electron beam processing, and electrohydrodynamic processing); with mechanical action because of shock waves or hydrodynamic effects (ultrasound and plasma); through radical formation (plasma, ozonation, ultrasound, UV light); and pressure-based technologies (using extremely high pressures). At the same time, Režek Jambrak et al. (2018) mentioned another categorizing in wave technologies (UV, plasma, and ultrasound); field technologies (i.e., electric and magnetic fields); and pressure technologies (e.g., high-pressure processing/homogenization). Depending on material influence principles, Gimenez-Escalante and Rahimifard (2019) classified novel nonthermal technologies in: physical (e.g., membrane technologies, nanotechnologies) and physicochemical (e.g., HHP, PEF, OMF, ultrasound, irradiation, UV and pulsed light, cold plasma technologies).

An accurate classification based on engineering principles is difficult because exist some NTP technologies, which can use a wide array of technical methods and operating conditions that can act in various ways on biological matrices. For instance, several technologies (e.g., plasma, ozonation, irradiation) are based on different mechanisms and methods, thus can be grouped at the same time in many classes. Ozonation technologies allow ozone formation through chemical or electrotechnologies. As Pankaj et al. (2018) reviewed, the plasma can be produced using various kinds of energy for the gas ionizing, namely, thermal, electrical, optical (UV light), X-ray electromagnetic radiation, and radioactive (gamma radiation), even if the electric or electromagnetic fields are widely used. Also, as Niemira (2012) pointed out, many cold plasma systems operate in a variety of conditions (e.g., pressure, temperature, and carrier gas) using different pieces of equipment (e.g., referring only plasma generator), and several mechanisms (e.g., chemical, radiation) by which cold plasma damage the microorganisms. On the other hand, because many of these emerging technologies are still under development, evaluation of their mechanisms, alternative issues, and other scientific-technical consequences are in progress and a complete understanding of these is not yet accomplished. For example, ohmic heating (OH) is considered a novel thermal technology, but a few research in pasteurization emphasized that OH has also a nonthermal, mild effect on microbial hazards due to the electric field (i.e., electroporation, electropermeabilization) (Sun et al. 2008; Park and Kang 2013). On the contrary, Schottroff et al. (2020), studying nonthermal effects against vegetative microbial populations by OH (of 12–300 kHz), concluded that, except for potential electrochemical reactions, pasteurization by ohmic heating basically follows the thermal process concepts. Consequently, research on assessment, differentiation, or experimental setup of the emerging nonthermal technologies proceeds further. Other matters in a systematic approach refer to their diversity and novelty. Nonetheless, the terminology on nonthermal technologies is also evolving, that is why some methods or technological variants are named differently without a unitary system of nomenclature (e.g., see HHP vs. HPP or UHP). In conclusion, at the moment, an in-depth taxonomy of emerging nonthermal technologies cannot be rigorously achieved because of manifold arisen questions that should be elucidated furthermore. However, based on literature review, a classification of these nonthermal technologies depending on engineering aspects on proposes, as follows:

- Electrotechnologies and electromagnetic technologies: pulsed electric fields (PEF), cold plasma technologies (CPT/e.g., atmospheric cold plasma CAP), magnetic fields (MF/e.g., OMF), electrohydrodynamic (EHD) processing, ionizing radiation (IOR/e.g., electron beam (EB) processing)
- Pressure-based technologies: high hydrostatic pressure (HHP/HPP), hydrodynamic pressure processing (HP homogenization – HPH), and carbon dioxide processes (e.g., supercritical fluid SCF processing)
- Mechanical technologies, caused by shock waves or hydrodynamic effects: ultrasound (US or ultrasonication), hydrodynamic cavitation
- Chemical technologies (e.g., through formation or interactions of radicals): ozonation
- Nanotechnologies

- Non-electro-technologies and others: UV light, pulsed light, photosensitization, membrane technologies (microfiltration, membrane separation, etc.), modified atmosphere, etc.

The electrotechnologies effectively act on biological matrices (i.e., microbial inactivation, biological cell electropermeabilization, etc.) due to matrix exposure to electrical fields, the operating conditions and experimental setup being of utmost importance in this regard (Toepfl et al. 2006; Niemira 2012; Schottroff et al. 2020); that is why different works propose treatment conditions in various range of electric field intensity and different energy input requirements. Pulsed electric fields consist of application the oscillating, bipolar, exponentially decaying, or electric pulses of high voltage (e.g., 20–80 kV/cm) to food placed between two electrodes (treatment time is from several nanoseconds to several milliseconds) (Alzamora et al. 2016; Zhang et al. 2019). Cold plasma (CPL) is a low-temperature non-thermal plasma (non-equilibrium plasma) generated at room temperature or nearly (Misra et al. 2016). Cold plasma technology (CPT) uses an ionized gas composed of a large number of charged species (i.e., ions, photons, electrons, free radicals, radiations within the UV wavelengths) (Priyadarshini et al. 2019). Plasma composition is influenced by methods used to ionize gases as regards the carrier gas, plasma generator, and operating conditions (Niemira 2012). For instance, electrical energy can ionize the gas for plasma production, gas being subjected to electric fields (with constant or alternating amplitude) existing between two electrodes (Sonawane et al. 2020). Depending on the pressure operating conditions, plasma can be grouped as high-pressure, atmospheric pressure (CAP), and low-pressure plasma. According to Pankaj et al. (2018), dielectric barrier discharge (DBD) and jet plasma devices are mostly compatible and commonly used with current food industry equipment. Electrohydrodynamic (EHD) caused by corona discharge refers to an airflow ionized locally near the corona electrode when a high voltage difference is produced between emitter and ground electrodes (Singh et al. 2012).

Magnetic fields (MF) are generated by either electric currents direct or alternating in conductors or permanent magnets (Miñano et al. 2020). They can be static magnetic field (SMF) or mobile/moderate magnetic field (MMF), oscillating magnetic field (OMF), or pulsed magnetic field (PMF) (Yıkmış 2016; Miñano et al. 2020).

Group of ionizing radiation (IOR) comprises electron beam (EB), X-rays, and gamma rays, in which EB uses high energy electrons unlike X- and gamma -rays (that employ energetic photons). Electron beam processing is generated electrically with an electron beam linear accelerator, in which electrons are accelerated to 99% of the speed of light and energies of up to 10 MeV (Hernández-Hernández et al. 2019). EB irradiation does not use a radioactive source (radioisotopes) as gamma rays radiation, although the process has a similar effect to that of gamma-ray irradiation. Irradiation refers to food exposed to radiation of various frequencies, ionizing energy being provided by any of the three sources: gamma-rays, X-rays, and EB. Food irradiation is performed with gamma rays generated using radionuclides such as Cobalt 60 or Cesium 137, and with X rays, which are produced by reflecting a high-energy stream of electrons off a target substance into food (Priyadarshini et al. 2019; Hernández-Hernández et al. 2019).

High hydrostatic pressure processing consists of the application of pressures spanning from 100 to 1000 MPa (typically, between 300 and 600 MPa) at a temperature in the range of 0–100 °C for a variable time (from seconds to about 20 min) (Barba et al. 2016; Alzamora et al. 2016). Dynamic high-pressure (i.e., high pressure homogenization HPH and (ultra)-high pressure homogenization UHPH) means processing with a pressurized liquid (obtained using a high-pressure generator (100–400 MPa) to be forced through a designed disruption valve (Zamora and Guamis 2015; Picart-Palmade et al. 2019). Carbon Dioxide Processes introduce pressurized carbon dioxide under fluid phase or supercritical state into a liquid or solid matrix and keep pressure/temperature ranges a sufficient period for treatment, according to its scope. Depending on operating conditions, these processes are known as Dense Phase Carbon Dioxide (DPCD), High Pressure Carbon Dioxide (HPCD), and Super Critical Carbon Dioxide (SC-CO_2) (Spilimbergo et al. 2013; Balaban and Duong 2014; Picart-Palmade et al. 2019; Yu et al. 2020). However, treatment of solid food with DPCD presents some difficulties (Balaban and Duong 2014).

Ultrasound (US) or ultrasonication are applied for fluid food processing with ultrasound waves at frequencies ranging from 18 to100 kHz, as low-intensity sonication and high-intensity (power ultrasound) depending on the ultrasound waves intensity (Yıkmış 2016; Hernández-Hernández et al. 2019). Ultrasounds may be generated using several methods (i.e., mechanically, electrically, optically, etc.) (Musielak et al. 2016). Hydrodynamic cavitation (HC) has been deemed as an alternative to acoustic cavitation (US) for exploring fluid food applications, as regards the intensification of several physical and chemical processes (Gogate 2011).

Ozonation represents a chemical treatment in which foodstuffs are exposed to ozone in a fluidic (gaseous or aqueous) state, ozone being produced by several methods (e.g., chemical, photochemical, electrical, etc.) (Brodowska et al. 2018).

Food nanotechnologies deal with size manipulation of the particles of cc. 1–100 nm to use these nanomaterials within the food matrix for any of enhancing food quality, safety, or shelf life, delivering nanoencapsulated food ingredients, and developing novel food with improved physicochemical, nutritive, etc., properties by matter manipulation at the nanoscale (Singh et al. 2017; Nazir and Azaz Ahmad Azad 2019).

Ultraviolet (UV) light technology works with radiations on the light spectrum (100–380 nm) of shortwaves UV-C, whereas pulsed light (PL) uses few flashes in the fraction of a second of intense pulses from a larger electromagnetic spectrum of 180–1100 nm (Priyadarshini et al. 2019). UV is mainly used for liquid food, whilst PL had better results on decontamination of food surfaces and packages.

Photosensitization is based on the usage of photosensitizers, light-sensitive molecules for microbial spoilage (Luksiene and Paskeviciute 2011).

Membrane technologies use a separation process with a selective barrier or interface between two phases; many advancements in design and membrane composition are in progress. Different membrane systems for ultrafiltration (UF), microfiltration (MF), reverse osmosis (RO), or nanofiltration (NF) are applied in nonthermal processing of liquid food (Echavarría et al. 2011; Picart-Palmade et al. 2019).

3 Prospects on Novel NTPs' Adoption Within the Sustainable Food Chain

Among emerging technologies, nonthermal ones appeared as an innovative alternative to traditional food processing for reducing harmful microbial populations and ensuring food preservation, with less damage on sensory attributes of the food products, and keep food nutritional value to a certain extent. During research on nonthermal technologies under development, the interest focused also on improving the physical and chemical properties of food, facilitate food processing and optimize processes, enhance quality attributes of food, keep as much as possible product integrity and stability, develop novel packaging systems for convenience foodstuffs, high valorization of bioactive compounds or other food ingredients, develop novel food, nutraceuticals, dietetic food, etc. (see also Table 2.1). High pressure treatment, gamma irradiation, high electric field pulses are a few nonthermal processing methods successfully used in minimal processing applications (Ohlsson 1994). Also, Barbosa-Cánovas et al. (2005) emphasized the use of nonthermal technologies in food processing in combinations based on the "hurdle" concept, starting from the idea that each emerging food processing technology has its own advantages and limitations, and some nonthermal technologies fit better or is applicable with one food type than another.

"Hurdle" approach materialized in combinations of nonthermal emerging technologies proved to have a double outcome, namely, on one hand, can have synergistic or additive operational effects related to scope or technological purpose, and, on the other hand, may counterbalance drawbacks of individual technologies applied. The combination of ultrasound with UV light (Antonio-Gutiérrez et al. 2017), a pulsed electric field with OMF (Mok et al. 2017), high-pressure homogenization with nanotechnologies (Ruiz-Montañez et al. 2017), high pressure (HHP), PEF, and sonication (Pyatkovskyy et al. 2018) are only a few examples of this view.

As shown, innovative sustainable food processing using nonthermal technologies tries to respond to social needs and solve manifold technical problems, known and unpredictable. The steps from scientific ideas to a novel technology going through all technology readiness levels until commercialization are numerous. As Picart-Palmade et al. (2019) observed, although implementation of NTPs and technology transfer from one field (e.g., membranes and PEF) to another is promising for future sustainable food applications, the progress registered in industrial scaling-up implementation of NTPs differs from one technology to another. Also, novel technologies that have the potential to be commercialized need capital for adequate industrial equipment, one of the actual limitations being high investment costs for nonthermal technologies implementation (Shabbir et al. 2021; Khouryieh 2020). Besides capital and operational costs, scalability and flexibility seem to be other criteria for the food manufacturers to adopt novel food processing technologies. Attention given to high-pressure processing industrial applications is obvious when HPP has had exponential growth in the food industry worldwide since the 1990s. For example, several authors commented on how food manufacturers commercialized high-pressure processing and mapped worldwide HPP applications and scalability by food sectors from 2008 to 2016 with a forecast until 2026 (Tonello 2010; Barba et al. 2016; Priyadarshini et al. 2019). In order to faster start-up and reduce long-term investment, toll-processing business models became attractive. For example, Hernández-Hernández et al. (2019) presented how food manufacturers accessed industrial HPP toll-processing in different countries and a few obstacles related to industrial processing PEF implementation. Nonetheless, economic reasons as the increase in revenue and productivity, and profitably placing the novel foodstuffs in the market are important drivers for the commercialization NTPs by food manufacturers. For a better illustration of the importance of innovative processing along the food chain and the adoption of emerging nonthermal technologies in the spirit of sustainability, a schematic representation is depicted in Fig. 2.1.

Increasing reliability and cost-efficiency become other challenges for scaling up NTPs at the industrial level, which concentrate various innovations and efforts to fine-tuning processing equipment, operations optimization, and suitable control of nonthermal processes implemented along the food supply chain. Reengineering and integrated automation systems using Information and Communication Technologies (ICTs) are drivers for the next generation of advanced agricultural and food industry systems that will allow modeling new food business using NTPs within Food industry 4.0 for lasting development and industrial competitiveness.

Nonthermal food technologies, either novel or emerging, under development, stand in a continuous process of evaluation for process optimization, productivity improvement, cut costs, and environmental assessment. Based on life-cycle assessment (LCA), many nonthermal emerging food technologies proved to be eco-friendly as regards the energy and water savings, emissions decrement, and other environmental indicators (Pereira and Vicente 2010). For example, Arnal et al. (2018) applied PEF technology as a pre-treatment for tomato peeling in the canning industry and determined an energy saving of about 20%, as well as the decreasing of fossil, water, and ozone depletion with cc. 18–20% in comparison with conventional tomato peeling pre-treatment.

Anyway, before 2016, many studies on the environmental impact of nonthermal technologies were performed at the laboratory or pilot scale, where different NTPs proved their added value in terms of processing, energy-saving, or waste decrement. Režek Jambrak et al. (2018) reviewed several difficulties in environmental impact assessment of NTP technologies, regarding inventory data, scope, scalability, and processing units, etc. But then, Barba et al. (2016) observed gaps in systematic cost-benefit analysis concerning the inputs for commercialization of emerging NTP technologies and industrial outputs. On a case study of orange juice pasteurization using nonthermal technologies of HHP and PEF compared to thermal pasteurization,

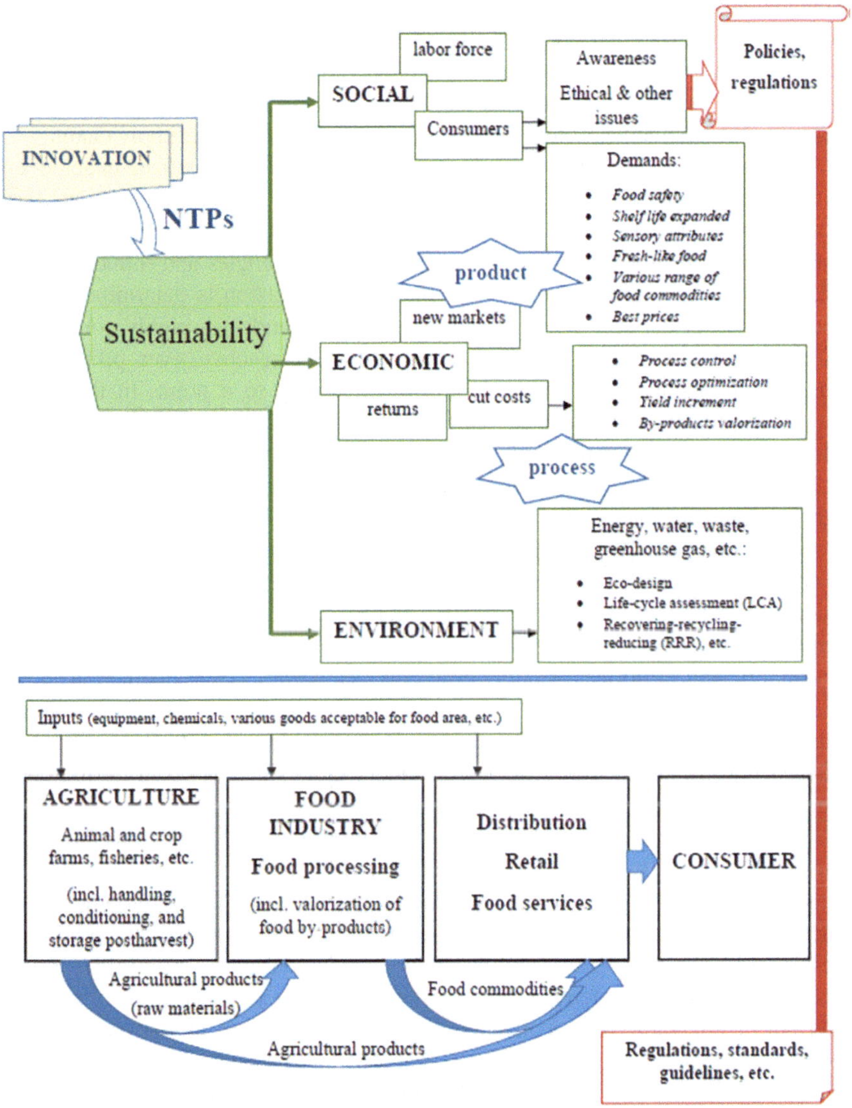

Fig. 2.1 Prospects on novel NTPs' adoption along the sustainable food supply chain

Barba et al. (2016) concluded that the overall cost of processing (calculated with capital, labor, and energy costs-based) was a few times-fold higher for NTPs than thermal pasteurization that means a higher price of end-product obtained by emerging technologies. Similar results were observed in a case study of milk processing between PEF and some thermal processes (Barba et al. 2016). In conclusion, at this stage, even environmentally attractive, in terms of efficiency, many emerging technologies are still expensive enough to be implemented in comparison with beneficial consequences of energetic optimization, waste management, or reduced costs by saving water, decrease waste, etc. But, as Picart-Palmade et al. (2019) emphasized, further holistic and cross-sectoral evaluations of novel NTPs should be considered to assess the potentialities and drawbacks of each emerging nonthermal technology implemented within the sustainable innovation food supply chain.

As shown in Fig. 2.1, the social dimension plays an important role in achieving sustainability. From a market perspective, stakeholders whose commitments have increased over the last two decades are final consumers, as their behavior determines whether or not they accept novel food or technologies and influences final consumption. According to marketing, purchasing decision is determined by the food product, price, convenience, and other factors, and in a highly competitive market as the food one is, food producers should be reactive, even proactive, to consumer demands, taking into account those for up to a point. In this sense, Khouryieh (2020) conducted a survey to food experts on the use of new NTP technologies in the US and concluded that food companies, motivated to meet consumer needs, chose to use new thermal technologies mainly for reasons, as follows: better nutritional and sensory quality (71.1%), improved shelf life (39.3%), and solutions for safety issues (25.4%). Khouryieh (2020) also pointed out that food manufacturers decide or prefer to implement a certain type of nonthermal technology in a direct relationship with these reasons (e.g., HHP or PEF for better nutritional and sensory quality, and increase the shelf life whilst pulsed light for a better nutrient and sensory quality, and solutions for safety problems). However, a clear distinction should be done between the acceptance of these novel technologies between food scientists and common consumers. Also, as Jaeger et al. (2015) pointed out, a difference between how scientists and regular consumers understand the NTP terminology (e.g., minimally processing, cold preservation, etc.) is.

As mentioned before, end-consumers have more demands related to food attributes and food safety, clean label, attractive and resilient food packaging, a wide array of foodstuffs through range diversification or new product development (e.g., functional food, chemical-free, dietetic food, etc.), food commodities with higher sensory characteristics, even food products customized according to their preferences. So, consumer preferences and food choices are expressed in favor of novel food but, at the same time, consumers seem to react differently when discusses novel technologies. For example, Sonne et al. (2012) performed qualitative research on consumer attitudes of HPP and PEF apple juices in four European countries and observed that although the subjects liked attributes of new products more than of conventional heat-processed ones, some consumers expressed concerns about PEF technology associated with potential food safety risks. Also, according to Hassoun et al. (2020), a major issue related to consumer perception and doubts toward novel technologies refer to food safety. Similarly, Hernández-Hernández et al. (2019) commented on consumer acceptance of food processed employing irradiation (IOR), although adequate food legislation was put in force to ensure food safety in this respect. Consequently, consumer awareness about emerging technologies may be perceived within market risks and a suitable and transparent communication about novel technologies towards end-consumers seems to be a key to increase consumer acceptance on the food market further (Barba et al. 2016). Nonetheless, besides these food concerns, there are many other triggers and obstacles that emerging food technologies face in gaining consumer acceptance (Ronteltap et al. 2007). And in purpose to anticipate consumer behavior towards future innovative processing, different tools as the framework of "distal" and "proximal" factors of Ronteltap et al. (2007) or Consumer Value (CV) framework (Perrea et al. 2015) were proposed for exploring the consumer acceptance, and ensure the marketability of the food product processed by emerging technologies. In conclusion, consumer acceptance is a criterion aiming to determine the social sustainability of the emerging nonthermal technologies onwards.

4 Conclusion

The lasting development is in-depth connected to the innovation the steps from eco-innovation to responsible innovation, being in progress of integration into the food supply chain for sustainability achieving nowadays. As innovation type, technological ones are in the top-4 of sustainable innovation with increasing applications in the agriculture and food industry. From those, innovative processing as a "hurdle" approach and nonthermal processing are key enable food technologies that can contribute to sustainable food chain value to a certain extent.

An overview of applications of the innovative nonthermal and minimal thermal processing technologies on the food supply chain has been introduced. Despite difficulties for an accurate categorizing, a classification of the emerging nonthermal processing food technologies is proposed. Also, several relations between innovative NTPs and social, economic, and environmental dimensions of the sustainable food supply chain are described. Some emerging nonthermal food processing technologies have already proved their efficiency, e.g., in terms of food safety and shelf life of agro-food commodities compared to conventional technologies. Nowadays, consumer-oriented innovation focuses on the "hurdle concept" and choosing green nonthermal technologies. Due to unknown adverse effects or unforeseen potential hazards of nonthermal emerging technologies in food processing, sometimes question marks about consumer acceptance have arisen, but through appropriate communication, marketability of the food products processed by nonthermal emerging technologies can be sustainably accomplished.

References

Alarcon-Rojo AD, Janacua H, Rodriguez JC, Paniwnyk L, Mason TJ (2015) Power ultrasound in meat processing. Meat Sci 107:86–93. https://doi.org/10.1016/j.meatsci.2015.04.015

Alzamora SM, López-Malo A, Tapia MS, Welti-Chanes J (2016) Minimally processed foods. In: Caballero B, Finglas PM, Toldrá F (eds) Encyclopedia of food and health. Academic Press, pp 767–771. https://doi.org/10.1016/B978-0-12-384947-2.00470-0

Antonio-Gutiérrez O, López-Malo A, Ramírez-Corona N, Palou E (2017) Enhancement of UVC-light treatment of tangerine and grapefruit juices through ultrasonic atomization. Innovative Food Sci Emerg Technol 39:7–12. https://doi.org/10.1016/j.ifset.2016.10.019

Arnal ÁJ, Royo P, Pataro G, Ferrari G, Ferreira VJ, López-Sabirón AM, Ferreira GA (2018) Implementation of PEF treatment at real-scale tomatoes processing considering LCA methodology as an innovation strategy in the agri-food sector. Sustainability 10(4):979–996. https://doi.org/10.3390/su10040979

Balaban MO, Duong T (2014) Dense phase carbon dioxide research: current focus and directions. Agric Agric Sci Procedia 2:2–9. https://doi.org/10.1016/j.aaspro.2014.11.002

Barba FJ, Orlien V, Mota MJ, Lopes RP, Pereira SA, Saraiva JA (2016) Chapter 7 – Implementation of emerging technologies. In: Galanakis CM (ed) Innovation strategies in the food industry. Academic Press, pp 117–148. https://doi.org/10.1016/B978-0-12-803751-5.00007-6

Barbosa-Cánovas GV, Góngora-Nieto MM, Rodriguez JJ, Swanson BG (2005) Nonthermal processing of foods and emerging technologies. In: Barbosa-Cánovas GV (ed) Food engineering. Encyclopedia of life support sciences. EOLSS Publishers/UNESCO, Paris, pp 575–593

Bautista-Ortín AB, Jiménez-Martínez MD, Jurado R, Iniesta JA, Terrades S, Andrés A et al (2017) Application of high-power ultrasounds during red wine vinification. Int J Food Sci Technol 52(6):1314–1323. https://doi.org/10.1111/ijfs.13411

Beske P, Land A, Seuring S (2014) Sustainable supply chain management practices and dynamic capabilities in the food industry: a critical analysis of the literature. Int J Prod Econ 152:131–143. https://doi.org/10.1016/j.ijpe.2013.12.026

Brodowska AJ, Nowak A, Śmigielski K (2018) Ozone in the food industry: principles of ozone treatment, mechanisms of action, and applications: an overview. Crit Rev Food Sci Nutr 58(13):2176–2201. https://doi.org/10.1080/10408398.2017.1308313

Chao D, He R, Jung S, Aluko RE (2013) Effect of pressure or temperature pretreatment of isolated pea protein on properties of the enzymatic hydrolysates. Food Res Int 54:1528–1534

Chávez-Martínez A, Reyes-Villagrana RA, Rentería-Monterrubio AL, Sánchez-Vega R, Tirado-Gallegos JM, Bolivar-Jacobo NA (2020) Low and high-intensity ultrasound in dairy products: applications and effects on physicochemical and microbiological quality. Foods 9(11):1688. https://doi.org/10.3390/foods9111688. PMCID: PMC7698897

Defraeye T, Martynenko A (2018) Electrohydrodynamic drying of food: new insights from conjugate modeling. J Clean Prod 198:269–284. https://doi.org/10.1016/j.jclepro.2018.06.250

Echavarría AP, Torras C, Pagán J, Ibarz A (2011) Fruit juice processing and membrane technology application. Food Eng Rev 3:136–158. https://doi.org/10.1007/s12393-011-9042-8

Farkas J (2006) Irradiation for better foods. Trends Food Sci Technol 17(4):148–152. https://doi.org/10.1016/j.tifs.2005.12.003

Gimenez-Escalante P, Rahimifard S (2019) A methodology to assess the suitability of food processing technologies for distributed localised manufacturing. Sustainability 11(12):3383. https://doi.org/10.3390/su11123383

Gogate PR (2011) Hydrodynamic cavitation for food and water processing. Food and Bioproc Technol 4(6):996–1011. https://doi.org/10.1007/s11947-010-0418-1

Gómez B, Munekata PES, Gavahian M, Barba FJ, Martí-Quijal FJ, Bolumar T et al (2019) Application of pulsed electric fields in meat and fish processing industries: an overview. Food Res Int 123:95–105. https://doi.org/10.1016/j.foodres.2019.04.047

Hassoun A, Ojha S, Tiwari B, Rustad T, Nilsen H, Heia K et al (2020) Monitoring thermal and non-thermal treatments during processing of muscle foods: a comprehensive review of recent technological advances. Appl Sci 10(19):6802. https://doi.org/10.3390/app10196802

Hausmann A, Sanciolo P, Vasiljevic T, Kulozik U, Duke M (2014) Performance assessment of membrane distillation for skim milk and whey processing. J Dairy Sci 97:56–71. https://doi.org/10.3168/jds.2013-7044

Hernández-Hernández HM, Moreno-Vilet L, Villanueva-Rodríguez SJ (2019) Current status of emerging food processing technologies in Latin America: novel non-thermal processing. Innovative Food Sci Emerg Technol 58:102233. https://doi.org/10.1016/j.ifset.2019.102233

Hertwig C, Meneses N, Mathys A (2018) Cold atmospheric pressure plasma and low energy electron beam as alternative nonthermal decontamination technologies for dry food surfaces: a review. Trends Food Sci Technol 77:131–142. https://doi.org/10.1016/j.tifs.2018.05.011

Jaeger H, Knorr D, Szabó E, Hámori J, Bánáti D (2015) Impact of terminology on consumer acceptance of emerging technologies through the example of PEF technology. Innovative Food Sci Emerg Technol 29:87–93. https://doi.org/10.1016/j.ifset.2014.12.004

Katariya P, Arya SS, Pandit AB (2020) Novel, non-thermal hydrodynamic cavitation of orange juice: effects on physical properties and stability of bioactive compounds. Innovative Food Sci Emerg Technol 62:102364. https://doi.org/10.1016/j.ifset.2020.102364

Keyser M, Műller IA, Cilliers FP, Nel W, Gouws PA (2008) Ultraviolet radiation as a non-thermal treatment for the inactivation of microorganisms in fruit juice. Innovative Food Sci Emerg Technol 9(3):348–354. https://doi.org/10.1016/j.ifset.2007.09.002

Khan I, Tango CN, Miskeen S, Lee BH, Oh D-H (2017) Hurdle technology: a novel approach for enhanced food quality and safety – a review. Food Control 73(B):1426–1444. https://doi.org/10.1016/j.foodcont.2016.11.010

Khouryieh HA (2020) Novel and emerging technologies used by the U.S. food processing industry. Innov Food Sci Emerg Technol 67:102559. https://doi.org/10.1016/j.ifset.2020.102559

Kneipp JM, Gomes CM, Bichueti RS, Frizzo K, Perlin AP (2019) Sustainable innovation practices and their relationship with the performance of industrial companies. Rev Gestão 26(2):94–111. https://doi.org/10.1108/REGE-01-2018-0005

Li G, Zhu F (2018) Effect of high pressure on rheological and thermal properties of quinoa and maize starches. Food Chem 241:380–386

López-Pedrouso M, Díaz-Reinoso B, Lorenzo JM, Cravotto G, Barba FJ, Moure A et al (2019) Chapter 3 – Green technologies for food processing: principal considerations. In: Barba FJ, Saraiva JMA, Cravotto G, Lorenzo JM (eds) Innovative thermal and non-thermal processing, bioaccessibility and bioavailability of nutrients and bioactive compounds, Woodhead Publishing series in food science, technology and nutrition. Woodhead Publishing, pp 55–103. https://doi.org/10.1016/B978-0-12-814174-8.00003-2

Luksiene Z, Brovko L (2013) Antibacterial photosensitization-based treatment for food safety. Food Eng Rev 5:185–199. https://doi.org/10.1007/s12393-013-9070-7

Luksiene Z, Paskeviciute E (2011) Novel approach to decontaminate food-packaging from pathogens in non-thermal and not chemical way: chlorophyllin-based photosensitization. J Food Eng 106(2):152–158. https://doi.org/10.1016/j.jfoodeng.2011.04.024

Miñano HLA, Silva ACS, Souto S, Costa EJX (2020) Magnetic fields in food processing perspectives, applications and action models. Processes 8(7):814. https://doi.org/10.3390/pr8070814

Misra NN, Pankaj SK, Segat A, Ishikawa K (2016) Cold plasma interactions with enzymes in foods and model systems. Trends Food Sci Technol 55:39–47. https://doi.org/10.1016/j.tifs.2016.07.001

Mok JH, Her JY, Kang T, Hoptowit R, Jun S (2017) Effects of pulsed electric field (PEF) and oscillating magnetic field (OMF) combination technology on the extension of supercooling for chicken breasts. J Food Eng 196:27–35. https://doi.org/10.1016/j.jfoodeng.2016.10.002

Musielak G, Mierzwa D, Kroehnke J (2016) Food drying enhancement by ultrasound – a review. Trends Food Sci Technol 56:126–141. https://doi.org/10.1016/j.tifs.2016.08.003

Nazir S, Azaz Ahmad Azad ZR (2019) Food nanotechnology: an emerging technology in food processing and preservation. In: Malik A, Erginkaya Z, Erten H (eds) Health and safety aspects of food processing technologies. Springer, Cham, pp 567–576. https://doi.org/10.1007/978-3-030-24903-8_20

Niemira BA (2012) Cold plasma decontamination of foods. Annu Rev Food Sci Technol 3:125–142. https://doi.org/10.1146/annurev-food-022811-101132

Ohlsson T (1994) Minimal processing-preservation methods of the future: an overview. Trends Food Sci Technol 5(11):341–344. https://doi.org/10.1016/0924-2244(94)90210-0
Pankaj SK, Wan Z, Keener KM (2018) Effects of cold plasma on food quality: a review. Foods 7(1):4. https://doi.org/10.3390/foods7010004. PMC5789267
Park IK, Kang DH (2013) Effect of electropermeabilization by ohmic heating for inactivation of *Escherichia coli* O157:H7, *Salmonella enterica* serovar Typhimurium, and *Listeria monocytogenes* in buffered peptone water and apple juice. Appl Environ Microbiol 79(23):7122–7129. https://doi.org/10.1128/AEM.01818-13. PMID: 23995939
Pereira RN, Vicente AA (2010) Environmental impact of novel thermal and non-thermal technologies in food processing. Food Res Int 43(7):1936–1943. https://doi.org/10.1016/j.foodres.2009.09.013
Perrea T, Grunert KG, Krystallis A (2015) Consumer value perceptions of food products from emerging processing technologies: a cross-cultural exploration. Food Qual Prefer 39:95–108. https://doi.org/10.1016/j.foodqual.2014.06.009
Picart-Palmade L, Cunault C, Chevalier LD, Belleville MP, Marchesseau S (2019) Potentialities and limits of some non-thermal technologies to improve sustainability of food processing. Front Nutr 5:130. https://doi.org/10.3389/fnut.2018.00130
Priyadarshini A, Rajauria G, O'Donnell CP, Tiwari BK (2019) Emerging food processing technologies and factors impacting their industrial adoption. Crit Rev Food Sci Nutr 59(19):3082–3101. https://doi.org/10.1080/10408398.2018.1483890
Pyatkovskyy TI, Shynkaryk MV, Mohamed HM, Yousef AE, Sastry SK (2018) Effects of combined high pressure (HPP), pulsed electric field (PEF) and sonication treatments on inactivation of *Listeria innocua*. J Food Eng 233:49–56. https://doi.org/10.1016/j.jfoodeng.2018.04.002
Rajeev A, Pati RK, Padhi SS, Govindan K (2017) Evolution of sustainability in supply chain management: a literature review. J Clean Prod 162:299–314. https://doi.org/10.1016/j.jclepro.2017.05.026
Režek Jambrak A, Vukušić T, Donsi F, Paniwnyk L, Djekic I (2018) Three pillars of novel nonthermal food technologies: food safety, quality, and environment. J Food Qual 2018:8619707, 18 p. https://doi.org/10.1155/2018/8619707
Ronteltap A, van Trijp JCM, Renes RJ, Frewer LJ (2007) Consumer acceptance of technology-based food innovations: lessons for the future of nutrigenomics. Appetite 49(1):1–17. https://doi.org/10.1016/j.appet.2007.02.002
Ruiz-Montañez G, Ragazzo-Sanchez JA, Picart-Palmade L, Calderón-Santoyo M, Chevalier-Lucia D (2017) Optimization of nanoemulsions processed by high-pressure homogenization to protect a bioactive extract of jackfruit (*Artocarpus heterophyllus* Lam). Innovative Food Sci Emerg Technol 40:35–41. https://doi.org/10.1016/j.ifset.2016.10.020
Sakdatorn V, Thavarungkul N, Srisukhumbowornchai N, Intipunya P (2018) Improvement of rheological and physicochemical properties of longan honey by non-thermal magnetic technique. Int J Food Sci Technol 53(7):1717–1725. https://doi.org/10.1111/ijfs.13756
Schottroff F, Biebl D, Gruber M, Burghardt N, Schelling J, Gratz M et al (2020) Inactivation of vegetative microorganisms by ohmic heating in the kilohertz range – evaluation of experimental setups and non-thermal effects. Innovative Food Sci Emerg Technol 63:102372. https://doi.org/10.1016/j.ifset.2020.102372
Shabbir, MA, Ahmed H, Maan AA, Rehman A, Afraz MT, Iqbal MW, et al (2021). Effect of non-thermal processing techniques on pathogenic and spoilage microorganisms of milk and milk products. Food Sci Technol (Campinas) 41(2):279–294. https://doi.org/10.1590/fst.05820
Singh A, Orsat V, Raghavan V (2012) A comprehensive review on electrohydrodynamic drying and high-voltage electric field in the context of food and bioprocessing. Dry Technol 30(16):1812–1820. https://doi.org/10.1080/07373937.2012.708912
Singh T, Shukla S, Kumar P, Wahla V, Bajpai VK, Rather IA (2017) Application of nanotechnology in food science: perception and overview. Front Microbiol 8:1501. https://doi.org/10.3389/fmicb.2017.01501

Sonawane SK, Marar T, Patil S (2020) Non-thermal plasma: an advanced technology for food industry. Food Sci Technol Int 26(8):727–740. https://doi.org/10.1177/1082013220929474. PMID: 32501116

Sonne A-M, Grunert KG, Veflen Olsen N, Granli B-S, Szabó E, Banati D (2012) Consumers' perceptions of HPP and PEF food products. Br Food J 114(1):85–107. https://doi.org/10.1108/00070701211197383

Spilimbergo S, Komes D, Vojvodic A, Levaj B, Ferrentino G (2013) High pressure carbon dioxide pasteurization of fresh-cut carrot. J Supercrit Fluids 79:92–100. https://doi.org/10.1016/j.supflu.2012.12.002

Stilgoe J, Owen R, Macnaghten P (2013) Developing a framework for responsible innovation. Res Policy 42(9):1568–1580. https://doi.org/10.1016/j.respol.2013.05.008

Sun H, Kawamura S, Himoto J-I, Itoh K, Wada T, Kimura T (2008) Effects of ohmic heating on microbial counts and denaturation of proteins in milk. Food Sci Technol Res 14(2):117–123. https://doi.org/10.3136/fstr.14.117

Sunil, Chauhan N, Singh J, Chandra S, Chaudhary V, Kumar V (2018) "Non-thermal techniques: application in food industries". A review. J Pharmacogn Phytochem 7(5):1507–1518. Available from: https://www.phytojournal.com/archives/2018/vol7issue5/PartZ/7-4-659-545.pdf. [cited 25 Mar 2021]

Tebaldi L, Bigliardi B, Bottani E (2018) Sustainable supply chain and innovation: a review of the recent literature. Sustainability 10(11):3946. https://doi.org/10.3390/su10113946

Toepfl S, Mathys A, Heinz V, Knorr D (2006) Review: potential of high hydrostatic pressure and pulsed electric fields for energy efficient and environmentally friendly food processing. Food Rev Intl 22(4):405–423. https://doi.org/10.1080/87559120600865164

Tonello C (2010) Chapter 4 – Case studies on high-pressure processing of foods. In: Zhang HQ, Barbosa-Cánovas GV, Balasubramaniam VM, Dunne CP, Farkas DF, Yuan JTC (eds) Nonthermal processing technologies for food. Wiley-Blackwell, pp 36–50. https://doi.org/10.1002/9780470958360.ch4

van den Oord A, van Witteloostuijn A (2018) A multi-level model of emerging technology: an empirical study of the evolution of biotechnology from 1976 to 2003. PLoS One 13(5):0197024. https://doi.org/10.1371/journal.pone.0197024

Yaldagard M, Mortazavi SA, Tabatabaie F (2008) The principles of ultra high pressure technology and its application in food processing/preservation: a review of microbiological and quality aspects. Afr J Biotechnol 7(16):2739–2276

Yıkmış S (2016) New approaches in non-thermal processes in the food industry. Int J Nutr Food Sci 5(5):344–351. https://doi.org/10.11648/j.ijnfs.20160505.15

Yu T, Niu L, Iwahashi H (2020) High-pressure carbon dioxide used for pasteurization in food industry. Food Eng Rev 12:364–380. https://doi.org/10.1007/s12393-020-09240-1

Zamora A, Guamis B (2015) Opportunities for ultra-high-pressure homogenisation (UHPH) for the food industry. Food Eng Rev 7:130–142. https://doi.org/10.1007/s12393-014-9097-4

Zhang Z-H, Wang L-H, Zeng X-A, Han Z, Brennan CS (2019) Non-thermal technologies and its current and future application in the food industry: a review. Int J Food Sci Technol 54(1):1–13. https://doi.org/10.1111/ijfs.13903

Part II
Mechanism of Action of Nonthermal Processing Technologies (NTP)

Chapter 3
Engineering and Nonthermal Technologies: Process Optimization Through Kinetic Modelling

George Katsaros, Varvara Andreou, and Marianna Giannoglou

1 Introduction

Food research and innovation during the last 50 years has significantly offered in the technological advancement. Significant effort has been targeted on the production of safer, of higher quality, of improved nutritional profile foods, as well as in the development of new novel food products that could not be produced, or their safety and stability could not be ensured by the conventional equipment and knowledge. Novel Nonthermal technologies in food processing are among the technology advancements. Their aim is mainly to substitute conventional thermal treatment applied for the pasteurization and sterilization of foods, while simultaneously retaining their quality and nutritional characteristics. High pressure (HP), Pulsed Electric fields (PEF), Pulsed Electromagnetic Fields (PEMF), Cold Atmospheric Plasma (CAP) and Osmotic Dehydration (OD) are the most important novel technologies studied and (some of them) efficiently applied in an industrial environment.

Although the research activity on novel Nonthermal technologies has been continuously growing the last decades (Fig. 3.1) resulting in production of data, thus continuously growing application of some of the novel technologies, some scientific, technological, and technical issues should be answered. Food industry demands fully documented and validated answers with regards to the applicability and the benefits of all these technologies, as physical processes of food preservation. Process optimization (selection of most appropriate process conditions) for the production of safer, high quality, nutritious, cost effective and consumer acceptable products is essential. The implementation of the scientific achievements for an effective process is based on the kinetic approach of the destructive reactions of several factors that

G. Katsaros (✉) · V. Andreou · M. Giannoglou
Laboratory of Food Processing & Engineering, Institute of Technology of Agricultural Products, Hellenic Agricultural Organization–DEMETER, Attica, Greece
e-mail: gkats@chemeng.ntua.gr; vandreou@chemeng.ntua.gr; giannoglou@chemeng.ntua.gr

A. Režek Jambrak (ed.), *Nonthermal Processing in Agri-Food-Bio Sciences*,
Food Engineering Series, https://doi.org/10.1007/978-3-030-92415-7_3

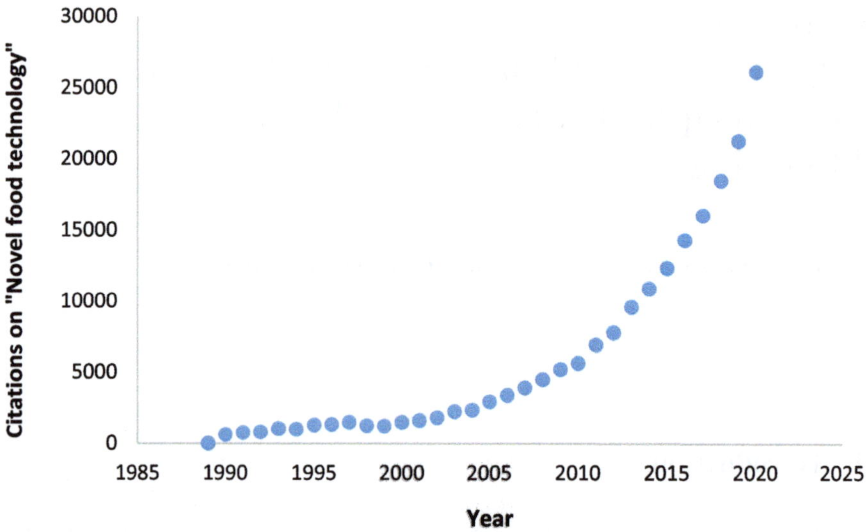

Fig. 3.1 Number of cited papers in the literature through years (keywords: novel and food and technology)

lead to spoilage or degradation of food, during the novel technologies' treatments. A systematic study of the technical parameters would allow for food engineers to predict and evaluate for the process applied, its intensity and efficiency, enabling the use of corrective actions if needed, for production of safe, sustainable, and cost-efficient final food products.

Although the novel technologies are aimed to substitute conventional thermal treatment, there is still research being conducted even for the well-studied thermal processing that has only two process parameters (temperature and time). For all the other Nonthermal technologies, that the number of their variables is higher (Table 3.1), it is evident that for the production of data, increased number of experiments has to be conducted.

Depending on the product to be treated, the dependent variables need to be named and studied as a function of the independent variables of each technology applied. The most important quality and safety parameters to be studied are the pathogens inactivation, the spoilage microflora inactivation, the endogenous deteriorative enzymes inactivation and the non-alteration of colour, texture, pH value, nutritional characteristics and finally organoleptic acceptability. The mathematical description of the effect of independent variables (process parameters) on the safety and quality indices of the treated foods is of high importance since allows for the food engineer to optimize the process; mathematical equations describe the effect of the process parameters values on selected quality indices. Thus, food engineers save time and money by not applying the trial and error approach for the selection of the appropriate conditions for the Nonthermal treatment.

Table 3.1 Number of independent variables for each process nonthermal technology

Technology	Number of variables	Independent variables
High pressure	3	P: Pressure T: Temperature t: Treatment time
Pulsed electric fields	7	E: Electric field strength t: Treatment time s: Shape of pulse w: Width of pulse f: Frequency Q: Specific energy T: Temperature
Pulsed electromagnetic fields	7	M: Magnetic field strength t: Treatment time s: Shape of pulse w: Width of pulse f: Frequency Q: Specific energy T: Temperature
Cold atmospheric plasma	9	F: Flow of the gas T: Temperature t: Treatment time l: Gap distance between electrode and food vol: Volume of food V: Voltage s: Shape of pulse w: Width of pulse f: Frequency
Osmotic dehydration	6	t: Treatment time T: Temperature $m_{solution}/m_{product}$: Solution-to-product mass ratio Type of osmotic solutes %C: Concentration of the osmotic solution Type and level of agitation

For all thermal and Nonthermal technologies, the mathematical description of the effect of process time on any food quality index is characterized as primary modelling. Within this approach, all other independent variables of each technology are kept constant, thus the effect on any chemical change, microbial population or enzyme activity is described by zero order, first order, second order or n^{th} order kinetics (Table 3.2) (Labuza 1984; Saguy and Karel 1980). First-order kinetics is the one mostly applied in nonthermal processing for the estimation of the rate constant of a quality index alteration vs process time and will further be discussed within this chapter.

The primary models are useful when the processing conditions (values of the independent variables for each Nonthermal technology) are kept constant. For processing conditions that are changed, new experiments should be performed to

Table 3.2 Reaction order kinetics applied for the mathematical description of the effect of process time on any microbial, enzymatic or chemical index

	Zero order	First order	Second order	nth order
Rate law	$-d[C]/dt = k$	$-d[C]/dt = k[C]$	$-d[C]/dt = k[C]^2$	$-d[C]/dt = k[C]^n$
Integrated rate law	$[C] = [C]_0 - kt$	$[C] = [C]_0\, e^{-kt}$	$1/[C] = 1/[C]_0 + kt$	$[C]^{n-1} = [C]_0^{n-1} - (n - 1)kt$

estimate the corresponding new primary model parameters. For a number of different process conditions studied with simultaneous application of primary models for each alteration of the process conditions, new mathematical equations known as secondary models can be applied to express the independent variables parameters effect on the predicted primary model parameters. These equations are the result of theoretical considerations or empirical observations and in most cases are nonlinear.

2 High Pressure Processing

The initial interest of the scientific community in the application of High pressure (HP) technology included inactivation of microorganisms such as bacterial stem cells, yeasts and fungi (Hoover et al. 1989; Patterson et al. 1997; Smelt 1998). The main target was, through HP processing optimization, to increase foods shelf-life by inactivating spoilage and pathogenic bacteria (Bayindirli et al. 2006; Donaghy et al. 2007; Jofré et al. 2008). HP causes microorganisms cellular membrane damage and is mainly reported for pathogenic and spoilage bacteria vegetative forms when food is treated with pressures higher than 200 MPa, where irreversible protein/enzyme denaturation and intracellular content leakage occur. The effects of HP on enzymes (mainly pectin methylesterase, polyphenoloxidase and peroxidases) has also been reported in a large number of studies cited in the literature (Irwe and Olsson 1994; Seyderhelm et al. 1996; Butz et al. 1996; Weemaes et al. 1997; Ludikhuyze et al. 1996, 1998, 2000; Indrawati et al. 2000; Nienaber and Shellhammer 2001; Moatsou et al. 2008a, b; Eylen et al. 2008; Katsaros et al. 2006, 2017; Boulekou et al. 2010). The conclusion of the studies is that HP significantly affects enzymes and enzymatic reactions, allowing for the development of an alternative to thermal treatment preservation technique that leads to reduced undesirable changes in the texture and sensory characteristics of a food product. The enzyme inactivation is mainly attributed to changes in the secondary and tertiary structure of enzymes occurring at specific pressures (Giannoglou et al. 2016). HP process of pressurization is based on the principle of "Le Chatelier," inducing a reduction in molecular volume and, consequently exponentially accelerating the occurrence of reactions favoured by pressure. Thus, the rates of the chemical or physical reactions resulting in lower volume products are accelerated by HP, whereas the reactions that result in an increase in the total volume are retarded. Based on the literature, the results show that different microorganisms and enzymes, exhibit different kinetic inactivation or activation with HP.

The observation from the researchers that the effect of HP processing, on a particular characteristic, over time follows a specific trend that could be kinetically expressed through mathematical modelling, led to the need mathematical models to be developed in order to predict the impact of HP processing on a specific characteristic (e.g., activity of enzymes, microorganisms etc). A mathematical model predicting the change of a target parameter as a function of time (primary model) is an essential tool when designing HP experiments and industrial processes. Furthermore, the application of a primary model can be extended if a secondary model describing the pressure dependence of the primary model parameters is available too. Below the most used primary and secondary mathematical models by various researchers in the field of HP processing are presented, to describe and predict the effect of HP on different microorganisms, enzymes and quality indices of various food systems.

2.1 Primary Mathematical Modeling

The main equations reported in the literature describing the effect of treatment time at constant pressure and temperature on the microbial, enzyme or any other quality index alteration are depicted in Table 3.3.

In Table 3.4, the effect of processing conditions range on the inactivation of indicative microorganisms and enzymes from various food products is depicted. The model used along with the model parameters estimated are also presented.

2.2 Secondary Mathematical Modeling

The primary models are useful when the processing conditions (pressure, temperature, pH, etc.) are kept constant. If any processing condition is changed, a new set of experiments must be performed to obtain new primary model parameters. To extend the application of primary models, mathematical expressions known as secondary models can be developed to estimate the pressure and/or temperature effect on the predicted primary model parameters. As in the case of primary models, secondary models can be obtained from theoretical considerations or empirical observations. Most of the secondary models here presented (Table 3.5) are nonlinear, reflecting complex biological behaviors under high-pressure/high-temperature conditions.

In most references cited in the literature, the Bigelow model is used to describe the effect of temperature and pressure on the reduction of microbial load. As in the case of thermal treatment, the thermal resistance constant z_T was developed is used, the analogous approach was established for pressure effect as well, estimating the pressure resistance constant z_P. Both these values may be used for the description of the effect of pressure (z_P) and temperature (z_T) on the decimal reduction time of microorganisms. The parameter z_P determines the pressure increase required to

Table 3.3 Mathematical equations cited in the literature and used for the description of the effect of treatment time at any combination of pressure and temperature on the value of microbial, enzymatic or chemical reactions

Model used	Observations
First order kinetics	
$\log_{10}(N/N_0) = -k\cdot t$ $\ln_{10}(N/N_0) = -(2.303/D)\cdot t$	A change in the initial concentration of an index, N_0 at $t = 0$, up to a value of the concentration equal to N after a process time, t, is described by an inactivation rate constant (k min^{-1}) under constant isobaric and isothermal conditions. Decimal reduction time, D (min) can be used for the microbial load reduction.
Fractional conversion model	
$C = C_\infty + (C_0 - C_\infty)\cdot \exp(-k\cdot t)$	Applied when part of a baroresistant or thermoresistant enzyme or isoenzyme or microbial load or component concentration with much higher resistance, C_∞ ($t = \infty$).
Multi phasic model	
$C = C_L\cdot \exp(-k_L\cdot t) + C_S\cdot \exp(-k_S\cdot t)$	The simplest form of the multiphasic model considers the presence of a labile fraction (C_L) that is inactivated more rapidly and a stable fraction (C_S) able to withstand longer treatment times. Each fraction is inactivated at a distinct rate, and the concentration (C) observed represents the sum of C_L and C_S at any given time
Weibull model	
$\log_{10}(N/N_0) = -b\, t^n$	The residual microbial/enzyme activity curve can be interpreted as a cumulative function of the distribution that dictates the treatment time at which the microorganism or enzyme will fail to resist and result in inactivation. The Weibull frequency distribution is applied, where N_0, the initial number of cells (CFU ml^{-1} or g^{-1}); N, the number of survivals after an exposure time t (CFU ml^{-1} or g^{-1}); t, the holding time (min) at pressure and b, n are the scale and shape factors respectively.

(continued)

Table 3.3 (continued)

Model used	Observations
Weibull biphasic model	
$N_{(t)} = N_0\left[f \cdot 10^{-(t/b'1)n_1} + (1-f) \cdot 10^{-(t/b'2)n_2}\right]$	When two bacterial subpopulations are present, the Weibull model is reparametrized as a function of the labile population fraction (f)
Log-logistic model	
$\log_{10}N = \alpha + (\omega - \alpha)/1 + \exp[4 \cdot \Omega\omega - \alpha(\tau - \log_{10}t)]$	The maximum inactivation rate (Ω), and the time at which Ω occurs (τ), along with the dependent variable microbial population logarithm ($y = \log_{10}N$) and the independent variable logarithm of time ($\log_{10}$ t) were incorporated into the Log-logistic equation. Parameter ω was defined as the difference between the lower and upper asymptotes ($\omega = \beta - \alpha$)
Modified Gompertz model	
$\text{Log}_{10}N/N_0 = A\ \exp \cdot [-\exp\ [(\mu_{max}\ \exp(1)/A) \cdot (\lambda - t) + 1]]$	Parameter A represents the difference between the lower and upper asymptotes of microbial survival curves ($\log_{10}N/N_0$ vs time), μ_{max} is the maximum inactivation rate, and parameter λ is the inflection point, or the time at which the linear portion of the curve starts
Baranyi-Roberts model	
$y(t) = \ln x(t) = y_0 + \mu_{max}A(t) - \ln(1 + e\mu_{max}A(t) - 1{*}e(y_{max} - y_0))$	The model encompasses both the lag to exponential and exponential to stationary transitions of bacterial growth. Letting the bacterial concentration at time t be given by x(t), where, $y_0 = \ln x_{(0)}$, $y_{max} = \ln x_{max}$ are the initial and maximum bacterial concentrations respectively, μ_{max} denotes the maximum specific growth rate.

Table 3.4 Effect of High pressure processing conditions range on the inactivation of indicative microorganisms, enzymes and substituents from various food products as cited in the literature

Microorganisms	Environment	Processing conditions	Kinetic model	References
Aerobic bacteria	Fresh, whole, raw milk pH = 6.64 Fresh, filtered orange juice pH = 3.35 Fresh, filtered peach juice pH = 5.21	300, 400, 600 MPa 100–200 MPa s^{-1} 0–105 min 25 °C	First-order	Dogan and Erkmen (2004)
Aerobic bacterial spores *B. amyloliquefaciens* TMW 2.479 Fad 82 *B. amyloliquefaciens* TMW 2.482 Fad 11/2 *B. sphaericus* NZ14 *B. amyloliquefaciens* ATCC 49763 Anaerobic bacterial spores *C. sporogenes* ATCC 7955 *C. tyrobutylicum* ATCC 27384 *T. thermosaccharolyticum* ATCC 27384	Deionized water	700 MPa 0–5 min 105, 121 °C	First-order	Ahn et al. (2007)
Bacillus stearothermophilus spores	Egg	400, 600, 700 MPa 0–16 min 105.8 ± 0.6 °C	First-order	Rajan et al. (2006)
Lactobacillus delbrueckii subsp. bulgaricus ACA-DC0105	Phosphate buffer 20 mM pH 7.0	100–700 MPa 20–40 °C	First-order	Katsaros et al. (2009a)
Lactobacillus delbrueckii subsp. bulgaricus ACA-DC0105 *Streptococcus thermophilus ACA-DC0022* *Lc. lactis ACA-DC 0049*	Reconstituted skimmed milk	100, 200, 450 MPa 0–40 min 20, 30, 40 °C	Baranyi model	Giannoglou et al. (2019)
Pediococcus spp.	Raw seabream (*Sparus aurata*) extract	150–600 MPa 20–40 °C	First-order	Tsironi et al. (2015)
Escherichia coli	Fresh extracted carrot juice pH = 6.6	200, 250, 300, 350, 400, 450, 500, 550, 600 MPa 100 MPa min^{-1} 0–60 min 5–45 °C	First-order	Van Opstal et al. (2005)

(continued)

Table 3.4 (continued)

Microorganisms	Environment	Processing conditions	Kinetic model	References
Listeria monocytogenes	Fresh, whole, raw milk pH = 6.6	300, 400, 600 MPa	First-order	Dogan and Erkmen (2004)
	Fresh, filtered orange juice pH = 6.64	100–200 MPa s^{-1} 0–105 min		
	Fresh, filtered peach juice pH = 3.35	25 °C		
Native microflora	Unpasteurized Hamlin variety orange juice	350, 400, 450, 500 MPa 1–300 s 25 ± 5 °C	First-order	Parish (1998)
Saccharomyces cerevisae ascospores *Saccharomyces cerevisae* vegetative cells	Commercial pasteurized orange juice			
Vibrio cholerae *Vibrio parahaemolyticus* *Vibrio vulnificus*	Phosphate-buffered saline	200–250 MPa 55–80 s 0–240 s <2 s 8–10 °C (initial)	First-order	Cook (2003)
Bacteriophage P008	Milk	0–800 MPa 10–70 °C 0–60 min	nth order n = 1.25	Müller-Merbach and Hinrichs (2006)

Enzymes	Environment	Processing conditions ranges	Kinetic model	References
a-Amylase (Barley malt (*Hordeum vulgare*))	0.1 M ACES buffer pH 5.6	0–800 MPa 30–90 °C	nth order n = 1.75	Buckow et al. (2007a)
	0.1 M ACES buffer + 3.8 M Ca2+ pH 5.6	0–1000 MPa 30–90 °C	nth order n = 2.1	
Alkaline phosphatase	Raw bovine milk	0–800 MPa 20–100 °C	First order	Ludikhuyze et al. (2000)
β-Amylase (Barley malt (*Hordeum vulgare*))	0.1 M ACES buffer pH 5.6	0–800 MPa 20–80 °C	nth order n = 1.4	Heinz et al. (2005)
β-Glucanase (*Bacillus subtilis*)	0.1 M ACES buffer, pH 5.6	0–1000 MPa 20–80 °C	nth order n = 1.8	Buckow et al. (2007b)
Lipoxigenase (Green pea)	Green peas	0–650 MPa −20–115 °C	First order	Indrawati et al. (2001)
	Supernatant of squeezed green peas			

(continued)

Table 3.4 (continued)

Enzymes	Environment	Processing conditions ranges	Kinetic model	References
Lipoxigenase (Tomato (*Lycopersicom esculentum, cv Malpica*))	Tomato puree	0–750 MPa 5–75 °C	Fractional conversion	Rodrigo et al. (2006)
X-prolyl dipeptidyl aminopeptidase (PepX)	Sodium phosphate buffer solution (pH 7.0)	100–450 MPa 20–40 °C	First order	Giannoglou et al. (2018)
Actinidin (Kiwi fruit)	Solution of phosphate buffer, L-cystein, EDTA and deionized water	0.1, 600, 750, 900 MPa 50–70 °C	First order	Alexandrakis et al. (2017)
Actinidin (Kiwi fruit)	Kiwi fruit juice	200–800 MPa 25–50 °C	First order	Katsaros et al. (2009b)
Ficin (EC 3.4.22.3)	Phosphate buffer 50 mM pH 7.0	500–900 MPa 50–80 °C	First order	Katsaros et al. (2009c)
Papain (EC 3.4.22.2)				
Pectinmethylesterase (Persimmon (*Hachiya cv.*))	Pulp pH 5.5	500–800 MPa 40–70 °C	First order	Katsaros et al. (2006)
Pectinmethylesterase (Sea buckthorn (*Golden sea berry cv.*))	Juice pH 2.8	200–600 MPa 25–35 °C	Fractional conversion	Alexandrakis et al. (2014a)
Pectinmethylesterase (Orange (*Navel cv.*))	Juice pH 3.4	100–800 MPa 30–60 °C	Fractional conversion	Polydera et al. (2004)
Pectinmethylesterase (Orange (*Valencia cv.*))	Juice pH 3.8	100–500 MPa 20–40 °C	Fractional conversion	Katsaros et al. (2010)
Pectinmethylesterase (Orange (*Valencia cv.*)) (Orange (*Navel cv.*))	Tris buffer pH 7.5	200–700 MPa 40–55 °C	First order	Alexandrakis et al. (2014b)
Pectinmethylesterase (Peach (*Everts cv.*))	Phosphate buffer pH 7.0	100–800 MPa 30–70 °C	First order	Boulekou et al. (2010)
Pectinmethylesterase (Carrot (*Daucus carota L. cv.*))	Tris buffer pH 7.0	100–825 MPa 10–65 °C	Fractional conversion	Ly-Nguyen et al. (2003a)
Pectinmethylesterase (Banana (*Cavendish cv.*))	Tris buffer pH 7.0	100–900 MPa 30–76 °C	Fractional conversion	Ly-Nguyen et al. (2003b)
Pectinmethylesterase (Carrot (*D. carota L. cv.*))	Citrate buffer pH 6.0	650–800 MPa 10–25 °C	First order	Balogh et al. (2004)
	Juice pH 6.0	700–800 MPa 10 °C		
	Pieces pH 6.0	700–800 MPa 40 °C		

(continued)

Table 3.4 (continued)

Enzymes	Environment	Processing conditions ranges	Kinetic model	References
Polyphenoloxidase (Pineapple (*Ananas comosus* L.))	Puree pH 3.48	0.1–600 MPa, 30–70 °C 0–20 min	nth order n = 0.991	Chakraborty et al. (2015)
Peroxidase (Pineapple (*Ananas comosus* L.))			nth order n = 0.995	

Food constituents	Environment	Processing conditions ranges	Kinetic model	References
Ascorbic acid	Fresh pineapple juice	0.1–600 MPa 30–95 °C	First order fractional conversion	Dhakal et al. (2018)
Chlorophyll (Broccoli (*Brassica oleracea L. italica*))	Broccoli juice	0–800 MPa 50–105 °C	First order	Van Loey et al. (1998)
Folate (5-methyl-tetrahydropholic acid, 5 CH3 H4 folate)	0.1 M phosphate buffer pH 7.0	0–700 MPa 20–90 °C	First order	Indrawati et al. (2005)
Starch (Normal maize)	5% w/w deionised water	0–700 MPa 20–75 °C	nth order n = 1.65	Buckow et al. (2007c)
Anthocyanins	Raspberry paste	200–700 MPa 90–115 °C	First order	Verbeyst et al. (2011)

Table 3.5 The most common kinetic models used for secondary mathematical modeling for high pressure processing as cited in the literature

Bigelow model	Observations
$Z_p = -P - P_{ref}/\log D_p - \log D_{Pref}$	**Zp or Z_T**: the inverse negative slope of $\log D_P$ or $\log D_T$ versus pressure or temperature level and determines the pressure or temperature increase required to achieve a tenfold increase in the inactivation rate
Model proposed by Santillana Farakos and Zwietering (2011) (based on Bigelow Model)	
$\log D = 1/z_P(P_{ref} - P) + 1/z_T(T_{ref} - T) + 1/z_{PT}[(T_{ref}P_{ref}) - (T_P)] + \log DP_{ref}T_{ref}$	Z_{PT} the inverse negative slope of $\log D_{PT}$ versus pressure•temperature level and represents the amount that the linear term P·T needs to increase for a tenfold decrease in D
Eyring-Arrhenius Model	
$k(P) = k_{refP}\cdot\exp[-\Delta V^{\neq}(T)/R\cdot(P - P_{ref})T]$ $\Delta V^{\neq}(T) = a\cdot(T - T_{ref}) + \Delta V_T^{\neq}$ $k(T) = k_{refT}\cdot\exp[-Ea(P)/R\cdot(1/T - 1/T_{ref})]$ $Ea(P) = Ea_P\cdot\exp[-g\cdot(P - P_{ref})]$ $k = k_{refP,T}\cdot\exp\{-E_{aP}/R\cdot\exp[-g\cdot(P - P_{ref})]\cdot(1/T - 1/T_{ref}) - a\cdot(T - T_{ref}) + \Delta V_T^{\neq}/R\cdot P - P_{ref}/T\}$	$\mathbf{E_a}$: the activation energy $\mathbf{\Delta V^{\neq}}$: the activation volume

(continued)

Table 3.5 (continued)

Bigelow model	Observations
Model proposed by Weemaes et al. (1998) (based on Eyring-Arrhenius Model)	
$\ln k_{ref}(P) = c_1 + c_2 \cdot P + c_3 \cdot P^2 + c_4 \cdot P^3$ Weemaes et al. (1998) $E_a(P) = E_{aP} \cdot [\exp(-c_5 \cdot P)]$ $k = \exp\{c_1 + c_2 \cdot P + c_3 \cdot P^2 + c_4 \cdot P^3 + [-E_{aP} \cdot [\exp(-c_5 \cdot P)]R(1/T - 1/T_{ref})]\}$ van den Broeck et al. (2000) $E_a(P) = c_5 - c_6 \cdot P$ $k = \exp\{c_1 + c_2 \cdot P + c_3 \cdot P^2 + c_4 \cdot P^3 + [-c_5 - c_6 \cdot PR(1/T - 1/T_{ref})]\}$	$\mathbf{c_1 - c_4}$: Empirical parameters describe the effect of pressure on k_{ref} Antagonistic pressure effects on k
Model proposed by Ludikhuyze et al. (1998) (based on Eyring-Arrhenius Model)	
$\ln kref(T) = c_1 + c_2 \cdot T + c_3 \cdot T^2$ $\Delta V \neq (T) = c_4 \cdot T \cdot [\exp(-c_5 \cdot T)]$ $\ln k = c_1 + c_2 \cdot T + c^2{}_3 \cdot T - \{c_4 \cdot T \cdot [\exp(-c_5 \cdot T)]/R \cdot T \cdot (P - P_{ref})\}$	Antagonistic effects for combined pressure–temperature treatments for the low-temperature (T < 40 °C) and high-pressure (P > 475 MPa) region
Model proposed by Katsaros et al. (2010) (based on Eyring-Arrhenius Model)	
$k/k_{Tref} = D_{Tref} D_T \cdot 10(T_{ref} - Tz_T)$ $kk_{Pref} = D_{Pref} DP \cdot 10(P_{ref} - Pz_P)$ $D = D_{PrefTref} \cdot \exp\{(P - P_{ref}) \cdot [\Delta V^{\neq}(T)/R \cdot T + 2.303/z_P] + 2.303(T - T_{ref})/z_T + E_a(P)/RT \cdot (1/T - 1/T_{ref})\}$	Proposed for microbial inactivation
Log-logistic model for Weibull distribution	
$\log_{10}(N/N_0) = -b' t^n$ $b'(T) = \ln\{1 + \exp[w_T(T - T_c)]\}^m$ $b'(P) = \ln\{1 + \exp[w_P(P - P_c)]\}^m$ $Pc(T) = P_{c0} \cdot \exp(-w_1 T)$ $Tc(P) = T_{c0} \cdot \exp(-w_2 P)$ $n(P) = d_0 \cdot \exp(-d_1 P)$	T_c: the temperature at which b′ (T) increases linearly for m = 1. If $T > T_c$, the parameter b′ (T) increases to the power $w_T \cdot (T - T_c)$ w_T determines the rate at which b′ (T) increases with temperature. If $T < T_c$, the exponential term tends to zero and b′ (T) is approximately ln (1) = 0

achieve a tenfold increase in the inactivation rate, while the z_T determines the temperature increase required to achieve the tenfold increase in the inactivation rate constant.

The Eyring equation was also developed to mathematically describe the effect of pressure (P) on the inactivation rate constant (k). This effect was correlated to the Activation Volume value (ΔV). Similarly to the Arrhenius equation that is widely used for the mathematical description of temperature on the inactivation rate constant of a quality index. In this case, the Activation Energy (E_a) is estimated and is correlated to the effect of temperature on the inactivation rate constant. In High pressure treatments that pressure and temperature are the main process parameters apart from treatment time, the combined effect of Eyring and Arrhenius equations may be used. Thus, the effects of pressure and temperature may be expressed through the Activation volume and Activation energy, respectively. For more than one different treatment temperatures, the Activation volume value is estimated for each temperature and the effect of temperature on the ΔV values may be estimated (by plotting ΔV values vs treatment temperatures). The same approach may be followed for the Activation energy as well, that expresses the effect of temperature on the inactivation rate constant of an index. For more than one pressures studied, the E_a value is determined for each pressure and then the effect of pressure on the E_a value can also be determined by plotting the E_a values vs treatment pressures.

Secondary models for the Weibull parameters include a logistic-exponential expression for parameter b', indicating that the inactivation rate constant will be close to zero until a critical pressure level (Pc) is reached. If the pressure level is increased beyond Pc, b′ will increase at a rate w_P. An exponential decay model has been used to predict pressure effects on the Weibull parameter "n", where the dimensionless parameter d_0 is the value of n at pressures P approaching 0, and d_1 is the rate at which n exponentially decays (Doona and Feeherry 2007; Peleg 2006; Doona et al. 2007).

3 Pulsed Electric Field Processing

For PEF treatment, the most critical processing factors are electric field strength, treatment time, shape and width of the pulse, frequency, specific energy, and pre-heating temperature (Barbosa-Canovas et al. 1999; Wouters et al. 2001). Depending on the conditions intensity, the treatment efficacy is affected leading to microbial population or enzyme activity decrease, causing cell disintegration. The development of mathematical models that can predict the death of microorganisms and inactivation levels of quality-related enzymes by PEF is a very useful tool to design safe and effective PEF processes.

3.1 Primary Mathematical Modeling

Many researchers have focused on developing mathematical models to understand the physiological mechanism of each quality or safety index alteration and how the treatment conditions affected microorganisms, enzymes, quality indices, retention of bioactive compounds and extraction of intracellular compounds from plant tissues. Kinetic models for studying the effects of PEF treatment on the activity of microorganisms or enzymes, have been extensively used in different process conditions and are presented in Table 3.6.

First-order kinetic model as a function of treatment time or electric field strength is the most widely mathematical equation used to describe successfully the effect of PEF treatment on inactivation rate of microorganism, enzymes or retention of health-related compounds in the foods. This model describes the dependency of residual activity of microorganisms or enzymes as a function of the PEF conditions intensity (Bendicho et al. 2002; Giner et al. 2001, 2002, 2003). In literature, the experimental data clearly show an exponentially decrease in residual activity as the

Table 3.6 Mathematical equations cited in the literature and used for the description of the effect of PEF treatment on the value of microbial, enzymatic or chemical reactions

Model	Mathematical equation	Where	References
Kinetic models for the inactivation of microorganisms and enzymes by PEF			
First order kinetic	$RA = e^{-k_P \bullet P}$	RA is the residual activity, P is the studied parameter (PEF treatment time (t, s), electric field strength (E, kV/cm), pulse frequency (f, Hz), the pulse width (τ, s), or the PEF energy input (Q, J/kg)), kP is the inactivation constant rate for the respective studied parameter	Giner et al. (2000, 2001, 2002, 2003)
Empirical Hulsheger's model	$RA = \left(\frac{t}{t_c}\right)^{-\frac{(E-E_c)}{k_c}}$	RA is the residual activity, t is PEF treatment time, E is the electric field strength, E_c, t_c, and k_c are proposed to be independently determined by the target microorganism or enzyme	Hülsheger et al. (1981)
Empirical Fermi model	$RA = \frac{1}{1+e^{\frac{(E-E_c(t))}{a_c(t)}}}$	RA is the residual activity, E is the electric field strength (kV/cm), $E_c(t)$ is the electric field strength (kV/cm) for RA equal to 50% and $a_c(t)$ is the parameter indicating the slope of the curve around E_c. E_c and k_c, are exponentially related to the PEF treatment time t.	Peleg (1995)
Weibull distribution model	$RA = e^{-\left(\frac{t}{a}\right)^{\gamma}}$	RA is the residual activity, P is the studied PEF parameter (treatment time (t, s) or PEF energy input (Q, J/kg)), a and γ are the scale and shape parameters, respectively. Apart from rare exceptions, an exponential relationship exists between the values "a" and electric field strength E	Weibull (1951)
Kinetic models for the extraction of intracellular compounds by PEF			
Empirical Peleg model	$C_t = \frac{t}{k_1 + k_2 \cdot t}$	C_t is the concentration of extracted compound (mg g^{-1}) at time t (s), k_1 is Peleg's rate constant (min g mg^{-1}) and k_2 is Peleg's capacity constant (g mg^{-1})	Peleg (1988)
First-order fractional model	$\frac{C_t - C_\infty}{C_0 - C_\infty} = -k \cdot t$	C_t is the concentration of extracted compound (mg g^{-1}) at time t (s), C_0 is the concentration of extracted compound (mg g^{-1}) at time t = 0, C_∞ is the concentration of extracted compound (mg g^{-1}) at time ∞ time, k is the constant rate of the concentration increase of extracted compound	Levenspiel (1972)

intensity of PEF conditions (electric field strength or treatment time) is increased (Bendicho et al. 2002; Giner et al. 2002). Moreover, the effect of PEF on microbial or enzyme inactivation could also be explained by empirical models such as those of Hülsheger, Fermi's and Weibull distribution model (Table 3.6). These models were first proposed for predicting microbial inactivation, and later to describe the destruction of enzymes by PEF (Giner et al. 2000; Min et al. 2003). Hülsheger's and Fermi's models describe the decrease of microbial or enzyme activity (RA) as a function of both electric field strength and treatment time.

PEF efficacy is crucial as a knowledge for accepting this technology as a pasteurization process equivalent of thermal treatment. Critical factors affecting microbial inactivation, enzymes inactivation as well as effect on other quality indices is necessary to be studied. The data received from these studies will establish optimized processes that are applicable under a wide range of conditions. Mathematical models applied target to fit microorganism and enzyme responses to processing factors and environmental variables. The fitting accuracy of the traditional first-order kinetics, Hulsheger's, Fermi's, and Weibull distribution models for microbial inactivation by PEF treatments has been reported in several studies (Table 3.7). Similarly to microbial inactivation, effort has been done on the mathematical description of PEF effect on enzymes inactivation. Indicative studies are presented, reporting the mathematical model they applied to describe their results in Table 3.8.

Numerous papers report studies of quality parameters of foods and how are they affected by PEF treatment (Table 3.9). The results obtained are promising, since the health-related compounds and quality indices are retained better when compared to conventional thermal treatment (Min et al. 2003; Odriozola-Serrano et al. 2008a, b).

PEF technology led to retention or enhancement (due to increased extractability offered by PEF technology) of health-related compounds in juices, such as lycopene, vitamin C and polyphenols. Generally, more intense process conditions (higher electric field strength and treatment time), resulted in lower vitamin C (in strawberry as reported by Odriozola-Serrano et al. 2008c, 2009b in tomato reported by Odriozola-Serrano et al. 2007, 2008b, d, 2009a and in watermelon juice cited by Oms-Oliu et al. 2009) and in higher lycopene contents (Odriozola-Serrano et al. 2007) up to 146.2% for tomato juices.

Mathematical description of data obtained by PEF treatment of foods may improve the prediction of the variation of the health-related compounds and antioxidants as affected by key parameters involved in PEF treatments.

PEF was also used to assist and accelerate extraction of high added value compounds from by products, such as tomato peels (Andreou et al. 2020a; Luengo et al. 2014; Pataro et al. 2020), olive pomace (Andreou et al. 2020b), potato peels (Frontuto et al. 2019), grapes by-products (Corrales et al. 2008), orange peels (Luengo et al. 2013), and sesame cake (Sarkis et al. 2015). PEF-assisted extraction could lead to cell alterations or disruption of cell membranes resulting in higher recovery yields of valuable compounds while decreasing the extraction time or solvent volume.

There is limited data in literature about the modeling description of the effect of PEF treatment on the enhancement of mass transfer phenomena. Several researchers have used PEF as pretreatment to conventional extraction procedure and several mathematical models have been used to optimize the PEF conditions' selection.

For instance, Andreou et al. (2020b) used a fractional fist order equation to describe the polyphenol and protein extraction assisted by PEF from olive pomace, and the extracted concentration of each compound was correlated with PEF energy

Table 3.7 Effect of PEF processing conditions range on the inactivation of indicative microorganisms from various food products as cited in the literature

Microorganism	Medium	Process conditions	Maximum reduction (logCFU/g)	Mathematical models	References
E. coli	Orange juice	15–40 kV/cm, pulse width 2.5 μs, treatment time 700 μs, flow rate 60 mL/min, <55 °C	3.83	First order kinetic model Hulsheger's model Weibull distribution model	Rivas et al. (2006)
E. coli O157:H7 Salmonella *Enteriditis*	Liquid egg yolk	30 kV/cm, pulse width 2 μs, treatment time 210 μs, flow rate 12 mL/min, 40 °C	4.9 4.8	First order kinetic model	Amiali et al. (2006)
E. coli CGMCC 1.90	Carrot juice	5–25 kV/cm, pulse width 1.5 μs, treatment time 207–1449 μs, coaxial treatment chamber, flow rate 52.5 mL/min, <40 °C	3.6	Hulsheger's model Fermi's model	Zhong et al. (2005)
Salmonella Dublin (ATCC 15480)	Skim milk	15–40 kV/cm, treatment time 12–127 μs, 10–50 °C	4	Hulsheger's model Fermi's model	Sensoy et al. (1997)
E. sakazakii CECT 858	Buffered peptone water Rehydrated infant formula milk	10–40 kV/cm, pulse width 2.5 μs, treatment time 360 μs, flow rate 1.8 L/h, 25 °C	2.7 1.7	First order kinetic model Weibull distribution model	Pérez et al. (2007)
Salmonella Senftenberg 775 W	Liquid whole egg	20–45 kV/cm, square pulse width 3 μs, treatment time 0–150 μs, 55 °C	3.3	Weibull distribution model	Monfort et al. (2010)
E. coli *L. monocytogenes*	Melon & watermelon juice	35 kV/cm, pulse width 4 μs, pulse frequency 217 Hz, treatment time 1440 μs, 40 °C	3.7 3.56 3.6 3.41	Quadratic response model	Mosqueda-Melgar et al. (2007)

(continued)

Table 3.7 (continued)

Microorganism	Medium	Process conditions	Maximum reduction (logCFU/g)	Mathematical models	References
E. coli O157:H7 *Salmonella Enteriditis*	Apple Pear Orange Strawberry juice	35 kV/cm, bipolar pulse width 4 μs, flow rate 80–110 mL/min, <40 °C	4.29 4.34 4.59 4.87 5.16 5.22 5.56 4.43	Quadratic response model	Mosqueda-Melgar et al. (2008)
G. oxydans *K. apiculata* *L. bacteria* *S. cerevisiae*	Grape juice	35 kV/cm, bipolar pulse width 5 μs, pulse frequency 303 Hz, flow rate 3.33 mL/s, inlet temperature 15 °C, maximum temperature <30.4 °C	2.24 3.88 3.54 3.90	Quadratic response model	Marsellés-Fontanet et al. (2009)

Table 3.8 Effect of PEF processing conditions range on the inactivation of indicative enzymes from various food products as cited in the literature

Enzyme	Medium	Process conditions	Residual activity (%)	Mathematical models	References
LOX	Tomato juice	35 kV/cm, pulse width 3 μs, 50 μs treatment time, flow rate 1 mL/s, 30 °C	20	First order kinetic model Hulsheger's model Fermi's model Quadratic response model	Min et al. (2003)
PME	Tomato juice	5–24 kV/cm, 10.9–108.0 MJ/m^3 energy input, pulse width 0.02–0.04 ms, 0–400 pulses	8	First order kinetic model Hulsheger's model Fermi's model	Giner et al. (2000)
LOX	Tomato juice	35 kV/cm, 250 Hz, bipolar pulse width 7 μs, treatment time, 1000 μs	81	Quadratic response model	Aguiló-Aguayo et al. (2009a)
PME PG	Tomato juice	5.5–12.5 kV/cm, 0–12 ms treatment time, pulse width 15 μs, 300 Hz, bipolar pulses	98 45	First order kinetic model	Andreou et al. (2016)

(continued)

Table 3.8 (continued)

Enzyme	Medium	Process conditions	Residual activity (%)	Mathematical models	References
PME PG POD	Tomato juice	35 kV/cm, 250 Hz bipolar pulse width 7 μs, flow rate 60 mL/min	10 45 0	Quadratic response model	Aguiló-Aguayo et al. (2008a, b, 2009a, b)
LOX	Soymilk	20–40 kV/cm, 400 Hz pulse width 2 μs, treatment time 1036 μs, 25 °C	12	First order kinetic model Fermi's model Weibull distribution model	Li et al. (2008)
PME	Orange juice	25 kV/cm, 700 Hz pulse width 2.0 ms, flow rate 0.31 mL/s, 50 °C	10	First order kinetic model	Yeom et al. (2002)
PME	Orange juice	5–35 kV/cm, 200 Hz bipolar and monopolar pulse width 4 μs, treatment time 1500 μs, 60 mL/min, 37.5 °C	20	First order kinetic model Hulsheger's model Fermi's model Weibull distribution model	Elez-Martinez et al. (2007)
PME	Fresh mixed orange and carrot juice	25–40 kV/cm, bipolar pulse length 2.5 μs, treatment time 340 μs, 60 mL/min	18.6	First order kinetic model Hulsheger's model Weibull distribution model	Rodrigo et al. (2003)
PME	Red grape juice	40 kV/cm, 15 Hz pulse width 1 μs, treatment time 100 μs	3.2	First order kinetic model	Riener et al. (2009)
POD, PPO	Grape juice	25–35 kV/cm, 600 Hz bipolar pulse width 4 μs, treatment time 5 ms, flow rate 7.8 mL/s, 40 °C	49.4 0	First order kinetic model	Marsellés-Fontanet and Martin-Belloso (2007)
POD, PPO	Grape juice	25–35 kV/cm, 600 Hz bipolar pulse width 4 μs, treatment time 5 ms, flow rate 7.8 mL/s, 40 °C	49.4 0	Quadratic response model	Marsellés-Fontanet and Martin-Belloso (2007)
POD, PPO	Apple juice	23–50 kV/cm, 15 Hz pulse width 1 μs, treatment time 100 μs, 50 °C	32, 29	First order kinetic model	Riener et al. (2008)

(continued)

Table 3.8 (continued)

Enzyme	Medium	Process conditions	Residual activity (%)	Mathematical models	References
PPO	Apple & pear	22.3–24.6 kV/cm, up to 6 ms treatment time, bipolar mode, exponential decay pulses, 0.02 ms pulse width	96.8 62.0	First order kinetic model	Giner et al. (2001)
PPO	Peach juice	24.3 kV/cm, bipolar pulse width 0.02 ms, 5 ms treatment time	30	First order kinetic model	Giner et al. (2002)
PME, PG	Strawberry juice	35 kV/cm, 100 Hz monopolar pulse width 1 μs, flow rate 60 mL/min	10, 75	Quadratic response model	Aguiló-Aguayo et al. (2009a)
LOX, POD	Watermelon juice	35 kV/cm, 50 Hz monopolar pulse width 1 μs, treatment time 1000 μs	112.25 15.25	Quadratic response model	Aguiló-Aguayo et al. (2010)
PME	Gazpacho	35 kV/cm, 200 Hz monopolar pulse width 4 μs, treatment time 1500 μs, flow rate 60 mL/min, 40 °C	3.8	Giner-Segui's model	Giner-Seguí et al. (2009)

input. Moreover, lycopene extraction kinetic from industrial tomato peels was well fitted by Peleg's model (Pataro et al. 2020), allowing to select the optimal PEF conditions with low energy consumption.

3.2 *Secondary Mathematical Modeling*

Except on empirical models that are used as a first step to obtain optimum values of each factor with maximal reduction for every microorganism or enzyme, there is no secondary modeling approach proposed in the literature. This is mainly attributed to the numerous process parameters that counteract between them not allowing for development of secondary models.

4 Pulsed Electromagnetic Fields Processing

Pulsed electromagnetic fields (PEMF) technology involves the generation and powerful direction of pulsed electromagnetic waves. The generated waves seem to react with the cells that come in contact changing the state of the electrons spin system

Table 3.9 Effect of PEF processing conditions range on the effect of quality indices of various food products as cited in the literature

Quality and bioactive compounds	Medium	Process conditions	RC	Mathematical models	References
Anthocyanin Vitamin C Antioxidant capacity	Strawberry juice	25–35 kV/cm, 232 Hz bipolar pulse width 1 μs, 100 μs treatment time, flow rate 60 mL/min, 40 °C	100.5 93 0	First order kinetic model Weibull distribution model	Odriozola-Serrano et al. (2008c)
Antioxidant capacity Vitamin C	Tomato juice	35 kV/cm, 100 Hz bipolar pulse width 4 μs, 1500 μs treatment time, flow rate 60 mL/min, 40 °C	0	First order kinetic model	Odriozola-Serrano et al. (2008b)
Antioxidant capacity Lycopene Vitamin C	Tomato juice	35 kV/cm, 250 Hz bipolar pulse width 1 μs, 500 μs treatment time, flow rate 60 mL/min, 40 °C	137.7 100 97	Weibull distribution model Fermi's model First order kinetic model	Odriozola-Serrano et al. (2008d)
Ascorbic acid	Milk	A static parallel plate treatment chamber, 27.1 kV/cm, 20–25 °C	93.4	First order kinetic model	Bendicho et al. (2002)
Anthocyanin Vitamin C	Strawberry juice	35 kV/cm 250 Hz, bipolar pulse width 1 μs, treatment time 1000 μs	101.9 100.3	Quadratic response model	Odriozola-Serrano et al. (2009b)
Antioxidant capacity Lycopene Vitamin C	Tomato juice	35 kV/cm, 150 Hz bipolar pulse width 4 μs, 1000 μs treatment time, flow rate 60 mL/min, 40 °C	92.3 146.2 99	Quadratic response model	Odriozola-Serrano et al. (2007)
Antioxidant capacity Lycopene Vitamin C	Watermelon juice	35 kV/cm 200 Hz, bipolar pulse width 7 μs, treatment time 50 μs	100 72 113	Quadratic response model	Oms-Oliu et al. (2009)

(Pawluk 2015). PEMF is mainly studied for its use for human therapeutic purposes, indicating significant effects on cells, tissues and biological processes such as embryogenesis, regeneration, wound healing (Hammerick et al. 2010), as well as in cell migration, DNA synthesis and gene expression (Tsai et al. 2009; Luo et al. 2012; Kang et al. 2013). PEMF technology has hardly been studied for its effect on food products, however it is considered to be a promising technology for microbial inactivation (Tadevosian et al. 2006; Torgomyan et al. 2011; Giannoglou et al. 2020a, 2021). Extensive work on the effect of PEMF on a food system (whole fresh strawberries) has been performed by Giannoglou et al. (2021). PEMF processing did not appear to have a significant impact on the weight loss, the color, the total anthocyanin content and on the pH-value of the strawberries, after processing and during storage. PEMF processing led to a 16% decrease in the firmness of the strawberries immediately after processing compared to Control samples, which was also maintained during storage. A significant increase in the total phenolic content and in the free radical scavenging activity was observed for PEMF processed fruits immediately after processing. PEMF processed strawberries also presented higher peak values in the total phenolic content during storage compared to the untreated ones. The PEMF strawberries presented the highest values in ascorbic acid content after processing and also during storage, compared to Control. PEMF technology could be used for enhancement of nutritional value of fruits in addition to the quality retention related with safety and consumer perceived traits. Nevertheless, there has not yet been described the effect of PEMF on any quality index by mathematical equations. There is a lot of space for work towards this direction, since the equations applied for PEF treatments could be modified and applied also for PEMF treated food products.

5 Cold Atmospheric Plasma Processing

Cold atmospheric plasma (CAP) is an emerging non-thermal processing method that attracts an ever increasing interest for future application in food industry (Schlüter et al. 2013; Niemira 2012; Giannoglou et al. 2020a, b). Plasma is a partially ionized gas consisting of a reactive mixture of charged particles, free radicals, excited species, and UV photons. Novel plasma reactor designs and electrical power supplies have enabled the generation of non-thermal, far from thermodynamic equilibrium, plasmas in atmospheric pressure (Pappas 2011; Brandenburg 2018). The high reactivity combined with the low temperature operation render plasma suitable for treatment of heat-sensitive food products, due to their ability to deactivate microorganisms (Kelly-Wintenberg et al. 1999; Moisan et al. 2002; Laroussi 2005; Puač et al. 2017; Dimitrakellis et al. 2021). In the field of fruits, the investigation mainly concerns the CAP effect on the microorganisms/enzymes and quality characteristics of fruit products such as orange juice (Xu et al. 2017), white grape juice (Pankaj et al. 2017), siriguela juice (Paixão et al. 2019), fresh-cut apples (Ramazzina et al. 2016; Tappi et al. 2014), fresh-cut melon (Tappi et al. 2016). A more limited number of studies concern the effect of this technology on whole fruit i.e. blueberries (Sarangapani et al. 2017) and cherry tomatoes (Misra et al. 2014). Studies on the effect of CAP processing on quality parameters of whole fruits during storage are limited to strawberries (Rana et al. 2020; Giannoglou et al. 2021) and mandarins (Won et al. 2017). Based on the reported results, CAP processing induced the inactivation of microorganisms mostly due to the interaction with reactive oxygen and nitrogen species (RONS) generated in the gas phase during processing. The effect of plasma treatment on the quality characteristics varied from significant to insignificant, depending on the investigated parameters, the product characteristics, as well as the intensity of the processing and the plasma source design. Indirect plasma treatment of strawberries through immersion in plasma activated water has also

Table 3.10 Indicative mathematical equations cited in the literature and used for the description of the effect of CAP treatment on the value of microbial and enzymatic reactions

Model	Mathematical equation	Where	References
First-order model	$\ln A/A_0 = -k \cdot \Delta t$	Where A_0 is the initial activity of PPO, A represents the PPO activity at time t, k is the first-order kinetic constant (min^{-1}), and t is the treatment time (min).	Dong et al. (2021); Liang et al. (2012)
Weibull model	$\ln A/A_0 = -(t/a)^{\beta}$	Where A and A_0 have the same meaning as in Eq. (2), t is the DBD plasma exposure time (min), α is the scale parameter (characteristic time, min), and β is the shape parameter. The β value denotes an idea of the form of the inactivation curve: upward concavity (β < 1), straight line (β = 1), or downward concavity (β > 1).	Dong et al. (2021); Liang et al. (2012)
Logistic model	$A = [(100 - A_{min})]/[1 + (t/t_{50})^p] + A_{min}$	Where A_{min} (≥0) is the minimum value attained by the logistic function, t_{50} is the time of half-maximal activity (min), and p is the power term.	Pankaj et al. (2013); Dong et al. (2021)

been proposed in the literature for efficient disinfection based on secondary RONS formed in liquid phase upon interaction with gas discharges (Ma et al. 2015).

In general, CAP can be used either as direct (direct processing of a food product by the ionized gas), semi-direct (surface dielectric barrier discharge used for the treatment of whole food products) or even indirect (production of "plasma activated water" and immersion of food products within this water rich in Reactive Oxygen Nitrogen Species with the antimicrobial effect) applications. In all cases, there is limited work done on the mathematical description of the data obtained. This is mainly attributed to the numerous process parameters for CAP treatment making it harder for the researchers to apply or develop appropriate mathematical equations. Nevertheless, the models applied in the limited works cited in the literature are depicted in Table 3.10. No secondary models have been applied or developed for the description of the effect of process parameters on the inactivation rate constants for microbial, enzymatic and chemical indices.

The works having included the kinetic approach in data obtained by applying the Cold Atmospheric Plasma technology on microbial, enzyme and chemical indices are very limited and some of them are depicted in Table 3.11. The Weibull model is mainly used by the researchers both for microbial and enzymatic inactivation.

6 Osmotic Dehydration

Osmotic dehydration (OD) has received greater attention in recent years as an important complementary treatment and food preservation technique in the processing of dehydrated foods. OD is a water removal process which is based on the implementation of foods, mainly fruits and vegetables, in a hypertonic solution,

Table 3.11 Effect of CAP processing conditions range on the effect of indicative microorganisms, enzymes and quality indices of various food products as cited in the literature

Model	Process conditions	Medium	Reference
Weibull model	Dielectric barrier discharge-atmospheric cold plasma treatment at high voltages (40, 50 and 60 kV) for durations ranging between 15 s and 5 min	Alkaline phosphatase enzyme	Segat et al. (2016)
Weibull model	Fixed distance of 35 mm, input power of 200 W for 30, 60, 120, 180, and 240 s	Salmonella and Escherichia coli in apples	Kilonzo-Nthenge et al. (2018)
Weibull model	Plasma exposure time varied between 0 and 480 s. The gas flow rate was 5 slm (standard litre per minute), and the power supply was operated at a voltage of 65 V and a resonance balancing of 0.05 A. The distance between nozzle outlet and sample was set to 10 mm.	*Citrobacter freundii* in apple juice	Surowsky et al. (2014)
Weibull model and a three-parameter logistic model	Different voltages (30, 40 and 50 kV) for different time intervals (15 s–5 min)	Tomato POD inactivation	Pankaj et al. (2013)
Weibull and logistic model	Different voltages (18, 23 and 28 kV) for different time intervals (15 s–5 min)	Peroxidase and polyphenol oxidase in tender coconut water	Chutia et al. (2019)
Weibull and logistic model	19.4, 26.4, and 32.6 W, for 3 min	PPO isolated from *Agaricus bisporus*	Dong et al. (2021)
Weibull model	50 kV (65–75 W @ 0.5–0.8 mA) with an electrode gap or depth of 2.5 cm. Samples exposures time to the cold plasma discharge were 0, 15, 30, 60, and 120 s.	B. subtilis spores	Mendes-Oliveira et al. (2019)

reducing its water content while increasing the soluble solid content. The raw material is placed in concentrated solutions of soluble solids with higher osmotic pressure and lower water activity. Diffusion phenomenon takes place with two countercurrent flows: a water flow from the food to the outer solution and a simultaneous flow of solute from the solution to the food (de Mello Jr et al. 2019), enriching its composition by enhancing its nutritional value. These mechanisms lead to water loss and solid gain in the food. OD process occurs at mild temperatures (<50 °C), thus low energy consumption and processing costs are required (Torreggiani 1993). Although OD process will not give a product of sufficiently low moisture content to be considered a shelf stable product and therefore, OD should be combined with other preservation techniques.

The kinetics of mass transfer is described using terms such as water loss (WL), solids or solutes gain (SG) and water activity (a_w) reduction (Pękosławska and Lenart 2009). The most important variable markedly affecting the kinetics of mass transfer during OD is temperature (Mokhtarian et al. 2014). This enhances the removal of water and uptake of solids.

The most widely used model for OD processes at atmospheric pressure is the Crank's model, which consists of a solution of non-steady Fick's law and represents the diffusional mechanism. Crank's model (Crank and Gupta 1975) consists of a group of analytical solutions of Fick's diffusion law that were obtained by Crank for various geometries and several initial and boundary conditions. Other models can also be found in the literature trying to describe the effect of process parameters on the mass transfer phenomena and are presented in Table 3.12.

Many researchers used the Crank's model in order to describe the mass transfer phenomena during OD treatment of various tissues such as apple (Sereno et al. 2001), strawberry (Dermesonlouoglou et al. 2017a), goji berry (Dermesonlouoglou et al. 2018), tomato (Dermesonlouoglou et al. 2017b), cucumber (Dermesonlouoglou et al. 2008), kiwi (Dermesonlouoglou et al. 2016), pumpkin (Dermesonlouoglou et al. 2020), apricot (Dermesonlouoglou et al. 2019a; Dermesonlouoglou and Giannakourou 2018), bananas (Mercali et al. 2010), etc. The effective diffusivities of water and solids were found to be in the range of $2.0 \pm 1.0 \times 10^{-7}$ to $0.7 \pm 0.2 \times 10^{-11}$ m^2/s, depending on the food tissue and the OD treatment conditions applied.

OD is the process that can be used as pre-treatment for conventional drying procedures, such as air-drying, microwave-drying and freeze drying. High temperature and long drying time in conventional drying may change flavor, color and rehydration capacity of dried products (Garcia et al. 2007). In all cases the drying efficiency and energy demand is associated with drying time, which is highly related with volume of moisture in a material to be removed or the rate at which drying can be accomplished. As a pretreatment it is necessary for the food engineer to effectively predict the effect of OD on quality indices, so as not to over process the food product. Thus, kinetic modeling is essential and can provide useful information on the necessary treatment conditions, depending on the tissue to be treated. Several researchers have studied and mathematically modeled the use of OD as a pretreatment to air-drying for fruits and vegetables such as tomato and cucumber (Dermesonlouoglou et al. 2019b), goji berry (Dermesonlouoglou et al. 2018), apple (Mandala et al. 2005), banana (Fernandes et al. 2006), pumpkin (Garcia et al. 2007) and melon (Teles et al. 2006). The beneficial use of OD has also been demonstrated for dairy (Giannoglou et al. 2020c) and meat products (Andreou et al. 2018).

7 Models Application, Process Parameters Estimation and Validation

The development and application of predictive models for the description of the effect of Nonthermal process conditions on safety and quality indices is a very useful tool for the food engineers and food scientists in general. These models already

Table 3.12 Mathematical equations cited in the literature and used for the description of the effect of OD treatment on the value of mass transfer phenomena

Model	Equation	Where
Crank's model: Semi-infinite plane	$M_{OD}=\frac{(m_t-m_\infty)}{(m_0-m_\infty)}=\frac{8}{\pi^2}\sum_{n=0}^{\infty}\frac{1}{(2n+1)^2}\exp\left[-\left(n+\frac{1}{2}\right)^2\pi^2 D_{ew}\frac{t}{l^2}\right]$ $S_{OD}=\frac{(s_t-s_\infty)}{(s_0-s_\infty)}=\frac{8}{\pi^2}\sum_{n=0}^{\infty}\frac{1}{(2n+1)^2}\exp\left[-\left(n+\frac{1}{2}\right)^2\pi^2 D_{es}\frac{t}{l^2}\right]$	M and S are the diffused moisture and solute ratio, respectively, m and s are the moisture and solute content, the subscripts o, t and ∞ represent the relevant values at time 0, t and at equilibrium, D_{ew} and D_{es} (m^2/s) are the effective coefficients of water and solute diffusivity, respectively, l (m) is the half thickness of the slab.
Crank's model: Sphere	$M_{OD}=\frac{(m_t-m_\infty)}{(m_0-m_\infty)}=\frac{6}{\pi^2}\sum_{n=0}^{\infty}\frac{1}{n^2}\exp\left[n^2\pi^2 D_{ew}\frac{t}{a^2}\right]$ $S_{OD}=\frac{(s_t-s_\infty)}{(s_0-s_\infty)}=\frac{6}{\pi^2}\sum_{n=0}^{\infty}\frac{1}{n^2}\exp\left[n^2\pi^2 D_{ew}\frac{t}{a^2}\right]$	M and S are the diffused moisture and solute ratio, respectively, m and s are the moisture and solute content, the subscripts o, t and ∞ represent the relevant values at time 0, t and at equilibrium, D_{ew} and D_{es} (m^2/s) are the effective coefficients of water and solute diffusivity, respectively, a (m) is the radius of the sphere.
Crank's model: Rectangular parallelepiped	$M=\frac{m_t-m_\infty}{m_0-m_\infty}=\sum_{n=1}^{\infty}C_n^3\exp\left[-D_{ew}q_n^2 t\left[\left(\frac{1}{a^2}\right)+\left(\frac{1}{b^2}\right)+\left(\frac{1}{c^2}\right)\right]\right]$ $S=\frac{S_t-S_\infty}{S-S_\infty}=\sum_{n=1}^{\infty}C_n^3\exp\left[-D_{es}q_n^2 t\left[\left(\frac{1}{a^2}\right)-\left(\frac{1}{b^2}\right)-\left(\frac{1}{c^2}\right)\right]\right]$	M and S are the diffused moisture and solute ratio, respectively, m and s are the moisture and solute content, the subscripts o, t and ∞ represent the relevant values at time 0, t and at equilibrium, D_{ew} and D_{es} (m^2/s) are the effective coefficients of water and solute diffusivity, and C_n was equal to 2a(1 + a)/(1 + a + a2qn2), where qn's were the positive roots other than zero of equation: tan(qn) = −αqn. α was the ratio of the volume of the osmotic solution to that of piece, a, b, c are the dimensions of parallelepiped.

(continued)

Table 3.12 (continued)

Model	Equation	Where
Crank's model: Cube	$M = \frac{m_t - m_\infty}{m_0 - m_\infty} = \sum_{n=1}^{\infty} \frac{2 \cdot a \cdot (1+a)}{1+a+a^2 \cdot q_n^2} \exp\left(\frac{D \cdot q_n^2 \cdot t}{l^2}\right)$ $S = \frac{S_t - S_\infty}{S - S_\infty} = \sum_{n=1}^{\infty} \frac{2 \cdot a \cdot (1+a)}{1+a+a^2 \cdot q_n^2} \exp\left(\frac{D \cdot q_n^2 \cdot t}{l^2}\right)$	M and S are the diffused moisture and solute ratio, respectively, m and s are the moisture and solute content, the subscripts o, t and ∞ represent the relevant values at time 0, t and at equilibrium, D_{ew} and D_{es} (m^2/s) are the effective coefficients of water and solute diffusivity, and C_n was equal to 2a(1 + a)/(1 + a + a2qn2), where qn's were the positive roots other than zero of equation: tan(qn) = −αqn. α was the ratio of the volume of the osmotic solution to that of piece, l is the edge of the cube
Magee's Model	WL or SG = $kt^{0.5}$ + k0	k and k0 are empirical kinetic parameters. k is associated with the transfer rates of water and solute that occur through the osmotic-diffusional mechanism, and k0 with the gain or loss of mass that occurs after very short processing times due to the action of the hydrodynamic mechanism promoted by imposed or capillary pressures.
Peleg's model	$WL_t = WL_0 + \frac{t}{k_1 + k_2}$ $SG_t = SG_0 + \frac{t}{k_1 + k_2}$	WL and SG are the amount of water loss or solids gain at time t, g; WL_0 and SG_0 are the initial amount of water or solids, g; k_1 and k_2: Peleg's constants; and t is the time, h.
Page's model	$M = \frac{m_t - m_\infty}{m_0 - m_\infty} = \exp\left(-At^B\right)$	m and s are the moisture and solute content, the subscripts o, t and ∞ represent the relevant values at time 0, t and at equilibrium, A and B are the Page's water loss or solid gain parameters
Newton model	$M = \frac{m_t - m_\infty}{m_0 - m_\infty} = \exp\left(-k_1 t\right)$	m and s are the moisture and solute content, the subscripts o, t and ∞ represent the relevant values at time 0, t and at equilibrium, k_1 is the Newton parameter

Model	Equation	Where
Henderson–Pabis	$M = \frac{m_t - m_\infty}{m_0 - m_\infty} = a \cdot \exp(-k_2 t)$	m and s are the moisture and solute content, the subscripts o, t and ∞ represent the relevant values at time 0, t and at equilibrium, k_2 and a are the Henderson–Pabis parameters
Azuara model	$\frac{t}{WL} = \frac{1}{s1(WL_\infty)} + \frac{t}{WL_\infty}$ $\frac{t}{SG} = \frac{1}{s1(SG_\infty)} + \frac{t}{SG_\infty}$	WL or SGt, ∞: Water loss or solid gain fraction at any time, t or at equilibrium s_1 and s_2 are parameters that can be defined as relative rate constants for moisture loss and solid gain, respectively.

developed by food scientists and researchers allow for the scaling up of Nonthermal processes. Industry food engineers can adopt these equations and effectively apply them for predicting the optimal process conditions for the products to be treated. This does not require any special skills from the food engineer, neither a special tool, apart from a common software (Excel, Sigmaplot, SYSTAT, Origin softwares etc.) already installed to most PCs.

The food engineer will have to transfer the equation to one of the softwares and by replacing the model parameters with the ones estimated by scientists after numerous experiments will be able to predict the total effect on microbial, enzymatic or chemical indices (depending on the dominant deterioration factor) allowing him to decide which process conditions are more appropriate and efficient for the production.

Of course, since the models are just equations, there is always an error in the predictions, thus the food engineer has to validate the results obtained in his/her PC by results obtained from the actual production considering the process parameters estimated by the models. The deviation between predicted and observed values (effect on safety and quality indices after the processing) must not be high otherwise the model cannot predict with high accuracy and cannot be used for the scale up of processes.

8 Application of Kinetic Modeling for Process Optimization and Case Studies

Two cases studies on how a food engineer could work and take advantage of the kinetic approach and the predictions by the kinetic models on selecting the optimal process conditions are presented below. Both case studies concern the application of High pressure technology on the cold pasteurization of orange juices of different varieties; the first one is Valencia var., while the second one is Navel var.

The dominant quality indices that were taken into consideration are the dominant microbial flora-Lactic acid bacteria, LAB (*L. plantarum* and *L. brevis*) and the endogenous enzyme pectinmethylesterase that causes cloud loss, thus quality degradation of the juices. For both cases, a wide range of experiments on inactivating these indices by high pressure was conducted. The data received were mathematically described by the combined Eyring-Arrhenius equation, used to predict the inactivation rate constants at any combination of pressure and temperature for both juices. By estimating the inactivation rate constants, the necessary processing time can be determined for achieving a stable final food product. Iso-reduction plots for achieving a seven microbial log reduction and 90% PME inactivation after High processing were developed.

8.1 Case Study 1: Valencia Orange Juice

The Fig. 3.2 depicts the necessary combination of pressure and temperature process conditions for achieving the seven microbial log reduction and 90% PME inactivation, after 2 and 5 min processing time, conducted by Katsaros et al. (2010). By the obtained results, the food engineer may select the necessary process conditions for optimal high pressure processing, avoiding over-processing and products degradation or not sufficient processing for pasteurization thus survival of degradation factors. Process conditions required for the simultaneous targeted inactivation of PME and LAB in 5 min are 325 MPa and 30 °C. More intense process conditions (360 MPa and 35 °C) are required for 2 min processing.

8.2 Case Study 2: Navel Orange Juice

Similarly to the case study 1, the authors of this current chapter have unpublished data showing the iso-reduction plots for achieving a seven microbial log reduction and 90% PME inactivation after processing for 3 and 5 min for another orange juice variety, the Navel cv. one. *Lactobacillus plantarum* appeared to be relatively sensitive to pressures above 300 MPa. Navel orange PME sensitivity is in accordance with the general statement that PMEs are generally more resistant than microorganisms and that treatment for PME inactivation is sufficient for juices pasteurization. For Navel orange juice pasteurization, inactivation of 90% of the pressure/temperature labile PME fraction was considered as process target. For LAB a 7D reduction of the most resistant strain was considered. The required processing times for pasteurization of Navel orange juice PME at different process pressures, at 25 and 30 °C are shown in Fig. 3.3. Processing at 25 and 30 °C requires longer times for the PME inactivation compared to the inactivation of LAB species. The necessary pressure and temperature process conditions for the inactivation of 90% PME and 7D LAB reduction were estimated for 5 min (milder conditions are required) and 3 min (more intense treatment conditions are required) processing time. The microbial and enzymatic iso-reduction contour plots for achieving 90% PME inactivation and seven microbial log reductions after processing for 5 and 3 min are depicted in Fig. 3.4.

According to the above results, the selection of HP processing conditions was based mainly on PME inactivation. A treatment of fresh Greek Navel orange juice at 600 MPa and 40 °C for 3 min can cause inactivation of the labile isoenzyme, leading to a remaining PME activity equal to approximately 10% of the initial activity of untreated juice. These conditions also exceeded process requirements for microbial stability of orange juice.

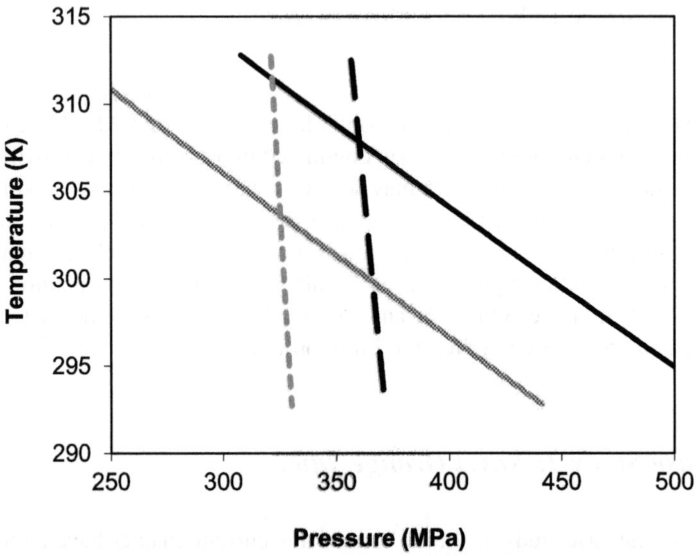

Fig. 3.2 Microbial (LAB) and enzymatic (PME) iso-reduction plots for achieving a seven microbial log reduction and 90% PME inactivation after processing for 2 and 5 min. Dashed lines represent a 7D LAB destruction and solid lines represent 90% PME inactivation. Black lines show processing for 2 min, while grey lines show processing for 5 min (own data, as published to Katsaros et al. 2010)

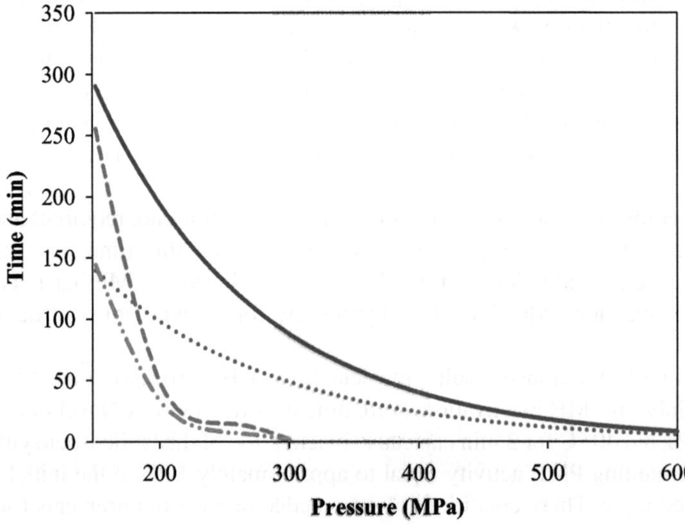

Fig. 3.3 Required processing time for the inactivation of PME and LAB as a function of pressure at 25 and 30 °C. Solid and dotted lines represent 90% PME inactivation at 25 °C and 30 °C, respectively. Dashed and dash-double dotted lines represent 7D LAB reduction at 25 °C and 30 °C, respectively

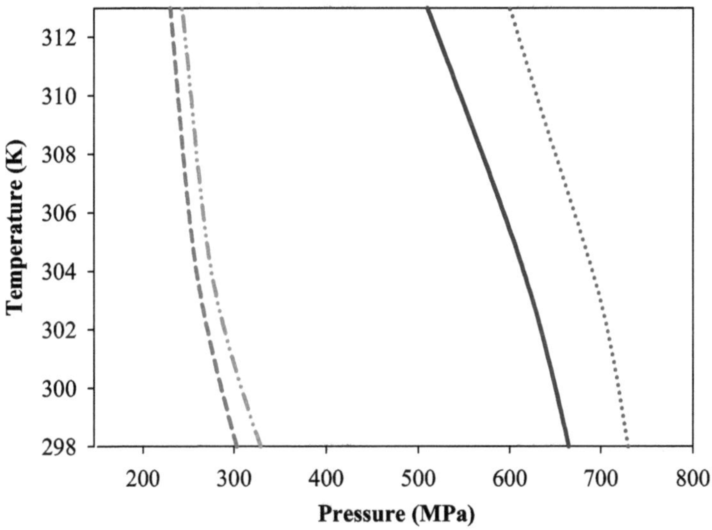

Fig. 3.4 Microbial (LAB) and enzymatic (PME) iso-reduction plots for achieving a seven microbial log reduction and 90% PME inactivation after processing for 3 and 5 min. Solid and dotted lines represent 90% PME inactivation after processing for 3 and 5 min, respectively. Dashed and dash-double dotted lines represent 7D LAB reduction after processing for 3 and 5 min, respectively

9 Conclusions and Future

In general, kinetic modeling is essential in nowadays for achieving an efficient processing in terms of producing food products of increased safety, quality, healthier profile while simultaneously being more cost effective. The scientists have done a lot of work towards producing data necessary for the development of mathematical equations for the description of the effect of process conditions with Nonthermal technologies on quality indices of new or improved food products. Nevertheless, there is space for more work on developing more reliable models with not significant errors, thus of higher accuracy when using them to scale up production from the labs to the industries. New measuring techniques providing more reliable values will boost the applicability and efficiency of kinetic models.

Nonthermal technologies are the future of food processing, but some issues related to products specificity have to be taken into consideration. Food engineers will have in their hands useful tools to predict optimal process conditions, thus evaluate for the process applied, its intensity and efficiency, enabling the use of corrective actions if needed, for production of safe, sustainable and cost-efficient final food products. Training and education has to be improved towards this direction enabling for the food engineers to efficiently take advantage of the tools-kinetic models available in the literature.

References

Aguiló-Aguayo I, Soliva-Fortuny R, Martín-Belloso O (2008a) Comparative study on color, viscosity and related enzymes of tomato juice treated by high-intensity pulsed electric fields or heat. Eur Food Res Technol 227:599–606

Aguiló-Aguayo I, Odriozola-Serrano I, Quintão-Teixeira LJ et al (2008b) Inactivation of tomato juice peroxidase by high-intensity pulsed electric fields as affected by process conditions. Food Chem 107:949–955

Aguiló-Aguayo I, Soliva-Fortuny R, Martín-Belloso O (2009a) Changes in viscosity and pectolytic enzymes of tomato and strawberry juices processed by high-intensity pulsed electric fields. Int J Food Sci Technol 44:2268–2277

Aguiló-Aguayo I, Soliva-Fortuny R, Martín-Belloso O (2009b) Avoiding non-enzymatic browning by high-intensity pulsed electric fields in strawberry, tomato and watermelon juices. J Food Eng 92:37–43

Aguiló-Aguayo I, Soliva-Fortuny R, Martín-Belloso O (2010) Impact of high-intensity pulsed electric field variables affecting peroxidase and lipoxygenase activities of watermelon juice. LWT-Food Sci Technol 43:897–902

Ahn J, Balasubramaniam VM, Yousef AE (2007) Inactivation kinetics of selected aerobic and anaerobic bacterial spores by pressure-assisted thermal processing. Int J Food Microbiol 113(3):321–329

Alexandrakis Z, Kyriakopoulou K, Katsaros G et al (2014a) Process condition optimization of high pressure pasteurized sea buckthorn juice with long shelf-life and antioxidant activity. Food Bioprocess Technol 7:3226–3234

Alexandrakis Z, Katsaros G, Stavros P et al (2014b) Comparative structural changes and inactivation kinetics of pectin methylesterases from different orange cultivars processed by high pressure. Food Bioprocess Technol 7:853–867

Alexandrakis Z, Katsaros G, Stavros P et al (2017) Inactivation kinetics and structural changes of high pressure treated actinidin. Int J Agric Sci Technol 5(1):18–29

Amiali M, Ngadi MO, Smith JP et al (2006) Inactivation of Escherichia coli O157: H7 and Salmonella enteritidis in liquid egg white using pulsed electric field. J Food Sci 71:M88–M94

Andreou V, Dimopoulos G, Katsaros G et al (2016) Comparison of the application of high pressure and pulsed electric fields technologies on the selective inactivation of endogenous enzymes in tomato products. Innov Food Sci Emerg Technol 38:349–355

Andreou V, Tsironi T, Dermesonlouoglou E et al (2018) Combinatory effect of osmotic and high pressure processing on shelf life extension of animal origin products–application to chilled chicken breast fillets. Food Packag Shelf Life 15:43–51

Andreou V, Dimopoulos G, Dermesonlouoglou E et al (2020a) Application of pulsed electric fields to improve product yield and waste valorization in industrial tomato processing. J Food Eng 270:109778

Andreou V, Psarianos M, Dimopoulos G et al (2020b) Effect of pulsed electric fields and high pressure on improved recovery of high-added-value compounds from olive pomace. J Food Sci 85:1500–1512

Balogh T, Smout C, Ly-Nguyen B et al (2004) Thermal and high pressure inactivation kinetics of carrot pectinmethylesterase (PME): from model systems to real foods. Innov Food Sci Emerg Technol 5:429–436

Barbosa-Canovas GV, Pothakamury UR, Gongora-Nieto MM et al (1999) Preservation of foods with pulsed electric fields. Academic Press, Elsevier, London

Bayindirli A, Alpas H, Bozoglu F et al (2006) Efficiency of high pressure treatment on inactivation of pathogenic microorganisms and enzymes in apple, orange, apricot and sour cherry. Food Control 17:52–58

Bendicho S, Barbosa-Cánovas GV, Martín O (2002) Milk processing by high intensity pulsed electric fields. Trends Food Sci Technol 13:195–204

Boulekou S, Katsaros G, Taoukis P (2010) Inactivation kinetics of peach pulp pectin methylesterase as a function of high hydrostatic pressure and temperature process conditions. Food Bioprocess Technol 3(5):699–706

Brandenburg R (2018) Dielectric barrier discharges: progress on plasma sources and on the understanding of regimes and single filaments. Plasma Sources Sci Technol 26:053001

Buckow R, Weiss U, Heinz V et al (2007a) Stability and catalytic activity of α-amylase from barley malt at different pressure–temperature conditions. Biotechnol Bioeng 97:1–11

Buckow R, Weiss U, Knorr D (2007b) Combined pressure and temperature effects on the catalytic activity of cellulase from Bacillus subtilis. In: Abe F, Suzuki A (eds) Proceedings of the 4th international conference on high pressure bioscience and biotechnology. J-Stage, Tsukuba

Buckow R, Heinz V, Knorr D (2007c) High pressure phase transition kinetics of maize starch. J Food Eng 81:469–475

Butz P, Fister H, Losch S et al (1996) Response of immobilized Bacillus subtilis a-amylase to high pressure inactivation. Food Biotechnol 10(2):93–103

Chakraborty S, Rao PS, Mishra HN (2015) Kinetic modeling of polyphenoloxidase and peroxidase inactivation in pineapple (Ananas comosus L.) puree during high-pressure and thermal treatments. Innov Food Sci Emerg Technol 27:57–68

Chutia H, Kalita D, Mahanta C et al (2019) Kinetics of inactivation of peroxidase and polyphenol oxidase intender coconut water by dielectric barrier discharge plasma. LWT-Food Sci Technol 101:625–629

Cook DW (2003) Sensitivity of vibrio species in phosphate-buffered saline and in oysters to high-pressure processing. J Food Prot 66(12):2276–2282

Corrales M, Toepfl S, Butz P et al (2008) Extraction of anthocyanins from grape by-products assisted by ultrasonics, high hydrostatic pressure or pulsed electric fields: a comparison. Innov Food Sci Emerg Technol 9:85–91

Crank J, Gupta RS (1975) Isotherm migration method in two dimensions. Int J Heat Mass Transf 18:1101–1107

de Mello RE Jr, Corrêa JLG, Lopes FJ et al (2019) Kinetics of the pulsed vacuum osmotic dehydration of green fig (Ficus carica L.). Int J Heat Mass Transf 55:1685–1691

Dermesonlouoglou EK, Giannakourou MC (2018) Modelling dehydration of apricot in a non-conventional multi-component osmotic solution: effect on mass transfer kinetics and quality characteristics. J Food Sci Technol 55:4079–4089

Dermesonlouoglou EK, Pourgouri S, Taoukis PS (2008) Kinetic study of the effect of the osmotic dehydration pre-treatment to the shelf life of frozen cucumber. Innov Food Sci Emerg Technol 9:542–549

Dermesonlouoglou E, Zachariou I, Andreou V et al (2016) Effect of pulsed electric fields on mass transfer and quality of osmotically dehydrated kiwifruit. Food Bioprod Process 100:535–544

Dermesonlouoglou EK, Bimpilas A, Andreou V et al (2017a) Process optimization and kinetic modeling of quality of fresh-cut strawberry cubes pretreated by high pressure and osmosis. J Food Process Preserv 41:e13137

Dermesonlouoglou EK, Andreou V, Alexandrakis Z et al (2017b) The hurdle effect of osmotic pretreatment and high-pressure cold pasteurisation on the shelf-life extension of fresh-cut tomatoes. Int J Food Sci Technol 52:916–926

Dermesonlouoglou E, Chalkia A, Taoukis P (2018) Application of osmotic dehydration to improve the quality of dried goji berry. J Food Eng 232:36–43

Dermesonlouoglou EK, Angelikaki F, Giannakourou MC et al (2019a) Minimally processed fresh-cut peach and apricot snacks of extended shelf-life by combined osmotic and high pressure processing. Food Bioprocess Technol 12:371–386

Dermesonlouoglou EK, Pantelaiaki K, Andreou V et al (2019b) Osmotic pretreatment for the production of novel dehydrated tomatoes and cucumbers. J Food Process Preserv 43:e13968

Dermesonlouoglou E, Paraskevopoulou E, Andreou V et al (2020) Osmotic dehydration for the production of novel pumpkin cut products of enhanced nutritional value and sustainability. Appl Sci 10:6225

Dhakal S, Balasubramaniam VM, Ayvaz H et al (2018) Kinetic modeling of ascorbic acid degradation of pineapple juice subjected to combined pressure-thermal treatment. J Food Eng 224:62–70

Dimitrakellis P, Giannoglou M, Zeniou A et al (2021) Food container employing a cold atmospheric plasma source for prolonged preservation of plant and animal origin food products. MethodsX 8:101177

Dogan C, Erkmen O (2004) High pressure inactivation kinetics of Listeria monocytogenes inactivation in broth, milk, and peach and orange juices. J Food Eng 62(1):47–52

Donaghy JA, Linton M, Patterson MF et al (2007) Effect of high pressure on Mycobacterium avium ssp. paratuberculosis in milk. Lett Appl Microbiol 45:154–159

Dong S, Fan L, Ma Y et al (2021) Inactivation of polyphenol oxidase by dielectric barrier discharge (DBD) plasma: kinetics and mechanisms. LWT-Food Sci Technol 145:111322

Doona CJ, Feeherry FE (2007) High pressure processing of foods. IFT Press & Blackwell Publishing: London, 2007

Doona CJ, Feeherry FE, Ross EW et al (2007) The quasi-chemical and Weibull distribution models of nonlinear inactivation kinetics of Escherichia coli ATCC 11229 by high pressure processing. In: Doona CJ, Feeherry FE (eds) High pressure processing of foods. IFT Press Blackwell Publishing, Ames, pp 115–114

Elez-Martinez P, Suarez-Recio M, Martin-Belloso O (2007) Modeling the reduction of pectin methyl esterase activity in orange juice by high intensity pulsed electric fields. J Food Eng 78:184–193

Eylen DV, Bellostas N, Strobel BW et al (2008) Influence of pressure/temperature treatments on glucosinolate• conversion in broccoli (Brassica oleraceae L. cv Italica) heads. Food Chem 112(3):646–653

Fernandes FA, Rodrigues S, Gaspareto OC et al (2006) Optimization of osmotic dehydration of bananas followed by air-drying. J Food Eng 77:188–193

Frontuto D, Carullo D, Harrison SM et al (2019) Optimization of pulsed electric fields-assisted extraction of polyphenols from potato peels using response surface methodology. Food Bioprocess Technol 12:1708–1720

Garcia CC, Mauro MA, Kimura M (2007) Kinetics of osmotic dehydration and air-drying of pumpkins (Cucurbita moschata). J Food Eng 82:284–291

Giannoglou M, Karra Z, Platakou E et al (2016) Effect of high pressure treatment applied on starter culture or on semi-ripened cheese in the quality and ripening of cheese in brine. Innov Food Sci Emerg Technol 38:312–320

Giannoglou M, Alexandrakis Z, Stavrou P et al (2018) Effect of high pressure on structural modifications and enzymatic activity of a purified X-prolyl dipeptidyl aminopeptidase from Streptococcus thermophilus. Food Chem 248:304–311

Giannoglou M, Katsaros G, Moatsou G et al (2019) Effect of high hydrostatic pressure treatment on the viability and acidification ability of lactic acid bacteria. Int Dairy J 96:50–57

Giannoglou M, Dimitrakellis P, Efthimiadou A et al (2020a) Comparative study on the effect of cold atmospheric plasma, ozonation, pulsed electromagnetic fields and high pressure technologies on sea-bream fillets quality indices and shelf-life. Food Eng Rev 13:175–184

Giannoglou M, Stergiou P, Dimitrakellis P et al (2020b) Effect of cold atmospheric plasma processing on quality and shelf life of ready-to-eat leafy salads. Innov Food Sci Emerg Technol 66:102502

Giannoglou M, Koumandraki H, Andreou V et al (2020c) Combined osmotic and air dehydration for the production of shelf-stable white cheese. Food Bioprocess Technol 13:1435–1446

Giannoglou M, Xanthou ZM, Chanioti S et al (2021) Effect of cold atmospheric plasma and pulsed electromagnetic fields on strawberry quality and shelf-life. Innov Food Sci Emerg Technol 68:102631

Giner J, Gimeno V, Espachs A et al (2000) Inhibition of tomato (Licopersicon esculentum Mill.) pectin methylesterase by pulsed electric fields. Innov Food Sci Emerg Technol 1:57–67

Giner J, Gimeno V, Barbosa-Cánovas GV et al (2001) Effects of pulsed electric field processing on apple and pear polyphenoloxidases. Food Sci Technol Int 7:339–345
Giner J, Ortega M, Mesegué M et al (2002) Inactivation of peach polyphenoloxidase by exposure to pulsed electric fields. J Food Sci 67:1467–1472
Giner J, Gimeno V, Palomes M et al (2003) Lessening polygalacturonase activity in a commercial enzyme preparation by exposure to pulsed electric fields. Eur Food Res Technol 217:43–48
Giner-Seguí J, Elez-Martínez P, Martín-Belloso O (2009) Modeling within the Bayesian framework, the inactivation of pectinesterase in gazpacho by pulsed electric fields. J Food Eng 95:446–452
Hammerick KE, Longake MT, Prinz FB (2010) In vitro effects of direct current electric fields on adipose-derived stromal cells. Biochem Biophys Res Commun 397:12–17
Heinz V, Buckow R, Knorr D (2005) Catalytic activity of b-amylase from barley in different pressure/temperature domains. Biotechnol Prog 21:1632–1638
Hoover DG, Metrick C, Papineau AM et al (1989) Biological effects of high hydrostatic pressure on food microorganisms. Food Technol 43:99–107
Hülsheger H, Potel J, Niemann EG (1981) Killing of bacteria with electric pulses of high field strength. Radiat Environ Biophys 20:53–65
Indrawati I, Van Loey AM, Ludikhuyze LR et al (2000) Kinetics of pressure inactivation at subzero and elevated temperatures of lipoxygenase in crude green beans (Phaseolous vulgaris) extract. Biotechnol Prog 16(1):109–115
Indrawati I, Van Loey AM, Ludikhuyze LR et al (2001) Pressure–temperature inactivation of lipoxygenase in green peas (Pisum sativum): a kinetic study. J Food Sci 66:686–693
Indrawati A, Van Loey A, Hendrickx M (2005) Pressure and temperature stability of 5-methyltetrahydrofolic acid: a kinetic study. J Agric Food Chem 53(8):3081–3087
Irwe S, Olsson I (1994) Reduction of pectinesterase activity in orange juice by high pressure treatment. In: Singh RP, Oliveira FAR (eds) Minimal processing of foods and process optimisation: an interface. CRC Press, Boca Raton, pp 35–42
Jofré A, Garriga M, Aymerich T (2008) Inhibition of Salmonella sp. Listeria monocytogenes and Staphylococcus aureus in cooked ham by combining antimicrobials, high hydrostatic pressure and refrigeration. Meat Sci 78:53–59
Kang KS, Hong JM, Kang JA et al (2013) Regulation of osteogenic differentiation of human adipose-derived stem cells by controlling electromagnetic field conditions. Exp Mol Med 45(1):e6
Katsaros GI, Apseridis I, Taoukis PS (2006) Modelling of high hydrostatic pressure inactivation of pectinmethylesterase from persimmon (Diospyros virginiana). IUFoST Edpsciences:1227–1238. https://doi.org/10.1051/IUFoST:20060753
Katsaros G, Giannoglou M, Taoukis P (2009a) Kinetic study of the combined effect of high hydrostatic pressure and temperature on the activity of Lactobacillus delbrueckii ssp. bulgaricus aminopeptidases. J Food Sci 74:219–225
Katsaros G, Katapodis P, Taoukis PS (2009b) Modeling the effect of temperature and high hydrostatic pressure on the proteolytic activity of kiwi fruit juice. J Food Eng 94(1):40–45
Katsaros G, Katapodis P, Taoukis P (2009c) High hydrostatic pressure inactivation kinetics of the plant proteases ficin and papain. J Food Eng 91(1):42–48
Katsaros GI, Tsevdou M, Panagiotou T et al (2010) Kinetic study of high pressure microbial and enzyme inactivation and selection of pasteurisation conditions for Valencia Orange Juice. Int J Food Sci Technol 45(6):1119–1129
Katsaros G, Alexandrakis Z, Taoukis P (2017) Kinetic assessment of high pressure inactivation of different plant origin pectinmethylesterase enzymes. Food Eng Rev 9:170–189
Kelly-Wintenberg K, Hodge A, Montie TC (1999) Use of a one atmosphere uniform glow discharge plasma to kill a broad spectrum of microorganisms. J Vac Sci Technol A17:1539
Kilonzo-Nthenge A, Liu S, Yannam S et al (2018) Atmospheric cold plasma inactivation of salmonella and Escherichia coli on the surface of golden delicious apples. Front Nutr 5:120

Labuza TP (1984) Application of chemical kinetics to deterioration of foods. J Chem Educ 61:4–348
Laroussi M (2005) Low temperature plasma-based sterilization: overview and state-of-the-art. Plasma Process Polym 2(5):391–400
Levenspiel O (1972) Interpretation of batch reactor data. Chemical reaction engineering. Wiley, New York, pp 41–47
Li YQ, Chen Q, Liu XH et al (2008) Inactivation of soybean lipoxygenase in soymilk by pulsed electric fields. Food Chem 109:408–414
Liang J, Zheng S, Ye S (2012) Inactivation of Penicillium aerosols by atmospheric positive corona discharge processing. J Aerosol Sci 54:103–112
Ludikhuyze L, De Cordt S, Weemaes C et al (1996) Kinetics for heat and pressure-temperature inactivation of *Bacillus subtilis* α-amylase. Food Biotechnol 10(2):105–129
Ludikhuyze L, Indrawati I, Van den Broeck I et al (1998) Effect of combined pressure and temperature on soybean lipoxygenase. 2. Modeling inactivation kinetics under static and dynamic conditions. J Agric Food Chem 46(10):4081–4086
Ludikhuyze L, Claeys W, Hendrickx M (2000) Combined pressure-temperature inactivation of alkaline phosphatase in bovine milk: a kinetic study. Food Eng Phys Prop 65(1):155–159
Luengo E, Álvarez I, Raso J (2013) Improving the pressing extraction of polyphenols of orange peel by pulsed electric fields. Innov Food Sci Emerg Technol 17:79–84
Luengo E, Álvarez I, Raso J (2014) Improving carotenoid extraction from tomato waste by pulsed electric fields. Front Nutr 1:12
Luo F, Hou T, Zhang Z et al (2012) Effects of pulsed electromagnetic field frequencies on the osteogenic differentiation of human mesenchymal stem cells. Orthopedics 35:e526–e531
Ly-Nguyen B, Van Loey AM, Smout C et al (2003a) Mild heat and high-pressure inactivation of carrot pectinmethylesterase: a kinetic study. J Food Sci 68:1377–1383
Ly-Nguyen B, Loey AMV, Smout C et al (2003b) Effect of mild-heat and high-pressure processing on banana pectin methylesterase: a kinetic study. J Agric Food Chem 51:7974–7979
Ma R, Wang G, Tian Y et al (2015) Non-thermal plasma-activated water inactivation of food-borne pathogen on fresh produce. J Hazard Mater 300:643–651
Mandala IG, Anagnostaras EF, Oikonomou CK (2005) Influence of osmotic dehydration conditions on apple air-drying kinetics and their quality characteristics. J Food Eng 69:307–316
Marsellés-Fontanet ÁR, Martin-Belloso O (2007) Optimization and validation of PEF processing conditions to inactivate oxidative enzymes of grape juice. J Food Eng 83:452–462
Marsellés-Fontanet ÀR, Puig A, Olmos P et al (2009) Optimising the inactivation of grape juice spoilage organisms by pulse electric fields. Int J Food Microbiol 130:159–165
Mendes-Oliveira G, Jensen JL, Keener KM et al (2019) Modeling the inactivation of Bacillus subtilis spores during cold plasma sterilization. Innov Food Sci Emerg Technol 52:334–342
Mercali GD, Tessaro IC, Noreña CP et al (2010) Mass transfer kinetics during osmotic dehydration of bananas (Musa sapientum, shum.). Int J Food Sci Technol 45:2281–2289
Min S, Min SK, Zhang QH (2003) Inactivation kinetics of tomato juice lipoxygenase by pulsed electric fields. J Food Sci 68:1995–2001
Misra NN, Keener KM, Mosnier JP et al (2014) Effect of in-package atmospheric pressure cold plasma treatment on quality of cherry tomatoes. J Biosci Bioeng 118(2):177–182
Moatsou G, Bakopanos C, Katharios D et al (2008a) Effect of high-pressure treatment at various temperatures on indigenous proteolytic enzymes and whey protein denaturation in bovine milk. J Dairy Res 75:262–269
Moatsou G, Katsaros G, Bakopanos C et al (2008b) Effect of high-pressure treatment at various temperatures on activity of indigenous proteolytic enzymes and denaturation of whey proteins in ovine milk. Int Dairy J 18:1119–1125
Moisan M, Barbeau J, Crevier MC et al (2002) Plasma sterilization, methods and mechanisms. Pure Appl Chem 74:349–358

Mokhtarian M, Majd MH, Koushki F et al (2014) Optimisation of pumpkin mass transfer kinetic during osmotic dehydration using artificial neural network and response surface methodology modelling. Qual Assur Saf Crops Foods 6:201–214

Monfort S, Gayán E, Saldaña G et al (2010) Inactivation of Salmonella Typhimurium and Staphylococcus aureus by pulsed electric fields in liquid whole egg. Innov Food Sci Emerg Technol 11:306–313

Mosqueda-Melgar J, Raybaudi-Massilia RM, Martín-Belloso O (2007) Influence of treatment time and pulse frequency on Salmonella Enteritidis, Escherichia coli and Listeria monocytogenes populations inoculated in melon and watermelon juices treated by pulsed electric fields. Int J Food Microbiol 117:192–200

Mosqueda-Melgar J, Raybaudi-Massilia RM, Martín-Belloso O (2008) Non-thermal pasteurization of fruit juices by combining high-intensity pulsed electric fields with natural antimicrobials. Innov Food Sci Emerg Technol 9:328–340

Müller-Merbach M, Hinrichs J (2006) Thermal and hydrostatic inactivation of bacteriophages. Chem Ing Tech 78(11):1723–1730

Niemira BA (2012) Cold plasma decontamination of foods. Annu Rev Food Sci Technol 3:125–142

Nienaber U, Shellhammer TH (2001) High-pressure processing of orange juice: kinetics of pectinmethylesterase inactivation. J Food Sci 66(2):328–331

Odriozola-Serrano I, Aguiló-Aguayo I, Soliva-Fortuny R et al (2007) Lycopene, vitamin C, and antioxidant capacity of tomato juice as affected by high-intensity pulsed electric fields critical parameters. J Agric Food Chem 55:9036–9042

Odriozola-Serrano I, Soliva-Fortuny R, Martín-Belloso O (2008a) Effect of minimal processing on bioactive compounds and color attributes of fresh-cut tomatoes. LWT-Food Sci Technol 41:217–226

Odriozola-Serrano I, Soliva-Fortuny R, Martín-Belloso O (2008b) Changes of health-related compounds throughout cold storage of tomato juice stabilized by thermal or high intensity pulsed electric field treatments. Innov Food Sci Emerg Technol 9:272–279

Odriozola-Serrano I, Soliva-Fortuny R, Martín-Belloso O (2008c) Phenolic acids, flavonoids, vitamin C and antioxidant capacity of strawberry juices processed by high-intensity pulsed electric fields or heat treatments. Eur Food Res Technol 228:239–248

Odriozola-Serrano I, Soliva-Fortuny R, Gimeno-Añó V et al (2008d) Modeling changes in health-related compounds of tomato juice treated by high-intensity pulsed electric fields. J Food Eng 89:210–216

Odriozola-Serrano I, Soliva-Fortuny R, Hernández Jover T et al (2009a) Carotenoid and phenolic profile of tomato juices processed by high intensity pulsed electric fields compared with conventional thermal treatments. Food Chem 112:258–266

Odriozola-Serrano I, Soliva-Fortuny R, Martín-Belloso O (2009b) Impact of high-intensity pulsed electric fields variables on vitamin C, anthocyanins and antioxidant capacity of strawberry juice. LWT-Food Sci Technol 42:93–100

Oms-Oliu G, Odriozola-Serrano I, Soliva-Fortuny R et al (2009) Use of Weibull distribution for describing kinetics of antioxidant potential changes in fresh-cut watermelon. J Food Eng 95:99–105

Paixão LMN, Fonteles TV, Oliveira VS et al (2019) Cold plasma effects on functional compounds of siriguela juice. Food Bioprocess Technol 12:110–121

Pankaj SK, Misra NN, Cullen PJ (2013) Kinetics of tomato peroxidase inactivation byatmospheric pressure cold plasma based on dielectric barrier discharge. Innov Food Sci Emerg Technol 19:153–157

Pankaj SK, Wan Z, Colonna W et al (2017) Effect of high voltage atmospheric cold plasma on white grape juice quality. J Sci Food Agric 97(12):4016–4021

Pappas D (2011) Status and potential of atmospheric plasma processing of materials. J Vac Sci Technol A29(2):020801

Parish ME (1998) High pressure inactivation of Saccharomyces cerevisiae, endogenous microflora and pectinmethylesterase in orange juice. J Food Saf 18:57–65

Pataro G, Carullo D, Falcone M et al (2020) Recovery of lycopene from industrially derived tomato processing by-products by pulsed electric fields-assisted extraction. Innov Food Sci Emerg Technol 63:102369

Patterson MF, Margey DM, Mills G et al (1997) The effect of high hydrostatic pressure treatment on microorganisms in foods. In: Heremans K (ed) High pressure research in the biosciences and biotechnology, Proceedings of the XXXIVth meeting of the European High Pressure Research Group. Leuven University Press, Leuven, pp 269–272

Pawluk W (2015) Magnetic fields for pain control. In: Markov M (ed) Electromagnetic fields in biology and medicine. CRC Press

Pękosławska A, Lenart A (2009) Osmotic dehydration of pumpkin in starch syrup. J Fruit Ornam Plant Res 2(17):107–113

Peleg M (1988) An empirical model for the description of moisture sorption curves. J Food Sci 53:1216–1217

Peleg M (1995) A model of microbial survival after exposure to pulsed electric fields. J Sci Food Agric 67:93–99

Peleg M (2006) Advanced quantitative microbiology for foods and biosystems. CRC Press, Boca Raton

Pérez MP, Aliaga DR, Bernat CF et al (2007) Inactivation of Enterobacter sakazakii by pulsed electric field in buffered peptone water and infant formula milk. Int Dairy J 17:1441–1449

Polydera A, Galanou E, Stoforos N et al (2004) Inactivation kinetics of pectin methylesterase of greek Navel orange juice as a function of high hydrostatic pressure and temperature process conditions. J Food Eng 62:291–298

Puač N, Škoro N, Spasić K et al (2017) Activity of catalase enzyme in Paulownia tomentosa seeds during the process of germination after treatments with low pressure plasma and plasma activated water. Plasma Process Polym 15(2):1700082

Rajan S, Pandrangi S, Balasubramaniam VM et al (2006) Inactivation of Bacillus stearothermophilus spores in egg patties by pressure assisted thermal processing. LWT-Food Sci Technol 39(8):844–851

Ramazzina I, Tappi S, Rocculi P et al (2016) Effect of cold plasma treatment on the functional properties of fresh-cut apples. J Agric Food Chem 64:8010–8018

Rana S, Mehta D, Bansal V et al (2020) Atmospheric cold plasma (ACP) treatment improved in-package shelf-life of strawberry fruit. J Food Sci Technol 57:102–112

Riener J, Noci F, Cronin DA et al (2008) Combined effect of temperature and pulsed electric fields on apple juice peroxidase and polyphenoloxidase inactivation. Food Chem 109:402–407

Riener J, Noci F, Cronin DA et al (2009) Combined effect of temperature and pulsed electric fields on pectin methyl esterase inactivation in red grapefruit juice (Citrus paradisi). Eur Food Res Technol 228:373–379

Rivas A, Rodrigo D, Martinez A et al (2006) Effect of PEF and heat pasteurization on the physical–chemical characteristics of blended orange and carrot juice. LWT-Food Sci Technol 39:1163–1170

Rodrigo D, Barbosa-Cánovas GV, Martinez A et al (2003) Pectin methyl esterase and natural microflora of fresh mixed orange and carrot juice treated with pulsed electric fields. J Food Prot 66:2336–2342

Rodrigo D, Jolie R, Van Loey A et al (2006) Combined thermal and high pressure inactivation kinetics of tomato lipoxygenase. Eur Food Res Technol 222:636–642

Saguy I, Karel M (1980) Modeling of quality deterioration during food processing and storage. Food Technol 37(2):78–85

Santillana Farakos SM, Zwietering MH (2011) Data analysis of the inactivation of foodborne microorganisms under high hydrostatic pressure to establish global kinetic parameters and influencing factors. J Food Prot 74(12):2097–2106

Sarangapani C, O'Toole G, Cullen PJ et al (2017) Atmospheric cold plasma dissipation efficiency of agrochemicals on blueberries. Innov Food Sci Emerg Technol 44:235–241

Sarkis JR, Boussetta N, Blouet C et al (2015) Effect of pulsed electric fields and high voltage electrical discharges on polyphenol and protein extraction from sesame cake. Innov Food Sci Emerg Technol 29:170–177

Schlüter O, Ehlbeck J, Hertel C et al (2013) Opinion on the use of plasma processes for treatment of foods. Mol Nutr Food Res 57:920–927

Segat A, Misra NN, Cullen PJ et al (2016) Effect of atmospheric pressure cold plasma (ACP) on activity and structure of alkaline phosphatase. Food Bioprod Process 98:181–188

Sensoy I, Zhang QH, Sastry SK (1997) Inactivation kinetics of Salmonella dublin by pulsed electric field. J Food Process Eng 20:367–381

Sereno AM, Moreira R, Martinez E (2001) Mass transfer coefficients during osmotic dehydration of apple in single and combined aqueous solutions of sugar and salt. J Food Eng 47:43–49

Seyderhelm I, Boguslawski S, Michaelis G et al (1996) Pressure induced inactivation of selected food enzymes. J Food Sci 61(2):308–310

Smelt JPPM (1998) Recent advances in the microbiology of high pressure processing. Trends Food Sci Technol 9:152–158

Surowsky B, Schlüter O, Knorr D (2014) Interactions of non-thermal atmospheric pressure plasma with solid and liquid food systems: a review. Food Eng Rev 7(2):82–108

Tadevosian A, Kalantarian V, Trchounian A (2006) The effects of electromagnetic radiation of extremely high frequency and low intensity on the growth rate of bacteria Escherichia coli and the role of medium pH. Biofizika 52(5):893–898

Tappi S, Berardinelli A, Ragni L et al (2014) Atmospheric gas plasma treatment of fresh-cut apples. Innov Food Sci Emerg Technol 21:114–122

Tappi S, Gozzi G, Vannini L et al (2016) Cold plasma treatment for fresh-cut melon stabilization. Innov Food Sci Emerg Technol 33:225–233

Teles UM, Fernandes FA, Rodrigues S et al (2006) Optimization of osmotic dehydration of melons followed by air-drying. Int J Food Sci Technol 41:674–680

Torgomyan H, Kalantaryan V, Trchounian A (2011) Low intensity electromagnetic irradiation with 70.6 and 73 GHz frequencies affects Escherichia coli growth and changes water properties. Cell Biochem Biophys 60:275–281

Torreggiani D (1993) Osmotic dehydration in fruit and vegetable processing. Food Res Int 26:59–68

Tsai MT, Li WJ, Tuan RS et al (2009) Modulation of osteogenesis in human mesenchymal stem cells by specific pulsed electromagnetic field stimulation. J Orthop Res 27:1169–1174

Tsironi T, Maltezou I, Tsevdou M, Katsaros G, Taoukis PS (2015) High pressure cold pasteurization of gilthead sea bream fillets: selection of process conditions and validation of shelf-life extension. Food Bioprocess Technol 8:681–690

Van den Broeck I, Ludikhuyze LR, Van Loey AM et al (2000) Inactivation of orange pectinesterase by combined high-pressure and temperature treatments: a kinetic study. J Agric Food Chem 48(5):1960–1970

Van Loey A, Ooms V, Weemaes C et al (1998) Thermal and pressure-temperature degradation of chlorophyll in broccoli (Brassica oleracea L. italica) juice: a kinetic study. J Agric Food Chem 46(12):5289–5294

Van Opstal I, Vanmuysen SCM, Wuytack EY et al (2005) Inactivation of Escherichia coli by high hydrostatic pressure at different temperatures in buffer and carrot juice. Int J Food Microbiol 98(2):179–191

Verbeyst L, Crombruggen, KV, Plancken, IV et al (2011) Anthocyanin degradation kinetics during thermal and high pressure treatments of raspberries. J Food Eng 105(3):513–521

Weemaes C, Rubens P, De Cordt S et al (1997) Temperature sensitivity and pressure resistance of mushroom polyphenoloxidase. J Food Sci 62(2):261–266

Weemaes CA, Ludikhuyze LR, Van den Broeck I et al (1998) Kinetics of combined pressure-temperature inactivation of avocado polyphenoloxidase. Biotechnol Bioeng 60(3):292–300

Weibull W (1951) A statistical distribution function of wide applicability. J Appl Mech 18:293–297

Won MY, Lee SJ, Min SC (2017) Mandarin preservation by microwave-powered cold plasma treatment. Innov Food Sci Emerg Technol 39:25–32

Wouters PC, Alvarez I, Raso J (2001) Critical factors determining inactivation kinetics by pulsed electric field food processing. Trends Food Sci Technol 12:112–121

Xu L, Garner AL, Tao B et al (2017) Microbial inactivation and quality changes in orange juice treated by high voltage atmospheric cold plasma. Food Bioprocess Technol 10:1778–1791

Yeom HW, Zhang QH, Chism GW (2002) Inactivation of pectin methyl esterase in orange juice by pulsed electric fields. J Food Sci 67:2154–2159

Zhong KUI, Chen F, Wu J et al (2005) Kinetics of inactivation of Escherichia coli in carrot juice by pulsed electric field. J Food Process Eng 28:595–609

Chapter 4
Electro – Technologies

Ilknur Ucak, Maliha Afreen, Evgenia Benova, Plamena Marinova, Todor Bogdanov, Maria Turtoi, Livia Patraşcu, and Iuliana Aprodu

1 Pulsed Electric Field

Ilknur Ucak and Maliha Afreen

1.1 Introduction to Pulsed Electric Field

Pulsed electric field (PEF) is a technique that includes the use of uninterrupted current voltage pulses for a very small duration, ranging from nanoseconds to milliseconds, from 100–300 V/cm to 20–80 kV/cm (Fincan and Dejmek 2002; Koubaa

I. Ucak · M. Afreen
Nigde Omer Halisdemir University, Faculty of Agricultural Sciences and Technologies, Nigde, Turkey
e-mail: ilknurrucak@nigde.edu.tr

E. Benova
Clean & Circle Center of Competence, Sofia University, Sofia, Bulgaria

P. Marinova
Faculty of Forest Industry, University of Forestry, Sofia, Bulgaria
e-mail: plamena_dragozova@ltu.bg

T. Bogdanov
Medical Faculty, Medical University, Sofia, Bulgaria
e-mail: tbogdanov@medfac.mu-sofia.bg

M. Turtoi (✉) · L. Patraşcu
Cross-Border Faculty, Dunarea de Jos University of Galati, Galati, Romania
e-mail: maria.turtoi@ugal.ro; livia.patrascu@ugal.ro

I. Aprodu
Faculty of Food Science and Engineering, Dunarea de Jos University of Galati, Galati, Romania
e-mail: iuliana.aprodu@ugal.ro

A. Režek Jambrak (ed.), *Nonthermal Processing in Agri-Food-Bio Sciences*, Food Engineering Series, https://doi.org/10.1007/978-3-030-92415-7_4

et al. 2015). The strength of PEF can rely on an object sited in the middle of two electrodes. This voltage produces an electric field, the strength of which can be determined by the distance between the electrodes and the voltage supplied. There is not a proper description of PEF, electric fields of low-intensity include field strengths of $E < 0.1$ kV/cm, electric fields of moderate intensity include the range of 0.1 to 1 kV/cm, and electric fields of high-intensity include the range of electric field greater than 1 kV/cm (Asavasanti et al. 2010).

Pulsed electric field is an eco-friendly procedure; that can be efficiently used in many food handling processes such as inactivation of enzyme/microorganism, freezing and dehydration and extraction of bioactive compounds etc. (Li and Farid 2016). Muscle foods, liquids, semi-liquid, and solid can be processed by PEF (Barba et al. 2016; Bhat et al. 2018). Electric fields greater than 20 kV/cm establish a substitute to traditional thermal techniques to eradicate pathogenic microorganisms and similar enzymes, with slight modification in nutritive, sensory and health promoting characteristics of liquid foodstuffs (Sánchez-Vega et al. 2014). Low electric field perforates the biological membrane and makes it semi-permeable for the time being or eternally (Barba et al. 2014; Deng et al. 2014), which can pass significant high-quality worth compounds from diverse mediums. When cytoplasmic membranes of cells become damaged due to deterioration, then objects can be transferred through the cell membrane by treating them with heat, grinding or through enzymatic procedures in many cases. Food materials are pretreated by mechanical grinding, heat, or enzymes to improve the rate of transferring objects by enhancing the penetrability of the cell membranes (Toepfl et al. 2006a). These procedures need a substantial quantity of mechanical or thermal energy. Pulsed electric fields (PEFs) is a replacement of conservative procedures. This technique has been verified as an effective operational procedure for permanent cell membranes permeabilisation in animal and plant tissues without enhancing temperature and at a low operational charge (Toepfl et al. 2006b). This technique was used in the middle of the past century by Russian scientists for the extraction of diverse internal compounds, like juice from vegetables and fruits and sugar from beets (Flaumenbaum 1949).

1.1.1 PEF Processing System

The general electrical system for pulse electric field processing comprises a pulsed power supply, treatment chamber, and governing system (Buchmann et al. 2018). High-voltage is delivered across the food containing treatment chamber through a pulse power generator (Sack and Mueller 2017). The pulse power generator contains distinct inert components, including (resistive, inductive and capacitive), power switches and transformers (Kumar et al. 2019). Power switches are used to transmit the energy deposited in the capacitors of inductors (Redondo 2017). Every component of the pulsed electric field system has many related constraints that control the circumstances of foodstuff and any required procedure. Some factors are essential for the efficient use of this technology to a profitable level. The significant factors of this procedure are pulse waveform, flow rate, pulse repetition rate, specific energy density, strength of the electric field, exposure time, and temperature deviation in the treated product (Amit et al. 2017). These factors can be evaluated and equated according to the desired energy requirement.

1.1.2 Apparatus of Pulsed Electric Field

Pulsed electric field apparatus contains two fundamental parts: treatment chamber and pulse generator (de Haan 2007). The pulse generator contains a charger that transforms the alternate current to direct current and charges an energy storage device, including an inductor or a capacitor. Electrical energy released in the treatment chamber is ordered by a switch that is the peak critical part of a pulse generator for industrial uses because it essentially turns off and, on a circuit, at large current and high voltages in microseconds. A pulse transformer is used to control the voltage supply in the capacitor. The treatment chamber that comprises the sample to be handled contains two electrodes alienated by a dividing material. Many new strategies of treatment chamber have been established in the last few years. The best significant designs used for a pulsed electric field are coaxial, colinear and parallel electrode patterns. The most typical factors considered for the PEF process are the strength of the electric field, pulse width, pulse shape, number of pulses, specific energy of pulses, and frequency (Barssoti et al. 1999).

1.1.3 Mechanism of Action

When an external electric field is not applied, a transmembrane potential of about 10 mV arises vertically in a cell due to the accumulation of opposed pole charges on the respective side of a membrane (Toepfl et al. 2014). Once the external electric field is applied, extra potential energy is produced, determined by the power of the applied field adjoining the cells. In PEF processing, the food is positioned between two electrodes and an external electric field is implemented, which produces the movement of the ions of the applied electric field alongside the way of force lines of both the outside and inside the cells. Due to these cells become polarized by gathering of ions on the membrane (Teissie et al. 1985), which leads to a decrease in the membrane thickness because of the electro compressive forces between oppositely charged ions on each side of the membrane. When the power of the electric field increases from the acute onset value of the transmembrane potential of about 1 V (Weaver 2000), these electro-compressive forces produce an electrical breakdown of the cell membrane, which conclude as pores in the membrane. This procedure is also known as electroporation, which raises membrane permeability.

1.2 Applications of PEF Technology

Pulsed electric field procedure has been used to deactivate diverse enzymes and microorganisms and decrease their actions in egg products, milk products, juice, and other food liquids to confirm harmless and suitable food quality according to consumer's requirements (Kempkes and Munderville 2017). Food preservation through PEF requires less energy, so the temperature of the preserved food sample does not increase as compared to conservative sterilization procedures (Timmermans et al. 2019). This technique is used to pretreat muscle foods and apples, eggshells and potatoes (Barba et al. 2016; Bhat et al. 2018). The efficiency of the extraction procedure can be enhanced through the PEF procedure, including extraction of bioactive compounds, extraction of juice from grapes or apple, and sugar from beetroot (Gómez et al. 2019; Redondo et al. 2018). Pretreatment of potatoes through PEF presents considerable effects on quality features of French fries throughout the industrial level production (Fauster et al. 2018). Processing of beef muscles through PEF decreases shear force and enhanced softness (Bekhit et al. 2014). Pulsed electric field treatment can influence the properties of dehydration and desiccating time can be reduced, which encourages the preservation of bio-compounds in dehydrated samples (Parniakov et al. 2016). The effectiveness of PEF depends on different factors, including pH and conductivity of handled food sample; electric field strength, specific energy, treatment time, pulse width, pulse shape, temperature and frequency for cell permeabilization; size, shape and membrane structure of treated cells. Industrial applications of PEF are the following.

1.2.1 Extraction by Diffusion

The resulted juices from cold diffusion had less amount of pectin, and the juice color was routinely 3 to 4 times less intense than the commercial juice color (Jemai and Vorobiev 2006). Furthermore, the juice quality was better than for raw (untreated) slices after thermal diffusion at around 70 °C for slices pretreated with PEF (Lebovka et al. 2007). The scale-up study has been reported on PEF-assisted aqueous sugar filtration that used a pilot counter current extraction system. Moreover, with electric field strength ranging from 100 to 600 V/cm, diffusion heat from 30–70 °C and drafts ranging from 120 to 90%, PEF processing of sugar beet cassettes was conducted. Sugar diffusivity was reported mostly the same in raw tissue, around 60 °C in sugar beet tissue, while pretreated with PEF at 30 °C (Jemai and Vorobiev 2003). The application of moderate heating at 50 °C and implementation of PEF is a viable technique for diffusion time reduction. In contrast to those industrial thermal diffusion techniques, PEF-assisted cold provided juices with fewer contaminants (Loginov et al. 2011; Loginova et al. 2011, 2012). The findings showed that using PEF technologies in sugar processing could lead to a more environmentally sustainable and effective process by reducing energy usage and cost-efficiency. Thus, PEF treatment permits the productive extract of raw collagen protein from bovine bone against the impurity of all other small proteins obtained from previous researches. With electrical field strength of 22 kV/cm, the maximum performance of soluble collagen (16.21 mg/ml) was attained. The results of PEF were analyzed mainly on extraction from saffron (*Crocus sativus*) of significant compounds (crocin, safranal and picrocrocin) (Pourzaki et al. 2013). Pulsed electric field therapy (5 kV/cm, pulsation period 35 μs and 100 pulses) has been shown to have induced a notable rise in the saffron-stigmatic and saffron-pomace recovery of crocin, saffron, and picrocrocin. Previous research developed a method of extraction of PEF from the Chinese medicinal herb *Aconitum coreano* to increase the yield of the alkaloids. A solid to solvent ratio of 20 kV/cm of electric field and a 90% ethanol-water solution around 1:12 solid-to-solvent ratio were used (Bai et al. 2013). It has been revealed from other experiments that a high intensity PEF-assisted extraction technique was used to enhance extractive conditions to retrieve Tibetan spiritual mushroom broth's exopolysaccharides (EPS) (Zhang et al. 2011). The findings show that such extraction strategies have an optimal 40 kV/cm electric field pressure at eight pulses with pH 7. The influence on exopolysaccharides (EPS) extraction of the electric field strength > pH > the number of pulses was identified by the order mentioned above. EPS extraction improved by 84.3% under ideal conditions. For oil extracted with PEF processed (3 KV/cm), rape seeds were obtained higher oil yields and a higher concentration of tocopherols, polyphenols, complete antioxidants, and phytosterols (Guderjan et al. 2007). Pulse electric field helped press enhanced Golden Delicious apple slices juice extraction (Bazhal and Vorobiev 2000). The pre-treatment of plant tissue by thermal and PEF resulted in an improved yield of juice during the further description. In yet another research, the method for handling apple tissue damage under simultaneous pressure and PEF field treatment was addressed. Compared to the treated sample, the yield of juice increased with an increase in field pressure. The whole structure, polyphenol content, and antioxidant ability for PEF-assisted and conventional pressing were identical.

1.2.2 Extraction by Pressing

Pressing is commonly used in the processing of fruit and vegetable juices. Throughout these operations, juices bound in plant materials are extracted by compression or pressing. Various devices (screw, belt, hydraulic, and or filter presses) and pre-treatment operations were used to promote the expression of juices. The severe mechanical, thermal or enzymatic treatment induces deterioration of plant tissues, juice contamination, and multi-stage juice clarity (Albagnac et al. 2002). Several studies have shown that PEF pre-treatment before pressing or a mixture of PEF and pressing permits significant increases in juice yield and consistency (Donsì et al. 2010; Grimi et al. 2014; Jaeger et al. 2012). Pulsed electric field assisted pressing is a laboratory device that, consisting of a treatment cell, is filled initially with slices, a moveable electrode is linked to a diaphragm, and a gauze electrode is straddling between a filter cloth and a layer of slices. The two electrodes are attached to the PEF generator. The strain of compressed air is applied to the sheet of slices employing a movable electrode and a flexible diaphragm in studies performed to confirm the findings of PEF treatment during the processing slices of sugar beet (Praporscic 2005). PEF-linked pressing greatly improved the juice yield, raising it to 43%, 68%, and 79%, accordingly, for the electrical field strength gradations of 215–427 V/cm. It was found that the initial pressing of slices was used to ensure strong electrical interaction. The energy intake for the raw resources was 0.6–1 Wh/kg and the juice generated was non-uniform, less flavored, and had a higher concentration of sugar. An output of 80% in juice per original cossette amount was placed adjacent to washing. The consistency of juice (96–98%) was significantly higher for PEF-assisted pressing relative to raw juice. Sugar crystals formed after evaporation and crystallization of PEF-treated juices have been less colored than sugar crystals collected from factory juices. As a result of PEF application, the amounts of pulp and juice obtained contained 3–5 fold more α-amino nitrogen and two to three-fold more Na and K than that derived from marketing sugar pulp.

1.2.3 Colorants

Nowadays, scientists focus on replacing artificial colorants with natural colorants in the food industry, which can be obtained from vegetable and fruit extracts. According to consumer desires, this approach can be accredited to healthy and safe food products with natural colorants (Cai et al. 2005). These natural colorants involve carotenoids, chlorophylls, betalains, flavonoids and anthocyanin. It was observed during extraction procedure that pre-treatment of fruits and vegetables through PEF enables the discharge of carotenoids (Grimi et al. 2007; Koehler et al. 2005), betalains (Chalermchat et al. 2004; Fincan et al. 2004; Lopéz et al. 2009), chlorophylls (Koehler et al. 2005), and anthocyanins (Corrales et al. 2008; Gachovska et al. 2010). Koehler et al. (2005) evaluated the use of PEF by treating *Chlorella vulgaris* with 100 kJ/kg at 15 kV/cm for the extraction of carotenoids and chlorophyll and attained higher production of 52.2% and 80% correspondingly. Lopéz et al. (2009) stated that the implementation of five pulses with an energy of 0.24 kJ/kg at 7 kV/cm produced four times the total quantity of betalain obtained from red beetroot. Gachovska et al. (2010) reported that PEF treatment of 16.63 kJ/kg at 2.5 kV/cm on red cabbage released anthocyanins 2.15 times more than the normal extraction process.

1.2.4 Microbial Inactivation through PEF

The potential of PEF for liquids food production has grown over the last centuries in terms of its potential to manufacture peaceful and sustainable products. Microbial inhibition in whole milk was caused by PEF therapy in conjunction with managed preheating (55 °C for 24 s) and stepwise moderate cooling in a survey performed by Sharma et al. (2014). Pulsed electric field therapy at 22–28 kV/cm, 20 μs pulse at 10–60 Hz, 17–101 μs, results in a 5–6 log phase reduction of certain bacteria's (*Pseudomonas aeruginosa, Escherichia coli, Staphylococcus aureus*, and *Listeria innocua*) to points far below detectable threshold. Sharma et al. (2014) stated cumulative plate count decreases to 2–3 log phases in bovine raw milk following PEF treatment (20.7–26.2 kV/cm, 20 μs pulse at 10–60 Hz, 17–101 μs). Such rate of inactivation of microorganisms was close to those obtained through thermal pasteurization (63 °C for 30 min or 73 °C for 15 s) (Amiali et al. 2004). The findings revealed that the inactivation of microorganisms improved with an increasing number of pulses, and a reduction of 1, 3, and 3.5 logs were attained from whole egg products.

The PEF treatment (35 kV/cm, 4 μs bipolar pulses, max. 40 °C) on *Salmonella enteritidis* and *E. coli O157:H7* populations inoculated in the treatment period (0–2000 μs) and pulse duration (100–250 Hz), for mango, pear, orange and strawberry juices have shown by Mosqueda-Melgar et al. (2008). Pulsed electric field blends against specific pathogens have been assessed through citric acid or cinnamon bark oil. In this analysis, only fruit juice was PEF-treated and it has been observed that microbes have been reduced by upwards of 5 log units. The mixture of PEF with citric acid at 0.5, 0.1 and 1.5%, or cinnamon bark oil at 0.05, 0.1, and 0.1% have been used to pasteurize the juices of the grapes, apples, and pears. Inhibition of *Salmonella panama, E. coli, Listeria monocytogenes,* and *Saccharomyces cerevisiae* with a steady flow of PEF (14 mL/min., 20 kV/cm, and 120–964 Hz variable frequencies) was studied in apple, watermelon and orange juice (Timmermans et al. 2014). After induction of cell membrane electroporation, materials from foreign sources are incorporated into the cell, such as DNA. When PEF is detached, the membranes are repaired and the electropores are sealed. A similar process can be used for fusing two microbial cells by the formation and reunification of electropores. The method must be carefully supervised in genetically engineering technology and biotechnology to ensure that pathogens are sustainable by using PEF. A specific theory refers to inhibition by increased medication time or severity of microbes, thus ensuing the irretrievable interruption of the cell membrane. Over the past decades, there has grown interest in applying PEF to fluid food processing as regards the potential to manufacture healthy, stable products. Pulsed electric field treatment combined with managed preheating (55 °C in 24 s) and moderate cooling were triggered by the microbial inactivation in whole milk (Sharma et al. 2014). The reduction of certain bacteria to levels below the detectable limit was obtained from an electric field of 22–28 kV/cm for 20 μs pulse at 10–60 Hz, for 17–101 μs, 5 to 6 log cycles. In another study, overall plate counting was reduced by PEF (20 μs pulse, 20.7–26.2 kV/cm, 10–60 Hz for 17–101 μs) in bovine whole milk to 2–3 log phases. These grades of pathogenic inactivation are similar to those that resulted in pasteurization (63 °C for 30 min or 73 °C for 15 °C) (Sharma et al. 2014).

Mechanisms of Inactivation by PEF

The microorganism subjected to PEF is inactivated because of the cell membrane's electromechanical instability (Castro et al. 1993; Zimmermann 1986a). The cell membrane defends the microorganisms from the atmosphere, which is an enduring obstacle. The cell membrane regulates the metabolic activity of the cell by establishing an appropriate osmotic border between both the cell and its surroundings. The microbial inhibition hypothesis by PEF has been established in 1968 (Sale and Hamilton 1968). A transmembrane potential (TMP) is generated through the cell membrane by the external electric field. Sale and Hamilton (1968) mentioned that cell lysis due to lack of membrane stability took place while the TMP was approximately 1 volt. Zimmermann (1986a) suggested the concept of a "dielectric-rupture theory," close to that of Sale and Hamilton (1968). According to this principle, the cell membrane is assumed to be a capacitor packed with a dielectric substance. The dielectric constant of most foodstuffs was around 60 and 80. Free charges at all membrane surfaces accumulate due to the disparity in dielectric constancies. The regular TMP is approximately 10 mV. A rise in TMP induces cell membrane susceptibility to an electric field. The growth of TMP decreases the thickness of the membrane. An electric pressure of the membrane is opposed by a viscoelastic restore force. As the electro-compressive force grows faster than the viscoelastic force at a TMP of about 1 V, local degradation of the membrane happens. Any local disruption in the cell membrane will develop abruptly in an electrical field with sufficient strength to overpower the opposite viscoelastic power. Having a standard value of 0.5 mm as the cell radius, it is appropriate to raise the TMP to the predefined threshold by 13.33 kV/cm and thus induce the forming of pores. When the electrical field intensity is further enhanced, large pores are formed and an inevitable deterioration happens.

The membrane has a well-ordered structure. As moderate potentials are used, dipolar reorientation happens in single layer phospholipids. Sale and Hamilton (1968) proposed that such a polarized realignment may well be an initiating mechanism leading to conformational changes in the membrane structure, contributing to a deterioration of the role of the membrane as a semi-permeable buffer, which in turn contributes to cell inactivation. Tsong (1991) suggested a related hypothesis, named the "electroporation theory". The electroporation process is frequently used for the genetic transformation of microbial species, and the technique may provide visibility into the inactivation of microorganisms under PEF treatment (Palaniappan and Sastry 1990). The cell membrane is sensitive to electrical fields due to the dipole existence of the lipid molecules and its limited permeability to ions (Tsong 1991). The introduction of an electrical field induces both electrical and thermal effects in cell membranes. The lipid bilayer is vulnerable to generated voltage fields due to its net electrical current (Tsong 1991). The introduction of an electrical field induces improvements in the validation of lipid molecules by widening internal pores or by developing new hydrophobic pores that eventually create hydrophilic pores that are more structurally stable.

1.2.5 Dehydration

Many previous studies presented that high permeability of cell membrane by PEF treatment increases conservative and osmotic dehydration. For instance, it has been stated that PEF treatment of vegetables and fruits such as red peppers, potatoes, and coconuts, decreased the conventional dehydration time by 20% to 30% without enhancing the temperature higher than 60 °C (Ade-Omowaye et al. 2000, 2003b). Researchers observed the rate of osmotic dehydration of vegetables, including carrots, apples, and peppers, and found that PEF pretreated samples have a higher release of water (up to 30%) and dehydration rate as compared with the untreated samples (Ade-Omowaye et al. 2002, 2003a; Amami et al. 2006, 2007; Taiwo et al. 2002, 2003).

1.2.6 Biorefinery

In recent research, scientists attempted to make valuable products from food processing wastes that appealed consumers (Parliament 2011; Roselló-Soto et al. 2015a). In this context, PEF related procedures can be used to produce high-added value compounds from food wastes and by-products, which can be used for the production of many useful products, including food additives, pharmaceuticals or nutraceuticals) (Roselló-Soto et al. 2015b). Scientists retrieved these valuable compounds from seeds, peels, and husks of grapes, kernels, marks and oil-cakes (Mahnič-Kalamiza et al. 2014). Pulsed electric field related procedures were also used for biorefinery of agricultural remains, including sawdust, leaves, stems, bark, vine shoots, energy crops, municipal wastes and debris, etc. (Mahnič-Kalamiza et al. 2014). These resources contain large quantities of bioactive compounds, particularly polyphenols (anthocyanins, flavonol glycosides, phenolic acids, catechins) including antibacterial, antifungal, antiviral and antioxidant properties. Moreover, PEF-assisted retrieval of proteins and trehalose can be a valuable tool in the food, medicine, and cosmetics industries. Pulsed electric field related treatment resulted in considerable enhancement in biogas production (Mahnič-Kalamiza et al. 2014).

1.2.7 Reduction of Food Contaminants

The European Food Safety Authority reports that acrylamide in foodstuff is a public well-being issue as it can be a cancer-causing substance. There is also a need to create alternative methods that can decrease the consumption of acrylamide in foodstuffs. Few findings have been reported to support the potential formulation or degradation of undesirable Maillard reaction related products, such as acrylamide, by PEF treatment. Previous researches have indicated that PEF processing decreases the accumulation of acrylamide in potato goods (Kalum and Hendriksen 2014). It has been suggested that PEF pre-treatment can facilitate the enzymatic action of glucose oxidase contained in membrane-derived vegetable cells, thereby decreasing the quantity of asparagine or glucose in potato goods prior to the frying process. Besides, Jaeger et al. (2010) contributed that the PEF inhibits the action of Maillard primarily owing to its capacity to facilitate the release of reduced potato sugars. Thus, PEF pretreatment can be used as a possible method to extract sugars before cooking.

1.3 Advantages of PEF

The three most critical specifications defined for PEF production are electric field power, temperature treatment, and energy supply (Toepfl et al. 2014; Amiali et al. 2007). As a nonthermal system, PEF processing produces less loss of the sensory and nutritional properties of food than conventional thermal production technology (Buckow et al. 2013; Walkling-Ribeiro et al. 2010). It has several benefits, such as lower treatment temperatures, faster processing cycles, and possible continuous flow, relative to conventional production technology rendering it a really promising technology for food producers (Walkling-Ribeiro et al. 2011; Puértolas et al. 2010a, b). Pulsed electric field may allow the food industry to manufacture high-quality foodstuffs based on enlisted points:

- Efficient protection of fluids of low ionic strength and low conduction
- Provides maximum returns and higher consistency during the processing of juices and oils
- Improves mass transfer in citrus, meat, or fish drying methods at lower temperatures
- At lower temperatures, pigment extraction can be achieved
- Good effect on quicker protein digestion
- Pulsed electric field decreases the energy required to cut by the softening effect (Ignat et al. 2015)
- Reducing temperature increase and processing times (Chotphruethipong et al. 2019)
- Eco-friendly, fewer production costs, energy-efficient (Jambrak et al. 2018)
- When linked with other methods, such as thermal, high hydrostatic pressure, and ultrasound, the pulsed electric field enhances the process efficiency to attain improved results (Katiyo et al. 2017).

1.4 Disadvantages of PEF

Among the advantages of PEF in the food sector, certain drawbacks hinder the implementation of the technology. For instance, the absence of efficient and more realistic electrical systems is recognized as the key constraint for the commercial adoption of the technology (Priyadarshini et al. 2019). However, other drawbacks for PEF-based food manufacturing are as follows:

- Substantial cost of capital (Priyadarshini et al. 2019)
- Inefficient for certain enzymes (Wang et al. 2020)
- Challenging for use with conductors
- The dissolving of gasses and bubble formation is an operating challenge that triggers a dielectric breakdown
- Insufficient economic and innovation studies for the high-quality process
- Fewer procedures for food production techniques.

2 Cold Plasma

Evgenia Benova, Plamena Marinova, and Todor Bogdanov

2.1 Cold Plasma Properties and Sources

Plasma is a partially or completely ionized quasi-neutral gas. It can be low-temperature plasma with a temperature of the ions $T_i < 10^5$ K and high-temperature plasma with $T_i > 10^5$ K (as is the Sun and stars plasma) which is out of our interest. The low-temperature plasma produced by various electrical discharges is partially ionized and consists of charged particles (electrons and ions), highly reactive radicals, excited atoms, molecules, UV radiation. It can be classified into two main categories: equilibrium (thermal) and non-equilibrium (nonthermal) plasma. All particles in thermal plasma are in thermal equilibrium and have the same (high) temperature, $T_e \approx T_i \approx T_g$. The electrons in the nonthermal plasma have temperature T_e much higher than the heavy particles (ions, atoms and molecules) temperature, $T_e >> T_i \geq T_g$, where the temperature of the neutral particles is denoted as T_g and is called gas temperature.

The nonthermal plasmas at atmospheric pressure usually have electron temperature of about 1–2 eV while the heavy particles temperatures can vary from room temperature to above 1000 K. For treatment of biological systems, food and thermosensitive materials, the thermal damage has to be eliminated. For such purposes, the temperature of the heavy particles needs to be low enough, and a "cold atmospheric plasma" (CAP) is used for direct plasma treatment in plasma food technology.

Various indirect plasma treating methods were developed in the past decade to eliminate the thermal damage on the plasma treated products. The plasma treated water (PTW), usually called plasma activated water (PAW), is a common name for the liquids (water and not only) treated with atmospheric low-temperature plasma. It is classified as an eco-friendly technique with minor changes in food products. Most of nowadays investigations of PAW applications have been focused on washing disinfection, surface decontamination and bacteria inactivation. PAW is successfully used not only in the food industry but in the agricultural and biomedical fields.

The temperature is not the only important parameter in plasma food technology. For each type of products, the plasma source and discharge conditions have to be optimized to achieve the required effects without any additional damage. Many parameters have to be considered and controlled during the plasma process, such as type of plasma source, geometry, working gas (gas mixture), degree of ionization, treatment time, pH changes, etc.

The main advantages of plasma food technology are the high level of inactivation of food pathogens (bacteria, viruses and fungi) and possible degradation of pesticides without using chemicals and eliminating any thermal damage of the products.

The plasma devices used in food technology can be separated into the following categories:

2.1.1 Gas Discharge Plasma Sources

Dielectric Barrier Discharge (DBD)

The various types of dielectric barrier discharges (DBD) are widely used and studied in food technology. The advantages are the simple construction, easy operation and geometry flexibility of the atmospheric DBD plasma sources.

In all geometrical configurations, the DBD consists of two electrodes with one or two dielectric layers between them (Fig. 4.1). One of the electrodes is grounded and the other is powered by AC or pulsed high voltage (~kV) with frequency varying from Hz to MHz (Guo et al. 2014). Most popular are planar (Fig. 4.1a–c), cylindrical (Fig. 4.1d–f), and plasma-jet (Fig. 4.1g) configurations.

Corona Discharge

Corona discharge occurs when high pulsed DC voltage is applied between two electrodes, one of which is in the form of wire. Its surface is small compared to the other electrode, usually having the form of a plate (Fig. 4.2). For surface treatment applications, the wire electrode is more often cathode (connected to the negative voltage) and the discharge appears in the air around it in the form of a crown called negative corona (Fig. 4.2). The positive corona also exists when the wire electrode is the

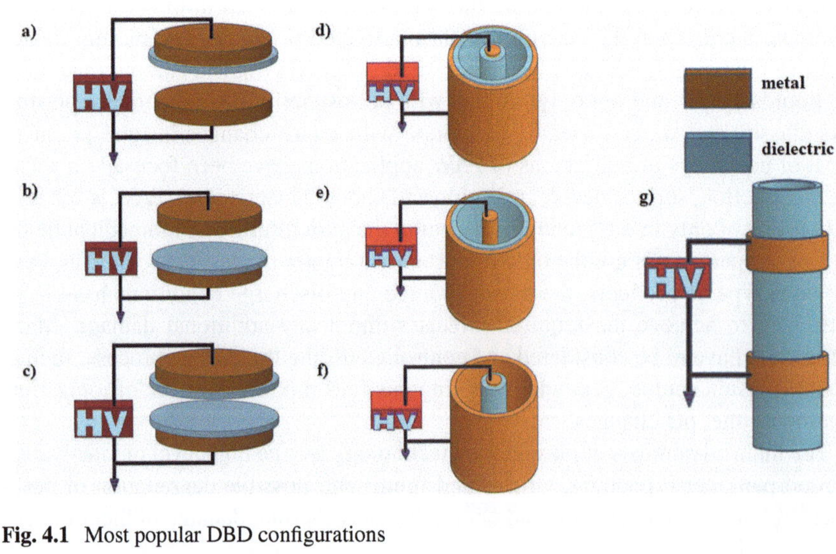

Fig. 4.1 Most popular DBD configurations

Fig. 4.2 Corona discharge

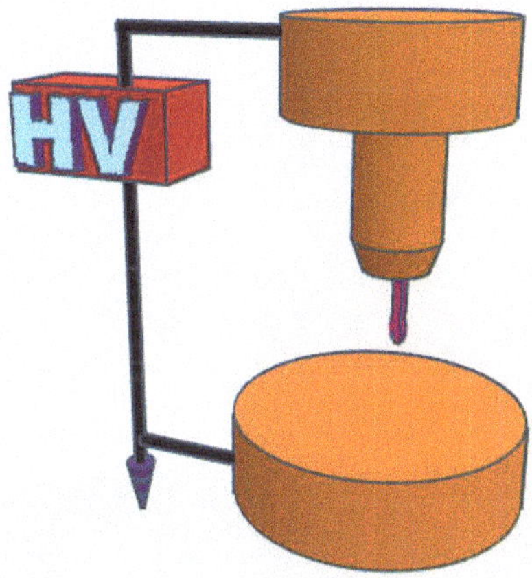

anode. The pulsed high voltage is used to prevent an increase in temperature and keep the discharge in non-equilibrium conditions important for food treatment. If not pulsed but continues high voltage is used, the temperature becomes very high and the arc discharge close to thermal equilibrium is produced.

The corona discharge has a simple construction and can operate in the open air without adding any noble gas, making it a cheap and easy operation device. It is widely used for ozone production for fruits and vegetables ozone treatment (Baggio et al. 2020).

Microwave Plasma Torch

Various microwave plasma torch systems are developed for different purposes: TIA (Torch á Injection Axiale) (Moisan et al. 1994), TIAGO (Torche à Injection Axiale sur Guide d'Onde) (Moisan and Nowakowska 2018), surface-wave discharges (SWD) with different wave-power applicators (surfaguide, surfatron, waveguide surfatron, Ro-Box) (Moisan and Nowakowska 2018; Moisan and Zakrzewski 1991). Operating at atmospheric pressure at high microwave power, they produce plasma with a gas temperature above 1000 K (and can reach 6000 K in some cases). Recently, atmospheric pressure SWD operating at low microwave power demonstrated the possibility to produce a plasma torch with gas temperature below 40 °C and was used for biological systems treatment, including fresh berries, without thermal damage (Bogdanov et al. 2018; Krčma et al. 2018) (Fig. 4.3).

2.1.2 Plasma-Liquid Systems

The plasmas sustained by various gas discharges interacting with liquids (mainly water) produce active particles in the air at the plasma-liquid interface and inside the liquid (Fig. 4.4). Other types of plasma-liquid systems are the plasma sustained

Fig. 4.3 Surface-wave-sustained plasma torch: (**a**) Surfatron discharge; (**b**) general view; (**c**) low temperature allows touching the torch

directly inside the liquid, the discharges in bubbles in the liquid and the gas discharges containing sprays. A wide variety of plasma reactors based on plasma-liquid interaction is built and used. Generally, as a result of the plasma-liquid interaction, a conductive liquid with reactive species, changed physicochemical properties, and oxidation-reduction potential (ORP) is produced, and that is called plasma treated water (PTW) or plasma activated water (PAW).

Classification of devices for PAW production could be done based on the working media: the working plasma gas, the surrounding gas or the activated liquid. They are determined by the direct (discharges in liquids or bubbles) or non-direct (discharges over the liquids and discharges with aerosol) contact of plasma and liquid.

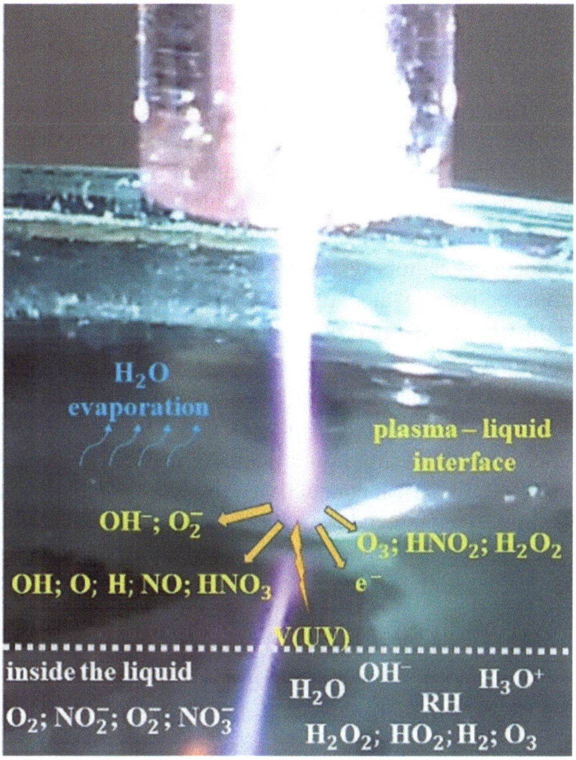

Fig. 4.4 Main processes for active particles production at the interface of microwave plasma torch and water

Depending on the plasma sustaining devices, the following systems for PAW production are used more often: (i) gas discharges in direct and non-direct contact with liquid, such as plasma jets in all of their variations (Joshi et al. 2018; Qi et al. 2018; Royintarat et al. 2019; Tian et al. 2015; Wu et al. 2017; Xiang et al. 2018; Ye et al. 2013; Zhang et al. 2013, 2016; Zhou et al. 2018) and dielectric barrier discharges (DBD) (H€ansch et al. 2015; Kojtari et al. 2013; Royintarat et al. 2019); (ii) inside liquid discharges as pinhole (Kamgang-Youbi et al. 2007, 2009; Lipovan et al. 2015; Naïtali et al. 2010) and gliding arc (Suganuma and Yasuoka 2018) discharges. Different working gases produce plasma in such devices: air (Kamgang-Youbi et al. 2007, 2009), noble gases as Ar or He (Oh et al. 2019), and gas mixtures as Ar + O_2 (Ye et al. 2013; Zhang et al. 2013) in different proportions (oxygen concentration varies from 1% to 3%). Details about such systems and the plasma properties can be found in (Bruggeman et al. 2016; Gorbanev et al. 2018; Vanraes and Bogaerts 2018; Xiang et al. 2020).

The long living active particles produced due to plasma interaction with the ambient air and the liquid systems are mainly reactive oxygen (ROS) and nitrogen species (RNS). The processes of reactive species production are highly dynamic and pass through various short living active particles production. An example of the main processes for active particles production at the interface of the microwave plasma torch and water is shown in Fig. 4.4. The chemical reactions pathway and particles concentration strongly depend on the type of the discharge system and the discharge conditions. The result of plasma interaction with the liquid (water or solution) is the production of PTW (or plasma activated water, PAW), which can be used further in food technology. Variations in the physicochemical properties of treated liquid due to the complex mechanism of interaction between plasma and solution have to be described in detail to understand PAW inactivation ability better. The oxidative stress and the physical effect of treated water are identified as leading influencing factors of PAW efficiency. A short review of the last published papers about activated water applications demonstrates that PAW could be used in the food industry, enhancing product quality even more than the classical thermal and chemical processes.

Complex chemical reactions between the liquid and the plasma occur at the interface region between plasma and treated substance. Reactive species are produced due to plasma-liquid interaction and physicochemical properties of PAW are changed. Electrical conductivity, oxidation-reduction potential (ORP) and pH of the liquids are changed during the activation process. Most commented chemical reactions between gaseous species and liquid molecules (Bruggeman et al. 2016; Gorbanev et al. 2018; Vanraes and Bogaerts 2018; Wende et al. 2018) are presented in Table 4.1.

Fast reactions between hydroxyl ions (OH^-) and solved electrons leads to the formation of stable reactive oxygen species (ROS) as superoxide (O_2^-), ozone (O_3) and H_2O_2. Production of ROS depends on the treatment time and increases with it, while the active time of species varies with the storage time and pH of PAW. Short living radicals (oxygen, ozone and OH*) could react with some organic admixtures in the liquid and form new radicals triggering new reactions. Relatively stable species as H_2O_2 could be identified in PAW a long time after treatment (Niquet et al. 2018). H_2O_2 is a biologically active agent with important antimicrobial and cytotoxic properties.

The formation of nitrite ions affects PAW in the reduction of pH and acidic environment. Variation of liquid pH accelerates nitrite degradation to nitrates. The short living time of reactive nitrogen species determines difficulties in the detection of the same. Long exposure time was reported (Chen et al. 2018) as the main mechanism

Table 4.1 Chemical reactions for ROS and RNS production at the plasma-liquid interaction (Khlyustova et al. 2019)

ROS production processes	RNS production processes
$H_2O + e^- \rightarrow OH^* + H^+ + e^-$	$NO_2^- + H^+ \rightarrow HNO_2$
$H_2O + e^- \rightarrow OH^* + H^+ + 2e^-$	$NO_2^* + OH^* \rightarrow HNO_3$
$OH^* + OH^* \rightarrow H_2O_2$	$3NO_2^- + 3H^+ \rightarrow 2NO + NO_3^- + H_3O^+$
$H^* + O_2 \rightarrow HO_2$	$2HNO_2 \rightarrow NO^* + NO_2^* + H_2O$
$OH^* + H_2O_2 \rightarrow OOH + H_2O$	$2NO_2 + H_2O \rightarrow NO_3^- + NO_2^- + 2H^+$
$O_2^- + H^+ \leftrightarrow OOH$	$NO_2^- + H_2O_2 + H^+ \rightarrow ONOOH + H_2O$
$O_2 + e^- \rightarrow O^+ + O^* + 2e^-$	$ONOOH \rightarrow NO_2^* + OH^*$
$O_2 + O^* \rightarrow O_3$	$ONOOH \rightarrow HNO_3 \rightarrow NO_3^- + H^+$

of copper, zinc or other metal ions dissipation into the PAW due to electrode erosion, especially in inside liquid discharges.

Plasma treatment time of PAW is inversely proportional to acidification of the treated liquid and pH decrease. The microbial inactivation ability of PAW is assumed to result from acidic pH (Naïtali et al. 2010). The electrical conductivity of PAW is assumed to increase with the activation period. The ability of a liquid to conduct electricity depends on the type of discharge, too. The conductivity of He and Ar discharges was shown (Vlad and Anghel 2017) as lower than conductivity of water activation by in-air discharges at similar working conditions. The treated liquid temperature was not reported to be dramatically changed (Tian et al. 2015), even for relatively long treatment times. A combination between "cold" PAW and the short living time of some reactive species determine the applicability of the activated liquid directly to heat-sensitive materials in a short time after the activation without the need for conservation for a long time.

2.2 *Cold Plasma in Food Technology*

Due to the reactive species, the plasma treatment has shown good antibacterial and antifungal effect and is used to decontaminate food and packages. The plasma technology in food, although relatively new field, uses a huge variety of plasma systems applied in (i) direct treatment of food, packages and instrumentation by the active gas-discharge region, by afterglow discharge region or by the air treated by the plasma; (ii) plasma treatment of water with food products immersed in it; (iii) direct or indirect plasma treatment of water for PTW production further used for microbial disinfection of food products (Thirumdas et al. 2018).

The different direct treatment technologies, plasma sources and treated food products are presented in Table 4.2. The main active particles and obtained results are also included.

The plasma treated water (PTW), often called plasma activated water (PAW), has also shown significant degradation of pesticides (Xiang et al. 2020) and is used for PTW washing and soaking of fresh fruits, vegetables and other products. Because of the wide use of PAW in food technology, the following important PAW features have to be taken into account:

2.2.1 Oxidation-Reduction Potential (ORP) of PAW

ORP has been pointed to as the main factor of disinfection due to its effect on cell membranes and integrity of the microbes (McFerson 1993; Suslow 2004). H_2O_2 concentration determines the ability of PAW to oxidize. The treatment time dependence of H_2O_2 production leads to a positive correlation of ORP from the same parameter. Depending on the plasma - liquid interface (plasma above the water or inside it), oxidation-reduction ability from the activated liquid has been reported to vary (Tian et al. 2015).

Table 4.2 Direct treatment

Method of treatment	Treated object	Plasma source	Working gas	Dominant active particles considered	Results reported	Article
Post discharge – Larger distance between plasma and substrate	Eggs	Atmospheric pressure plasma source	Compressed air	Antimicrobial effect is caused due to irradiation, long-living radicals and metastable and inhibitory substances	A statistically significant reduction of *S. enteritidis* (4.1 log CFU) was achieved	Moritz et al. (2020)
Exposure to plasma afterglow – in-package plasma treatments	Strawberries	DBD plasma (two aluminium circular and thick polypropylene dielectric sheets)	65% O_2 + 16% N_2 + 19% CO_2 (G1), 90% N_2 + 10% O_2	Ozone, excited nitrogen species	Background microflora of the strawberries was reduced by an average of ~3.0 log cycles from the initial levels of 5 log10 CFU/g in 300 s	Misra et al. (2014)
Direct treatment – Zein powder samples were onto bottom dielectric barriers	Zein powders	DBD plasma reactor		Ultraviolet photons, electrons, positive and negative ions, free radicals, and excited or non-excited molecules and atoms.	The structure of zein underwent reorganizations, corresponding to the modifications of the morphology and other physicochemical properties. The increase of CP treatment time could reinforce the tensile strength and surface hydrophilicity of zein films when the time ranged from 0 to 10 min.	Dong et al. (2017)
Direct treatment – samples placed in polypropylene boxes onto bottom dielectric barriers	Wheat flour, whole grains	High voltage atmospheric air plasma reactor, based on a dielectric barrier discharge (DBD)			Plasma processes may be tailored to regulate flour functionality and it is associated with partial disorganization of the structure of starch in wheat flour. Changes in the flours and the flour hydration properties of flour.	Chaple et al. (2020)

Method of treatment	Treated object	Plasma source	Working gas	Dominant active particles considered	Results reported	Article
Treatment of *Salmonella* typhimurium and *L. monocytogenes* inoculated surfaces – effect of treatment time and distance from the CAP source	Decontamination of stainless-steel food processing surfaces	Surface barrier discharge	Ambient air	OH● and NO●, UV	The most important factor affecting inactivation of both microorganisms was the distance from the CAP source, followed by treatment time and then interactions, intermediate-lived species and UV photons may play a key role, activities should focus on optimizing the transport of the desired reactive species to the target surface	Katsigiannis et al. (2021)
Direct treatment-juice samples placed in the center of the enclosure between the electrodes	Orange and carrot juice blend	Dielectric barrier discharge	Atmospheric air	Radicals, heat and UV light	There are no changes in regard to pH, changes in sugar content are statistically significant, most significant changes in color. Samples of treated juice were lighter in color and paler. Atmospheric air DBD plasma is a strong oxidizing agent and enzymatic browning is an oxidative reaction process.	Vukić et al. (2017)
Direct treatment-in sealed box to prevent leakage of the plasma species that were generated	White wheat flour	SDBD	Air	Atomic oxygen, ozone (O_3), hydrogen peroxides (H_2O_2), hydroxyl radicals (•OH), and nitrogen oxides	It is expected that the degradation of *Alternaria* toxins was through combined effects of the oxidative species: OH•, H_2O_2, and O_3. It can be stated that both investigated factors (time of exposure and distance from discharge) affect the degradation efficiency of *Alternaria* toxins in the wheat flour matrix. S, degradation efficiency increases with increasing treatment time	Janić Hajnal et al. (2019)

(continued)

Fig 4.2 (continued)

Method of treatment	Treated object	Plasma source	Working gas	Dominant active particles considered	Results reported	Article
Direct-plasma torch tip in contact with the fruit's surface	Small berries	Microwave plasma torch	Argon	Reactive oxygen and nitrogen species (ROS and RNS), ozone, oxides, peroxides, monoxides and ultraviolet (UV)	There is no significant variation in the number of colony-forming microorganism units between the groups for strawberries. In contrast to strawberries, for both cherries and blueberries, a decrease in the number of colonies is observed. The inhibiting impact of the plasma torch is determined by the surface characteristics of the tested fruits and the kind of microbial contamination.	Bogdanov et al. (2018)
Direct-both sides of the meat exposed to plasma	Inactivation of *S. typhimurium* in chicken breast and pork loins	Atmospheric pressure plasma jet based on an arc plasma	N_2 and O_2	Atoms, molecules, and radicals, as well as excited and charged particles	APP jet treatment conditions, including distance, time, and direction, may affect the efficiency of inactivating *S.* typhimurium in chicken breast and pork loin	Kim et al. (2013)

2.2.2 Storage of PAW

Antimicrobial potential and bactericidal efficiency of PAW decrease with the storage time and temperature (Shen et al. 2016). The main reason for such behavior of PAW could reduce the long-lived reactive species as H_2O_2 and NO_3. The lower storage temperature of the PAW could slightly enhance the antimicrobial potential of the liquid.

2.2.3 Combination of PAW with Other Technologies

Last few years, a set of cross-methods of treatment have been reported. PAW was combined with heat treatment and ultrasound cleaning (Royintarat et al. 2020; Zhang et al. 2020). Synergistic treatment of PAW and mild heating (50–55 °C) strongly inactivate some bacteria in food products, while treatment with PAW only is not providing satisfying results. A combination of PAW and ultrasound has been presented as a promising methodology for cleaning ready-to-consume plants and seeds. All investigated cross-treatments almost have not been reported any observable adverse effect on the firmness, color, total soluble solids, total phenolics, vitamin C, and antioxidant properties of samples.

The various applications of indirect treatment (mainly washing and soaking) by PTW are included in Table 4.3.

Table 4.3 Indirect treatment

Method	Treated object	Plasma source	Working gas	Dominant active particles considered	Results reported	Article
Washing – plasma processed air (PPA) was used to treat distilled water	Fresh-cut lettuce	Microwave-driven device (2.45 GHz)	Air	Nitrogen monoxide radical (·NO), nitronium cation (NO^+_2), and hydroxide anion (OH^-)	High concentrations of nitrite and nitrate in PTW did not affect the nitrate concentration of the lettuce	Schnabel et al. (2021)
Washing – plasma processed air (PPA) was used to treat distilled water	Fresh-cut lettuce	Microwave-based	Air	H_2O_2, NO_2^-, and NO_3^- and the acidification by low pH values	Cause only minor cell membrane damage, the metabolic activity could be significantly affected	Schnabel et al. (2020)

(continued)

Table 4.3 (continued)

Method	Treated object	Plasma source	Working gas	Dominant active particles considered	Results reported	Article
Immersed in PAW – plasma discharge over the sterile distilled water surface, no contact	Strawberries	Single electrode nonthermal atmospheric pressure plasma jet	Ar/O2	Hydroxyl radical (OH), singlet oxygen ($^1O_2$), superoxide anion (O_2), and hydrogen peroxide (H_2O_2)	Treatment achieved a higher bacterial reduction of S. Aureus. After PAW treatment, the S. Aureus cells exhibited different degrees of damage.	Ma et al. (2015)
Soaking – plasma discharge over the sterile distilled water surface, no contact	Button mushrooms	Single-electrode, nonthermal, atmospheric-pressure plasma jet	Ar/O2	Hydroxyl radical (OH), singlet oxygen ($^1O_2$), superoxide anion (O_2), and hydrogen peroxide (H_2O_2)	PAW treatment was more effective in reducing microorganisms than water treatment. The antimicrobial effects are well established and the mechanism of action has been related to membrane damage.	Xu et al. (2016)
Exposure to plasma afterglow – distance between the end of the plasma nozzle and the jujube surface was 5 cm	Jujube	The plasma system consisted of electrodes and a dielectric. A copper cylinder was the cathode and a metal rod made of tungsten alloy the anode. The plasma region was at the end of the inner anode.	Air	N_2*, N_2^+, N^+, O_2^+, and O	Cold plasma pretreatment significantly changed the moisture content of jujube samples, which reduced to 61.80%, and 59.29% after 30 and 60 s pretreatment on each side. Cold plasma pretreatment can effectively inhibit the degradation of antioxidants.	Bao et al. (2021)

2.3 *Conclusion*

Plasma technologies offer new approaches in the food industry. Producing reactive oxygen and nitrogen species like ozone, singlet oxygen, H_2O_2, OH^-, NO_2^-, NO_3^-, N_2^*, N_2^+, N^+, and many others during the treatment process or in the plasma treated liquid (water) without additional chemicals makes the plasma technology clean, green and environmentally friendly. The synergetic action of many active plasma components leads to high efficiency and effectiveness of plasma decontamination processes, reducing the pathogens (bacteria, viruses, fungi) and possibly degrading pesticides in food. This is why using the plasma or plasma treated water (PTW) in a broad spectrum of food processes leads to disinfection, sterilization and increased storage time and quality of fruits, juices, fresh products, and other foods. The plasma technology is also applied to disinfect instrumentation surfaces in food production and food packaging without using heat and chemicals.

Acknowledgments
This work is partially supported by the Operational programme "Science and Education for Smart Growth" 2014-2020, co-funded by EU: Project BG05M2OP001-1.002-0019 "Clean technologies for a sustainable environment – water, waste, energy for a circular economy" (Clean&Circle CoC).

3 Radio-Frequency

Maria Turtoi, Livia Patraşcu, and Iuliana Aprodu

3.1 *Introduction*

Radio-frequency (RF) represents the oscillation rate of an alternating electric current or an electric, magnetic, or electromagnetic field that consists of a band of frequencies that range from around 20 kHz to about 300 GHz. This interval is approximately between the upper limit of audio frequencies and the lower limit of infrared frequencies (NASEM 2015; IEEE 521-2019 2021). At these frequencies, the energy from an alternating current goes through a conductive material and generates radio waves into space. RF includes the frequencies used for communications signals (radio and television broadcasting and cell-phone and satellite transmissions) or radar signals.

Therefore, according to the Federal Communications Commission (FCC 1988) and International Telecommunication Unit Radio Regulations (ITU 2012), there are only a few permitted frequencies for industrial, scientific and medical (ISM)

Table 4.4 Federal Communications Commission (FCC) allocated RF bands designated for industrial, scientific and medical (ISM) applications (FCC 1988)

Frequency range, MHz		Centre frequency, MHz	Bandwidth, kHz	Frequency tolerance, ± %	Free space wavelength, m
13,553	13,567	13.56	14	0.052	22.1
26,957	27,283	27.12	326	0.600	11.1
40,660	40,700	40.68	40	0.005	7.4

applications. The frequencies reserved internationally for ISM purposes other than telecommunications are between 10 and 50 MHz (Table 4.4).

There are other several frequency bands admitted for ISM applications such as: 6765–6795 kHz (centre frequency 6780 kHz), 433.05–434.79 MHz (centre frequency 433.92 MHz), 61–61.5 GHz (centre frequency 61.25 GHz), 122–123 GHz (centre frequency 122.5 GHz), 244–246 GHz (centre frequency 245 GHz) (FCC 1988, art. 5.138), and 902–928 MHz (centre frequency 915 MHz), 2400–2500 MHz (centre frequency 2450 MHz), 5725–5875 MHz (centre frequency 5800 MHz), 24–24.25 GHz (centre frequency 24.125 GHz) (FCC 1988, art. 5.150). The last four bands are used for microwave ovens.

The microbicidal effect of the electric current had been known since the early twentieth century when milk started being electrically pasteurized. The lethality of milk microorganisms in the electrical pasteurization was due to the thermal effects of the direct or alternating electric current applied to the milk (Anderson and Finkelsein 1919; Beattie and Lewis 1913; Beatie 1914; Stabler 1931).

RF is also referred to as dielectric heating (RF-H) and dielectric loss heating and is considered a novel thermal processing method in food engineering (Jiao et al. 2011). There are many studies on RF-H in the literature, both research articles and reviews (Altemimi et al. 2019; Komarov 2021; Marra et al. 2009; Piyasena et al. 2003). RF-H is considered an eco-friendly technology, better than microwave heating, and has many applications for food pasteurization (Gao et al. 2011; Geveke et al. 2017; Liu et al. 2018; Uemura et al. 2009, 2010a, b, 2012; Xu et al. 2020; Wang et al. 2003), blanching (Jiang et al. 2020; Zhang et al. 2019), thawing (Llave et al. 2015; Uyar et al. 2015), meat processing (McKenna et al. 2005; Rincon et al. 2015), and drying (Alfaifi et al. 2014; Ozturk et al. 2016).

Several studies related to the use of RF electric fields reported either lethal effects on different microorganisms or even nonthermal effects of the treatment. Thus, a long debate over whether a nonthermal effect can be associated with the electromagnetic field has existed for over seven decades. Among the first studies reporting results that suggested nonthermal effects of the electromagnetic field are Nyrop (1946), Sale and Hamilton (c), Webb and Booth (1969) and Webb and Dodds (1968).

For instance, Nyrop (1946) applied a 0.23 kV/cm electric field at 20 MHz for a total of 7 s and reported the destruction of 99.5% of *Escherichia coli* in liquid medium at temperatures even lower than 40 °C. The author mentioned no difference in the results whether the treatment took place between 12 °C and 40 °C, or between

40 °C and 60 °C. If the strength of the electric field was increased to 0.288 kV/cm, the time was reduced to 4 s for the same killing effect. Because the information on the experimental setup and procedures did not contain enough details, the experiments could not be repeated or reproduced.

A few years later, aware of the work of Nyrop, Ingram and Page (1953) studied the effect of RF electric field on suspensions of baker's yeast, *E. coli*, the mosaic virus of tobacco and bacteriophages. The temperature of the samples was maintained below 30 °C with adequate and sufficient cooling. The samples were exposed to a high-frequency electric field of a) 10 MHz modulated with half-sine waves of 50 Hz, and b) 20 MHz modulated with 10 to 120 kHz, with rectangular pulses varied from 1/3 to 1/15 of the complete cycle. The electric field strength was up to 2 kV/cm, applied in periods of up to 12 min. However, no significant lethal effect was observed under these conditions. Similar conclusions are presented in other studies such as Carroll and Lopez (1969) and Kou et al. (2018).

Carroll and Lopez (1969) aimed to determine if 60-MHz RF energy has a selective effect on microorganisms other than that attributable to heat. They suspended *S. cerevisiae*, *E. coli* and *Bacillus subtilis* in an aqueous buffer medium of various pH values (7.0, 5.0 and 3.0 for *S. cerevisiae*, and 7.0 and 5.0 for *E. coli* and *B. subtilis*) and other complex food liquids such as tomato juice, orange juice, and milk, alcoholic buffer with 2.0%, 5.0%, and 12% alcohol, beer and wine. No lethal effect of the RF energy and no visible synergistic killing effect of RF energy and heat on microorganisms existed after 46 min of treatment, except a synergistic effect of ethanol and heat at 48.8 °C for *S. cerevisiae*.

Recently, in a study by Kou et al. (2018), *E. coli* and *Staphylococcus aureus* in apple juice and mashed potato were exposed to an RF electric field of 27.12 MHz and less than 0.3 kV/cm simulated by Huang et al. (2016) and, in parallel, to a conventional heating block system to compare the inactivation of their populations. The results were similar in terms of uniform temperature distribution, inactivation rates and survival patterns for RF and conventional heating. The statistical analysis indicated no significant difference ($p > 0.05$) in inactivating bacteria between the two systems. Therefore, the conclusion was that no nonthermal effect of RF energy at 27.12 MHz existed (Kou et al. 2018).

Several studies disseminated in the last two decades have taken up the idea of establishing whether there are nonthermal effects of RF treatment (Brunkhorst et al. 2000; Geveke and Brunkhorst 2003, 2004, 2008; Geveke et al. 2002, 2007; Geveke 2020; Masood et al. 2017; Uemura and Isobe 2002). Some of the results confirmed the nonthermal effects of RF treatment for field strengths higher than 5 kV/cm and frequency lower than 70 kHz, as discussed below.

Also, several reviews (Fleming 1944; Kotnik and Miklavčič 2006; Miyakoshi 2013; Weaver and Chizmadzhev 1996; Weaver 2000; Wust et al. 2020; Zimmermann 1986b) and a few book chapters (Geveke 2005, 2011; Geveke et al. 2006; Trujillo and Geveke 2014) related to nonthermal RF electric field treatments were published.

3.2 Mechanism of Inactivation

Processing of liquid foods in RF electric field is similar in many respects to pulsed electric field (PEF) processing. Both methods have been extensively investigated in the last three decades. Also, the inactivation of microorganisms is thought to occur by electroporation in both RF and PEF methods (Chang et al. 1991).

When a cell is exposed to an electric field, a voltage is formed across the cell membrane. Cell membranes are composed of different molecules with positive and negative electrical charges, thus having specific electrical properties such as electrical resistance and capacitance. The opposite charges on either side of the membrane attract to each other. The attraction takes place with greater force as the electric field is more robust and the membrane becomes thinner. At a sufficiently high voltage, pores are formed in the membrane and the cell ruptures (Coster and Zimmermann 1975a, b; Zimmermann et al. 1973, 1974, 1975; Zimmermann 1986b).

Kotnik et al. (1998) used a general method to analyze time courses of transmembrane voltage induced by time-varying electric fields. They applied the technique to different shapes of the electric field used for electroporation and electrofusion such as rectangular, trapezoidal, exponential and sine-modulated pulses with different durations, each for both the normal physiological and the low-conductivity medium. An important conclusion of the study was the need to measure the conductivity of the medium due to its influence on the dynamics of induced transmembrane voltage, imposing the range of pulse duration. Thus, the process of transmembrane voltage inducement is slowed down as the conductivity of the extracellular medium is reduced for all the pulse shaped investigated. Therefore, the authors stated that it is generally tough to predict the peak value of the induced transmembrane voltage because the shape of the response strongly depends on the conductivity of the medium, and it is influenced by other parameters as well. The results analysis shows that an increase of the frequency from 100 kHz to 1 MHz leads to a significant reduction of the voltage.

While most electroporation applications aim to transport extracellular ions or molecules into cells, the primary purpose of PEF and RF food applications are related to the dielectric breakdown of the cells of microorganisms (Weaver 2000). The breakdown can be reversible or irreversible, usually requiring voltage much higher than 1 V.

Electroporation involves more than electro-permeabilization of cells membrane because, during the application of the electromagnetic field, there is a rapid evolution of pore population, which causes the fast membrane discharge than rupture (Weaver and Chizmadzhev 1996; Weaver 2003).

A better understanding of the electroporation phenomenon can be obtained with advanced investigation. Geveke et al. (2006) got scanning electron microscope (SEM) images of *E. coli* K12 that were untreated, thermally treated, and RF nonthermally treated. The untreated cells (control) were usually individual, rod-shaped with smooth surfaces. The thermally and RF nonthermally treated cells were mostly agglomerated in small groups of three to ten cells with different irregularities of the surface of almost every cell. The shapes of thermally treated cells were distorted by large, irregular depressions and evaginations of their membranes. The profiles of RF nonthermally treated cells were less distorted than previous ones and had at least a few small vesicles attached to the visible surfaces. The SEM images showed remarkable differences indicating that thermal and RF nonthermal inactivation mechanisms are different (Geveke et al. 2006).

According to Ukuku et al. (2008), the bacterial cells having ruptures in the membrane structure can lose through leakage intracellular substances such as ATP and UV-absorbing substances. Increasing the RF treatment temperature of apple juice containing *E. coli* K-12 from 23 °C to 27 °C, and 45 °C led to an increased extracellular ATP concentration indicating the damage of the cytoplasmic membrane (Ukuku et al. 2008). A few years later, Ukuku et al. (2012) detected the injuries produced to the cell membrane with Transmission Electron Microscopy (TEM), then determined the leakage of intracellular materials with an ATP luminometer (20 D) and characterised changes in bacterial cell surfaces with electrostatic and hydrophobic interaction chromatography. The hydrophobicity of the bacterial cell surface decreased significantly, and relative negative ions were lost due to RF treatment compared to heat treatment alone at 55 °C and 75 °C (Ukuku et al. 2012). The results of these studies suggest that the mechanism of inactivation of RF treatment consists in the disruption of the cell membrane with the hydrophobicity decrease and negative ions loss, which led to injury and leakage of the intracellular biological active compounds, and if the membrane structure of the bacteria is damaged beyond self-repair, loss of biological activities and death (Geveke et al. 2006; Ukuku et al. 2008, 2012).

Saulis (2010) discussed in a review the electroporation of cell membranes in food processing and considered that the whole process could be divided into four main stages: (1) increasing the transmembrane potential because of charging the cell membrane by the applied external electric field; (2) creation of tiny metastable hydrophilic pores after the transmembrane potential has been built up; (3) evolution of pore population, meaning changes in the number and sizes of pores during an electric treatment; (4) post-treatment stage consisting of the processes that take place after the electric treatment, such as leakage of intracellular compounds, the entrance of extracellular substances, pore resealing, etc. The duration of the first three stages is from nanoseconds to milliseconds, while the fourth is from milliseconds to hours, or even days (Saulis 2010).

Although the majority of the scientific community assumes, during the last decades, that the inactivation mechanism in the electromagnetic field applications is irreversible electroporation of the cell membrane, because a few studies consider the statement as being still controversial (Schoenbach et al. 2000), further investigations are needed (Geveke 2011; Trujillo and Geveke 2014).

3.3 RF Electric Field Equipment

When a high voltage is applied to two electrodes that are close to each other and have a liquid placed between them, a high electric field is produced. The process is similar to the pulsed electric field (PEF) process, except that the power supply is continuous and not pulsed. In the case of RF electric fields, an alternating current (AC) generator supplies the power.

Several types of equipment and AC power generators for nonthermal RF electric field processing have been reported in the literature. Detailed descriptions, diagrams, and photographs are presented in numerous articles (Brunkhorst et al. 2000; Carroll and Lopez 1969; Geveke and Brunkhorst 2003, 2004, 2008; Geveke et al. 2002, 2007; Geveke 2020; Ingram and Page 1953; Kou et al. 2018; Masood et al. 2017; Sale and Hamilton 1967; Uemura and Isobe 2002) and book chapters (Geveke 2005, 2011; Geveke et al. 2006; Trujillo and Geveke 2014), so that only a few experimental setups with certain features are highlighted below.

To isolate thermal and nonthermal effects of RF energy on microorganisms in liquid foods, Brunkhorst et al. (2000) developed a modified RF dielectric heater as a component of a continuous process. The concept combines instantaneous input of RF energy to the food liquid with fast removal of heat resulting from RF energy conversion. It accurately and precisely controlled the temperature. The authors integrated two concentric pipes into the RF heater. The inner pipe was stainless steel with a 2.5 outer diameter, and it was grounded in the RF circuit. The outer tube was polypropylene (PP) with a 3.8 cm internal diameter. The tested liquid food flowed through the annular region between the two concentric pipes and absorbed the RF energy. The inner tube carried cooling water to remove the thermal energy from the processed fluid, controlling its temperature this way. This assembly was flanked on either side by concave aluminium capacitor plates 7.0 cm wide and 76.2 cm long, which transmitted the RF energy through the polypropylene tube wall to the processed fluid. The first tests were performed using water from a storage tank with a small pump via polyvinyl chloride (PCV) piping. Subsequently, the experimental setup was improved by replacing the outer tube of PP with Teflon that is transparent to RF and with a recycling pump to achieve turbulent flow and cumulative treatment time necessary for the inactivation of microorganisms (Brunkhorst et al. 2000; Geveke et al. 2002).

The RF power supply system consisted of four 1 kW RF amplifiers and four step-up transformers connected in series to produce a voltage of 4.0 kV peak over a frequency range of 20 to 100 kHz. It also contained a generator that drove the amplifiers (Geveke and Brunkhorst 2003).

In addition to the basic RF power supply system, the experimental setup generally included a stainless-steel feed tank, a peristaltic pump for product feeding, a treatment chamber, controllers and monitoring equipment. A stainless-steel heat exchanger and a temperature controller were inserted before the treatment chamber to control the inlet temperature of the fluid. Also, for rapidly cooling the treated liquid to less than 25 °C, another stainless-steel heat exchanger as a cooling coil submerged in a water bath was used immediately after the treatment chamber (Geveke and Brunkhorst 2003).

A continuous RF electric field experimental setup was designed and used by Geveke et al. (2006, 2007, 2009), which contained three co-linear treatment chambers connected in series and four heat exchangers. The first heat exchanger controlled the inlet temperature of the fluid entering the first treatment chamber, the last one was used for the final cooling of the processed fluid, and the other two were used for cooling between the treatment chambers. Uemura and Isobe (2002) used four parallel plate treatment chambers connected in series. The flow rates of processed liquid varied from 150 mL/min (Uemura and Isobe 2002) to 1.400 mL/min (Geveke and Brunkhorst 2008).

Apart from the experimental setups mentioned above, Masood et al. (2017) proposed a novel design called Steinmetz for the RF electric field treatment chamber to inactivate the microorganisms. Steinmetz chamber is a geometrical configuration obtained by intersecting two perpendicular cylinders of equal radius, the horizontal one as the flow channel and the vertical one representing the two electrodes, the ground electrode situated below and the high voltage electrode situated above Masood et al. (2017). After the construction based on simulated optimal dimensions, the Steinmetz chamber was validated by treating *E. coli* in water for microbial inactivation.

3.4 Processing Parameters

The RF electric field processing technology is influenced by several variables, e.g., factors depending on the medium, microorganism, equipment, operating parameters such as RF electric field strength, frequency, time, temperature, inoculum, etc. Different research groups have tested such variables for the RF treatment; however, the medium, microorganisms and equipment were quite different, that comparison may not always be practical. Therefore, the main processing parameters influencing the nonthermal RF processing are the electric field strength, frequency, treatment time and treatment temperature.

3.4.1 Electric Field Strength

The RF electric field strength has an essential role in the inactivation of microorganisms. In most studies, the electric field used was up to 25–30 kV/cm, leading to the conclusion that increasing the electric field increased the inactivation (Geveke and Brunkhorst 2003, 2004, 2008; Geveke et al. 2009). However, there is an upper limit to the electric field strength harmful to the treatment chamber and the liquid foods. This limit is situated at electric field strength higher than 30–40 kV/cm, for which arcing across the electrodes occurs (Geveke 2011; Masood et al. 2017).

After applying a series of direct current pulses to suspensions of vegetative bacteria and yeast cells, Sale and Hamilton (1967) observed that a minimum electric field strength of 5 kV/cm is necessary to accomplish inactivation.

Geveke et al. (2002) applied an approximately 0.5 kV/cm electric field strength to different liquids, i.e., deionized water and apple cider inoculated with *L. inocua*, beer with naturally occurring microorganisms, and liquid whole egg and tomato juice inoculated with *E. coli* K-12 and thermal effects were observed. Because the electric field strength was relatively low, no nonthermal effects of RF energy were detected, and no synergistic effects of RF energy with heat was obtained.

Increasing the electric field strength from 20 kV/cm to 30 kV/cm has been shown to increase the inactivation of *S. cerevisiae* in water (Geveke and Brunkhorst 2003). Also, intensifying the field strength up to 16 kV/cm increased the inactivation of *E. coli* K-12 in apple juice; however, for a further increase up to 26 kV/cm, the inactivation remained constant (Geveke and Brunkhorst 2004).

Similar results were obtained when *E. coli* in saline water was treated with RF electric field. Thus, increasing the electric field strength above 7 kV/cm, the number of *E. coli* decreased from 1.3×10^6 to one log order per 1 kV/cm and with 14 kV/cm below 10 CFU/mL (Uemura and Isobe 2002).

Investigating the inactivation of *Lactobacillus plantarum* ATCC 49445, a gram-positive bacterium, in apple cider, Geveke et al. (2009) applied field strength 0.15 to 15 kV/cm and obtained a steadily increase of the inactivation from 7.5 kV/cm to 15 kV/cm. The minimum field required to irreversibly rupture the cell membrane, E_c, calculated in this study, was 7.1 kV/cm. It is situated between the reported values for *E. coli* in apple cider, which are 13.5 and 4.0 kV/cm at 55 °C and 60 °C, respectively (Geveke and Brunkhorst 2008).

3.4.2 Frequency

The frequency used in those studies reporting that there is no nonthermal effect of RF electrical field was of the order of MHz, e.g., 20 MHz and 0.23 kV/cm field strength (Nyrop 1946), 10 and 20 MHz at up to 2 kV/cm field strength (Ingram and Page 1953), 60 MHz (Carroll and Lopez 1969), and 27.12 MHz at less than 0.3 kV/cm field strength (Kou et al. 2018).

Brunkhorst et al. (2000) used a frequency of 18 MHz both for water and fluid tests in their attempts to develop new experimental setups with which the thermal and nonthermal effects produced by the RF electric field can be separated. The equipment was used to study the inactivation of *E. coli* K-12, *L. inocua* and yeast in apple cider, beer, deionised water, liquid whole egg, and tomato juice with RF electric field treatment at 18 MHz and 0.5 kV/cm. Even though the treatment was capable of pasteurizing the liquids, no nonthermal effects of RF energy were reported (Geveke et al. 2002).

In subsequent studies, frequencies of the order of kHz were used, e.g., 20 kHz (Geveke and Brunkhorst 2003, 2004; Geveke et al. 2009; Geveke 2020; Masood et al. 2017; Ukuku et al. 2008, 2012; Ukuku and Geveke 2010), 21.3 kHz (Geveke and Brunkhorst 2008). Also, RF electric field strengths between 15 and 30 kV/cm were used, higher than a minimum of 5 kV/cm; that is the value hypothesised as the one above that nonthermal inactivation might be achieved (Geveke et al. 2002; Sale and Hamilton 1967).

According to (Geveke and Brunkhorst 2003), the frequency of a nonthermal RF electric field treatment did not influence the inactivation of microorganisms in the range of 20 to 60 kHz. In a series of experiments performed at 20, 40, and 60 kHz, a 20-kV/cm electric field at a temperature of 50 °C was applied to *S. cerevisiae*. The reductions achieved ranged from 1.8 to 2.0 log CFU/mL and were not significantly different across the limited range of frequencies ($P > 0.05$).

However, the exposure of *E. coli* K-12 in apple juice to an RF electric field strength of 20 kV/cm at a temperature of 50 °C, over a frequency range of 15 to 70 kHz led to significantly greater inactivation at frequencies less or equal to 20 kHz ($P < 0.01$) compared to frequencies of 30 to 70 kHz (Geveke and Brunkhorst 2004).

Also, Geveke et al. (2007) stated that increasing the treatment time and temperature and decreasing the frequency enhanced inactivation. *E. coli* in orange juice was exposed to an RF electric field strength of 15 and 20 kV/cm, at frequencies of 21.1, 29.5, and 40.1 kHz. The inactivation of *E. coli* decreased as the frequency was increased from 21.1 kHz to 41.1 kHz. The authors considered the results interesting because the RF treatment could be more efficient at even lower frequencies. Also, the equipment costs would be significantly lower if the industrially usual frequency of 60 Hz could be used. There is a disadvantage of 60 Hz because electricity can penetrate the human body deeper at lower frequencies, but enough safety features can be designed to prevent injury.

3.4.3 Temperature

The temperature of the processed fluid increases during the RF electric field treatment as a function of the RF field strength and the duration of the treatment. It is said that the effects of temperature and RF electric field strengths on the inactivation of microorganisms are synergistic (Geveke and Brunkhorst 2004; Geveke et al. 2009).

Based on several data presented and discussed in the works of Geveke and Brunkhorst (2003, 2004, 2008) and Geveke et al. (2009), Table 4.5 has been built to show how increasing the temperature for a specific electric field strength increased the inactivation, and how a specific thermal treatment in identical conditions except for the absence of RF electric field determines a very small or even no inactivation of microorganisms.

Geveke and Brunkhorst (2003) performed a series of experiments at 20 kHz, a 30-kV/cm RF electric field strength, and different outlet temperatures (Table 4.5). The inactivation of *S. cerevisiae* increased with the increase of the outlet temperature from 40 °C to 55 °C. When the inlet temperature was increased at 45 °C, the same value as the outlet temperature and the electric field was eliminated, a reduction of only 0.3 ± 0.2 log CFU/mL was obtained. That also proved the nonthermal inactivation.

Similar effects were obtained when *E. coli* in apple juice or apple cider was exposed to RF electric field at varying temperatures and specific field strengths, e.g., 24 kV/cm (Geveke and Brunkhorst 2004) and 20 kV/cm (Geveke and Brunkhorst 2008). This synergism of temperature and RF electric field was also shown in the work of Geveke et al. (2009) during the RF electric field processing of apple cider inoculated with *L. plantarum* (Table 4.5).

Table 4.5 Effect of electric field strength and treatment temperature on the inactivation of several microorganisms in different media subjected to RF electric field treatment

Microorganism/ medium	Frequency, kHz	Time, ms	Flow rate, L/min	Electric field strength, kV/cm	Inlet temperature, °C	Outlet temperature, °C	Inactivation, log CFU/mL	References
S. cerevisiae/ water	20	10	1.2	30	26	40	2.1 ± 0.1	Geveke and Brunkhorst (2003)
						45	3.1 ± 0.1	
						55	4.7 ± 0.5	
				<<1 (control)	45	45	0.3 ± 0.2	
E. coli/apple juice	20	0.17	0.55	24	10	45	1.4 ± 0.1	Geveke and Brunkhorst (2004)
						50	1.9 ± 0.1	
				<<1 (control)	45	45	<0.1	
E. coli/apple cider	21.3	0.42	1.5	20	Ambient	50	1.3 ± 0.2	Geveke and Brunkhorst (2008)
						55	2.4 ± 0.7	
						60	5.0 ± 0.1	
				<<1 (control)	Ambient	50	No inactivation	
						55	No inactivation	
						60	0.2 ± 0.1	
L. plantarum/ apple cider	20	0.17	0.55	15	Ambient	45	0.74 ± 0.42	Geveke et al. (2009)
						50	1.04 ± 0.23	
						55	2.51 ± 0.36	
				<<1 (control)	Ambient	45	No inactivation	
						50	No inactivation	
						55	0.23 ± 0.06	

3.4.4 Time

The inactivation effect on the cells of microorganisms increases with the increase of the treatment time. Treatment time can be easily increased by using treatment chambers in series, with heat exchangers for cooling between them, as described above.

Geveke and Brunkhorst (2003) observed that a single treatment with a 30-kV/cm RF electric field at 20 kHz and 35 °C reduced the population of *S. cerevisiae* in water by 0.8 ± 0.1 log CFU/mL. In contrast, three treatments led to a reduction of 3.8 ± 0.4 log units. Similar behaviour was obtained for RF electric field treatment of *E coli* in apple juice and apple cider. The treatment with 24-kV/cm RF electric field at 20 kHz and 50 °C reduced the population of *E. coli* in apple juice by 1.8 ± 0.3 log CFU/mL for 170 µs duration and by 3.0 ± 0.5 log CFU/ml for 510 µs (Geveke and Brunkhorst 2004). The RF electric field processing of *E. coli* in apple cider was applied at 20-kV/cm field strength, two outlet temperatures, 55 and 60 °C, and varying treatment times, 140, 280 and 420 µs obtained by using either one, two or three treatment chambers in series (Geveke and Brunkhorst 2008). The inactivation of *E. coli* at 55 °C was 1.0 log CFU/mL for 140 µs, 1.5 log CFU/mL for 280 µs, and 2.4 log CFU/mL for 420 µs treatment time, whereas at 60 °C the values were 1.3, 3.6 and 5.0 log CFU/mL for the exact treatment times. One could observe that the increase of the inactivation of *E. coli* as the treatment time increases. The authors revealed that data followed first-order kinetics with a multiple regression correlation coefficient (r^2) of 0.972 and 0.986 for 55 and 60 °C, respectively. Also, they calculated D values for the studied outlet temperatures and obtained 194 µs for 55 °C and 74 µs for 60 °C.

The holding time after reaching the outlet temperature contributes to increasing the inactivation. For instance, in the RF electric field processing of *L. plantarum* in apple cider at 15 kV/cm field strength, 20 kHz frequency, and 170 µs treatment time, holding cider at 55 °C for 5 to 50 s resulted in 2.5- and 3.1-log reductions, respectively (Geveke et al. 2009).

3.5 *Applications of Nonthermal RF Electric Field*

There are not so many studies investigating nonthermal RF electric field processing. As discussed before, the first attempts to show nonthermal effects of the electromagnetic fields were not very trustworthy because the experiments could not be repeated (Nyrop 1946; Ingram and Page 1953). However, one of the first reliable studies suggesting the nonthermal effect of the RF electric field is that of Sale and Hamilton (1967). The study investigated a high electric field effect on several species of vegetative bacteria and yeasts in saline solution (0.1% NaCl). An electric field of 10 to 25 kV/h passed in pulses of 2–20 µs, in steps of 2 µs, between two electrodes immersed in the liquid. Sale and Hamilton (1967) observed that the electric field controlled the degree of microorganisms killing whether the temperature of suspension at the start of the treatment was 20 °C or 40 °C, and the electrolysis that occurs near each electrode did not influence the lethal effect. Based on their results, Geveke et al. (2002) hypothesized that a minimum of 5-kV/cm field strength is necessary to achieve nonthermal inactivation.

After Brunkhorst et al. (2000) developed a modified RF dielectric heater to isolate thermal and nonthermal effects of RF energy on microorganisms in liquid foods, reliable studies followed. Table 4.6 summarises the main findings of the works related to nonthermal RF electric field treatments used to inactivate microorganisms in liquid foods.

3.6 Concluding Remarks

RF processing uses relatively simple equipment to generate high electric fields at reasonable strengths and frequency and a very short time of food exposure. The essential processing parameters are field strength, frequency, temperature, and treatment time. The electric field strength should be higher than 5–10 kV/cm to inactivate microorganisms and lower than 30–40 kV/cm to avoid arcing between the electrodes. The frequency used in nonthermal RF processing ranged between 10 to 70 kHz; however, significantly higher inactivation was obtained at frequencies less or equal to 20 kHz. The RF electric field strength and the duration of the treatment increase the temperature of the processed fluid. To preserve the freshness of the product and inactivate the microorganisms, the processing temperature ranged between 35 and 60 °C. These values are relatively mild, lower than the average values encountered in thermal pasteurization. The effects of temperature and RF field strength on microbial inactivation are synergistic. The RF electric fields are applied to food for a very short time, less than 500 μs.

Several research groups demonstrated the inactivation of different microorganisms in model systems and liquid food products. Thus, nonthermal RF electric field processing has been effective at inactivating yeasts, *e.g.*, *S. cerevisiae* and *C. utilis* and bacteria, *e.g.*, *E. coli*, *L. inocua*, and *L. plantarum*. Nonthermal RF electric field processing is similar in many respects to PEF. However, PEF technology has various industrial applications, while RF does not. That means further researches are necessary.

Table 4.6 Summary of the study results related to nonthermal RF electric field treatments used to inactivate microorganisms in liquid food and other liquid media

Product/medium	Microorganism(s)	RF system/ conditions	Results	References
Apple cider	*L. innocua*	Power supply up to 19 kW Electric field strength: 0.5 kV/cm Exposure time: 9 s or 4.2 min Frequency: 18 MHz Flow rate: 9 kg/min Temperature: 50.5 °C Outer pipe d_i = 3.8 cm Inner pipe *de* = 2.5 cm Length of RF treatment zone 86 cm	Reduction of *L. inocua* from 5.9 to 4.5 log CFU/mL in RF treatment Reduction of *L. inocua* from 5.8 to 4.7 log CFU/mL in RF treatment No nonthermal effects of RF energy were detected no synergistic effects of RF energy with heat	Geveke et al. (2002)
	E. coli K-12	Power supply: 80 kW Electric field strength: 20–30 kV/cm Exposure time: 140–420 μs Frequency: 21, 30, and 4[illegible] kHz Flow rates: 1.5 and 1.9 L/min Temperature: 60 °C	The population of *E. coli* was reduced by 4.8 log at an outlet temperature of 60 °C The treatment at 20 kV/cm and 60 °C allows a calculated D value of 74 μs and E_c = 4.0 kV/cm Electric energy density: 260 J/mL	Geveke and Brunkhorst (2008)
	L. plantarum ATCC 49445	Electric field strength: 0.[illegible]5 to 15 kV/cm Frequency: 5 to 65 kHz Treatment time: 170 μs Holding time: 5 to 50 s Temperature: 45 to 55 °C	The population of *L. plantarum* was reduced by 1.0 log at 15 kV/cm, 20 kHz, and 50 °C, with a 5-s holding time Inactivation of *L. plantarum* increased steadily from 0.28 ± 0.12 log reduction at 7.5 kV/cm to 2.51 ± 0.36 log at 15 kV/cm Synergistic effect between RFEF and heat above 50 °C: Holding cider at 55 °C after RFEF exposure for 5 and 50 s resulted in 2.5- and 3.1-log reductions E_c = 7.1 kV/cm (calculated) Electric energy density: 51 J/mL	Geveke et al. (2009)

(continued)

Table 4.6 (continued)

Product/medium	Microorganism(s)	RF system/ conditions	Results	References
Apple juice	*E. coli* K12	Power supply: 4 kW Electric field strength: Up to 26 kV/cm Exposure time: 0.17 ms Frequency: 15–70 kHz Temperature: 50 °C	The population of *E. coli* was reduced by 1.8 log following exposure to 18 kV/cm at an outlet temperature of 50 °C Repeated three treatment stages at 50 °C increased inactivation to 3 log Electric energy density: 300 J/mL	Geveke and Brunkhorst (2004)
	E. coli K-12, ATCC 23716, initial 7.8 log CFU/mL	RFEF chamber: 0.1 cm diameter, 0.2 cm gap Electric field strength: 15 kV/cm Exposure time: 170 µs Frequency: 20 kHz Flow rate: 540 mL/min Outlet temperatures: 40, 45, 50, 55, and 60 °C	RFEF treatment at 23 °C cause insignificant change in the population of *E. coli* in apple juice Significant changes in the population of *E. coli* were observed at 40 °C and above: 3-log reduction at 45 °C and until 55 °C and 4-log at 55 °C A healthy population of *E. coli* cells recovered in RFEF-treated apple juice at 55 °C averaged <1.5 log CFU/mL on selective agar plates RFEF treatment deformed the bacterial surface structure at all tested temperatures, leading to the leakage of intracellular substances (SEM analysis) A decrease in *E. coli* population and a slight increase in extracellular ATP was obtained at 60 °C RFEF treatment Separate treatment at 55 °C without RFEF caused a viability loss of 1.5 log and a 27.8% injury on *E. coli* cells in apple juice Treatment at 27 °C without RFEF led to no significant differences in viability loss and cells injury than the control	Ukuku et al. (2008)

Product/medium	Microorganism(s)	RF system/ conditions	Results	References
	E. coli K-12, ATCC 23716, initial 7.8 log CFU/mL	(1) RF treatment RF electric field chamber 0.1 cm diameter, 0.2 cm length (gap between electrodes) Electric field strength: 15 kV/cm Exposure time: 170 μs Frequency: 20 kHz Flow rate: 540 mL/min Outlet temperatures: 25, 30, and 40 °C (2) UV-light treatment 30 W UV bulb with 90% energy at the UV wavelength 254 nm UV transparent Chemfluor tubing wrapped around the UV bulb: 3.3 cm internal diameter, 1.6 mm thickness and 14 m length Exposure time: 12 s Temperature: 25, 30 and 40 °C Feed rate: 540 mL/min	At 40 °C, UV-light treatment alone caused 5.8 log reduction of E. coli in apple juice while RFEF caused only 2.8 log reduction. A combination of the two processing treatments did not increase cell injury or leakage of intracellular bacterial UV-substances more than that from the UV-light treatment. Viability loss was not significantly ($P < 0.05$) different than UV-light treatment alone UV-substances determined in apple juice treated with RF electric field were significantly ($P > 0.05$) different than UV-light treated samples RF treatment causes more injury to the bacterial cells leading to more leakage of intracellular UV-substances than cells treated with UV-light alone The effect of the two processing treatment combination on bacterial inactivation was not additive	Ukuku and Geveke (2010)
	E. coli K-12, ATCC 23716, initial 7.8 log CFU/mL	RF treatment chamber: 0.1 cm diameter, 0.2 cm gap Electric field strength: 25 kV/cm Exposure time: 3.4 ms Frequency: 20 kHz Flow rate: 540 mL/min Outlet temperatures: 25, 55, and 75 °C	RF treatment caused a significant decrease in bacterial cell surface hydrophobicity and loss of relative negative ions compared to heat treatment alone at 55 °C and 75 °C Leakage of cellular materials into the media indicated cell damage and TEM observation showed altered intracellular membrane structure in RF treated *E. coli* cells The mechanism of inactivation of RF is by disruption of the bacterial cell surface hydrophobicity and loss of a relative negative ions, which led to injury and leakage of cellular materials and death	Ukuku et al. (2012)

(continued)

Table 4.6 (continued)

Product/medium	Microorganism(s)	RF system/ conditions	Results	References
Beer	Naturally occurring microorganisms	Power supply up to 19 kW Electric field strength: 0.5 kV/cm Exposure time: 9 s or 4.2 min Frequency: 18 MHz Flow rate: 9 kg/min Temperature: 50 °C Outer pipe d_i = 3.8 cm Inner pipe *de* = 2.5 cm Length of RF treatment zone 86 cm	Reduction of naturally occurring microorganisms from 6.3 to 4.4 log CFU/mL in thermal treatment reduction of naturally occurring microorganisms from 6.3 to 4.3 log CFU/mL in RF treatment No nonthermal effects of RF energy were detected no synergistic effects of RF energy with heat	Geveke et al. (2002)
Deionized water	*L. innocua*	Power supply up to 19 kW Electric field strength: 0.5 kV/cm Exposure time: 9 s or 4.2 min Frequency: 18 MHz Flow rate: 9 kg/min Temperature: 55.5 °C Outer pipe d_i = 3.8 cm Inner pipe *de* = 2.5 cm Length of RF treatment zone 86 cm	Reduction of *L. inocua* from 5.9 to 4.4 log CFU/mL in thermal treatment Reduction from 5.9 to 4.3 log CFU/mL in RF treatment No nonthermal effects of RF energy were detected no synergistic effects of RF energy with heat	Geveke et al. (2002)
NaCl solution, 0.1%	Several species of vegetative bacteria: *E. coli*, *S. aureus*, *Micrococcus lysodeikticus*, *Sarcina lutea*, *B. subtilis*, *B. cereus*, *B. megaterium*, *Clostridium welchii*, Pseudomonad, and yeasts: *S. cerevisiae*, *Candida utilis*	Electric field strength: Up to 25 kV/cm Pulse voltage: Up to 10 kV Pulse length: 2–20 μs in steps of 2 μs Temperature control: 20 °C water circulation through the brass blocks that support the carbon electrodes	*E. coli* was used as the test organism to study the parameters likely to influence the inactivation Electrolysis occurring near each electrode does not influence the lethal effect The electric field controlled the degree of microorganisms killing whether the temperature of suspension at the start of the treatment was 20 °C or 40 °C Yeasts need lower electric field strength: Up to 7.3 kV/cm for *S. cerevisiae* and 12 kV/cm for *C. utilis* than bacteria: Up to 17 kV/cm for *E. coli*, and up to 22 kV/cm for motile pseudomonad to be destroyed until the same level around 1% of survivors	Sale and Hamilton (1967)

Product/medium	Microorganism(s)	RF system/ conditions	Results	References
Orange juice	*E. coli* K-12	Power supply: 80 kW Electric field strength: 15 and 20 kV/cm Exposure time: 140–420 µs Frequency: 21, 30, and 40 kHz Flow rates: 1.0 and 1.4 L/min Temperature: 65 °C	The population of *E. coli* was reduced by 2.1 ± 0.4 log after exposure to a 15 kV/cm peak electric field, 270 µs, hold time of 3 s and outlet temperature of 60 °C The population of *E. coli* was reduced by 3.9 ± 0.1 log after exposure to 20 kV/cm and outlet temperature of 65 °C Electric energy density: 180 J/mL	Geveke et al. (2006)
Saline water	*E. coli*	Steinmetz treatment chamber Power: Up to 2 kW Frequency: 20 kHz Treatment times: 900, 675, and 540 µs Outlet temperature: 25, 35, and 45 °C Inlet temperature: 20 °C	Maximum inactivation of 3.6 log CFU/mL at 26.5 kV/cm, 45 °C, and 900 µs The energy demand of 63 kJ/kg for 1 log reduction at 25 °C for foods having 0.1 S/m conductivity at 20 °C	Masood et al. (2017)
Water	*S. cerevisiae*	Electric field strength: 20 and 30 kV/cm Temperature: 35...55 °C Flow rate: 1.2 L/min	The population of *S. cerevisiae* was reduced by 2.1 ± 0.1 log units following exposure to a 30-kV/cm field at 40 °C	Geveke and Brunkhorst (2003)

4 Oscillating Magnetic Fields

Livia Patraşcu, Iuliana Aprodu, and Maria Turtoi

4.1 Introduction

It is well known that a magnetic field is a vector field surrounding magnetized materials being created by electric currents. Starting from the well-known fact that biological substances are mostly diamagnetic, and can subsequently be permeated by magnetic fields, to either static (SMF), pulsed (PMF) or oscillatory magnetic fields (OMF), applications in agricultural and food sciences were found. Researches were done in the areas of microorganisms' inactivation with the help of OMF, food preservation by MF assisted freezing, or plant growth stimulation. Also, MF of various intensities led to untargeted cell mutations, both in microorganisms and somatic cells (Valiron et al. 2005). The same phenomenon was used to increase some compounds production by microorganisms (Al-Hawash et al. 2018). If subjected to a short description, one could differentiate that SMF is a magnetic field applied with a constant intensity in time, while OMF is characterized by a sinusoidal constant or decreasing wave amplitude with constant frequency.

4.2 Magnetic Field Assisted Inactivation of Microorganisms

The most studied possibilities of using OMF in agricultural and food sciences is based on the theory according to which a single pulse of the magnetic field at frequencies between 5 to 500 kHz and intensities above 5 to 50 T can decrease the microorganism population from food products by at least two log cycles (Hofmann 1985). However, fewer intensities were also stated to be effective (Table 4.7). According to Hofmann (1985) theory, complete sterility could be attained after subjecting it to additional pulses. More recent studies found inhibitory effects also at much lower intensities. When comparing the efficiency of different magnetic fields, researchers reported that OMF was more efficient in reducing the growth rate of both bacteria and fungi compared with SMF or PMF (Abdelhameed 2014; Masood et al. 2020).

The temperature rising in the treated food products with OMF is avoided because microorganisms' destruction is accomplished within a brief time. The impulse duration ranges between 10 μs and several milliseconds, with the total exposure time varying from 25 to 100 μs (Barbosa-Canovas et al. 2000; Butz and Tauscher 2002). For instance, after 100 such pulses, the core temperature of the food rises by only 2 °C (Hofmann 1985). Frequencies above 500 kHz were found to be less effective for microorganisms tending also to overheat the treated product.

Table 4.7 Summary of several reported effects of oscillating magnetic fields on microorganisms

Experiment conditions	Microorganism	Source of microorganisms	Reported effect	References
OMF (a) 5 pulses × 5 T (b) 1 pulse × 8 T	Saprophytic bacteria and fungi (c.f.u.)	Oat sprouts	(a) Halving bacteria c.f.u., fungi c.f.u.decrease by a factor of 10 (b) no significant differences in c.f.u. number	Lipiec et al. (2005)
PMF 8 T 10–50 pulses 50 Hz	*L. monocytogenes*	Pure culture	9.6% survival rate at 20 pulses. Up to ~50% survival rate at higher pulse number	Qian et al. (2020)
PMF 1.21–9.48 T 20 pulses	*E. coli* AS1.129, *S. aureus* AS1.89, *S. cerevisiae* ATTC 7552 *B. subtilis* AS 1.921	Pure culture	Varying survival rates "window effect". Min. survival rates reached: *S. cerevisiae* - 6.7% at 5.07 T. At 6.33 T *E. coli* - 2.25%, *S. aureus* 3.8% and *B. subtilis* - 22.5%.	Haile et al. (2008)
PMF 6, 8, or 10 T for 1, 3, and 6 pulses with 10 ms/pulse and time interval of 5 s between two pulses	*Cunninghamella echinulata*	Soil	Strain mutation with a significant increase of biomass, lipid content and yield, γ-linolenic acid content and yield	Al-Hawash et al. (2018)
SMF Two fixed magnets of 0.365 T each	Total bacteria counts	Milk	Total bacteria counts decreased from $250 \cdot 10^4$ for the control sample to $29 \cdot 10^2$ after 60 minutes of exposure	Ali et al. (2015)
OMF of 0.25 T and SMF of 0,4 T 50/60 Hz with 24, 48, 72, 96 and 120 h exposure time	*Plasmopara viticola, Rhizopus stolonifer, Asperguillus niger* and *Rhizopus nigricans*	Cheese, apple, and bread	OMF: Best results - 72% inhibition for *P. viticola,* less effective for *Rh. nigricans* with 53% inhibition SMF: Best results – 57% inhibition for *P. viticola,* less effective for *Rh. nigricans* with 47% inhibition	Abdelhameed (2014)

The primary requirement for OMF applications for microorganisms' inactivation is a low electrical conductivity of the treated material (electrical resistivity higher than 10 to 25 Ωcm), such as to avoid the shielding phenomenon of the interior regions by induced eddy currents from the coil-generated magnetic field (Hofmann 1985; Barbosa-Canovas et al. 2000).

The initial theory on the OMF mechanism of action on microorganisms was based on the OMF coupling energy into DNA resulting in the covalent bonds breakdown and growth inhibition (Hofmann 1985). Takao et al. (1995) revealed the genotoxic effect of a homogeneous static magnetic field of 0.6 T on *Drosophila melanogaster* somatic cells. The mechanism of action of low static magnetic fields on somatic cells was similar to that of a UV light of 0.14 mJ m^{-2} s^{-1}. However, Zhang et al. (2011) concluded that there isn't a lasting effect on biological systems within achievable magnetic fields.

Pothakamury et al. (1993) theory stipulate that weak OMF may loosen the bonds between ions and proteins. It has been observed that OMF give responses, especially at the cyclotron resonance frequencies of ions (a phenomenon related to the movement of ions in a magnetic field) that are of the maximum importance for biological systems such as the case of Ca^{2+} (Pazur et al. 2007), in particular for calcium-binding proteins. Barbosa-Canovas et al. (2000) stated that cyclotron resonance is attained when the frequency at which the ions revolve in a magnetic field equals that of the magnetic field itself. For Ca^{2+} at 50 μT, the resonance frequency was reported to be 38.7 Hz, while for most ions of biological importance, the ion cyclotron resonance frequency was calculated in the range of 2–100 Hz (Grigelmo-Miguel et al. 2011). Also, Abdelhameed (2014) stated that the inhibition of microorganisms when subjected to the magnetic field might be attributed to energy transfer to cell ions such as Ca^{2+} increasing its velocity across the cell membrane, resulting in metabolic changes of the cells.

The sterilization efficiency of magnetic field treatment depends on the intensity and species of the bacteria (Haile et al. 2008). Studies revealed an inconsistency in the sterilization effect of OMF, as shown in Table 4.7. Different microorganisms presented different survival rates when treated under magnetic fields. Moreover, the same bacteria strain could have a lower survival rate when subjected to a lower number of pulses. In comparison, at a higher number of pulses of similar intensity, the same strain could present a higher survival rate (Qian et al. 2020, data presented in Table 4.7). The phenomenon was explained by a so-called "*window effect*", "*frequency window*", or "*vale value*". It is described as one of the fundamental characteristics of electromagnetic field, indicating that "*targets inside bio-system only respond to the electromagnetic waves with some discrete frequency or intensity range*" (Haile et al. 2008; Qian et al. 2020).

The use of OMF in microorganism inactivation is still in its laboratory research phase. In addition to inconsistent results reported, there are also many factors to be considered, such as magnetic field intensity, number of pulses, frequency, and product characteristics like resistivity, electrical conductivity, and thickness (Zhang et al.

2011). Researches on magnetic fields effect on microorganisms found that it could inhibit and stimulate bacterial growth. Ahmad et al. (2013) reported that total aflatoxin increased when *Aspergillus flavus* was exposed to the magnetic field. Another concern is the safety of the technology for humans. While magnetic fields up to 0.5 T are considered safe for humans in European Union, Yadollahpour et al. (2014) concluded that the mechanism by which magnetic fields influence biological cells is still purely understood. In comparison, the vertical component of Earth's geomagnetic field is ~67 μT at the magnetic pole and 0 T at the equator (Kobayashi et al. 2004).

4.3 Magnetic Field Assisted Freezing

More successful use of the magnetic field in food processing is represented by magnetic fields assisted freezing (MFAF). This technology is currently used for commercial freezers, like Cells Alive System (CAS) offered by ABI Corporation, Ltd., equipment that can generate weak OMFs in a rapid freezer, or "Proton freezers"by Ryoho Freeze Systems Co., Ltd., that use static magnetic fields and electromagnetic waves (Otero et al. 2017). Magnetic fields and electromagnetic waves freezing are used in the food industry and biology to increase tissue survival. The apparatus consists of a magnetic field generator attached to a conventional quick-freezing unit.

Magnetic field assisted freezing technology is based on the fact that it can affect water properties, leading to changes in freezing kinetics. As an essential component of the food matrix, understanding the effect of magnetic field on water molecules is decisive in studying the MFAF mechanism. First, it is necessary to start with the science of food freezing. In short, the size of ice crystals formed during the freezing process is the topmost factor to determine the sensory quality of frozen foods, and it depends on the rate of heat extraction from the product, together with the nucleation rate. A low nucleation rate means fewer ice crystals but large-sized damaging products texture and leading to high drip loss when thawing. In contrast, a high nucleation rate means many tiny ice crystals causing minor damage to the quality of frozen foodstuff.

Because water molecule has no intrinsic magnetic dipole moment, its arrangement in the food matrix is scattered. The leading theory states that OMF determines reorientation of the electronic and nuclear spin of water molecules leading to a higher number of tiny ice crystals than conventional freezing (Rodríguez et al. 2017b). In response to an applied MF, the orbital motion of electrons can be altered, developing an opposite moment (Tang et al. 2019). That may increase the difficulty for the water molecules to bind together, avoiding ice crystals formation, which may lead to lower nucleation temperature. It was observed that pulsed magnetic field (PMF) treatment decreased the initial nucleation temperature, shortening thus water-tempering time (Zhang et al. 2021). This lower nucleation temperature was attributed to MF's ability to vibrate water molecules, inhibiting ice crystals nucleation.

Various physical changes were observed to be induced by MFs to water molecules, like hydrogen bonding, refractive index, supercooling, conductivity and vaporization rate (Zhang et al. 2021 and cited literature therein). MF is thought to determine displacement and polarization of molecules and atoms and modification of hydrogen bonding, thus leading to changes in water structure. Cai et al. (2009) found that magnetic field leads to decreased water surface tension while increasing its viscosity. Okuda et al. (2020a, b) stated that OMFs inhibit latent heat by suppressing the hydrogen bonding between water molecules during air-blast freezing. Another advantage of MFAF is the significant shortening of the cooling time needed to acquire freezing temperature (Okuda et al. 2020b).

Another theory claims that MFs influence the freezing process through Lorentz force. As it is known that Lorentz forces can produce small currents in liquids, producing the same effect as water electrolysis (Colic and Morse 1999), it is considered that MF can act on the charged solid particles and biologically precipitated magnetite in food tissue, increasing the cooling efficiency and the extent of supercooling (Cai et al. 2009).

MF was shown to significantly influence the zeta potential and particle size distribution in solutions (Higashitani et al. 1993). Chang and Weng (2006) stated that contrary to electric fields, which break water hydrogen-bond network, magnetic fields enhance the hydrogen-bonding ability, improving the stability of liquid water. However, the precise mechanisms that cause the effects observed in water exposed to MFs are not yet completely understood (Otero et al. 2017).

Considering that the magnetic sensibility of water is low, it would take strong magnetic fields (>10 T) for macroscopic observations. However, field intensities used in MFAF are much lower than for microorganism inactivation, ranging from 0.1 mT up to 100 mT (Table 4.8), while the MF frequency ranges between 50 Hz and 10 MHz. Magnetic flux densities from 1 to 10 T were effective in increasing the number of hydrogen bonds from water (Cai et al. 2009).

Although the efficiency of OMF in preserving frozen food quality is pretty recognized, research results are still varying, as can be seen from Table 4.8. The difficulty in understanding the MF action on the freezing process and the different results reported can be attributed to the different design of freezers and sample location inside, leading to other MF vectors (Rodríguez et al. 2017b).

The thawing process might significantly influence the quality characteristics of food products frozen under OMFs. During the thawing process, hydrogen bonds of ice crystals are broken due to an occurring endothermic reaction. Thus, if heat exchange efficiency is low, there is an increased probability of repeated recrystallisation and re-thawing, leading to severe tissue damage (Okuda et al. 2020a).

Table 4.8 Summary of several reported effects of oscillating magnetic fields assisted freezing

Experiment conditions	Product	Reported effect	Reference
OMF <2 mT 6–59 Hz	Crab sticks	No effect on the drip loss, water-holding capacity, toughness, and whiteness was observed	Otero et al. (2017)
OMF – CAS 1.66 mT 6 Hz	Egg white	MFAF significantly reduced the foamability of egg whites. No differences in DSC profiles were observed.	Fernández-Martín et al. (2017)
OMF 50–100 mT 1 Hz	Chicken breasts	Microstructure, drip loss, pH, colour and texture of chicken breasts maintained the original qualities without any quality deteriorations.	Mok et al. (2017)
OMF - CAS 0.1–1 mT	Mackerel	Very slight tissue damage due to ice crystal formation. OMF inhibited histological damage in mackerel samples after freezing and thawing.	Okuda et al. (2020a)
OMF: 0.02–1.74 mT, PMF: 0.00–16 mT 50 Hz	Pork meat	The initial nucleation temperature decreased significantly, with a higher supercooling degree in comparison to control for PMF. PMF is considered a more promising method to be used in the food freezing industry than AMF.	Tang et al. (2019)
OMF-CAS 0.04–0.5 3mT	Pork loin	OMF conditions had no significant effect on the freezing characteristics of pork loin compared with conventional freezing	Rodríguez et al. (2017a)

CAS cells alive system, *PMF* permanent magnetic field

5 Electrohydrodynamic Processing

Iuliana Aprodu, Livia Patraşcu, and Maria Turtoi

5.1 Introduction

The electrohydrodynamic processing includes simple and cost-effective voltage-driven technologies, such as electrospinning and electrospraying, which are used for the fabrication of nano- to micrometers sized fibers and particles, respectively. In addition, electrohydrodynamic drying consists of the induction of electric wind (corona wind) generated by gaseous ions under the influence of a high-voltage electric field applied to an electrode with a small radius of curvature (Singh et al. 2012).

Electrospinning and electrospraying are perceived as 'sister' technologies (Bhushani and Anandharamakrishnan 2014). The mechanism behind these two electrohydrodynamic processing techniques relies on the electro-hydrodynamic force that acts uniformly on a liquid sample and breaks it up into fine jets after passing through a spinneret. The main difference between electrospinning and electrospraying techniques is the concentration of the polymer solution subjected to processing. In the case of electrospinning, the polymer concentration is chosen such as to allow the formation of a charged jet with whipping instability, which is finally distributed over a grounded collector. In contrast, in the case of electrospraying, a lower concentration of polymer solution is used, such as to allow the formation of fine highly charged droplets prior to the deposition on the collector because the varicose instability driven jet destabilization (Bhushani and Anandharamakrishnan 2014).

Electrospun and electrosprayed products have particular structural advantages and functional features. The most important structural benefits of these products are nano- to micrometer size, which can be controlled by manipulating the processing, solution and ambient related parameters, tailored morphology with high porosity which can go higher than 90%, high surface to volume ratio and intertwined fibrous structure (Elias et al. 2011; Bhushani and Anandharamakrishnan 2014). The functional features responsible for the wide applications of the electrospun and electrosprayed products are: allow efficient encapsulation, enhanced stability and further controlled release of different compounds such as bioactives, are obtained through nonthermal processing, and are prone to food related applications when processing food grade (bio)polymers (Bhushani and Anandharamakrishnan 2014). Therefore, the fibers and particles obtained through electrohydrodynamic processing have several important applications in various fields, such as encapsulation and packaging in the food industry, tissue engineering in biomedical applications, or water and air nanofiltration for environment protection etc. (Echegoyen et al. 2017; Haider et al. 2019). The scale-up of the laboratory electrohydrodynamic processing technique for food application was achieved recently (Echegoyen et al. 2017).

5.2 *Electrospinning*

Although the first patents of Morton (1902) and Cooley (1902), dealing with the method and apparatus for electrically dispersing fluids, date back to more than a century, significant progress in the field of nanotechnology involving the electrospinning technology was registered only in the 1990s (Haider et al. 2019).

Electrospinning is suitable for preparing fibrous materials with diameters in the 10–500 nm range and small inter-fibrous pore size, having a high surface area to volume ratio, which allows desired properties in terms of permeability and porosity (Haider et al. 2019).

Prior to electrospinning, a sol-gel is prepared out of one or more polymers and suitable solvents. A large variety of stable and biodegradable materials ranging from natural polymers to complex composites can be processed through electrospinning for preparing nanomaterials.

Electrospinning has several advantages over other techniques used to obtain nanofibers, such as the laboratory scale drawing, template synthesis, phase separation and self-assembly. These advantages are related to the process itself, which is simple, flexible and cost-effective, and allows controlling the diameter fibers (Leidy and Ximena 2019). The main challenge in the case of electrospinning refers to the difficulties in handling the instability of the jet.

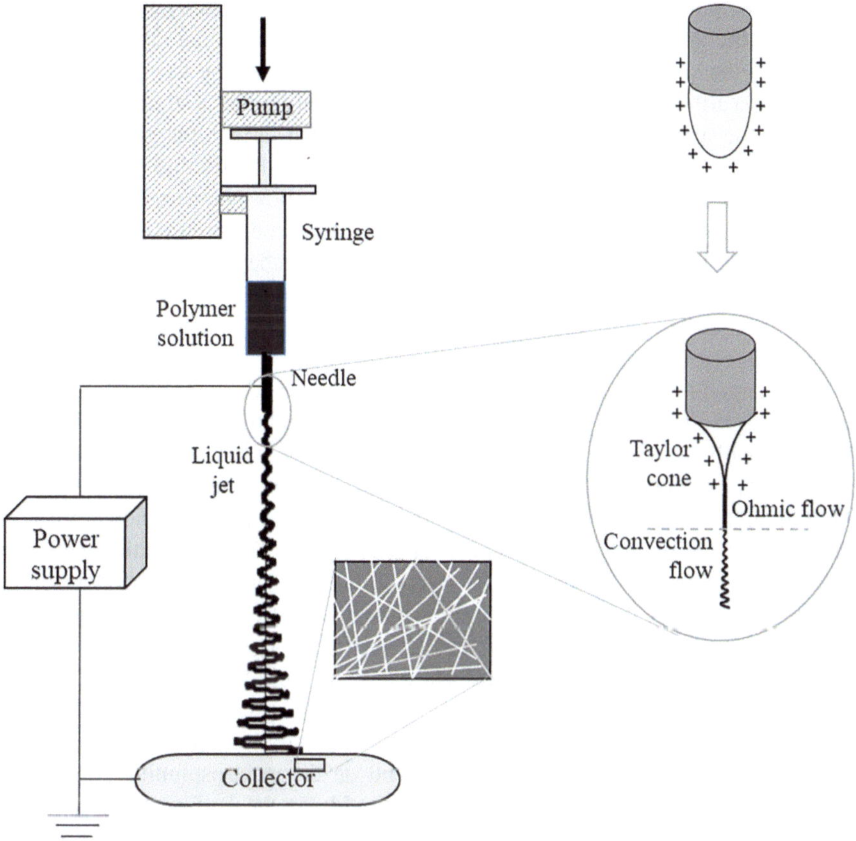

Fig. 4.5 Schematic representation of the electrospinning set up. (Adapted from Xue et al. 2019)

Typically, the fabrication method of the electrospun materials requires the use of a simple set-up involving a high-voltage power supply of either direct current (DC) or alternating current (AC), a syringe pump fitted with a capillary tube (syringe) with a needle or capillary tip (spinneret) and a conductive collector (Fig. 4.5) (Faridi-Esfanjani and Jafari 2016; Xue et al. 2019). Considering that all these features are accessible, the system scale-up is relatively simple (Nagam Hanumantharao and Rao 2019).

5.2.1 Principle of Electrospinning

Upon pushing at a constant rate, the liquid with the syringe pump through the spinneret, pendant droplets are produced due to the surface tension. When applying a potential difference between the spinneret and the collector, the electric charges will migrate in the liquid, producing excess changes on the surface of the liquid, that resemble the polarity of the spinneret. The density of the surface charges is correlated with the applied voltage. Up to a certain electric field, the droplet shape is favored by the surface tension, which tends to compensate the free energy of the surface. At a critical voltage, the electrostatic repulsion reaches a strength enough to overcome the surface tension. Therefore, because of the surface charges of the same sign induced through electrification, the droplets further deform into a cone-shaped jet, termed the Taylor cone (Fig. 4.5). This new shape allows increased surface area, and consequently, the attenuation of the electrostatic repulsion and the minimization of the surface free energy. The critical voltage needed for getting the Taylor cone is a function of the viscosity of the liquid subjected to electrospinning. Higher voltage is required when processing viscous liquids, such as compensating the viscoelastic force of the material.

The charged jet ejected from the Taylor cone has a particular trajectory that starts with a straight line moving toward the collector. The length of the straight segment decreases with the voltage. To prevent the disintegration of the jet into tiny droplets, the viscoelastic properties of the samples should overcome the Rayleigh instability (Duft et al. 2003; Xue et al. 2019). At the end of the straight line, which delineates the near-field region, the velocity of the jet is 1–15 m/s. The acceleration of the jet is gradually attenuated because of the viscoelastic force and the surface tension of the jet. After passing the straight line, the jet begins to stretch, and when the acceleration drops to zero, the movement gets disturbed by any small perturbation (He et al. 2005). This behavior is noticed while moving from the near-field to the far-field region, when the trajectory continues with whipping motions, caused by electrical bending instability. In fact, the electrically charged jet can be affected by the following types of instabilities, which depend on the electrospinning parameters and the physicochemical properties of the liquid: (i) the surface tension driven Rayleigh instability, which is axisymmetric and may result in splitting the jet into small droplets; the Rayleigh instability can be controlled through a strong electric field; (ii) the axisymmetric instability associated to stronger electric fields; (iii) the non-axisymmetric whipping or bending instability, when the movement of the jet in a radial direction is governed by the lateral electrostatic force (Reneker and Yarin 2008). Achieving rapid whipping instability, with the trajectory of the jet consisting of loops and conical shaped coils around the original direction, is crucial for obtaining ultrathin electrospun nanofibers (Xue et al. 2019). With the gradual jet bending, elongation and subsequent diameter stretching, the second bending instability stage is reached. The reduction of the charges density on the jet that appeared through elongation might result in capillary instability, which finally leads to beaded fibers because of the jet disruption into tiny droplets. Taking into account that whipping and bending instability, accompanied by jet elongation by up to 10,000 times over max. 0.05 s, is essential for obtaining fibers with diameters in the sub-micrometer to nanometer scale, the evaporation of the solvent used to prepare the liquid subjected to electrospinning should be strictly controlled, as can modulate the viscoelastic parameters of the jet. Solvent evaporation or cooling of the melt are the phenomena that allow quick solidification and final deposition of solid fibers on the grounded collector (Xue et al. 2019). Thinner fibers are obtained when the evaporation is slow. Although the surface of the fibers is still residually charged, any kind of instability disappears once the solidification process is completed.

The morphology of the fibers deposited in the ground collector depends on the final bending instability stage before deposition. In the loop like region of the jet trajectory, corresponding to the first bending instability, the fibers are collected as a "nonwoven mat", whereas in the small coiled region, corresponding to the second and the third bending instabilities, fibers with straight or waxy morphology are obtained. The thickness of the collected electrospun fibers is restricted to 0.5–1 mm by the residual charges of the solidified layers, which incline to prevent the building-up events with additional similarly charged jet.

5.2.2 Factors Affecting the Electrospinning Process

The quality of the final electrospun fibers can be modulated by controlling both the parameters of the electrospinning process and the characteristics of the working solution.

Among the processing parameters used to control the diameter of the fibers, the applied electric field, the flow rate of the solution subjected to electrospinning, the distance between the tip of the needle (spinneret) and the collector, and the diameter of the needle are the most important (Xue et al. 2019; Haider et al. 2019; Castro Coelho et al. 2021).

To generate the electric field required for carrying out the electrospinning, both DC and AC can be applied to the spinneret. Still, different behaviors of the jet have been reported for the two cases (Xue et al. 2019). The density of the charges on the surface of the jet, the intensity of the repulsive electrostatic forces acting between charges of the same type on the surface of the jet and the overall interaction between the external electric field and the jet are influenced by the applied electric field. The distribution of the charges on the surface of the jet will depend on the polarity of the applied voltage. In particular, the electrospinning capacity of the electrolytes relies on the polarity of the applied voltages (Terada et al. 2012; Xue et al. 2019).

The morphology of the nanofibers depends on the velocity of the solution subjected to electrospinning when passing through the needle. The flow rate of the liquid will directly influence the diameter of the fibers. Thus, low flow rates of the liquid are required to obtained narrow fibers. The diameter of the fibers will increase with the flow rate of the liquid up to a critical value, which depends on the polymer system, over which beads will be formed instead of nanofibers (Haider et al. 2019). Moreover, Megelski et al. (2002) indicated that the pore size of the polystyrene fibers increased with the flow rate.

The thickness of the fibers is also influenced by the type of instability characterizing the jet before deposition on the grounded collector. Efficient elongation and solidification of the jet are usually achieved at relatively long working distances between the tip of the spinneret and the collector. As detailed before, the instability of the jet is a factor of this distance. Within a certain range, the increase of the working distance will result in thinner fibers. In fact, managing the morphology and diameter of the fibers is possible only by simultaneously controlling all processing parameters When increasing the flow rate of the solution, one should also increase the working distance and the critical voltage, such as to get total elongation of the jet prior to deposition of the fiber (Xue et al. 2019).

As Leidy and Ximena (2019) reviewed, most electrospinning studies involving various biopolymers (gelatin, zein, amaranth proteins, pullulan, dextran, cyclodextrins, hydroxypropyl-γ-cyclodextrin, ovalbumin, whey and soy proteins) used flow rates ranging from 0.01 mL/h to 1 mL/h (most frequently 0.5 mL/h), a voltage of 10–35 kV (most often 15 kV), and distance between the tip of the needle and the collector of 7.5–20 cm (most frequently 10 cm).

The success of the process is as well influenced by factors related to the materials for electrospinning. The common materials used for generating nanofibers through electrospinning are solutions or melts of organic polymers. The solution electrospinning is the most studied electrospinning method and was employed for processing over 100 types of organic polymers to produce different end-use products, such as biocompatible and biodegradable nanofibers suitable for obtaining scaffolds for biomedical applications (poly(ε-caprolactone) (PCL), poly(lactic acid) (PLA), and poly(lactic-co-glycolic acid) (PLGA)) or nanofibers with applications in environmental protection (polystyrene (PS) and poly(vinyl chloride) (PVC)) (Xue et al. 2019). Other natural polymers (dextran, chitin, chitosan, alginate, collagen, gelatin etc.) have been successfully used to generate nanofibers with various applications in the food industry (Leidy and Ximena 2019; Xue et al. 2019). In addition, small molecules capable to self-assemble and generate entanglements are suitable for obtaining nanofibers. Finally, different research groups reported on the possibility of integrating nanometers sized components into the polymer solutions prior to electrospinning. The solution parameters that affect the electrospinning process are related to the polymer and the solvent. The molecular weight and concentration of the polymer in the solution exert a high influence on the viscosity of the solution, which is an important parameter for the success of the electrospinning process. Choosing the appropriate solvent for preparing a homogeneous polymer solution is essential (Luo et al. 2012). Anyway, the solidification rate of the jet, which is directly related to the evaporation rate of the solvent, should be carefully factored in when deciding on the solvent for preparing the polymer solution. The use of solvents with very high volatility results in fast solidification of the jet generated through the spinneret, whereas solvents with very low volatility lead to wet fibers deposited on the collector. The electrical properties of the solutions are influenced by the molecular weight of a polymer, while the intensity of the electrostatic repulsion among the charges located on the surface of the jet is controlled by the dielectric constant of the solvent (Xue et al. 2019). Higher voltage should be applied to stabilize the jet when processing solutions with increasing dielectric constant (Luo et al. 2012). The spinnability of the polymer solutions depends on the concentration and electrical conductivity. Since the polymer concentration affects the surface tension and viscosity of the working solution, which further impact the diameter and morphology of the nanofibers, the concentration of the polymer solution should be high enough to overcome the Rayleigh instability, such as to avoid splitting the jet into small droplets, and should not be too high to prevent difficulties in being ejected from the spinneret and in further handling the viscoelastic force which impedes jet formation (Xue et al. 2019). The importance of the electrical conductivity of the polymer solution in producing thin nanofibers relies on the involvement in jet elongation and bending. It has been demonstrated that polymer solutions having high conductivity raise difficulties in generating the Taylor cone or activating bending instability of the jet. One way to control the electrical conductivity is to add different ionic compounds to the polymer solution subjected to electrospinning.

Finally, the solution electrospinning is influenced by environmental conditions. In particular, the relative humidity, the ambient temperature and the air flow impact the solvent evaporation rate and further on the solidification of the jet (Xue et al. 2019; Castro Coelho et al. 2021).

5.3 *Electrospraying*

Electrospraying is a process of liquid atomization by electrical forces, also known as electrohydrodynamic atomization (Bhushani and Anandharamakrishnan 2014). Electrospraying provides several advantages over the conventional processes, such as limiting the polydispersity of the particles and improving the encapsulation efficiency. Moreover, this technique allows easier scale up of the atomization process, simply and effectively, generating high quality products (Castro Coelho et al. 2021).

As in the case of electrospinning, a typical unit for electrospraying includes four major components: a high-voltage power supply operated more often in DC mode and only rarely in AC, a syringe pump, a stainless-steel needle or capillary (spinneret), and a grounded collector, which can be a rotating drum or a flat plate (Fig. 4.6).

5.3.1 Principle of Electrospraying

Electrosprayed nano- or microparticles are obtained at mild temperature upon subjecting a polymer solution to flow under high voltage in a one-step process. The main difference between electrospinning and electrospraying relies on the concentration of the polymer solution, which is lower in electrospraying. The concentration of the electrospraying solution has to be low enough to avoid getting the elongation of stable Taylor cone and further whipping instability. Instead, varicose instability of the polymer jet is desired, as it ensures the self-dispersion of the highly charged droplets. When the polymer solution is pushed out of the needle, the surface will become charged because of the existence of the external electric field, and the density and distribution of the charges will depend on the electrical conductivity and dielectric constant. Because of the electrostatic attraction, the charged jet will get stretched toward the collector. When the charges density reaches 50–80% of the Rayleigh limit, the fission of the charged droplets out of the Taylor cone will occur

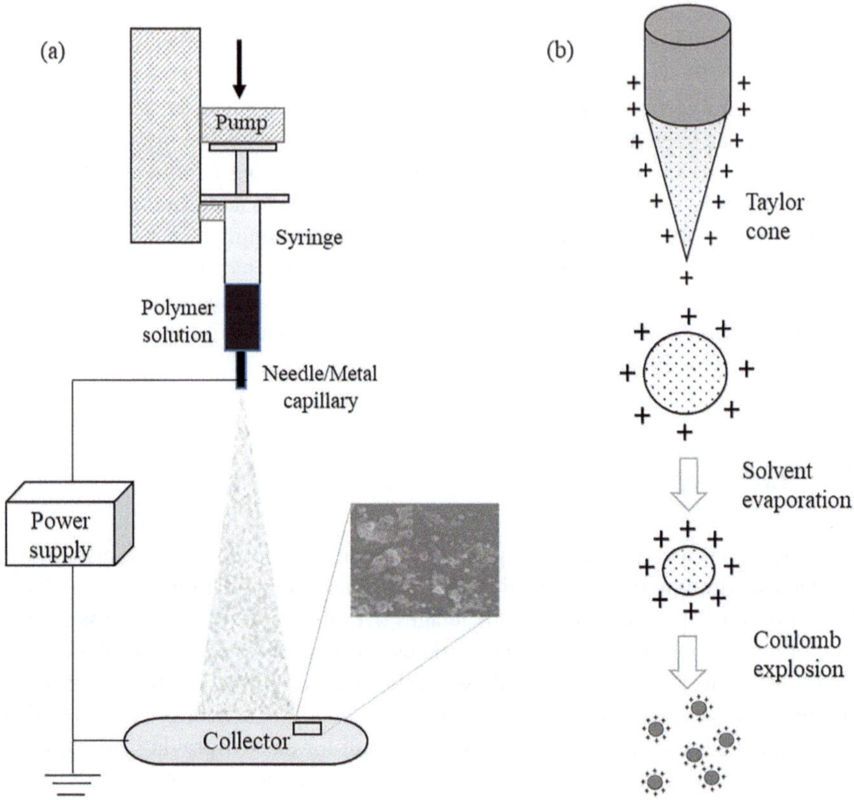

Fig. 4.6 Schematic representation of the electrospraying set-up (**a**) and mechanism (**b**). (Adapted from Haider et al. 2019)

due to the Coulomb repulsions (Haider et al. 2019). As a result of rapid solvent evaporation and solidification of the droplets, solid particles are deposited on the collector (Bhushani and Anandharamakrishnan 2014; Alehosseini et al. 2019). Droplets stretching and atomization are influenced by the electrostatic force due to the electric field between the spinneret and the collector, the repulsive Coulomb forces acting between droplets carrying the same type of charge on the surface, the viscoelastic behavior of the polymer solution and the surface tension (Haider et al. 2019).

5.3.2 Factors Affecting the Electrospraying Process

As in the case of electrospinning, the electrospraying process and properties of the final products are influenced by factors related to the solution subjected to atomization, processing and environment parameters. Regarding the polymer solutions, the solvent used as well as the molecular weight and polymer concentration exert a high influence on the pH, viscosity, conductivity and surface tension. As Haider et al. (2019) reviewed, a critical range of viscosity of 1.5–5500 mPa·s and surface tension usually lower than 50 mN/m are required to get a stable Taylor cone. As discussed in the case of electrospinning, the processing parameters used to control the size of the final product include the applied voltage, the distance between the tip of the needle and the collector, and the flow rate (Bhushani and Anandharamakrishnan 2014). Finally, environmental parameters like temperature, relative humidity and air flow are highly important because of the impact on the solvent evaporation rate.

5.4 Applications of Electrospinning and Electrospraying in the Food Industry

Electrospun nanofibers and the micro- to nanoscale size particles obtained through electrospraying have multiple applications in various fields, such as environment protection, medical, pharmaceutical, textile and electrical industries. When targeting edible applications of electrospinning and electrospraying, only non-toxic food grade compounds approved by authorities, such as the Food and Drug Administration (FDA) or the European Food Safety Authority (EFSA), should be used for preparing nano- to micrometer sized fibers and particles. The polymers must be degraded during human digestion and the concentration used should ensure no toxicity. Several biopolymers, including proteins like whey proteins, soy proteins, casein, gelatin, zein, and protein concentrate from *Spirulina* and polysaccharides like pullulan, alginates, dextrans, and cellulose (Table 4.9), have been tested as materials for electrospinning and electrospraying, and blending with synthetic polymers and crosslinking method were considered for improving the morphology of the final products. In particular, taking into account that the linear structure to random coil conformation of biopolymers in dissolution is essential for obtaining fibers (Nieuwland et al. 2013), one important issue when subjecting protein solutions to electrospinning to obtain nanofibers consists in the appropriate unfolding of the molecules to ensure droplet stretching and jet formation (Mendes et al. 2017). Different strategies have been tested to improve the electrospinnability of globular proteins: applying thermal treatments, using denaturing agents, choosing particular solvents or blending wheat proteins or casein with other polymers (Zhang et al. 2006; Ratanavaraporn et al. 2010; Mascheroni et al. 2013; Leidy and Ximena 2019). On the other hand, chain entanglements occurring in electrospinning solution in case of some proteins or polysaccharides ensure avoiding the breakup of the jet during electrospinning (Leidy and Ximena 2019).

Because of the safety issues related to the use of toxic solvents and crosslinking treatment which are not approved by the FDA, food applications of electrospinning and electrospraying are still limited mainly to packaging, encapsulation of different types of compounds or enzymes and development of halochromic nanosensors and for real-time monitoring of fish quality (Table 4.9) (Echegoyen et al. 2017; Leidy and Ximena 2019; Aghaei et al. 2020). A less explored food-based application of the electrospinning consists in producing nanofibrous membranes for the pressure-driven liquids filtration (Bhushani and Anandharamakrishnan 2014). On the other hand, the most important applications of electrospraying in the food industry refer to the encapsulation of different biologically active compounds which are otherwise prone to degradation and obtaining edible films with actives properties (Torres-Giner et al. 2010; Alehosseini et al. 2019; Rostami et al. 2019; Soleimanifar et al. 2020). Finally, combining electrospinning and electrospraying was proposed by Schmatz et al. (2019) for developing an innovative active food packaging material. They developed nanocomposite materials by incorporating nanoparticles that stabilize sensitive bioactive compounds such as phycocyanin into the nanospun fibers.

Successful applications of electrospun fibers for producing innovative and biodegradable food packaging materials have been recently reviewed by Castro Coelho et al. (2021). The shelf-life and stability of different food products can be improved by incorporating different bioactives in the electrospun materials used as active packages, which allow further controlled release of the biomolecules exhibiting antioxidant and antimicrobial properties. In addition to the food simulants, these applications targeted various types of food products, such as bananas, bread and cheese (Yao et al. 2016; Altan et al. 2018; Göksen et al. 2020).

Promising results on the use of electrospinning to improve the thermal stability of different enzymes used in the food industry were reported by Baştürk et al. (2013) and Işik et al. (2019). A blend of poly(vinyl alcohol) (PVA) and poly(acrylic acid) (PAA) was used to fabricate thermally crosslinked nanofibers suitable for covalent immobilization of α-amylase (Table 4.9) upon activation of the amine reactive imidazolyl-carbamate group through 1,1′-carbonyldiimidazole (Baştürk et al. 2013). On the other hand, the interaction between PVA and ionic metals (Zn^{2+}) was exploited for obtaining electrospun nanofibers, suitable for immobilization of lipase upon cross-linking with glutaraldehyde (Işik et al. 2019). The immobilized α-amylase and lipase exhibited enhanced storage and thermal stability and could be reused over 14 times (Baştürk et al. 2013; Işik et al. 2019).

5.5 *Final Remarks*

The electrohydrodynamic processes are emerging technologies suitable for obtaining fibers and particles in the nano- to micrometer size range. The most promising applications of electrospinning and electrospraying in the field of the food industry consist mainly in producing food-grade delivery vehicles for different bioactive compounds and active nanofiber materials for packaging or coatings for different food systems.

Table 4.9 Examples of main food applications of electrospinning and electrospraying

Application	(Bio)polymer solution	Process parameters	Reference
Encapsulation of bioactive compounds			
Encapsulation of docosahexaenoic acid (DHA) using zein by electrospraying. The resulting beads exhibited good stability	Zein and ω-3/zein solutions (20%) prepared in 85 wt% aqueous ethanol. The zein:DHA ranged from 1: 1 to 3: 1	Flow rate: 0.20 mL/h Applied voltage: 12 kV internal diameter of the needle: 0.9 mm Tip-to-collector distance: 15 cm	Torres-Giner et al. (2010)
Encapsulation of volatile aroma compound in nanofibrous electrospun membranes made of pullulan and β-cyclodextrin (β-CD)	The polymer solution consisted of 20% pullulan and eventually 5% β-CD. The volatiles, rich in perillaldehyde, were incorporated after precipitation with β-CD	Flow rate: 0.5 mL/h, Applied voltage: 15 kV Tip-to-collector distance: 12 cm	Mascheroni et al. (2013)
Highly efficient encapsulation of polyphenols (epigallocatechin gallate) into gelatin micro- and submicroparticles through electrospraying	Gelatin solutions (5, 8, 10 and 20%, w/v) prepared by dissolving in acetic acid 20% (v/v)	Flow rate: 0.15–0.5 mL/h Applied voltage: 15–28 kV Tip-to-collector distance: 10 cm	Gómez-Mascaraque et al. (2015)
Electrospraying encapsulation of the green tea catechins in two different protein matrices: Zein and gelatin. Because of the catechins instability in aqueous solutions, zein allowed better encapsulation efficiency. The process requires further optimization before applying to food products that require backing.	Gelatin (8%, w/v) and zein (12%, w/v) solutions prepared in acetic acid (20%, v/v) and aqueous ethanol (85%, v/v), respectively. After proteins solubilization, the green tea extract was incorporated at a concentration of 20 g/100 g dw.	Flow rate: 0.2 and 0.15 mL/h for gelatin and zein solutions, respectively Applied voltage: 15 and 13 kV for gelatin and zein solutions, respectively Inner diameter of the needle: 0.9 mm Tip-to-collector distance: 10 cm	Gómez-Mascaraque et al. (2017)
Encapsulation of polyphenols from *Rosmarinus officinalis* in polyvinyl alcohol nanofibers to ensure the controlled release of the bioactive	Poly(vinyl alcohol) solution (12%) prepared in water or ethanol 70%, supplemented with 5% citric acid in respect to the polymer	Flow rate: 2.2 mL/h Applied voltage: 30 kV Tip-to-collector distance: 20 cm Inner diameter of the needle: 0.8 mm	Estevez-Areco et al. (2018)

(continued)

Table 4.9 (continued)

Application	(Bio)polymer solution	Process parameters	Reference
Encapsulation of bioactive compounds from saffron extract (picrocrocin, safranal, and crocin) by electrospinning	Zein solution (20%) prepared in aqueous ethanol 80%	Flow rate: 0.15–0.20 mL/h Applied voltage: 15 kV, Tip-to-collector distance: 10 cm	Alehosseini et al. (2019)
Encapsulation of bioactive compounds from saffron extract (picrocrocin, safranal, and crocin) by electrospraying	Zein solution (10%) prepared in 80% aqueous ethanol	Flow rate: 0.15–0.20 mL/h Applied voltage: 15 kV Tip-to-collector distance: 10 cm	Alehosseini et al. (2019)
Nanoencapsulation of the phenolics from olive leaf. Different whey proteins-based particles were loaded with varying amounts of oleuropein, caffeic acid or hydroxytyrosol	Whey protein concentrate solutions (15, 20, and 30% w/v) were prepared in distilled water. The pH was adjusted to pH and tween 20 was used as surfactant	Flow rate: 0.5 mL/h applied voltage: 18 kV internal diameter of the needle: 0.9 mm Tip-to-collector distance: 14 cm	Soleimanifar et al. (2020)
Food coating or packaging applications			
Preparing starch films and coatings by electrospraying modified maize starch	Gelatinization of the 5% (w/v) starch dispersion in de-ionized water was carried out at 120 °C for 30 min. The ultrasonic treatment was then applied to the ethanolic mixture to efficiently disrupt and stabilize the modified starch solution	Flow rate: 5×10^{-11} m^3/s Applied voltage: 7 kV Tip-to-collector distance: 20 mm Inner diameter: 0.25 mm	Pareta and Edirisinghe (2006)
Preparing hydrophobic edible films by electrospraying water-in-oil emulsions	Water-in-oil emulsions were prepared by high shear mixing while gradually adding the water (0.5–40%, w/w) to the sunflower oil with 5% (w/w) surfactant (Span80, lecithin or polyglycerol polyricinoleate)	Flow rate: 15 mL/h Applied voltage: 9–14 kV	Khan et al. (2013)

(continued)

Table 4.9 (continued)

Application	(Bio)polymer solution	Process parameters	Reference
Enhancement of O_2 barrier properties of the multilayer food packaging films based on polyhydroxybutyrate containing 12% valerate, with an interlayer of electrospun zein nanofibers	Zein solution (33 wt.%) prepared in ethanol 85% v/v	Flow rate: 0.3 mL/h Applied voltage: 12–14 kV Tip-to-collector distance: 10 cm Inner diameter: 0.9 mm	Fabra et al. (2013)
Electrospun fibre mats based on loaded zein functionalized with gallic acid as food contact active packaging material	Zein solution (25%, w/w) prepared in 80% aqueous ethanol were supplemented with 5%, 10% and 20% (w/w) of gallic acid	Feed rate: 0.8 ml/h Applied voltage: 16 kV tip-to-collector distance: 13 cm	Neo et al. (2013)
Improvement of oxygen and water barrier properties of food packaging films based on polyhydroxyalkanoates with 3% valerate by the use of adhesive electrospun interlayers. Whey protein isolate (WPI), zein, pullulan and blends of zeine:WPI or zeine:Pullulan (50:50) were used to prepare electrospun interlayers	Aqueous solutions of 30 wt.% of WPI (with 10 wt.% of glycerol and 5 wt.% of tween 80) or 15 wt.% of pullulan (the electrospinning results: Nano- or microparticles) Zein solution (33 wt.%) prepared in ethanol 85% v/v	Flow rate: 0.15 mL/h for WPI and 0.3 mL/h for zein and pullulan Applied voltage: 12–14 kV	Fabra et al. (2014)
Coating the poly(3-hydroxybutyrate-co-3-hydroxyvalerate) films for getting effective bactericide systems used for food packaging and food contact surfaces	PHBV18/ZnO solution (10%) consisting of PHBV18 (18 mol% valerate content) mixed with ZnO nanoparticles	Flow rate: 5 mL/h Applied voltage: 10–12 kV Tip-to-collector distance: 17 cm. Inner diameter of the needle: 0.9 mm	Castro-Mayorga et al. (2017)
Development of ultrafine fibers for active packaging, based on protein concentrate from spirulina sp. LEB 18/ poly(ethylene oxide) (PEO), having antioxidant activity due to the entrapped phycocyanin	Polymeric solutions consisted of 5–10% (w/w) protein concentrate from spirulina sp. LEB 18 and 0.5% (w/w) PEO, dissolved in acetic acid: Water (3:1, w/w). The mixture was further supplemented with phycocyanin (2%, w/w)	Diameter of the whole for releasing the polymer solution: 0.6 mm Distance between grounded and wire electrodes: 180 mm	Moreira et al. (2019)

(continued)

Table 4.9 (continued)

Application	(Bio)polymer solution	Process parameters	Reference
Development of electrospun nanofiber functionalized with cinnamon essential oil (CEO) nanophytosomes, used in active packaging of shrimp for shelf life extension	Polymeric solutions consisted of 7% PVA, eventually cross-linked with boric acid. The CEO was added to the PVA solution to a final concentration of ~5% (w/w)	Applied voltage:17–19 kV Flow rate: 1.5 mL/h Tip-to-collector distance: 18 cm Blunt-ended stainless-steel needle no. 18	Nazari et al. (2019)
Combining electrospinning and electrospraying for producing novel nanocomposite with high antioxidant activity and good thermal and mechanical properties, suitable for active food packaging	Electrospinning: 12% PCL solution in chloroform or 13% PLA solution in chloroform and acetone, with 1.4% NaCl (w/v). Electrospraying: 11% PVA solution with 2% phycocyanin (w/v) prepared in distilled water	For the PCL/PLA *electrospun nanofibers*: Flow rate: 2/1 mL/h, Voltage: 25/20 kV Capillary diameter: 0.80/0.45 mm Capillary to collector distance: 12/15 cm For *electrospraying*: Flow rate: 0.05 mL/h, Applied voltage: 20 kV Capillary diameter: 0.45 mm Capillary to collector distance: 18 cm	Schmatz et al. (2019)
Development of edible electrospun films loaded with phytochemicals (essential oils rich in 1,8 cineole) from spices (*Laurus nobilis* and *Rosmarinus officinalis*) with antimicrobial properties	Zein solutions (25%, w/v) prepared in a mixture of glacial acetic acid and ethanol (30:70). The biopolymer solution was loaded with essential oils at 1, 5 and 10% (w/w) in respect to the zein powder	Flow rate: 0.04 mL/h Applied voltage: 18 kV Tip-to-collector distance: 11 cm	Göksen et al. (2020)
Other food related applications			
Covalent immobilization of α-amylase on thermally crosslinked PVA/PAA nanofibers	Mixture of 5% PVA and 10% PAA aqueous solutions with 10:1 weight ratio	Flow rate rate: 1 mL/h Applied voltage: 25 kV Tip-to-collector distance: 15 cm	Baştürk et al. (2013)

(continued)

Table 4.9 (continued)

Application	(Bio)polymer solution	Process parameters	Reference
Modulation of the physicochemical properties of the WPI-guar gum mixtures	Solution subjected to electrospinning consisted of WPI (0–12% wt) and guar gum (0.7–0.9% wt)	Flow rate: 0.6 ml/min Applied voltage: 20 kV Tip-to-collector distance: 8 cm	Aman Mohammadi et al. (2019)
Producing gelatin nanofiber sheets with good solubility in cold water	Gelatin solution of 10–25% (w/v) concentration prepared in a 3:1 (v/v) solvent mixture of acetic acid: Water	Flow rate: 0.5 ml/h Applied voltage: 25 kV Tip-to-collector distance: 15 cm Inner diameter of the needle: 0.84 mm	Ghorani et al. (2020)
Halochromic nanosensor based on zein nanofibers incorporated with alizarin for real time checking the quality parameters of rainbow trout fillets	Zein solutions (10, 15 and 20% w/v) prepared in ethanol 70% v/v, incorporated with 4% alizarin and 1% glycerol	Flow rate: About 1 ml/h, Applied voltages: 20, 25 and 30 kV Tip-to-collector distance: About 15 cm Inner diameter: 0.51 mm	Aghaei et al. (2020)

6 Electron Beam Processing

Ilknur Ucak and Maliha Afreen

6.1 *Introduction to Electron Beam Processing*

Electron beam processing is a chemical process that produced irradiation. The electron beam is the movement of electrons with energy. When an electron beam hits an object, a chemical process occurs in which ionization and excitation of electrons arise due to the contact between the object and the ray. The process is known as "radiation chemical response".

Electron beam processing is a nonthermal intrusion that applying to food in the appropriate quantity at which no modifications occur in the physical features of foods. Electron beam processing is also recognized as "cold pasteurization" or "electronic pasteurization". The results of electron beam processing interference in food safety are reflected by water chlorination, milk pasteurization, and the protection of the public's health by boosting immunity against chronic diseases. The effectiveness of energy consumption through electron beam processing is very high because it is directly inserted through electron irradiation. On the other side, the thermochemical reaction required heating energy to trigger molecules by indirect insertion of energy. Absorbed radiation of 10 kGy is equivalent to a *2.4* caloric energy for 1 gram of water. Radiation energy ranges from 10 kGy to 200 kGy, is used for less heat resistance materials, including plastics and papers. No facilitator is needed in this process. The energy is rightly inserted into the object, and the chemical process is prompted. The electron beam has excellent performing efficiency because of its direct insertion. The level of reported electron beam energy (corpuscular beam) is much higher than that of other electromagnetic emissions, including electric wave, visible light, ultraviolet light, X-ray, and Gamma ray (γ-ray). Currently, radiation chemistry mainly involves electron beam energy.

Electron beam technology produced from commercial voltage, so it is known as a on-off technology. Although the complete antimicrobial properties of gamma radiation and electron beam are analogous, electron beam allows application of high frequency (e-beam, 10^3–10^5 Gy/second; gamma, 0.01–1 Gy/second) within a shorter treating time. The electron beam has a less penetrating ability as compared to gamma radiation. Those electrons with high energy can efficiently penetrate around 8–10 cm in usual foodstuffs to inactivate foodborne pathogens (Jaczynski and Park 2003). Thus, the dimensions, size and particular compactness of foods should be sensibly measured before treating with an electron beam. The electron beam dose is the amount of radiation immersed by the food when exposed to the electron beam. The dose is usually considered in Grays (G) or more suitable kilo Grays (kGy), where 1 Gray = 0.001 kGy = 1 Joule (J) of energy immersed in each kilogram of treated food.

The apparatus that produces electron beams is usually known as the "linear accelerator". There are diverse types of accelerators liable on the potential treating line speeds, the energy, the electrical effectiveness, etc. (Brown 2015). According to commercial electron beam accelerators, there are three categories: DC accelerators (direct current), CW accelerators (continuous wave), and pulsed accelerators. Meanwhile, the high piercing potential of electrons is always required; the 10 MeV, LINAC-style accelerators are resulting in enlarged uses in food irradiation. A LINAC-style accelerator with speed up structure, the sub-components, and the electron beam alert above the conveyor belt that takes the product below the electron beam. In the US, food irradiation is controlled by the "Food and Drug Administration" and the USDA-FSIS (United States Department of Agriculture-Food Safety Inspection).

6.2 Applications of Electron Beam Processing

Food materials can be decontaminated when photons and electrons are discharged from the accelerator and inserted in foodstuffs to initiate a sequence of ionization processes with attached atoms of the molecules, which come to be agitated by the hit of an emitted electron from the adjacent orbitals. These electrons convert solid and liquid atoms into ions, but water ionization is the main feature in decontamination. Decontamination of foodstuffs from biological objects occurs in two ways one is "direct DNA damage", and the other is "indirect DNA damage". The formation of free radicals converted the DNA molecules of biological objects that are a symptom of "indirect DNA damage". Though, direct interruption of DNA bands in molecules is known as "direct DNA damage". Microbial cells can be destroyed by the permanent breakdown of DNA strands (Lung et al. 2015). The potential of electrons of living cell destruction can be influenced by the trembling energy of electrons, and the quantity of irradiation detects this rate of energy. Less energy stops DNA duplication and increment in the amount of irradiation can disturb RNA structure and conformation, while higher energy ions damage the cell membrane and speed up the demolition rate of microorganisms (Schmidt et al. 2018). Concerning the effects of irradiation on mycotoxins, ionization of water molecules produced hydrogen radicals that react with the hydrocarbon chain of the mycotoxin's molecules and form new free radicals that have the ability of other methods of destruction. It should be noted that food configuration, water portion, microbial species and toxins are the main aspects associated with the efficiency of electron beam ionization and the appropriately selected quantity for treatment. Many factors that affect food irradiation are categorized into groups according to the following standards:

1. Irradiation dose: The quantity of irradiation is generally directly proportional to the rate of microbial destruction.
2. Food composition: Those factors which control food maintenance are density, pH, and temperature and gas configuration. When these factors are not suitable for microbial development, their confrontation with the electron beam is decreased and destroyed by less irradiation.
3. Microbial species: Microorganisms show diverse acceptance levels concerning particular quantities of irradiation. In normal circumstances, gram-positive bacteria show higher resistance than gram-negative bacteria, even though prokaryotic microorganisms are also more resistant to electron beam radiation than eukaryotic microorganisms (Moosekian et al. 2012).

6.2.1 Microbial Inactivation by Electron Beam Processing

Electron beam processing deactivates microorganisms. Microbial deactivation rises steadily with the growing quantity of electron beam that is usually used for food handling. The DNA of microorganisms is the acute goal of ionizing radiation. Though the cellular modifications produced by radiation, and these modifications differ in different microbes. Generally, as the main point of radiation is the genetic material, the connection of radiation sensitivity is contrariwise to the size and complication of an organism and its DNA (Urbain 1996). The antimicrobial characteristics of the electron beam can be categorized into direct and indirect special effects. Thus, from the appropriate point of view, it is essential to remember that the direct and indirect special effects produce the complete deactivation of microbes. The direct effects conclude from the non-specific collision of radiating photons with the atoms in the molecules of the microbes. The electron beam produces a breakdown of important biomolecules, including DNA, RNA, enzymes and membrane proteins of microbes. Each DNA strand of the microbial DNA contains four bases guanine, cytosine, thymine, and adenine. The bases from two distinct strands bond with each other, generating distinctive base pairs "guanine with cytosine" and "thymine with adenine" (G-C and T-A). The electron beam can disrupt these G-C and T-A bonds, consequently breaking the double helix strand. The breakdown of the DNA reduces microbial cells that are unable to divide, which is frequently known as cellular propagative death. This is the key mechanism of deactivation of microbes by direct effects of the electron beam. The indirect effect of electron beam for deactivation of microbes relies on availability or activity of water in a foodstuff because indirect effects are produced with free radicals produced during ionization of water through radiations (Miller 2005). Radiating photons inserted in foodstuff and come upon water molecules, this interaction providing many radiolysis products even some types of unpaired electrons known as free radicals (Miller 2005). The free radicals are much responsive and pursue the development of constant products by joining with each other or with oxygen; thus, generating oxidizing mediators. These oxidizing mediators and free radicals are not particular only for microbial cells. If they come upon a microbial cell, then oxidizing mediator's and free radicals damaged the cell membrane. If the destruction is enough, then cell leakage leads to cell death and, in the end, complete cell burst.

6.2.2 Electron Beam Processing for Agriculture and Food Products

Attentive combination of research methodologies is inevitable for an effective overview of seed processing procedures before sowing involving several components and several functions, including Nanosystems originating, preferably, from recyclable assets, involving agricultural remnants (Revina 1998, 2009; Ruban et al. 2012). When Biologically active Nanochips (BAN chips) are used to treat seeds of agricultural plants, permit implementation of preferred ultra-small quantities of active ingredients for the objective of simultaneous attainment of decontamination of plants, for the nourishment with microelements or nutrients, enrichment of seed germination, and crop produce rise under conflicting situations. The BAN chip commonly contains substances of permeable material and nanoparticles of biologically active and phytosanitary materials and nutrients. Treatment of BAN chips with Electron beam (EB) processing (Pikaev et al. 1993; Revina 2003; Revina and Magomedbekov 2010; Pavlov et al. 2014; Pavlov 2015) can propose many solutions proficient for enhancing the efficacy of BAN chips.

6.3 Advantages of Electron Beam Processing

Electron-beam irradiation (EBI) is a beneficial technique that directs the proficiency of usage in a two-direction method with a minute time to eradicate the microbial burden in foodstuffs (Jeyakumari et al. 2019; Schoeller et al. 2002). Moreover, due to less harmful special effects on the nutritional and chemical configuration of foods, specifically cereal-based foodstuffs, electron beam irradiation can be an encouraging technique in their sanitization (Carocho et al. 2012). Electron beam irradiation is a beneficial obstacle technique, joint with other conservative or evolving food handling processes, more efficient food protection can be attained (Yang et al. 2020). The usage of combining electron beam irradiation and probiotic *Listeria rhamnosus* on pathogen *Klebsiella pneumonia* showed as an efficient method for the reticence of infectious microorganisms (Balayan et al. 2019).

6.4 Disadvantages of Electron Beam Processing

Less penetrability of the electron beam is the main disadvantage of consuming electron beam ionization. The sanitization effect of Electron beam irradiation may be affected by the size, thickness, single or double direction, and packaging of the food. Scientists evaluated the influence of electron beam irradiation on antimicrobial mediator covered with low-density polyethylene or polyamide films and oriented polypropylene. Radiation quantity of 3 kGy was perceived to produce slight or insignificant modifications in the film functions but did not change the inhibitory outcome on *E. coli* and *L. inocua* (Han et al. 2007). Additionally, some customers still have a negative opinion about "radioactivity" and "radiation". Therefore, consumer training via reports, database, and seminars is necessary.

7 Ionizing Radiation

Ilknur Ucak and Maliha Afreen

7.1 Introduction of Ionizing Radiation

Treatment of any material with radiation is a chemical process that produced irradiation. Mainly gamma rays (γ-rays) are used in these reactions. The γ-ray is an electromagnetic beam as with light and it is an energy stream. Irradiation processing includes implementing ionizing or non-ionizing emissions on foodstuffs (raw or processed) for the long-term storage and conservation of food. Ionizing emissions can be produced from high-energy electrons, including gamma rays produced from (Cesium-137 or Cobalt-60) or X-rays (produced from a machine). Non-ionizing emission is an electromagnetic emission that does not have sufficient energy to convert atoms or molecules into ions, characterized mainly through visible light, microwaves, infrared and ultraviolet rays (UV-A, B, and C).

Applying radiations on food is well recognized, nonthermal method of food preservation (cold- pasteurization) at surrounding temperature. Treatment of food products with radiations can produce slight changes in the nutrients, color, taste, and other quality parameters of food. These changes mainly depend on the nature of the object, the quantity of radiation, and which source of radiation is used, either X-ray, γ-ray and UV (Bhat and Sridhar 2008; Bhat et al. 2007b; Mexis et al. 2009). Previous studies prove that treatment of foodstuffs with ionizing radiation, including gamma rays, is efficient in controlling confinement barricades for international trading, as a source of disinfestation, decontamination, and for enhancement of nutritional characteristics and shelf life of food (Hong et al. 2008; Lacroix and Ouattara 2000; Teets et al. 2008). Some studies also show the application of non-ionizing radiation, including ultraviolet, for food decontamination (Selma et al. 2008; Walkling-Ribeiro et al. 2008). Some people have negative concepts, but it is just a movement of energy in the form of light, γ-ray, X-ray, electric wave and corpuscular beam. The radiation can be allocated into two categories: the corpuscular beam, including electron beam and α-ray, while X-ray and γ-ray fall into the second category.

Irradiation involves Gamma irradiation, X-ray irradiation which are basic types of food irradiation process used broadly with the capability to "kick" electrons out of their orbital shells on atoms, resulting in "ionization of the atoms" due to which it is known as "ionizing radiation "(Pillai and Shayanfar 2017; Simas et al. 2010). The food irradiation process is a portion of the similar electromagnetic spectrum which comprises incandescent lights, radio waves, TV broadcasts, UV radiation, microwaves, and cosmic radiation (Chmielewski et al. 2006). Ionizing and non-ionizing radiation frequencies collectively made electromagnetic spectrum. Food irradiation depends on both X-rays (from X-ray tubes) and gamma irradiation (from radioactive isotopes such as cobalt-60 or cesium-137). According to their energy sources, there are significant variations between the three irradiation methods of production, their particular protective necessities, and the control setting for each of these techniques. Gamma irradiation is mainly produced in nuclear reactors from cesium-137 or cobalt-60. Gamma irradiation is principally photons and they do not have any physique. Gamma ray (γ-ray) released by the breakdown of radioactive isotope cobalt-60, which has a high penetrating ability, so they can completely penetrate in objects of variable bulk compactness.

7.2 *Applications of Ionizing Radiation*

It has been proved that when less heat or irradiation is applied to food collectively with the other food preservation methods, it will give purified food and effectively keep the food fresh for a long time. First time in the 1950s, heat and irradiation were collectively used for decontamination and preservation of food (Farkas 1990). It was stated that less quantity irradiation (4.1 kJ m^2) of ultraviolet-C with the usage of 45 °C heat for 3 hours in the air, before storage of fruit at 20 °C increased advantage as compared to applying these actions independently, and could be beneficial to improve the shelf life of strawberries, even though the phenolic and anthocyanins gathering ratio will be reduced as compared to untreated fruits (Pan et al. 2004). One scientist observed that vegetables and fruits have phytochemical abstracts which have resilient anti-proliferative and antioxidant actions (Liu 2004). Irradiation can affect the ability of a particular plant to yield phytochemicals/ antioxidants in different quantities at different stages. It has been stated that under definite advantageous situations, the amount of plant phytochemicals may be increased. These situations involve contact to radiation sources, cutting, contact to higher temperatures and storing at low temperatures (Zobel 1997). The quantity of radiation supplied to a product plays an important role in the determining antioxidants level. For example, it was observed that antioxidant action of the methanol abstract of mushroom (*Agaricus blazei*) (Huang and Mau 2006) though prior, it was found that hot water and ethanolic abstracts of the similar mushroom (*A. blazei*) contained greater antioxidant actions (101–104% at 5–10 mg/ml) (Tsai et al. 2007). In the same way, Pérez et al. (2007) evaluated the influence of gamma-irradiation (30 kGy) on the dry rosemary leaves by extraction with ethanol, methanol and water. Consequently, they observed that the gamma radiation enhanced antioxidant action in water and ethanol abstracts with a consistent escalation in the whole phenolic part of the rosemary leaf. At the same time, the implementation of irradiation had no considerable effect on the methanol abstracts.

Radiation processing has been presented to might enhance or reduce the antioxidant part of fresh plant products, which relies on the supply rate, contact time, and the usage of a fresh food product. The improved antioxidant actions of a plant after irradiation are chiefly accredited to improved enzyme activity (peroxidase activity and phenylalanine ammonia-lyase activity) or enhanced extraction ability from the tissues (Bhat et al. 2007a; González-Aguilar et al. 2007a, b; Tomás-Barberán and Espin 2001). Some scientists worked on soybeans and observed that the implementation of gamma irradiation enriched their potential of antioxidant, which was accredited to enhanced amount of genistein (an isoflavone) and a minimum degree on the antioxidant actions of daidzein degradative foodstuffs. They perceived that at a greater quantity of 5 kGy, antioxidant activity moved toward the typical 39.16% Trolox. This increment was associated with the higher degrees of free isoflavones in products treated with radiation quantity less than 1 kGy (Variyar et al. 2004). Commonly, the reduction in antioxidants levels is accredited to free radicals or the production of degradative products by implementing radiation (Sajilata and Singhal 2006; Wong and Kitts 2001). Scientists stated that gamma irradiation treatment of strawberries ranges from 1–10 kGy causes the breakdown of phenolic acids like p-coumaric, gallic, cinnamic and hydroxybenzoic acids (Breitfellner et al. 2002). The hydroxylation of these phenolic acids has been accredited to free hydroxyl (OH) radicals during the treatment. Irradiation quantity ranged from 0.25–1.00 kGy held out on cashew nuts has been presented to reduce the anti-oxidative activity, due to the breakdown of vitamin E (tocopherols) existed in the cashew nuts after the radiation treatment, which was decreased additionally during storage time (Sajilata and Singhal 2006).

7.2.1 Microbial Decontamination by Food Irradiation

Irradiation is the only interference existing that will constantly decrease foodborne pathogens and contaminants to untraceable points when the quantity of irradiation is correctly and consistently applied based on food producers. Irradiation, within appropriate amounts, has been presented to abolish at least *99.9%* of communal foodborne pathogens, such as *E. coli O157:H7, Salmonella, L. monocytogenes* and *Campylobacter jejuni*, which are associated with poultry and meat (GAO 2000). It was summarized that the quantity of irradiation between *25* and 60 kGy would not produce any possible health hazards while keeping tolerable criteria of nutritional value and sensory features of food. Thus, this irradiation technology improves food protection and decreases crop-associated economic damages (Bhat et al. 2012). Mostly used types of radiation handlings can be approximately divided into three levels based on quantity: (a) low-rate treatment (<1 kGy) for insect-killing in cereals, spices and dried fruits, postponement in fruit ripeness such as in bananas and inhibition of propagation such as in onions, potatoes, ginger and garlic; (b) central-rate treatment (1–10 kGy) to eradicate microbial pathogens and prolong the shelf life of foodstuffs such as fruits/vegetables, spices, coffee beans, seafood, and poultry; (c) high-rate treatment (10–60 kGy) applied for irradiation of food equipped for low immune patients and astronauts (Fan et al. 2012).

7.2.2 Phytosanitation

Ionizing radiation processing is used worldwide for handling agricultural products to eradicate insects and pests. Some strong international principles control the use of various techniques (e.g., hot water treatment, methyl bromide, ionizing radiation) for handling agricultural products. There is an increasing mandate for the usage of ionizing radiation to treat plants from insects, even at the less quantity minimally at 3–4 log deactivation of significant bacterial pathogens can happen (Shayanfar et al. 2016). Irradiation processing in higher amounts is used worldwide to ensure food safety by eradicating pathogens. Ionizing radiation is allowed for protecting animal diets, including packaged feeds, bagged complete diets, bulk feeds, feed elements, and animal treats in the US. Though, the quantity cannot go above 50 kGy. For whole poultry diets and feed ingredients, the amount of radiation cannot be less than 2 kGy and cannot go above 25 kGy. The upper quantity radiating boundary is founded on the hypothesis that the irradiation treatment is actually used to govern the *Salmonella* species.

7.3 *Advantages of Ionizing Radiation*

- Ionizing radiation techniques are applied to minimize the risks of food-related diseases caused by microorganisms such as *Salmonella*. Radiation is used to sterilize food from pathogens.

- Ionizing radiation is also used to increase the storage time of food by killing those pathogens, which can cause food deterioration.
- According to the United State Department of Agriculture and Word Health Organization, irradiation is nontoxic and completely harmless for food practising.
- Irradiation has a more negligible effect on the nutritious food value than conventional preservation methods like freezing and boiling.
- Those Foods that have been processed through radiation have proper labelling. This Radura label helps customers to decide whether they want to buy irradiated food or not.

7.4 *Disadvantages of Ionizing Radiation*

- There is no guarantee of complete eradication of viruses and toxins from food. To avoid this problem, an appropriate dosage of radiation should be used.
- Irradiated food products can be risky for mammal's health in some exceptional circumstances. These health problems can be premature death, chromosomal abnormalities and cancer.
- Ionizing radiation is a costly procedure because of its facilitation and due to the requirement of a wholly sterilized environment

References

Pulsed Electric Field

Ade-Omowaye BIO, Angersbach A, Eshtiaghi NM, Knorr D (2000) Impact of high intensity electric field pulses on cell permeabilisation and as pre-processing step in coconut processing. Innov Food Sci Emerg Technol 1:203–209

Ade-Omowaye BIO, Rastogi NK, Angersbach A, Knorr D (2002) Osmotic dehydration of bell peppers: influence of high intensity electric field pulses and elevated temperature treatment. J Food Eng 54:35–43

Ade-Omowaye BIO, Rastogi NK, Angersbach A, Knorr D (2003a) Combined effects of pulsed electric field pre-treatment and partial osmotic dehydration on air drying behavior of red bell pepper. J Food Eng 60:89–98

Ade-Omowaye BIO, Taiwo KA, Eshtiaghi NM, Angersbach A, Knorr D (2003b) Comparative evaluation of the effects of pulsed electric field and freezing on cell membrane permeabilisation and mass transfer during dehydration of red bell peppers. Innov Food Sci Emerg Technol 4:177–188

Albagnac G, Varoquaux P, Montigaux JC (2002) Technologies de transformation des fruits. Lavoisier, Paris

Amami E, Vorobiev E, Kechaou N (2006) Modelling of mass transfer during osmotic dehydration of apple tissue pretreated by pulsed electric field. LWT-Food Sci Technol 39:1014–1021

Amami E, Fersi A, Vorobiev E, Kechaou N (2007) Osmotic dehydration of carrot tissue enhanced by pulsed electric field, salt and centrifugal force. J Food Eng 83:605–613

Amiali M, Ngadi MO, Raghavan VGS, Smith JP (2004) Inactivation of Escherichia coli O157: H7 in liquid dialyzed egg using pulsed electric fields. Food Bioprod Process 82:151–156
Amiali M, Ngadi MO, Smith JP, Raghavan GSV (2007) Synergistic effect of temperature and pulsed electric field on inactivation of Escherichia coli O157:H7 and Salmonella enteritidis in liquid egg yolk. J Food Eng 79:689–694
Amit SK, Uddin MM, Rahman R, Islam SMR, Khan MS (2017) A review on mechanisms and commercial aspects of food preservation and processing. Agric Food Secur 6:51
Asavasanti S, Ersus S, Ristenpart W, Strove P, Barret DM (2010) Critical field strength of onions tissues treated by pulsed electric fields. J Food Sci 75:433–443
Bai Y, Li C, Zhao J, Zheng P, Li Y, Pan Y, Wang Y (2013) A high yield method of extracting alkaloid from Aconitum coreanum by pulsed electric field. Chromatographia 76:635–642
Barba FJ, Grimi N, Vorobiev E (2014) New approaches for the use of non-conventional cell disruption technologies to extract potential food additives and nutraceuticals from microalgae. Food Eng Rev 7(1):45–62
Barba FJ, Zhu Z, Koubaa M, Sant'Ana AS, Orlien V (2016) Green alternative methods for the extraction of antioxidant bioactive compounds from winery wastes and by-products. Trends Food Sci Technol 49:96–109
Barssoti P, Merle P, Cheftel JC (1999) Food processing by pulsed electric fields. I Phys Aspects Food Rev Int 15:163–220
Bazhal M, Vorobiev E (2000) Electrical treatment of apple cossettes for intensifying juice pressing. J Sci Food Agric 80:1668–1674
Bekhit AEDA, van de Ven R, Suwandy V, Fahri F, Hopkins DL (2014) Effect of pulsed electric field treatment on cold-boned muscles of different potential tenderness. Food Bioprocess Technol 7:3136–3146
Bhat ZF, Morton JD, Mason SL, Bekhit AEDA (2018) Applied and emerging methods for meat tenderization: a comparative perspective. Compr Rev Food Sci Food Saf 17:841–859
Buchmann L, Bloch R, Mathys A (2018) Comprehensive pulsed electric field (PEF) system analysis for microalgae processing. Bioresour Technol 265:268–274
Buckow R, Ng S, Toepfl S (2013) Pulsed electric field processing of orange juice: a review on microbial, enzymatic, nutritional, and sensory quality and stability. Compr Rev Food Sci Food Saf 12:455–467
Cai YZ, Sun M, Corke H (2005) Characterization and application of betalain pigments from plants of the Amaranthaceae. Trends Food Sci Technol 16:370–376
Castro AJ, Barbosa-Canovas GV, Swanson BG (1993) Microbial inactivation of foods by pulsed electric fields. J Food Process Preserv 19:47–73
Chalermchat Y, Fincan M, Djmek P (2004) Pulsed electric field treatment for solid-liquid extraction of red beetroot pigment: mathematical modelling of mass transfer. J Food Eng 64:229–236
Chotphruethipong L, Aluko RE, Benjakul S (2019) Effect of pulsed electric field assisted process in combination with porcine lipase on defatting of seabass skin. J Food Sci 84:1799–1805
Corrales M, Toepfl S, Butz P, Knorr D, Tauscher B (2008) Extraction of anthocyanins from grape by-products assisted by ultrasonics, high hydrostatic pressure or pulsed electric fields: a comparison. Innov Food Sci Emerg Technol 9:85–91
de Haan SWH (2007) Circuitry and pulse shapes in pulsed electric field treatment of food. In: Notermans S, de Haan SWH (eds) Lelieveld HLM. Food Preservation by Pulsed Electric Fields, Elsevier, pp 43–69
Deng Q, Zinoviadou KG, Galanakis CM, Orlien V, Grimi N, Vorobiev E, Lebovka N, Barba FJ (2014) The effects of conventional and non-conventional processing on glucosinolates and its derived forms, isothiocyanates: extraction, degradation, and applications. Food Eng Rev 7(3):357–381
Donsì F, Ferrari G, Pataro G (2010) Applications of pulsed electric field treatments for the enhancement of mass transfer from vegetable tissue. Food Eng Rev 2(2):109–130

Fauster T, Schlossnikl D, Rath F, Ostermeier R, Teufel F, Toepfl S, Jaeger H (2018) Impact of pulsed electric field (PEF) pretreatment on process performance of industrial French fries production. J Food Eng 235:16–22
Fincan M, Dejmek P (2002) In situ visualization of the effect of a pulsed electric field on plant tissue. J Food Eng 55(3):223–230
Fincan M, DeVito F, Dejmek P (2004) Pulsed electric field treatment for solid-liquid extraction of red beetroot pigment. J Food Eng 64:381–388
Flaumenbaum BL (1949) Electrical treatment of fruits and vegetables before extraction of juice. Trudy OTIKP 3:15–20
Gachovska T, Cassada D, Subbiah J, Hanna M, Thippareddi H, Snow D (2010) Enhanced anthocyanin extraction from red cabbage using pulsed electric field processing. J Food Sci 75:323–329
Gómez B, Munekata PE, Gavahian M, Barba FJ, Martí-Quijal FJ, Bolumar T, Campagnol PCB, Tomasevic I, Lorenzo JM (2019) Application of pulsed electric fields in meat and fish processing industries: an overview. Int Food Res J 123:95–105
Grimi N, Praporscic I, Lebovka N, Vorobiev E (2007) Selective extraction from carrot slices by pressing and washing enhanced by pulsed electric fields. Sep Purif Technol 58:267–273
Grimi N, Dubois A, Marchal L, Jubeau S, Lebovka NI, Vorobiev E (2014) Selective extraction from microalgae Nannochloropsis sp. using different methods of cell disruption. Bioresour Technol 153:254–259
Guderjan M, Elez-Martínez P, Knorr D (2007) Application of pulsed electric fields at oil yield and content of functional food ingredients at the production of rapeseed oil. Innov Food Sci Emerg Technol 8:55–62
Ignat A, Manzocco L, Brunton NP, Nicoli MC, Lyng JG (2015) The effect of pulsed electric field pretreatments prior to deep-fat frying on quality aspects of potato fries. Innov Food Sci Emerg Technol 29:65–69
Jaeger H, Janositz A, Knorr D (2010) The Maillard reaction and its control during food processing. The potential of emerging technologies [La reaction de Maillard et son contrôle pendant la fabrication des aliments. Le potentiel des nouvelles technologies]. Pathol Biol 58(3):207–213
Jaeger H, Schulz M, Lu P, Knorr D (2012) Adjustment of milling, mash electroporation and pressing for the development of a PEF assisted juice production in industrial scale. Innov Food Sci Emerg Technol 14:46–60
Jambrak A, Djekić I, Van Impe J (2018) Non-thermal food processing: Modelling of processes towards safety, quality and sustainability. In: 10th international conference on simulation and modelling in the food and bio-industry 2018. FOODSIM, vol 2018, pp 19–21
Jemai AB, Vorobiev E (2003) Enhancing leaching from sugar beet cossettes by pulsed electric field. J Food Eng 59:405–412
Jemai AB, Vorobiev E (2006) Pulsed electric field assisted pressing of sugar beet slices: towards a novel process of cold juice extraction. Biosyst Eng 93:57–68
Kalum L, Hendriksen HV (2014) Process for treating vegetable material with an enzyme. US Patent 20080241315 A1
Katiyo W, Yang R, Zhao W (2017) Effects of combined pulsed electric fields and mild temperature pasteurization on microbial inactivation and physicochemical properties of cloudy red apple juice (Malus pumila Niedzwetzkyana (Dieck)). J Food Saf 37(4):12369
Kempkes M, Munderville M (2017) Pulsed electric fields (PEF) processing of fruit and vegetables. In: IEEE 21st International Conference on Pulsed Power (PPC) 2017, pp 1–7
Koehler E, Toepfl S, Knorr D, Pulz O (2005) Unconventional procedures for the production and stabilization of extracts with active agents. Eur. Work. Microalgal Biotechnol., 6th, Potsdam
Koubaa M, Roselló-Soto E, Šic-Žlabur J, Režek-Jambrak A, Brnčic M, Grimi N, Boussetta N, Barba FJ (2015) Current and new insights in the sustainable and green recovery of nutritionally valuable compounds from Stevia rebaudiana Bertoni. J Agric Food Chem 63(31):6835–6846
Kumar R, Vijayalakshmi S, Kathiravan T, Nadanasabapathi S (2019) PEF processing of fruits, vegetables, and their products. In: Chauhan OP (ed) Nonthermal processing of foods. Taylor & Francis Group, Boca Raton, pp 104–124

Lebovka NI, Shynkaryk MV, El-Belghiti K, Benjelloun H, Vorobiev E (2007) Plasmolysis of sugarbeet: pulsed electric fields and thermal treatment. J Food Eng 80(2):639–644
Li X, Farid M (2016) A review on recent development in non-conventional food sterilization technologies. J Food Eng 182:33–45
Loginov M, Loginova K, Lebovka N, Vorobiev E (2011) Comparison of dead-end ultrafiltration behavior and filtrate quality of sugar beet juices obtained by conventional and "cold" PEF-assisted diffusion. J Membr Sci 377(1–2):273–283
Loginova K, Loginov M, Vorobiev E, Lebovka NI (2011) Quality and filtration characteristics of sugar beet juice obtained by "cold" extraction assisted by pulsed electric field. J Food Eng 106(2):144–151
Loginova K, Loginov M, Vorobiev E, Lebovka NI (2012) Better lime purification of sugar beet juice obtained by low temperature aqueous extraction assisted by pulsed electric field. LWT – Food Sci Technol 46(1):371–374
Lopéz N, Puértolas E, Condóon S, Raso J, Álvarez I (2009) Enhancement of the extraction of betanine from red beetroot by pulsed electric fields. J Food Eng 90:60–66
Mahnič-Kalamiza S, Vorobiev E, Miklavčič D (2014) Electroporation in food processing and biorefinery. J Membr Biol 247:1279–1304
Mosqueda-Melgar J, Raybaudi-Massilia RM, Martín-Belloso O (2008) Non-thermal pasteurization of fruit juices by combining high-intensity pulsed electric fields with natural antimicrobials. Innov Food Sci Emerg Technol 9(3):328–340
Palaniappan S, Sastry SK (1990) Effects of electricity on microorganisms. J Food Process Preserv 14:393–414
Parliament E (2011) European Parliament resolution of 19 January 2012 on how to avoid food wastage: Strategies for a more efficient food chain in the EU (2011/2175(INI)
Parniakov O, Bals O, Lebovka N, Vorobiev E (2016) Pulsed electric field assisted vacuum freeze-drying of apple tissue. Innov Food Sci Emerg Technol 35:52–57
Pourzaki A, Mirzaee H, Hemmati Kakhki A (2013) Using pulsed electric field for improvement of components extraction of saffron (Crocus sativus) stigma and its pomace. J Food Process Preserv 37(5):1008–1013
Praporscic I (2005) Influence du traitement combiné par champ électrique pulsé et chauffage modéré sur les propriétés physiques et sur le comportement au pressage de produits végétaux. (PhD Thesis). Université de Technologie de Compiègne, France
Priyadarshini A, Rajauria G, O'Donnell CP, Tiwari BK (2019) Emerging food processing technologies and factors impacting their industrial adoption. Crit Rev Food Sci Nutr 59(19):3082–3101
Puértolas E, Hernández-Orte P, Sladaña G, Álvarez I, Raso J (2010a) Improvement of winemaking process using pulsed electric fields at pilot-plant scale. Evolution of chromatic parameters and phenolic content of Cabernet Sauvignon red wines. Food Res Int 43:761–766
Puértolas E, López N, Condón C, Álvarez I, Raso J (2010b) Potential applications of PEF to improve red wine quality. Trends Food Sci Technol 21:247–255
Redondo LM (2017) Basic concepts of high-voltage pulse generation. In: Miklavčič D (ed) Handbook of electroporation. Springer, Cham (CH). pp 859–879
Redondo D, Venturini ME, Luengo E, Raso J, Arias E (2018) Pulsed electric fields as a green technology for the extraction of bioactive compounds from thinned peach by-products. Innov Food Sci Emerg Technol 45:335–343
Roselló-Soto E, Galanakis CM, Brnčić M, Orlien V, Trujillo FJ, Mawson R, Knoerzer K, Tiwari BK, Barba FJ (2015a) Clean recovery of antioxidant compounds from plant foods, by-products and algae assisted by ultrasounds processing. Modeling approaches to optimize processing conditions. Trends Food Sci Technol 42(2):134–149
Roselló-Soto E, Koubaa M, Moubarik A, Lopes RP, Saraiva JA, Boussetta N, Grimi N, Barba FJ (2015b) Emerging opportunities for the effective valorization of wastes and byproducts generated during olive oil production process: non-conventional methods for the recovery of high-added value compounds. Trends Food Sci Technol 45(2):296–310

Sack M, Mueller G (2017) Design considerations for electroporation reactors. IEEE Trans Dielectr Electr Insul 24:1992–2000

Sale AJH, Hamilton WA (1968) Effects of high electric fields on microorganisms. III. Lysis of erythrocytes and protoplasts. Biochim Biophys Acta 163:37–43

Sánchez-Vega R, Elez-Martínez P, Martín-Belloso O (2014) Influence of high-intensity pulsed electric field processing parameters on antioxidant compounds of broccoli juice. Innov Food Sci Emerg Technol 29:70–77

Sharma P, Oey I, Bremer P, Everett DW (2014) Reduction of bacterial counts and inactivation of enzymes in bovine whole milk using pulsed electric fields. Int Dairy J 39(1):146–156

Taiwo KA, Angersbach A, Knorr D (2002) Influence of high intensity electric field pulses and osmotic dehydration on the rehydration characteristics of apple slices at different temperatures. J Food Eng 52:185–192

Taiwo KA, Angersbach A, Knorr D (2003) Effects of pulsed electric field on quality factors and mass transfer during osmotic dehydration of apples. J Food Process Eng 26:31–48

Teissie J, Prats M, Soucaille P, Tocanne JF (1985) Evidence for conduction of protons along the interface between water and a polar lipid monolayer. Proc Natl Acad Sci U S A 82:3217–3221

Timmermans RAH, Nierop Groot MN, Nederhoff AL, van Boekel MAJS, Matser AM, Mastwijk HC (2014) Pulsed electric field processing of different fruit juices: impact of pH and temperature on inactivation of spoilage and pathogenic micro-organisms. Int J Food Microbiol 173:105–111

Timmermans R, Mastwijk H, Berendsen L, Nederhoff A, Matser A, Van Boekel M, Nierop Groot MN (2019) Moderate intensity Pulsed Electric Fields (PEF) as alternative mild preservation technology for fruit juice. Int J Food Microbiol 298:63–73

Toepfl S, Heinz V, Knorr D (2006a) Application of pulsed electric field technology for the food industry. In: Raso J, Heinz V (eds) Pulsed electric fields technology for the food industry, Food engineering series. Springer, Boston, pp 197–221

Toepfl S, Mathys A, Heinz V, Knorr D (2006b) Potential of high hydrostatic pressure and pulsed electric fields for energy efficient and environmentally friendly food processing. Food Rev Int 22:405–423

Toepfl S, Siemer C, Saldaña-Navarro G, Heinz V (2014) Overview of pulsed electric fields processing for food. In: Sun D-W (ed) Emerging technologies for food processing, 2nd edn. Elsevier/Academic Press, Amsterdam/Boston, pp 93–114

Tsong TY (1991) Electroporation of cell membranes. Biophys J 60:297–306

Walkling-Ribeiro M, Noci F, Cronin DA, Lyng JG, Morgan DJ (2010) Shelf life and sensory attributes of a fruit smoothie-type beverage processed with moderate heat and pulsed electric fields. LWT – Food Sci Technol 43:1067–1073

Walkling-Ribeiro M, Rodrıguez-Gonzalez O, Jayaram SH, Griffiths MW et al (2011) Processing temperature, alcohol and carbonation levels and their impact on pulsed electric fields (PEF) mitigation of selected characteristic microorganisms in beer. Int Food Res J 44:2524–2533

Wang LH, Pyatkovskyy T, Yousef A, Zeng XA, Sastry SK (2020) Mechanism of Bacillus subtilis spore inactivation induced by moderate electric fields. Innov Food Sci Emerg Technol 62:102349

Weaver JC (2000a) Electroporation of cells and tissues. IEEE Trans Plasma Sci 28(1):24–33

Zhang TH, Wang SJ, Liu DR, Yuan Y, Yu YL, Yin YG (2011) Optimization of exopolysaccharide extraction process from Tibetan spiritual mushroom by pulsed electric fields. J Jilin Uni 41(3):882–886

Zimmermann U (1986a) Electric breakdown, electro permeabilization and electrofusion. Rev Phys Biochem Pharmacol 105:196–256

Cold Plasma

Baggio A, Marino M, Innocente N, Celotto M, Maifreni M (2020) Antimicrobial effect of oxidative technologies in food processing: an overview. Eur Food Res Technol Springer. https://doi.org/10.1007/s00217-020-03447-6

Bao T, Hao X, Shishir MRI, Karim N, Chen W (2021) Cold plasma: an emerging pretreatment technology for the drying of jujube slices. Food Chem 337:127783

Bogdanov T, Tsonev I, Marinova P, Benova E, Rusanov K, Rusanova M, Atanassov I, Kozáková Z, Krčma F (2018) Microwave plasma torch generated in argon for small berries surface treatment. Appl Sci 8:1870

Bruggeman PJ, Kushner MJ, Locke BR, Gardeniers JE, Graham WG, Graves DB, Hofman-Caris RCHM, Maric D, Reid JP, Ceriani E, Fernandez Rivas D, Foster JE, Garrick SC, Gorbanev Y, Hamaguchi S, Iza F, Jablonowski H, Klimova E, Kolb J, Krcma F, Lukes P, Machala Z, Marinov I, Mariotti D, Mededovic Thagard S, Minakata D, Neyts EC, Pawlat J, Lj Petrovic Z, Pflieger R, Reuter S, Schram DC, Schröter S, Shiraiwa M, Tarabová B, Tsai PA, Verlet JRR, von Woedtke T, Wilson KR, Yasui K, Zvereva G (2016) Plasma–liquid interactions: a review and roadmap. Plasma Sources Sci Technol 25(5):053002

Chaple S, Sarangapani C, Jones J, Carey E, Causeret L, Genson A, Duffy B, Bourke P (2020) Effect of atmospheric cold plasma on the functional properties of whole wheat (Triticum aestivum L.) grain and wheat flour. Innov Food Sci Emerg Technol 66:102529

Chen TP, Liang JF, Su TL (2018) Plasma-activated water: antibacterial activity and artifacts? Environ Sci Pollut Res Int 25(27):26699–26706

Dong S, Gao A, Xu H, Chen Y (2017) Effects of dielectric barrier discharges (DBD) cold plasma treatment on physico-chemical and structural properties of zein powders. Food Bioprocess Technol 10:434–444

Gorbanev Y, Privat-Maldonado A, Bogaerts A (2018) Analysis of short-lived reactive species in plasma-air-water systems: the do's and the don'ts. Anal Chem 90(22):13151–13158

Guo C, Tang F, Chen J, Wang X, Zhang S, Zhang X (2014) Development of dielectric-barrier-discharge ionization. Anal Bioanal Chem 407(9):2345–2364

H€ansch MAC, Mann M, Weltmann KD, Woedtke T (2015) Analysis of antibacterial efficacy of plasma-treated sodium chloride solutions. J Phys D Appl Phys 48(45):454001

Janić Hajnal E, Vukić M, Pezo L, Orčić D, Puač N, Škoro N, Mi-lidrag A, Šoronja Simović D (2019) Effect of atmospheric cold plasma treatments on reduction of Alternaria toxins content in wheat flour. Toxins 11:704

Joshi I, Salvi D, Schaffner DW, Karwe MV (2018) Characterization of microbial inactivation using plasma-activated water and plasma-activated acidified buffer. J Food Prot 81(9):1472–1480

Kamgang-Youbi G, Herry JM, Bellon-Fontaine MN, Brisset JL, Doubla A, Naïtali M (2007) Evidence of temporal postdischarge decontamination of bacteria by gliding electric discharges: application to Hafnia alvei. Appl Environ Microb 73(15):4791–4796

Kamgang-Youbi G, Herry JM, Meylheuc T, Brisset JL, Bellon-Fontaine MN, Doubla A, Naïtali M (2009) Microbial inactivation using plasma-activated water obtained by gliding electric discharges. Lett Appl Microbiol 48(1):13–18

Katsigiannis AS, Bayliss DL, Walsh JL (2021) Cold plasma decontamination of stainless steel food processing surfaces assessed using an industrial disinfection protocol. Food Control 121:107543

Khlyustova A, Labay C, Machala Z, Ginebra M-P, Canal C (2019) Important parameters in plasma jets for the production of RONS in liquids for plasma medicine: a brief review. Front Chem Sci Eng 13(2):238–252

Kim H-J, Yong HI, Park S, Kim K, Bae Y, Choe W, Oh M, Jo C (2013) Effect of inactivating salmonella typhimurium in raw chicken breast and pork loin using an atmospheric pressure plasma jet. J Anim Sci Technol 55(6):545–549

Kojtari A, Ercan UK, Smith J, Friedman G, Sensenig RB, Tyagi S, Joshi SG, Ji HF, Brooks AD (2013) Chemistry for antimicrobial properties of water treated with nonequilibrium plasma. J Nanomed Biother Discov 4(1):1000120

Krčma F, Tsonev I, Smejkalová K, Truchlá D, Kozáková Z, Zhekova M, Marinova P, Bogdanov T, Benova E (2018) Microwave micro torch generated in argon based mixtures for biomedical applications. J Phys D Appl Phys 51:414001

Lipovan I, Bostanaru AC, Nastasa V, Hnatiuc E, Vuple V, Mares M (2015) Assessment of the antimicrobial effect of nonthermal plasma activated water against coagulase positive Staphylococci. Bull Univ Agric Sci Vet Med Cluj Napoca 72(2):363–367

Ma R, Wang G, Tian Y, Wang K, Zhang J, Fang J (2015) Nonthermal plasma activated water inactivation of food-borne pathogen on fresh produce. J Hazard Mater 300:643–651

McFerson L (1993) Understanding ORP's role in the disinfection process. Water Eng Manag 140:29–31

Misra NN, Moiseev T, Patil S, Pankaj SK, Bourke P, Mosnier JP, Keener KM, Cullen PJ (2014) Cold plasma in modified atmospheres for post-harvest treatment of strawberries. Food Bioprocess Technol 7:3045–3054

Moisan M, Nowakowska H (2018) Surface-wave (SW) sustained plasma columns: their contribution to the modeling of RF and microwave discharges with new insight into some of their features. A survey of other types of SW discharges. Plasma Sources Sci Technol 27(7):073001

Moisan M, Zakrzewski Z (1991) Plasma sources based on the propagation of electromagnetic surface waves. J Phys D Appl Phys 24:1025–1048

Moisan M, Sauvé G, Zakrzewski Z, Hubert J (1994) An atmospheric pressure waveguide-fed microwave plasma torch: the TIA design. Plasma Sources Sci Technol 3(4):584–592

Moritz M, Wiacek C, Weihe T, Ehlbeck J, Weltmann K-D, Braun PG (2020) Effect of cold atmospheric pressure plasma treatment of eggshells on the total bacterial count inoculated salmonella Enteritidis and selected quality parameters. Plasma Process Polym:e2000061

Naïtali M, Kamgang-Youbi G, Herry JM, Bellon-Fontaine MN, Brisset JL (2010) Combined effects of long-living chemical species during microbial inactivation using atmospheric plasma-treated water. Appl Environ Microb 76(22):7662–7664

Niquet R, Boehm D, Schnabel U, Cullen P, Bourke P, Ehlbeck J (2018) Characterising the impact of post-treatment storage on chemistry and antimicrobial properties of plasma treated water derived from microwave and DBD sources. Plasma Process Polym 15(3):1700127

Oh J-S, Szili EJ, Hatta A, Ito M, Shirafuji T (2019) Tailoring the chemistry of plasma-activated water using a DC-pulse-driven nonthermal atmospheric-pressure helium plasma jet. Plasma 2(2):127–137

Qi ZH, Tian EQ, Song Y, Sosnin EA, Skakun VS, Li TT, Xia Y, Zhao Y, Lin XS, Liu DP (2018) Inactivation of Shewanella putrefaciens by plasma activated water. Plasma Chem Plasma Process 38(5):1035–1050

Royintarat T, Seesuriyachan P, Boonyawan D, Choi EH, Wattanutchariya W (2019) Mechanism and optimization of nonthermal plasma-activated water for bacterial inactivation by underwater plasma jet and delivery of reactive species underwater by cylindrical DBD plasma. Curr Appl Phys 19(9):1006–1014

Royintarat T, Choi EH, Boonyawan D, Seesuriyachan P, Wattanutchariya W (2020) Chemical-free and synergistic interaction of ultrasound combined with plasma-activated water (PAW) to enhance microbial inactivation in chicken meat and skin. Sci Rep 10(1):1559

Schnabel U, Handorf O, Stachowiak J, Boehm D, Weit C, Weihe T, Schafer J, Below H, Bouke P, Ehlbeck J (2020) Plasma-functionalized water: from bench to prototype for fresh-cut lettuce. Food Eng Rev 13:115–135

Schnabel U, Handorf O, Winter H, Weihe T, Weit C, Schäfer J, Stachowiak J, Boehm D, Be-low H, Bourke P, Ehlbeck J (2021) The efffect of plasma treated water unit processes on the food quality characteristics of fresh-cut endive. Front Nutr 7:627483

Shen J, Tian Y, Li YL, Ma RN, Zhang Q, Zhang J, Fang J (2016) Bactericidal effects against S. aureus and physicochemical properties of plasma activated water stored at different temperatures. Sci Rep 6:28505

Suganuma R, Yasuoka K (2018) Air-supplied pinhole discharge in aqueous solution for the inactivation of Escherichia coli. Jpn J Appl Phys 57(4):046202

Suslow T (2004) Oxidation-reduction potential (ORP) for water disinfection monitoring. Control and Documentation, University of California, Division of Agriculture and Natural Resources. https://doi.org/10.3733/ucanr.8149

Thirumdas R, Kothakota A, Annapure U, Siliveru K, Blundell R, Gatt R, Valdramidis VP (2018) Plasma activated water (PAW): chemistry, physico-chemical properties, applications in food and agriculture. Trends Food Sci Tech 77:21–31

Tian Y, Ma RN, Zhang Q, Feng HQ, Liang YD, Zhang J, Fang J (2015) Assessment of the physicochemical properties and biological effects of water activated by nonthermal plasma above and beneath the water surface. Plasma Process Polym 12(5):439–449

Vanraes P, Bogaerts A (2018) Plasma physics of liquids—a focused review. Appl Phys Rev 5(3):031103

Vlad IE, Anghel SD (2017) Time stability of water activated by different on-liquid atmospheric pressure plasmas. J Electrost 87:284–292

Vukić M, Vujadinović D, Ivanović M, Gojković V, Grujić R (2017) Color change of orange and carrot juice blend treated by nonthermal atmospheric plasma. J Food Process Preserv 42(2):e13525

Wende K, von Woedtke T, Weltmann K-D, Bekeschus S (2018) Chemistry and biochemistry of cold physical plasma derived reactive species in liquids. Biol Chem 400(1):19–38

Wu SJ, Zhang Q, Ma RN, Yu S, Wang KL, Zhang J, Fang J (2017) Reactive radical-driven bacterial inactivation by hydrogen- peroxide-enhanced plasma-activated-water. Eur Phys J Spec Top 226(13):2887–2899

Xiang QS, Kang CD, Niu LY, Zhao DB, Li K, Bai YH (2018) Antibacterial activity and a membrane damage mechanism of plasma-activated water against Pseudomonas deceptionensis CM2. LWT – Food Sci Technol 96:395–401

Xiang Q, Fan L, Li Y, Dong S, Li K, Bai Y (2020) A review on recent advances in plasma-activated water for food safety: current applications and future trends. Crit Rev Food Sci Nutr. https://doi.org/10.1080/10408398.2020.1852173

Xu Y, Tian Y, Ma R, Liu Q, Zhang J (2016) Effect of plasma activated water on the postharvest quality of button mushrooms, Agaricus bisporus. Food Chem 197(Part A):436-444

Ye GP, Zhang Q, Pan H, Sun K, Pan J, Zhang J, Wang J (2013) Efficiency of pathogenic bacteria inactivation by nonthermal plasma activated water. Scientia Sinica Vitae 43(8):679–684

Zhang Q, Liang YD, Feng HQ, Ma RN, Tian Y, Zhang J, Fang J (2013) A study of oxidative stress induced by nonthermal plasma-activated water for bacterial damage. Appl Phys Lett 102(20):203701

Zhang Q, Ma RN, Tian Y, Su B, Wang KL, Yu S, Zhang J, Fang J (2016) Sterilization efficiency of a novel electrochemical disinfectant against Staphylococcus aureus. Environ Sci Technol 50(6):3184–3192

Zhang R, Ma YF, Wu D, Fan LM, Bai YH, Xiang QS (2020) Synergistic inactivation mechanism of combined plasma-activated water and mild heat against Saccharomyces cerevisiae. J Food Prot 83(8):1307–1314

Zhou RW, Zhou RS, Prasad K, Fang Z, Speight R, Bazaka K, Ostrikov K (2018) Cold atmospheric plasma activated water as a prospective disinfectant: the crucial role of peroxynitrite. Green Chem 20(23):5276–5284

Radio-Frequency

Alfaifi B, Tang J, Jiao Y, Wang S, Rasco B, Jiao S, Sablani S (2014) Radio frequency disinfestation treatments for dried fruit: model development and validation. J Food Eng 120:268–276
Altemimi A, Aziz SN, Al-Hilphy ARS, Lakhssassi N, Watson DW, Ibrahim SA (2019) Critical review of radio-frequency (RF) heating applications in food processing. Food Qual Saf 3(2):81–91
Anderson AK, Finkelsein R (1919) A study of the Electropure process of treating milk. J Dairy Sci 2:374–406
Beatie JM (1914) Electrical treatment of milk. Brit Med J 1(2791):1433
Beattie JM, Lewis FC (1913) Utilization of electricity in the continuous sterilization of milk. J Pathol Bacteriol 18:120
Brunkhorst C, Ciotti D, Fredd E, Wilson JR, Geveke DJ, Kozempel M (2000) Development of process equipment to separate nonthermal and thermal effects of RF energy on microorganisms. J Microw Power Electromagn Energy 35(1):44–50
Carroll DE, Lopez A (1969) Lethality of radio-frequency energy upon microorganisms in liquid, buffered, and alcoholic food systems. J Food Sci 34(4):320–324
Chang DC, Gao PQ, Maxwell BL (1991) High efficiency gene transfection by electroporation using a radio-frequency electric field. Biochim Biophys Acta 1092(2):153–160
Coster HG, Zimmermann U (1975a) The mechanism of electrical breakdown in the membranes of *Valonia utricularis*. J Membr Biol 22(1):73–90
Coster HG, Zimmermann U (1975b) Dielectric breakdown in the membranes of *Valonia utricularis*. The role of energy dissipation. Biochim Biophys Acta Biomembr 382(3):410–418
FCC (1988) Federal Communications Commission Online table of frequency allocations. 47 C.F.R. § 2.106, revised on February 1, 2021, https://www.fcc.gov/engineering-technology/policy-and-rules-division/general/radio-spectrum-allocation. Accessed 11 March 2021
Fleming H (1944) Effect of high frequency fields on micro-organisms. Electr Eng 63(1):18–21
Gao M, Tang J, Villa-Rojas R, Wang Y, Wang S (2011) Pasteurization process development for controlling Salmonella in in-shell almonds using radio frequency energy. J Food Eng 104(2):299–306
Geveke DJ (2005) Non-thermal processing by radio frequency electric field. In: Sun D-W (ed) Emerging technologies for food processing. Elsevier, Academic Press, London/San Diego, pp 307–322
Geveke DJ (2011) Radio frequency electric field as a nonthermal processing. In: Zhang HQ, Barbosa-Canovas GV, Balasubramaniam VM, Dunne CP, Farkas DF, Yuan JTC (eds) Nonthermal processing technologies for food. Wiley-Blackwell IFT Press, Ames, pp 213–221
Geveke DJ (2020) Inactivation of yeast and bacteria using combinations of radio frequency electric fields and ultraviolet light. J Food Process Preserv 44:e14385
Geveke DJ, Brunkhorst C (2003) Inactivation of Saccharomyces cerevisiae using radio frequency electric fields. J Food Prot 66(9):1712–1715
Geveke DJ, Brunkhorst C (2004) Inactivation of Escherichia coli in apple juice by radio frequency electric fields. J Food Sci 69(3):134–138
Geveke DJ, Brunkhorst C, Fan X (2007) Radio frequency electric fields processing of orange juice. Innov Food Sci Emerg Technol 8(4):549–554
Geveke DJ, Brunkhorst C (2008) Radio frequency electric fields inactivation of Escherichia coli in apple cider. J Food Eng 85(2):215–221
Geveke DJ, Kozempel M, Scullen OJ, Brunkhorst C (2002) Radio frequency energy effects on microorganisms in foods. Innov Food Sci Emerg 3(2):133–138
Geveke DJ, Brunkhorst C, Cooke P, Fan X (2006) Nonthermal inactivation of *E. coli* in fruit juices using radio frequency electric fields. In: Juneja VK, Cherry JP, Tunick MH (eds) Advances in microbial food safety. American Chemical Society, Washington, DC, pp 121–139
Geveke DJ, Gurtler J, Zhang HQ (2009) Inactivation of Lactobacillus plantarum in apple cider, using radio frequency electric fields. J Food Prot 72(3):656–661

Geveke DJ, Bigley ABW, Brunkhorst CD (2017) Pasteurization of shell eggs using radio frequency heating. J Food Eng 193:53–57

Huang Z, Marra F, Wang S (2016) A novel strategy for improving radio frequency heating uniformity of dry food products using computational modeling. Innov Food Sci Emerg 34:100–111

IEEE 521-2019 (2021) IEEE Standard Letter Designations for Radar-Frequency Bands. Revision of IEEE Std 521-2002, https://standards.ieee.org/standard/521-2019.html. Accessed 11 March 2021

Ingram M, Page LJ (1953) The survival of microbes in modulated high-frequency voltage fields. Proc Soc Appl Bacteriol 16:69–87

ITU (2012) Radio regulations articles. Article 1. Terms and definitions, Section I – General terms, Article 1.15, definition: ISM application, http://search.itu.int/history/HistoryDigitalCollectionDocLibrary/1.41.48.en.101.pdf. Accessed 11 March 2021

Jiang H, Ling B, Zhou X, Wang S (2020) Effects of combined radio frequency with hot water blanching on enzyme inactivation, color and texture of sweet potato. Innov Food Sci Emerg 66:102513

Jiao S, Johnson JA, Tang J, Tiwari G, Wang S (2011) Dielectric properties of cowpea weevil, black-eyed peas and mung beans with respect to the development of radio frequency heat treatments. Biosyst Eng 108(3):280–291

Komarov VV (2021) A review of radio frequency and microwave sustainability-oriented technologies. Sustain Mater Technol 28:e00234

Kotnik T, Miklavčič D (2006) Theoretical evaluation of voltage inducement on internal membranes of biological cells exposed to electric fields. Biophys J 90(2):480–491

Kotnik T, Miklavcic D, Slivnik T (1998) Time course of transmembrane voltage induced by time-varying electric fields e a method for theoretical analysis and its application. Bioelectrochem Bioenerg 45(1):3–16

Kou X, Li R, Wang S (2018) Identifying possible non-thermal effects of radio frequency energy on inactivating food microorganisms. Int J Food Microbiol 269:89–97

Liu S, Ozturk S, Xu J, Kong F, Gray P, Zhu M-J, Sablani SS, Tang J (2018) Microbial validation of radio frequency pasteurization of wheat flour by inoculated pack studies. J Food Eng 217:68–74

Llave Y, Liu S, Fukuoka M, Sakai N (2015) Computer simulation of radiofrequency defrosting of frozen foods. J Food Eng 152:32–42

Marra F, Zhang L, Lyng JG (2009) Radio frequency treatment of foods: review of recent advances. J Food Eng 91(4):497–508

Masood H, Razaeimotlagh A, Cullen PJ, Trujilo FJ (2017) Numerical and experimental studies on a novel Steinmetz treatment chamber for inactivation of Escherichia coli by radio frequency electric fields. Innov Food Sci Emerg 41:337–347

McKenna BM, Lyng J, Brunton N, Shirsat N (2005) Advances in radio frequency and ohmic heating of meats. J Food Eng 77(2):215–229

Miyakoshi J (2013) Cellular and molecular responses to radio-frequency electromagnetic field. Proc IEEE 101(6):1494–1502

National Academies of Sciences, Engineering, and Medicine (NASEM) (2015) Handbook of frequency allocations and spectrum protection for scientific uses, 2nd edn. The National Academies Press, Washington, DC. https://www.nap.edu/catalog/21774/handbook-of-frequency-allocations-and-spectrum-protection-for-scientific-uses. Accessed 11 March 2021

Nyrop JE (1946) A specific effect of high-frequency electric currents on biological objects. Nature 157(3976):51

Ozturk S, Kong F, Trabelsi S, Singh RK (2016) Dielectric properties of dried vegetable powders and their temperature profile during radio frequency heating. J Food Eng 169:91–100

Piyasena P, Dussault C, Koutchma T, Ramaswamy H, Awuah G (2003) Radio frequency heating of foods: principles, applications and related properties—a review. Crit Rev Food Sci 43(6):587–606

Rincon AM, Singh RK, Stelzleni AM (2015) Effects of endpoint temperature and thickness on quality of whole muscle non-intact steaks cooked in a radio frequency oven. LWT-Food Sci Technol 64(2):1323–1328

Sale AJH, Hamilton WA (1967) Effects of high electric fields on microorganisms. I. Killing of bacteria and yeasts. Biochim Biophys Acta 148:781–788

Saulis G (2010) Electroporation of cell membranes: the fundamental effects of pulsed electric fields in food processing. Food Eng Rev 2:52–73

Schoenbach KH, Joshi RP, Stark RH, Dobbs FC, Beebe SJ (2000) Bacterial decontamination of liquids with pulsed electric fields. IEEE Trans Dielectr Electr Insul 7(5):637–645

Stabler SH (1931) The electrical process of milk pasteurization. Am J Epidemiol 14(2):433–452

Trujillo FJ, Geveke DJ (2014) Nonthermal processing by radio frequency electric fields. In: Sun D-W (ed) Emerging technologies for food processing, 2nd edn. Elsevier Academic Press, Oxford, UK, pp 259–269

Uemura K, Isobe S (2002) Developing a new apparatus for inactivating Escherichia coli in saline water with high electric field AC. J Food Eng 53(3):203–207

Uemura K, Kobayashi I, Inoue T (2009) Inactivation of Alicyclobacillus acidoterrestris in orange juice by high electric field alternating current. Food Sci Technol Res 15:211–216

Uemura K, Kobayashi I, Inoue T (2010a) Inactivation of Bacillus subtilis spores in orange juice and the quality change by high electric field alternating current. Jap Agr Res Q 44(1):61–66

Uemura K, Takahashi C, Kobayashi I (2010b) Inactivation of Bacillus subtilis spores in soybean milk by radio-frequency flash heating. J Food Eng 100(4):622–626

Uemura K, Takahashi C, Kobayashi I (2012) Inactivation of lactobacillus brevis in liquid egg white by radio-frequency flash heating. Food Sci Technol Res 18:357–362

Ukuku DO, Geveke DJ (2010) A combined treatment of UV-light and radio frequency electric field for the inactivation of Escherichia coli K-12 in apple juice. Int J Food Microbiol 138(1–2):50–55

Ukuku DO, Geveke DJ, Cooke P, Zhang HQ (2008) Membrane damage and viability loss of Escherichia coli K-12 in apple juice treated with radio frequency electric field. J Food Prot 71(4):684–690

Ukuku DO, Geveke DJ, Cooke PH (2012) Effect of thermal and radio frequency electric fields treatments on Escherichia coli bacteria in apple juice. J Microb Biochem Technol 4(3):76–81

Uyar R, Bedane TF, Erdogdu F, Palazoglu TK, Farag KW, Marra F (2015) Radio-frequency thawing of food products – a computational study. J Food Eng 146:163–171

Wang Y, Wig T, Tang J, Hallberg L (2003) Sterilization of foodstuffs using radio frequency heating. J Food Sci 68(2):539–544

Weaver JC (2000b) Electroporation of cells and tissues. IEEE T Plasma Sci 28:24–33

Weaver JC (2003) Electroporation of biological membranes from multicellular to nano scales. IEEE Trans Dielectr Electr Insul 10(5):754–768

Weaver JC, Chizmadzhev YA (1996) Theory of electroporation: a review. Bioelectrochem Bioenerg 41(2):135–160

Webb SJ, Booth AD (1969) Absorption of microwaves by microorganisms. Nature 222:1199–1120

Webb SJ, Dodds DD (1968) Inhibition of bacterial cell growth by 136 gc microwaves. Nature 218:374–375

Wust P, Kortüm B, Strauss U, Nadobny J, Zschaeck S, Beck M, Stein U, Ghadjar P (2020) Nonthermal effects of radiofrequency electromagnetic fields. Sci Rep 10:e13488

Xu J, Yang R, Jin Y, Barnett G, Tang J (2020) Modeling the temperature-dependent microbial reduction of Enterococcus faecium NRRL B-2354 in radio-frequency pasteurized wheat flour. Food Control 107:106778

Zhang L, Lyng JG, Xu R, Zhang S, Zhou X, Wang S (2019) Influence of radio frequency treatment on in-shell walnut quality and *Staphylococcus aureus* ATCC 25923 survival. Food Control 102:197–205

Zimmermann U (1986b) Electrical breakdown, electropermeabilization and electrofusion. Rev Physiol Bioch Pharmacol 105:176–256

Zimmermann U, Schulz J, Pilwat G (1973) Transcellular ion flow *in Escherichia coli* B and electrical sizing of bacterias. Biophys J 13(10):1005–1013
Zimmermann U, Pilwat G, Riemann F (1974) Dielectric breakdown of cell membranes. Biophys J 14(11):881–899
Zimmermann U, Pilwat G, Riemann F (1975) Preparation of erythrocye ghosts by dielectric breakdown of the cell membrane. Biochim Biophys Acta Biomembr 375(2):209–219

Oscillating Magnetic Fields

Abdelhameed AE (2014) Growth rate inhibition of some spoilage fungi of food by magnetic field. Misr J Agric Eng 31(1):299–308
Ahmad AM, Yahya AGI, Jabir AWS (2013) Effect of magnetic field energy on growth of Aspergillus flavus and aflatoxins production. Al-Nahrain J Sci 16(2):180–187
Al-Hawash AB, Li S, Zhang X, Zhang X, Ma F (2018) Productivity of γ-linoleic acid by oleaginous fungus Cunninghamella echinulata using a pulsed high magnetic field. Food Biosci 21:1–7
Ali HI, Al-Hilphy AR, Al-Darwash AK (2015) The effect of magnetic field treatment on the characteristics and yield of Iraqi local white cheese. IOSR J Agric Vet Sci 8(9):63–69
Barbosa-Canovas GV, Schaffner DW, Pierson MD, Zhang QH (2000) Oscillating magnetic fields. J Food Sci 65:86–89
Butz P, Tauscher B (2002) Emerging technologies: chemical aspects. Food Res Int 35(2–3):279–284
Cai R, Yang H, He J, Zhu W (2009) The effects of magnetic fields on water molecular hydrogen bonds. J Molec Struct 938(1–3):15–19
Chang KT, Weng CI (2006) The effect of an external magnetic field on the structure of liquid water using molecular dynamics simulation. J Appl Phys 100(4):043917
Colic M, Morse D (1999) The elusive mechanism of the magnetic 'memory' of water. Colloids Surf A Physicochem Eng Asp 154(1–2):167–174
Fernández-Martín F, Pérez-Mateos M, Dadashi S, Gómez-Guillén CM, Sanz PD (2017) Impact of magnetic assisted freezing in the physicochemical and functional properties of egg components. Part 1: egg white. Innov Food Sci Emerg Technol 44:131–138
Grigelmo-Miguel N, Soliva-Fortuny R, Barbosa-Cánovas GV, Martín-Belloso O (2011) Use of oscillating magnetic fields in food preservation. In: Zhang HQ, Barbosa-Cánovas GV, Balasubramaniam VM, Dunne CP, Farkas DF, Yuan JTC (eds) Nonthermal processing technologies for food. Wiley Blackwell, Oxford, pp 222–243
Haile M, Pan Z, Gao M, Luo L (2008) Efficacy in microbial sterilisation of pulsed magnetic field treatment. Int J Food Eng 4(4):Article 15
Higashitani K, Kage A, Katamura S, Imai K, Hatade S (1993) Effects of a magnetic field on the formation of $CaCO_3$ particles. J Colloid Interface Sci 156(1):90–95
Hofmann GA (1985) Deactivation of microorganisms by an oscillating magnetic field. U.S. Patent No. 4,524,079. U.S. Patent and Trademark Office, Washington, DC
Kobayashi M, Soda N, Miyo T, Ueda Y (2004) Effects of combined DC and AC magnetic fields on germination of hornwort seeds. Bioelectromagnetics 25:552–559
Lipiec J, Janas P, Barabasz W, Pysz M, Pisulewski P (2005) Effects of oscillating magnetic field pulses on selected oat sprouts used for food purposes. Acta Agrophysica 5(2):357–365
Masood S, Saleem I, Smith D, Chu WK (2020) Growth pattern of magnetic field-treated bacteria. Curr Microbiol 77(2):194–203
Mok JH, Her JY, Kang T, Hoptowit R, Jun S (2017) Effects of pulsed electric field (PEF) and oscillating magnetic field (OMF) combination technology on the extension of supercooling for chicken breasts. J Food Eng 196:27–35
Okuda K, Kawauchi A, Yomogida K (2020a) Quality improvements to mackerel (Scomber japonicus) muscle tissue frozen using a rapid freezer with the weak oscillating magnetic fields. Cryobiology 95:130–137

Okuda K, Kawauchi A, Yomogida K (2020b) Data of the freezing curves of tuna blocks with or without the weak oscillating magnetic fields. Data Brief 105852
Otero L, Pérez-Mateos M, Rodríguez AC, Sanz PD (2017) Electromagnetic freezing: effects of weak oscillating magnetic fields on crab sticks. J Food Eng 200:87–94
Pazur A, Schimek C, Galland P (2007) Magnetoreception in microorganisms and fungi. Open Life Sci 2(4):597–659
Pothakamury UR, Barbosa-Canovas GV, Swanson BG (1993) Magnetic-field inactivation of microorganisms and generation of biological changes. Food Technol 47(12):85–93
Qian J, Zhang M, Dai C, Huo S, Ma H (2020) Transcriptomic analysis of Listeria monocytogenes under pulsed magnetic field treatment. Food Res Int 109195
Rodríguez AC, James C, James SJ (2017a) Effects of weak oscillating magnetic fields on the freezing of pork loin. Food Bioprocess Technol 10(9):1615–1621
Rodríguez AC, Sánchez-Benítez J, Sanz PD (2017b) Simulation of the magnetic freezing process applied to foods. Food Eng Rev 9(4):271–294
Takao K, Ikehata M, Nakagawa M (1995) Estimation of genetic effects of a static magnetic field by a somatic cell test using mutagen-sensitive mutants of Drosophila melanogaster. Bioelectrochem Bioenerg 36(2):95–100
Tang J, Shao S, Tian C (2019) Effects of the magnetic field on the freezing parameters of the pork. Int J Refrig 107:31–38
Valiron O, Peris L, Rikken G, Schweitzer A, Saoudi Y, Remy C, Job D (2005) Cellular disorders induced by high magnetic fields. J Magn Reson Imaging 22(3):334–340
Yadollahpour A, Jalilifar M, Rashidi S (2014) Antimicrobial effects of electromagnetic fields: a review of current techniques and mechanisms of action. J Pure Appl Microbiol 8(5):4031–4043
Zhang L, Yang Z, Deng Q (2021) Effects of pulsed magnetic field on freezing kinetics and physical properties of water and cucumber tissue fluid. J Food Eng 288:110149

Electrohydrodynamic Processing

Aghaei Z, Ghorani B, Emadzadeh B, Kadkhodaee R, Tucker N (2020) Protein-based halochromic electrospun nanosensor for monitoring trout fish freshness. Food Control 111:107065
Alehosseini A, Gómez-Mascaraque LG, Ghorani B, López-Rubio A (2019) Stabilization of a saffron extract through its encapsulation within electrospun/electrosprayed zein structures. LWT – Food Sci Technol 113:108280
Altan A, Aytac Z, Uyar T (2018) Carvacrol loaded electrospun fibrous films from zein and poly (lactic acid) for active food packaging. Food Hydrocoll 81:48–59
Aman Mohammadi M, Ramazani S, Rostami M, Raeisi M, Tabibiazar M, Ghorbani M (2019) Fabrication of food-grade nanofibers of whey protein isolate–guar gum using the electrospinning method. Food Hydrocoll 90:99–104
Baştürk E, Demir S, Danış Ö, Kahraman MV (2013) Covalent immobilization of α-amylase onto thermally crosslinked electrospun PVA/PAA nanofibrous hybrid membranes. J Appl Polym Sci 127(1):349–355
Bhushani AJ, Anandharamakrishnan C (2014) Electrospinning and electrospraying techniques: potential food based applications. Trends Food Sci Technol 38(1):21–33
Castro Coelho S, Nogueiro Estevinho B, Rocha F (2021) Encapsulation in food industry with emerging electrohydrodynamic techniques: electrospinning and electrospraying – a review. Food Chem 339:127850
Castro-Mayorga JL, Fabra MJ, Pourrahimi AM, Olsson RT, Lagaron JM (2017) The impact of zinc oxide particle morphology as an antimicrobial and when incorporated in poly (3-hydroxybut yrate-co-3-hydroxyvalerate) films for food packaging and food contact surfaces applications. Food Bioprod Process 101:32–44
Cooley JF (1902) Apparatus for electrically dispersing fluids. US Patent 692631

Duft D, Achtzehn T, Müller R, Huber BA, Leisner T (2003) Rayleigh jets from levitated microdroplets. Nature 421(6919):128–128

Echegoyen Y, Fabra MJ, Castro-Mayorga JL, Cherpinski A, Lagaron JM (2017) High throughput electro-hydrodynamic processing in food encapsulation and food packaging applications. Trends Food Sci Technol 60:71–79

Elias AL, Nearingburg B, Zahorodny-Burke M (2011) Microfluidic cell culture devices. In: Mitra SK, Chakraborty S (eds) Microfluidics and nanofluidics handbook: chemistry, physics, and life science principles. CRC Press, Boca Raton, pp 951–1016

Estevez-Areco S, Guz L, Candal R, Goyanes S (2018) Release kinetics of rosemary (Rosmarinus officinalis) polyphenols from polyvinyl alcohol (PVA) electrospun nanofibers in several food simulants. Food Packag Shelf Life 18:42–50

Fabra MJ, Lopez-Rubio A, Lagaron JM (2013) High barrier polyhydroxyalcanoate food packaging film by means of nanostructured electrospun interlayers of zein. Food Hydrocoll 32(1):106–114

Fabra MJ, López-Rubio A, Lagaron JM (2014) On the use of different hydrocolloids as electrospun adhesive interlayers to enhance the barrier properties of polyhydroxyalkanoates of interest in fully renewable food packaging concepts. Food Hydrocoll 39:77–84

Faridi-Esfanjani A, Jafari SM (2016) Biopolymer nano-particles and natural nanocarriers for nano-encapsulation of phenolic compounds. Colloid Surface B 146:532–543

Ghorani B, Emadzadeh B, Rezaeinia H, Russell SJ (2020) Improvements in gelatin cold water solubility after electrospinning and associated physicochemical, functional and rheological properties. Food Hydrocoll 104:105740

Göksen G, Fabra MJ, Ekiz HI, López-Rubio A (2020) Phytochemical-loaded electrospun nanofibers as novel active edible films: characterization and antibacterial efficiency in cheese slices. Food Control 112:107133

Gómez-Mascaraque LG, Lagarón JM, López-Rubio A (2015) Electrosprayed gelatin submicroparticles as edible carriers for the encapsulation of polyphenols of interest in functional foods. Food Hydrocoll 49:42–52

Gómez-Mascaraque LG, Hernández-Rojas M, Tarancón P, Tenon M, Feuillère N, Ruiz JFV, Fiszman S, López-Rubio A (2017) Impact of microencapsulation within electrosprayed proteins on the formulation of green tea extract-enriched biscuits. LWT – Food Sci Technol 81:77–86

Haider S, Haider A, Alghyamah AA, Khan R, Almasry WA, Khan N (2019) Electrohydrodynamic processes and their affecting parameters. In: Daider S (ed) Electrospinning and electrospraying-techniques and applications. IntechOpen, London/Rijeka, pp 1–25

He JH, Wu Y, Zuo WW (2005) Critical length of straight jet in electrospinning. Polymers 46(26):12637–12640

Işik C, Arabaci G, Ispirli Doğac Y, Deveci İ, Teke M (2019) Synthesis and characterization of electrospun PVA/Zn2+ metal composite nanofibers for lipase immobilization with effective thermal, pH stabilities and reusability. Mat Sci Eng 99:1226–1235

Khan MKI, Maan AA, Schutyser M, Schroën K, Boom R (2013) Electrospraying of water in oil emulsions for thin film coating. J Food Eng 119(4):776–780

Leidy R, Ximena QCM (2019) Use of electrospinning technique to produce nanofibres for food industries: a perspective from regulations to characterisations. Trends Food Sci Technol 85:92–106

Luo CJ, Stride E, Edirisinghe M (2012) Mapping the influence of solubility and dielectric constant on electrospinning polycaprolactone solutions. Macromolecules 45(11):4669–4680

Mascheroni E, Fuenmayor CA, Cosio MS, Di Silvestro G, Piergiovanni L, Mannino S, Schiraldi A (2013) Encapsulation of volatiles in nanofibrous polysaccharide membranes for humidity-triggered release. Carbohydr Polym 98(1):17–25

Megelski S, Stephens JS, Bruce Chase D, Rabolt JF (2002) Micro- and nanostructured surface morphology on electrospun polymer fibers. Macromolecules 35:8456–8466

Mendes AC, Stephansen K, Chronakis IS (2017) Electrospinning of food proteins and polysaccharides. Food Hydrocoll 68:53–68

Moreira JB, Lim LT, da Rosa Zavareze E, Dias ARG, Costa JAV, de Morais MG (2019) Antioxidant ultrafine fibers developed with microalga compounds using a free surface electrospinning. Food Hydrocoll 93:131–136
Morton WJ (1902) Method of dispersing fluids. US Patent 705691
Nagam Hanumantharao S, Rao S (2019) Multi-functional electrospun nanofibers from polymer blends for scaffold tissue engineering. Fiber 7(7):66
Nazari M, Majdi H, Milani M, Abbaspour-Ravasjani S, Hamishehkar H, Lim LT (2019) Cinnamon nanophytosomes embedded electrospun nanofiber: its effects on microbial quality and shelf-life of shrimp as a novel packaging. Food Packag Shelf Life 21:100349
Neo YP, Swift S, Ray S, Gizdavic-Nikolaidis M, Jin J, Perera CO (2013) Evaluation of gallic acid loaded zein sub-micron electrospun fibre mats as novel active packaging materials. Food Chem 141(3):3192–3200
Nieuwland M, Geerdink P, Brier P, van den Eijnden P, Henket JTMM, Langelaan MLP, Stroeks N, van Deventer HC, Martin AH (2013) Food-grade electrospinning of proteins. Innov Food Sci Emerg Techn 20:269–275
Pareta R, Edirisinghe MJ (2006) A novel method for the preparation of starch films and coatings. Carbohydr Polym 63(3):425–431
Ratanavaraporn J, Rangkupan R, Jeeratawatchai H, Kanokpanont S, Damrongsakkul S (2010) Influences of physical and chemical crosslinking techniques on electrospun type A and B gelatin fiber mats. Int J Biol Macromol 47(4):431–438
Reneker DH, Yarin AL (2008) Electrospinning jets and polymer nanofibers. Polymers 49:2387–2425
Rostami M, Yousefi M, Khezerlou A, Mohammadi MA, Jafari SM (2019) Application of different biopolymers for nanoencapsulation of antioxidants via electrohydrodynamic processes. Food Hydrocoll 97:105170
Schmatz DA, Costa JAV, de Morais MG (2019) A novel nanocomposite for food packaging developed by electrospinning and electrospraying. Food Packag Shelf Life 20:100314
Singh A, Orsat V, Raghavan V (2012) A comprehensive review on electrohydrodynamic drying and high-voltage electric field in the context of food and bioprocessing. Dry Technol 30(16):1812–1820
Soleimanifar M, Jafari SM, Assadpour E (2020) Encapsulation of olive leaf phenolics within electrosprayed whey protein nanoparticles; production and characterization. Food Hydrocoll 101:105572
Terada D, Kobayashi H, Zhang K, Tiwari A, Yoshikawa C, Hanagata N (2012) Transient charge-masking effect of applied voltage on electrospinning of pure chitosan nanofibers from aqueous solutions. Sci Technol Adv Mat 13(1):015003
Torres-Giner S, Martinez-Abad A, Ocio MJ, Lagaron JM (2010) Stabilization of a nutraceutical omega-3 fatty acid by encapsulation in ultrathin electrosprayed zein prolamine. J Food Sci 75(6):N69–N79
Xue J, Wu T, Dai Y, Xia Y (2019) Electrospinning and electrospun nanofibers: methods, materials, and applications. Chem Rev 119(8):5298–5415
Yao ZC, Chang MW, Ahmad Z, Li JS (2016) Encapsulation of rose hip seed oil into fibrous zein films for ambient and on demand food preservation via coaxial electrospinning. J Food Eng 191:115–123
Zhang Y, Venugopal J, Huang ZM, Lim CT, Ramakrishna S (2006) Crosslinking of the electrospun gelatin nanofibers. Polymers 47(8):2911–2917

Electron Beam Processing

Balayan MH, Pepoyan AZ, Manvelyan AM, Tsaturyan VV, Grigoryan B, Abrahamyan A, Chikindas ML (2019) Combined use of eBeam irradiation and the potential probiotic Lactobacillus rhamnosus Vahe for control of foodborne pathogen Klebsiella pneumoniae. Ann Microbiol 69:1579–1582

Brown D (2015) Integrating electron beam equipment into food processing facilities: strategies and design consideration. In: Pillai SD, Shayanfar S (eds) Electron beam pasteurization and complimentary food processing technologies chapter 3. Woodhead Publishing, Oxford

Carocho MR, Barreira JOC, Antonio AL, Bento A, Kaluska I, Ferreira IC (2012) Effects of electron-beam radiation on nutritional parameters of Portuguese chestnuts (Castanea sativa Mill.). J Agric Food Chem 60:7754–7760

Han J, Castell-Perez ME, Moreira RG (2007) The influence of electron beam irradiation of antimicrobial-coated LDPE/ polyamide films on antimicrobial activity and film properties. LWT- Food Sci Technol 40:1545–1554

Jaczynski J, Park JW (2003) Microbial inactivation and electron penetration in surimi seafood during electron beam processing. J Food Sci 68(5):1788–1792

Jeyakumari A, Narasimha Murthy L, Visnuvinayagam S, Rawat K, Khader SA (2019) Electron beam irradiated Tilapia fish chunk: quality and shelf life under chilled storage. FishTech Rep 5(1):8–9

Lung HM, Cheng YC, Chang YH, Huang HW, Yang BB, Wang CY (2015) Microbial decontamination of food by electron beam irradiation. Trends Food Sci Technol 44:66–78

Miller RB (2005) Electronic irradiation of foods: an introduction to the technology. Springer, New York, p 350

Moosekian SR, Jeong S, Marks BP, Ryser ET (2012) X-Ray irradiation as a microbial intervention strategy for food. Annu Rev Food Sci Technol 3:493–510

Pavlov Yu S (2015) VI Russian conference (with participation of CIS specialists) "Actual problems of high energy chemistry" Industrial electronic-beam technologies realized by accelerator UELV-10-20-S-70-2 Moscow, pp 268–278

Pavlov Yu S, Dobrohotov BB, Pavlov BA, Nepomnyaschy ON, Danilichev VA (2014) Magnetic buncher accelerator UELV-10-10-T-1 for studying fluorescence and radiation physical researches Proceedings of XXIV Russian Particle Accelerators Conference (RuPAC'2014) Obninsk, Russia, pp 259–261

Pikaev AK, Glazunov PY, Pavlov Yu S (1993) Radiation Center of Institute-of-Physical Chemistry in Moscow. Radiat Phys Chem 42:887–890

Revina AA (1998) Patent № 2104927, The method for producing a modified activated carbon

Revina AA (2003) Patent RF № 2212268, System of modification of objects with nanoparticles Bull. № 26

Revina AA (2009) Adsorbtion and oxidation processes in modern nanotechnology. Prot Met Phys Chem Surf 45(1):58–63

Revina AA, Magomedbekov EP (2010) V All-Russian conference "Physicochemical processes in condensed media and intermediate boundaries FAGRAN-2010" Role of radiation chemistry in creation of functional nanomaterials based on metallic and bimetallic nanoparticles Voronezh, Russia, p 145

Ruban IN, Voropaeva NL, Figovsky OL, Sharipov MD, Dadajanov TK (2012) Biologically active multifunctional nanochips and method of application for production of high-quality seed. U.S. Patent № 8209902

Schmidt M, Zannini E, Arendt EK (2018) Recent advances in physical post-harvest treatments for shelf-life extension of cereal crops. Foods 7:45

Schoeller NP, Ingham SC, Ingham BH (2002) Assessment of the potential for Listeria monocytogenes survival and growth during alfalfa sprout production and use of ionizing radiation as a potential intervention treatment. J Food Prot 65(8):1259–1266

Urbain WM (1996) Food irradiation. Academic Press, Orlando, p 351

Yang K, Li K, Pan L, Luo X, Xing J, Wang J, Wang L, Wang R, Zhai Y, Chen Z (2020) Effect of ozone and electron beam irradiation on degradation of zearalenone and ochratoxin A. Toxins 12:138

Ionizing Radiation

Bhat R, Sridhar KR (2008) Nutritional quality evaluation of electron beam irradiated (Nelumbo nucifera) seeds. Food Chem 107:174–184
Bhat R, Sridhar KR, Bhushan B (2007a) Free radicals in velvet bean seeds (Mucuna pruriens L. DC.) and their status after gamma irradiation and conventional processing. LWT-Food Sci Technol 40:1570–1577
Bhat R, Sridhar KR, Yokotani KT (2007b) Effect of ionizing radiation on antinutritional features of velvet bean seeds (Mucuna pruriens). Food Chem 103:860–866
Bhat R, Alias AK, Paliyath G (2012) Use of electron beams in food preservation. In: Bhat R, Alias AK, Paliyath G (eds) Progress in food preservation. Wiley, Blackwell, UK, pp 343–372
Breitfellner F, Solar S, Sontag G (2002) Effect of g-irradiation on phenolic acids in strawberries. J Food Sci 67:517–521
Chmielewski AG, Kang CM, Kang CS, Vujic JL (2006) Radiation technology. Introduction to industrial and environmental applications. Seoul National University Press, Seoul, p 274
Fan X, Sommers CH, Marshall RC (2012) Advances in electron beam and X-ray technologies for food irradiation. In: Fan X, Sommers CH (eds) Food irradiation research and technology, 2nd edn. Wiley, Hoboken, pp 9–28
Farkas J (1990) Combination of irradiation with mild heat treatment. Food Control 1:223–229
GAO (2000) US General Accountability Office, Report to Congressional Requesters 24 August 2000, RCED-00-217. Food Irradiation: Available Research Indicates That Benefits Outweigh Risks, GAO, and Washington, DC, USA, p 3 and 14
González-Aguilar GA, Villegas-Ochoa MA, Martínez-Téllez MA, Gardea AA, Ayala-Zavala JF (2007a) Improving antioxidant capacity of fresh-cut mangoes treated with UV-C. J Food Sci 72:197–202
González-Aguilar GA, Zavaleta-Gatica R, Tiznado-Hernández ME (2007b) Improving postharvest quality of mango 'Haden' by UV-C treatment. Postharvest Biol Tech 45:108–116
Hong YH, Park JY, Park JH, Chung MS, Kwon KS, Chung K, Won M, Song KB (2008) Inactivation of Enterobacter sakazakii, Bacillus cereus, and Salmonella typhimurium in powdered weaning food by electron-beam irradiation. Rad Phys Chem 77:1097–1100
Huang SJ, Mau JL (2006) Antioxidant properties of methanolic extracts from Agaricus blazei with various doses of g-irradiation. LWT- Food Sci Technol 39:707–716
Lacroix M, Ouattara B (2000) Combined industrial processes with irradiation to assure innocuity and preservation of food products. Int Food Res J 33:719–724
Liu RH (2004) Potential synergy of phytochemicals in cancer prevention: mechanism of action. J Nutr 134(12 Suppl):3479S–3485S
Mexis SF, Badeka AV, Chouliara E, Riganakos KA, Kontominas MG (2009) Effect of g-irradiation on the physicochemical and sensory properties of raw unpeeled almond kernels (Prunus dulcis). Innov Food Sci Emerg Technol 10:87–92
Pan J, Vicente AR, Martínez GA, Chaves AR, Civello PM (2004) Combined use of UV-C irradiation and heat treatment to improve postharvest life of strawberry fruit. J Sci Food Agric 84:1831–1838
Pérez MB, Calderón NL, Croci CA (2007) Radiation-induced enhancement of antioxidant activity in extracts of rosemary (Rosmarinus officinalis L.). Food Chem 104:585–592
Pillai SD, Shayanfar S (2017) Electron beam technology and other irradiation technology applications in the food industry. In: Applications of radiation chemistry in the fields of industry, biotechnology and environment. Springer, Cambridge, pp 249–268

Sajilata MG, Singhal RS (2006) Effect of irradiation and storage on the antioxidative activity of cashew nuts. Radiat Phys Chem 75:297–300

Selma MV, Allende A, López-Gálvez F, Conesa MA, Gil MI (2008) Disinfection potential of ozone, ultraviolet C, and their combination in wash waster for the fresh-cut vegetable industry. Food Microbiol 25:809–814

Shayanfar S, Mena K, Pillai SD (2016) Quantifying the reduction in potential infection risks from non-O157 Shiga toxin producing E. coli in strawberries by low dose electron beam processing. Food Control 72(B):324–327

Simas MM, Albuquerque R, Oliveira CA, Rottinghaus GE, Correa B (2010) Influence of gamma radiation on productivity parameters of chicken fed mycotoxin contaminated corn. Appl Radiat Isot 68:1903–1908

Teets AS, Sundararaman M, Were LM (2008) Electron beam irradiated almond skin powder inhibition of lipid oxidation in coked salted ground chicken breast. Food Chem 111:934–941

Tomas-Barberan FA, Espin JC (2001) Phenolic compounds and related enzymes as determinants of quality in fruit and vegetables. J Sci Food Agric 81:853–876

Tsai SY, Tsai HL, Mau JL (2007) Antioxidant properties of Agaricus blazei, Agrocybe cylindracea, and Boletus edulis. LWT – Food Sci Technol 40:1392–1402

Variyar PS, Limaye A, Sharma A (2004) Radiation-induced enhancement of antioxidant contents of soybean (Glycine max Merrill). J Agric Food Chem 52:3385–3388

Walkling-Ribeiro M, Noci F, Cronin DA, Riener J, Lyng JG, Morgan DJ (2008) Reduction of Staphylococcus aureus and quality changes in apple juice processed by ultraviolet irradiation, pre-heating and pulsed electric fields. J Food Eng 89:267–273

Wong PYY, Kitts DD (2001) Factors influencing ultraviolet and electron beam irradiation-induced free radical damage of ascorbic acid. Food Chem 74:75–84

Zobel AM (1997) Coumarins in fruit and vegetables. In: Tomás Barberán FA, Robbins RJ (eds) Photochemistry of fruit and vegetables. Clanderon Press, Oxford, pp 173–204

Chapter 5
Pressure-Based Technologies: High Pressure Processing; Supercritical and Subcritical Fluid Processing

Zoya Okun and Avi Shpigelman

1 Introduction

High pressure-based technologies are a cluster of unit operations that involve elevated pressure to obtain a desired effect and can be extremely useful in the field of agri-food processing. Similar to heat, pressure is an extensive thermodynamic parameter, reactions such as phase transition and molecular reorientation depend on both temperature and pressure and cannot be treated separately (Martínez-Monteagudo and Balasubramaniam 2016). While temperature is commonly the main thermodynamic parameter used to ensure safety or to manipulate the structure and techno-functionalities of biological systems, the application of pressure often brings important technological advantages. Pressure is the ratio of a force divided by the area on which that force acts. While pressure is increasing during several processes commonly occurring in the food industry, it is not the main parameter responsible for the effect (like microbial inactivation or texture formation), as will be presented for the technologies that will be briefly introduced in this chapter. The relatively new use of high pressure may assist in the manufacture of improved or entirely new products with special properties and functionalities. The application of pressure for the production of bioactive compounds is widely explored and results in efficient and green extraction procedures for ingredients that are hard to economically obtain conventionally.

In addition to high hydrostatic pressure processing (HPP), which is the most widely adopted non-thermal processing technology for food so far (mostly aimed at shelf life extension) (Bolumar et al. 2021), other non-thermal technologies exist, some of them also combine pressure application for induction of microbial and enzymatic inactivation (Yu et al. 2020). Another extremely promising industrial

Z. Okun · A. Shpigelman (✉)
Faculty of Biotechnology and Food Engineering, Technion, Haifa, Israel
e-mail: avis@bfe.technion.ac.il

A. Režek Jambrak (ed.), *Nonthermal Processing in Agri-Food-Bio Sciences*,
Food Engineering Series, https://doi.org/10.1007/978-3-030-92415-7_5

direction for the application and advantages of high pressure is the extraction of valuable compounds from plant materials or sidestreams and their formulation (Knez et al. 2014). Such high pressure technologies, utilizing sub and supercritical fluids, have the potential of the creation of new products with unique properties or assist in the creation of new environmentally friendly and sustainable processes (Knez et al. 2014).

Mechanistically, pressure changes the distance between molecules, thus having a major impact on distance-dependent interactions, such as van der Waals, hydrogen bonds, electrostatic and hydrophobic interactions. On the other hand, covalent bonds are hardly affected (Martínez-Monteagudo and Balasubramaniam 2016). An additional governing principle is the principle of Le Chatelier, stating, that for a system in equilibrium, due to the application of pressure, the equilibrium will shift to a direction that tends to decrease the volume, thus any phenomenon that will be accompanied by a volume decrease will be enhanced (Martínez-Monteagudo and Balasubramaniam 2016; Tauscher 1995). Furthermore to its effect on reactions in equilibrium, pressure also affects the reaction rate constant, k. The reaction rate, k, depends on the activation volume $\Delta V^{\neq}$, while the pressure effect on the equilibrium constant K depends on the reaction volume ΔV^0 as described in Eqs. (5.1 and 5.2) (Martinez-Monteagudo and Saldaña 2014; Ramirez et al. 2009).

$$\left(\frac{\partial lnK}{\partial P}\right)_T = -\frac{\Delta V^0}{RT} \tag{5.1}$$

$$\left(\frac{\partial lnk}{\partial P}\right)_T = -\frac{\Delta V^{\neq}}{RT} \tag{5.2}$$

Where R is the gas constant, T is the absolute temperature, and P is the applied pressure. $\Delta V^{\neq}$ is composed of changes in molar volume due to molecular reorganization, alteration of bond angles and length, and those changes due to interactions of the reactants with the solvent (Martinez-Monteagudo and Saldaña 2014).

2 High Hydrostatic Pressure

To obtain microbial inactivation at pasteurization level high hydrostatic pressure (HHP) or high pressure processing (HPP) is well studied and already industrially applied with several hundred industrial units (Bolumar et al. 2021) manufactured in U.S.A, Spain, U.K, Japan, and China (Huang et al. 2017). In this technology, packaged (in most cases) foods are pressurized via water, commonly used as a pressure transmitting medium, at pressure levels of up to 600 MPa and temperatures that usually will not pass 35 °C. Those conditions can, when compared to thermal treatment of a similar inactivation capability, better preserve some of the sensorial and nutritional aspects of processed products (Yu et al. 2020). Over the past decade, the technology has been evolving with manufacturers releasing HP vessels with larger capacity (i.e., 35 up to 525 L) and are gradually automating the processing lines, which enables improvements in productivity, energy consumptions, reduced processing costs and times, which are directly related to the processing vessel volume (Bolumar et al. 2021). The most recent commercial system by Hiperbaric (Burgos, Spain) the "Hiperbaric bulk" having a volume of 1050 L allows a continuous treatment of liquid foods by filling a recyclable plastic 'bladder' that occupies 90% of the inside vessel.

The idea and concept for the beneficial effect of pressure to ensure microbial stability are not new and were introduced in 1899 with the work of Hite on milk (Hite 1899). One of the most important principles of HHP is the isostatic rule: a force transported to the surface of a fluid is equally transmitted through the contact surface. In food processing, it means that when a packed product is submerged in a pressurized fluid, the pressure effects are distributed within the food quasi-instantaneously and uniformly, regardless of the food's geometry and size (Martínez-Monteagudo and Balasubramaniam 2016). Furthermore, the pressurized material, while is reduced in volume due to the application of pressure, retains the original shape. By subjecting the food to pressures between 400–600 MPa for a short holding time (mostly 3–5 min for food preservation), pathogens and spoilage bacteria can be inactivated, while effects on low-molecular-weight compounds (e.g. sugar, vitamins, pigments, flavor compounds) are mostly avoided (Cheftel 1995; Farkas and Hoover 2000). Bacterial spores are resistant in the industrially relevant pressure levels at ambient temperature (Balasubramaniam, Martínez-Monteagudo,, and Gupta 2015), while foodborne viruses presented a pH and virus type dependent inactivation (Govaris and Pexara 2021), therefore requiring significant additional research. At high pressure, various mechanisms were suggested to explain the microbial inactivation: denaturation of membrane proteins, damage to cell membrane due to phase transition and changes in its fluidity, enzyme inactivation, and the disintegration of ribosomes that can lead to cell death. Importantly, in real food products, one should also consider the impact of the presence of solutes in solutions that can infer baroprotection to some extent (Georget et al. 2015). Moderate pressures below 180 MPa can result in sublethal cellular damage and decrease the rate of microbial growth (Bolumar et al. 2021). Furthermore, such pressure levels (<200 MPa) were suggested to expand the application of pressure beyond a processing step to replace thermal pasteurization towards a novel method of storage, replacing refrigeration (termed hyperbaric storage). It is reported that the application of pressure at such moderate (compared to the pressure during HPP) levels has the potential to replace refrigeration by controlling microbial growth at ambient temperatures. The potential advantage of such technology for energy conservation stems from the fact that for hyperbaric storage energy is only required during the compression and decompression phases, and not during the whole storage period as in refrigeration (Fernandes et al. 2019; Santos et al. 2015). In literature, pressure was also suggested to be combined with elevated temperatures to ensure spore inactivation and/or better enzyme inhibition. The combination of pressure with heat, often termed as high-pressure thermal sterilization (HPTS) or as pressure-assisted thermal sterilization (PATS), was suggested to be able to lead to benefits in terms of food safety and food quality when compared to conventional retorting. The suggested benefits stem from shorter dwell times in comparison to the needed for achieving a similar sterilization effect in standard retorting, as well as due to a reduction in post-processing levels of unwanted potentially carcinogenic food processing contaminants (FPCs), such as furan (Al-Ghamdi et al. 2020; Sevenich et al. 2014), acrylamide, 3-MCPD-esters, HMF (Sevenich et al. 2020). One of the benefits of HPTS over regular retorting is the utilization of the adiabatic heat originating from the compression work that results in a temperature increase ranging from 3 to 9 °C per 100 MPa and helps to heat the product to the required temperatures (i.e., from 70 °C to 90 °C achieved by regular heating followed by an increase to 90 °C–130 °C due to the adiabatic heating) (Sevenich et al. 2014). The gained temperature will also drop during decompression (Bolumar et al. 2016). Several studies suggested that further to the US FDA certification for the technology that only considers the thermal contribution to the lethality, pressure and temperature may act synergistically on spores inactivation (Sevenich et al. 2020).

HPP is suitable for a large variety of foods, as long as the liquid content is sufficient to transmit the pressure, and the amount of air voids is low. Noncovalent molecular interactions such as ionic, hydrophobic, and hydrogen bonding are affected by pressure, and previous works suggested that the solvation of charged groups is accompanied by volume reduction, while the formation of coulombic interactions is accompanied by positive changes in volume (Balny et al. 1997). Dissociation of ionic interactions is mostly favored by pressure, as well as the exposure of hydrophobic residues during protein unfolding (Gross and Jaenicke 1994). Equilibrium thermodynamics concerning the structure and stability of biological macromolecules depend mainly on such noncovalent interactions (Gross and Jaenicke 1994), and therefore proteins and other macromolecules whose structure is often stabilized by different inter- and intra-molecular interactions are usually affected by HPP even at the short times required to achieve pasteurization level microbial inactivation.

The pressure effect on proteins is governed by Le Chatelier's principle, shifting the equilibrium towards the state that occupies a smaller volume (Balny et al. 1997). The resulting effects such as unfolding, disassociation, denaturation, and aggregation are often described and reviewed in the literature (Bolumar et al. 2016). For comprehensive reviews on the impact of pressure on proteins, the readers are referred to several reviews on the topic (Balny et al. 1997; Heremans 1982; Knorr et al. 2006). Recent works suggest that hydrostatic pressurization can be a valuable tool to tailor protein properties, especially of importance for plant-based proteins, as it can impact solubility, aggregation, digestability, and other properties (Queirós et al. 2018).

As preservation without application of high temperatures reduces the occurrence of Maillard and caramelization reactions, less or no new aromatic compounds are likely to be formed during such preservation process, conserving the color and flavor close to the original product (Huang et al. 2020). Starch swelling can be induced by HPP, similarly to thermal gelation, and pressure can also cause lipid crystallization (Huang et al. 2020).

In addition to the main beneficial application for food preservation, additional functionalities attributed to high hydrostatic pressure were suggested such as the utilization for protein gelation with reduced thermal load (Alvarez et al. 2008; Katzav et al. 2020), an environmentally friendly method to enhance mass transfer for bioactive compounds recovery (Barba et al. 2015), improvement (compared a preservation similar condition) of the bioaccessibility of small bioactives (Barba et al. 2015; Eran Nagar et al. 2021) although contradicting reports were also published, changes in the allergenicity, and higher saltiness perception by HPP allowing lower salt usage (Barba et al. 2015).

Pressure-induced pH shift is an additional important parameter that is often overlooked with respect to the impact of pressure. Application of pressure results in the displacement of the equilibrium associated with food pH, typically towards more acid values (Mathys et al. 2008; Samaranayake and Sastry 2013). The pH shifts can affect the microbial inactivation by HPP (Mújica-Paz et al. 2011) and also the reaction rates (Shkolnikov et al. 2020). It is important to note that due to the relatively short processing times, combined with the complexity of food matrices, reaching equilibrium is difficult, therefore the rate at which a reaction occurs, that can decrease, increase, or not to be affected by pressure, is often more important than the equilibrium (Martinez-Monteagudo and Saldaña 2014).

2.1 *Supercritical Fluid Extraction*

Compressing a gas or liquid past its critical point will cause it to enter a phase known as the supercritical phase, and the extraction medium in this phase is called a "supercritical fluid" (Shi et al. 2012). With a combination of temperature and pressure, certain chemicals become excellent solvents for particular solutes, allowing improved extraction (Rozzi and Singh 2002), therefore pressure can also assist in liquid-solid extractions. The supercritical solvent extraction process refers to the actions of liquids at or near supercritical conditions for separation or extraction processes. Over the past few years, several applications of this technology have been proposed, covering industries such as food, pharmaceuticals, chemical, and oil processing.

The specific temperature (Tc) and pressure (Pc) for such a critical point are unique for pure substances. Critical pressure is defined as the highest pressure at which a liquid can be converted to gas by increasing its temperature, while the critical temperature is the highest temperature at which a gas can be converted to a liquid by increasing the pressure (Hedrick et al. 1992). At conditions above their critical temperature and pressure, a supercritical fluid (SCF) is a gas-like fluid that takes the shape of its container, fills it, and has a liquid-like density (0.1–1 g/ml) providing the high solvent power, while the gas like viscosity and diffusivity, enhance the mass transfer properties during extraction (Demirbaş 2001; Shi et al.

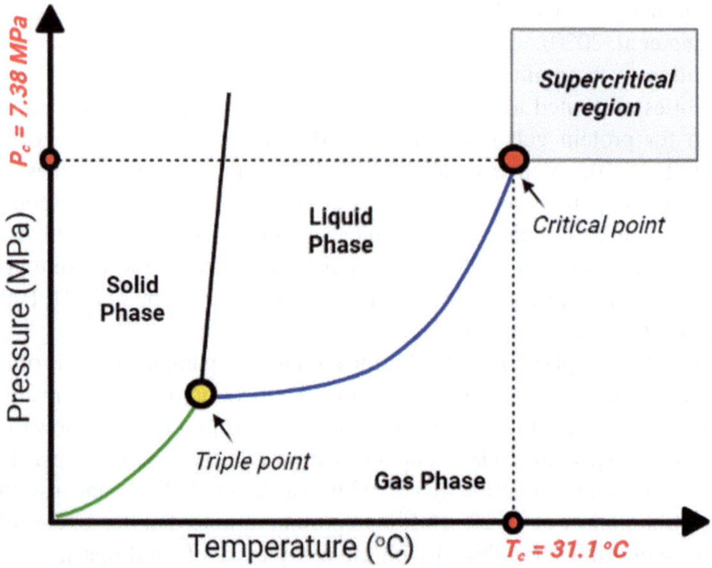

Fig. 5.1 Schematic pressure-temperature phase diagram for carbon dioxide

2012). A schematic supercritical pressure-temperature phase diagram for carbon dioxide a representative of SCFs is presented in Fig. 5.1. Parameters such as density, viscosity, diffusivity, heat capacity, and thermal conductivity are affected by changing pressure and temperature beyond the critical points, altering the extraction process (Shi et al. 2012).

Due to the fact that SCF's have both gas and liquid-like properties, they have the advantage of having solvation characteristics of liquid while the mass transfer properties of gas, allowing high diffusivity. SFCs have no surface tension and can be compressed, attenuating their density and their solubilization properties. The variable fluid density can provide an additional parameter for extraction control. Typically, SCFs are maintained at a temperature close to their critical value and at a pressure high enough for their density to become more than their critical density.

The following compounds were previously suggested as SCF's: CO_2, ethane, propane, hexane, ethylene, nitrous oxide, methanol, ethanol. Since CO_2 is nonflammable, harmless, noncorrosive, inexpensive, nontoxic, highly available, odorless, and tasteless it is probably the most studied SCF. In addition, the critical pressure and temperature (7.38 MPa, 31.1 °C) are relatively mild contributing to its desirability for extraction. Many supercritical- CO_2 applications utilize pressures and temperatures ranging from 20 to 50 MPa, and between 40 °C and 80 °C respectively, to extract natural products or bioactive compounds.

Pressure reduction to atmospheric pressure allows elimination of the CO_2 yielding a solvent-free extract, and as the extracted product is no longer soluble in the gas, it precipitates, removing the need for an additional separation step. In some cases, limited quantities of added polar or nonpolar co-solvents are used to increase the extractability of polar components or components with relatively higher molecular weight (Shi et al. 2012). The addition of a small quantity of co-solutes, usually 3–20%, like ethanol may also result in improved selectivity. Current literature requires a better understanding of the phase behavior and solubility of multi-bioactive component combinations in supercritical- CO_2. Solvent power is related to the density of the supercritical fluid. At constant pressure, usually density decreases with increasing temperature. Higher density was previously correlated with increased solvating power (Shi et al. 2007). Elevation in temperature usually leads to increased solubility of targeted compounds in supercritical CO_2.

It has been established that supercritical fluid- CO_2 is suitable for the extraction of essential oils, functional fatty acids, antioxidants, and other bioactive compounds. High-pressure carbon dioxide (HPCD) was also proposed as an alternative for cold pasteurization techniques for a wide range of food products without the application of high temperatures (Ferrentino and Spilimbergo 2011; Silva et al. 2020). HPCD involves contact of food with sub-or supercritical (pressurized at $\geq$ 0.1 MPa (1 bar)) CO_2 in a continuous, semi-continuous, or batch process, at a relatively low temperature (lower than thermal pasteurization) (Yu et al. 2020). Its milder conditions (in terms of pressure level) give HPCD some advantages over HHP. Several steps in a complex mechanism were suggested to explain how pressurized CO_2 can induce bacterial inactivation including: modification of cell membrane by the solubilized CO_2, intracellular pH decrease, enzyme inactivation due to the pH change, direct inhibitory effect on metabolism, and removal of vital components from cells. For a detailed review on the impact of such technology on microbial inactivation, the readers are referred to a previously published article (Garcia-Gonzalez et al. 2007). In addition, HPCD was shown to inactivate enzymes such as Polyphenol Oxidase (PPO) and Pectin Methylesterase (PME) in pressure, temperature, and CO_2/enzyme concentration-dependent ratios. Different mechanisms were suggested to explain the enzymatic inactivation, including pH reduction, conformational changes of the secondary and tertiary structure, and direct inhibitory effects of CO_2 on enzyme activity, yet the underlying mechanism is not clear (Benito-Román et al. 2020). N_2 was also suggested for a similar application, yet it showed no or little effect for microbial reduction. An increase in CO_2 pressure generally accelerates microbial inactivation, allowing to decrease exposure time, yet a maximal pressure where no further improvement occurs. Also, an increase in temperature enhances microbial inactivation (when all other conditions are equal). The supercritical CO_2 conditions were found to be more effective in inactivating microbial cells than subcritical conditions (liquid state under pressure and temperature levels above the boiling point at atmospheric pressure), likely stemming from the physico-chemical properties of supercritical CO_2 like the liquid-like density in combination with gas like mass transport properties as described above. In addition cell rupture due to rapid release of pressure, and fast cooling due to Joule-Thomson effect occurring when pressurized CO_2 expands during the decompression may also play a role in the inactivation.

2.2 Subcritical Extraction

While the utilization of water for extraction is preferred in terms of practically all aspects, as it a ubiquitous, non-toxic and with minimal costs, the high polarity of water at ambient conditions limits its capacity to solubilize and extract compounds with moderate and low polarity. In order to extract moderately polar, and low polarity compounds, water can be used at elevated pressures that keep it in a liquid state at high temperatures (Shi et al. 2012). The pressurized low polarity extraction, mostly relevant for water, is known as subcritical water extraction (SWE) and is based on applying sufficiently high pressures to keep water, at temperatures between 100 °C and 374 °C, in a liquid state. This allows the utilization of water as a safe and efficient extraction solvent for compounds from different sources such as plant and algae matrices (Zakaria and Kamal 2016). The pressure levels between 1 and 22.1 MPa are below the critical pressure (Zhang et al. 2020). Such conditions increase extraction efficiency by improving mass transfer and altering the polarity of water (Shi et al. 2012). Water at room temperature and atmospheric pressure is a highly polar solvent with a high dielectric constant (ε) due to hydrogen bonding. Increasing the temperature diminishes those electrostatic interactions between the water molecules and also between water molecules and the surrounding ions and solute molecules. Thus, intermolecular interactions involving hydrogen bonding become less pronounced, making water a poorer solvent for salts, favoring London dispersion forces (Plaza and Turner 2015). Specifically, two aspects were suggested to explain the extraction efficiency of subcritical water, improvement in solubility and mass transfer effects, and increased damage to surface balance (Ong et al. 2006). As the polarity decreases when temperature increases, keeping the water at a liquid state can allow the extraction of lower polarity compounds. The dielectric constant (or relative permittivity) is easily manipulated to a range of values from ~80 to ~20 by temperature under those moderate pressures allowing to dissolve low and medium polarity compounds (Zhang et al. 2020). As an example, subcritical water at 250 °C and a $P > 40$ bar has a dielectric constant of 37 compared to 80 at regular ambient conditions. Such a dielectric constant is similar to the one of ethanol, which is known to be used for the extraction of low polarity compounds. Such decrease in the dielectric constant is only possible when the temperature increase is sufficient to induce a substantial hydrogen bond break down, while the liquid state needed for the extraction is maintained by pressurization. Hydrogen bonds in water are self associating, with the strength of one hydrogen bond being affected by additional hydrogen bonds around it. Therefore a small change to one hydrogen bond affects the entire water volume, affecting the dielectric constant. The increased thermal agitation with increasing temperature decreases the strength of each hydrogen bond leading to the amplification in the reduction in the dielectric constant (Carr et al. 2011).

In addition, in such conditions, the viscosity and surface tension will decrease steadily (Zhang et al. 2020). Unlike the crucial effect of temperature in modifying the dielectric constant and therefore the capacity to efficiently extract compounds with low polarity, the pressure level has no major effect on the recovery by SWE as long as the pressure is sufficiently high to keep the liquid state of water. Other factors that must be controlled during process optimization is particle size (usually smaller particles improve yield), solid to liquid ratio and flow rate, extraction time (considering also the impact of the high temperature on degradation), the addition of modifiers, type of extractor (continuous or batch) (Zhang et al. 2020). The effect on the solubility of non-polar compounds is not always directly proportional to the dielectric constant values, and/or complex molecular interactions can take place when a co-solute is added to the water system (Carr et al. 2011), therefore specific verification for extraction depending on the matrix and target molecule is required. It is also important to take into account that additional processes will likely be needed to remove moisture after extraction. A possible downside of the technology, stemming from the high temperatures, could be enhanced rates of unwanted reactions such as oxidation and hydrolysis, therefore in such a process, still, there is an advantage for minimization of processing time and temperature for thermally liable compounds. In addition, it should be noted that with the increasing solubility of less polar compounds with increasing temeperature, the co-extraction of polar compounds may decrease. An additional possible advantage for the extraction process is that the energy provided by the subcritical water can disturb interactions between the target solute and the matrix or the target solute and other solutes, therefore, reducing the activation energy required for the desorption process (Zhang et al. 2020). Successful extraction by SWE was reported for a range of polysaccharides aiming at immunological, antioxidant, and anti-tumor activity. The efficacy of SWE extraction for flavonoids, as well as the optimal processing conditions, were found to be dependent on flavonoid structure (Ko et al. 2014). The structure-dependent solubility improvement by SWE requires further work, despite some generalities on the effect of structure on solubility were already reported (Carr et al. 2011).

3 Future Considerations

The agri-food industry is one of the most principal industries directly and strongly related to consumer health and well-being. Pressure is an important thermodynamic parameter that can help in extending the range and diversity of processed foods and extracted ingredients while minimizing negative thermal effects and the need for solvents, negatively affecting the environment. The utilization of pressure-based technologies, due to the advantages in terms of the preservation of original compounds and reducing the need for solvents can play a major role in the growing trends of sustainable and health-promoting food production and agri-produce processing. Pressure can assist in both preserving higher content of beneficial components and in reducing the need for compounds like preservatives and additives often considered having a negative health effect, thus promoting a cleaner product label. When considering the application of pressure in processing one should clearly consider the costs, often stemming from the investment in equipment and personnel, but also should take into account less-discussed impacts such as the effect of pressure on reaction kinetics.

References

Al-Ghamdi S, Sonar CR, Patel J, Albahr Z, Sablani SS (2020) High pressure-assisted thermal sterilization of low-acid fruit and vegetable purees: microbial safety, nutrient, quality, and packaging evaluation. Food Control 114:107233. https://doi.org/10.1016/j.foodcont.2020.107233

Alvarez PA, Ramaswamy HS, Ismail AA (2008) High pressure gelation of soy proteins: effect of concentration, pH and additives. J Food Eng 88(3):331–340. https://doi.org/10.1016/J.JFOODENG.2008.02.018

Balasubramaniam (Bala) VM, Martínez-Monteagudo SI, Gupta R (2015) Principles and application of high pressure–based technologies in the food industry. Annu Rev Food Sci Technol 6(1):435–462. https://doi.org/10.1146/annurev-food-022814-015539

Balny C, Mozhaev VV, Lange R (1997) Hydrostatic pressure and proteins: basic concepts and new data. Comp Biochem Physiol A Physiol 116(4):299–304. https://doi.org/10.1016/S0300-9629(96)00355-6

Barba FJ, Terefe NS, Buckow R, Knorr D, Orlien V (2015) New opportunities and perspectives of high pressure treatment to improve health and safety attributes of foods. A review. Food Res Int 77:725–742. https://doi.org/10.1016/j.foodres.2015.05.015

Benito-Román Ó, Sanz MT, Illera AE, Melgosa R, Beltrán S (2020) Polyphenol oxidase (PPO) and pectin methylesterase (PME) inactivation by high pressure carbon dioxide (HPCD) and its applicability to liquid and solid natural products. Catal Today 346:112–120. https://doi.org/10.1016/j.cattod.2018.12.051

Bolumar, T., Middendorf, D., Toepfl, S., & Heinz, V. (2016). Structural changes in foods caused by high-pressure processing. In V. M. Balasubramaniam, G. V Barbosa-Cánovas, & H. L. M. Lelieveld (Eds.), High pressure processing of food: principles, technology and applications (pp. 509–537). New York: Springer New York. https://doi.org/10.1007/978-1-4939-3234-4_23

Bolumar T, Orlien V, Sikes A, Aganovic K, Bak KH, Guyon C et al (2021) High-pressure processing of meat: molecular impacts and industrial applications. Compr Rev Food Sci Food Saf 20(1):332–368. https://doi.org/10.1111/1541-4337.12670

Carr AG, Mammucari R, Foster NR (2011) A review of subcritical water as a solvent and its utilisation for the processing of hydrophobic organic compounds. Chem Eng J 172(1):1–17. https://doi.org/10.1016/j.cej.2011.06.007

Cheftel JC (1995) Review: high-pressure, microbial inactivation and food preservation/revision: alta-presion, inactivacion microbiologica y conservacion de alimentos. Food Sci Technol Int 1(2–3):75–90. https://doi.org/10.1177/108201329500100203

Demirbaş A (2001) Supercritical fluid extraction and chemicals from biomass with supercritical fluids. Energy Convers Manag 42(3):279–294. https://doi.org/10.1016/S0196-8904(00)00059-5

Eran Nagar E, Berenshtein L, Katz IH, Lesmes U, Okun Z, Shpigelman A (2021) The impact of chemical structure on polyphenol bioaccessibility, as a function of processing, cell wall material and pH: a model system. J Food Eng 289:110304. https://doi.org/10.1016/j.jfoodeng.2020.110304

Farkas DF, Hoover DG (2000) High pressure processing. J Food Sci 65:47–64. https://doi.org/10.1111/j.1750-3841.2000.tb00618.x

Fernandes PAR, Moreira SA, Santos MD, Duarte RV, Santos DI, Inácio RS et al (2019) Hyperbaric storage at variable room temperature – a new preservation methodology for minced meat compared to refrigeration. J Sci Food Agric 99(7):3276–3282. https://doi.org/10.1002/jsfa.9540

Ferrentino G, Spilimbergo S (2011) High pressure carbon dioxide pasteurization of solid foods: current knowledge and future outlooks. Trends Food Sci Technol 22(8):427–441. https://doi.org/10.1016/j.tifs.2011.04.009

Garcia-Gonzalez L, Geeraerd AH, Spilimbergo S, Elst K, Van Ginneken L, Debevere J et al (2007) High pressure carbon dioxide inactivation of microorganisms in foods: the past, the present and the future. Int J Food Microbiol 117(1):1–28. https://doi.org/10.1016/j.ijfoodmicro.2007.02.018

Georget E, Sevenich R, Reineke K, Mathys A, Heinz V, Callanan M et al (2015) Inactivation of microorganisms by high isostatic pressure processing in complex matrices: a review. Innovative Food Sci Emerg Technol 27:1–14. https://doi.org/10.1016/j.ifset.2014.10.015

Govaris A, Pexara A (2021) Inactivation of foodborne viruses by high-pressure processing (HPP). Foods. https://doi.org/10.3390/foods10020215

Gross M, Jaenicke R (1994) Proteins under pressure. Eur J Biochem 221(2):617–630. https://doi.org/10.1111/j.1432-1033.1994.tb18774.x

Hedrick JL, Mulcahey LJ, Taylor LT (1992) Supercritical fluid extraction. Microchim Acta 108(3):115–132. https://doi.org/10.1007/BF01242421

Heremans K (1982) High pressure effects on proteins and other biomolecules. Annu Rev Biophys Bioeng 11(1):1–21. https://doi.org/10.1146/annurev.bb.11.060182.000245

Hite BH (1899) The effect of pressure in the preservation of milk: a preliminary report, vol 58. West Virginia Agricultural Experiment Station

Huang H-W, Wu S-J, Lu J-K, Shyu Y-T, Wang C-Y (2017) Current status and future trends of high-pressure processing in food industry. Food Control 72:1–8. https://doi.org/10.1016/j.foodcont.2016.07.019

Huang H-W, Hsu C-P, Wang C-Y (2020) Healthy expectations of high hydrostatic pressure treatment in food processing industry. J Food Drug Anal 28(1):1–13. https://doi.org/10.1016/j.jfda.2019.10.002

Katzav H, Chirug L, Okun Z, Davidovich-Pinhas M, Shpigelman A (2020) Comparison of thermal and high-pressure gelation of potato protein isolates. Foods 9(8):1041

Knez Ž, Markočič E, Leitgeb M, Primožič M, Knez Hrnčič M, Škerget M (2014) Industrial applications of supercritical fluids: a review. Energy 77:235–243. https://doi.org/10.1016/j.energy.2014.07.044

Knorr D, Heinz V, Buckow R (2006) High pressure application for food biopolymers. Biochim Biophys Acta 1764(3):619–631. https://doi.org/10.1016/j.bbapap.2006.01.017

Ko M-J, Cheigh C-I, Chung M-S (2014) Relationship analysis between flavonoids structure and subcritical water extraction (SWE). Food Chem 143:147–155. https://doi.org/10.1016/j.foodchem.2013.07.104

Martínez-Monteagudo SI, Balasubramaniam VM (2016) Fundamentals and applications of high-pressure processing technology. In: Balasubramaniam VM, Barbosa-Cánovas GV, Lelieveld HLM (eds) High pressure processing of food: principles, technology and applications. Springer New York, New York, pp 3–17. https://doi.org/10.1007/978-1-4939-3234-4_1

Martinez-Monteagudo SI, Saldaña MDA (2014) Chemical reactions in food systems at high hydrostatic pressure. Food Eng Rev 6(4):105–127. https://doi.org/10.1007/s12393-014-9087-6

Mathys A, Kallmeyer R, Heinz V, Knorr D (2008) Impact of dissociation equilibrium shift on bacterial spore inactivation by heat and pressure. Food Control 19(12):1165–1173. https://doi.org/10.1016/j.foodcont.2008.01.003

Mújica-Paz H, Valdez-Fragoso A, Samson CT, Welti-Chanes J, Torres JA (2011) High-pressure processing technologies for the pasteurization and sterilization of foods. Food Bioprocess Technol 4(6):969–985. https://doi.org/10.1007/s11947-011-0543-5

Ong ES, Cheong JSH, Goh D (2006) Pressurized hot water extraction of bioactive or marker compounds in botanicals and medicinal plant materials. J Chromatogr A 1112(1):92–102. https://doi.org/10.1016/j.chroma.2005.12.052

Plaza M, Turner C (2015) Pressurized hot water extraction of bioactives. TrAC Trends Anal Chem 71:39–54. https://doi.org/10.1016/j.trac.2015.02.022

Queirós RP, Saraiva JA, da Silva JAL (2018) Tailoring structure and technological properties of plant proteins using high hydrostatic pressure. Crit Rev Food Sci Nutr 58(9):1538–1556. https://doi.org/10.1080/10408398.2016.1271770
Ramirez R, Saraiva J, Pérez Lamela C, Torres JA (2009) Reaction kinetics analysis of chemical changes in pressure-assisted thermal processing. Food Eng Rev 1(1):16–30. https://doi.org/10.1007/s12393-009-9002-8
Rozzi NL, Singh RK (2002) Supercritical fluids and the food industry. Compr Rev Food Sci Food Saf 1(1):33–44. https://doi.org/10.1111/j.1541-4337.2002.tb00005.x
Samaranayake CP, Sastry SK (2013) In-situ pH measurement of selected liquid foods under high pressure. Innovative Food Sci Emerg Technol 17:22–26. https://doi.org/10.1016/j.ifset.2012.09.006
Santos MD, Queirós RP, Fidalgo LG, Inácio RS, Lopes RP, Mota MJ et al (2015) Preservation of a highly perishable food, watermelon juice, at and above room temperature under mild pressure (hyperbaric storage) as an alternative to refrigeration. LWT Food Sci Technol 62(1):901–905. https://doi.org/10.1016/j.lwt.2014.06.055
Sevenich R, Kleinstueck E, Crews C, Anderson W, Pye C, Riddellova K et al (2014) High-pressure thermal sterilization: food safety and food quality of baby food puree. J Food Sci 79(2):M230–M237. https://doi.org/10.1111/1750-3841.12345
Sevenich R, Rauh C, Belkova B, Hajslova J (2020) Chapter 6 – effect of high-pressure thermal sterilization (HPTS) on the reduction of food processing contaminants (e.g., furan, acrylamide, 3-MCPD-esters, HMF). In: Barba FJ, Tonello-Samson C, Puértolas E, Lavilla M (eds) Present and future of high pressure processing. Elsevier, pp 139–172. https://doi.org/10.1016/B978-0-12-816405-1.00006-6
Shi J, Mittal G, Kim E, Xue SJ (2007) Solubility of carotenoids in supercritical CO_2. Food Rev Intl 23(4):341–371. https://doi.org/10.1080/87559120701593806
Shi J, Xue SJ, Ma Y, Jiang Y, Ye X, Yu D (2012) Green separation technologies in food processing: supercritical-CO_2 fluid and subcritical water extraction. In: Boye JI, Arcand Y (eds) Green technologies in food production and processing. Springer US, Boston, pp 273–294. https://doi.org/10.1007/978-1-4614-1587-9_11
Shkolnikov H, Belochvostov V, Okun Z, Shpigelman A (2020) The effect of pressure on the kinetics of polyphenolics degradation – implications to hyperbaric storage using epigallocatechin-gallate as a model. Innovative Food Sci Emerg Technol 59:102273. https://doi.org/10.1016/J.IFSET.2019.102273
Silva EK, Meireles MAA, Saldaña MDA (2020) Supercritical carbon dioxide technology: a promising technique for the non-thermal processing of freshly fruit and vegetable juices. Trends Food Sci Technol 97:381–390. https://doi.org/10.1016/j.tifs.2020.01.025
Tauscher B (1995) Pasteurization of food by hydrostatic high pressure: chemical aspects. Z Lebensm Unters Forsch 200(1):3–13. https://doi.org/10.1007/BF01192901
Yu T, Niu L, Iwahashi H (2020) High-pressure carbon dioxide used for pasteurization in food industry. Food Eng Rev 12(3):364–380. https://doi.org/10.1007/s12393-020-09240-1
Zakaria SM, Kamal SMM (2016) Subcritical water extraction of bioactive compounds from plants and algae: applications in pharmaceutical and food ingredients. Food Eng Rev 8(1):23–34. https://doi.org/10.1007/s12393-015-9119-x
Zhang J, Wen C, Zhang H, Duan Y, Ma H (2020) Recent advances in the extraction of bioactive compounds with subcritical water: a review. Trends Food Sci Technol 95:183–195. https://doi.org/10.1016/j.tifs.2019.11.018

Chapter 6
Mechanical Technologies: Ultrasound and Cavitation in Food Processing

Kaouther Kerboua, Djihane Mazouz, Imen Hasaounia, and Oualid Hamdaoui

1 Introduction

Food processing can be simply defined as the effective conversion of agricultural products into more appealing, shelf-stable, compact and value-added consumer ready to use products (Feng et al. 2011).

Conserving the natural flavor and taste of foods and reducing or eliminating the use of chemical additives and preservatives increased the interest in the use of non-thermal technologies in food processing. Ultrasound assisted techniques have emerged as one of the promising non thermal solutions in agri-food sector involving less time, less water and less energy than conventional techniques. These techniques in the field of food industry are well known for significantly influencing the processing rate of different used processes (Bhargava et al. 2021; Dastkhoon et al. 2018). Sono-food processing has been a subject of research and development for many years (Gogate and Pandit 2011; Demirdöven and Baysal 2009; Mason et al. 1996), and is still in the surge of technologists' attention owing to the variety of possible applications ranging from cutting, dehydration, extraction, freezing, crystallization, filtration to preservation and more.

The ultrasound assisted techniques in food processing rely principally on the mechanical effects induced by the propagation of ultrasonic waves in fluids, but also all the physical and chemical effects induced by acoustic cavitation bubble. In the present chapter, the fundamentals of acoustic cavitation and its physical and chemical effects will be presented, with the objective of linking the microscopic events

K. Kerboua · D. Mazouz · I. Hasaounia
Higher School of Industrial Technologies, Department of Engineering, Annaba, Algeria

O. Hamdaoui (✉)
Chemical Engineering Department, College of Engineering, King Saud University, Riyadh, Saudi Arabia

A. Režek Jambrak (ed.), *Nonthermal Processing in Agri-Food-Bio Sciences*, Food Engineering Series, https://doi.org/10.1007/978-3-030-92415-7_6

occurring when irradiating a fluid with ultrasound, to the macroscopic applications (Bratsikhin et al. 2019) of ultrasound and acoustic cavitation in food industry.

The chapter will also cover selected applications of ultrasound in food-processing, with reference to experimental and theoretical works retrieved in the literature and dealing with those application at laboratory and industrial scales.

Finally, some perspectives are given to sono-processing and ultrasound assisted technologies in agri-food sector, while highlighting the main challenges facing this field.

2 Fundamentals of Acoustic Cavitation Bubble

2.1 Bubble Inception, Growth and Dynamics of Oscillation

An acoustic wave is a propagation of pressure oscillation with sound velocity in a medium such as liquid, gas, or solid. The acoustic pressure amplitude P_a is defined as the amplitude for pressure oscillation. The wavelength λ of an acoustic wave is defined as the length for one pressure oscillation. The acoustic period T is defined as time for one pressure oscillation. The frequency f of an acoustic wave is defined as the number of pressure oscillations per unit time and is then the reverse of the period. The sound velocity (or sound speed) c is defined as the distance for a pressure disturbance propagating per unit time. The sound velocity in liquid water at room temperature is about 1500 m/s, i.e., almost 4.5-fold higher than in dry air, where it is estimated at 340 m/s. The sound velocity in liquid water increases with the increase of the emperature and has a maximum value of about 1555 m/s at around 74 °C.

Owing to the high oscillating frequency of liquid molecules in an ultrasonic field and the high viscosity of the liquid medium compared with a gas, liquid molecules absorb a large amount of energy from the propagation of ultrasonic wave. The high energy absorption of liquid molecules and high acoustic pressure in the presence of gaseous germs make it possible for ultrasonic field to overcome intermolecular interactions in a solvent, and produce numerous cavities, which is referred to as acoustic cavitation.

In practice, weaknesses in liquid medium can typically occur in two forms. (i) Homogeneous nucleation, induced by the thermal motions within the liquid, which form temporary microscopic voids that constitute nuclei. (ii) Heterogeneous nucleation that explains the major weaknesses occurring in liquid medium, and manifesting at the boundary between the liquid and a solid phase or between the liquid and small particles suspended in the liquid (Jena 1965).

2.1.1 Homogeneous Nucleation Theory

We present here a brief and simplified version of homogeneous nucleation theory. Acoustic cavitation, concisely defined as the formation, growth and collapse of microbubbles within an aqueous solution, results from pressure fluctuations that occur in the liquid medium irradiated by an acoustic field in the range of ultrasound, with appropriate value of frequency and amplitude. The formed cavities drain and accumulate ultrasonic energy, and explosively release their energy by the collapse of cavities (Luo et al. 2015). The event of a collapsing bubble is a microscopic implosion accompanied of a significant increase of temperature and pressure of up to thousand Kelvin and hundred Bar (Yasui et al. 2007; Kerboua and Hamdaoui 2017), a release of heat energy and high local turbulence. The elevated temperatures in the vicinity of collapsing bubble, known as "hot spots", can activate chemical mechanisms particularly characterized by the formation of free radicals (Leong et al. 2011; Kerboua and Hamdaoui 2019a), while the local turbulence and the abrupt release of concentrated localized energy induces several physical effects of mechanical nature such as mass transfer and surface erosion (Fu et al. 2020).

The ultrasonic wave propagates through ultrasonic generator, transducer, horn and radiation rod to liquid, causing medium molecules vibration. The average distance between molecules decreases in the acoustic compression phase, while increases in the rarefaction phase. In pure liquid, cavitation will occur when the liquid additional negative pressure reaches a critical value, whose magnitude is called cavitation threshold (Ye and Zhu 2017). In fact, cavities are formed when adequately large negative pressure is applied to the liquid so that the average distance between the molecules exceeds the critical molecular distance required to hold the liquid intact. However, in order to attain this critical distance, a huge energy is needed, to illustrate, the creation of a microscopic cavity-void with a radius of barely 0.4 nm in water, the minimum required negative pressure is estimated at 1400 atm.

However, what makes it easier in the case of acoustic cavitation is the presence of gaseous impurities, which dramatically reduces the tensile strength and consequently the required negative pressure. This has been proved by experiments on tensile strength (Brennen 2013).

2.1.2 Heterogeneous Nucleation Theory

Acoustic cavitation is observed with a pressure amplitude of about 1 atm, implying that pre-existing nuclei or sites for nucleation assist the growth of acoustic cavitation bubbles. Acoustic cavitation event is then explained by a sound wave imposing a sinusoidally varying pressure upon existing cavities in solution, namely gaseous impurities. In the rarefaction phase of the ultrasonic wave, instantaneous local pressures in liquid become negative when the acoustic pressure amplitude is larger than the ambient pressure. Negative pressure is the result of the "force" to expand a liquid element and is possible only in liquids or solids, and impossible in gases.

Consequently, gases dissolved in the liquid appear as gas bubbles because gases can no longer be dissolved in the liquid under negative pressures (Yasui 2018). During the negative pressure cycle, the liquid is pulled apart at sites containing such a gaseous impurity, which are known as “weak spots” in the fluid. The number of bubbles that are produced during this rarefaction cycle is proportional to the density of such weak spots present in the fluid (Leong et al. 2011).

2.1.3 Cavitation Types and Expected Effects

During its lifetime, acoustic cavitation bubble can be either vaporous or gaseous. Vapor of liquid is introduced to the bulk of the bubble through the non-equilibrium of evaporation and condensation process (Fujikawa and Akamatsu 1980; Yasui 1998; Kerboua and Hamdaoui 2018a), while gases are introduced through a diffusion process, much slower than the evaporation process. Physical effects are more pronounced when acoustic cavitation bubbles are of vaporous nature (Yasui 2018). The reason is that acoustic cavitation bubbles under low frequency ultrasound tend to grow to larger size than under mid to high frequencies (Carrillo-Lopez et al. 2017). During this growth process, vapors of the liquid accumulate in the volume of the bubble and give it a vaporous nature. These bubbles, of bigger size, are characterized by a harsh collapse accompanied of strong cavitation microstreaming, shock waves and microjets. This observation is summarized in Table 6.1.

2.1.4 Growth and Oscillation of Acoustic Cavitation Bubble

There are two mechanisms in growth of a bubble in acoustic cavitation. The first mechanism is coalescence of bubbles. The coalescence of bubbles is driven by the two processes. One is the attractive radiation force between bubbles called secondary Bjerknes force. The other is the radiation force called the primary Bjerknes force which drives active bubbles to the pressure antinode of a standing wave field (Sunartio et al. 2007).

The second mechanism for bubble growth is known as rectified diffusion and is defined as the gas diffusion into a bubble due to the area and shell effects. The bubble growth rate due to rectified diffusion strongly depends on acoustic

Table 6.1 Broad classification of acoustic frequencies and their major effects

Acoustic conditions	Frequency range	Major effect	Nature of the bubble
Low frequency, high power	20–100 kHz	Physical effects	Vaporous bubble
Intermediate frequency, medium power	100 kHz–1 MHz	Sonochemistry	Gaseous bubble
High frequency, low power	1–10 MHz	Diagnostic applications	–

conditions. This rate decreases as ultrasonic frequency increases for the same acoustic pressure amplitude (Crum 1980; Lee et al. 2005).

At the initial development of acoustic cavitation, rectified diffusion may be the main mechanism of bubble growth, whilst when acoustic cavitation is fully started, coalescence of bubbles may be the main mechanism of the bubble growth as the rate of coalescence is proportional to the square of the number density of bubbles which should be small at the initial stage of acoustic cavitation (Pankaj and Ashokkumar 2011).

Within the bubble population, if the bubbles are small and occur below the cavitation threshold, they will simply dissolve away, they are known as dissolving bubbles (Yasui 2002; Kerboua et al. 2021a).

Bubbles formed through cavitation will begin to expand and collapse under the influence of the acoustic field, if the expansion-collapse cycle is sinusoidal, mimicking that of the acoustic wave, the bubbles are qualified as linearly oscillating bubbles (Louisnard and González-García 2011).

Alternatively, for certain bubble sizes and acoustic pressures, the bubble expansion phase is extended and is followed by a violent collapse back to a smaller bubble size. Bubbles occurring such oscillation dynamics are known as inertial cavitation (Johansen et al. 2017).

If inertial acoustic cavitation bubbles persist for many hundreds of acoustic cycles, they are referred to as stable, or repetitive transient cavitation (Kerboua and Hamdaoui 2018b). Furthermore, if stable or repetitive transient cavitation are active in sonochemistry, they are qualified as high energy stable cavitation (Yasui 2011; Kerboua et al. 2021b). An example of the dynamics of oscillation of a repetitive transient acoustic cavitation bubble is given in Fig. 6.1.

Alternatively, if the acoustic amplitude is higher, the inertial acoustic cavitation bubble will grow and collapse spectacularly within a very few acoustic cycles and the collapsed bubble then disintegrates into a mass of smaller daughter bubbles, which are often small and themselves collapse in chain so that complete annihilation of the original bubble occurs. These bubbles are known as unstable or transient cavitation (Yasui 2011; Kerboua and Hamdaoui 2020). An example of the dynamics of a transient bubble oscillation is illustrated in Fig. 6.2.

Finally, it is worthy to mention that higher yield of gases diffuses into the bubble during the expansion phase than leaks out during collapse. This causes larger bubbles to grow by rectified diffusion over a very large number of acoustic cycles. Large bubbles also tend to form coalesced bubbles. By both mechanisms, the formed big bubbles will attain a size where they will simply float away to the liquid surface. These bubbles are known as degassing bubbles (Leighton 1995).

2.2 *Physical Effects of Acoustic Cavitation Bubble*

This overview focuses on the physical effects induced by the passage of ultrasound through a liquid medium and the formed transient acoustic cavitation bubble. The physical effects may vary from the mechanical turbulence created due to the

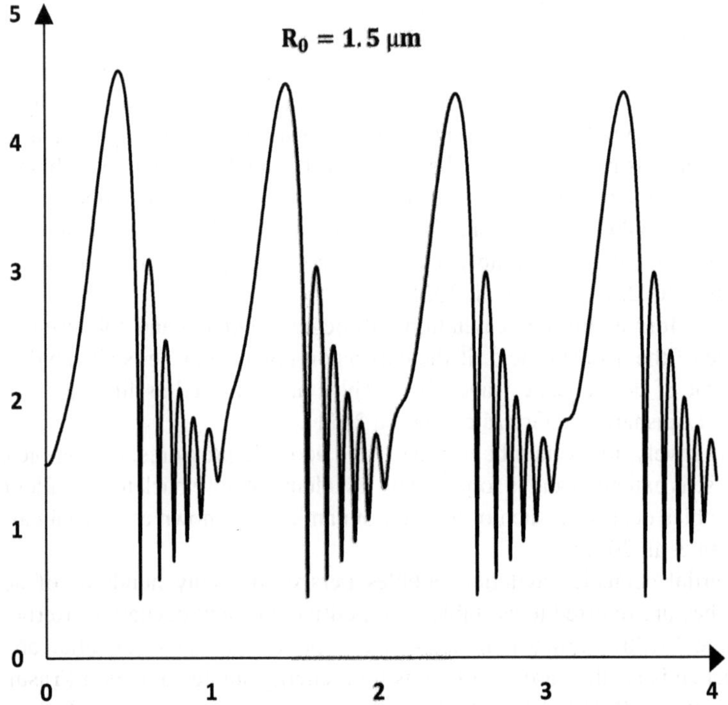

Fig. 6.1 Simulated stable or repetitive transient bubble radius over four acoustic cycles for acoustic bubbles of 1.5 μm of ambient radius under an exciting wave of 300 kHz and 1.5 atm. The bubbles are supposed to oscillate in water under oxygen atmosphere

passage of ultrasound through water, namely vibration and acoustic streaming, to the physical consequences of acoustic cavitation, namely streamers, cavitation microstreaming, microjets and shock waves (Yusof et al. 2016). All these effects are reported in Fig. 6.3.

2.2.1 Acoustic Streaming

When a liquid medium is irradiated with ultrasound, the immediate consequence is acoustic streaming. This large-scale streaming refers to the regular oscillatory motion of the fluid volume elements with a time independent flow velocity due to the second-order nonlinear propagation of the acoustic wave occurring when the acoustic wave propagates through the liquid medium. Acoustic streaming can be viewed as physical forces of the sound waves that provide a driving force capable of displacing ions and small molecules (Marshall and Wu 2015). Acoustic streaming can have extremely complex pattern, with linear and nonlinear interactions between the cavitation bubbles and with the microstreaming associated with the microbubbles occurring at the small scale. Wu (2018) exposed in a recent review of acoustic

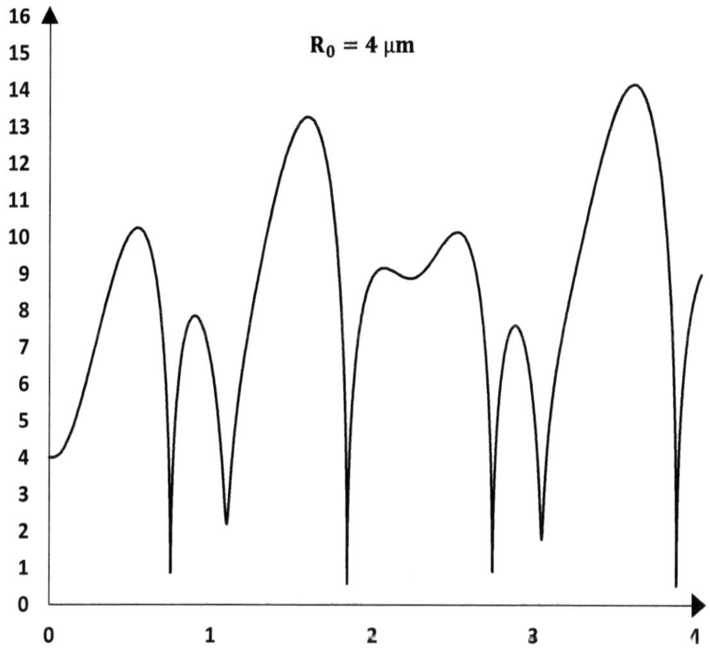

Fig. 6.2 Simulated unstable or transient bubble radius over four acoustic cycles for acoustic bubbles of 4 μm of ambient radius under an exciting wave of 300 kHz and 1.5 atm. The bubbles are supposed to oscillate in water under oxygen atmosphere

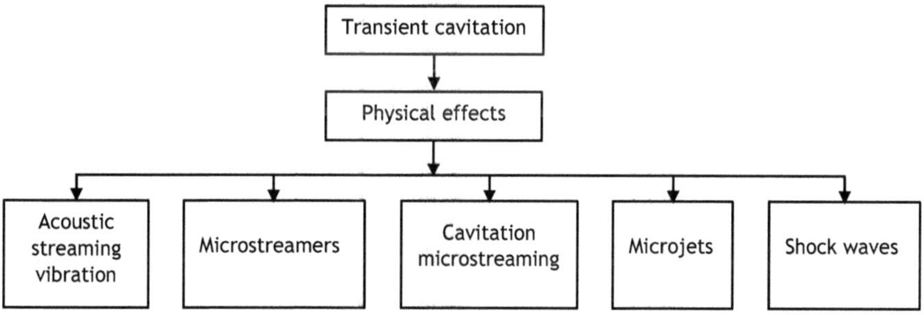

Fig. 6.3 Physical effects accompanying the oscillation and collapse of transient acoustic cavitation bubble

streaming and its applications the detailed theoretical background behind the acoustic streaming.

There are mainly two types of acoustic streaming, namely, Eckart streaming and Rayleigh streaming (Luo et al. 2018). One is accelerating direct current fluid flow in the direction of the acoustic traveling wave propagation. This is caused by the attenuation of the acoustic traveling wave mainly resulting of the viscosity of the fluid.

Due to attenuation, an unbalance of radiation forces is created since the radiation pressure pushing a fluid parcel becomes stronger than that pulling it. This causes the accelerating fluid flow called Eckart streaming (Bjørnø 2017). The second type of acoustic streaming is a vortex-like streaming caused by viscous stress at the boundary layer near a wall or an object, regardless of the situation of a traveling or standing wave. This type is known as Rayleigh streaming (Yasui 2017).

Large scale acoustic streaming plays a significant role in enhancing heat and mass transfer processes in the liquid but also around solid surfaces. Indeed, the acoustic streaming patterns around solid objects are believed to be the strongest within a few millimeters of the solid surface. This boundary layer effect around a solid surface is important and quite interesting in several applications, because it is usually such boundary layers that offer the greatest resistance to both heat and mass transfer. The surface effects of large-scale acoustic streaming are also important in emulsification, where interfacial turbulence is associated with droplet formation and nebulization. The acoustic streaming effects cause a fountain structure to form at the air/water interface from which microdroplets are ejected (Luo 2020).

2.2.2 Microstreamers

Acoustic standing waves can result from the reflection of sound from a solid surface or an air–liquid interface back into the solution at the same time that a wave is generated at the transducer. At the pressure antinode of such a standing wave pattern, the pressure fluctuates from a maximum to a minimum amplitude with time. Conversely, at the pressure node, the acoustic pressure is invariant and close to zero.

A phenomenon referred to as Bjerknes forces causes smaller bubbles to accumulate at the antinode, while larger bubbles accumulate at the node. In moving to the antinodes, the cavitation bubbles travel in ribbon-like structures referred to as "streamers" coalescing as they collide (Bai et al. 2014). A filamentary structure, referred to as an acoustic "Lichtenberg" figure is then created (Parlitz et al. 1999).

To illustrate, at 20 kHz, the bubbles forming the streamers are typically less than 10 μm in size, about a millimeter apart and traveling at less than 1 μm/s. This bubble translation is known to dislodge particles from fouled surfaces, in cases where the surface itself acts as the pressure antinode. Movement of bubbles toward a solid surface acting as a pressure antinode within an acoustic standing wave pattern results in increased turbulence within this same zone (Kentish and Ashokkumar 2010).

2.2.3 Cavitation Microstreaming

Acoustic cavitation microstreaming refers to the microscopic streaming most often created by a microbubble. Acoustic microstreaming do not rely at all on the presence of microbubbles, it can also occur around a solid particle. However, it is much more significant around a bubble because the speed of microstreaming is proportional to the square of vibration speed of an object. The speed of microstreaming around a pulsating acoustic cavitation bubble is 10^2 to 10^6 times larger than that around a solid particle. Thus, the term "microstreaming" is usually used for liquid streaming around a pulsating bubble and is preferentially called "cavitation microstreaming". This is consistent with the early observations of Kolb and Nyborg (1956), who allowed cavitation bubbles to form under the influence of various frequencies in the audible range up to 11.4 kHz. They observed that streaming is most pronounced when the bubble is oscillating in its volumetric mode or when the bubble is on a solid boundary.

Cavitation microstreaming was noted to have detrimental biological effects. For instance, ultrasonic destruction of Paramecium cells was observed to be caused specifically by microstreaming flows. Marmottant and Hilgenfeldt (2003) also reported that microstreaming from a microbubble was also observed to disrupt an artificial vesicle, which is a model biological cell.

2.2.4 Microjets

Microjetting is a microscopic flow of liquid emitted by the bubble and occurring during asymmetric bubble collapse. Near an extended solid surface, the collapse of an acoustic cavitation bubble is non-spherical and drives high-speed jet of liquid to the surface known as microjet. This microjet is directed toward the solid surface and this can lead to pitting and erosion. The surface action can also dislodge particles attached to the surface and break down large aggregates into smaller particles (Yuan and Prosperetti 1997; Peters et al. 2015). The mechanism of microjets is attributed to the difference in pressures applied on the bubble wall from the solid and liquid sides, respectively. Just before the bubble collapses near a solid surface, the liquid pressure on the bubble surface near a solid boundary becomes much lower than that on the other side of the bubble surface. As a result, a jet penetrates into the bubble and finally hits the solid surface.

However, jet penetration also occurs when two neighboring bubbles collapse simultaneously. Another situation of jet penetration into a bubble can be encountered when the bubble collapses in a traveling ultrasonic wave (Yasui 2018).

Kornfeld and Suvorov (1944) were the first researchers to propose the microjet theory, they indicated that when a bubble collapses near a solid wall, the bubble deforms into flat-shape or ingot-shape and finally collapses producing a microscopic jet flow impingement on that wall. Crum (1995) and Gregorčič et al. (2007) subsequently observed the microjet impact produced by bubble collapse on a wall, using high-speed photography. Plesset and Chapman (1970) reported that the microjet velocity is about 130 m/s when the bubble collapses attached to the wall and is about 170 m/s when the bubble collapses near the wall. This result is coherent with the findings of Brujan (2004) who observed that microjet velocity produced by the collapse of a bubble having a maximum radius of 150 μm ranges from 80 to 130 m/s. Sato et al. (2016) used the impact effect of bubble collapse to deal with the surface of a hard aluminum alloy plate and found that the fatigue of the material increased more than 10 times (Yasui 2018), microjets are then believed to be the key phenomenon responsible of material erosion under the effect of ultrasound and acoustic cavitation.

2.2.5 Shock Wave

The collapse of a bubble during transient or repetitive transient cavitation induces extremely high pressures that result in outward propagating shock waves. This suddenly released pressure into the liquid medium generates shock waves that are helpful to enhance mass transfer and shear induced processes. This cataclysmic event, i.e., shock waves, can also cause severe turbulence within the immediate surroundings and produce high velocity interparticle collisions, the impact of which is sufficient to melt most metals (Jena 1965), break polymer chains or destroy the cell walls of plant and animals. The shock wave approach considers bubbles to remain spherical during collapse, and the collapse causes a shock wave with an order of magnitude of several megapascals at a final stage.

The energy released from a single transient collapse is extremely small, but millions of bubbles collapse every second and the cumulative effect is large. Fortes Patella and Reboud (Fortes Patella and Reboud 2004) investigated numerically the interaction between the shock wave emitted by a collapsing acoustic cavitation bubble and the material, they concluded that cavitation damage is directly related to the shock wave and material characteristics. Brujan (Brujan 2004) reported that at a distance of 68 μm from the bubble wall, the shock wave pressure is 1.3 ± 0.3 GPa (Ye and Zhu 2017).

2.3 *Chemical Effects of Acoustic Cavitation Bubble*

The violent collapse events that occur during transient and repetitive transient cavitation can also generate enormous temperatures at a localized level, exceeding 5000 K (Kerboua and Hamdaoui 2018c). These high temperatures and the violent pressure changes occurring simultaneously can cause a number of chemical changes to occur within both the gaseous phase inside the cavitation bubble and in the immediate fluid surrounding it (Pankaj and Ashokkumar 2011).

Primary radicals are formed as a direct result of the high temperatures inside a collapsing bubble. In the heated interior of the bubble, water vapor and gas molecules, if reactive, are dissociated and oxidants such as OH radicals, H_2O_2, O atoms, and O_3 are formed. Table 6.2 illustrates the possible reactions occurring in the bulk volume of the bubble under oxygen atmosphere.

The oxidants diffuse out of the bulk volume of the bubble, represented in Fig. 6.4, into the surrounding liquid, and chemically react with solutes if present. Such

Table 6.2 Possible reactions occurring inside an O_2/H_2O collapsing bubble

No.	Reactions	A_{fi}	b_{fi}	E_{afi}/R_g (K)	A_{ri}	b_{ri}	E_{ari}/R_g (K)	ΔH_i (kJ/mol)
1	$H + O_2 \rightleftharpoons O + {}^\bullet OH$	1.92×10^8	0	8270	7.18×10^5	0.36	−342	6917
2	$O + H_2 \rightleftharpoons H^\bullet + {}^\bullet OH$	5.08×10^{-2}	2.67	3166	2.64×10^{-2}	2.65	2245	823
3	${}^\bullet OH + H_2 \rightleftharpoons H^\bullet + H_2O$	2.18×10^2	1.51	1726	1.02×10^3	1.51	9370	−6435
4	${}^\bullet OH + {}^\bullet OH \rightleftharpoons H_2O + O$	2.1×10^2	1.4	200	2.21×10^3	1.4	8368	−7259
5	$H_2 + M \rightleftharpoons H^\bullet + H^\bullet + M$ Coef. H_2: 2.5, H_2O: 16.0	4.58×10^{13}	−1.4	52,500	2.45×10^8	−1.78	480	44,447
6	$O + O + M \rightleftharpoons O_2 + M$ Coef. H_2: 2.5, H_2O: 16.0	6.17×10^3	−0.5	0	1.58×10^{11}	−0.5	59,472	−5054
7	$O + H^\bullet + M \rightleftharpoons {}^\bullet OH + M$ Coef. H_2O: 5.0	4.72×10^5	−1.0	0	4.66×10^{11}	−0.65	51,200	−43,623
8	$H^\bullet + {}^\bullet OH + M \rightleftharpoons H_2O + M$ Coef. H_2: 2.5, H_2O: 16.0	2.25×10^{10}	−2.0	0	1.96×10^{16}	−1.62	59,700	−50,882
9	$H^\bullet + O_2 + M \rightleftharpoons HO_2^\bullet + M$ Coef. H_2: 2.5, H_2O: 16.0	2.00×10^3	0	−500	2.46×10^9	0	24,300	−2048
10	$H^\bullet + HO_2^\bullet \rightleftharpoons O_2 + H_2$	6.63×10^7	0	1070	2.19×10^7	0.28	28,390	−23,967
11	$H^\bullet + HO_2^\bullet \rightleftharpoons {}^\bullet OH + {}^\bullet OH$	1.69×10^8	0	440	1.08×10^5	0.61	18,230	−16,226
12	$O + HO_2^\bullet \rightleftharpoons O_2 + {}^\bullet OH$	1.81×10^7	0	−200	3.1×10^6	0.26	26,083	−23,185
13	${}^\bullet OH + HO_2^\bullet \rightleftharpoons O_2 + H_2O$	1.45×10^{10}	−1.0	0	2.18×10^{10}	−0.72	34,813	−30,444
14	$HO_2^\bullet + HO_2^\bullet \rightleftharpoons O_2 + H_2O_2$	3.0×10^6	0	700	4.53×10^8	−0.39	19,700	−17,535
15	$H_2O_2 + M \rightleftharpoons {}^\bullet OH + {}^\bullet OH + M$ Coef. H_2: 2.5, H_2O: 16.0	1.2×10^{11}	0	22,900	9.0×10^{-1}	0.90	−3050	21,789
16	$H_2O_2 + H^\bullet \rightleftharpoons H_2O + {}^\bullet OH$	3.2×10^8	0	4510	1.14×10^3	1.36	38,180	−29,093
17	$H_2O_2 + H^\bullet \rightleftharpoons H_2 + HO_2^\bullet$	4.82×10^7	0	4000	1.41×10^5	0.66	12,320	−6432
18	$H_2O_2 + O \rightleftharpoons {}^\bullet OH + HO_2^\bullet$	9.55	2	2000	4.62×10^{-3}	2.75	9277	−5608
19	$H_2O_2 + {}^\bullet OH \rightleftharpoons H_2O + HO_2^\bullet$	1.00×10^7	0	900	2.8×10^7	0	16,500	−12,867
20	$O_3 + M \rightleftharpoons O_2 + O + M$ Coef. O_2: 1.64 Coef. O_2: 1.63, H_2O: 15	2.48×10^8	0	11,430	4.1	0	−1057	10,927
21	$O_3 + O \rightleftharpoons O_2 + O_2$	5.2×10^6	0	2090	0	0	0	−39,614
22	$O_3 + {}^\bullet OH \rightleftharpoons O_2 + HO_2^\bullet$	7.8×10^5	0	960	0	0	0	−16,492
23	$O_3 + HO_2^\bullet \rightleftharpoons O_2 + O_2 + {}^\bullet OH$	1×10^5	0	1410	0	0	0	−12,192
24	$O_3 + H^\bullet \rightleftharpoons HO_2^\bullet + O$	9×10^6	0.5	2010	0	0	0	−13,565
25	$O_3 + H^\bullet \rightleftharpoons O_2 + {}^\bullet OH$	1.6×10^7	0	0	0	0	0	−962

M is the third body. Subscript "f" denotes the forward reaction and "r" denotes the reverse reaction. A is in (m^3/mol s) for two body reaction and in (m^6/mol^2 s) for a three body reaction (Yasui 1997)

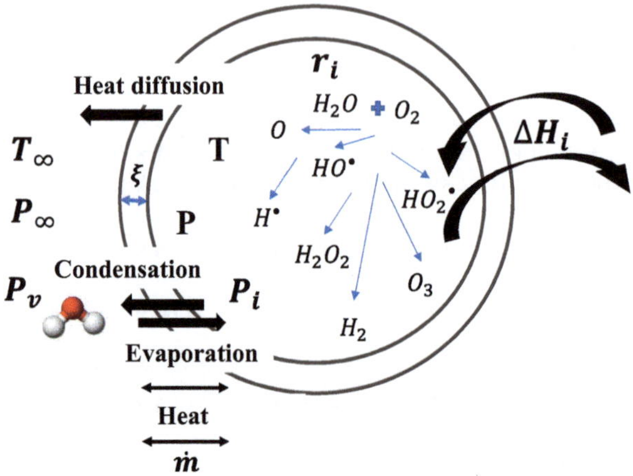

Fig. 6.4 Sonochemical activity in the bulk volume of an acoustic cavitation bubble

chemical reactions are called sonochemical reactions, and chemistry associated with acoustic cavitation is called sonochemistry.

There are three sites for chemical reactions for a cavitation bubble. (i) The interior of the bubble, where volatile solutes which evaporate into this region are dissociated by high temperature. (ii) The gas-liquid interface region, where the temperature dramatically increases due to the thermal conduction from the heated interior of a bubble where radicals with a relatively short lifetime such as OH and O react with solutes or radicals themselves and surfactants adsorbed at the bubble surface can dissociate at the interface region due to both heat and radical attack. (iii) The liquid region outside the interface region which is at the ambient temperature and where chemical species with a relatively long lifetime such as H_2O_2 diffusing out of the interface region chemically react with solutes (Jena 1965).

Sonochemistry is rigorously concerned by multibubble system, which is a complex system owing to bubble-bubble interactions, changes in local acoustic field due to attenuation and scattering by moving bubbles. Moreover, bubbles frequently coalesce each other, fragment into daughter bubbles, grow by rectified diffusion, shrink by gas diffusion, and spatially move in a complex pattern due to radiation forces. In addition, new bubbles are permanently created and the bubble pulsation as well as the acoustic wave propagation are strongly nonlinear. For all these reasons, most of researchers consider first a simpler system, which is the single acoustic cavitation bubble.

When a single bubble is considered, the number of generated free radicals abruptly increases when the temperature inside the collapsing bubble is at its maximum. This temperature can be increased by increasing the sonication power, increasing the static pressure, or decreasing the liquid temperature. The nature of the saturating gas plays a role as well, the heavier inert gases have a lower thermal conductivity and hence are less efficient at transferring heat away from the bubble to the surrounding fluid, they then increase the temperature at the bulk volume of the bubble. The amount of generated heat depends also on the size of the cavitation bubble, the bigger the transient cavitation, the higher the amount of heat.

In a multibubble field, the total yield of generated free radicals is governed not only by the bubble temperature, but also by the number of active bubbles. In fact, it has been shown that the number of bubbles generated is the predominant factor in controlling the radical yield. Thus, for a given liquid volume and acoustic power, a greater number of radicals are generated at higher frequencies (Kerboua and Hamdaoui 2019b) and this can dominate over the radical production per bubble. For this reason, sonochemical effects are generally the most dominant at intermediate frequencies generally ranging from 200 to 500 kHz (Kerboua et al. 2021c; Bermúdez-aguirre et al. 2011).

The sonochemical effect accompanying the oscillation of an acoustic cavitation bubble, when submitted to appropriate acoustic conditions, knows several applications in wastewater treatment, food and beverages industry, and biological processes (Leong et al. 2011; Yusof et al. 2016).

3 Sono-processing in Agri-food Applications

The effects induced by ultrasound and acoustic cavitation bubbles, particularly those of physical nature, know several applications in agri-food sector. Table 6.3 reports most of the applications involving ultrasound and/or acoustic cavitation bubble in food processing. Some of these applications are discussed in detail in the next section.

3.1 Sono-extraction

Extraction is a major step in the development of natural functional food. The conventional extraction technique is solvent extraction that has been associated with a large amount of organic solvent and heating. The development of extraction technologies has placed the ultrasound-assisted process as an emerging solution for improved recovery of food compounds with higher quality, lower extraction times, lower temperatures, less solvent volumes and lower energy consumption (Fu et al. 2020).

Sono-extraction is mainly explained by cell disruption caused by the implosion of acoustic cavitation bubble. The increase of the number of disrupted cells participate in weakening the limitation of cell structure on mass transfer from the solid matrix to the liquid phase (Yang et al. 2017).

Microstreaming accompanying the collapse of acoustic cavitation bubble causes violent turbulence that disrupts the surfaces of solid foodstuffs and facilitates the mixing of food materials and components in the medium. This also decreases diffusion boundary layers (Bhaskaracharya et al. 2009), which consequently accelerates

Table 6.3 Ultrasound assisted processes applied to food industry and their basic ultrasound mechanism (Bhargava et al. 2021; Fu et al. 2020; Singla and Sit 2021; Chemat et al. 2011).

Application	Definition	Ultrasound mechanism
Cutting	Size reduction of large to medium sized parts of food material.	Cavitation phenomenon
Cooking	Transferring heat to food component to bring about some physical/chemical transformations.	Uniform heat transfer
Picking and marinating	Soaking foods in a seasoned liquid before cooking.	Increased mass transfer
Filtration	Separation of solids from a suspension in a liquid by means of a porous medium which retains the solids and allows the liquid to pass through.	Vibration
Extraction	Recover valuable soluble components from raw materials by primarily dissolving them in a liquid solvent, so that the components can be separated and recovered later from the liquid.	Increased diffusion
Drying	The removal of solvent from a solid, semi-solid or a liquid.	Uniform heat transfer
Dehydration	Removal of water from the water-containing compound until there is not enough moisture to support microbial activity.	Uniform heat transfer
Rehydration	Imbibition of water into the dried material, swelling, and leaching of soluble materials.	Increased absorption
Freezing	Lowering the temperature to inhibit microorganism growth.	Uniform heat transfer
Thawing	Taking a frozen product from frozen to a temperature (usually above 0 °C) where there is no residual ice.	Uniform heat transfer
Crystallization	The process of forming crystals from solution.	Uniform heat transfer
Foaming	Forming small bubbles in or on liquids.	Dispersion of gas bubbles
Defoaming	Removing bubbles and air from the surface of liquids.	Cavitation phenomenon
Aeration/deaeration	Adding/removing very tiny pockets of air to something.	Agitation
Degassing	Removing dissolved gas from a liquid.	Compression-rarefaction
Fermentation	Converting carbohydrates to alcohol or organic acids using microorganisms like yeasts or bacteria, under anaerobic conditions.	Cavitation phenomenon
Homogenization	Reducing a substance to extremely small particles and distributing it uniformly throughout a fluid.	Cavitation phenomenon
Emulsification	Breaking up the dispersed phase into small droplets.	Cavitation phenomenon

(continued)

Table 6.3 (continued)

Application	Definition	Ultrasound mechanism
Pasteurization-microbial inactivation	Destroying pathogenic microorganisms in certain foods and beverages.	Uniform heat transfer
Sterilization	Complete destruction or elimination of all viable organisms in/on object.	Uniform heat transfer
Preservation-enzyme inactivation	Inhibiting the activity of enzymes present in the food that can lead to undesirable changes in flavor, texture, or color.	Uniform heat transfer
Encapsulation	Incorporation of food ingredients, enzymes, cells or other materials in small capsules. Fats, starches, dextrins, alginates, protein and lipid materials can be employed as encapsulating materials.	Cavitation phenomenon
Packaging	Enclosing food to protect it from tampering or contamination from physical, chemical, and biological sources.	Cavitation phenomenon
Gelation	Converting liquid food into solid using viscoelastic substances.	Cavitation phenomenon Uniform heat transfer
Demolding	Separation of the set food from the mold substrate.	Vibration
Oxidation	Chemical reactions in the presence of oxygen affecting the food components.	Cavitation phenomenon

mass and heat transfers (Tao and Sun 2015). The oscillation of the acoustic cavitation bubble, ending by its implosion can induce shear forces, fragmentation, sonoporation erosion, capillary effect, turbulence and mixing, which also contribute to mass-transfer enhancement (Rodrigues and Fernandes 2017).

Ultrasound also contributes to the reduction of the size of solid materials, which increases the contact surface area between the solid phase and the solvent. Many microscopic channels in tissue can appear as well due to alternative compression and rarefaction of the ultrasonic wave, inducing the so called “sponge effect”. This effect facilitates the penetration of the solvent into the solid matrix (Fernandes et al. 2008; Fernandes et al. 2009; Nowacka et al. 2014).

Literature counts a significant number of works that investigated the efficiency of sono-extraction in food processing. For instance, Vilkhu et al. (2008) studied the sono-extraction of bioactives such as polyphenols, anthocyanins and aroma compounds.

Panchev et al. (1988) examined the extraction of pectin from apple pomace in concentrated nitric acid using 22 kHz intermittent ultrasonication during 10–60 min at 1–1.2 W/cm^2 followed by coagulation with 95% ethanol and separation by filtration. Panchev et al. (1988) revealed that ultrasound-assisted extraction reduced the extraction time by more than 33% for similar yields and improved the pectin yield by 22% for similar time.

Tang (2007) performed ultrasound assisted extraction of solids from buckwheat flour at 15 kHz and 500 W for 30 min. They demonstrated that the technique allowed the extraction of lipids and proteins in 87.5% shorter time compared to conventional method. This observation was explained by increased mass transfer.

Albu et al. (2004) carried out the extraction of an antioxidant, namely carnosic acid, from dried and fresh leaves of rosemary, they demonstrated improved yields and higher extraction rates using ultrasound and solvent at 50 °C.

Other researchers examined the possibility of coupling ultrasound with other conventional and innovative techniques in order to achieve synergetic effects. For instance, Lianfu and Zelong (2008) optimized the extraction of lycopene from fresh tomatoes at 86.4 °C to theoretical values of 97% using coupled ultrasound and microwave. At this temperature, cavitation would be unlikely to occur, which explains the ultrasound effect by acoustic streaming. Another example is that of Cravotto et al. (2008) who extracted vegetable oils from soybean germ and sea weed using a solvent with a combination of ultrasound and microwave. The innovative extraction technique allowed the reduction of the extraction time by 8-fold compared to separated techniques.

Some other studies combined Soxhlet extraction (Alara et al. 2018; Luque de Castro and Priego-Capote 2010; López-Bascón-Bascon and Luque de Castro 2019) with ultrasound to quicken the extraction of fat and oil (Chemat et al. 2017). This strategy has been adopted for instance by Djenni et al. (2013) who developed a novel design of sono-Soxhlet extraction of oils from crushed dried olives using an ultrasonic horn in situ to provide rapid and complete recovery of analytes from the matrix. The authors proved through physicochemical characterization and gravimetric analysis that extraction time was substantially reduced compared to conventional Soxhlet extraction, without affecting the composition and quality of target extracts. Luque de Castro and coworkers (2004) used similar procedure as Djenni et al. (2013) and proved that ultrasound boosts mass transfer, the specimen is then collected completely, effortlessly and precisely.

3.2 Sono-emulsification

An emulsion consist of two naturally immiscible liquids and is thermodynamically unstable due to the interfacial tension between the continuous and dispersed phases (Ashokkumar et al. 2010). Ultrasound has been introduced in the last decades as efficient emulsification technique, particularly in food industry.

Two mechanisms occurring in parallel explain the ultrasound-assisted emulsification, namely the development of instabilities at the oil/water interface and the turbulence induced by acoustic cavitation bubble (Bhaskaracharya et al. 2009).

low frequency and high intensity ultrasound is usually used in the food sector for the preparation of mayonnaise, emulsions fruit juices, ketchup, and homogenized milk (Hongyu et al. 2000). Several research groups studied the ultrasound assisted emulsification of protein. For instance, Li et al. (2020) used ultrasound at 20 kHz and 450 W to improve the emulsification of chicken myofibrillar protein. Their experiments showed that better oil/water emulsion gel elasticity was obtained after 6 min of sonication.

3.3 Sono-homogenization

Sono-homogenization is particularly applied in the dairy industry. Indeed, milk is an oil in water emulsion. During storage, in order to avoid the creaming of the natural full cream milk, the size milk fat particles should be reduced below 0.8 μm (Silva and Chandrapala 2021; Villamiel and De Jong 2000). Ultrasound is considered as an alternative method for reducing the size of fat globules and can be applied effectively to homogenize milk (Bosiljkov et al. 2011). For instance, Vercet et al. (2002) demonstrated that the firmness and texture of a yogurt made from milk exposed to sonication are considerably improved.

Karlović (2015) exposed milk to high pressure ultrasound of up to 600 MPa and treatment time up to 9 min. They reported an improvement in the homogenization of milk fat globules, which were significantly reduced in size by the high pressure (Carrillo-Lopez et al. 2017).

Bermúdez-Aguirre et al. (2008) exposed whole milk to thermosonication treatment of 24 kHz and 400 W over a duration of 30 min and studied the changes to the microstructure of fat globules. The sonication of whole milk at 63 °C resulted in fat globules of less than 1 μm with more binding sites on the fat globule membrane enhancing the amalgamation of casein and serum proteins, which produces an ideal ingredient for cheese making. The authors attributed the observed effect to cavitation since heat treatment alone did not show similar changes to the fat globules (Ashokkumar et al. 2010).

Ertugay et al. (2004) proved that the size range of fat globules in ultrasonically homogenized milk samples at 20 kHz was much smaller than that obtained with conventional homogenizer (2–5 μm). The mean size and size distribution of the fat globules were dependent on the ultrasonic power and duration of sonication. The authors explained the observed effect by the physical effects generated during acoustic cavitation.

In addition to the decrease of the diameter of fat globules (Nguyen and Anema 2010), several authors demonstrated that high pressure ultrasound induces alterations in the secondary structure of proteins in milk, aggregation of protein particles, and denaturation (Chandrapala et al. 2011) depolarizes the particles of gamma-carrageenan and reduces their size, allowing for better homogenization of nanoparticles in a dispersion mixed with beta-lactoglobulin. Consequently, ultrasound participate significantly in the enrichment of acidified milk drinks (Hosseini et al. 2013). Power ultrasound is also proved to cause alterations in the composition and structure of the membrane of fat globules, which improves the efficiency of homogenization in casein gel (Chandrapala et al. 2013).

3.4 Sono-foaming/defoaming

The formation of foams, which passes through three main steps (i) transportation (ii) penetration and (iii) reorganization of the molecules at the gas-liquid interface, is influenced by multiple factors such as particle size, structural flexibility of surfactant molecules and surface hydrophobicity (Wilde and Clark 1993). Application of ultrasound technologies in foaming of food processes has emerged significantly in dairy industry. For instance, Jambrak et al. (2008) sonicated whey protein isolate, whey protein concentrate and whey protein hydrolysate solutions using low-intensity ultrasound through sonication bath at 500 kHz and high-intensity ultrasound through sonication probe at 20 kHz and sonication bath at 40 kHz. Significant increase in foaming capacities and foam stabilities were observed at 20 and 40 kHz. This observation was explained by the homogenization effect of ultrasound which disperses the protein and fat particles enhancing the foaming properties, and eventually by the partial unfolding of proteins during sonication.

In the other hand, Tan et al. (2015) observed that the foaming properties of whey protein concentrates improved with increasing ultrasound amplitude and time. To illustrate, at 60% amplitude and 25 min treatment time, foam drainage decreased by 35% while foaming capacity, viscosity and consistency increased significantly. In another study, Tan et al. (2016) demonstrated that sonication of whey protein suspensions at 20%, 40% and 60% amplitude, 20 kHz and 400 W for 5, 15 and 25 min significantly enhanced the foaming properties. Better foaming properties can then be obtained using higher ultrasound amplitudes and longer treatment times. The observed effect was attributed to the acoustic cavitation that induces the fracturing of bubbles in the protein suspension leading to the formation of small-sized bubbles. Due to the denaturation of proteins, gel film is created on the surface of bubbles which further improve the stability.

Intensive foaming in several technological food processes negatively impacts the useful volume of the technological equipment and the conditions of sterility in biotechnological processes (Chemat et al. 2011). Conventional techniques rely on mechanical breakers, decrease of temperature and addition of anti-foaming chemical reagents in order to control excessive foaming (Gallo et al. 2018; Morey et al. 1999). Ultrasonic defoaming has been developed and successfully applied to control the excessive foaming, particularly in beverages industry (Riera et al. 2006).

Rodríguez et al. (2010) achieved the scale-up of ultrasonic defoaming systems by mounting a focused airborne ultrasonic emitter on an electronically controlled rotating system. When the ultrasonic transducer rotates, it creates a complex movement covering a large defoaming area at different rotation speeds for sufficient duration. The authors demonstrated that most of the bubbles break almost instantaneously under the acoustic beam.

3.5 Ultrasound-assisted Fermentation

Ultrasound has been introduced in food processing, particularly in dairy industry, as an accelerator for the fermentation process. Low intensity ultrasound is believed to increase mass transfer into fermentation chamber, which enhances fermentation rate. It was proved by Sakakibara et al. (1994) that ultrasound accelerates lactose hydrolysis in milk at starter *Lactobacillus delbrueckii* and could promote the fermentation process. Several researchers demonstrated through experimental studies that fermentation time for milk can be significantly decreased, along with the syneresis reduction and viscous increasing of yogurt because of enzyme activity enhancement under ultrasonic treatment (Bratsikhin et al. 2019). Indeed, Vercet et al. (2002) studied the manothermosonication at 20 kHz and 40 °C and demonstrated that it produces low syneresis in yoghurts prepared from cow milk (Ashokkumar et al. 2010). Moreover, Riener et al. (2009) experienced the thermosonication of milk at 25 kHz and 45 or 75 °C and proved that the obtained yoghurts had higher water holding capacities and viscosities. Obtained yoghurts also displayed improvements in microstructural properties containing honeycomb-like network with a more porous nature. Nguyen and Anema (2010) showed that yoghurt prepared from skim milk treated with ultrasound at 22 kHz and 50 W and up to 30 min had higher viscosity. In the other hand, Masuzawa and Ohdaira (2002) proved that ultrasonic treatment at 20 kHz reduced ripening time for yogurt production (Ashokkumar et al. 2010). Sfakianakis et al. (2015) demonstrated significant improvements in water holding capacity and viscosity of yoghurts with reduced syneresis can be achieved using ultrasound treatment and the effect has been shown to be more pronounced with the increases in the ultrasound amplitude. Hence, high intensity ultrasound provides several benefits over conventional fermentation in milk gels and yoghurts. All of the mentioned effects were justified by water binding capacity enhancing because of fat globe surface enlargement (Ashokkumar et al. 2010; Hongyu et al. 2000).

3.6 Sono-crystallization

In order to obtain appropriate food structure, texture and consumer appeal, crystallization is a key factor to control in the food industry. With conventional techniques of crystallization, it is difficult to get a uniform crystal size due to non-uniform nucleation, inefficient cooling caused by surface encrustation of cooling coils and non-uniform growth of crystals due to uneven mixing. With ultrasound, it is possible to initiate the nucleation process at a higher temperature or in a shorter time, smaller and more uniform crystals are then obtained (Mason 2007). The use of power ultrasound provides a useful approach to producing crystals with desired properties. Sono-crystallization facilitates process control by modulating crystal size distribution and morphology for several food products such as chocolate, honey, fats and frozen foods (Deora et al. 2013).

Bund and Pandit (2007) examined the sonication of a lactose solution using a 22 kHz ultrasonic bath, with 85% v/v ethanol as an anti-solvent. They demonstrated that the ultrasound assisted process led to 92% recovery of lactose as compared to 15% recovery for mechanically stirred samples. They also proved that the size and shape characteristics of the lactose crystals were enhanced. The observed improvements were explained by rapid mixing of the anti-solvent into the solution and eventually by either cavitation bubbles acting as crystal nucleation sites or solvent depletion in the zone of a cavitation bubble. These hypotheses are confirmed by the observations of Chow et al. (2005) who confirmed through direct imaging of a 15% sucrose solution that cavitational micro-bubbles can indeed act as nucleation sites. Kougoulos et al. (2010) employed power ultrasound to crystallize lactose and showed that a longer duration time resulted in a smaller crystal size, the minimum crystal size can reach 10 μm. They also demonstrated that ultrasound plays a positive role in increasing nucleation rate and yield.

In the other hand, Luque De Castro and Priego-Capote (2007) indicated that the results obtained so far in the literature make foreseeable that crystal size distribution and crystal shape can be tailored by appropriate selection of the sonication conditions. Patrick et al. (2004) indicated that the optimum acoustic conditions in order to obtain small crystals in the shortest time should be selected just below the cavitation intensity threshold.

3.7 *Sono-osmo-dehydration*

Osmo-dehydration is a technique used for dehydration of fruits and vegetables. Employing ultrasound with osmotic dehydration results in higher wastage of water and solute at a lesser temperature of the solution, in addition to protecting the attributes like flavor, color and temperature-sensitive nutrients (Rahaman et al. 2019; Prithani and Dash 2020). Deng and Zhao (2008a) studied the osmo-dehydration and rehydration of apple cylinders using an ultrasonic bath at 50/60 kHz and a pulsed vacuum chamber with 13 MPa vacuum applied twice for 5 min intervals. They demonstrated that ultrasound treatment increased the mass transfer of water from the substrate without heating the fruit pieces to very high temperatures. Such mass transfer enhancement is attributed to the turbulence and physical shear induced both by the acoustic field and by the acoustic cavitating bubbles (Bhaskaracharya et al. 2009). In this context, Simal et al. (1998) also affirmed that the sono-osmo-dehydratation to dry out the apples has been found to accelerate the mass transfer rate of "water out" and "solute in". Cárcel et al. (2007) indicated that diffusion coefficients of water and solute were found to increase around 117% and 137% separately due to ultrasound.

Similar observations were made by Stojanovic and Silva when treating blue berries with ultrasound (Stojanovic and Silva 2006), they showed that the diffusion rates of moisture throughout osmotic dehydration of rabbit eye blueberries are notably enhanced with application of ultrasound though, this is attributed to the loss of anthocyanins and phenolics. Sánchez et al. (1999) also confirmed the increase of the rate of expulsion of water and sodium chloride in the case of cheese brining, when ultrasound are used.

Moreover, Bchir et al. (2020) showed that ultrasound pretreatment of the pomegranate seeds induces a reduction in the osmotic solution temperature and leads to more hardness of the samples microstructure compared to the fresh and osmotic dehydrated samples. Deng and Zhao (2008a) compared ultrasound assisted dehydratation with pulsed vacuum technique, and proved that ultrasound treatment resulted in higher glass transition temperature, lower water activity, reduction in rehydration rate and water content, a more extreme rupturing of the structure, and fewer crevices and calcium take-up, as compared to pulsed vacuum. However, it is important to notice that the same research group (Deng and Zhao 2008b) observed that rehydration rates were slower for apples pretreated with ultrasound, which would be explained by cell deformation and severe tissue damage at the surface. Deng and Zhao reported an increased hardness and crispness of apples pre-treated with osmo-dehydration assisted by ultrasound upon air drying or freeze drying.

3.8 Ultrasound Assisted Freezing

Freezing is an important unit operation in food industry intended to produce food products in frozen state, which extends their shelf life. Freezing passes generally by two phases: (i) initial ice nucleation and (ii) crystal growth towards the liquid phase (Tao and Sun 2015). The speed of ice nucleation is often slower than the growth of ice crystals. In conventional freezing, the crystals will rapidly grow to large size which may destroy the cell tissue instead of forming small and uniform ice crystals.

When using ultrasound, acoustic cavitation bubbles constitute nucleation sites for ice, which results in generating a mass of ice nuclei. Simultaneously, heat and mass transfer are enhanced due to the turbulence created by acoustic waves and acoustic cavitation (Dalvi-Isfahan et al. 2017). During primary nucleation, it is believed that the extreme conditions of pressure created during the collapse of acoustic cavitation bubble contribute to the decrease of the supercooling degree driving the process of nucleation (Kiani et al. 2011). During secondary nucleation, the collapse of cavitation bubbles and the shear forces accompanying it can break up the dendrites of the pre-existing crystal ice into many smaller fragments, producing more nucleation sites (Fu et al. 2020).

Delgado et al. (2009) performed the ultrasound assisted freezing of apple cylinders. They demonstrated that the application of ultrasonic irradiation using ultrasonic bath working intermittently at 40 kHz and 131.3 W at 0 or −1 °C for 120 s with 30 s intervals highly improved the freezing rate. Besides, Islam et al. (2014) studied the ultrasound-assisted freezing of three types of mushroom, namely *Lentinula edodes*, *Agaricus bisporus* and *Pleurotus eryngii*. The times for nucleation formation during sonication for *Lentinula edodes*, *Agaricus bisporus* and *Pleurotus eryngii* were reduced by 24%, 53%, and 34%, respectively, while the drip loss during the thawing process was decreased by an order of 10%. Power ultrasound proved to inhibit polyphenol oxidase and peroxidase enzyme activities by generating fine crystals with even distribution, which leads to high quality of mushrooms during long term storage. It is worthy to mention that power ultrasound results in the sponge effect which retains some residual water, the microstructures of frozen samples were better in comparison with samples obtained using conventional methods.

Kiani and Sun (2018) were able to accurately predict the ultrasound-assisted freezing of various shapes of potatoes and suggested the existence of optimum condition of ultrasonic intensity to reduce freezing time. The same research group (Kiani et al. 2015) assessed the ultrasound-assisted freezing of potato spheres using experimental, numerical and analytical approaches aiming to predict temperature distribution, phase ratios and process time. The author demonstrated that the shortest freezing time could be achieved only within the range of 30–70% duty cycles. This was attributed to enhanced heat transfer and the thermal effect at the sample surface. Numerical simulations proposed by Kiani et al. (2015) provided the temperature and water fraction profiles in potatoes under different conditions.

Cheng et al. (2014) also employed ultrasound to assist the immersion freezing of strawberries. The authors proved that usage of ultrasound resulted in nucleation at a lower supercooling degree, they also revealed a linear correlation of the supercooling with the ultrasonic temperature.

Comandini et al. (2013) investigated the ultrasound-assisted freezing of potato cubes using a sonotrode at 35 kHz. The sonication was activated when the temperature of the geometrical center of the sonotrode was within the range of −0.1 to −3 °C. The authors reported that nucleation occurred significantly earlier when applying ultrasound. However, the authors mentioned that a significant reduction in freezing time was observed only when ultrasound was applied at −2 °C.

3.9 Ultrasound Assisted Filtration

The ultrasound-assisted filtration depends on several factors such as feed solution characteristics, cleaning solution characteristics, membrane properties and ultrasound parameters (Ashokkumar et al. 2010). The technique takes advantage of four mechanisms induced by ultrasound, namely (i) the bulk water movement toward and away from the membrane cake layer due to turbulence generated by the acoustic streaming, which removes particles from the membrane surface. (ii) the mechanical vibrational energy generated through sonication, which keeps the particles partly suspended in the solution and provides more free channels for the solvent elution, (iii) the cleaning of the membrane crevices and pores by acoustic cavitation bubbles, and (iv) the agglomeration of fine particles and thereby the reduction of the pore blockage of the membranes and compaction of cake on the membrane (Muthukumaran et al. 2005a, b).

Koh et al. (2014) observed a small but significant decrease in membrane fouling at solid concentrations higher than 10% w/w when applying ultrasound irradiation at 20 kHz to the whey protein concentration solution prior to filtration. This observation is explained by the ultrasound induced reduction of the solution viscosity and the size of the aggregates leading to the reduced cake growth and pore blockage.

3.10 Sono-preservation (Enzyme Inactivation)

Ultrasound can be used for food preservation in combination with other treatments by improving its inactivation efficacy. There have been many studies combining ultrasound with either pressure, temperature, or pressure and temperature (Ercan and Soysal 2012). When ultrasound is used alone, long time of ultrasonic treatment is required to achieve efficient inactivation of enzymes and microorganisms (Zheng and Sun 2006). When ultrasonic irradiated is combined to moderated heating, the technique is known as thermosonication. If thermosonication is employed for pasteurization or sterilization purpose, it is observed that lower process temperatures and times are required to achieve the same lethality values as compared to conventional processes (Mason et al. 1996). If pressure and sonication are combined, the technique is then called manosonication, its use in enzyme inactivation showed higher efficiency than ultrasound alone at the same temperature (Demirdöven and Baysal 2009). Finally, if ultrasound is coupled to both temperature and pressure effects, it results in manothermosonication. This technique is believed to inactivate more enzymes at lower temperatures and in a shorter time than thermal treatments at the same temperatures (Chemat et al. 2011).

First enzyme inactivation by ultrasound was applied to pure pepsin almost 60 years ago (Ercan and Soysal 2012) and since then, several researchers performed enzyme inactivation using ultrasound alone or with heating and pressure. For instance, Raso and Barbosa-Cánovas (2003) studied the ultrasound-assisted inactivation of polyphenol oxidase. Ercan and Soysal (2011) performed the therrmosonication technique to inactivate tomato peroxidase, while De Gennaro et al. (1999) explored the inactivation of peroxidase by thermosonication. Besides, Guiseppi-Elie et al. (2009) inspected the effect of ultrasonic processing on enzymatic activity of glucose oxidase. Raviyan et al. (2005) carried out both ultrasonication and thermosonication to inactivate tomato pectin methyl esterase, they examined the effect of cavitation intensity and temperature. Vercet et al. (2001) also studied the inactivation of proteases and lipases using ultrasound, while Cruz et al. (2006) investigated the effect of thermal and thermosonication treatments on inactivation kinetics of watercress peroxidase.

3.11 Sono-pasteurization (Microbial Inactivation)

Conventional thermal techniques of pasteurization result in unwanted flavors and loss of nutrients. These undesirable effects can be overcome using ultrasound (Ercan and Soysal 2012). Ultrasound is believed to inactivate microbial activity in liquid food. The ultrasound-assisted inactivation of microbial activity is explained by four mechanisms. (i) Ultrasound induces intracellular cavitation in the bacterial cells, which destroys the structural and functional components of the bacterial cells up to cell lysis. (ii) The collapse of acoustic cavitation bubbles on hydrophobic surfaces of micro-organism results in severe damage to the microbial cell walls. (iii) the microstreaming generated by the passage of the ultrasonic wave causes thinning of the cell membranes. (iv) The generated hot spot and the formation of free radicals can damage the DNA of the microorganism (Ashokkumar et al. 2010).

Cameron et al. (2009) studied the ultrasound assisted pasteurization of milk and proved that ultrasound eliminates spoilage and potential pathogens to zero or to levels acceptable by legislation, even when initial inoculum loads were 5 times higher than permitted. According to their study, viable cell counts of *E. coli* were reduced by 100% after 10 min of ultrasonication, while viable counts of *Pseudomonas fluorescens* were reduced by 100% after 6 min and *Listeria monocytogenes* were reduced by 99% after 10 min.

Besides, Saikia et al. (2016) performed ultrasound pasteurization of juices and demonstrated that sonication has a positive effect on the total flavonoid content in carambola, black jamun, watermelon, pineapple and litchi juice samples followed by microwave treatment with exceptions in some cases.

Salleh-Mack and Roberts (2007) carried out a parametric study of ultrasound pasteurization by examining the effects of temperature, sugar, organic acids and pH on ultrasound inactivation of *Escherichia coli* ATCC 25922. Overall, ultrasound increased the sensitivity of *Escherichia coli* to thermal inactivation. The presence of soluble solids had a protective effect where the sonication time requirement increased. Similar to heat sensitivity, the lower pH environment resulted in *Escherichia coli* having less resistance to sonication. In another study, Wordon et al. (2012) examined the effect of the acoustic frequency and suggested that high frequency ultrasound was more effective in inactivation of microorganisms.

4 Ultrasound in Agri-food Application: Challenges

Although ultrasound assisted processes in food industry have known a rapid development and even actual implementation at large scale, the literature and industrial documentation seriously lack standardized documentation of methodology and control parameters until the widespread implementation of the technology. The inexistence of such documentation prevents reproducibility of results and questions the liability of the use of ultrasound assisted processes in large-scale applications (Rastogi 2011).

Joint efforts of fundamental studies and technical research should be gathered in order to define parametric standards in terms of sono-food processing technologies. Efforts should also be invested in the security of ultrasound technologies applied to food processing, notably in order to prevent undesirable effects reported by some researchers at laboratory scale and design reliable and viable industrial solutions (Singla and Sit 2021).

References

Alara OR, Abdurahman NH, Ukaegbu CI (2018) Soxhlet extraction of phenolic compounds from Vernonia cinerea leaves and its antioxidant activity. J Appl Res Med Aromat Plants 11:12–17. https://doi.org/10.1016/j.jarmap.2018.07.003

Albu S, Joyce E, Paniwnyk L, Lorimer JP, Mason TJ (2004) Potential for the use of ultrasound in the extraction of antioxidants from Rosmarinus officinalis for the food and pharmaceutical industry. Ultrason Sonochem 11:261–265. https://doi.org/10.1016/j.ultsonch.2004.01.015

Ashokkumar M, Bhaskaracharya R, Kentish S, Lee J, Palmer M, Zisu B (2010) The ultrasonic processing of dairy products – an overview, dairy. Sci Technol 90:147–168. https://doi.org/10.1051/dst/2009044

Bai L, Xu W, Deng J, Li C, Xu D, Gao Y (2014) Generation and control of acoustic cavitation structure. Ultrason Sonochem 21:1696–1706. https://doi.org/10.1016/j.ultsonch.2014.02.027

Bchir B, Bouaziz MA, Ettaib R, Sebii H, Danthine S, Blecker C, Besbes S, Attia H (2020) Optimization of ultrasound-assisted osmotic dehydration of pomegranate seeds (Punica granatum L.) using response surface methodology. J Food Process Preserv 44:1–17. https://doi.org/10.1111/jfpp.14657

Bermúdez-Aguirre D, Mawson R, Barbosa-Cánovas GV (2008) Microstructure of fat globules in whole milk after thermosonication treatment. J Food Sci 73:325–332. https://doi.org/10.1111/j.1750-3841.2008.00875.x

Bermúdez-aguirre D, Mobbs T, Barbosa-cánovas GV (2011) Ultrasound technologies for food and bioprocessing. https://doi.org/10.1007/978-1-4419-7472-3

Bhargava N, Mor RS, Kumar K, Sharanagat VS (2021) Advances in application of ultrasound in food processing: a review. Ultrason Sonochem 70:105293. https://doi.org/10.1016/j.ultsonch.2020.105293

Bhaskaracharya RK, Kentish S, Ashokkumar M (2009) Selected applications of ultrasonics in food processing. Food Eng Rev 1:31–49. https://doi.org/10.1007/s12393-009-9003-7

Bjørnø L (2017) Finite-amplitude waves. In: Applied underwater acoustics. Elsevier, pp 857–888. https://doi.org/10.1016/B978-0-12-811240-3.00013-8

Bosiljkov T, Tripalo B, Brnčić M, Ježek D, Karlovi S, Jagušt I (2011) Influence of high intensity ultrasound with different probe diameter on the degree of homogenization (variance) and physical properties of cow milk. Afr J Biotechnol 10:34–41. https://doi.org/10.4314/ajb.v10i1

Bratsikhin A, Leschenko E, Kostenko K (2019) Influence of cavitation disintegration on dairy foods production. J Hyg Eng Des 27:173–177

Brennen CE (2013) Cavitation and bubble dynamics. https://doi.org/10.1017/CBO9781107338760

Brujan EA (2004) The role of cavitation microjets in the therapeutic applications of ultrasound. Ultrasound Med Biol 30:381–387. https://doi.org/10.1016/j.ultrasmedbio.2003.10.019

Bund RK, Pandit AB (2007) Sonocrystallization: effect on lactose recovery and crystal habit. Ultrason Sonochem 14:143–152. https://doi.org/10.1016/j.ultsonch.2006.06.003

Cameron M, McMaster LD, Britz TJ (2009) Impact of ultrasound on dairy spoilage microbes and milk components. Dairy Sci Technol 89:83–98. https://doi.org/10.1051/dst/2008037

Cárcel JA, Benedito J, Bon J, Mulet A (2007) High intensity ultrasound effects on meat brining. Meat Sci 76:611–619. https://doi.org/10.1016/j.meatsci.2007.01.022

Carrillo-Lopez LM, Alarcon-Rojo AD, Luna-Rodriguez L, Reyes-Villagrana R (2017) Modification of food systems by ultrasound. J Food Qual 2017. https://doi.org/10.1155/2017/5794931

Chandrapala J, Zisu B, Palmer M, Kentish S, Ashokkumar M (2011) Effects of ultrasound on the thermal and structural characteristics of proteins in reconstituted whey protein concentrate. Ultrason Sonochem 18:951–957. https://doi.org/10.1016/j.ultsonch.2010.12.016

Chandrapala J, Zisu B, Kentish S, Ashokkumar M (2013) Influence of ultrasound on chemically induced gelation of micellar casein systems. J Dairy Res 80:138–143. https://doi.org/10.1017/S0022029912000696

Chemat F, Zill-E-Huma, Khan MK (2011) Applications of ultrasound in food technology: processing, preservation and extraction. Ultrason Sonochem 18:813–835. https://doi.org/10.1016/j.ultsonch.2010.11.023

Chemat F, Rombaut N, Sicaire AG, Meullemiestre A, Fabiano-Tixier AS, Abert-Vian M (2017) Ultrasound assisted extraction of food and natural products. Mechanisms, techniques, combinations, protocols and applications. A review. Ultrason Sonochem 34:540–560. https://doi.org/10.1016/j.ultsonch.2016.06.035

Cheng XF, Zhang M, Adhikari B, Islam MN, Xu BG (2014) Effect of ultrasound irradiation on some freezing parameters of ultrasound-assisted immersion freezing of strawberries. Int J Refrig 44:49–55. https://doi.org/10.1016/j.ijrefrig.2014.04.017

Chow R, Blindt R, Chivers R, Povey M (2005) A study on the primary and secondary nucleation of ice by power ultrasound. Ultrasonics 43:227–230. https://doi.org/10.1016/j.ultras.2004.06.006

Comandini P, Blanda G, Soto-Caballero MC, Sala V, Tylewicz U, Mujica-Paz H, Valdez Fragoso A, Gallina Toschi T (2013) Effects of power ultrasound on immersion freezing parameters of potatoes. Innov Food Sci Emerg Technol 18:120–125. https://doi.org/10.1016/j.ifset.2013.01.009

Cravotto G, Boffa L, Mantegna S, Perego P, Avogadro M, Cintas P (2008) Improved extraction of vegetable oils under high-intensity ultrasound and/or microwaves. Ultrason Sonochem 15:898–902. https://doi.org/10.1016/j.ultsonch.2007.10.009

Crum LA (1980) Measurements of the growth of air bubbles by rectified diffusion. J Acoust Soc Am 68:203–211. https://doi.org/10.1121/1.384624

Crum LA (1995) Comments on the evolving field of sonochemistry by a cavitation physicist. Ultrason Sonochem 2:147–152. https://doi.org/10.1016/1350-4177(95)00018-2

Cruz RMS, Vieira MC, Silva CLM (2006) Effect of heat and thermosonication treatments on peroxidase inactivation kinetics in watercress (Nasturtium officinale). J Food Eng 72:8–15. https://doi.org/10.1016/j.jfoodeng.2004.11.007

Dalvi-Isfahan M, Hamdami N, Xanthakis E, Le-Bail A (2017) Review on the control of ice nucleation by ultrasound waves, electric and magnetic fields. J Food Eng 195:222–234. https://doi.org/10.1016/j.jfoodeng.2016.10.001

Dastkhoon M, Ghaedi M, Asfaram A, Jannesar R, Sadeghfar F (2018) Magnetic based nanocomposite sorbent combination with ultrasound assisted for solid-phase microextraction of Azure II in water samples prior to its determination spectrophotometric. J Colloid Interface Sci 513:240–250. https://doi.org/10.1016/j.jcis.2017.11.031

De Gennaro L, Cavella S, Romano R, Masi P (1999) Use of ultrasound in food technology I: inactivation of peroxidase by thermosonication. J Food Eng 39:401–407. https://doi.org/10.1016/S0260-8774(99)00028-X

Delgado AE, Zheng L, Sun DW (2009) Influence of ultrasound on freezing rate of immersion-frozen apples. Food Bioprocess Technol 2:263–270. https://doi.org/10.1007/s11947-008-0111-9

Demirdöven A, Baysal T (2009) The use of ultrasound and combined technologies in food preservation. Food Rev Int 25:1–11. https://doi.org/10.1080/87559120802306157

Deng Y, Zhao Y (2008a) Effects of pulsed-vacuum and ultrasound on the osmodehydration kinetics and microstructure of apples (Fuji). J Food Eng 85:84–93. https://doi.org/10.1016/j.jfoodeng.2007.07.016

Deng Y, Zhao Y (2008b) Effect of pulsed vacuum and ultrasound osmopretreatments on glass transition temperature, texture, microstructure and calcium penetration of dried apples (Fuji). LWT – Food Sci Technol 41:1575–1585. https://doi.org/10.1016/j.lwt.2007.10.018

Deora NS, Misra NN, Deswal A, Mishra HN, Cullen PJ, Tiwari BK (2013) Ultrasound for improved crystallisation in food processing. Food Eng Rev 5:36–44. https://doi.org/10.1007/s12393-012-9061-0

Djenni Z, Pingret D, Mason TJ, Chemat F (2013) Sono-soxhlet: in situ ultrasound-assisted extraction of food products. Food Anal Methods 6:1229–1233. https://doi.org/10.1007/s12161-012-9531-2

Ercan SŞ, Soysal Ç (2011) Effect of ultrasound and temperature on tomato peroxidase. Ultrason Sonochem 18:689–695. https://doi.org/10.1016/j.ultsonch.2010.09.014

Ercan SŞ, Soysal Ç (2012) Ultrasound in food preservation. Nat Sci 5:251–262. https://doi.org/10.1007/978-1-4614-2038-5_11

Ertugay MF, Şengül M, Şengül M (2004) Effect of ultrasound treatment on milk homogenisation and particle size distribution of fat. Turk J Vet Anim Sci 28:303–308

Feng H, Barbosa-Cánovas GV, Weiss J (2011) Ultrasound technologies for food and bioprocessing. Springer, New York

Fernandes FAN, Gallão MI, Rodrigues S (2008) Effect of osmotic dehydration and ultrasound pretreatment on cell structure: melon dehydration. LWT – Food Sci Technol 41:604–610. https://doi.org/10.1016/j.lwt.2007.05.007

Fernandes FAN, Gallão MI, Rodrigues S (2009) Effect of osmosis and ultrasound on pineapple cell tissue structure during dehydration. J Food Eng 90:186–190. https://doi.org/10.1016/j.jfoodeng.2008.06.021

Fortes Patella R, Reboud J-L, Briancon Marjollet L (2004) A phenomenological and numerical model for scaling the flow agressiveness in cavitation erosion. In: EROCAV workshop. https://www.researchgate.net/publication/281921326

Fu X, Belwal T, Cravotto G, Luo Z (2020) Sono-physical and sono-chemical effects of ultrasound: primary applications in extraction and freezing operations and influence on food components. Ultrason Sonochem 60:104726. https://doi.org/10.1016/j.ultsonch.2019.104726

Fujikawa S, Akamatsu T (1980) Effects of the non-equilibrium condensation of vapour on the pressure wave produced by the collapse of a bubble in a liquid. J Fluid Mech 97:481–512. https://doi.org/10.1017/S0022112080002662

Gallo M, Ferrara L, Naviglio D (2018) Application of ultrasound in food science and technology: a perspective. Foods 7:1–19. https://doi.org/10.3390/foods7100164

Gogate PR, Pandit AB (2011) Sonocrystallization and its application in food and bioprocessing. In: Ultrasound technologies for food and bioprocessing, pp 65–105. https://doi.org/10.1007/978-1-4419-7472-3

Gregorčič P, Petkovšek R, Možina J (2007) Investigation of a cavitation bubble between a rigid boundary and a free surface. J Appl Phys 102. https://doi.org/10.1063/1.2805645

Guiseppi-Elie A, Choi SH, Geckeler KE (2009) Ultrasonic processing of enzymes: effect on enzymatic activity of glucose oxidase. J Mol Catal B Enzym 58:118–123. https://doi.org/10.1016/j.molcatb.2008.12.005

Hongyu W, Hulbert GJ, Mount JR (2000) Effects of ultrasound on milk homogenization and fermentation with yogurt starter. Innov Food Sci Emerg Technol 1:211–218. https://doi.org/10.1016/S1466-8564(00)00020-5

Hosseini SMH, Emam-Djomeh Z, Razavi SH, Moosavi-Movahedi AA, Saboury AA, Mohammadifar MA, Farahnaky A, Atri MS, Van Der Meeren P (2013) Complex coacervation of β-lactoglobulin - κ-carrageenan aqueous mixtures as affected by polysaccharide sonication. Food Chem 141:215–222. https://doi.org/10.1016/j.foodchem.2013.02.090

Islam MN, Zhang M, Adhikari B, Xinfeng C, Xu BG (2014) The effect of ultrasound-assisted immersion freezing on selected physicochemical properties of mushrooms. Int J Refrig 42:121–133. https://doi.org/10.1016/j.ijrefrig.2014.02.012

Jambrak AR, Mason TJ, Lelas V, Herceg Z, Herceg IL (2008) Effect of ultrasound treatment on solubility and foaming properties of whey protein suspensions. J Food Eng 86:281–287. https://doi.org/10.1016/j.jfoodeng.2007.10.004

Jena F (1965) Sonochemistry of selected hydrocarbons, sulfur-containing and nitrogen-containing organic compounds in aqueous solutions and nonaqueous liquids. Dissertation

Johansen K, Song JH, Prentice P (2017) Validity of the Keller-Miksis equation for "non-stable" cavitation and the acoustic emissions generated. In: 2017 IEEE International Ultrasonics Symposium (IUS). https://doi.org/10.1109/ULTSYM.2017.8091852

Karlović S (2015) Reducing fat globules particle-size in goat milk: ultrasound and high hydrostatic pressures approach. Chem Biochem Eng Q J 28:499–507. https://doi.org/10.15255/cabeq.2014.19400

Kentish S, Ashokkumar M (2010) The physical and chemical effects of ultrasound. In: Ultrasound technologies for food and bioprocessing. Springer, New York, pp 1–19

Kerboua K, Hamdaoui O (2017) Computational study of state equation effect on single acoustic cavitation bubble's phenomenon. Ultrason Sonochem 38. https://doi.org/10.1016/j.ultsonch.2017.03.005

Kerboua K, Hamdaoui O (2018a) Ultrasonic waveform upshot on mass variation within single cavitation bubble: investigation of physical and chemical transformations. Ultrason Sonochem 42:508–516. https://doi.org/10.1016/j.ultsonch.2017.12.015

Kerboua K, Hamdaoui O (2018b) Numerical investigation of the effect of dual frequency sonication on stable bubble dynamics. Ultrason Sonochem 49. https://doi.org/10.1016/j.ultsonch.2018.08.025

Kerboua K, Hamdaoui O (2018c) Influence of reactions heats on variation of radius, temperature, pressure and chemical species amounts within a single acoustic cavitation bubble. Ultrason Sonochem 41. https://doi.org/10.1016/j.ultsonch.2017.10.001

Kerboua K, Hamdaoui O (2019a) Oxidants emergence under dual-frequency sonication within single acoustic bubble: effects of frequency combinations. Iran J Chem Chem Eng. https://doi.org/10.30492/ijcce.2019.36705

Kerboua K, Hamdaoui O (2019b) Void fraction, number density of acoustic cavitation bubbles, and acoustic frequency: a numerical investigation. J Acoust Soc Am 146. https://doi.org/10.1121/1.5126865

Kerboua K, Hamdaoui O (2020) Sonochemistry in green processes: modeling, experiments, and technology. In: Sustainable green chemical processes and their allied applications, p 603. https://doi.org/10.1007/978-3-030-42284-4

Kerboua K, Merouani S, Hamdaoui O, Alghyamah A, Islam MH, Hansen HE, Pollet BG (2021a) How do dissolved gases affect the sonochemical process of hydrogen production?An overview of thermodynamic and mechanistic effects – on the "hot spot theory". Ultrason Sonochem 72:105422. https://doi.org/10.1016/j.ultsonch.2020.105422

Kerboua K, Hamdaoui O, Alghyamah A (2021b) Energy balance of high-energy stable acoustic cavitation within dual-frequency sonochemical reactor. Ultrason Sonochem 73. https://doi.org/10.1016/j.ultsonch.2021.105471

Kerboua K, Hamdaoui O, Alghyamah A (2021c) Acoustic frequency and optimum sonochemical production at single and multi-bubble scales: a modeling answer to the scaling dilemma. Ultrason Sonochem 70:105341. https://doi.org/10.1016/j.ultsonch.2020.105341

Kiani H, Sun DW (2018) Numerical simulation of heat transfer and phase change during freezing of potatoes with different shapes at the presence or absence of ultrasound irradiation. Heat Mass Transf Und Stoffuebertragung 54:885–894. https://doi.org/10.1007/s00231-017-2190-5

Kiani H, Zhang Z, Delgado A, Sun DW (2011) Ultrasound assisted nucleation of some liquid and solid model foods during freezing. Food Res Int 44:2915–2921. https://doi.org/10.1016/j.foodres.2011.06.051

Kiani H, Zhang Z, Sun DW (2015) Experimental analysis and modeling of ultrasound assisted freezing of potato spheres. Ultrason Sonochem 26:321–331. https://doi.org/10.1016/j.ultsonch.2015.02.015

Koh LLA, Nguyen HTH, Chandrapala J, Zisu B, Ashokkumar M, Kentish SE (2014) The use of ultrasonic feed pre-treatment to reduce membrane fouling in whey ultrafiltration. J Memb Sci 453:230–239. https://doi.org/10.1016/j.memsci.2013.11.006

Kolb J, Nyborg WL (1956) Small-scale acoustic streaming in liquids. J Acoust Soc Am 28:1237–1242

Kornfeld M, Suvorov L (1944) On the destructive action of cavitation. J Appl Phys 15:495–506. https://doi.org/10.1063/1.1707461

Kougoulos E, Marziano I, Miller PR (2010) Lactose particle engineering: influence of ultrasound and anti-solvent on crystal habit and particle size. J Cryst Growth 312:3509–3520. https://doi.org/10.1016/j.jcrysgro.2010.09.022

Lee J, Kentish S, Ashokkumar M (2005) Effect of surfactants on the rate of growth of an air bubble by rectified diffusion. J Phys Chem B 109:14595–14598. https://doi.org/10.1021/jp051758d

Leighton TG (1995) Bubble population phenomena in acoustic cavitation. Ultrason Sonochem 2:123–136. https://doi.org/10.1016/1350-4177(95)00021-W

Leong T, Ashokkumar M, Sandra K (2011) The fundamentals of power ultrasound – a review. Acoust Aust 39:54–63

Li K, Fu L, Zhao YY, Xue SW, Wang P, Xu XL, Bai YH (2020) Use of high-intensity ultrasound to improve emulsifying properties of chicken myofibrillar protein and enhance the rheological properties and stability of the emulsion. Food Hydrocoll 98. https://doi.org/10.1016/j.foodhyd.2019.105275

Lianfu Z, Zelong L (2008) Optimization and comparison of ultrasound/microwave assisted extraction (UMAE) and ultrasonic assisted extraction (UAE) of lycopene from tomatoes. Ultrason Sonochem 15:731–737. https://doi.org/10.1016/j.ultsonch.2007.12.001

López-Bascón-Bascon MA, Luque de Castro MD (2019) Soxhlet extraction. Liq Extr:327–354. https://doi.org/10.1016/B978-0-12-816911-7.00011-6

Louisnard O, González-García J (2011) Acoustic cavitation. Food Eng Ser:13–64. https://doi.org/10.1007/978-1-4419-7472-3_2

Luo MR (2020) Encyclopedia of color science and technology. https://doi.org/10.1007/978-3-642-27851-8

Luo J, Fang Z, Smith RL, Qi X (2015) Fundamentals of acoustic cavitation in sonochemistry. https://doi.org/10.1007/978-94-017-9624-8_1

Luo X, Cao J, Gong H, Yan H, He L (2018) Phase separation technology based on ultrasonic standing waves: a review. Ultrason Sonochem 48:287–298. https://doi.org/10.1016/j.ultsonch.2018.06.006

Luque de Castro MD, Priego-Capote F (2007) Ultrasound-assisted crystallization (sonocrystallization). Ultrason Sonochem 14:717–724. https://doi.org/10.1016/j.ultsonch.2006.12.004

Luque de Castro MD, Priego-Capote F (2010) Soxhlet extraction: past and present panacea. J Chromatogr A 1217:2383–2389. https://doi.org/10.1016/j.chroma.2009.11.027

Luque-García JL, Luque De Castro MD (2004) Ultrasound-assisted Soxhlet extraction: an expeditive approach for solid sample treatment – application to the extraction of total fat from oleaginous seeds. J Chromatogr A 1034:237–242. https://doi.org/10.1016/j.chroma.2004.02.020

Marmottant P, Hilgenfeldt S (2003) Controlled vesicle deformation and lysis by single oscillating bubbles. Nature 423:153–156. https://doi.org/10.1038/nature01613

Marshall JS, Wu J (2015) Acoustic streaming, fluid mixing, and particle transport by a Gaussian ultrasound beam in a cylindrical container. Phys Fluids 27. https://doi.org/10.1063/1.4932232

Mason TJ (2007) Developments in ultrasound-non-medical. Prog Biophys Mol Biol 93:166–175. https://doi.org/10.1016/j.pbiomolbio.2006.07.007

Mason TJ, Paniwnyk L, Lorimer JP (1996) The uses of ultrasound in food technology. Ultrason Sonochem 3. https://doi.org/10.1016/S1350-4177(96)00034-X

Masuzawa N, Ohdaira E (2002) Attempts to shorten the time of lactic fermentation by ultrasonic irradiation. Jpn J Appl Physics Part 1 Regul Pap Short Notes Rev Pap 41:3277–3278. https://doi.org/10.1143/jjap.41.3277

Milton RBC, Plesset S, Office of Naval Research (1970) Collapse of an initially spherical vapor cavity in the neighborhood of a solid boundary. J Fluid Mech 47:49–85

Morey MD, Deshpande NS, Barigou M (1999) Foam destabilization by mechanical and ultrasonic vibrations. J Colloid Interface Sci 219:90–98. https://doi.org/10.1006/jcis.1999.6451

Muthukumaran S, Kentish S, Lalchandani S, Ashokkumar M, Mawson R, Stevens GW, Grieser F (2005a) The optimisation of ultrasonic cleaning procedures for dairy fouled ultrafiltration membranes. Ultrason Sonochem 12:29–35. https://doi.org/10.1016/j.ultsonch.2004.05.007

Muthukumaran S, Kentish SE, Ashokkumar M, Stevens GW (2005b) Mechanisms for the ultrasonic enhancement of dairy whey ultrafiltration. J Memb Sci 258:106–114. https://doi.org/10.1016/j.memsci.2005.03.001

Nguyen NHA, Anema SG (2010) Effect of ultrasonication on the properties of skim milk used in the formation of acid gels. Innov Food Sci Emerg Technol 11:616–622. https://doi.org/10.1016/j.ifset.2010.05.006

Nowacka M, Tylewicz U, Laghi L, Dalla Rosa M, Witrowa-Rajchert D (2014) Effect of ultrasound treatment on the water state in kiwifruit during osmotic dehydration. Food Chem 144:18–25. https://doi.org/10.1016/j.foodchem.2013.05.129

Panchev I, Kirchev N, Kratchanov C (1988) Improving pectin technology. II. Extraction using ultrasonic treatment. Int J Food Sci Technol 23:337–341. https://doi.org/10.1111/j.1365-2621.1988.tb00587.x

Pankaj, Ashokkumar M (2011) Theoretical and experimental sonochemistry involving inorganic systems. Theor Exp Sonochem Involv Inorg Syst:1–404. https://doi.org/10.1007/978-90-481-3887-6

Parlitz U, Mettin R, Luther S, Akhatov I, Voss M, Lauterborn W (1999) Spatio-temporal dynamics of acoustic cavitation bubble clouds. Philos Trans R Soc A Math Phys Eng Sci 357:313–334. https://doi.org/10.1098/rsta.1999.0329

Patrick M, Blindt R, Janssen J (2004) The effect of ultrasonic intensity on the crystal structure of palm oil. Ultrason Sonochem 11:251–255. https://doi.org/10.1016/j.ultsonch.2004.01.017

Peters A, Sagar H, Lantermann U, el Moctar O (2015) Numerical modelling and prediction of cavitation erosion. Wear 338–339:189–201. https://doi.org/10.1016/j.wear.2015.06.009

Prithani R, Dash KK (2020) Mass transfer modelling in ultrasound assisted osmotic dehydration of kiwi fruit. Innov Food Sci Emerg Technol 64:102407. https://doi.org/10.1016/j.ifset.2020.102407

Rahaman A, Zeng XA, Kumari A, Rafiq M, Siddeeg A, Manzoor MF, Baloch Z, Ahmed Z (2019) Influence of ultrasound-assisted osmotic dehydration on texture, bioactive compounds and metabolites analysis of plum. Ultrason Sonochem 58. https://doi.org/10.1016/j.ultsonch.2019.104643

Raso J, Barbosa-Cánovas GV (2003) Nonthermal preservation of foods using combined processing techniques. Crit Rev Food Sci Nutr 43:265–285. https://doi.org/10.1080/10408690390826527

Rastogi NK (2011) Opportunities and challenges in application of ultrasound in food processing. Crit Rev Food Sci Nutr 51:705–722. https://doi.org/10.1080/10408391003770583

Raviyan P, Zhang Z, Feng H (2005) Ultrasonication for tomato pectinmethylesterase inactivation: effect of cavitation intensity and temperature on inactivation. J Food Eng 70:189–196. https://doi.org/10.1016/j.jfoodeng.2004.09.028

Riener J, Noci F, Cronin DA, Morgan DJ, Lyng JG (2009) The effect of thermosonication of milk on selected physicochemical and microstructural properties of yoghurt gels during fermentation. Food Chem 114:905–911. https://doi.org/10.1016/j.foodchem.2008.10.037

Riera E, Gallego-Juárez JA, Mason TJ (2006) Airborne ultrasound for the precipitation of smokes and powders and the destruction of foams. Ultrason Sonochem 13:107–116. https://doi.org/10.1016/j.ultsonch.2005.04.001

Rodrigues S, Fernandes FAN (2017) Extraction processes assisted by ultrasound. Elsevier Inc. https://doi.org/10.1016/B978-0-12-804581-7.00014-2

Rodríguez G, Riera E, Gallego-Juárez JA, Acosta VM, Pinto A, Martínez I, Blanco A (2010) Experimental study of defoaming by air-borne power ultrasonic technology. Phys Procedia 3:135–139. https://doi.org/10.1016/j.phpro.2010.01.019

Saikia S, Mahnot NK, Mahanta CL (2016) A comparative study on the effect of conventional thermal pasteurisation, microwave and ultrasound treatments on the antioxidant activity of five fruit juices. Food Sci Technol Int 22:288–301. https://doi.org/10.1177/1082013215596466

Sakakibara M, Wang D, Ikeda K, Suzuki K (1994) Effect of ultrasonic irradiation on production of fermented milk with Lactobacillus delbrueckii. Ultrason Sonochem 1:107–110. https://doi.org/10.1016/b0-12-227235-8/00243-1

Salleh-Mack SZ, Roberts JS (2007) Ultrasound pasteurization: the effects of temperature, soluble solids, organic acids and pH on the inactivation of Escherichia coli ATCC 25922. Ultrason Sonochem 14:323–329. https://doi.org/10.1016/j.ultsonch.2006.07.004

Sánchez ES, Simal S, Femenia A, Benedito J, Rosselló C (1999) Influence of ultrasound on mass transport during cheese brining. Eur Food Res Technol 209:215–219. https://doi.org/10.1007/s002170050483

Sato M, Takakuwa O, Nakai M, Niinomi M, Takeo F, Soyama H (2016) Using cavitation peening to improve the fatigue life of titanium alloy Ti-6Al-4V manufactured by electron beam melting. Mater Sci Appl 07:181–191. https://doi.org/10.4236/msa.2016.74010

Sfakianakis P, Topakas E, Tzia C (2015) Comparative study on high-intensity ultrasound and pressure milk homogenization: effect on the kinetics of yogurt fermentation process. Food Bioprocess Technol 8:548–557. https://doi.org/10.1007/s11947-014-1412-9

Silva M, Chandrapala J (2021) Ultrasound in dairy processing. In: Innovative food processing technologies a comprehensive review. Elsevier, pp 439–464. https://doi.org/10.1016/b978-0-08-100596-5.22955-7

Simal S, Benedito J, Sánchez ES, Rosselló C (1998) Use of ultrasound to increase mass transport rates during osmotic dehydration. J Food Eng 36:323–336. https://doi.org/10.1016/S0260-8774(98)00053-3

Singla M, Sit N (2021) Application of ultrasound in combination with other technologies in food processing : a review. Ultrason Sonochem 73:105506. https://doi.org/10.1016/j.ultsonch.2021.105506

Stojanovic J, Silva JL (2006) Influence of osmoconcentration, continuous high-frequency ultrasound and dehydration on properties and microstructure of rabbiteye blueberries. Dry Technol 24:165–171. https://doi.org/10.1080/07373930600558995

Sunartio D, Ashokkumar M, Grieser F (2007) Study of the coalescence of acoustic bubbles as a function of frequency, power, and water-soluble additives. J Am Chem Soc 129:6031–6036. https://doi.org/10.1021/ja068980w

Tan MC, Chin NL, Yusof YA, Taip FS, Abdullah J (2015) Characterisation of improved foam aeration and rheological properties of ultrasonically treated whey protein suspension. Int Dairy J 43:7–14. https://doi.org/10.1016/j.idairyj.2014.09.013

Tan MC, Chin NL, Yusof YA, Abdullah J (2016) Effect of high power ultrasonic treatment on whey protein foaming quality. Int J Food Sci Technol 51:617–624. https://doi.org/10.1111/ijfs.13013

Tang CH (2007) Functional properties and in vitro digestibility of buckwheat protein products: influence of processing. J Food Eng 82:568–576. https://doi.org/10.1016/j.jfoodeng.2007.01.029

Tao Y, Sun DW (2015) Enhancement of food processes by ultrasound: a review. Crit Rev Food Sci Nutr 55:570–594. https://doi.org/10.1080/10408398.2012.667849

Vercet A, Burgos J, Crelier S, Lopez-Buesa P (2001) Inactivation of proteases and lipases by ultrasound. Innov Food Sci Emerg Technol 2:139–150. https://doi.org/10.1016/S1466-8564(00)00037-0

Vercet A, Oria R, Marquina P, Crelier S, Lopez-Buesa P (2002) Rheological properties of yoghurt made with milk submitted to manothermosonication. J Agric Food Chem 50:6165–6171. https://doi.org/10.1021/jf0204654

Vilkhu K, Mawson R, Simons L, Bates D (2008) Applications and opportunities for ultrasound assisted extraction in the food industry – a review. Innov Food Sci Emerg Technol 9:161–169. https://doi.org/10.1016/j.ifset.2007.04.014

Villamiel M, De Jong P (2000) Influence of high-intensity ultrasound and heat treatment in continuous flow on fat, proteins, and native enzymes of milk. J Agric Food Chem 48:472–478. https://doi.org/10.1021/jf990181s

Wilde PJ, Clark DC (1993) The competitive displacement of β-lactoglobulin by tween 20 from oil-water and air-water interfaces. J Colloid Interface Sci 155:48–54. https://doi.org/10.1006/jcis.1993.1008

Wordon BA, Mortimer B, McMaster LD (2012) Comparative real-time analysis of Saccharomyces cerevisiae cell viability, injury and death induced by ultrasound (20 kHz) and heat for the application of hurdle technology. Food Res Int 47:134–139. https://doi.org/10.1016/j.foodres.2011.04.038

Wu J (2018) Acoustic streaming and its applications. Fluids 3. https://doi.org/10.3390/fluids3040108

Yang RF, Geng LL, Lu HQ, Fan XD (2017) Ultrasound-synergized electrostatic field extraction of total flavonoids from Hemerocallis citrina baroni. Ultrason Sonochem 34:571–579. https://doi.org/10.1016/j.ultsonch.2016.06.037

Yasui K (1997) Alternative model of single bubble sonoluminescence. Phys Rev E 56:6750–6760. https://doi.org/10.1103/PhysRevA.65.054304

Yasui K (1998) Effect of non-equilibrium evaporation and condensation on bubble dynamics near the sonoluminescence threshold. Ultrasonics 36:575–580. https://doi.org/10.1016/S0041-624X(97)00107-8

Yasui K (2002) Influence of ultrasonic frequency on multibubble sonoluminescence. J Acoust Soc Am 112:1405–1413. https://doi.org/10.1121/1.1502898

Yasui K (2011) Fundamentals of acoustic cavitation and sonochemistry. In: Theoretical and experimental sonochemistry involving inorganic systems. National Institute of Advanced Industrial Science and Technology, Anagahora-Japan, pp 1–29. https://doi.org/10.1007/978-90-481-3887-6

Yasui K (2017) Acoustic cavitation and bubble dynamics. Springer Briefs in Molecular Science, Nagoya

Yasui K (2018) Acoustic cavitation. In: Acoustic cavitation and bubble dynamics, pp 1–35, Springer, Cham, Switzerland

Yasui K, Tuziuti T, Kozuka T, Towata A, Iida Y (2007) Relationship between the bubble temperature and main oxidant created inside an air bubble under ultrasound. J Chem Phys 127:1–9. https://doi.org/10.1063/1.2790420

Ye L, Zhu X (2017) Analysis of the effect of impact of near-wall acoustic bubble collapse micro-jet on Al 1060. Ultrason Sonochem 36:507–516. https://doi.org/10.1016/j.ultsonch.2016.12.030

Yuan H, Prosperetti A (1997) Gas-liquid heat transfer in a bubble collapsing near a wall. Phys Fluids 9:127–142. https://doi.org/10.1063/1.869153

Yusof NSM, Babgi B, Alghamdi Y, Aksu M, Madhavan J, Ashokkumar M (2016) Physical and chemical effects of acoustic cavitation in selected ultrasonic cleaning applications. Ultrason Sonochem 29:568–576. https://doi.org/10.1016/j.ultsonch.2015.06.013

Zheng L, Sun DW (2006) Innovative applications of power ultrasound during food freezing processes – a review. Trends Food Sci Technol 17:16–23. https://doi.org/10.1016/j.tifs.2005.08.010

Chapter 7
Non-electro-Technologies: Pulsed Light

Gianpiero Pataro and Giovanna Ferrari

1 Fundamentals of Pulsed Light Application

Pulsed light (PL), also known by several other names such as pulsed UV light, high intensity pulsed light, pulsed white light, or intense light pulses, is one of the emerging non-thermal technologies that has gained increasing interest, from both the research world and food processing industry, as a rapid, effective and mild alternative to conventional chemical and thermal treatment to decrease microbial count in foods, food contact surfaces and equipment (Gómez-López et al. 2007; Mahedran et al. 2019).

In its basic, PL consists of the exposure of foods or non-food material to a successive repetition of short duration (100 ns-1 ms), high-intensity pulses (flashes) of polychromatic light (200 nm – 1100 nm) including UV (100–400 nm), visible (400–700 nm) and near infrared region (700–1100 nm) emitted by an inert-gas lamps (Pirozzi et al. 2020). The UV portion of the electromagnetic spectrum includes long-wave UV-A (315–400 nm), medium-wave UV-B (280–315 nm), and short-wave UV-C (200–280 nm) (Wekhof et al. 2001). The light used for food processing applications is typically pulsed at 1 to 20 flashes per second at an energy density in the range of about 0.01 to 50 J/cm^2 at the surface (Oms-Oliu et al. 2010).

The efficacy of PL has been successfully tested against a great variety of pathogenic and spoilage microorganisms, including bacterial species (both as vegetative cells or spores), yeast, fungi and viruses spread on agar, or contaminating packed

G. Pataro (✉)
Department of Industrial Engineering, University of Salerno, Fisciano (SA), Italy
e-mail: gpataro@unisa.it

G. Ferrari
Department of Industrial Engineering, University of Salerno, Fisciano (SA), Italy

ProdAl Scarl – University of Salerno, Fisciano (SA), Italy

A. Režek Jambrak (ed.), *Nonthermal Processing in Agri-Food-Bio Sciences*,
Food Engineering Series, https://doi.org/10.1007/978-3-030-92415-7_7

and unpackaged foods or food contact surfaces (packaging material), or suspended in transparent beverages (e.g., drinking water, fruit juices) or in fluid involved in the production processes (air, water, etc.) (Chen et al. 2015a, b; Heinrich et al. 2015; Mahedran et al. 2019; Manzocco et al. 2015).

PL treatment can allow to achieve anywhere from 0.5 to 8 log reduction (Mahedran et al. 2019). The wide range in levels of inactivation achieved indicates that on one hand, PL can be an effective treatment against food borne pathogens in certain substrates, but also that probably microbes exhibit variations in PL susceptibility (Chen et al. 2015a, b). On the other hand, it also indicate that an optimization of equipment design as well as of the main factors that influence the effectiveness of PL treatment, is necessary (Pataro et al. 2016).

The effect of PL on the microbial inactivation is commonly ascribed to the effects of UV component of the broad spectrum of the flash and impacts of the high peak power (Oms-Oliu et al. 2010), which result in the coexistence of different inactivation mechanisms. In particular, the UV component of PL can be absorbed by proteins, DNA and RNA, thereby causing photochemical damage, which kills microorganisms (Wang et al. 2005). On the other hand, PL spectrum includes visible and especially near-infrared regions, which convey heat to the surface of the processed substrate, inducing a local instantaneous increase of the temperature of only a thin (few μm thick) surface layer without considerably increasing the internal temperature of the irradiated substrates. In microbial cells this localized overheating may cause rapid vaporization of intracellular liquid, generating a small steam flow that induces disruption of cell membrane and wall (photothermal effect) (Wekhof 2000; Wekhof et al. 2001). Additionally, structural cell damages caused by the high power pulsing effect (photophysical effect) have been also detected through microscopic observation as well as quantification of intracellular matter leakage, such as proteins, which was not observed when the same microbial cells were exposed to continuous wave UV light treatment (Krishnamurthy et al. 2007; Pataro et al. 2011; Takeshita et al. 2003). The relative importance of each mechanism may depend on the peak power of the light pulses, the composition of the emitted light spectrum, the type of microorganism as well as the physical and optical properties of the target substrate, among other.

The superimposition of these different microbial inactivation mechanisms acting in parallel or in sequence, along with the high emission power, which likely increases the capability of PL to penetrate the treated substrates, can explain the generally reported higher decontamination effectiveness of PL as compared with continuous wave UV light treatment (Dunn et al. 1989).

Moreover, PL decontamination technology possesses the characteristics of fast and easy operation, residue-free method, and low energy consumption. It has also an easy integration in the current industrial processing lines, alone or in combination with other disinfection methods or preservation technologies (Artíguez and Marañon 2015; Donsi et al. 2015; Kramer et al., 2017; Ferrario et al. 2015; Muñoz et al. 2011), while enabling high product throughput.

2 Technical Aspect and PL Systems

2.1 Definitions and Terminology

The exposure conditions of a given substrate to PL irradiation are commonly expressed by several parameters and units (Gómez-López and Bolton 2016; Gómez-López et al. 2007, 2021; Pirozzi et al. 2020), whose formal definitions can be found in IUPAC (1996).

- Pulse duration (or pulse width): time interval (ns - ms) during which the light energy is delivered;
- Number of pulses (or flashes): total number of pulses or flashes of light delivered to the target;
- Pulse repetition frequency: number of pulses of light delivered per second (Hertz [Hz]) or commonly expressed as pps (pulses per second). Due to the design features of the lamp, in PL systems, the pulse frequency is typically limited to a few Hertz. However, higher pulse frequency can be achieved by using two or more lamps, placed and flashed in sequence;
- Exposure time: actual time (in seconds) of exposure of the substrate to the flash of light, and is calculated as the number of pulses times the pulse width;
- Peak power, ϕ: is measured in Watt (W) and is the pulse energy divided by the pulse duration.
- Fluence rate (F_o): is measured in Watt/meter^2 (W/m^2) and is the energy received from the lamp by the sample per unit area per second.
- Fluence (F) (or PL dose): is the energy received from the lamp by the sample per unit area during the treatment. The fluence is the fluence rate multiplied by the exposure time, and its IS unit is J/m^2, even though in PL technology it is often given as J/cm^2. However, when the substrate exposed to PL treatment is a given volume of a liquid containing a certain microbial load expressing the PL dose as total radiant energy per unit area is not correct. Thus, several authors have proposed that the fluence (in J/L or J/mL) is correctly defined, according to Eq. (7.1), as the PL energy output (radiant power, Φ, in W) delivered to a volumetric flow (Q, in L/s, or mL/s):

$$F = \frac{\varnothing}{Q} \tag{7.1}$$

PL treatments expressed as volumetric electrical energy input (J/L or J/mL) allows comparing the results of PL treatments carried out with different PL units, as well as with those obtained utilizing other inactivation technologies (Gouma et al. 2016).

Proper measurement of fluence received by the treated surface of solid substrate or within the liquid substrate, which is substantially different from the energy delivered by the light source due to light absorption and scattering phenomena (Artíguez and Marañón 2015; Hsu and Moraru 2011; Gómez-López et al. 2021; Pirozzi et al.

2021), is therefore fundamental to characterize a PL treatment. However, fluence determination can be complex, requiring a good knowledge of light properties. Additionally, it also must be kept in mind the lamp emission decreases with lamp use (Schaefer et al. 2007), therefore, fluence must be measured frequently and do not rely in a single initial measurement.

It is also worth noting that fluence alone does not give information about how much radiation of effective wavelengths (for example, germicidal UV-C) is reaching the target substrate (Gómez-López et al. 2021). In this line, the light spectrum emitted by the flash lamp represents a relevant feature. However, the spectral distribution of the emitted light differs among PL systems and, in some instances, the same PL system can generate different emission spectra depending on the operating conditions (Schaefer et al. 2007). Moreover, reporting emission spectra is not easy since its measurement requires specialized equipment (a spectroradiometer), which is uncommon in food laboratories. (Gómez-López et al. 2021). Recommendations on fluence determinations and measurement of emission spectra can be found in Bolton and Linden (2003), Bolton et al. (2015), Gómez-López and Bolton (2016), Gómez-López et al. (2021).

In addition to the above parameters, other key parameters characterize a PL treatment and should be included in the overall analyses of the process. For instance, since the assessment of fluence distribution inside liquid substrrates is a difficult task, liquid samples should be also characterized for the different factors that attenuate light propagation, namely, suspended solids, absorption spectrum and sample thickness (Gómez-López et al. 2021; Pataro et al. 2016).

Furthermore, temperature evolution or the highest temperature increment of the treated substrate should be indicate. In fact, even though PL is considered a non-thermal technology, a very high increase in temperature either caused by the absorption of light by the treated substrate or by lamp heating can be observed during PL treatment (Pataro et al. 2011, 2016). This especially occurs at higher energy doses and long operation time that could strongly compromise the quality of the treated substrate. The temperature increase in the substrate exposed to PL treatment is mainly a function of the total amount of energy delivered by the light source to the target, as well as of the pulse frequency, distance from the light source, and composition of the PL spectrum. Pulses of equivalent fluence and similar spectra generate the same amount of heat. However, a slow pulse repetition rate and an appropriate distance between the product and the lamp source may lower sample heating (Gómez-López et al. 2021; Pataro et al. 2011, 2016).

2.2 PL Systems

PL is generated using pulsed power technology that involves the generation of high-power electrical pulses and their transformation into high-power light pulses. Although PL equipment differ according to manufacturer, a typical PL system include a power/control module, a lamp unit, a treatment chamber (Fig. 7.1), as well

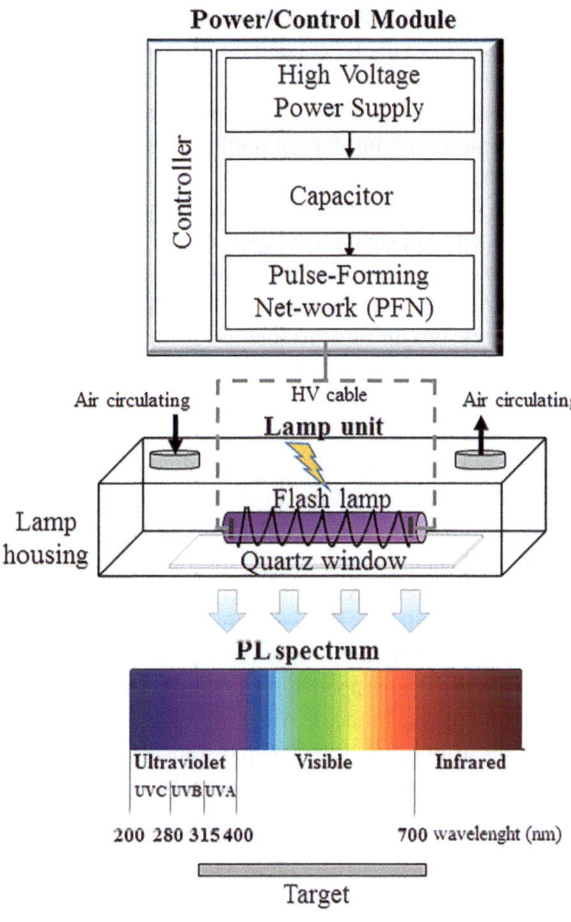

Fig. 7.1 Schematics of the operative principle of PL generation

as auxiliary devices, such as pump or conveyor belt for moving the product through the irradiated zone, cooling systems, and devices for measurements of temperature, fluence rate or fluence (Pirozzi et al. 2020).

The power/control module consists of high-voltage power supply, a storage capacitor, a pulse-forming net-work (PNF), and a controller (Fig. 7.1). It is used to manage the process, start the flashes, control the treatment time, set the mode of pulse deliver such as single pulse, burst of pulses (timed mode), or continuous array of pulses (continuous mode), and modulate the generation of high-power electrical pulses to obtain the desired configuration of the electric pulse energy and rate (Pirozzi et al. 2020). In particular, during PL treatment, the high voltage power supply converts alternating current (AC) into direct current (DC), which is temporarily accumulated in the capacitor bank. When the capacitor reaches a pre-established level, the controller releases the stored electrical energy through the PNF that determines the shape of the electric pulse, which is delivered as a high voltage and high-energy electric current to the flash lamp through a coaxial cable (Fig. 7.1).

The lamp unit consists of a lamp housing with one or more flash lamps made of quartz envelops, each equipped with two electrodes and typically filled with inert gases (e.g., xenon, krypton) or a mixture of noble gases due to their high-efficiency electrical-to-optical energy conversion (Franco-Vega et al. 2021). When a high-energy pulsed electric current passes through the lamp, the gases are ionized and intense light is produced. Approximately 25% of the wavelength lies in the UV range, 45% in the visible range and 30% in the infrared range (Chen et al. 2015a, b). The shape (linear, spiral, etc.) and the size of the lamp can be customized for the specific application to ensure uniform irradiation of the target surface (Chen et al. 2015a, b). For the dissipation of the heat generated by the lamp inside the lamp housing, a cooling system of the lamp is also provided through a water or filtered air circulation (Pataro et al. 2011, 2016). Moreover, the lamp unit is also equipped with fans and/or a vacuum pump to purge the ozone produced by lamp ignition and connected to an ozone decontamination unit (Franco-Vega et al. 2021).

Several PL light systems are currently available especially for laboratory scale applications by different manufacturers, with the most commonly used being the RS-3000C SteriPulse-XL system (Xenon Corp., Wilmington, Mass., USA), and the XeMaticA-2L System (SteriBeam Systems GmbH, Germany), equipped with a single linear and Xenon flash lamp, and the PL mobile decontamination unit (Claranor, Rouaine, France) equipped with 4 xenon lamps (JA series, Verre et Quartz, Bussy Saint Georges, France). All these equipment differ for the wavelength distribution, pulse duration (50–360 μs), pulse repetition rate (0–5-5 Hz), input voltage (100–3800 V), the cooling system of the lamp (forced air or circulating water). Moreover, most of them operate in batch mode, while for continuous flow treatment of liquid, solid, or powder products, only a few laboratory-scale PL systems, which include in-house developed equipment (Chen et al. 2018; Krishnamurthy et al. 2007; Ferrario and Guerrero 2016; Muñoz et al. 2012; Pataro et al. 2011; Xu et al. 2019), or commercial dynamic flow-through pilot unit (Maria PUD system, Claranor, Manosque, France) (Artíguez and Marañón 2015), are currently available. Figure 7.2 shows simplified schematics of typical laboratory scale batch and continuous flow PL treatment chamber configurations.

Batch systems are the most widespread and used especially for preliminary investigations at laboratory scale on the effect of the main PL treatment parameters on the decontamination of liquid and solid products. During processing the solid or liquid sample contained in a Petri dish is placed inside the treatment chamber on an adjustable tray that allows regulating the distance between the sample and light source (Fig. 7.2a, b). A housing lamp with a xenon lamp is generally mounted on the top of the tray. The power/control module allows setting the treatment time or number of flashes to be delivered to the treated sample.

Improvements of PL efficiency in batch operations have been achieved by adding devices to the pulsed-light system to create multidirectional PL, the use of more than one lamp (Collazo et al. 2019), or enable manual or automatic rotation or

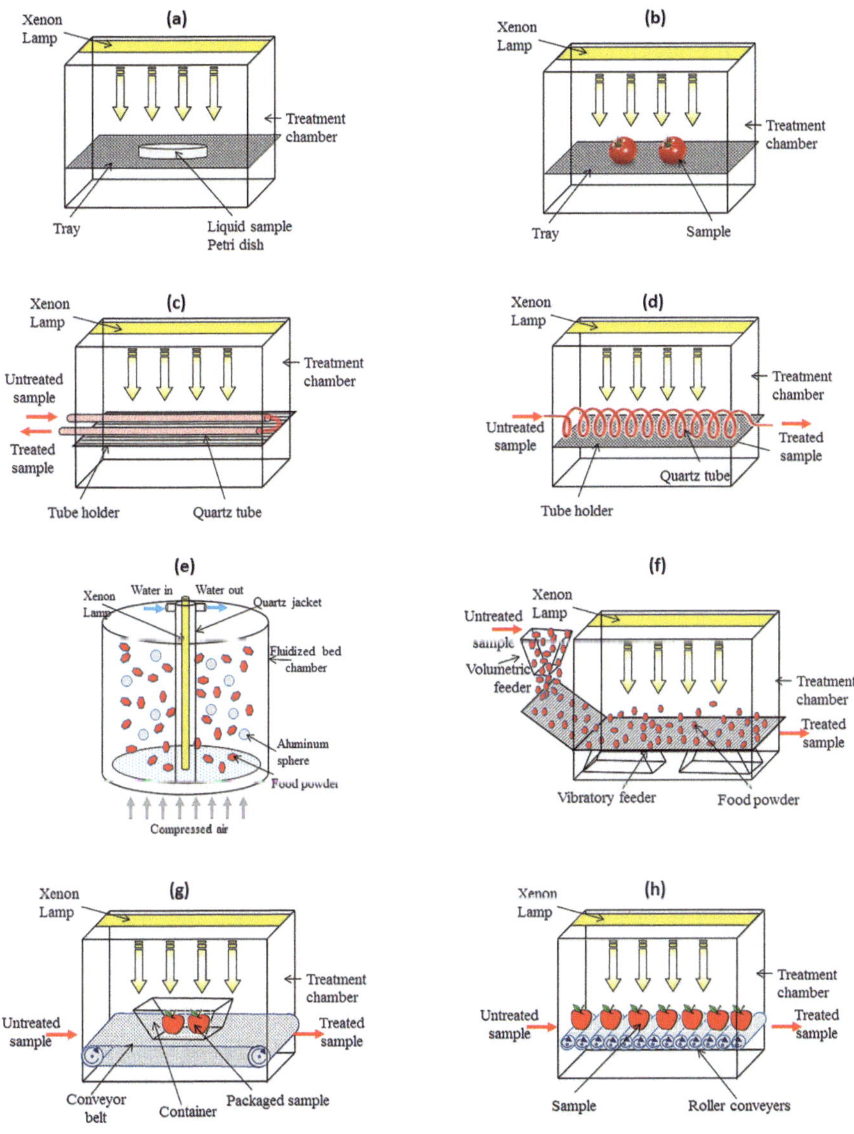

Fig. 7.2 Schematic of operating principle of typical batch or continuous flow PL treatment chamber. (**a**) batch chamber for solid products. (Adapted from Pirozzi et al. 2020); (**b**) batch chamber for liquid products. (Adapted from Pirozzi et al. 2020); (**c**) continuous flow chamber with linear quartz tubes. (Adapted from Pirozzi et al. 2020); (**d**) continuous flow chamber with spiral quartz tubes (adapted from Xu et al. 2019); (**e**) fluidized bed chamber for powdered or granular products (adapted from Fine and Gervais 2004); (**f**) vibrating treatment chamber for granular foods (adapted from Chen et al. 2018, 2019); (**g**) conveyor belt chamber for unpacked or packed solid products, food contact surface and packaging material. (Adapted from Pirozzi et al. 2020); (**h**) roller conveyors chamber for unpacked or packed solid products. (Adapted from Pirozzi et al. 2020)

mixing of products such that all surfaces or volume of the treated material could be irradiated uniformly (Franco-Vega et al. 2021; Pataro et al. 2015b).

Continuous flow systems can be designed to process either liquid or solid products, prior to or after packaging. Their use can result in a better control of processing temperature, the capability to process a large amount of product, and higher efficiency in microbial inactivation due to the improved light exposure of the product that would enhance the treatment uniformity, as compared with the discontinuous systems (Pataro et al. 2011). In continuous flow systems, the exposure time and, consequently, the total number of flashes or fluence delivered to the treated product depends on the pulse repetition frequency and residence time of the treated product in the irradiated zone, which in turn is given by the ratio between the volume of the treated product and its flow rate.

For the treatment of liquid products, the treatment chamber is typically equipped with linear (Krishnamurthy et al. 2007; Muñoz et al. 2012; Pataro et al. 2011) or spiral (Xu et al. 2019) quartz tubes through which the liquid (e.g., water, fruit juices, milk) is pumped, while it is exposed to the flash of light (Fig. 7.2c, d). Optimization of the geometry and number of quartz tubes along with their relative position to the lamp source is required to ensure the maximum disinfection efficacy with the minimum energy consumption.

For processing of granular or powder food products, well designed PL fluidized bed system could improve treatment uniformity by reducing or eliminating shadowed area (Fig. 7.2e, f). In this line, in 2004, Fine and Gervais (2004) tested a 3-L fluidized bed PL unit labeled as a One-Shot EN2/2143–1 developed by Solsys (Marseille, France) to evaluate the efficiency of PL on the destruction of dried microorganisms. Fluidized bead of aluminum sphere or cylinder where used to maximize light reflection. A quartz jacket around the lamp provided water circulation to minimize overheating. However, microbial decontamination levels remained insufficient being around 0.5 log-cycles after 31 J/cm^2 PL treatment, thus suggesting the need for further improvement of the design of the PL unit. More recently, Chen et al. (2018, 2019) developed a vibrating treatment chamber including a vibrating feeder, and automatic controls for temperature and humidity to facilitate the tumbling of nonfat dry milk particles and barley kernels in an effort to enable exposure of the entire product surface to PL for the inactivation of *Cronobacter sakazakii* and the destruction of deoxynivalenol produced by Fusarium. Results indicated that significantly higher inactivation and destruction were observed when PL was combined with the vibratory feeder.

For the PL treatment of packaged or unpacked solid products, as well as for food contact surface or packaging material (e.g., polymeric film, cap, cup, lid, bottle), the treatment chamber can be equipped with a moving platform (e.g., conveyor belts, roller conveyors), to move the product through the irradiated zone (Fig. 7.2g, h) at the velocity required for achieving the desired exposure time (or fluence) (Pirozzi et al. 2020).

This scenario illustrates the important advances made in only two decades of constant evolution and application of PL technology in the development of treatment devices designed to achieve high treatment efficiency during processing of specific substrates in view of industrial level application of the technology.

Nevertheless, commercial-scale PL systems are currently successfully used in the food industry only for the decontamination of packaging material (e.g., caps, cups, trays, and steel cans, bottles and lids). These PL units are easily integrated into the existent continuous processing line to decontaminate up to 4000–90,000 items/hour and can provide decontamination levels with a 3–5 log reduction in the reference microorganisms (http://www.claranor.com/).

3 Factors Determining the Effectiveness of the PL Process

PL process depends on many factors, which are interrelated to each other, and that are critical to the outcome of the process as they may affect the treatment uniformity and the level of energy dose than ultimately reaches the target (Pataro et al. 2016). In general, these factors can be classified into three groups, namely process set up factors, product properties and microbial factors (Gómez-López et al. 2005, 2007; Pataro et al. 2016).

3.1 Process Set Up Factors

Process set up parameters markedly affect the effectiveness of PL treatment for microbial inactivation. Among these parameters, the most important factor determining the inactivation efficacy of PL is the total fluence incident on the sample. Therefore, to achieve the required decontamination effect, the parameters affecting the total fluence, such as the distance of the target substrate from the lamp source (or pulse fluence), the number and duration of pulses, and the input voltage have to be properly optimized. In general, increasing the intensity of these parameters or decreasing the distance from the lamp, increase the total fluence and, hence, enhances microbial inactivation (Gómez-López et al. 2005, 2007; Mahendran et al. 2019; Pataro et al. 2011, 2016; Pirozzi et al. 2020). Moreover, the shorter the distance is, the narrower lamp foot print becomes, thus reducing the treatment uniformity (Gómez-López et al. 2005). For globular bodies, multidirectional and uniform illumination of all surfaces present another challenge, which can be overcome by increasing the distance to the light source, the treatment time, the relative movement of the body or by using a conveyor with transparent sections or reflectors (Heinrich et al. 2015).

The composition of the PL spectrum is another important process parameter determining the antimicrobial action of PL technology. Despite the fact that the whole spectrum can contribute to the lethal effect accomplished by PL against a wide range of microbial cells, the fluence of wavelength below 300 nm (i.e., UVC) seems play a key role in the inactivation effect of PL technology. To this regards, several scientists observed that the use UV-blocking filters led to a drastically reduction in microbial lethality of different vegetative or spores cells of bacterial and fungi species, like *Escheriachia coli, Listeria innocua, Bacillus subtilis and Aspergillus niger* (Levy et al. 2012; Ramos-Villaroel et al. 2012a, b; Wang et al. 2005; Woodling and Moraru 2007). Therefore, the use of wavelength filters, such as solid filters (e.g., glass UV filters) or liquid filters (e.g., CuSO4 solution) (Chen et al. 2015a, b; Ramos-Villaroel et al. 2012b), or the adjustment of electrical conditions (peak power or voltage) (Keklik et al. 2012), as well as the proper selection of the gas lamp, enables, for each application, the most suitable wavelength ranges emitted by the lamp to achieve the desired effect, while avoiding or minimizing alterations of the substrate properties.

Expectedly, also design parameters such as geometry and setup of the treatment chamber, type and shape of the lamp, and lamp housing configuration, as well as the number of lamps and their location in the treatment chamber, significantly influence the treatment uniformity and, hence, the efficiency of the PL process and product quality (Pataro et al. 2016; Pirozzi et al. 2020).

The temperature increase in a given substrate exposed to PL treatment is mainly a function of the total fluence delivered by the light source to the target, as well as of the pulse frequency, the distance from lamp, and composition of the PL spectrum. Thus, special attention should be taken to prevent overheating of the treated substrate through an efficient cooling system incorporated in the equipment (Pataro et al. 2011), as well as appropriate selection of pulse repetition rate and distance between the treated matrix and the lamp source, since it may seriously compromise product quality especially during long processing time (Dunn et al. 1989; Gómez-López et al. 2005).

3.2 *Product Properties*

The physical properties, surface characteristics and nutrient composition of the substrate exposed to PL treatment are also important factors, since they can enhance absorption, reflection or scattering phenomena of the incident light or induce shading effect, thus reducing the efficacy and uniformity of the PL treatment. In general, high treatment efficiency is achieved when the processed product has a low reflection-, high absorption- and transmission coefficient (Heinrich et al. 2015).

To this regard, in the case of liquid products, aside from transparent liquids (e.g., drinking water and clear fruit juices), properties such as opacity, turbidity, coloring compounds, sample thickness, viscosity and suspending particles may significantly affect the penetration of the light pulses into the depth of the liquid and, therefore, decrease the efficiency and uniformity of PL treatment (Artíguez and Marañón 2015; Mandal et al. 2020; Pataro et al. 2011). Therefore, in liquid with more limited transparency, such orange juice and milk, the effect of PL might be limited to only a superficial layer of the treated substrate. In this line, Palgan et al. (2011) evaluated the feasibility of using PL (0–8 s) to reduce *E. coli* and *L. innocua* in apple juice, orange juice, and milk, which exhibit the following transparencies trend (in decreasing order): apple juice, orange juice, and milk. The authors observed that microbial inactivation of *E. coli* and *L. innocua* dropped with reducing transparency of the medium from 4.7 and 1.93 $\log_{10}$ CFU/mL in apple juice (the most transparent media) up to 1.06 and 0.84 $\log_{10}$ CFU/mL in milk (the opaquest medium), respectively.

In such a case, optimal design of PL reactors along with the adoption of proper flow conditions leading to enhanced turbulent mixing of the fluid might be necessary to ensure a suitable intensity of PL throughout the entire volume (Dunn et al. 1989; Pataro et al. 2011).

On the other hand, when flashing solid products, properties of the treated surface such as topography, reflectivity, and hydrophobicity, as well as its color, are expected to greatly affect microbial inactivation (Chen et al. 2015a, b). For example, the target surface should be as smooth as possible since the presence of irregularities, such as surface roughness, crevices, or pores, constitute obstacle for the incident light and may provide opportunities for microbial cell shading or hiding during treatment (Heinrich et al. 2015; Gómez-López et al. 2007). The hydrophobic of the treated surface may affect the distribution of microbial cells on the surface promoting the formation of cells cluster that may negatively affect the efficacy of PL inactivation (Koch et al. 2019). Furthermore, a high reflectivity surface might decrease the light absorption of the microbial cells leading to poor inactivation (Woodling and Moraru 2005; Ringus and Moraru 2013).

The challenge is, therefore, represented by engineering solutions that might include the re-design of the equipment to promote more uniform surface irradiation even on non-smoothed surfaces (Oms-Oliu et al. 2010; Gómez-López et al. 2007). Additionally, the use of PL in combination with other disinfection methods might present significant potential benefits to food preservation, within the hurdle technology approach, to attain mild but reliable preservation effects (John and Ramaswamy 2018; Mahendran et al. 2019; Pirozzi et al. 2020).

Furthermore, the matrix should contain only low quantities of substances, such as fat and protein, able to competitively absorb light, thus decreasing the efficacy of PL inactivation. Carbohydrates, however, do not show this pronounced light-absorbing effect (Gómez-López et al. 2007). Therefore, high protein and fat-containing food products have little potential to be efficiently decontaminated by PL, while vegetables and fruit appear more suitable for PL treatment (Gómez-López et al. 2007).

3.3 *Microbial Factors*

Microbial contamination also influences the efficacy of a PL treatment in respect of the type microorganism, its physiological constitution, growth phase and population density.

For examples, the different type of microorganisms seem exhibit different PL susceptibility according to the following trend (in decreasing order): Gram-negative bacteria, Gram-positive bacteria, yeasts, fungal and bacterial spores, molds, viruses (Oms-Oliu et al. 2010). The growth phase is another microbial factor which may affect the light sensitivity of microorganisms. In general, microbial cells in stationary growth phase seem to show greater resistance to PL than cells in either lag or exponential phase (Gómez-López et al. 2005). The decontamination efficacy of PL treatment also decreased at high initial contamination levels; this fact could be due to light attenuation phenomena due to shadow effects (Gómez-López et al. 2007) when high population densities are found, thus preventing the PL incidence on microorganism placed in the lower layers (Maftei et al. 2014). This effect generates nonuniformity of the treatment and a reduction in terms of efficacy of the PL inactivation.

In conclusion, in order to optimize a PL treatment for its application in food industry, the influence of these factors on microbial killing and their interactions must be deeply understood. Moreover, optimal values of these parameters should be defined in order to maximize the microbial inactivation and treatment uniformity, thus increasing the process efficiency while minimizing any thermal effect and product impairment (Pataro et al. 2016).

4 Main Applications to Food Products

PL is a rapid, environmentally friendly, non-thermal technology with many potential applications in the food industry for the processing of foods and food contact surfaces. In this framework, PL treatment can be conducted at various stages of the food processing plant such as for the decontamination of incoming goods, or in-process treatment or treatment of the final product prior or post-packaging (Heinrich et al. 2015). More specifically, the technology appears especially interesting for the disinfection of liquid products, the sterilization of food contact surfaces and packaging material, and the decontamination solid foods such as fresh and fresh-cut produce, meat and fish and their products, prior to or after packaging. Moreover, PL appears effective in modulating the metabolic activity of fresh produce that might be useful to extend the shelf-life and produce high-added value products, as well as to induce targeted structural modification of proteins molecules.

Since 1996, PL irradiation has been approved by the U.S. Food and Drug Administration to decontaminate food or food contact surfaces, provided that the treatment uses a xenon lamp with an emission wavelength between 200–1000 nm, with a pulse duration not exceeding 2 ms and the cumulative energy level of the treatment not exceeding 12 J/cm^2 (Code of Federal Regulation, CFR: 21CFR179.41). In the case of the European Union, instead, PL is not explicitly regulated. A possible food-ingredient orientated approach is somehow covered by the regulation 258/97 "novel foods and novel food ingredients" (article 1, item f), but it is questionable if this technology can be allocated to this area. Based on the statements above, specific foods but not the treatment intrinsically can be approved in the EU at the moment (EU 1997).

However, despite this and of the recent advancements in PL technology, the increasing number of PL manufactures, and the number of studies focused on decontamination food and non-food materials, currently, only the decontamination of food packaging material has been successfully applied at industrial level (http://www.claranor.com/), while there is a lack in the application at the pilot or industrial scale on the use of PL to extend the shelf-life and improve quality of food products.

The main applications of PL technology in food processing, with the most important findings, advantages and limitations, are described in the following sections, and summarized in Tables 7.1.

4.1 PL Disinfection of Liquid Products

The use of PL as a non-thermal technology to disinfect liquids, especially beverages, is a new tendency in liquid food processing research. Although the application of PL in beverages has been reported to significantly reduce the initial number of microorganisms, which dwell in the whole volume of liquids, this reduction is highly variable and depends on the transparency and turbidity of the beverages (Mahendran et al. 2019).

For instance, highly transparent fluids like water (2.96–7.0 log-cycles) and clear apple juice (1.0–7 log-cycles) appear to be products, which can be efficiently disinfected by PL. However, in case of opaques beverages, such as orange juice, grape juice, strawberry juice or milk (0.3–2 log-cycles), PL treatments mostly show lower disinfection efficiencies, independently of the targeted microorganism (Artíguez and Marañón 2015; Kramer et al. 2017; Mahendran et al. 2019; Pataro et al. 2011; Yi et al. 2016).

Nevertheless, improvement of PL disinfention efficiency seems to some extent also possible in opaque drinks under the right conditions. An appropriate design of continuous flow PL reactors operating with thin layer of the liquid (Hillegas and Demirci 2003), as well as under flow conditions generating high turbulence intensity (Franco-Vega et al. 2021), appear crucial to improve the treatment homogeneity by ensuring that all the fluid elements receive the same energy dose, thus increasing the effectiveness of the PL process. Increased microbicidal effects, as well as minimally processed foods with improved nutritional and sensorial profiles and lower energy expenditure may be also attained by using PL in a hurdle approach in combination with other factors or decontamination technologies. Heat (Artıguez and de Marañón 2015), ultrasound (Ferrario et al. 2015; Ferrario and Guerrero 2016, 2017), thermosonication (Muñoz et al. 2011, 2012), and storage at low temperatures (<20 °C) (Ferrario et al. 2013), are the most common and reliable strategies to be used combined with PL treatments.

Regarding the impact of PL on quality attributes of beverages, relatively few data are available, with most of them related to the impact of PL on quality attributes of

Table 7.1 Summary of the main applications of PL technology in food processing, with the most important findings, advantages and limitations.

Application	Type of matrices	Total Fluence (J/cm^2)	Main findings	Advantages	Limitation Drawbacks	Refernces
Fluid disinfention	Air Buffer solutions Water Fruit juices Milk Whey Wastewater	0.5–70	0.5–7 Log reduction in microbial population	Short processing time High disinfection efficiency for light transparent, thin layer fluids Little changes in quality attributes Reduced energy consumption Reduced water footprint No residual toxic chemicals Use in a hurdle approach	Low disinfection efficiency for opaque fluids Shadow effect Sample heating	Artíguez and de Marañón (2015) Mahendran et al. (2019) Manzocco et al. (2015) Pataro et al. (2011) Yi et al. (2016)
Decontamination of solid foods before packaging	Fruit Vegetable Meat Fish	1–56 (for fruit and vegetables) 1–20 (for meat, fish and their products)	1–4 Log reduction in microbial population (for fruit and vegetables) 1–2.5 Log reduction in microbial population (for meat, fish and their products)	Short processing time Shelf-life extension No or reduced amount of chemical sanitizers Preservation of quality attributes at low or moderate intensity Use in a hurdle approach Minimally processed foods Reduced water footprint	No complete inactivation Uneven exposure Shadowing effect Sample heating Negative impact on quality attributes at high intensity Post-treatment recontamination	Kramer et al, (2017) Mahendran et al. (2019) Manzocco et al. (2015)

Application	Type of matrices	Total Fluence (J/cm^2)	Main findings	Advantages	Limitation Drawbacks	Refernces
Decontamination of in-packaged foods	Fruit Vegetables Meat	0.2–12	0.3–3 Log reduction in microbial population	Short processing time No chemicals No post-treatment recontamination	Uneven exposure Screening effect of packaging Changes in mechanical and barrier properties of packaging Possible process-product-packaging interactions Overheating of product and packaging material	Chen et al. (2015b) Haughton et al. (2011) Hierro et al. (2012) Keklik et al. (2009) Ramos-Villarroel et al. (2012a) Ramos-Villarroel et al. (2014)
Sterilization of food contact surfaces and packaging materials	Stainless steel Aluminum, Polyethylene (HDPE, LDPE) Polyethylene terephthalate (PET, PETE) Polystyrene (PS) Polypropylene (PP) Polyvinyl chloride	0.7–8	1–7 Log reduction in microbial population	Short processing time No contamination transferred from the food contact surfaces and packaging materials to the food Shelf-life extension No or reduced amount of chemical sanitizers	Limited effect of PL for material with high roughness values, water contact angle and reflectivity Possible undesired change of mechanical and barrier properties	Chen et al. (2015b) Haughton et al. (2011) Jean et al. (2011); Ringus and Moraru (2013) Woodling and Moraru (2005)

(continued)

Table 7.1 (continued)

Application	Type of matrices	Total Fluence (J/cm^2)	Main findings	Advantages	Limitation Drawbacks	Refernces
Enhancement of functionally and health properties of fresh produce	Tomatoes Apples Persimmons Figs Mangoes Mushrooms	0.2–10 J/cm^2	Delaying senescence and fruit ripening Induction of natural defenses against fungi and bacteria Enhanced amount of total carotenoids, lycopene, total anthocyanins and phenolic compounds and the vitamin D2 Increased antioxidant activity	Shelf-life extension Preservation of quality attributes Health benefits High-added value product	Definition of optimal conditions highly dependent on variability of fresh produce properties Sample heating	Pataro et al., (2015a, b) Denoya et al. (2020) Rodov et al. (2012) Lopes et al. (2016) Koyyalamudi et al. (2011), Kalaras and Beelman (2008)
Structural modification and allergens reduction	Milk proteins β-lactoglobulin Egg whites Whey protein isolate Gluten Whole peanut Almond protein extracts Peanut allergens in extracts Soybean extracts	1–30	Unfolding, aggregation, cross-linking between ingredients, oxidation, or glycosylation	Enhanced functional properties (solubility, foaming capacity) Improves the digestibility Reduced toxicity	Uneven treatment Sample heating	Elmnasser et al. (2008) Fernández et al. (2012) Manzocco et al. (2013) Siddique et al. (2016, 2017) John and Ramaswamy (2018)

fruit juices. For example, it appears that PL induces only little changes in colour (Aron-Maftei et al. 2014; Muñoz et al. 2012; Ferrario and Guerrero 2016) as well as some sensory deviations (Palgan et al. 2011), but no appreciable changes to pH, Brix, nonenzymatic browning index or total antioxidant activity (Caminiti et al. 2011; Palgan et al. 2011; Muñoz et al. 2012). Neverthless, a product-specific quality assessment as well as optimized treatment conditions are necessary in case of beverage treatment also to control the increase in product temperature that has to be considered when treating beverages. Moreover, with the successful decontamination of clear liquids by PL, more research has to be extended towards turbid and opaque liquid foods.

Besides the decontamination of beverages, PL may also be used for disinfection in municipal wastewater, which has been proposed by Uslu et al. (2014) and Garvey et al. (2014). Furthermore the application of PL has been proposed as an efficient measure to decontaminate wastewater deriving from salad washing, which could decrease the water footprint of fresh-cut vegetables by minimizing the overall requirement for water in industrial plants. In addition, it would decrease the risk for residuals of toxic chemicals in fresh-cut vegetables by avoiding the use of sanitizers, such as chlorine and its related compounds. (Manzocco et al. 2015).

4.2 PL Decontamination of Unpackaged Solid Foods

The capability of PL to reduce microbial counts on the surfaces of different commodities of vegetable and animal origin has been extensively investigated (Mahendran et al. 2019).

In this framework, several studies have been performed to assess the suitability of PL for the inactivation of either artificially inoculated or native microbial cells on the surface of fresh produce prior to packaging, and especially to extend the shelf-life of fresh cut fruits and vegetables due to the worldwide rising popularity of fresh-cut produce (Kramer et al. 2017; Pirozzi et al. 2020).

Many different kinds of vegetables have been tested, so far, including spinach, celeriac, green bell pepper, soybean sprouts, radicchio, carrot, iceberg lettuce, white cabbage, tomato, fresh-cut mushrooms, while data for PL treatments of fresh and fresh cut fruits are available for apple, mango, melon and some sorts of berries, such as blueberries, raspberries, and strawberries (Bialka and Demirci 2008; Kramer et al. 2017). The obtained microbial data show that the inactivation efficiency of PL is higher against artificially inoculated test bacteria than native microflora with a defined test matrix. Moreover, although PL treatment cannot enable complete inactivation, it allows to achieve anywhere from 1 to 4 log reduction (Kramer et al. 2017). Therefore, as the occurrence of residual populations on the product surface is usually inevitable, their growth during storage after PL exposure of the product surface needs to be considered.

Differing results have been obtained regarding the impact of PL on food quality of fresh vegetables or fruits. In some studies, positive effects of PL on apparent quality attributes were found. The colour and firmness of PL-treated mung bean sprouts (Kramer et al. 2015) as well as fresh-cut mangoes (Charles et al. 2013) were better maintained compared to control samples, while the colour of avocado was stabilized after PL exposure (Aguilo-Aguayo et al. 2014). However, in many studies it was reported that an intense PL treatment of plant-based foods may negatively affect several quality parameters like, for example, the texture of mushrooms (Oms-Oliu et al. 2010a), the appearance (wrinkled skin and softening) and weight loss of tomato (Aguayo et al. 2013), the colour of salad (Kramer et al. 2015) or sensory attributes of cut apple (Ignat et al. 2014). Significant deterioration is mostly not found immediately after PL exposure, but during storage. Storage trials of PL-exposed products are therefore not only important regarding the maintenance of microbial reductions but also to investigate the impact on shelf-life. As the effect on quality parameters varies significantly among products and mostly seems to be dose dependent, it is hence of great importance to find optimal process parameters, which guarantee a mild treatment while providing substantial microbial inactivation.

Several studies have been also performed in order to assess the suitability of PL for decontamination of fresh meat, fish or products of them. Similar to the results of fresh produce, it is generally clear that the count reduction of native microflora on the treated products is distinctively lower compared to in vitro studies and the count reductions is anyware usually limited to a maximum inactivation of about 1–2.5 log-cycles (Kramer et al. 2017; Mahendran et al. 2019). In most studies, a dose-dependent negative impact on key quality attributes has been found. For example, discolorations have been reported at high energy doses in case of raw salmon (Ozer and Demirci 2006), chicken frankfurter (Keklik et al. 2009), chicken meat (Haughton et al. 2011; Keklik et al. 2011), beef and tuna carpaccio (Hierro et al. 2012) or pork meat (Nicorescu et al. 2014). Intense treatments may lead to immediate quality changes due to overheating of the product surface (Ozer and Demirci 2006). This has to be considered especially when raw fish or meat is exposed to PL. Lipid oxidation as well as sensory deviations need to be addressed as well (Hierro et al. 2011, 2012; Ganan et al. 2013; Nicorescu et al. 2014).

Similarly to liquid product disinfection, the use of PL in a hurdle approach with other preservation factors has been also suggested for the decontamination of solid foods, in order to enhance the efficiency of the PL treatment, while enabling the minimal processing of food with reduced costs and energy expenditure. In this line, successful applications have been found for the decontamination of fresh and fresh cut produce like green onions, green beans, raspberries, strawberries and blueberries, fresh-cut cantaloupe and fresh-cut apples, among others, when PL has been combined with disinfectant products (Xu et al. 2013), advanced oxidation processes using hydrogen peroxide (Huang and Chen 2015), and edible coating (Donsi et al. 2015; Koh et al. 2017; Moreira et al. 2016; Tastan et al. 2017; Pirozzi et al. 2020).

4.3 PL Decontamination of In-Packaged Food Products

The in-package decontamination of food products with PL is also of great interest for the future commercial application of this technology. Despite the screening effects of the packaging could reduce PL treatment effectiveness, the exposure of foods to light flashes after packaging is particularly advantageous to extend the shelf-life of products, since it allows the simultaneous reduction of microbial count spread both on foods and package surface, while avoiding undesirable post-treatment recontaminations (Heinrich et al. 2015; John and Ramaswamy 2018; Pirozzi et al. 2020).

Nevertheless, only a few published works have focused, so far, with in-package decontamination of food products, with most of them related to meat and meat products (e.g., chicken breast, frankfurters, and skin, and beef carpaccio), and only very few studies deal with fish (e.g., tuna carpaccio and catfish fillets), and fruits and vegetables (e.g., fresh cut watermelon, mushroom and avocados) (Haughton et al. 2011; Hierro et al. 2012; Keklik et al. 2009; Ramos-Villarroel et al. 2012a; Ramos-Villarroel et al. 2014). A maximum inactivation level of about 0.7–3.0 log-cycles was achieved depending on the fluence, target microorganism, type of matrices and packaging material, with more limited effect being detected in the case of meat and fish products. Therefore, more effort is required to assess the feasibility of the in-packaging PL process of food products. Specifically, it is essential that future studies will take into account not only the properties of the food matrix, the processing conditions, the susceptibility of the native microflora to PL exposure, but especially the composition, as well as the physical (e.g., thickness) and optical properties (e.g., UV transmissivity) of the packaging films.

Particular attention should be paid to evaluate the structural and barrier property changes of packaging and migration of compounds from packing materials to foods induced by PL (Heinrich et al. 2015). Similarly, scientific investigation of process-product-packaging interactions is also required for successful PL in-package application (Heinrich et al. 2015). Finally, although preliminary results did not reveal drastic alterations of the polymeric packaging materials (Keklik et al. 2009, 2010; Ringus and Moraru 2013), PL treatment should be performed within reasonable limits to avoid excessive temperature rise and overheating of packaging material and product during treatment. The qualities of food, such as sensory properties and shelf-life, need to be considered when using PL for packaged food sterilization (Chen et al. 2015b).

4.4 Sterilization of Food Contact Surfaces and Packaging Materials

During processing and packaging steps of food manufacturing, contamination can be transferred from the food contact surfaces and packaging materials to the food itself and cause spoilage, quality deterioration, and even public health issues by pathogenic bacteria (Chen et al. 2015b). Therefore, there is a need to develop effective and sustainable methods and technologies, alternative to chemical sanitizers and thermal treatment, able to ensure food safety and elongate the shelf-life of food. In this framework, PL is emerging as a valuable rapid, physical and sustainable method for the disinfection and sterilization of food contact surfaces and packaging materials, which enable to reduce the need for chemical sanitizers and sterilants, water, and leaves no residues.

However, so far, only a few scientist have investigated the sterilization efficiency of PL of different food contact surfaces and packaging materials, which include stainless steel and aluminum, low density polyethylene (LDPE), high density polyethylene (HDPE), polyethylene-laminated ultra-metalized polyethylene terephthalate (MET), polyethylene-coated paper-board (TR), polyethylene-coated aluminum foil paperboard laminate (EP), polystyrene (PS), polypropylene (PP), polyvinyl chloride (Haughton et al. 2011; Jean et al. 2011; Ringus and Moraru 2013; Turtoi and Nicolau 2007; Woodling and Moraru 2005). A maximum inactivation level of about 1–7 log-cycles was achieved depending on the fluence, target microorganism and inoculum size, and especially type of material. More specifically, the outcomes of these studies highlighted the key role played by surface properties of the material, such as roughness, hydrophobicity and reflectivity, on the sterilization efficiency of PL. Limited effect of PL were detected in the case of materials characterized by higher roughness values, water contact angle and reflectivity, which seem to induce shading effect, cell cluster formation and decrease light absorption, respectively. However, additional investigation are necessary to gain more insight on the role played by the different properties and especially to better understand how their interaction affect the sterilization efficiency.

4.5 Enhancement of Functionally and Health Properties of Fresh Produce

In several recent studies, it has been shown that the post-harvest exposure of fresh produce to light-based technologies, like continuous UV (0.1–8 J/cm^2) and PL (0.2–10 J/cm^2) irradiation, may trigger a series of biochemical events within the plant tissue, which stimulate the biosynthesis of defensive secondary metabolites with antimicrobial activity. These compounds are highly desirable as they can contribute to prolong the life and maintain the quality of vegetable and fruits by delaying senescence and fruit ripening, and induction of natural defences against fungi and bacteria (Ribeiro et al. 2012; Urban et al. 2018).

Interestingly, the activation of plant defence mechanisms might also stimulate the formation of bioactive compounds, which exhibit antioxidant potential, increasing the nutritional value and health beneficial properties of light-treated products (Ribeiro et al. 2012). However, literature data on the use of UV light, and especially PL, to enhance the nutraceutical properties of fresh produce are relatively recent and in progress, thus, conclusions cannot yet be drawn.

Nevertheless, an enhanced formation of total carotenoid compounds, lycopene, and total phenolic compounds has been found during storage of post-harvest PL-treated green tomato, thereby increasing the nutritional value (Pataro et al. 2015a, b). Lopes et al. (2016) showed that PL treatments increase the concentrations of phytochemicals in mangoes without negative effects on quality criteria. Rodov et al. (2012) showed that PL treatments caused an increase in the anthocyanins and total phenolic content of fig fruit. An enhanced formation of total phenolics and consequent increase in antioxidant activity was also found in post-harvest PL treated Annurca apples and persimonns (Pataro et al. 2015a; Denoya et al. 2020). In addition, several studies have reported that PL treatment provides a highly effective way for increasing Vitamin D2 content in mushrooms with little to no discoloration depending on the energy dose. In an earlier study by Kalaras and Beelman (2008), vitamin D2 was enhanced up to 824% daily value (DV-400 IU/84 g being 100% DV) per 84 g brown and white button mushrooms by application of 3.5 J/cm^2. Kalaras et al. (2011) in a separate study, also observed Vitamin D2 content enhancement in white button mushrooms (Agaricus bisporus) was up to 27 μg/g dry weight using 18 pulses of pulsed UV light of 0.791 J/cm2 per pulse for 1 s treatment. In another study, Koyyalamudi et al. (2011) treated the white button mushrooms (Agaricus bisporus) passed in conveyor belts and enhanced Vitamin D2 by 22.5 μg/g dry weight using nine pulses at 1.15 J/cm^2 per pulse. Chen et al. (2015a) treated Pleurotus mushrooms with nine pulses of 1.15 J/cm^2 per pulse and enhanced vitamin D2 by up to 2.78 μg/g fresh weight. Overall, the vitamin D2 enhancement varies proportionally with energy delivered during PL treatment. It is particularly noteworthy that the vitamin D2 enhancement in mushrooms by PL was also patented (Chalupa and Schroeder 2014; Schroeder 2013) and recently, the industrial implementation of this kind of PL treatment already took place (Xenon 2020).

Despite these very promising results and preliminary industrial applications, further research is needed since, in some cases, investigations have resulted in different conclusions regarding the appropriate energy doses, the optimal ripening stage of fresh produce, and the storage conditions.

4.6 Reduction of Allergens in Foods and Improvement of Functional and Technological Properties

Non-thermal technologies like PL are gaining increasing interest by food scientists and manufacturers, also because they can be employed, among other, to induce targeted structural modifications of food protein chromospheres, which could allow producing different formulated foods with appropriate functional properties, (Siddique et al. 2016, 2017), as well as to design low-allergy food products (del Castillo-Santaella et al. 2014).

Proteins are the main structural and functional components of many food systems, e.g., meat, cheese, gelatine, egg white and most of the cereals. In addition, proteins are being increasingly used to facilitate the engineering and fabricate new food products, such as protein beverages and extruded foods. The effectiveness of the use of proteins in food processing depends on their functional and technological characteristics, which, in turn, are strictly related to their different structural conformations (Siddique et al. 2016). Moreover, proteins are also among the major causative agent of allergenic response in human (Mandal et al. 2020).

PL techniques has the potential to induce changes in the structural conformation of proteins either via unfolding, aggregation, cross-linking between ingredients, oxidation, or glycosylation (Okolie et al. 2018; Siddique et al. 2016, 2017), thus affecting their functional properties and allergenic power. However, the energy levels required to achieve the desired effect is peculiar to each protein, its matrix composition, and environmental conditions (Ekezie et al. 2018).

A close examination of the scientific literature shows that, until now, this new application of PL is more or less the stage of a feasibility study, with only very few data being published on the matter.

Specifically, it has been found that PL treatments cause some aggregation in milk proteins by disulphide bonds, although no changes in amino-acid composition was observed (Elmnasser et al. 2008). Fernández et al. (2009) observed conformational changes of β-lactoglobulin related with PL fluence, which led to the formation of high molecular proteins species improving surface and foaming properties of β-lactoglobulin solutions. Manzocco et al. (2013) shown that the exposure to PL treatments can modify the functional properties of egg white proteins. Later on, Siddique et al. (2016, 2017) observed a dose-dependent structural changes (unfolding) and aggregates formation in whey protein isolate (WPI) solutions upon PL exposure. The partial unfolding of whey proteins wee reported to increase the functional properties of WPI, such as solubility and foaming capacity (Siddique et al. 2016). Additionally, in a subsequent work the same authors hypothesized that partial denaturation and the formation of a small fraction of soluble aggregates might improve the functional and technological properties of WPI. In this contest, PL may be exploited to induce desirable changes in protein rich food ingredients.

On the other hand, PL treatment is proposed to mitigate the allergenicity and toxicity of foods, which may extends the applicability of PL treatment in food industry (Mandal et al. 2020). To this regard, it has been shown that PL-induced structural rearrangement can decrease the gluten immunoreactivity (Panozzo et al. 2016). Increasing fluence value modifies the conformation of β-lactoglobulin protein, which improves the digestibility of this modified protein and hence can be used to design low-allergy food products (del Castillo-Santaella et al. 2014). Pulsed UV light treatment reduced tropomyosin levels and IgE-binding capacity of shrimp extracts (Shriver Yang 2011), as well as the level of allergens and also decreased IgE-binding capacity of peanut extracts and liquid peanut butter (Chung et al. 2008; Zhao et al. 2014), almonds (Li et al. 2013), and wheat (Nooji 2011). Moreover, pulsed-UV light treatment of peanut extracts and peanut butter can deactivate Ara h1, Ara h2 and Ara h3, known to be the most potent allergens present in peanuts (Yang et al. 2012) and decrease the levels of soy allergens like glycinin and b-conglycinin in soy extracts (Yang et al. 2010).

Further research is needed in order investigate the effects of PL on proteins in complex food systems, as well as to better elucidate the role played by different process parameter, including wavelengths ranges of the spectrum, on protein denaturation. This will allow to select the most suitable fluence and wavelength ranges emitted by the lamp in order to develop tailored proteins with specific structural conformation and functionality, while reducing allergenic power and avoiding or minimizing alterations of the treated product.

5 Perspective and Final Remarks

Pulsed light (PL) is a fast, environmental friendly, non-thermal technology with many potential applications in different stages of the food processing industry, and especially for the production of safe, no toxic, and high quality foods, as well as for the decontamination food-contact surface and packaging materials. The latter, represent till now, the most concrete application of PL technology at industrial level. The successful exploitation of PL technology as a "dry" decontamination technique of packaging materials, is explained by the reduced operational costs, no or reduced water consumption and chemical sanitizers, and, consequently, reduced environmental impact and cost derived for wastewater treatment.

However, although the direct application of PL to food products open exciting new and sustainable opportunities to the food industry to increase the safety, shelf-life and functionality of certain of certain food products, with no or limited toxic or detrimental effects on their overall quality and sensory properties, it still require more efforts to achieve industrial scale.

In most food applications, the shadow effect, which prevent the irradiated solid surface or treated liquid volume from a uniform light exposure and reduce treatment efficiency, as well as sample heating, remain as the main concerns. To this regard, while transparent drinks and solid foods, which are smooth and possess low reflection coefficient, are efficiently treated, opaque or turbid liquids, and surfaces of solid food presenting irregular and complex microstructure may render PL scarcely effective. The challenge consists in the optimization of the processing conditions and the adoption of technological solutions enabling to improve treatment uniformity, while reducing the heating effects. In this line, different approach continue to be developed involving improvements in treatment chamber design, which integrate proper cooling system, for efficient treatment of a wide range of liquid and solid foods, as well as the combination of PL with other conventional or innovative food preservation techniques to enhance their antimicrobial effect, while reducing the energy and water consumption.

Finally, more studies on the effect of PL on microbial inactivation, food quality attributes, and allergen level, are necessary at laboratory and, especially, at pilot scale before the full exploitation of PL at industrial level.

References

Aguiló-Aguayo I, Charles F, Renard CMGC, Page D, Carlin F (2013) Pulsed light effects on surface decontamination, physical qualities and nutritional composition of tomato fruit. Postharvest Biol Technol 86:29–36

Aguilo-Aguayo I, Oms-Oliu G, Martın-Belloso O, Soliva-Fortuny R (2014) Impact of pulsed light treatments on quality characteristics and oxidative stability of fresh-cut avocado. LWT-Food Sci Technol 59:320–326

Aron-Maftei N, Ramos-Villarroel AY, Nicolau AI, Martín-Belloso O, Soliva-Fortuny R (2014) Influence of processing parameters on the pulsed-light inactivation of Penicillium expansum in apple juice. Food Control 41:27–31

Artíguez ML, Martínez de Marañón I (2015) Improved process for decontamination of whey by a continuous flow-through pulsed light system. Food Control 47:599–605

Bakar Siddique A, Maresca P, Pataro G, Ferrari G (2017) Influence of pulsed light treatment on the aggregation of whey protein isolate. Food Res Int 99:419–425

Bialka KL, Demirci A (2008) Efficacy of pulsed UV-light for the decontamination of Escherichia coli O157:H7 and Salmonella spp. on Raspberries and Strawberries. J Food Sci 73:201–207

Bolton JR, Linden KG (2003) Standardization of methods for fluence (UV dose) determination in bench-scale UV experiments. J Environ Eng 129(3):209–215. https://doi.org/10.1061/(ASCE)0733-9372(2003)129:3(209)

Bolton JR, Mayor-Smith I, Linden KG (2015) Rethinking the concepts of fluence (UV dose) and fluence rate: the importance of photon-based units - a systemic review. Photochem Photobiol 91(6):1252–1262. https://doi.org/10.1111/php.12512

Caminiti IM, Palgan I, Noci F, Muñoz A, Whyte P, Cronin DA, Morgan DJ, Lyng JG (2011) The effect of pulsed electric fields (PEF) in combination with high intensity light pulses (HILP) on Escherichia coli inactivation and quality attributes in apple juice. Innov Food Sci Emerg Technol 12:118–123

Chalupa WF, Schroeder GM (2014) Method and apparatus for vitamin D enhancement in dried mushroom powder. U.S. Patent Application No. 13/469,168, 4 September 2014

Charles F, Vidal V, Olive F, Filgueiras H, Sallanon H (2013) Pulsed light treatment as new method to maintain physical and nutritional quality of fresh-cut mangoes. Innov Food Sci Emerg Technol 18:190–195

Chen SY, Huang SJ, Cheng MC, Chen YK, Yang SC, Mau JL (2015a) Enhancement of vitamin D2 content in Pleurotus mushrooms using pulsed light. J Food Process Pres 39:2027–2034

Chen BY, Lung HM, Yang BB, Wang CY (2015b) Pulsed light sterilization of packagingmaterials. Food Packag Shelf-life 5:1–9

Chen D, Wiertzema J, Peng P, Cheng Y, Liu J, Mao Q, Ma Y, Andreson E, Chen P, Baumler D, Chen C, Vickers Z, Feirtag J, Lee L, Ruan R (2018) Effects of intense pulsed light on Cronobacter sakazakii inoculated in non-fat dry milk. J Food Eng 238:178–187

Chen D, Chen P, Cheng Y, Peng P, Liu J, Ma Y, Liu Y, Ruan R (2019) Deoxynivalenol decontamination in raw and germinating barley treated by plasma-activated water and intense pulsed light. Food Bioprocess Technol 12:246–254

Chung SY, Yang W, Krishnamurthy K (2008) Effects of pulsed UV-light on peanut allergens in extracts and liquid peanut butter. J Food Sci 73:400–404

Collazo C, Charles F, Aguiló-Aguayo I, Marín-Sáez J, Lafarga T, Abadias M, Viñas I (2019) Decontamination of Listeria innocua from fresh-cut broccoli using UV-C applied in water or peroxyacetic acid, and dry-pulsed light. Innov Food Sci Emerg Technol 52:438–449

del Castillo-Santaella T, Sanmartin E, Cabrerizo-Vilchez MA, Arboleya JC, Maldonado-Valderrama J (2014) Improved digestibility of b-lactoglobulin by pulsed light processing: a dilatational and shear study. Soft Matter 10:9702–9714

Denoya GI, Pataro G, Ferrari G (2020) Effects of postharvest pulsed light treatments on the quality and antioxidant properties of persimmons during storage. Postharvest Biol Technol 160:111055. https://doi.org/10.1016/j.postharvbio.2019.111055

Donsì F, Marchese E, Maresca P, Pataro G, Vu KD, Salmieri S, Lacroix M, Ferrari G (2015) Green beans preservation by combination of a modified chitosan based-coating containing nanoemulsion of mandarin essential oil with high pressure or pulsed light processing. Postharvest Biol Technol 106:21–32

Dunn JE, Clark RW, Asmus JF, Pearlman JS, Boyer K, Painchaud F, Hofmann GA (1989) Methods for preservation of foodstuffs. Maxwell Laboratories Inc., San Diego, USA. US Patent 4871559

Ekezie FC, Cheng J, Sun D (2018) Effects of nonthermal food processing technologies on food allergens: a review of recent research advances. Trends Food Sci Technol 74:12–25

Elmnasser N, Dalgalarrondo M, Orange N, Bakhrouf A, Haertlé T, Federighi M, Chobert JM (2008) Effect of pulsed-light treatment on milk proteins and lipids. J Agric Food Chem 56:1984–1991

EU (1997) Regulation EC No 258/97 of the European Parliaments and the Council of 27 January 1997 concerning novel foods and novel food ingredients. EU: European Union

Fernández M, Manzano S, De La Hoz L, Ordóñez JA, Hierro E (2009) Pulsed light inactivation of Listeria monocytogenes through different plastic films. Foodborne Pathog Dis 6:1265–1267

Fernández E, Artiguez ML, de Marañón IM, Villate M, Blanco FJ, Arboleya JC (2012) Effect of pulsed-light processing on the surface and foaming properties of β-lactoglobulin. Food Hydrocoll 27:154–160

Ferrario M, Alzamora SM, Guerrero S (2013) Inactivation kinetics of some microorganisms in apple, melon, orange and strawberry juices by high intensity light pulses. J Food Eng 118:302–311

Ferrario M, Alzamora SM, Guerrero S (2015) Study of the inactivation of spoilage microorganisms in apple juice by pulsed light and ultrasound. Food Microbiol 46:635–642

Ferrario M, Guerrero S (2016) Effect of a continuous flow-through pulsed light system combined with ultrasound on microbial survivability, color and sensory shelf-life of apple juice. Innov Food Sci Emerg Technol 34:214–224

Ferrario M, Guerrero S (2017) Impact of a combined processing technology involving ultrasound and pulsed light on structural and physiological changes of *Saccharomyces cerevisiae* KE 162 in apple juice. Food Microbiol 65:83–94

Fine F, Gervais P (2004) Efficiency of pulsed UV light for microbial decontamination of food powders. J Food Prot 67:787–792

Franco-Vega A, Reyes-Jurado F, González-Albarrán D, Ramírez-Corona N, Palou E, López Malo A (2021) Developments and advances of high intensity pulsed light and its combination with other treatments for microbial inactivation in food products. Food Eng Rev, https://doi.org/10.1007/s12393-021-09280-1

Ganan M, Hierro E, Hospital XF, Barroso E, Fernandez M (2013) Use of pulsed light to increase the safety of ready-to-eat cured meat products. Food Control 32:512–517

Garvey M, Thokala N, Rowan N (2014) A comparative study on the pulsed UV and the low-pressure UV inactivation of a range of microbial species in water. Water Environ Res 86:2317–2324
Gómez-López et al (2021) Guidelines on reporting treatment conditions for emerging technologies in food processing. Crit Rev Food Sci Nutr. https://doi.org/10.1080/10408398.2021.1895058
Gómez-López VM, Bolton JR (2016) An approach to standardize methods for fluence determination in bench-scale pulsed light experiments. Food Bioprocess Technol 9:1040–1048
Gómez-López VM, Devlieghere F, Bonduelle V, Debevere J (2005) Factors affecting the inactivation of micro-organisms by intense light pulses. J Appl Microbiol 99:460–470
Gómez-López VM, Ragaert P, Debevere J, Devlieghere F (2007) Pulsed light for food decontamination: a review. Trends Food Sci Technol 18:464–473
Gouma M, Condón S, Mañas P, Álvarez I, Gayán E (2016) The use of UV light for food preservation: an overview. In: Pataro G, Lyng JG (eds) High intensity pulsed light in processing and preservation of foods. Nova Science Publishers, Inc, pp 1–39
Haughton PN, Lyng JG, Morgan DJ, Cronin DA, Fanning S, Whyte P (2011) Efficacy of high intensity pulsed light for the microbiological decontamination of chicken, associated packaging, and contact surfaces. Foodborne Pathog Dis 8:109–117
Heinrich V, Zunabovic M, Bergmair, Kneifel W, Jäger H (2015) Post-packaging application of pulsed light for microbial decontamination of solid foods: a review. Innov Food Sci Emerg Technol 30:145–156
Hierro E, Barroso E, De la Hoz L, Ordo~nez JA, Manzano S, Fernandez M (2011) Efficacy of pulsed light for shelf-life extension and inactivation of Listeria monocytogenes on ready-to-eat cooked meat products. Innov Food Sci Emerg Technol 12:275–281
Hierro E, Ganan M, Barroso E, Fernandez M (2012) Pulsed light treatment for the inactivation of selected pathogens and the shelf-life extension of beef and tuna carpaccio. Int J Food Microbiol 158:42–48
Hillegas SL, Demirci A (2003) Inactivation of Clostridium sporogenes in Clover Honey by Pulsed UV-Light Treatment; V. Manuscript FP 03 009; American Society of Agricultural and Biological Engineers: St. Joseph, MI, USA
Hsu L, Moraru CI (2011) Quantifying and mapping the spatial distribution of fluence inside a pulsed light treatment chamber and various liquid substrates. J Food Eng 103:84–91
Huang Y, Chen H (2015) Inactivation of Escherichia coli O157:H7, Salmonella and human norovirus surrogate on artificially contaminated strawberries and raspberries by water-assisted pulsed light treatment. Food Res Int 72:1–7
Ignat A, Manzocco L, Maifreni M, Bartolomeoli I, Nicoli MC (2014) Surface decontamination of fresh-cut apple by pulsed light: effects on structure, colour and sensory properties. Postharvest Biol Technol 91:122–127
IUPAC (1996) Glossary of terms used in photochemistry. Pure Appl Chem 12:2223–2286
Jean J, Morales-Rayas R, Anoman MN, Lamhoujeb S (2011) Inactivation of hepatitis A virus and norovirus surrogate in suspension and on food-contact surfaces using pulsed UV light (pulsed light inactivation of food-borne viruses). Food Microbiol 28:568–572
John D, Ramaswamy HS (2018) Pulsed light technology to enhance food safety and quality: a mini-review. Curr Opin Food Sci 23:70–79
Kalaras MD, Beelman RB (2008) Vitamin D2 Enrichment in Fresh Mushrooms Using Pulsed UV Light. Foodscience. Available online: psu.edu/directory/rbb6/VitaminDEnrichment.pdf
Kalaras MD, Beelman RB, Elias RJ (2011) Effects of postharvest pulsed UV light treatment of white button mushrooms (Agaricus bisporus) on vitamin D2 content and quality attributes. J Agric Food Chem 60:220–225
Keklik NM, Demirci A, Puri VM (2009) Inactivation of listeria monocytogenes on unpackaged and vacuum-packaged chicken frankfurters using pulsed UV-light. J Food Sci 74:M431–M439
Keklik NM, Demirci A, Puri VM (2010) Decontamination of unpackaged and vacuum-packaged boneless chicken breast with pulsed ultraviolet light. Poult Sci 89:570–581
Keklik NM, Krishnamurthy K, Demirci A (2012) Microbial decontamination of food by ultraviolet (UV) and pulsedUVlight. In: Demirci A, Ngadi M (eds) Microbial decontamination in the

food industry. Novel methods and applications. Woodhead Publishing, pp 344–369. https://doi.org/10.1533/9780857095756.2.344
Keklik NM, Demirci A, Bock RG (2011) Surface decontamination of whole chicken carcasses using a pilotscale pulsed UV light system. T ASABE 54:993–1000
Koh PC, NoranizanMA HZAN, Karim R, Rosli SZ (2017) Application of edible coatings and repetitive pulsed light far shelf-life extension of fresh-cut cantaloupe (Cucumismelo L. reticulatus cv. Glamour). Postharvest Biol Technol 129:64–78
Koyyalamudi SR, Jeong SC, Pang G, Teal A, Biggs T (2011) Concentration of vitamin D2 in white button mushrooms (Agaricus bisporus) exposed to pulsed UV light. J Food Compos Anal 24:976–979
Kramer B, Wunderlich J, Muranyi P (2015) Pulsed light decontamination of endive salad and mung bean sprouts and impact on color and respiration activity. J Food Prot 78:340–348
Kramer B, Wunderlich J, Muranyi P (2017) Recent findings in pulsed light disinfection. J Appl Microbiol 122:830–856
Krishnamurthy K, Demirci A, Irudayaraj JM (2007) Inactivation of Staphylococcus aureus in milk using flow-through pulsed UV light treatment system. J Food Sci 72(7):M233–M239
Levy C, Aubert X, Lacour B, Carlin F (2012) Relevant factors affecting microbial surface decontamination by pulsed light. Int J Food Microbiol 152:168–174
Li Y, Yang W, Chung SY, Chen H, Ye M, Teixeira AA, Gregory JF, Welt BA, Shriver S (2013) Effect of pulsed ultraviolet light and high hydrostatic pressure on the antigenicity of almond protein extracts. Food Bioprocess Technol 6:431–440
Lopes MM, Silva EO, Canuto KM, Silva LM, Gallão MI, Urban L, Fernando Ayala-Zavala J, Miranda MRA (2016) Low fluence pulsed light enhanced phytochemical content and antioxidant potential of 'Tommy Atkins' mango peel and pulp. Innov Food Sci Emerg Technol 33:216–224
Maftei NA, Ramos-villarroel AY, Nicolau AI, Martin-Belloso O, Soliva-Fortuny R (2014) Influence of processing parameters on the pulsed - light inactivation of Penicillium expansum in apple juice. Food Control 41:27–31
Mahendran R, Ramanan KR, Barba FJ et al (2019) Recent advances in the application of pulsed light processing for improving food safety and increasing shelf-life. Trends Food Sci Technol 88:67–79. https://doi.org/10.1016/j.tifs.2019.03.010
Mandal R, Mohammadi X, Wiktor A, Singh A, Singh Anubhav P (2020) Applications of pulsed light decontamination technology in food processing: an overview. Appl Sci 10:3606. https://doi.org/10.3390/app10103606
Manzocco L, Panozzo A, Nicoli MC (2013) Effect of pulsed light on selected properties of egg white. Innov. Food Sci Emerg Technol 18:183
Manzocco L, Ignat A, Bartolomeoli I, Maifreni M, Nicoli MC (2015) Water saving in fresh-cut salad washing by pulsed light. Innov Food Sci Emerg Technol 28:47–51
Moreira MR, Álvarez MV, Martín-Belloso O, Soliva-Fortuny R (2016) Effects of pulsed light treatments and pectin edible coatings on the quality of fresh-cut apples: a hurdle technology approach. J Sci Food Agric 97:261–268
Muñoz A, Palgan I, Noci, F, Morgan, DJ, Cronin, DA, Whyte P, Lyng, JG (2011) Combinations of high intensity light pulses and thermosonication for the inactivation of Escherichia coli in orange juice. Food Microbiol 28:1200–1204
Muñoz A, Caminiti IM, Palgan I, Pataro G, Noci F, Morgan DJ, Cronin DA, Whyte P, Ferrari G, Lyng JG (2012) Effects on Escherichia coli inactivation and quality attributes in apple juice treated by combinations of pulsed light and thermosonication. Food Res Int 45:299–305
Nicorescu I, Nguyen B, Chevalier S, Orange N (2014) Effects of pulsed light on the organoleptic properties and shelf-life extension of pork and salmon. Food Control 44:138–145
Nooji JK (2011) Reduction of Wheat Allergen Potency by Pulsed Ultraviolet Light, High Hydrostatic Pressure and Nontherma Plasma (MSc thesis). University of Florida
Okolie CL, Aryee ANA, Udenigwe CC (2018) Detection and deactivation of allergens in food. In: Yada RY (ed) Proteins in food processing. Woodhead Publishing Series, pp 367–387. 71

Oms-Oliu G, Martín-Belloso O, Soliva-Fortuny R (2010) Pulsed light treatments for food preservation. A review. Food Bioprocess Technol 3:13–23
Oms-Oliu G, Aguilo-Aguayo I, Martın-Belloso O, Soliva-Fortuny R (2010a) Effects of pulsed light treatments on quality and antioxidant properties of freshcut mushrooms (Agaricus bisporus). Postharvest Biol Technol 56:216–222
Ozer NP, Demirci A (2006) Inactivation of Escherichia coli O157: H7 and Listeria monocytogenes inoculated on raw salmon fillets by pulsed UV-light treatment. Int J Food Sci Tech 41:354–360
Palgan I, Caminiti IM, Muñoz A, Noci F, Whyte P, Mogan DJ et al (2011) Effectiveness of high intensity light pulse (HILP) treatments for the control of Escherichia coli and Listeria innocua in apple juice, orange juice and milk. Food Microbiol 28:14–20
Panozzo A, Manzocco L, Lippe G, Nicoli MC (2016) Effect of pulsed light on structure and immunoreactivity of gluten. Food Chem 194:366–372
Pataro G, De Maria S, Lyng JG (2016) Factors determining the efficacy of a pulsed light treatment. In: Pataro G, Lyng JG (eds) High intensity pulsed light in processing and preservation of foods. Nova Science Publishers, Inc, pp 85–108
Pataro G, Donsi G, Ferrari G (2015a) Post-harvest UV-C and PL irradiation of fruits and vegetables. Chem Eng Trans 44:31–36. https://doi.org/10.3303/CET1544006
Pataro G, Muñoz A, Palgan I, Noci F, Ferrari G, Lyng JG (2011) Bacterial inactivation in fruit juices using a continuous flowpulsed light (PL) system. Food Res Int 44:1642–1648
Pataro G, Sinik M, Capitoli MM, Donsì G, Ferrari G (2015b) The influence of post-harvest UV-C and pulsed light treatments on quality and antioxidant properties of tomato fruits during storage. Innov Food Sci Emerg Technol 30:103–111
Pirozzi A, Pataro G, Donsì F, Ferrari G (2020) Edible coating and pulsed light to increase the shelf life of food products. Food Eng Rev. https://doi.org/10.1007/s12393-020-09245-w
Ramos-Villarroel AY, Aron-Maftei N, Martín-Belloso O, Soliva-Fortuny R (2012a) Influence of spectral distribution on bacterial inactivation and quality changes of fresh-cut watermelon treated with intense light pulses. Postharvest Biol Technol 69:32–39
Ramos-Villarroel AY, Aron-Maftei N, Martín-Belloso O, Soliva-Fortuny R (2012b) The role of pulsed light spectral distribution in the inactivation of *Escherichia coli* and *Listeria innocua* on fresh-cut mushrooms. Food Control 24:206–213
Ramos-Villarroel A, Aron-Maftei N, Martín-Belloso O, Soliva-Fortuny R (2014) Bacterial inactivation and quality changes of fresh-cut avocados as affected by intense light pulses of specific spectra. Int J Food Sci Technol 49:128–136
Ribeiro C, Canada J, Alvarenga B (2012) Prospects of UV radiation for application in postharvest technology. Emirates J Food Agric 24:586–597
Ringus DL, Moraru CI (2013) Pulsed light inactivation of Listeria innocua on food packaging materials of different surface roughness and reflectivity. J Food Eng 114:331–337
Rodov V, Vinokur Y, Horev B (2012) Brief postharvest exposure to pulsed light stimulates coloration and anthocyanin accumulation in fig fruit (Ficus carica L.). Postharvest Biol Technol 68:43–46
Schaefer R, Grapperhaus M, Schaefer I, Linden K (2007) Pulsed UV lamp performance and comparison with UV mercury lamps. J Environ Eng Sci 6(3):303–310. https://doi.org/10.1139/s06-068
Schroeder GM (2013) Method and Apparatus for Vitamin D Enhancement in Mushrooms. U.S. Patent No. 8,545,915, 1 October 2013
Shriver SK, Yang WW (2011) Thermal and nonthermal methods for food allergen control. Food Eng Rev 3:26–43
Siddique AB, Maresca P, Pataro G, Ferrari G (2016) Effect of pulsed light treatment on structural and functional properties of whey protein isolate. Food Res Int 87:189–196. http://hdl.handle.net/11386/4669309
Takeshita K, Shibato J, Sameshima T, Fukunaga S, Isobe S, Arihara K, Itoh M (2003) Damage of yeast cells induced by pulsed light irradiation. Int J Food Microbiol 85:151–158

Taştan Ö, Pataro G, Donsì F, Ferrari G, Baysal T (2017) Decontamination of fresh-cut cucumber slices by a combination of a modified chitosan coating containing carvacrol nanoemulsions and pulsed light. Int J Food Microbiol 260:75–80

Turtoi M, Nicolau A (2007) Intense light pulse treatment as alternative method for mould spores destruction on paperpolyethylene packaging material. J Food Eng 83:47–53

Urban L, Chabane Sari D, Orsal B, Lopes M, Miranda R, Aarrouf J (2018) UV-C light and pulsed light as alternatives to chemical and biological elicitors for stimulating plant natural defenses against fungal diseases. Sci Hortic (Amsterdam) 235:452–459

Uslu G, Demirci, A, Regan, JM (2014) Efficacy of pulsed UV-light treatment on wastewater effluent disinfection and suspended solid reduction. J Environ Eng-ASCE 141:04014090

Wang T, MacGregor SJ, Anderson JG, Woolsey GA (2005) Pulsed ultra-violet inactivation spectrum of Escherichia coli. Water Res 39:2921–2925

Wekhof A (2000) Disinfection with flash lamps. PDA J PharmSci Technol 54:264–276

Wekhof A, Trompeter F-J, Franken O (2001) Pulsed UV disintegration (PUVD): a new sterilisationmechanism for packaging and broad medical-hospital applications. In: Proceedings of the the first international conference on ultraviolet technologies, June14–16, 2001, Washington D.C, pp 1–15

Woodling SE, Moraru CI (2005) Influence of surface topography on the effectiveness of pulsed light treatment for the inactivation of listeria innocua on stainless-steel surfaces. J Food Sci 70:m345–m351. https://doi.org/10.1111/j.1365-2621.2005.tb11478.x

Woodling SE, Moraru CI (2007) Effect of spectral range in surface inactivation of Listeria innocua using broad-spectrum pulsed light. J Food Prot 70:909–916

Xenon food safety and enhancement with pulsed UV light. Available online: https://xenoncorp.com/food-safety/other-foodapplications/. Accessed 16 Apr 2020

Xu, W, Chen H, Huang Y, Wu C (2013) Decontamination of Escherichia coli O157: H7 on green onions using pulsed light (PL) and PL–surfactant–sanitizer combinations. Int J Food Microbiol 166:102–108

Xu F, Wang B, Hong C, Telebielaigen S, Nsor-Atindana J, Duan Y, Zhong F (2019) Optimization of spiral continuous flowthrough pulse light sterilization for Escherichia coli in red grape juice by response surface methodology. Food Control 105:8–12

Yang WW, Chung SY, Ajayi O, Krishnamurthy K, Konan K, Goodrich-Schneider R (2010) Use of pulsed ultraviolet light to reduce the allergenic potency of soybean extracts. Int J Food Eng 6:3

Yang WW, Mwakatage NR, Goodrich-Schneider R, Krishnamurthy K, Rababah TM (2012) Mitigation of major peanut allergens by pulsed ultraviolet light. Food Bioprocess Technol 5:2728–2738

Yi JY, Lee NH, Chung MS (2016) Inactivation of bacteria and murine norovirus in untreated groundwater using a pilot-scale continuous-flow intense pulsed light (IPL) system. LWT-Food Sci Technol 66:108–113

Zhao XY, Yang WD, Chung SY, Sims CA, Otwell SW, Rababah TM (2014) Reduction of IgE immunoreactivity of whole peanut (Arachis hypogaea L.) after pulsed light illumination. Food Bioprocess Technol 7:2637–2645

Chapter 8
Non-electro-Technologies: Gamma Rays, UV Light, Ozone, Photodynamic and Membrane Processing

S. Padma Ishwarya, Muhammad Haseeb Ahmad, A. M. Nandu Lal, V. Silpa, and T. Venkatesh

1 Radiation Processing

Radiation processing of foods began soon after the discovery of radioactivity in 1896. A pioneering application of irradiation was microbial inactivation in foods (Appleby and Banks 1905). Since then, the concept of food irradiation has evolved across years (Fig. 8.1) to encompass a wide range of applications, at present. Formally, food irradiation can be defined as a physical process in which food and agricultural commodities, either in bulk or pre-packed form, are exposed to a controlled dose of radiation energy to achieve various desirable effects (BARC n.d.) (Fig. 8.2). It is also referred to as *'cold pasteurization'* since the food remains isothermal as it is not heated up during the process (Moreira and Castell-Perez 2021). Further, irradiation is a validated substitute to the chemical treatment of agro and food produces. Regardless of the periodic concerns raised about the safety of irradiated foods, radiation processing has continued to be a favorable methodology which assures the safety of food produces (IFST 2006; Verma and Gautam 2015). The United States Food and Drug Administration (USFDA) has confirmed the safety of the food irradiation process after assessing a range of irradiated foods for more than

S. Padma Ishwarya
Department of Chemical Engineering, Indian Institute of Technology Madras, Chennai, Tamil Nadu, India

M. H. Ahmad
Department of Food Science, Government College University, Faisalabad, Pakistan

A. M. Nandu Lal · V. Silpa
Agro Processing and Technology Division, CSIR-NIIST, Thiruvananthapuram, Kerala, India

T. Venkatesh (✉)
Agro Processing and Technology Division, CSIR-NIIST, Thiruvananthapuram, Kerala, India

Academy of Scientific and Innovative Research, Ghaziabad, Uttar Pradesh, India

A. Režek Jambrak (ed.), *Nonthermal Processing in Agri-Food-Bio Sciences*, Food Engineering Series, https://doi.org/10.1007/978-3-030-92415-7_8

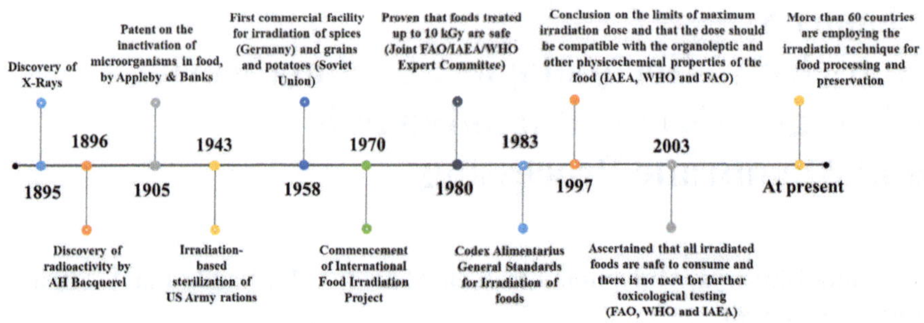

Fig. 8.1 Radiation processing of foods: a timeline of advancements

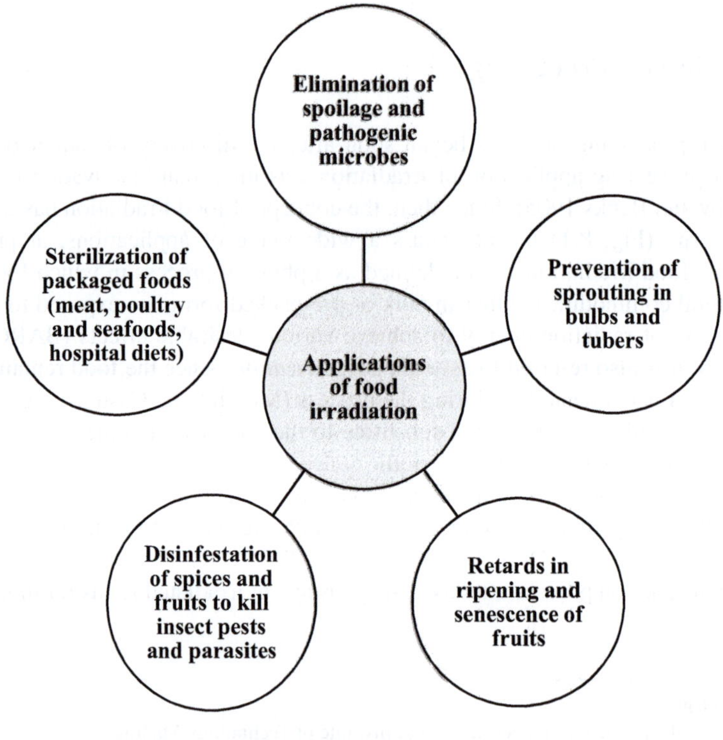

Fig. 8.2 Wide-ranging applications of food irradiation

30 years. Similarly, concern regarding safe dosage of radiation in food products has also been validated by the World Health Organization (WHO), the Center for Disease Control and Prevention (CDC), and the U.S. Department of Agriculture (USDA) (FDA 2018).

This section of the chapter will present an overview of the principle, classifications, and applications of food irradiation with relevant case studies.

1.1 *Principle of Food Irradiation*

According to the USFDA (2016), three categories of ionizing radiations have been approved for food processing applications (Bahrami et al. 2020):

1. gamma-ray (γ-rays)
2. electron beam of energy order 10 MeV, maximum), and
3. Roentgen ray (X-ray; high energy of 5 MeV, maximum)

When radio-active isotopes cobalt (^{60}Co) and cesium (^{137}Cs) gets decayed, gamma rays are generated. Whereas, electron beams are generated from an equipment called "electron gun" which operates under vacuum conditions. On the other hand, X-rays are generated from accelerators that can be of the linear or Rhodotron-type (Pillai and Shayanfar 2017).

The food preservation effect of these ionizing radiations is attributed to their interactions with the food substance, which can be classified as physical and chemical. Molecular ions of high energy, which gets generated due to physical interactions between UV and food results in electron capture, dissociation, and a quick reorganization via ion-molecule reactions. Alternatively, the ions might disintegrate with time, based on their convolution. Nevertheless, the impact of radiation on the food matrix is determined by several factors such as physical state, temperature, atmospheric conditions, radiation type and energy intensity. Apart from physical changes, the radiolysis-mediated chemical changes in treated foods arise as a consequence of the energy absorbed by the absorbing matter, since the UV radiation can excite the molecules and produce high reactive free radicals. The highly reactive intermediates thus formed can undergo several reactions to form stable chemical products known as the *'radiolysis products'*. If the targets of irradiation are living organisms such as microbes or insects, these chemical modifications will eventually have biological consequences (Morehouse and Komolprasert 2004). The specific governing mechanisms by which the ionizing radiations achieve each of the intended applications mentioned in Fig. 8.2 are explained in the subsequent subsections.

1.1.1 Mechanism of Microbial Inactivation and Sterilization by Radiation Processing

The mode of action by which the ionizing radiations inactivate food-borne pathogens and spoilage microbes is by triggering direct damage to their genomic material or by impacting the cell membrane. In the former mechanism, the ionizing radiation targets the DNA and RNA of microbes/pathogens and disrupts the composition and structure thereby causing the reproductive death of pathogens (Lung et al. 2015). When exposed to ionizing radiation, lesions are formed in the bacterial DNA. The single-stranded lesions do not harm the bacterial cell significantly, as the cell overcomes them by its DNA repair mechanism. Nevertheless, double-stranded lesions and multiple single-stranded DNA lesions are beyond the scope of repair and certainly lead to cell death (Ravindran and Jaiswal 2019) (Fig. 8.3a). Alternatively, ionizing radiations may also inactivate microorganisms by the generation of free radicals such and reactive oxygen species that are released upon the radiolysis of water molecules. These reactive species are capable of damaging the microbial cell membrane (Lung et al. 2015) (Fig. 8.3b). Free radicals emanating from the radiolysis of water and small molecules are effective in degrading mycotoxins by direct destruction of the organic molecules (Pankaj et al. 2018; Wang et al. 2015).

The effectiveness of microbial inactivation by ionizing radiations is determined by various factors such as source and dosage of radiation, microorganism and the characteristics of the food matrix (Deng et al. 2020b). For instance, microbial load reduction and radiation dose level are directly related. For example, diderm bacteria is more vulnerable to irradiation when compared to monoderm bacteria and prokaryotes are highly resilient than eukaryotes (Deng et al. 2020a). Since the penetration capability of γ-ray is substantially deeper than other radiation forms, its dosage is considerably lower than the X-ray and e-beam. However, X-rays have low energy efficiency (3–5%), and e-beams exhibit low penetration ability (Calado et al. 2014). With respect to the influence of food matrices, microorganisms in buffer solutions are more sensitive to radiation, rather than those present in solutions containing organic constituents such as proteins. The latter exhibit protective effects against radiations. Conversely, the presence of antimicrobial agents (ex. nitrite) in foods enhances the sensitivity of microorganisms to radiation (Erkmen and Bozoglu 2016).

1.1.2 Mechanism of Sprout Inhibition by Radiation Processing

Ionizing radiations inhibit the sprouting of tubers, bulbs, rhizomes, and corms by suppressing nucleic acid synthesis (DNA and RNA) in the meristem (bud). Consequently, germination is prevented and sprout formation is inhibited. But the extent of sprout inhibition depends on the dose of absorbed radiation (Farkas 2006). While dose less than the prescribed limit is not effective at inhibiting sprouting,

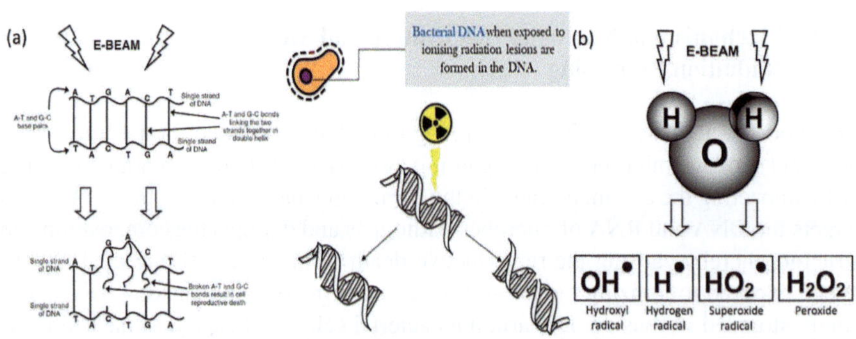

Fig. 8.3 Direct and indirect modes of microbial inactivation during food irradiation. (Modified from Lung et al. 2015; Ravindran and Jaiswal 2019)

higher doses lead to undesirable effects such as blackening of bud tissue (Mahto and Das 2014) (Fig. 8.4a), excessive browning, storage rot, and spoilage, reduced vitamin content and wound healing ability, besides sweetening and variations in chemical composition that are irreversible during the following storage period (Tripathi et al. 2015).

Fig. 8.4 Effect of irradiation on (**a**) sprouting in potatoes (exposure with 1 kGy resulted in blackening of the bud tissue). (Mahto and Das 2014); (**b**) physiological maturity of 'Dushehri' mangoes after upon 7 days storage period. (Mahto and Das 2013)

1.1.3 Mechanism of Delayed Ripening and Senescence of Fruits by Radiation Processing

Irradiation is known to induce modifications in the enzyme-mediated metabolic pathways that are accountable for the ripening of fruits and thereby delay ripening and senescence (Thomas 1988). According to another school of thought, irradiation slows down the production of hormones in fruits and vegetables, thereby delaying the fruit maturation (Agbaka and Ibrahim 2020). Other mechanisms by which irradiation postpones ripening include: (i) reducing sensitivity to ripening action of ethylene; (ii) altered carbohydrates metabolism by regulating certain vital enzymes that hinder the production of Adenosine Triphosphate (ATP), which is essential for several synthetic processes during ripening (Udipi and Ghurge 2010). As discussed in the earlier section, the effect of irradiation on delayed ripening of fruits is dose-dependent and excessive dosage can lead to deteriorative effects. For example, when irradiation was applied to delay the ripening of *'Dushehri'* mangoes, the peel color of the control fruit changed to yellowish-green within 4 days of storage and turned fully yellow at the end of the seventh day, relative to the irradiated fruits that were still green on the fourth day and yellowish-green on seventh day of storage. While the fruit treated with a low dose, i.e., 0.7 kGy had firmer flesh and looked more acceptable, that treated with a higher dose (6–10 kGy) showed brown patches on the peel and pulp, separation of cells, and development of fissures and cavities in the pulp. Also, treatment of fruit with 10 kGy dosage resulted in an excessive extrusion of sap (Mahto and Das 2013) (Fig. 8.4b).

1.1.4 Mechanism of Disinfestation by Radiation Processing

Similar to the mode of action involved in microbial inactivation, ionizing radiations destroy insects and pests by a sequence of oxidative reactions and the development of free peroxy radicals, that damage the organic molecules irreversibly. At the cytological level, ionization causes lethal mutations, translocations, and other aberrations in the germ cell chromosome. The result is the formation of imbalanced gametes and the consequent inhibition of mitosis and death of fertilized eggs (Espo et al. 2015). Irradiation also causes the breakage of chemical bonds, which at the DNA level retards the reproductive mechanism of the organism (Abbas et al. 2011). Nevertheless, the lethal dosage of radiation for disinfestation is determined by the type and life cycle stage of the insect. For example, beetles and mites are highly affected by radiation than the more resistant moths. Usually, a dosage of as low as 0.5 kGy is known to control the most resistant beetles and immature moths, and the progenies of the moths would be sterile. Similarly, insects in the egg stage are more susceptible to irradiation than those in their adult phase. A maximum radiation dose of up to 3 kGy is required to kill insects at any developmental stage. But, a lower dose might prevent progressive maturation or lead to sterilization (World Health Organization 1994).

1.2 Applications of Radiation Processing in Food Products: Pertinent Case Studies

The food applications of radiation processing can be broadly categorized according to the radiation dose used, which is measured in units of kilogray (kGy).

1. low dose (<1 kGy);
2. medium dose (1–10 kGy) &
3. high dose (>10 kGy).

This section would provide insights into the different applications of food irradiation under each of the above categories.

1.2.1 Low-Dose Applications

In low-dose applications, irradiation is employed as a preventive measure rather than for achieving sterility. These are mainly employed for insect disinfestation, phytosanitary irradiation, delaying of ripening, and sprout inhibition. Low-dose is frequently used for disinfection of fresh agro produces and decontamination of poultry products such as egg (Alvarez et al. 2007; Kim et al. 2010; Niemira and Fan 2006 & Kalyani and Manjula 2014). Table 8.1 lists the commonly encountered low dose irradiation applications and the corresponding dosage.

1.2.2 Medium Dose Applications

Medium-dose irradiation (1–10 kGy) for microbial inactivation of food products is termed as *'radurization'* or *Radiopasteurization* (due to its similar effects as heat pasteurization). *Radurization* involves treating the food with a dose of ionizing radiation that is adequate to improve its shelf-life by bringing about a significant reduction in the count of viable spoilage microorganisms. It is aimed at controlling pathogenic microbes (except viruses) and extending the shelf-life of fresh agro produces, spices and also wide range of dairy and poultry produces (Hallman 2011; Miller 2015). After radurization, the count of viable non-spore-forming pathogenic microbes is reduced to an undetectable level in the treated product. However, it is recommended that the treated products are stored under refrigeration conditions (Kalyani and Manjula 2014). The established medium-dose applications of irradiation and the corresponding dosage are mentioned in Table 8.1.

Table 8.1 An overview of the applications of food irradiation

Application	Radiation dosage	Products	Reference
Low-dose (up to 1 kGy)			
Sprout Inhibition	0.05–0.15 kGy	Onions & tubers	Kalyani and Manjula (2014)
Delaying of physiological processes (ex. ripening of fresh fruits)	0.5–1.0 kGy	Fresh fruits and vegetables	
Insects and parasite deactivation	0.15–0.5 kGy	Cereals and pulses, flours, cocoa beans, fresh and dried fruits, dried fish and meat, fresh pork	
Inactivation of pathogenic parasites like tapeworm and trichina	0.3–1 kGy	Meat products	
Quarantine treatment against		Fruits and vegetables	Hallman and Loaharanu (2016)
(i) fruit flies	150 Gy (minimum)		
(ii) insects of other species	300–500 Gy		
Medium-dose (1–10 kGy)			
Shelf-life enhancement of fresh produce	1.0–3.0 kGy	Fresh fish, strawberries, mushrooms, meat	Kalyani and Manjula (2014)
Spoilage prevention	1.0–7.0 kGy	Fresh and frozen seafood, raw or frozen poultry and meat	
Improving the properties of food	2.0–7.0 kGy	Grapes (to increase juice yield), dehydrated vegetables (to shorten the cooking time)	
Delay of fruit maturation	2–5 kGy	Strawberries	Erkmen and Bozoglu (2016)
Control of molds for prolonged storage life	2.5 kGy	Fresh fruits	Mongpraneet et al. (2002)
Sterilization of foods	7.0–10.0 kGy	Herbs, spices, meat, poultry, seafood	Antonelli et al. (1998); Kim et al. (2014)
Reduction of aerobic bacteria	e-beam 2.5 kGy (resulted in a 3.2 log CFU/g reduction)	Bay leaves	Gryczka et al. (2018)
Treatment for: Reduction of *E.coli* O157:H7 Reduction of *Salmonella typhimurium*	γ-ray (5 kGy) >4.4 log CFU/g reduction >5.2 log CFU/g reduction	Black pepper	Song et al. (2014)

(continued)

Table 8.1 (continued)

Application	Radiation dosage	Products	Reference
Reduction of *Salmonella*	X-ray 1.13–2.28 kGy 5-log reduction in Salmonella	Almonds	Jeong et al. (2012)
Inhibition of aflatoxicogenic molds (*Aspergillus* sp.)	γ-ray: 5 kGy Reduction in aflatoxin colony growth by 48–63%	Contaminated maize samples	Markov et al. (2015)
High-dose (10–50 kGy)			
Integrated sterilization and heat treatment	30–50 kGy	Meat, poultry, seafood, prepared foods, and hospital diets	Kalyani and Manjula (2014)
Decontamination of food additives	10–50 kGy	Spices, enzyme consortia, and natural gums	
Sterilization of packaging materials	10–25 kGy	Wine corks	Corsi et al. (2015); Davis et al. (1982)
Sterilized hospital diets	30–50 kGy	Ready meals	Ihsanullah and Rashid (2017)
Virus inactivation	20–50 kGy	Food products	Erkmen and Bozoglu (2016)
Radappertization	25–50 kGy	Canned foods	

1.2.3 High Dose Applications

High-dose radiation (10–50 kGy) treatment is usually applied for achieving sterilization and shelf-stability of packaged meat products and foods with multiple components (Table 8.1). High dose is effective for inactivating microbes in wide range of dried agro produces. It could also be considered as a promising substitute for chemical fumigation (Kalyani and Manjula 2014). Notably, irradiation with high dose γ-rays significantly reduces wide range of mycotoxins in dry pepper by 45–50% (Jalili et al. 2012). On the other hand, at a further higher dosage between 25 and 70 kGy, irradiation imparts a sterilization effect in enzyme-inactivated food products in sealed containers. The high radiation dose reduces both the number and activity of spoilage organisms and their spores below a detectable limit in the irradiated food product. This leads to an indefinite shelf-life extension of the treated products, irrespective of their subsequent storage conditions, provided the package is intact.

With respect to attaining shelf-stability, i.e., long-term storage without refrigeration, the high-dose irradiation process is also referred to as *'radappertization'* (Kalyani and Manjula 2014). Radappertization is the treatment of food products with a sufficient dose of ionizing energy to prevent spoilage or toxicity of microbial origin, irrespective of the storage time or conditions under which the treated food is stored, as long as it is not decontaminated (WHO 1994). The main application of radappertization is the sterilization of astronaut foods, for which the approved dosage is 44 kGy (FAO/IAEA/WHO 1999; FDA 2015). Besides, vacuum-packed muscle foods are sterilized using a dosage of 25–45 kGy and then frozen to prevent lipid oxidation (Miller 2015; Cabo Verde 2018). A list of high-dose applications of food irradiation is given in Table 8.1.

1.3 Advantages and Limitations of Food Irradiation

The major competitive edge of ionizing radiation over the conventional thermal preservation techniques is that it inactivates microbes and their spores and toxins, without causing predominant quality changes to the treated foods. The other advantages of food irradiation technology (Khan et al. 2018; Pan et al. 2017) are depicted in Fig. 8.5.

Despite the above advantages, the food irradiation process poses certain limitations. Free radicals formed from the interaction between ionizing radiation and food substances can oxidize lipids and vitamins and lead to off-flavors and undesirable color changes in the treated product. For instance, γ-ray processing of peanuts at a dosage of 10 kGy speeded the degree of lipid oxidation in the treated product (Liu et al. 2018). Also, it was detected that the ionizing radiations induced mutations in

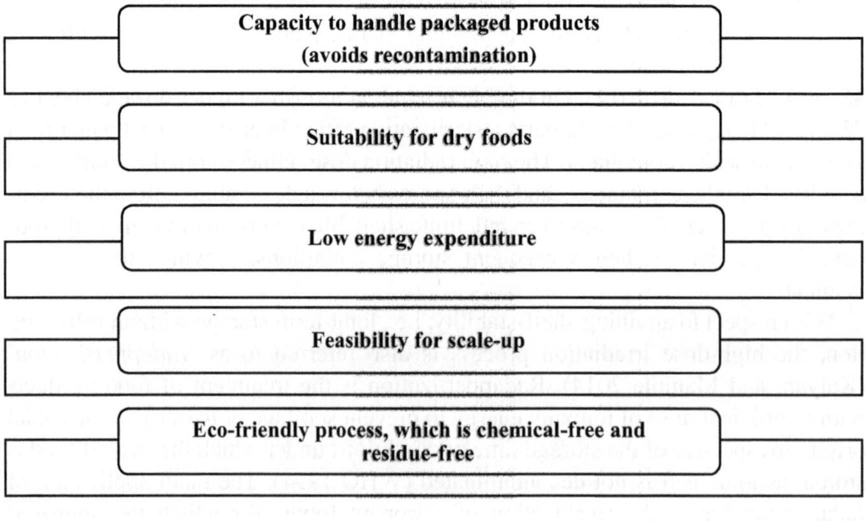

Fig. 8.5 Advantages of food irradiation

the pathogenic microorganisms to result in newer resistant strains (Calado et al. 2014). Strikingly, sub-inhibitory doses of radiation promoted sporulation, growth, and secondary metabolism of fungi. Consequently, generation of aflatoxin and ochratoxin was increased by two-fold in some irradiated commodities (Ribeiro et al. 2011). The commercial applications of food irradiation are limited by the capital investment involved and the adverse opinion of consumers about the health and safety concerns of irradiated foods (Deng et al. 2020a).

1.4 Safety Regulations of Irradiation Technology in the Food Processing Sector

According to the Food Additives Amendment to the Federal Food, Drug and Cosmetic Act, 1958, irradiation is classified as an 'additive', rather than a process. Therefore, the concern of misuse or safety arises only when the food is irradiated intentionally without following the prescribed regulations for safe use (Grégoire et al. 2003). Given the different schools of thought and ambiguities that prevail around the safety of irradiated foods, the US FDA has aligned many safety regulations in place. Accordingly, on the label of any package containing irradiated foods, it is mandatory to include the international symbol for irradiation – *"Radura"* (Fig. 8.6), along with the statement *"Treated with radiation"* or *"Treated by irradiation"*. In the case of bulk foods, the label has to be put up either on individual units or next to the sale container. However, the FDA does not demand to label the individual ingredients in multicomponent foods such as spices. It is imperative to follow the prescribed storage, handling, and cooking conditions for the irradiated foods in the same manner as non-irradiated foods. There is every chance that irradiated foods can still be contaminated with microbes when the fundamental rules of food safety are breached (FDA 2018).

Further, recent investigations have shown that not all radiolytic products are harmful to health. Their consumption levels are at low a concentration to have an impact on biological systems. Since both the food being treated and source are directly exposed to each other, possibilities of food turning into a radioactive

Fig. 8.6 International Radura symbol for irradiation. (IAEA 2012)

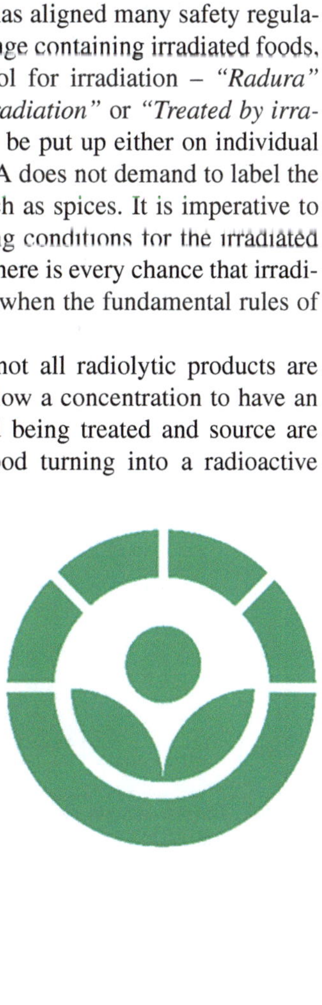

substance is nil. Consequently, the US FDA asserts that irradiated foods are not a thread for human use and that they do not contain any hazardous microorganisms and pathogens. The loss of nutrients that are prone to occur during food irradiation is similar to those caused by cooking or freezing. A positive sign is that in recent times, consumers are aware of the safety of irradiated foods and as a result, irradiated foods are gaining public acceptance in many countries. Irradiation technology is a boon to the developing countries that have abundant production of tropical fruits such as mangoes. It has opened up their export avenues to the United States, which once had prohibited the import of agro produces owing to the concern of pests. Particularly, specialty mangoes from Indian subcontinent cannot withstand hot water treatment, for which irradiation serves to be a one-stop solution for the elimination of pests (Ravindran and Jaiswal 2019).

2 Ultraviolet Light

The COVID-19 pandemic situation has made the safety of both food handlers and the food products that they produce on the shop floor, questionable. A critical bottleneck is the sanitation of food contact surfaces, which when contaminated can lead to loss of quality in food produces and also cause hazards to the operators. In this background, the ultraviolet (UV) light treatment is an effective alternative to the chemical disinfection methods. Ultraviolet radiation is a part of the electromagnetic spectrum, which falls in the wavelength range of 100–400 nm, which in turn comprises four sub-regions, namely, vacuum-UV (100–200 nm), UV-A (315–400 nm; long-wave black light), UV-B (280–315 nm; medium wave) and UV-C (200–280 nm; Short-wave, germicidal) (Fig. 8.7). Of the above, UV-C lighting is appropriate for the inactivation of harmful bacteria viruses, yeast, and molds, by damaging their genetic material (nucleic acids/DNA) and preventing their vital cellular functions. A broad range of applications is possible with UV light in the food industry including, cleaning, disinfection, and treatment of equipment, water, air, and surfaces. Not

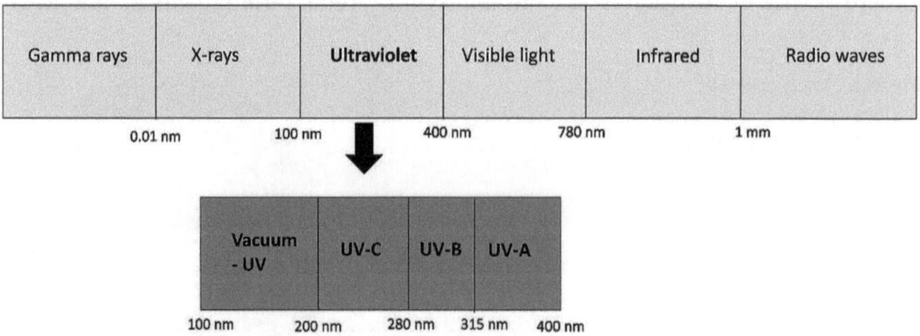

Fig. 8.7 The wavelength ranges of UV light in the electromagnetic spectrum. (Modified from Delorme et al. 2020)

limited to the above, a relatively recent interest of the food industry is on using UV radiation for the sanitization of solid surfaces and liquid foods such as milk products and fruit concentrates (Gayán et al. 2014). This section of the chapter would elaborate on the mechanism, applications, advantages, and limitations of the UV-light processing of foods.

2.1 Mechanism of Microbial Inactivation by Ultraviolet Light

As mentioned above, UV-C light exerts a germicidal effect on most microorganisms owing to its ability to cause photochemical modifications in the pyrimidine bases of nucleic acids (Fig. 8.8). Pyrimidine is an unsaturated organic compound that is essential for the reproduction and metabolism of cells. It forms the basic structure of the nucleobases including uracil (a constituent of tRNA), thymine (a part of DNA and tRNA), and cytosine (a constituent of DNA and RNA) (Cutler and Zimmerman 2011). The predominantly induced DNA lesion by UV light is the cyclobutane pyrimidine dimers (CPDs). Simultaneously, (6-4) photoproducts (6-4PPs) are also generated on about twenty-five percentages of CPDs (Sinha and Häder 2002). These lesions obstruct the normal replication and transcription of DNA, thereby leading to mutagenesis and eventual cell death (Friedberg et al. 2006). Nevertheless, the extent of lethality is determined by the radiation dose and the cells' capability to recover from and repair the damage. The latter point is relevant as microbes have several enzymatic repair pathways before replication to restore the replication errors in DNA molecules and recover them from the stroke of DNA-damaging agents (Friedberg et al. 2006, Sinha and Häder 2002). Moreover, in the case of extensive damage caused to DNA, the alternate repair mechanisms are activated such as that regulated by the SOS regulon. Thus, it is evident that the efficiency of UV light can be enhanced by weakening the repair mechanisms of bacterial DNA (Gayán et al. 2013). The other influential parameters of relevance to UV light processing can be classified under four categories, as shown in Fig. 8.9.

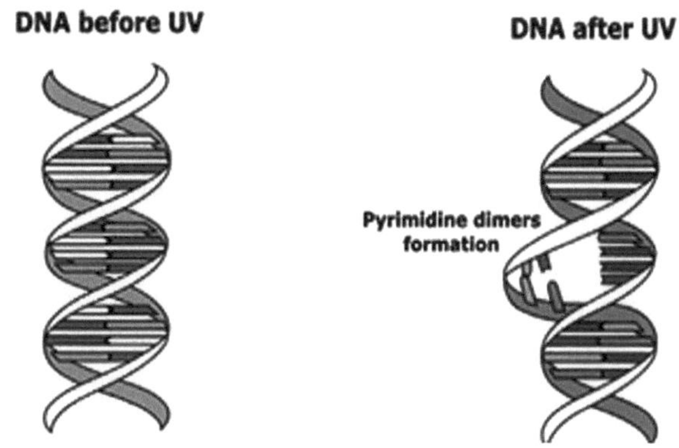

Fig. 8.8 Mechanism of microbial inactivation by UV light. (Modified from Delorme et al. 2020)

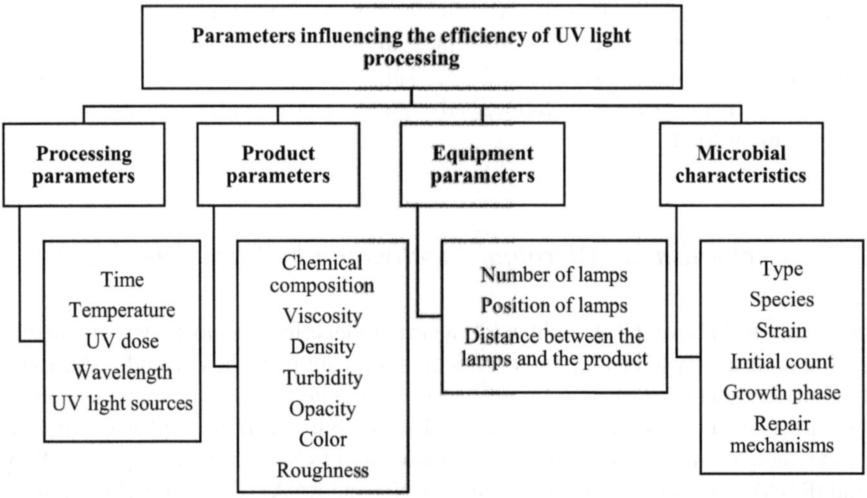

Fig. 8.9 Parameters influencing the efficiency of UV light processing. (Adapted from Delorme et al. 2020)

2.2 Sources of UV Light and Design of a UV Light Cabinet

Choosing the right light source and design of UV light device plays a major role in determining its microbial inactivation efficiency. Even though sun is a primary source of UV radiation, the intensity of radiation entering the earth is prevented due to the presence of ozone layer in the earth's atmosphere. This necessitates the use of synthetic sources such as black lights, certain types of lasers, fluorescent and incandescent sources, and mercury vapor lamps. Germicidal lamps are specifically employed for the emission of UV-C radiation, for reducing the microbial load in food products (FDA 2021). Other readily available methods are mercury lamps, pulsed light, and light-emitting diodes (LEDs). The appropriate UV light source is selected depending on its penetration depth and the end application. While low-pressure sources are preferred for the majority of the applications, mercury vapor discharges are specifically used for germicidal applications (Reed 2010). The pulsed UV light is an improved form of UV-C light, in which high power UV light (200–1100 nm) is emitted by lamps at regular interval of short times (Gómez-López et al. 2007).

The design of the UV light device depends on whether the product treated is solid or liquid. While a reactor (Figs. 8.10a, b) is used for processing fresh juices and other liquids, a chamber is developed for the treatment of solid surfaces (Fig. 8.10c) (Ha et al. 2016). The 'CiderSure' UV reactor shown in Fig. 8.10a is equipped with a low-pressure mercury lamp (LPM), housed inside a quartz sleeve that passes through the center of the reactor. The liquid product to be treated, say juice, is pumped as a thin film (Fig. 8.10a) from a storage tank via a diameter of 0.08 cm that is available as a hollow space (annulus) between the chamber and the

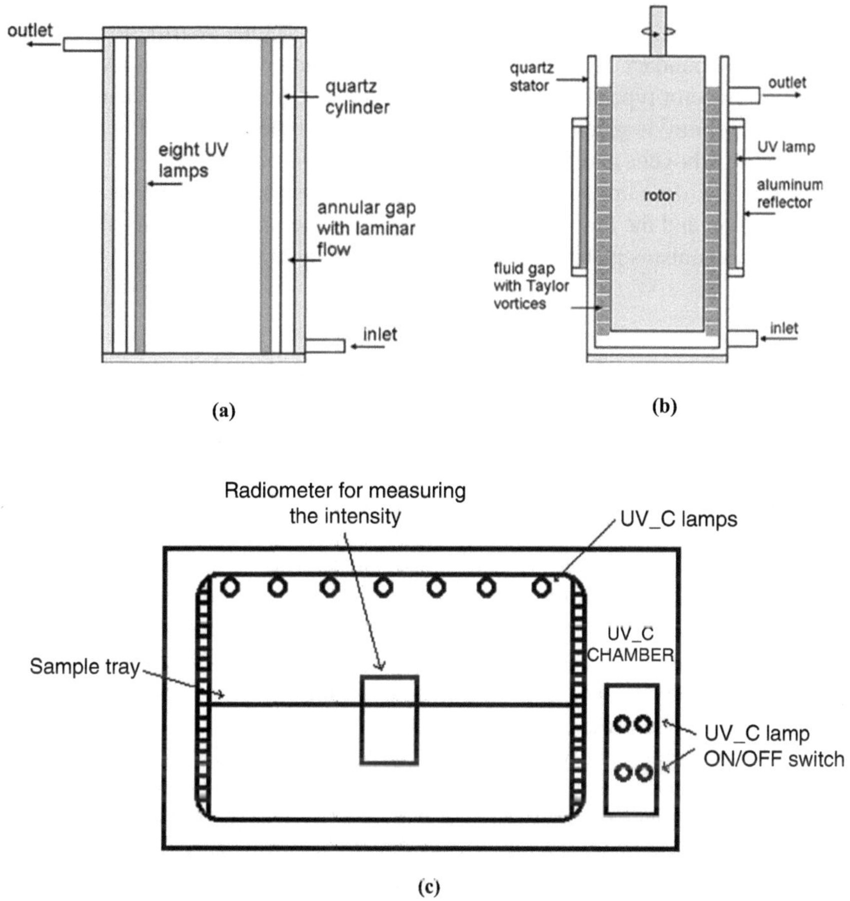

Fig. 8.10 Schematic representation of UV radiation devices currently in use: (**a**) laminar thin film reactor (CiderSure); (**b**) laminar Taylor–Couette UV reactor. (Koutchma 2008); (**c**) chamber for treatment of solid surfaces. (Ha et al. 2016)

quartz sleeve. The flow rate of juice is regulated by a processing sensors connected to UV source that monitor the UV transmission. Similar set-ups can be arranged in series connection depending upon the need and amount to be processed. A single pass of apple cider through the reactor ensued in reduction of *E. coli* by 1,00,000 folds (Worobo 1999).

Alternatively, the liquid may be pumped by a UV reactor via the annular space of a hollow cylinders, of which the outer one is static and the inner cylinder rotates (Fig. 8.10b. In this reactor type, the UV light source is located within the outer cylinder that is static, to offer adequate exposure and to decrease the boundary layer thickness of the fluid. As the smaller inner cylinder rotates at a lower rate, a vortexing laminar flow is established of the liquid inside the annular gap of width in millimeter-scale. The above type of flow is referred to as Taylor-Couette flow, which approaches ideal plug flow reactor where the residence time distribution does not dependent on boundary layer characteristics. The commonly treated liquid products in the above reactor type are the commercial apple and grape juices. When a dosage of around 9 mJ/cm^2 is given in a Taylor-Couette flow type reactor, 4-log & 3–5 log reduction of pathogens and E. *coli* is observed (Forney et al. 2004).

The total UV dose applied is strongly influenced by the flow patterns present in the UV reactor, and the flow patterns strongly alters the location and residence time of the microorganisms inside the reactor (Koutchma et al. 2009).

2.3 Applications of UV Light Processing in Food Systems

2.3.1 Disinfection of Food Contact Surfaces

The term 'food contact surface' refers to the exteriors and tops of bakery equipment, machinery in cheese and meat processing units, conveyor belts, product surface, packaging materials such as cartons, lids, cups, bottles, laminates, foils, and pouches, and tools of the equipment that come into direct contact with the product (Koutchma 2008). As mentioned above, UV light is a promising alternative to the conventionally employed chemical-based disinfection systems. UV light treatment of packaging materials prevents microbial contamination and aids in shelf life enhancement of food products.

The critical factors involved in the UV light treatment of food contact surfaces include the characteristics of the surface, matrix of the material being processed and the nature of microbes to be eliminated. For instance, when applied for the post-packaging treatment of finished packaged foods, the opaqueness of the packaging material inhibits the (Fig. 8.11) microbial inactivation (Tarek et al. 2015). Materials such as glass and polystyrene are opaque to UV light and hence not useful for UV light-based microbial inactivation of the contained food. Materials such as polyethylene film, polypropylene, and oriented polypropylene films are highly transparent to UV light, with 76%, 59%, and 57% transparency, correspondingly (Ha et al. 2016). Packaging materials that are treated by UV systems are categorized as "low-germ packaging materials", owing to their high purity. Further, UV light treatment is highly effective on plastic surfaces for the inactivation of Gram-negative bacteria, while the Gram-positive counterparts are less UV-sensitive. Similarly, vegetative forms (of bacteria and yeast) are more sensitive to UV than their spore forms (Koutchma 2016).

Similar to that of packaging materials, sanitation of conveyor belts in food manufacturing plants is central to the quality of products. As a conveyor belt carrying food products passes through the UV light module, up to 99% of bacteria on its surface are killed upon its exposure. In this manner, the UV light disinfects the surfaces of both the conveyor belt and the products that are transported on it (Morey et al. 2010). In certain cases, the UV light module is designed to be water-proof such

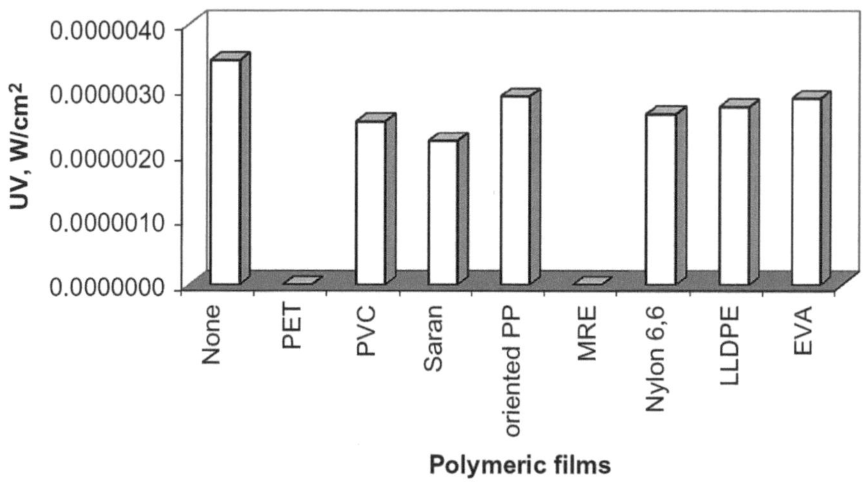

Fig. 8.11 UV transparency of polymeric films

that it can be incorporated in cleaning-in-place systems. Exposure of conveyor belts to UV light helps in reducing the downtime involved in complete wash-downs carried out during processing. This leads to the conservation of water resources and reduction of energy expenditure (Heraeus Amba Australia Pty Ltd. 2012).

2.3.2 Disinfection of Air and Water in the Food Industry

Worldwide, the mandate for clean water and air is on the rise. Moreover, germ-free air and water are the fundamental requirements in any food production unit. UV systems is found to be highly effective at the disinfection of water and air. This is because, UV light of short wavelength generates ozone from the oxygen present in the air, which in turn initiates the oxidation process. Subsequently, UV oxidation acts upon the pollutants in the exhaust air and breaks it down. Air can be disinfected and rendered free from bacteria, viruses, and molds by UV light, either in the state of moving stream or when it is stationary as in the drain pans of air conditioners and filter surfaces. Reflection of UV radiation from the material is an important parameter for microbial inactivation and hence UV lamps are coated with suitable materials such as stainless steel and anodized aluminum, to maximize the reflection of up to 80% of the emitted radiation. The sensitivity of airborne microbes to UV light depends on the temperature and humidity of the room. The efficiency of UV light decreases with an increase in relative humidity (Cutler and Zimmerman 2011). Besides, the positioning of the lamp and airflow patterns within a room also impact the effectiveness of UV disinfection. The efficiency of UV lamps has been observed to increase further in synergism with high-efficiency air filters. The above combination is effective for the disinfection of air in ducts, storerooms, and during seasoning, chilling, and drying, during which the food products, say, for instance, cheese, salami, and ham, cannot be removed intermittently during the unit operation (Stanga 2010). Attributed to its effectiveness in microbial inactivation and ecofriendly nature, UV light treatment is widely used in the public health sector comprising of health care centers, public shelters, hospitals etc., apart from the food and pharmaceutical industry (Lee 2011).

2.3.3 Liquids Handling: Disinfection and Pasteurization of Juices and Milk

The capability of UV light to decrease microbial contamination in liquid foods and beverages is well-established. The critical factors concerning the UV light processing of juices and pulp are the absorption coefficient (a measure of absorbance), turbidity, Brix, and viscosity of the liquid products. The above parameters vary substantially between fresh liquids and liquids containing pulp. For instance, fresh apple juice has a low absorption coefficient (11 cm^{-1}) than orange juice (50 cm^{-1}) (Koutchma et al. 2004). The processing conditions and outcomes of UV light processing for fruit juices are compiled in Table 8.2a.

In the case of milk and dairy products processing, UV treatment has been found to convert 7-dehydrocholesterol to vitamin D3 and thereby enhance the levels of the latter. No other significant changes have been detected other physical and sensory attributes (EFSA 2016; Orlowska et al. 2013). Similar observations have been reported with the UV-light treated cheese varieties (Can et al. 2014; Keklik et al. 2019). Nevertheless, depending on the processing conditions, protein precipitation (Orlowska et al. 2013), increase in the thiobarbituric acid reactive substances (Matak et al. 2007), and certain changes in the profile of fatty acid profile may occur. The plausible effects of UV-C on the quality attributes of dairy products are summarized in Fig. 8.12. A summary of the processing conditions and effects of UV light processing on the quality of milk and dairy products is presented in Table 8.2b.

2.4 Advantages and Limitations of the UV Light-Processing of Foods

As evident from the above discussions, UV light processing offers a multitude of applications and benefits to the food industry, owing to its non-thermal, non-chemical, non-toxic, and ecofriendly nature. It is an approved technology by the US Food and Drug Administration (FDA) and the European Food Safety Authority (EFSA). According to a survey conducted among the stakeholders from industry, government, and academic fraternities of North America and Europe, UV radiation turned out to be one of the most acceptable technology with high commercial viability and potential for food production. Besides, the UV-treated products were found to be safer (92%) with higher quality (94%) and enhanced shelf life (91%) (Jermann

Table 8.2 (**a**) Applications of UV light processing in the preservation of fruit juices. (**b**) Applications of UV light processing in the preservation of milk and dairy products

S.No.	Product	Target organism	Source of UV light	Processing conditions	Findings	Reference
(a)						
1.	Clear and cloudy apple juice	*E. coli* K12 (ATCC 25253)	UV LEDs emitting wavelengths 254, 280, 365 and 405 nm	Number of LEDs: 4 Working distance: 1 cm Sample volume: 3 ml	~ 2 log CFU/ml reduction (280 nm and 280/365 nm LEDs) 4.4 log reduction (280 nm LED) In clear apple juice, the polyphenoloxidase (PPO) enzyme was highly inactivated by the 280/365 and 280/405 nm LED treatment The lowest color difference was observed with 280/365 nm LED UV LEDs in combination were found to be effective	Akgün and Ünlütürk (2017)
2.	Orange juice	Cocktail of *Salmonella enterica* serovars, *Gaminara,* Montevideo, Newport, Typhimurium, and Saintpaul	Blue (460 nm) LED	Intensity: 92, 147.7, and 254.7 mW/cm^2 Illumination temperatures: 4, 12, and 20 °C	2–5 log reduction in *Salmonella* cocktail Best treatment conditions: 92 mW/cm^2 for 13.58 h at a dose of 4500 J/cm^2 at 12 °C	Ghate et al. (2016)

(continued)

Table 8.2 (continued)

S.No.	Product	Target organism	Source of UV light	Processing conditions	Findings	Reference
3.	Colored beverages and two readily available orange juices (A and B)	*E. coli* DH5α	UV-A LED (365 nm)	Intensity: 70 mW/cm^2 Coloring pigment concentrations: 0.001, 0.01, 0.1, and 1.0% Treatment time: 30 min; Dose: 126 J/cm^2	1.75 log CFU/ml reduction in the beverage containing 0.001% β carotene Orange juices (A and B) showed 0.35 and 1.58 log reduction, respectively An upsurge in the concentration of coloring agents reduced the antibacterial effect	Lian et al. (2010)
4.	Clear apple juice (pH 3.68, 12°Brix)	Spoilage yeasts: *Debaryomyces hansenii* *Clavispora lusitaniae* *Torulaspora delbrueckii* *Pichia fermentans* *Saccharomyces cerevisiae*	UV-C wavelength of 254 nm	Treatment time: 8 min	*D*-value = 8.27 min *D*-value = 9.78 min *D*-value = 9.39 min *D*-value = 11.04 min *D*-value = 6.38 min	Gabriel (2012)
5.	Clear apple juice (pH 3.68, 12°Brix)	*E. coli O157:H7*	UV-C wavelength of 254 nm	Treatment time: 1.5 min	*D*-value = 2.64–2.76 min	
6.	Grape juice	*S. cerevisiae* Yeasts Lactic acid bacteria	UV-C	138 mJ/cm^2, 9 min 280 mJ/cm^2, 24 min 280 mJ/cm^2, 24 min	5-log reduction 3-log reduction 4.3-log reduction	Kaya and Unluturk (2016)
(b)						
1.	Sliced camembert cheese	*Escherichia coli O157:H7* *Salmonella typhimurium* *Listeria monocytogenes*	UV emitting peak wavelengths 266, 270, 275, and 279 nm	Dose: 1 mJ/cm^2 (*E. coli*), 2 mJ/cm^2 (*S. typhimurium*), and 3 mJ/cm^2 (*L. monocytogenes*) Radiation intensity: 4 W/cm^2	4–5 log reduction in *E. coli*, *S. typhimurium* & *L. monocytogenes* were observed at a dose of 3 mJ/cm^2	Kim et al. (2017)

S.No.	Product	Target organism	Source of UV light	Processing conditions	Findings	Reference
2.	Sliced cheddar cheese packaged in Polypropylene (PP) and Polyethylene (PE)	*Escherichia coli O157:H7* *Salmonella typhimurium* *Listeria monocytogenes*	UV-C germicidal emitting lamps at 254 nm	No. of lamps: 5 Distance between lamp and product: 10 cm Intensity: 3.04 mW/cm², 182.4 mJ/cm² Treatment time: 1 min	3.36 and 3.12-log reductions	Ha et al. (2016)
3.	Unpackaged sliced cheese	*Escherichia coli O157:H7* *Salmonella typhimurium* *Listeria monocytogenes*	UV LEDs at 266, 270, 275, 279 nm	4 mJ/cm² 0.5 mJ/cm² 0.7 mJ/cm² 0.7 mJ/cm²	5-log reduction, irrespective of the target organism	Kim et al. (2015)
4.	Unpackaged white American cheese slices	*Listeria monocytogenes*	Pulsed ultraviolet light 3800 V of input at distance 13 cm and 8 cm	1270 mJ/cm² per pulse 3 pulses in 5–40 s	1.1–3.08-log	Can et al. (2014)
5.	Fresh Kashar cheese	*Staphylococcus aureus* *Escherichia coli O157:H7*	Pulsed ultraviolet light	Pulse rate: 3 pulses/s Treatment time: 5 s, 15 s, 30 s, 45 s, 60 s Distance between lamp and product: 5, 8, and 13 cm	1.62 log (cfu/cm²) reduction for *S. aureus* 3.02 log (cfu/cm²) reduction for *E. coli* O157:H7.	Keklik et al. (2019)
6.	Raw milk	*Staphylococcus aureus*	Pulsed ultraviolet light	Pulse rate: 3 pulses/s Distance between lamp and product: 8 cm Number of passes through the system: 1–3 Flow rate: 20–40 mL/min	Reduction of 0.55–7.26 log CFU/mL, depending on the sample, distance, number of pass and flow rate	Krishnamurthy et al. (2007)

(continued)

Table 8.2 (continued)

S.No.	Product	Target organism	Source of UV light	Processing conditions	Findings	Reference
7.	Fiordilatte cheese	*Pseudomonas* spp. *Enterobacteriaceae*	UV-C lamps	No. of lamps: 4 Distance between lamp and product: 2 cm Intensity: 20 W/m^2 Treatment time: 5, 30, 60, 150, 300, 450 and 750 s	1–2 log cycle reduction Sensory attributes were unaffected	Lacivita et al. (2016)
8.	White American Cheese	*Penicillium roqueforti* *Listeria monocytogenes*	Pulsed UV light	Pulse rate: 3 pulses/s Distance between lamp and product: 5. 8 and 13 cm Treatment time: 5, 30, and 60 s	Maximum reduction was obtained after 40 s at 5 cm Decrease ranged from 1.32 to 1.24 log CFU/cm^2	Can et al. (2014)
9.	Skim cow milk, whole cow milk, and whole sheep milk	*Bacillus subtilis* spores	UV-C pretreatment 254 nm	Intensity: 2.37 ± 0.126 J/mL combined with thermal treatment at 110 °C for 30 s	Reduction level: 6 log CFU/mL in bovine skim milk 2.90 log CFU/mL in bovine whole milk 1.1 log CFU/mL in, and ovine milk	Ansari et al. (2019)
10.	Goat's milk	*Listeria monocytogenes*	UV apparatus designed to increase UV penetration	15.8 ± 1.6 mJ/cm^2	>5-log reduction	Matak et al. (2005)
11.	Cottage cheese	*Pseudomonas* spp.	Pulsed UV light	Intensity: 16 J/cm^2 Pulse duration: 0.5 ms	1.5-log reduction No effect on sensory attributes (taste)	Dunn et al. (1991)

S.No.	Product	Target organism	Source of UV light	Processing conditions	Findings	Reference
12.	UHT skim milk (<0.5% fat)	*E. coli* ATCC 25922	405 nm (Near UV-Vis), 433 nm, and 460 nm (blue) LEDs	Illumination temperature: 5–15 °C Treatment time: 0–90 min	Highest inactivation at higher temperature and lower wavelengths; 406 nm LED treatment at 13.8 °C for 37.83 min can yield a 5-log reduction with The color change was minimal	Srimagal et al. (2016)

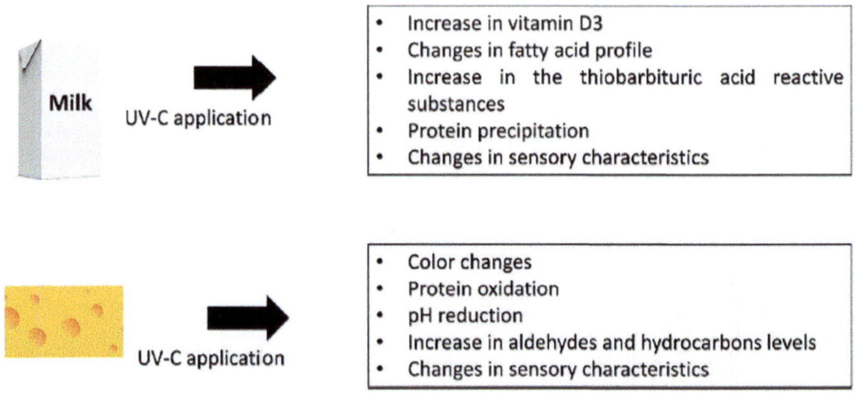

Fig. 8.12 Summary of the effects of UV light processing on milk and dairy products. (Delorme et al. 2020)

et al. 2015). The claim of high quality as mentioned above is justified, as UV light processing retains the sensory properties, texture, and nutritional aspects of the treated food product. Further, UV light can be used in conjunction with other non-thermal technologies for enhanced microbicidal and preservative effects (Datta et al. 2015; Delorme et al. 2020; Fan et al. 2017; Guneser and Karagul Yuceer 2012). Also, the energy consumption of the UV system is 10^4-fold lesser than thermal pasteurization. It consumes lesser specific energy than the other unconventional food processing techniques. However, the energy efficiency of UV systems is based on several factors such as the effectiveness of UV source, number, and type of UV light sources, flow rate, mixing proficiency within the UV reactor and matrix and properties of the material being processed (Rodriguez-Gonzalez et al. 2015). The other advantages of UV light processing are depicted in Fig. 8.13.

Regardless of its numerous advantages, UV light processing suffers from certain limitations. Since fruit juices contain organic compounds and suspended solids, efficiency of UV is sometimes reduced. It also depends on liquid parameters like viscosity, opaqueness and cloudiness. Likewise, the uniform and consistent distribution of ultraviolet dosage to large volumes of air are limited (Masotti et al. 2019). If the liquid is highly viscous, high power agitations are required to maintain the flow within the reactors (Koutchma 2019). Further, its commercial applications are limited by its lower penetration ability, which imposes an impractical requirement that food surfaces and the microorganisms on them are directly exposed to achieve the intended microbicidal effect (Fan et al. 2017; Koutchma 2009). Specifically, for milk processing, apart from the low penetration ability, the other disadvantages of UV light processing are modifications in the taste of the treated product, failure to inactivate bacterial spores, and the resistance of certain pathogenic organisms. Moreover, unlike the alkaline phosphatase test to determine the efficiency of milk pasteurization by conventional heat processing, there is no reliable indicator to confirm the efficacy of UV pasteurization (Datta and Tomasula 2015).

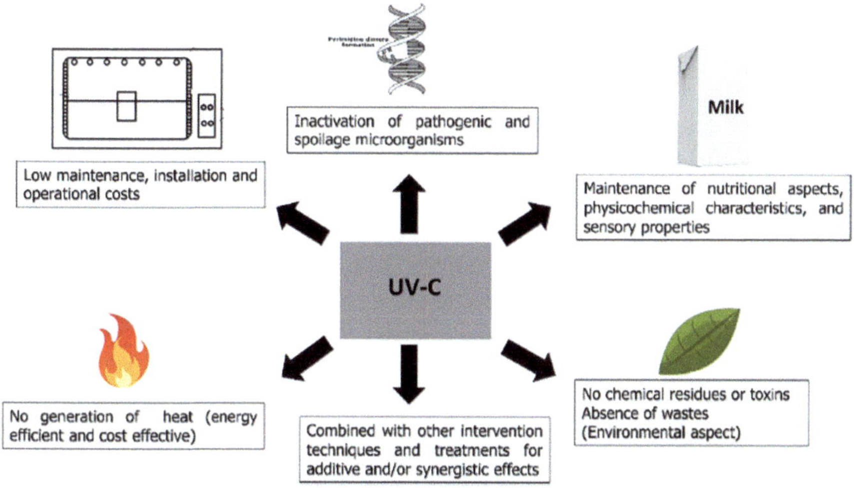

Fig. 8.13 Advantages of UV light processing. (Delorme et al. 2020)

Later studies have demonstrated the demerits of higher UV dosage, including, loss of quality and nutrients, changes in the physical (ex. color), chemical and sensory features of products, formation of undesirable compounds, and protein oxidation (Can et al. 2014; Orlowska et al. 2013). The aforesaid negative effects occurs when UV light gets absorbed due to the presence of amino acids and certain vitamins like vitamin B (Fernández et al. 2014). Other investigations have reported on the surge in levels of volatile aldehydes and hydrocarbons (Fernández et al. 2016). Thus, alleviating these negative effects of UV light processing presents an interest ing scope for future research.

3 Introduction to Ozone

Ozone is an inorganic molecule, which is a soft blue gas with a unique smell of pungency. Its chemical formula is O_3, having a less stability than that of O_2 (dioxygen). Ozone exists naturally in atmosphere and can also be produced in a laboratory via a synthetic route. When ultraviolet (UV) lights and electrical discharges from sun are exposed to diatomic oxygen, Ozone gas is produced. Most of the UV radiation from sun is absorbed in the stratosphere and hence the concentration of ozone is higher in stratosphere than that of Earth's atmosphere. Ozone, being highly unstable, it gets disintegrated into O_2 and O^- quickly and O^- acts as a powerful disinfection agent. Due to its powerful oxidizing effect, ozone is primarily considered among most potent disinfectant agents. U.S. Food and Drug Administration, in the year 1997 recognized Ozone as GRAS (Generally Recognized as Safe) element, permitting its usage as disinfectant and for sanitization in foods and food processing.

Since ozone is highly reactive and unstable making it unable to store, instantaneous generation of ozone is required (Fig. 8.14).

3.1 Principle of Ozone Generation

Generation of ozone requires splitting of oxygen molecule. High energy is required to breakdown the O–O bond. Free radicals, thus produced, interacts with oxygen molecule and forms ozone O3. Splitting of oxygen molecule can be cone via the action of electric discharges. Electric discharges for the production of ozone can be done with the aid of ultraviolet radiation, electrochemical, or electrical (corona) discharge. Once produced, ozone can be injected into water for decontamination and disinfection applications or directly in gaseous form into an atmosphere (Perry and Yousef 2011).

3.2 Ozone Generation Techniques

3.2.1 Ultraviolet Radiation

UV radiations of wavelength range less than 240 nm can be employed to generate ozone gas (Velasco et al. 2008). In this method, ozone is formed by exposing O_2 molecules to UV radiation in the wavelength range of 140–190 nm. For better yield of ozone, instead of air, oxygen must be used as feed gas (Miller et al. 2013). Via this method, small concentrations (0.03–0.05 ppm) of ozone can be generated by exposing the air to radiation using UV lamps at a wavelength of 185 nm. UV based ozone generators are easy to construct, maintance is easier and produces less amount of nitric acid as a by-product. However, low ozone output, higher production costs for fresh oxygen feed are certain limitations (Fig. 8.15).

3.2.2 Electrolysis

When high current is applied between two electrodes in an electrolytic system comprises of water and highly electronegative anionic solution, ozone gas is produced at the anode with oxygen (Mahapatra et al. (2005).

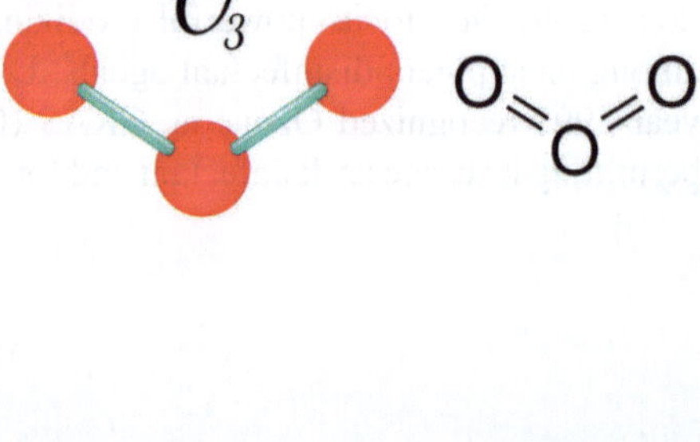

Fig. 8.14 Representation of Ozone molecule

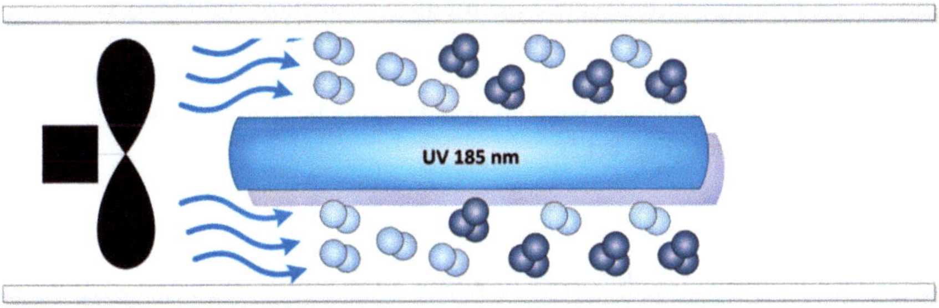

Fig. 8.15 Schematic representation of O_3 molecule formation

Fig. 8.16 Schematic representation of Electrolytic Ozone Generator

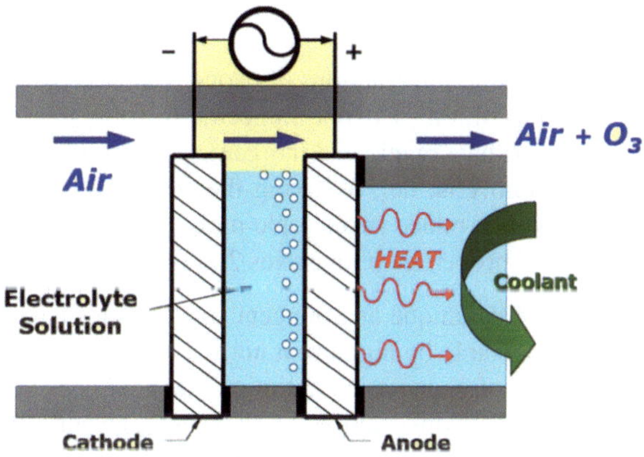

Reported electrochemical procedure for effective ozone generation involving electrolysis of water. Electrolysis produces hydrogen and oxygen, wherein hydrogen gas is vented from gas-water mixture whereas the oxygen atom get recombined in an oxygen mix, producing ozone. The ozone thus generated may be self- pressurized (≤20 psig), and can have about 12–14% (mass basis, in oxygen) ozone concentration. Since ozone is produced directly in the water, it does not require additional ozone contacting equipment. Its design and sizing can be made as compact as possible. However, few limitations include, higher energy consumption and shorter life time of anode and cathode (Fig. 8.16).

3.2.3 Corona Discharge

When two electrodes are separated by a dielectric material and Dry air/O_2 gas is passed over it, produces ozone due to the strong electric discharge. Electric discharge created produces free electrons, which in turn produces atomic oxygen. Atomic oxygen, then combines with available molecular oxygen to form ozone gas. Factors that decide the amount of ozone includes Voltage, current, frequency, type of dielectric material, discharge gap and absolute pressure (Horvath et al. 1985).

Though this method consumes huge electricity, it is the most commonly used, since it can able to produce ozone in commercial level production capacity. Dry air, gaseous mixture or pure oxygen can be used as a feed gas in this method. However, certain limitations include heat generation and requirement of pure feed without any presence of impurities (Fig. 8.17).

3.3 Theory and Mechanism of Food Preservation by Ozone Gas

de Souza et al. (2018) stated that there are three main pathways that describes the action of ozone.

1. Direct oxidation reactions of ozone where oxygen atom (O^-) reacts.
2. Decomposition of ozone molecule to form free radicals, which in turn reacts with various organic and inorganic compounds.
3. By ozonolysis, i.e., by fixing the complete molecule on double linked atoms. This will produce two simple molecules with dissimilar molecular characteristics and properties (Gonçalves 2009)

Because of its unique and pungent smell, even small quantities as low as 0.02 ppm can be detectable by humans at normal temperatures (Horvath et al. 1985).

Efficacy of ozone as an antimicrobial agent has been considered for wide range of micro-organisms such as yeast, molds, bacteria and even virus. Antimicrobial behavior of ozone is a complicated process, since it attacks and interacts differently with different cellular constituents (Khadre et al. 2001).

Komanapalli and Lau (1996) explained the process of deactivation of bacteria by the action of Ozone. Primary target of ozonation is cell surface, where unsaturated lipids on the cell envelope gets degraded, due to which vast part of the membrane barrier gets wrecked resulting in disruption of the cell. This in turn leads to leakage of cellular contents and bacterial cells. In addition, ozone is also capable of

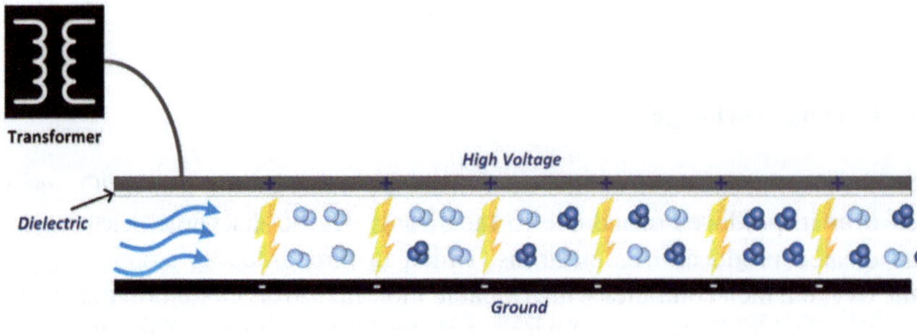

Fig. 8.17 Schematic representation of corona discharge

penetrating deep inside the bacterium thereby oxidizing the essential components like proteins, enzymes and nucleic acids.

3.4 Applications of Ozone Gas in Food Processing

Ozone plays an important role in food industries for food plant sanitation and also as a disinfection agent. It can be used as such in gaseous form or purged into water and used as a bubbling system in aqueous solution (Çatal and İbanoğlu 2012). Ozone has been reported to have improved the shelf life of cucumber (Gorzelany et al. 2017; Migut et al. 2019), blueberry (Pandiselvam et al. 2019), apples (Antos et al. 2018), fresh cut lettuce (Yucel Sengun and Kendirci 2018), melon juices (Miller et al. 2018) and sugarcane concentrates (Garud et al. 2018) etc., Other than food and agro produces, it has also been used as an antifungal effect in dog foods (Silva et al. 2018a), degradation of pesticides in chillies (Karadurmuş et al. 2019) and in removal of carcinogenic substances from water (Sintuya et al. 2018). Ozone is also extensively used for fruit juice processing. Since fruit juices are liquids or semi-solids, the process of ozonation can be done by injecting ozone gas into the container handling juices (Miller et al. 2013).

It has been evaluated that ozone efficacy as low as 0.4 ppm can able to reduce wide range of bacteria. Upon higher dosage and prolonged exposure, it can also disintegrate the structure of protein, carbohydrate and fat sources (Guzel-Seydim et al. 2004). Freitas-Silva and Venâncio (2010) found out that the inactivation of fungi by ozone is due to their implications in cell membranes of the microbes. Palou et al. (2001) detected that, even the development of Penicillium italicum was inhibited by ozone gas at desired concentrations. Majority of the applications of ozone are in destruction of fungi and mycotoxins in food produces (Table 8.3).

3.4.1 Treatment of Solid Food Items

Ozone has been extensively used in various solid/semi-solid food products for inactivation of pathogens and microorganisms. It is also used for the dissipation of pesticide and chemical residues in agro produces like fruits and vegetables (De Souza et al. 2018; Kim et al. 1999).

Ozone treatments on grapes and blackberries is found to enhance the shelf life by decreasing the fungal infection (Beuchat 1992). With the oxidation of ethylene (ripening hormone) and removal of highly toxic metabolic products, ozone helps in increasing shelf life of fruits (Horvath et al. 1985).

Ozone treatment at a rate of 0.2 μg L^{-1} for 8 h in potatoes and onions were found to significantly lower the surface microbial infection and also decreased antioxidant enzymes (catalase and peroxidase) and oxygen intake.

Table 8.3 A time line of Ozone discovery and its application in Food Processing

1801	**Cruickshank's** while performing electrolysis of water observed a distinct odour.	**1942**	Storage for eggs and cheese was done using Ozone in US.
1840	German Scientist **Christian Friedrich Schönbein** officially discovered ozone gas	**1965**	Decolouration of water was done using Ozone in Ireland and UK.
1857	First ozone generator was discovered by **Werner Von Siemens** now called by the name "Siemens Type" ozone generator	**1973**	For the first time, **IOA (International Ozone Association)** was created to provide ideas, innovation and consultancy for the industries utilizing ozone for disinfection and decontamination purposes.
1870	**Therapeutic** applications of ozone was introduced by C. Lender, Germany	**1976**	Approval for ozone as an antimicrobial agent was given by EPA (Environmental Protection Agency), USA.
1873	**C. B. Fox** observed and found that the ozone possesses the ability to act as anti-microbial agent.	**1977**	Ozone is found to reduce **Salmonella on eggshells** in Russia.
1893	First water treatment plant using ozone was first used by **Ousbaden** in Holland. Ozone for the first time used in Algae control also.	**1982**	**GRAS (Generally Recognised As Safe)** status was given to ozone by FDA (Food and Drug Administration) in the US
1896	Patent for first ozone generator was done by Nikola Tesla.	**1997**	A panel of experts concluded that ozone could be declared GRAS for food processing sectors in the United States.
1909	Ozone is used for the first time in the **preservation of frozen meat** in Germany.	**2001**	Ozone is approved as a food additive and anti-microbial agent by FDA.
1936	Ozone is used for **washing and storing fish** in France.	**2001**	FSIS (Food Safety and Inspection Services) formulated and standardized acceptable limits of ozone use in meat and poultry.
1939	Ozone is used for preventing yeast and mould growth in fruits.	**2004**	Industrial recommendations on safe usage and limits of ozone was given by FDA.

3.4.2 Treatment of Fruit Juices Via Ozone

After the approval of FDA to use ozone as a food additive, it has become extensively used in treatment of fruit juices for pathogen reduction. Ozone significantly reduces the population of wide range of microorganisms upto the level of five log reductions (100,000 fold). Ozone has also been used for processing of juices of various fruits such as apple cidar (Steenstrup and Floros 2004), orange (Tiwari et al. 2008), blackberry (Tiwari et al. 2009b), tomato (Tiwari et al. 2009a), apple (Choi et al. 2012), peach (Jaramillo-Sánchez et al. 2018), melon (Fundo et al. 2018) sugarcane (Garud et al. 2018), and lime purification of sugar beet juice (Gharib-Bibalan et al. 2018). In all fruit juices, ozone was injected as a stream of gas inside the juice solution and process was optimized. Based on experimental results and process optimization, the maximum permissible limit of ozone in bottled is fixed at 0.4 mg.L^{-1}.

3.4.3 Applications of Ozone on Pest Control

When gases like ozone enters the respiratory system of insects will lead to increase in temperature which in turn results in the increment in respiration rate causing higher gas exchange and leading to overall increase in the rate of metabolism and respiration (Pandiselvam et al. 2017). Since ozone toxicity will cause oxidative tissue damage, the insect tend to breath discontinuously (Hetz and Bradley 2005). Due to which, ozone gas will enter into the breathing system of insects (Lu et al. 2009), leading to sharp upsurge in temperature, respiration rate and gas exchange (Rozado et al. 2008). The oxidative tissue damage caused by ozone (Pimentel et al. 2007) results in alteration of pulmonary functions, breakage in DNA strands, bronchial responsiveness and membrane oxidation, which ultimately leads to the death of the pests (Ballinger et al. 2005).

Various reports of ozone being used in pest management of stored product such as grains and pulses has been reported (Pandiselvam et al. 2017, 2019). Ozone is considered as a safe potential substitute to present fumigants (Kells et al. 2001). One of the most striking advantage of ozone in fumigation is ozonation requires very less energy compared to other methods (Khadre et al. 2001).

Laboratory as well as field studies confirms that injecting ozone gas into stored grains shows promising effect in controlling the insects. Depending on the amount and development stage of insects present in the grains, amount of ozone required for disinfection varies. For example, larval and pupal stages of T. castaneum were sensitive to ozone and this sensitivity decreases with age (Erdman 1980).

Kells et al. (2001) observed a high mortality for the T. castaneum and P. interpunctella, upon exposure of 50 ppm ozone gas at an interval of 3 days. The eggs of P. interpunctella were destructed by injecting around 1500–2000 ppm ozone for a period of 180 minutes (McDonough et al. 2011).

However, at prolonged exposure of ozone leads to damage of storage structures and rubber couplings via corrosion. Majority of the researchers proposed a upper limit of 50 ppm of ozone to control the insect growth in grains. Though, higher concentrations of ozone are needed for complete killing of the insects, due to high cost and material damage, it is seldom practised. Considering these facts, researchers recommended the upper limit of 50 ppm of ozone should be used for storage of grains.

However, in certain cases, ozone treatment was found to be ineffective and needs further investigation. For example, Faroni et al. (2007) reported the effect of ozone on corn grains at various concentrations and revealed that there is no change in lipids and concluded that there is no significant change in the quality of corn oil for both treated and untreated sample. White et al. (2010) observed that ozone effectively decreases the dry matter loss in ozone.

Since ozone is highly reactive, development of suitable storage systems for effective diffusion of ozone becomes challenging. The concentration of ozone level for fumigation must be adjusted according to the moisture content of grains. The efficiency of fumigation in grains also depends on diffusion of ozone gas, time of exposure and concentration (Shunmugam et al. 2005).

3.5 *Advances, Challenges and Recent Advancements in Ozone Food Processing*

There has been growing advancements in the usage of Ozone gas in pesticide reduction in agro produces. According to Food and Agriculture Organization (FAO), pesticide is defined as:

> Any substance or mixture of substances intended for preventing, destroying, or controlling any pest, including vectors of human or animal disease, unwanted species of plants or animals, causing harm during or otherwise interfering with the production, processing, storage, transport, or marketing of food, agricultural commodities, wood and wood products or animal feedstuffs, or substances that may be administered to animals for the control of insects, arachnids, or other pests in or on their bodies. The term includes substances intended for use as a plant growth regulator, defoliant, desiccant, or agent for thinning fruit or preventing the premature fall of fruit. Also used as substances applied to crops either before or after harvest to protect the commodity from deterioration during storage and transport.

Thus, pesticides are mainly projected to kill the pests or to control the insects and growth of fungus present in the agro produces during the growing stage (González-Rodríguez et al. 2011). However, presence of residual pesticides in the agro produces after the post-harvest processing poses a huge thread to human health (Damalas and Eleftherohorinos 2011). Chemical reagents have been frequently used as a medium for eliminating the residual pesticides from agro produces, whereas Ozone has been growingly used for the destruction of pesticides from agro produces. Using ozone gas is highly advantageous since it is effective both in aqueous and gaseous state (Wang et al. 2019). Chemical methods are much costlier though they are efficient. For example, Hydrogen peroxide always require special conditions for their conveyance and stowage, thereby increasing the cost of maintenance (Zohair 2001; Wang et al. 2019). Ozone can change a variety of organic constituents into their simpler forms that are decomposable (Pandiselvam et al. 2017; Pandiselvam et al. 2019). Ozone can quickly decompose, thereby leaving no harmful residues after usage. Since ozone leaves no residues, ozone treated agro produces can be tagged as organic food (Selma et al. 2008).

Ong et al. (1996) proved that various pesticide residues on the surface of apple can be broken down using ozonated water. It has been documented by Hwang et al. (2001) that around 60–95% reduction in pesticide residues was detected in apples treated with ozone at a rate of 1–3 ppm. Osman (2015) reported that percentage of pesticides removed from many varieties of dates fruits with ozonated water in at a concentration of 2 mg/L were notably greater than immersing fruits in distilled water alone (Osman 2015).

Antos et al. (2018) showed that utilization of aqueous and gaseous ozone has more effect in degradation of dithiocarbamate residues present in blackcurrant. Karaca (2019) proved that six pesticides were removed from grape by applying ozone enriched air at a concentration of 0.64 mg/m^3 and exposing fruits to that air.

Whangchai et al. (2011) reported that degradation of chlorpyrifos was effective in gaseous ozone than ozone injected in aqueous solution. Metzger et al. (2007) proved that levels of pesticides on citrus fruit that were stored in ozonated conditions decreased to minimum levels. Ozone treatment also reduced the difenoconazole residue by 97% in strawberries. Ozone dosage of around 0.3–0.8 mg/L of solution is effective in reducing difenoconazole content in strawberries by 97% (Heleno et al. 2014).

Organophosphorus (OP) insecticides being widely used in vegetable crops for controlling the soil and foliage pests are effectively removed by ozone gas at a concentration of 2 mg. L − 1 of ozone (Khaled et al. 2017).

Al-Antary et al. (2018) demonstrated that ozone can effectively reduce the concentration of chlorfenapyr residues upto 95% from the surface of tomatoes at a concentration of 2 ppm for 15 minutes' exposure. Hence they concluded that ozonation for about 15 minutes at 5 ppm plays a significant decrease in myclobutanil residue in lettuce leaves.

From studies it has been observed pesticide removal largely depend on the ozone concentration, time of treatment and temperature and hence effective optimization will lead to more effective ozone treatment (de Souza et al. 2018).

4 Photodynamic Inactivation Treatment

Food safety helps to ensure that food, which is being consumed, is safe and free from any type of contamination. Conventionally, thermal methods are being applied for reducing and killing the microbial cells to make food fit for the consumption. But these techniques may have negative impact like loss of nutrition and sensory attributes. Therefore, the researchers and scientists are always interested to develop some new technologies that can serve the purpose with minimal damage to nutritional and sensory parameters of the food. There are various methods, which are adopted by the food industry to decontaminate the food items for assuring the food safety issues. In this regard, photodynamic inactivation (PDI) is an innovate technique to deactivate the propagation of different infectious microbes which causes diseases and results in wastage of food and creates a burden on economy due to food borne illness (Xuan et al. 2018). This method is considered favourable for inactivation of microbes and decontamination of food.

4.1 Principle of Photodynamic Inactivation

The light of specific wavelength is required to activate the photodynamic inactivation treatment. The light of specific wavelength starts the oxidative reactions in the microbial cells. The light sensitive components (photosensitizers) help to start the series of these oxidative reactions. These components may be naturally present or supplied from external sources (Ghate et al. 2019). The exposure of cell with incident light of certain wavelength results in the excitation of the photosensitizers by the absorption of the energy. After that when the photosensitizer comes back to the ground state result in the production of energy by collision of the cytoplasmic molecules and reactive oxygen species (ROS) are produced. ROS have high energy and include singlet oxygen, hydroxyl radical and hydrogen peroxide. ROS result in the oxidation of the various components of the cell, which ultimately results in death of microbial cells (Ghate et al. 2019; Luksiene and Brovko 2013).

4.2 Mechanism of Photodynamic Inactivation

In photodynamic inactivation there is a role of selectivity, speed of inactivation and invasiveness. The inactivation procedure starts when the photosensitizers absorb the light of specific wavelength and it goes to the excited state, which leads to the generation of singlet oxygen. After the production of singlet oxygen, the photosensitizer (PS) comes back to the ground state and at the same time emission of light and heat transfer occurs (Kashef et al. 2017). As mentioned above the photosensitizers (PS) go to the excited state by the incident light. There are two different modes of activation of photosensitizers called Type I and Type II mechanism.

In the former, the molecular oxygen which produces further hydroxyl radicals like the reactive oxygen radicals are considered as a main element in the whole mechanism because this molecular oxygen is very much important in the Type II mechanism of photodynamic inactivation treatment. The collaboration of molecular oxygen with the photosensitizer (PS), which is in the excited state in the Type II mechanism, leads the process of microbial cell inactivation and at last the microbial cell death. That is why molecular oxygen is the main element in this process. This collaboration in the mechanism lead to the production of singlet oxygen. The photodynamic inactivation proves to be very useful as compared to other techniques and very effective in inactivation of viruses, fungi and bacteria. It is now considered as an effective substitute because it does not cause an effect to the color, taste and the nutritional quality of that food to which this methodology is being applied (Cunha et al. 2017; Tim 2015; Nakonechny and Nisnevitch 2019; Durantini 2006).

4.3 *Factors Affecting the Photodynamic Inactivation*

There are various internal and external factors that affect the photodynamic inactivation process (Ghate et al. 2019). The internal factors mean the factors which can have an effect inside the bacterial cell and the environmental factors or environmental conditions are considered to be the external factors which are also equally as important as an internal factor. Wavelength, time and dose are some external factors, which are important and affect the photodynamic inactivation process (Angarano et al. 2020).

4.3.1 Wavelength

The most critical factor that affect the mechanism of photodynamic inactivation is the wavelength and have been discussed widely in the literature. Specific wavelength is required for every food, as single wavelength cannot be applied to all the foods. The two phases or modes of light are considered in this process. In the first phase, light gets absorbed by the photosensitizers (PS), whereas in the second phase, emission occurs. But there is a problem in it because the light sources are not available at every wavelength (Ghate et al. 2019).

The specific wavelengths are required but in this study there is not aim on this tat why specific wavelengths are required. In this study basically the mechanism and applications of photodynamic inactivation are discussed. The Flavin and Porphyrin type photosensitizers (PS) are applied at bacterial cell inactivation. These are the main Photosensitizers which can be used in the bacterial cell inactivation but some other photosensitizers can also be used in case of bacterial cell inactivation. The light of different wavelengths are absorbed by the photosensitizers means it has the capability to absorb lights in different ranges. The violet blue range of light is considered to be peak. The microbial species containing oxygen when absorb light produces reactive oxygen species (Angarano et al. 2020).

4.3.2 Time

The time of treatment is another factor that is important for photodynamic inactivation process. The selection of light also determines that how much time is required for the treatment. From 2 min to 7 days is considered to be appropriate time for such treatments. The time of 72 h is applicable to the fresh cut products according to the study. The shorter treatment time is required if the powerful beam of light is used. There is no need to use the so much expensive light source. The inexpensive light source having a shorter treatment is also applicable and appropriate. Although this method is new in quality and safety of food but proves to be very useful (Angarano et al. 2020).

4.3.3 Temperature

Temperature is also considered very important and influential factor in the photodynamic inactivation treatment. The temperature during illumination is very important. The physiology of the microbial cell plays a crucial role in determining the ambient temperature required for the inactivation. The other factor which has an effect the temperature is stress response. The stress response on the other hand also has an influence on the physiology (Ghate et al. 2019). The fatty acid and the protein composition of the membrane are adjusted by the tolerance level of the bacterial cells, which are in the vegetative phase after the temperature is being applied on the microbial cells especially bacterial cells. The permeability properties changes with the change in the ambient temperature. The changes occur according to the heat tolerance level of the microbial cells. The protein profile also shows changes along with the change in the temperature being applied. The cells, which grow at higher temperatures shows a high content in the heat shock proteins (Cebrián et al. 2017).

4.3.4 Dose

There is a direct relationship between dose and the illumination time. The microbial inactivation mechanism occurs after the regulation and control of the dose. After wavelength and the temperature, the dose applied is also very important factor for the inactivation of the microbes. Some studies have been conducted on this topic suggests that the dose requires is low comparatively. The different models like log linear model, Weibull model and Gompertz model are being used mostly. Although all the models can be used but the Gompertz model are considered to be most effective and appropriate model (Ghate et al. 2016; Angarano et al. 2020).

4.3.5 Water Activity (A_w)

Water activity is also very significant factor in the photodynamic inactivation treatment. As water activity is different for various types of microbes, therefore, it depends on the species and the type of microorganisms, which we want to deactivate like viruses, fungi and bacteria. Bacteria have a water activity of 0.9 while molds have a water activity 0.7 having the lowest water activity. There are many factors the controls and regulates the water activity. These factors include the stress response and the osmotic stress of microorganisms. The stress response shows its effect and increases the effect of the photodynamic inactivation method. The effect of water activity also increases or decreases by the type of microbes present in the matrix (Ghate et al. 2016; Cebrián et al. 2017).

4.3.6 Other Factors

Factors such as acidity, surface properties, irradiance, flux distribution and oxygen also play a crucial role in photodynamic activation. Surface Properties is very important among these minor factors and plays very significant role in the microbial cells inactivation (Ghate et al. 2016; Angarano et al. 2020).

4.4 Applications of Photodynamic Inactivation in Food Processing

The photodynamic inactivation treatment can be applied in almost all of the food sectors. It has proven to be playing an important role in the decontamination of the food matrices and ultimately ensures quality and shelf life of the agro and food produces.

These food sectors include fruits, vegetables, dairy products, poultry, meat and seafood as well. But the studies which have been done on the food sector is of fruits, vegetables and dairy products (Ghate et al. 2016).

4.4.1 Fruits and Vegetables

Fruits and vegetables are rich in minerals and vitamins and hence application of photodynamic inactivation becomes necessary for ensuring healthy and decontaminated agro produces. For this purpose we can apply the photodynamic inactivation treatment on the fruits and vegetables by which they can be decontaminated and safe to eat and their nutrients should also not disturbed.

It has been observed there is more and prominent transmission of many microbes which causes foodborne illness diseases from the raw fruits. Because the fruits and vegetables are being exported globally from all over the world and therefore very prominent route from the transmission of microbial activity. It has therefore caused great financial losses to the US economy of about 6.9 Billion. The losses are in terms of production and medical areas (Luksiene and Brovko 2013). During the past decade the biological and physical methods are being implemented instead of the chemical use. The major physical methods includes controlled atmosphere and irradiation. These methods are proven to be useful and safe as compared to chemical methods (Bhavya and Hebbar 2019).

The photodynamic treatment of strawberries with Listeria shows positive result and decrease in microbial activity. There has been shown a decrease of 1.8 log in microbial load in strawberry. The decrease sustained to a time period of 2 days in the berries. Hence it was proven by this activity that the photodynamic inactivation treatment (PDI) has a useful impact on the fruits and vegetables (Luksiene and Brovko 2013).

4.4.2 Dairy Products

In dairy products milk is an essential ingredient. Milk is also included in daily servings and has many important ingredients. Milk is considered as a main diet of children so it is very important that milk should be microbiologically safe. For this purpose many processes have been applied to make milk microbiologically safe. These treatments include sterilization, pasteurization and UHT. By using these treatments, the milk become safe but many of its nutritional ingredients have been lost so to overcome these nutritional loses, the photodynamic inactivation treatment has applied to milk which is very useful technique. It makes milk microbiologically safe as well as its nutritional loses are reduced.

In photodynamic treatment of milk, the blue LEDs at three different wavelengths and at different temperatures were applied. By this treatment the extinction of *E.coli* was observed. By this treatment the milk becomes microbiologically safe as well as the shelf life of milk was also increased. Its shelf life was observed to be increased 4 days more than the milk, which was treated, by the conventional methods or techniques (Galstyan and Dobrindt 2019).

4.4.3 Beverages

Photodynamic inactivation treatment (PDI) was also applied to different beverages like fruit juices. In orange juice the *Salmonella* was inactivated by applying LEDs of a specific wavelength. The wavelength which was applied was 460 nm. This method was proved to be successful but there was a slight problem which was observed. Some bleaching was observed but we can reduce or minimize this effect by proper selection of temperature. In short the photodynamic inactivation treatment can be also used in beverages to inactivate microbial activity (Ghate et al. 2016).

4.4.4 Seafood

There are many pathogens which are related with the seafood and that cause different diseases. For example Vibrio parahaemolyticus is a pathogen related to sea food which causes gastroenteritis. Similarly there are many other pathogens which causes different diseases which are harmful for human health. Photodynamic Inactivation is not so much old technique. It is a recent technique or approach, which is being applied in different food sectors to ensure the availability of microbiologically safe food (Deng et al. 2016).

4.4.5 Poultry and Meat

Poultry and meat are the also important food sectors. Meat is present in most of the servings of people. Therefore for the availability of microbiologically safe, different techniques were used. Photodynamic inactivation is also used now. The blue light of specific wavelength is used for the photodynamic inactivation of meat and poultry. Curcumin combination with blue light of specific wavelength is being applied for the inactivation of microbial activity. *S. aureus* in poultry that is in the chicken meat can be detected by using the combination discussed above. The use of this combination is necessary because the blue light alone has no impact on the inactivation of microbial activity in meat (Hadi et al. 2020).

4.4.6 Other Applications

The photodynamic therapy has been employed for treatment of cancers, inactivation of bacteria, inactivation of viruses and new selective media. Photodynamic therapy includes the usage of specialized photosensitizers which are directed intravenously to the patient. The norms for a successful photosensitizer are very strict due to the nature of its functionality. Present clinical trials to eliminate tumors or to relieve the symptoms use haematoporphyrin derivative (HpD). A purified fraction of HpD is accessible commercially as a sensitizer called DHE, which has been one photosensitizer extensively, used on clinical trials (Benov 2015).

4.5 *Future Prospects of Photodynamic Inactivation for Ensuring Food Safety*

Food Safety is a global issue. Millions of lives lost every year all over the world due to the consumption of unhealthy and unsafe food. Loses not only occurs in developing nations but also in the developed nations. It means this issue is equally dangerous or critical for the developing as well as the developed nations. It means it is present all over the world and should be tackled by the whole world and by the collaboration of all the countries. There are different types of loses which occurs due to the unsafe food. Obviously the loss of lives is the main loss and after that the financial loses. Financial loses are also much important to tackle it by using different approaches. Especially for the developing nations, it could be a disaster for the whole nation. The developed nations are also not safe from these loses as there are examples of it that even the countries like US bears the heavy loses (Silva et al. 2018a, 2018b).

Various methods have been used to tackle these problems and for the availability of microbiologically safe food. These conventional techniques were useful but at the same time the organoleptic properties and the nutritional loses has been occurred by using these techniques. So for the reduction or minimization of the nutritional loses some non- destructive and innovative techniques were developed. Photodynamic inactivation (PDI) is one of those techniques, which is being used for the microbial activity inactivation due to its effectiveness (Silva et al. 2018a, 2018b). Photodynamic inactivation is not only applied in the food sector but also in the other fields such as medicinal and clinical treatment of different diseases (Alves et al. 2014). The use of the food grade photosensitizers with its increased effectiveness to inactivate all the microbial activity like of bacteria, fungi and viruses is coming to happen soon. Likewise such different approaches can ensure the availability of microbiologically safe food to all the people easily (Cunha et al. 2017).

5 Membrane Processing Technology

Membrane processing is a simple and chemical free technique that can be applied for separation of impurities and useful substances. All the molecules and substances cannot pass through the membrane as some of the molecules are caught, as the membrane does not allow them to pass through due to the selectivity, thereby making the separation possible. Both factors are membrane dependent. Membrane processing technology is used at small and large-scale applications. The domestic as well as the treatment of water in different industries comes in the range of application of membrane processing technology. Similarly it is also used in chemical, pharmaceutical industries as well (Saleh and Gupta 2016; McKenna 1978).

5.1 Classification of Membrane Separation Technique

The membrane separation techniques have many types based on physical membrane separation process, diffusional membrane separation, heat separation and electric separation process. These all the types are further classified into further categories. Ultrafiltration, nanofiltration, micro filtration and reverse osmosis are basically the physical membrane separation procedures. These all the four types are also known as pressure driven membrane types because the pressure is the main driver in these all the four types and these forces have a special applications in the food sector. Apart from these four types the diffusional process is also included in the physical pressure driven membrane forces. The heat separation process includes membrane distillation and vacuum membrane distillation. Moreover, the electric process includes electro dialysis and pseudo liquid membrane (Moskvin 2016).

The most widely used membrane processing technologies are based on pressure driven membrane forces and are discussed here in detail. These forces are used in food processing. All these types have their specific uses in specific types of food industries (Nissar et al. 2018).

5.1.1 Microfiltration

Microfiltration (MF) is the oldest type of pressure driven membrane forces, which have been practiced, commercially in wide range of applications. It can remove micrometer-sized matter through its membrane. The major suspended particles are the harmful and beneficial microorganisms, macromolecules such as proteins. The membranes of MF can not allow to pass through the harmful contaminates which makes it a versatile type of pressure driven membrane process (Anis et al. 2019).

The range of the pore diameter of Microfiltration membrane is 0.1–5 micrometer. The pore size and diameter of MF determines its separation characteristics. Particles greater than the pore size is retained, whereas less porous particles are allowed to pass through (Davis 2019). The pressure applied usually in MF is 50–660 pounds per square inch (KPa). The macromolecules and micro molecules are separated from the complex bonding assemblies using MF processing technique (Uragami 2017; Su 2018).

Applications of MF Processing

Application of this technology in food industry is growing with every-day at global marketplace since 1960s. Innovative characteristics like as high separation accuracy, good selectivity, processing at room temperature, no elevated temperature or chemical damage, high automation, easy operation, economic friendly energy, reduced cost, complete consumption of resource and reduced pollution are the main advantages of the membrane technology in respective field (Nissar et al. 2018; Amaresh and Soumen 2018).

Microfiltration membranes are used for the separation, purification and the concentration of macromolecules. It is being employed in various industries for the treatment of wastewater, which can be further utilized in various applications for having good quality end products (Singh and Purkait 2019). The pore size of MF membrane is usually in the range of 0.1–1 micrometer that makes it compatible for the treatment of wastewater. There are many researches made in the past decade on the treatment of the wastewater by the microfiltration process. Some studies are also been done on the treatment of wastewater by combining the microfiltration process with other technologies to make it potable or safe for drinking purposes. This trend has been begun due to the growth rate of urbanization due to which the pollution of water also increases and as a result the demand of microbiologically safe water also increases. In 2003 Singapore has launched the project for water treatment. This project has three purification stages. In this three-stage project, microfiltration is an initial step and after that reverse osmosis method and ultra violet disinfection step was applied for sterilization to make it fit for drinking purpose (Anis et al. 2019).

Pasteurization is an important technique for milk processing to improve the shelf stability of various dairy based products. However, due to heat treatment there are chances of certain nutritional loss and decay in sensory attributes. Therefore, MF has been employed for the pasteurization of the milk and other liquid foods to maintain the nutritional and organoleptic parameters of the final product. Using this approach, skim milk and cream can be separated by utilizing the membranes of ceramic pore size of about 1.4 micrometer at constant temperature. The retentate, which is left behind, has nearly all the bacteria and spores. This retentate can be mixed and homogenized with the standardized cream (Kulozik 2019). Apart from milk the MF can also applied to the other dairy products for the removal of bacteria and spores. In cheese making the usage of milk having low bacterial count improves the quality of cheese due to the removal of spores by microfiltration, which ultimately results in elimination of additives for quality and safety enhancement of the cheese. Similarly, removal of bacterial spores can also be achieved for the production of whey protein concentrates and isolates using microfiltration process (Dhineshkumar and Ramasamy 2017).

In beverage industry the major application of the microfiltration process is the clarification of fruit juices. The clarification of fruit juices is mostly done by microfiltration and ultrafiltration. Microfiltration is used for the clarification and concentration of pineapple, kiwi, apricot, peach, pear, apple and many other juices. Various studies are done to analyse the effect of membrane processing on fruit juice clarification. The results exhibits that the juice which is clarified from microfiltration was in much better and improved quality as compared to the juices obtained by the treatment of conventional technologies. The juices obtained after the clarification by microfiltration and are microbiologically safe for drinking. Similarly in beverage industry the beer and wine is also clarified by the MF membrane technologies. The wine clarification by MF is effective in decreasing the turbidity and elimination of additives for improving the shelf stability (Urošević et al. 2017).

5.1.2 Ultrafiltration

Ultrafiltration is a type of membrane processing in which pressure may be applied as a driving force for the separation of solvent from solute. The selection of the membrane, its pore size and applied pressure are the most important factors for the separation of macromolecules from the liquid. The average diameter of the pores used for UF process is about 2–10 nm with applied pressure (2–5 bar) (Pal 2020). UF can retain macromolecules (greater than 5 KDa). UF technique eliminates the need of filter aids and flocculants and provides good quality of filtrate with no suspended particles as compared to conventional filtration process. The pores of UF may block that hinders its utilization and adoption by the various industrial applications. However, the development of membrane fouling treatment can open the block pores that may increase its utilization for industrial applications.

Applications of UF

The UF has been employed in various food industrial applications. The milk based products that are separated with UF process can be further used in various food items. It can be used for filtration of milk and whey proteins at temperature (50 °C) with greater efficiency as compared to the higher temperature due to denaturation of proteins that ultimately reduce the filtration efficiency. The skim milk that pass through the UF process helps to evaluate the proteins before the production of cheese as the quality and characteristics of these macromolecules determine their further applications in product development.

UF have been widely used in fruit juice processing plants where the main aim is to clarify the juice. The new development in the UF processing results in enhancement of filtration efficiency and increases its resistance to chemicals.

Water safety and quality is important for various industrial applications. It has been used for the elimination of various types of bacteria and viruses laving behind the safe and pure water for the utilization in various application. It is the main agent in the beverage industry and elimination of viruses and bacteria through this technology provide safe water for beverage production and consumer trust. Effective elimination of microorganisms can be obtained by careful selection of UF membranes. The types of membranes, which have been successfully used for the removal of pathogen microorganism, include cellulose, polyvinylidine and polyethersulfone. UF has been also applied to ensure the sterilization of heat sensitive ingredients as thermal processing may change the organoleptic and nutritional properties of the food (Pal 2020).

5.1.3 Reverse Osmosis (RO)

Reverse Osmosis is very important type of pressure driven membrane forces. The RO membranes are denser as compared to the other pressure driven membrane forces with predefined pore size. Due to this characteristic, the penetration of permeate through the membranes becomes slower and the residues on the RO membrane are disposed of due to solution-diffusion mechanism. The denser membrane reduce the permeability of the RO that results in application of high operating pressures and high energy as compared to other processes (Van der Bruggen 2018; Giorno et al. 2015).

It is mainly used for the treatment of water in small scale as well as on the large scale. It also controls the NaCl level in the water. The pressure which is the main driver in this type of membrane processing technology is applied in the range of about a few hundred to thousands square per inch. In the unit of bar, it is about 25 to 68 bar. Apart from the pressure the hydraulic force is also applied in reverse osmosis. Hydraulic force also acts as a driving force in this process (Uragami 2017; Su 2018).

Applications of Reverse Osmosis

Reverse Osmosis is well renowned and widely accepted technique in the beverage industry especially for the fruit juice concentration for attaining the required Brix° (Peyravi et al. 2020; Ganorkar et al. 2012). Reverse Osmosis has been extensively used for the waste water treatment, softening of hard water, removal of salts, removal of organics and for the clarification of water. In short Reverse Osmosis membrane technology is mainly used for the treatment of water bodies on a large scale (Criscuoli and Figoli 2019).

In the past years the reverse osmosis technology was considered an expensive method than other traditional technologies due to utilization of higher energy cost. It was also considered that this technology is only limited in some sectors and cannot be used in the treatment of water. However, the compactness, simple and good effluent quality, it has widely used for the waste water treatment, softening of hard water, removal of salts, removal of organics and for the clarification of water (Jiang et al. 2018). It also controls the NaCl level in the water. By removing the salts and demineralizing of the water through RO processing technology make it a strong candidate for various industrial applications. The presence of the minerals in water is important for providing various health benefits. But the seawater and brackish water are not healthy and potable that is why the reverse osmosis technology has been used for the desalination of water to make it potable (Wenten 2016).

Wastewater contains different contaminants including organic contaminants, pathogens and pesticides. These all contaminants have good water solubility due to which these contaminants are not removed in the sludge and will cause hurdles in reclamation of water. Therefore the RO has been applied for the water reclamation of water. Recently the reclamation is done in GWR facility in Orange County for making water potable. RO has an important role in the innovative and new treatment process used at this plant. Low pressure and high removal ESPA2 membranes are being applied which reduce the TDS less than 50 mg permeate that make the water safe and potable (Garud et al. 2011).

5.1.4 Nanofiltration

Nano Filtration (NF) is considered as one of the important process in pressure driven membrane forces. It was developed as a type of pressure driven membrane forces in 1985. It has the characteristics in between UF and RO. It requires low energy as compared to the other pressure driven membrane processes for its functioning (Agboola et al. 2014).

The pressure applied in this type of pressure driven membrane force is in the range of one to few hundred pounds per square inch (KPa) or 7–40 bar. It is the most open filtration type among all the types. In NF, the ion exchange occurs. The monovalent ions pass through the NF membrane but the other types of ions are being rejected. The divalent and multivalent ions may not be allowed to pass through these most open type of membrane (Uragami 2017; Su 2018).

Applications of NF

NF can perform at different and a wide range of application due to which it is also known as multipurpose filtration pressure driven membrane force. This process also shows its applications in the separation of particle process. It is also used for the softening of the hard water. It is used for the purification of water. It is used to remove the organic matter, different contaminants and disinfectants from the water which can harm the human health by causing different diseases and make it potable (Ahsan and Imteaz 2019). It is used in food processing and for controlling the concentration in the beverages area especially in the fruit juices (Nissar et al. 2018; Amaresh and Soumen 2018).

The NF processing technology shows a powerful impact on the dairy industry so it has been used on a wider range in the dairy industry. Among all the food industries the development of the NF process mainly occurs in the dairy industry as well as in the milk processing industries. The reason of it is mainly its characteristics and performance as well as its molecular weight that is 200–1000 Da between UF and RO processes. In dairy industry the NF is used in different applications such as in cheese making process, the concentration of whey protein, fractionation process and waste water purification. The fractionation process is applied here to control the level of concentration in the milk. Fractionation of milk makes it safe microbiologically by removing the bacterial spores and fractionation of whey proteins. Some of these processes, which are mentioned above, are performed in combination of NF with other pressure driven membrane forces. In the past when the NF process was not developed, the Electro dialysis process was used for cheese making, concentration of whey protein and wastewater purification (Nath et al. 2018; Marella et al. 2013).

Another very important use of the NF process is in the beverage sector. In beverage this process is utilized mainly in the processing of fruit juices. It has also a wide range of application in the fermented products like beer and wine. For the concentration of different fruits and vegetable juices the NF can be used for concentrating the different juices like grape, apple and pineapple. Ferrarini et al. (2001) concentrated the grape juice by using different pressure driven membrane forces like NF and RO and then differentiate them by comparing them and find that low pressure is required for the grape juice concentration by NF as compared to the RO. The results were same while concentrating other fruit juices. It is therefore clear by the results that NF can be operated at lower energy for getting the effective outcomes and this characteristic make it cost effective and environment friendly tool (Mohammad et al. 2019).

Beside the juice concentration NF has been used in the concentration of fermented products like wine and beer. It is used in maintaining the level of alcohol in the fermented products. It is also used for the partially removal of sugar contents from beer and wine. From the last decade the low alcoholic drinks are in demand because of the health issues therefore NF is used which perfuse the ethanol and water and keep the bioactive components in the alcoholic beverages at low operating pressure and temperature and therefore considered as an efficient method as compared to other pressure driven membrane forces (Conidi et al. 2020).

Besides all the uses or functions of NF in the different food processing industries the most important is the treatment of wastewater. In the dairy industry large amount of water is used and as well as produced as a waste. Therefore, treatment of wastewater produced during various processing operations using NF in combination with RO make the treated water reused for cleaning, heating, cooling and many other operations.

Rotating disc Membrane and NF 270 (Dow-Filmtec) treatment shows good results in dairy wastewater treatment. Moreover the energy can also save by this NF technique. The treatment of industrial wastewater by membrane bio-reactor and NF was also studied. The combining treatment of both these methodologies exhibits removal of 99.9% for COD and 93.1% for solids. This treated water by the combination of these methods can be used for good manufacturing practices also besides from the cooling and heating (Mohammad et al. 2019).

5.2 Future Opportunities for Membrane Processing Treatment for Food Industry

There are also many other separation techniques apart from the membrane processing treatment. But the membrane processing technology creates a powerful and useful impact than other separation techniques as compared to other methodologies. The membrane processing technologies can easily adopted at domestic scale but need proper infrastructure for adoption at industrial level. Now days we are using the reverse osmosis process as a powerful substitute to all the other separation technologies. Most of the membrane technologies which have been used on the small scale not on the industrial scale because of the difficulties and problems are using on the last scale in recent years and in future it will have more wider range of applications in small or lab scale as well as on the large or industrial scale. Membrane processing technology also have a wider range of uses in the juices area for the concentration transfer (Strathmann 2001; Baker 2006).

Membrane processing technology fall under the umbrella of non-thermal methodologies which require no heat, hence nutrient losses and organoleptic properties can be maintained using this technology. It can work on the different sources of energy. It is also capable of using the alternate sources of energy, which does not create bad impact on the environmental conditions. The capability of the membrane processing technology at low heat and less temperature make very useful for applications at various industrial level without wasting the resources and generate less harmful impact to the environment. More efforts are being needed to make the membrane processing technology much better and in some years it will create more powerful impact in the food-processing sector (Onsekizoglu 2012).

References

Abbas H, Nouraddin S, Reza ZH, Iraj B, Mohammad B, Hasan Z, Hadi F (2011) Effect of gamma radiation on different stages of Indian meal moth Plodia interpunctella Hübner (Lepidoptera: Pyralidae). Afr J Biotechnol 10(20):4259–4264

Agbaka JI, Ibrahim AN (2020) Irradiation: utilization, advances, safety, acceptance, future trends, and a means to enhance food security. Adv Appl Sci Res 11(3:1):12

Agboola O, Maree J, Mbaya R (2014) Characterization and performance of nanofiltration membranes. Environ Chem Lett 12(2):241–255

Ahsan A, Imteaz M (2019) Nanofiltration membrane technology providing quality drinking water. In: Nanotechnology in water and wastewater treatment. Elsevier, pp 291–295

Akgün MP, Ünlütürk S (2017) Effects of ultraviolet light emitting diodes (LEDs) on microbial and enzyme inactivation of apple juice. Int J Food Microbiol 260:65–74

Al-Antary TM, Shaderma AM, Al-Dabbas MB (2018) Effect of Ozonation treatment on spiked myclobutanil pesticide on tomato fruits. Feb Fresenius Environ Bulletin:8574

Alvarez I, Niemira BA, Fan X, Sommers CH (2007) Inactivation of Salmonella enteritidis and Salmonella senftenberg in liquid whole egg using generally recognized as safe additives, ionizing radiation, and heat. J Food Prot 70(6):1402–1409

Alves E, Faustino MA, Neves MG, Cunha A, Tome J, Almeida A (2014) An insight on bacterial cellular targets of photodynamic inactivation. Future Med Chem 6(2):141–164

Amaresh P, Soumen G (2018) Applications of membrane separation technology in food industry. Trends Prospects Food Technol Process Preserv 4:43–55

Angarano V, Akkermans S, Smet C, Chieffi A, Van Impe JF (2020) The potential of violet, blue, green and red light for the inactivation of P. fluorescens as planktonic cells, individual cells on a surface and biofilms. Food Bioprod Process 124:184–195

Anis SF, Hashaikeh R, Hilal N (2019) Microfiltration membrane processes: a review of research trends over the past decade. J Water Process Eng 32:100941

Ansari JA, Ismail M, Farid M (2019) Investigate the efficacy of UV pretreatment on thermal inactivation of *Bacillus subtilis* spores in different types of milk. Innovative Food Sci Emerg Technol 52:387–393

Antonelli A, Fabbri C, Boselli E (1998) Modifications of dried basil (*Ocimum basilicum*) leaf oil by gamma and microwave irradiation. Food Chem 63:485–489

Antos P, Piechowicz B, Gorzelany J, Matłok N, Migut D, Józefczyk R, Balawejder M (2018) Effect of ozone on fruit quality and fungicide residue degradation in apples during cold storage. Ozone Sci Eng 40(6):482–486. https://doi.org/10.1080/01919512.2018.1471389

Appleby J, Banks AJ (1905) Improvements in or relating to the treatment of foodstuffs, more especially cereals and their products. British Patent number 1609

Bahrami S, Amiri-Yekta A, Daneshipour A, Jazayeri SH, Mozdziak PE, Sanati MH et al (2020) Designing a transgenic chicken: applying new approaches toward a promising bioreactor. Cell J 22:133–139

Baker RW (2006) Membrane technology in the chemical industry: future directions. Memb Technol Chem Ind 7:305–335

Ballinger CA, Cueto R, Squadrito G, Coffin JF, Velsor LW, Pryor WA, Postlethwait EM (2005) Antioxidant-mediated augmentation of ozone-induced membrane oxidation. Free Radic Biol Med 38(4):515–526. https://doi.org/10.1016/j.freeradbiomed.2004.11.009

BARC (n.d.) Radiation processing for food preservation. Available from: https://mofpi.nic.in/sites/default/files/RadiationProcessingforFoodPreservation.pdf.pdf. Accessed 04 April 2021

Benov L (2015) Photodynamic therapy: current status and future directions. Med Princ Pract 24(1):14–28

Beuchat LR (1992) Surface disinfection of raw produce. Dairy Food Environ Sanit 12(1):6–9

Bhavya ML, Hebbar HU (2019) Sono-photodynamic inactivation of Escherichia coli and Staphylococcus aureus in orange juice. Ultrason Sonochem 57:108–115

Cabo Verde S (2018) Food irradiation as sanitary treatment. In: Ferreira ICFR, Antonio AL, Cabo Verde S (eds) Food irradiation technologies: concepts, applications and outcomes, food chemistry, functions and analysis, vol 4. The Royal Society of Chemistry, UK, pp 183–209

Calado T, Venâncio A, Abrunhosa L (2014) Irradiation for mold and mycotoxin control: A review. Compr Rev Food Sci Food Saf 13:1049–1061

Can FO, Demirci A, Puri VM, Gourama H (2014) Decontamination of hard cheeses by pulsed UV light. J Food Prot 77:1723–1731

Çatal H, İbanoğlu Ş (2012) Ozonation of corn and potato starch in aqueous solution: effects on the thermal, pasting and structural properties. Int J Food Sci Technol 47(9):1958–1963. https://doi.org/10.1111/j.1365-2621.2012.03056.x

Cebrián G, Condón S, Mañas P (2017) Physiology of the inactivation of vegetative bacteria by thermal treatments: mode of action, influence of environmental factors and inactivation kinetics. Foods 6(12):107

Choi MR, Liu Q, Lee SY, Jin JH, Ryu S, Kang DH (2012) Inactivation of Escherichia coli O157: H7, Salmonella typhimurium and listeria monocytogenes in apple juice with gaseous ozone. Food Microbiol 32(1):191–195. https://doi.org/10.1016/j.fm.2012.03.002

Conidi C, Castro-Muñoz R, Cassano A (2020) Nanofiltration in beverage industry. In: Nanotechnology in the beverage industry. Elsevier, pp 525–548

Corsi AJ, Robles FC, Corduay CG, Neal JA (2015) The effectiveness of electron beam irradiation to reduce or eliminate mould in cork stoppers. Int J Food Sci Technol 51:389–395

Criscuoli A, Figoli A (2019) Pressure-driven and thermally-driven membrane operations for the treatment of arsenic-contaminated waters: a comparison. J Hazard Mater 370:147–155

Cunha A, Couceiro J, Bonifácio D, Martins C, Almeida A, Neves MGPMS, Faustino MAF, Saraiva JA (2017) Inactivation of pathogenic bacteria in food matrices: high pressure processing, photodynamic inactivation and pressure-assisted photodynamic inactivation. In: IOP Conference Series: Earth and Environmental Science, vol 85, No. 1. IOP Publishing, p 012016

Cutler TD, Zimmerman JJ (2011) Ultraviolet irradiation and the mechanisms underlying its inactivation of infectious agents. Anim Health Res Rev 12(1):15–23

Damalas CA, Eleftherohorinos IG (2011) Pesticide exposure, safety issues, and risk assessment indicators. Int J Environ Res Public Health 8(5):1402–1419. https://doi.org/10.3390/ijerph8051402

Datta N, Tomasula PM (2015) Emerging dairy processing technologies: opportunities for the dairy industry. Wiley, Chichester, UK

Datta N, Harimurugan P, Palombo E (2015) Ultraviolet and pulsed light technologies in dairy processing. In: Datta N, Tomasula PM (eds) Emerging dairy processing technologies: opportunities for the dairy industry. Wiley, West Sussex, UK

Davis CR, Fleet GH, Lee TH (1982) Inactivation of wine cork microflora by a commercial sulphur dioxide treatment. Am J Enol Vitic 33:124–127

Davis RH (2019) Microfiltration in pharmaceutics and biotechnology. Current trends and future developments on (bio-) membranes, pp 29–67

de Souza LP, Faroni LRDA, Heleno FF, Cecon PR, Gonçalves TDC, da Silva GJ, Prates LHF (2018) Effects of ozone treatment on postharvest carrot quality. LWT 90:53–60. https://doi.org/10.1016/j.lwt.2017.11.057

Delorme MM, Guimarães JT, Coutinho NM, Balthazar CF, Rocha RS, Silva R, Margalho LP, Pimentel TC, Silva MC, Freitas MQ, Granato D, Sant'Ana AS, Duart MCKH, Cruz AG (2020) Ultraviolet radiation: an interesting technology to preserve quality and safety of milk and dairy foods. Trends Food Sci Technol 102:146–154

Deng LZ, Mujumdar AS, Pan Z, Vidyarthi SK, Xu J, Zielinska M, Xiao HW (2020a) Emerging chemical and physical disinfection technologies of fruits and vegetables: a comprehensive review. Crit Rev Food Sci Nutr 60(15):2481–2508

Deng L-Z, Tao Y, Mujumdar AS, Pan Z, Chen C, Yang X-H, Liu Z-L, Wang H, Xiao H-W (2020b) Recent advances in non-thermal decontamination technologies for microorganisms and mycotoxins in low-moisture foods. Trends Food Sci Technol 106:104–112

Deng X, Tang S, Wu Q, Tian J, Riley WW, Chen Z (2016) Inactivation of Vibrio parahaemolyticus by antimicrobial photodynamic technology using methylene blue. J Sci Food Agric 96(5):1601–1608

Dhineshkumar V, Ramasamy D (2017) Review on membrane technology applications in food and dairy processing. J Appl Biotechnol Bioeng 3(5-2017)

Dunn JE, Clark RW, Asmus JF, Pearlman JS, Boyer K, Painchaud F, et al (1991) U.S. Patent No. 5,034. Washington, DC: U.S. Patent and Trademark Office 235

Durantini EN (2006) Photodynamic inactivation of bacteria. Curr Bioact Comp 2(2):127–142

EFSA (2016) Safety of UV-treated milk as a novel food pursuant to Regulation (EC) No 258/97. EFSA J 14(1):4370

Erdman HE (1980) Ozone toxicity during ontogeny of two species of flour beetles, Tribolium confusum and T. castaneum. Environ Entomol 9(1):16–17. https://doi.org/10.1093/ee/9.1.16

Erkmen O, Bozoglu TF (2016) Food preservation by irradiation. Food microbiology: principles into practice. Wiley, West Sussex, UK

Espo E, Eyidozehi K, Ravan S (2015) Influence of gamma and ultraviolet irradiation on pest control. Magnt Res Rep 3(2):319–326

Fan X, Huang R, Chen H (2017) Application of ultraviolet C technology for surface decontamination of fresh produce. Trends Food Sci Technol 70:9–19

FAO/IAEA/WHO (1999) World Health Organization. High-Dose Irradiation: Wholesomeness of Food Irradiated with Doses above 10 kGy. Report of a Joint FAO/IAEA/WHO Study Group. Geneva, Switzerland: World Health Organization, WHO Technical Report Series No. 890

Farkas J (2006) Irradiation for better foods. Trends Food Sci Technol 17:148–152

Faroni LRD, Pereira AM, Sousa AH, Silva MTC, Urrichi WI (2007) Influence of corn grain mass temperature on ozone toxicity to Sitophilus zeamais (Coleoptera: Curculionidae) and quality of oil extracted from ozonized grains. In: IOA conference and exhibition, vol 1. IOA, Valência, pp 1–6

FDA (2018) Food irradiation: what you need to know. Available from: https://www.fda.gov/food/buy-store-serve-safe-food/food-irradiation-what-you-need-know. Accessed 04 April 2021

FDA (2021) UV lights and lamps: ultraviolet-C radiation, disinfection, and coronavirus. Available from: https://www.fda.gov/medical-devices/coronavirus-covid-19-and-medical-devices/uv-lights-and-lamps-ultraviolet-c-radiation-disinfection-and-coronavirus. Accessed 04 April 2021

Fernández M, Arias K, Hierro E (2016) Application of pulsed light to sliced cheese: effect on listeria inactivation, sensory quality and volatile profile. Food Bioprocess Technol 9:1335–1344

Fernández M, Ganan M, Guerra C, Hierro E (2014) Protein oxidation in processed cheese slices treated with pulsed light technology. Food Chem 159:388–390

Ferrarini R, Versari A, Galassi S (2001) A preliminary comparison between nanofiltration and reverse osmosis membranes for grape juice treatment. J Food Eng 50(2):113–116

Food and Drug Administration (2015) Federal Register, CFR 179.26. http://www.ecfr.gov/cgi-bin/textidx?SID=2c4e58481ca8b6ba26e94304f940150c&node=pt21.3.179&rgn=div5. Accessed 04 April 2021

Forney LJ, Pierson JA, Ye Z (2004) Juice irradiation with TaylorCouette flow: UV inactivation of Escherichia Coli. J Food Prot 67:2410–2415

Freitas-Silva O, Venâncio A (2010) Ozone applications to prevent and degrade mycotoxins: a review. Drug Metab Rev 42(4):612–620. https://doi.org/10.3109/03602532.2010.484461

Friedberg EC, Walker GC, Siede W, Wood RD, Schultz RA et al (2006) DNA repair and mutagenesis. ASM Press, Washington, DC

Fundo JF, Miller FA, Tremarin A, Garcia E, Brandão TR, Silva CL (2018) Quality assessment of cantaloupe melon juice under ozone processing. Innovative Food Sci Emerg Technol 47:461–466. https://doi.org/10.1016/j.ifset.2018.04.016

Gabriel AA (2012) Inactivation of Escherichia coli O157:H7 and spoilage yeasts in germicidal UV-C irradiated and heat-treated clear apple juice. Food Control 25(4):425–432

Galstyan A, Dobrindt U (2019) Determining and unravelling origins of reduced photoinactivation efficacy of bacteria in milk. J Photochem Photobiol B Biol 197:111554

Ganorkar P, Nandane A, Tapre A (2012) Reverse osmosis for fruit juice concentration–a review. Res Rev 1(1):23–36

Garud RM, Kore SV, Kore VS, Kulkarni GS (2011) A short review on process and applications of reverse osmosis. Universal Journal of Environmental Research & Technology 1(3)

Garud SR, Priyanka BS, Rastogi NK, Prakash M, Negi PS (2018) Efficacy of ozone and lactic acid as nonthermal hurdles for preservation of sugarcane juice. Ozone Sci Eng 40(3):198–208. https://doi.org/10.1080/01919512.2017.1415802

Gayán E, Condón S, Álvarez I (2014) Biological aspects in food preservation by ultraviolet light: a review. Food Bioprocess Technol 7:1–20

Gayán E, Mañas P, Álvarez I, Condón S (2013) Mechanism of the synergistic inactivation of Escherichia coli by UV-C light at mild temperatures. Appl Environ Microbiol 79(14):4465–4473

Gharib-Bibalan S, Keramat J, Hamdami N (2018) Better lime purification of raw sugar beet juice by advanced Fenton oxidation process. Ozone Sci Eng 40(1):54–63. https://doi.org/10.1080/01919512.2017.1345617

Ghate V, Kumar A, Zhou W, Yuk H (2016) Irradiance and temperature influence the bactericidal effect of 460-nanometer light-emitting diodes on salmonella in orange juice. J Food Prot 79(4):553–560

Ghate VS, Zhou W, Yuk HG (2019) Perspectives and trends in the application of photodynamic inactivation for microbiological food safety. Compr Rev Food Sci Food Saf 18(2):402–424

Giorno L, Drioli E, Strathmann H (2015) The principle of reverse osmosis (RO). In: Encyclopedia of membranes, pp 1–5

Gómez-López VM, Ragaert P, Debevere J, Devlieghere F (2007) Pulsed light for food decontamination: a review. Trends Food Sci Technol 18(9):464–473

Gonçalves AA (2009) Ozone: an emerging technology for the seafood industry. Braz Arch Biol Technol 52(6):1527–1539. https://doi.org/10.1590/S1516-89132009000600025

González-Rodríguez RM, Rial-Otero R, Cancho-Grande B, Gonzalez-Barreiro C, Simal-Gándara J (2011) A review on the fate of pesticides during the processes within the food-production chain. Crit Rev Food Sci Nutr 51(2):99–114. https://doi.org/10.1080/10408390903432625

Gorzelany J, Migut D, Matłok N, Antos P, Kuźniar P, Balawejder M (2017) Impact of pre-ozonation on mechanical properties of selected genotypes of cucumber fruits during the souring process. Ozone Sci Eng 39(3):188–195. https://doi.org/10.1080/01919512.2016.1273756

Grégoire O, Cleland MR, Mittendorfer J, Dababneh S, Ehlermann DAE, Fan X, Käppeler F, Logar J, Meissner J, Mullier B, Stichelbaut F, Thayer DW (2003) Radiological safety of food irradiation with high energy Xrays: theoretical expectations and experimental evidence. Radiat Phys Chem 67(2):169–183

Gryczka U, Migdał W, Bułka S (2018) The effectiveness of the microbiological radiation decontamination process of agricultural products with the use of low energy electron beam. Radiat Phys Chem 143:59–62

Guneser O, Karagul Yuceer Y (2012) Effect of ultraviolet light on water- and fat-soluble vitamins in cow and goat milk. J Dairy Sci 95(11):6230–6241

Guzel-Seydim ZB, Greene AK, Seydim AC (2004) Use of ozone in the food industry. LWT Food Sci Technol 37(4):453–460. https://doi.org/10.1016/j.lwt.2003.10.014

Ha JW, Back KH, Kim YH, Kang DH (2016) Efficacy of UV-C irradiation for inactivation of food-borne pathogens on sliced cheese packaged with different types and thicknesses of plastic films. Food Microbiol 57:172–177

Hadi J, Wu S, Brightwell G (2020) Antimicrobial blue light versus pathogenic bacteria: mechanism, application in the food industry, hurdle technologies and potential resistance. Foods 9(12):1895

Hallman GJ (2011) Phytosanitary applications of irradiation. Compr Rev Food Sci Food Saf 10:143–151

Hallman GJ, Loaharanu P (2016) Phytosanitary irradiation – development and application. Radiat Phys Chem 129:39–45

Heleno FF, De Queiroz MEL, Neves AA, Freitas RS, Faroni LRA, De Oliveira AF (2014) Effects of ozone fumigation treatment on the removal of residual difenoconazole from strawberries and on their quality. J Environ Sci Health B 49(2):94–101. https://doi.org/10.1080/03601234.2014.846736

Heraeus Amba Australia Pty Ltd (2012) Applications for UV light in the food industry. Available from: https://www.foodprocessing.com.au/content/processing/article/applications-for-uv-light-in-the-food-industry-1291409884. Accessed 04 April 2021

Hetz SK, Bradley TJ (2005) Insects breathe discontinuously to avoid oxygen toxicity. Nature 433(7025):516–519. https://doi.org/10.1038/nature03106

Horvath M, Bilitzky L, Huttner J (1985) Fields of utilization of ozone. Ozone:257–316

Hwang ES, Cash JN, Zabik MJ (2001) Postharvest treatments for the reduction of mancozeb in fresh apples. J Agri Food Chem *49*(6):3127–3132

IAEA (2012) International atomic energy agency. Food irradiation treatment facilities database. http://nucleus.iaea.org/fitf/. Accessed 04 April 2021

IFST (2006) Institute of Food Science and Technology. The use of irradiation for foodquality and safety. Available from: http://www.ifst.org/document.aspx?id=12. Accessed 04 April 2021

Ihsanullah I, Rashid A (2017) Current activities in food irradiation as a sanitary and phytosanitary treatment in the Asian and the Pacific Region and a comparison with advanced countries. Food Control 72:345–359

Jalili M, Jinap S, Noranizan MA (2012) Aflatoxins and ochratoxin a reduction in black and white pepper by gamma radiation. Radiat Phys Chem 81:1786–1788

Jaramillo-Sánchez GM, Garcia Loredo AB, Gómez PL, Alzamora SM (2018) Ozone processing of peach juice: impact on physicochemical parameters, color, and viscosity. Ozone Sci Eng 40(4):305–312. https://doi.org/10.1080/01919512.2017.1417111

Jeong S, Marks BP, Ryser ET, Harte JB (2012) The effect of X-ray irradiation on Salmonella inactivation and sensory quality of almonds and walnuts as a function of water activity. Int J Food Microbiol 153(3):365–371

Jermann C, Koutchma T, Margas E, Leadley C, Ros-Polski V (2015) Mapping trends in novel and emerging food processing technologies around the world. Innovative Food Sci Emerg Technol 31:14–27

Jiang L, Tu Y, Li X, Li H (2018) Application of reverse osmosis in purifying drinking water. In: E3S web of conferences, vol 38. EDP Sciences, p 01037

Kalyani B, Manjula K (2014) Food irradiation – technology and application. Int J Curr Microbiol App Sci 3(4):549–555

Karaca H (2019) The effects of ozone-enriched storage atmosphere on pesticide residues and physicochemical properties of table grapes. Ozone Sci Eng 41(5):404–414. https://doi.org/10.1080/01919512.2018.1555449

Karadurmuş E, Taşkın N, Göz E, Yüceer M (2019) Prediction of bromate removal in drinking water using artificial neural networks. Ozone Sci Eng 41(2):118–127. https://doi.org/10.1080/01919512.2018.1510763

Kashef N, Huang YY, Hamblin MR (2017) Advances in antimicrobial photodynamic inactivation at the nanoscale. Nano 6(5):853–879

Kaya Z, Unluturk S (2016) Processing of clear and turbid grape juice by a continuous flow UV system. Innovative Food Sci Emerg Technol 33:282–288

Keklik NM, Elik A, Salgin U, Demirci A, Koçer G (2019) Inactivation of *Staphylococcus aureus* and *Escherichia coli* O157: H7 on fresh kashar cheese with pulsed ultraviolet light. Food Sci Technol Int 25:680–691

Kells SA, Mason LJ, Maier DE, Woloshuk CP (2001) Efficacy and fumigation characteristics of ozone in stored maize. J Stored Prod Res 37(4):371–382. https://doi.org/10.1016/S0022-474X(00)00040-0

Khadre MA, Yousef AE, Kim JG (2001) Microbiological aspects of ozone applications in food: a review. J Food Sci 66(9):1242–1252. https://doi.org/10.1111/j.1365-2621.2001.tb15196.x

Khaled AO, Fahad B, Abdullah A (2017) Ozone as a safety post-harvest treatment for chlorpyrifos removal from vegetables and its effects on vegetable quality. Int J Food Nutr Sci 4(1):38–48
Khan QU, Mohammadzai I, Shah Z, Ullah I, Khattak TN, Noreen H, Hassan W (2018) Effect of gamma irradiation on nutrients and shelf life of peach stored at ambient temperature. Open Conf Proc J 09(1):8–15
Kim DK, Kim SJ, Kang DH (2017) Bactericidal effect of 266 to 279 nm wavelength UVC-LEDs for inactivation of Gram positive and Gram negative foodborne pathogenic bacteria and yeasts. Food Res Int 97:280–287
Kim HY, Ahn JJ, Shahbaz HM, Park KH, Kwon JH (2014) Physical, chemical and microbiological based identification of electron beam and γ-irradiated frozen crushed garlic. J Agric Food Chem 62:7920–7926
Kim JG, Yousef AE, Dave S (1999) Application of ozone for enhancing the microbiological safety and quality of foods: a review. J Food Prot 62(9):1071–1087. https://doi.org/10.4315/0362-028X-62.9.1071
Kim J, Moreira RG, Castell-Perez ME (2010) Optimizing irradiation treatment of shell eggs using simulation. J Food Sci 76(1):E173–E177
Kim S, Kim D, Kang D (2015) Using UVC light-emitting diodes at wavelengths of 266 to 279 nanometers to inactivate foodborne pathogens and pasteurize sliced cheese. Appl Environ Microbiol 82(1):11–17
Komanapalli IR, Lau BHS (1996) Ozone-induced damage of Escherichia coli K-12. Appl Microbiol Biotechnol 46(5):610–614. https://doi.org/10.1007/s002530050869
Koutchma T (2008) UV light for processing foods. Ozone Sci Eng 30:1–6
Koutchma T (2009) Advances in ultraviolet light technology for non-thermal processing of liquid foods. Food Bioprocess Technol 2(2):138–155
Koutchma T (2016) UV and light Technologies for disinfection of food contact and food surfaces. Ref Mod Food Sci. https://doi.org/10.1016/b978-0-08-100596-5.21201-8
Koutchma T (2019) Ultraviolet light in food technology principles and applications. CRC Press, Boca Raton, FL
Koutchma T, Forney LJ, Moraru CL (2009) Ultraviolet light in food technology. CRC Press, Boca Raton, FL
Koutchma T, Keller S, Chirtel S, Parisi B (2004) Ultraviolet disinfection of juice products in laminar and turbulent flow reactors. Innovative Food Sci Emerg Technol 5:179–189
Krishnamurthy K, Demirci A, Irudayaraj JM (2007) Inactivation of *Staphylococcus aureus* in milk using flow-through pulsed UV-light treatment system. J Food Sci 72:M233–M239
Kulozik U (2019) Ultra-and microfiltration in dairy technology. In: Current trends and future developments on (bio-) membranes. Elsevier, pp 1–28
Lacivita V, Conte A, Manzocco L, Plazzotta S, Zambrini VA, Del Nobile MA, Nicoli MC (2016) Surface UV-C light treatments to prolong the shelf-life of Fiordilatte cheese. Innovative Food Sci Emerg Technol 36:150–155
Lee BU (2011) Life comes from the air: a short review on bioaerosol control. Aerosol Air Qual Res 11:921–927
Lian X, Tetsutani K, Katayama M, Nakano M, Mawatari K, Harada N, Hamamoto A, Yamato M, Akutagawa M, Kinouchi Y, Nakaya Y, Takahashi A (2010) A new colored beverage disinfection system using UV-A light-emitting diodes. Biocontrol Sci 15(1):33–37
Liu K, Liu Y, Chen F (2018) Effect of gamma irradiation on the physicochemical properties and nutrient contents of peanut. LWT 96:535–542
Lu B, Ren Y, Du YZ, Fu Y, Gu J (2009) Effect of ozone on respiration of adult Sitophilus oryzae (L.), Tribolium castaneum (Herbst) and Rhyzopertha dominica (F.). J Insect Physiol 55(10):885–889. https://doi.org/10.1016/j.jinsphys.2009.05.014
Luksiene Z, Brovko L (2013) Antibacterial photosensitization-based treatment for food safety. Food Eng Rev 5(4):185–199

Lung HM, Cheng YC, Chang YH, Huang HW, Yang BB, Wang CY (2015) Microbial decontamination of food by electron beam irradiation. Trends Food Sci Technol 44:66–78. https://doi.org/10.1016/j.tifs.2015.03.005

Mahapatra AK, Muthukumarappan K, Julson JL (2005) Applications of ozone, bacteriocins and irradiation in food processing: a review. Crit Rev Food Sci Nutr 45(6):447–461. https://doi.org/10.1080/10408390591034454

Mahto R, Das M (2013) Effect of gamma irradiation on the physico-chemical and visual properties of mango (Mangifera indica L.), cv. 'Dushehri' and 'Fazli' stored at 20°C. Postharvest Biol Technol 86:447–455

Mahto R, Das M (2014) Effect of gamma irradiation on the physico-mechanical and chemical properties of potato (Solanum tuberosum L.), cv. 'Kufri Sindhuri', in non-refrigerated storage conditions. Postharvest Biol Technol 92:37–45

Marella C, Muthukumarappan K, Metzger LE (2013) Application of membrane separation technology for developing novel dairy food ingredients. J Food J Process Technol 4(9):1–5

Markov K, Mihaljević B, Domijan A, Pleadin J, Delaš F, Frece J (2015) Inactivation of aflatoxigenic fungi and the reduction of aflatoxin B1 in vitro and in situ using gamma irradiation. Food Control 54:79–85

Masotti F, Cattaneo S, Stuknytė M, Noni ID (2019) Airborne contamination in the food industry: an update on monitoring and disinfection techniques of air. Trends Food Sci Technol 90:147–156

Matak KE, Churey JJ, Worobo RW, Sumner SS, Hovingh E, Hackney CR, Pierson MD (2005) Efficacy of UV light for the reduction of *listeria monocytogenes* in goat's milk. J Food Prot 68:2212–2216

Matak KE, Sumner SS, Duncan SE, Hovingh E, Worobo RW, Hackney CR, Pierson MD (2007) Effects of ultraviolet irradiation on chemical and sensory properties of goat milk. J Dairy Sci 90:3178–3186

McDonough MX, Mason LJ, Woloshuk CP (2011) Susceptibility of stored product insects to high concentrations of ozone at different exposure intervals. J Stored Prod Res 47(4):306–310. https://doi.org/10.1016/j.jspr.2011.04.003

McKenna B (1978) Membrane processing of foods: a review of the concepts involved. Irish J Food Sci Technol:45–58

Metzger C, Barnes JD, Singleton I, Andrews P (2007) Effect of low level ozone-enrichment on the quality and condition of citrus fruit under semi-commercial conditions. In: IOA conference and exhibition, pp 29–31

Migut D, Gorzelany J, Antos P, Balawejder M (2019) Postharvest ozone treatment of cucumber as a method for prolonging the suitability of the fruit for processing. Ozone Sci Eng 41(3):261–264. https://doi.org/10.1080/01919512.2018.1525277

Miller FA, Fundo JF, Silva CL, Brandão TR (2018) Physicochemical and bioactive compounds of 'Cantaloupe' melon: effect of ozone processing on pulp and seeds. Ozone Sci Eng 40(3):209–215. https://doi.org/10.1080/01919512.2017.1414582

Miller FA, Silva CL, Brandão TR (2013) A review on ozone-based treatments for fruit and vegetables preservation. Food Eng Rev 5(2):77–106. https://doi.org/10.1007/s12393-013-9064-5

Miller RB (2015) Electronic irradiation of foods. An introduction to the technology. In: Food engineering series. Springer, New York

Mohammad AW, Teow YH, Ho KC, Rosnan NA (2019) Recent developments in nanofiltration for food applications. In: Nanomaterials for food applications. Elsevier, pp 101–120

Mongpraneet S, Abe T, Tsurusaki T (2002) Accelerated drying of welsh onion by far-infrared radiation under vacuum conditions. J Food Eng 55:147–156

Morehouse KM, Komolprasert V (2004) Irradiation of food and packaging: an overview. Available from: http://www.fda.gov/Food/FoodIngredientsPackaging/IrradiatedFoodPackaging/ucm081050.htm. Accessed 04 April 2021

Moreira RG, Castell-Perez E (2021) Fundamentals of food irradiation. In: Knoerzer K, Muthukumarappan K (eds) Innovative food processing technologies. Elsevier, pp 1–18

Morey A, McKee SR, Dickson JS, Singh M (2010) Efficacy of ultraviolet light exposure against survival of Listeria monocytogenes on conveyor belts. Foodborne Pathog Dis 7(6):737–740

Moskvin LN (2016) A classification of separation methods. Sep Purif Rev 45(1):1–27

Nakonechny F, Nisnevitch M (2019) Aspects of photodynamic inactivation of bacteria. In: Microorganisms. IntechOpen

Nath K, Dave HK, Patel TM (2018) Revisiting the recent applications of nanofiltration in food processing industries: Progress and prognosis. Trends Food Sci Technol 73:12–24

Niemira BA, Fan X (2006) Low-dose irradiation of fresh-cut produce: safety, sensory, and shelf life. In: Sommers HS, Fan X (eds) Food irradiation research and technology. Blackwell Publishing, USA, pp 169–184

Nissar N, Hameed OV, Nazir F (2018) Application of membrane technology in food processing industries: a review. Int J Adv Res Sci Eng 7(4):2016–2114

Ong KC, Cash JN, Zabik MJ, Siddiq M, Jones AL (1996) Chlorine and ozone washes for pesticide removal from apples and processed apple sauce. Food Chem 55(2):153–160. https://doi.org/10.1016/0308-8146(95)00097-6

Onsekizoglu P (2012) Membrane distillation: principle, advances, limitations and future prospects in food industry. Distill Adv Model Appl 11:233–265

Orlowska M, Koutchma T, Grapperhaus M, Gallagher J, Schaefer R, Defelice C (2013) Continuous and pulsed ultraviolet light for nonthermal treatment of liquid foods. Part 1: effects on quality of fructose solution, apple juice, and milk. Food Bioprocess Technol 6:1580–1592

Osman KA (2015) Production of date palm fruits free of acaricides residues by ozone technology as post-harvest treatment. J Food Sci Technol 52(6):3322–3335. https://doi.org/10.1007/s13197-014-1398-3

Pal P (2020) Membrane-based technologies for environmental pollution control. Butterworth-Heinemann

Palou L, Smilanick JL, Crisosto CH, Mansour M (2001) Effect of gaseous ozone exposure on the development of green and blue molds on cold stored citrus fruit. Plant Dis 85(6):632–638. https://doi.org/10.1094/PDIS.2001.85.6.632

Pan Y, Sun D-W, Han Z (2017) Applications of electromagnetic fields for nonthermal inactivation of microorganisms in foods: an overview. Trends Food Sci Technol 64:13–22

Pandiselvam R, Kothakota A, Thirupathi V, Anandakumar S, Krishnakumar P (2017) Numerical simulation and validation of ozone concentration profile in green gram (Vigna radiate) bulks. Ozone Sci Eng 39(1):54–60. https://doi.org/10.1080/01919512.2016.1244641

Pandiselvam R, Subhashini S, Banuu Priya EP, Kothakota A, Ramesh SV, Shahir S (2019) Ozone based food preservation: a promising green technology for enhanced food safety. Ozone Sci Eng 41(1):17–34. https://doi.org/10.1080/01919512.2018.1490636

Pankaj SK, Shi H, Keener KM (2018) A review of novel physical and chemical decontamination technologies for aflatoxin in food. Trends Food Sci Technol 71:73–83

Perry JJ, Yousef AE (2011) Decontamination of raw foods using ozone-based sanitization techniques. Annu Rev Food Sci Technol 2:281–298. https://doi.org/10.1146/annurev-food-022510-133637

Peyravi M, Jahanshahi M, Banafti S (2020) Application of membrane technology in beverage production and safety. In: Safety issues in bverage production. Academic, pp 271–308

Pillai SD, Shayanfar S (2017) Electron beam technology and other irradiation technology applications in the food industry. Top Curr Chem 375:6

Pimentel MAG, Faroni LRDA, Tótola MR, Guedes RNC (2007) Phosphine resistance, respiration rate and fitness consequences in stored-product insects. Pest Manag Sci 63(9):876–881. https://doi.org/10.1002/ps.1416

Ravindran R, Jaiswal AK (2019) Wholesomeness and safety aspects of irradiated foods. Food Chem 285:363–368

Reed NG (2010) The history of ultraviolet germicidal irradiation for air disinfection. Public Health Rep (Washington, DC: 1974) 125(1):15–27

Ribeiro J, Cavaglieri L, Vital H, Cristofolini A, Merkis C, Astoreca A, Orlando J, Carú M, Dalcero A, Rosa CAR (2011) Effect of gamma radiation on Aspergillus flavus and Aspergillus ochraceus ultrastructure and mycotoxin production. Radiat Phys Chem 80(5):658–663

Rodriguez-Gonzalez O, Buckow R, Koutchma T, Balasubramaniam VM (2015) Energy requirements for alternative food processing technologies-principles, assumptions, and evaluation of efficiency. Compr Rev Food Sci Food Saf 14(5):536–554

Rozado AF, Faroni LR, Urruchi WM, Guedes RN, Paes JL (2008) Ozone application against Sitophilus zeamais and Tribolium castaneum on stored maize. Rev Bras Eng Agríc Ambient 12(3):282–285. https://doi.org/10.1590/S1415-43662008000300009

Saleh TA, Gupta VK (2016) An overview of membrane science and technology. In: Nanomaterial and polymer membranes, pp 1–23

Selma MV, Ibáñez AM, Cantwell M, Suslow T (2008) Reduction by gaseous ozone of Salmonella and microbial flora associated with fresh-cut cantaloupe. Food Microbiol 25(4):558–565. https://doi.org/10.1016/j.fm.2008.02.006

Shunmugam G, Jayas DS, White NDG, Muir WE (2005) Diffusion of carbon dioxide through grain bulks. J Stored Prod Res 41(2):131–144. https://doi.org/10.1016/j.jspr.2003.09.005

Silva AF, Borges A, Giaouris E, Graton Mikcha JM, Simões M (2018a) Photodynamic inactivation as an emergent strategy against foodborne pathogenic bacteria in planktonic and sessile states. Crit Rev Microbiol 44(6):667–684

Silva J, Pereira MN, Scussel VM (2018b) Ozone gas antifungal effect on extruded dog food contaminated with Aspergillus flavus. Ozone Sci Eng 40(6):487–493. https://doi.org/10.1080/01919512.2018.1481361

Singh R, Purkait MK (2019) Microfiltration membranes. In: Membrane separation principles and applications. Elsevier, pp 111–146

Sinha RP, Häder DP (2002) UV-induced DNA damage and repair: a review. Photochem Photobiol Sci 1:225–236

Sintuya P, Narkprasom K, Jaturonglumlert S, Whangchai N, Peng-Ont D, Varith J (2018) Effect of gaseous ozone fumigation on organophosphate pesticide degradation of dried chilies. Ozone Sci Eng 40(6):473–481. https://doi.org/10.1080/01919512.2018.1466690

Song WJ, Sung HJ, Kim SY, Kim KP, Ryu S, Kang DH (2014) Inactivation of Escherichia coli O157:H7 and *Salmonella typhimurium* in black pepper and red pepper by gamma irradiation. Int J Food Microbiol 172:125–129

Srimagal A, Ramesh T, Sahu J (2016) Effect of light emitting diode treatment on inactivation of Escherichia coli in milk. LWT Food Sci Technol 71:378–385

Stanga M (2010) Sanitation: cleaning and disinfection in the food industry. Wiley-VCHVerlag, Weinheim

Steenstrup LD, Floros JD (2004) Inactivation of E. coli 0157: H7 in apple cider by ozone at various temperatures and concentrations. J Food Process Preserv 28(2):103–116. https://doi.org/10.1111/j.1745-4549.2004.tb00814.x

Strathmann H (2001) Membrane separation processes: current relevance and future opportunities. AICHE J 47(5):1077–1087

Su Y (2018) Current state-of-the-art membrane based filtration and separation technologies, vol 1, pp 1–13

Tarek AR, Rasco BA, Sablani SS (2015) Ultraviolet-C light inactivation kinetics of E. coli on bologna beef packaged in plastic films. Food Bioprocess Technol 8(6):1267–1280

Thomas P (1988) Radiation preservation of food of plants origin. Part 6. Mushrooms, tomatoes, minor fruits and vegetables, dried fruits and nuts. Crit Rev Food Sci Nutr 24:313–358

Tim M (2015) Strategies to optimize photosensitizers for photodynamic inactivation of bacteria. J Photochem Photobiol B Biol 150:2–10

Tiwari BK, Muthukumarappan K, O'Donnell CP, Cullen PJ (2008) Kinetics of freshly squeezed orange juice quality changes during ozone processing. J Agric Food Chem 56(15):6416–6422. https://doi.org/10.1021/jf800515e

Tiwari BK, O'Donnell CP, Brunton NP, Cullen PJ (2009a) Degradation kinetics of tomato juice quality parameters by ozonation. Int J Food Sci Technol 44(6):1199–1205. https://doi.org/10.1111/j.1365-2621.2009.01946.x

Tiwari BK, O'donnell CP, Muthukumarappan K, Cullen PJ (2009b) Anthocyanin and colour degradation in ozone treated blackberry juice. Innovative Food Sci Emerg Technol 10(1):70–75. https://doi.org/10.1016/j.ifset.2008.08.002

Tripathi J, Variyar PS, Singhal RS, Sharma A (2015) Radiation processing for sprout inhibition of stored potatoes and mitigation of acrylamide in fries and chips. In: Preedy VR (ed) Processing and impact on active components in food. Academic Press, Elsevier, London, UK, pp 89–96

Udipi SA, Ghurge PS (2010) Applications of food irradiation. In: Udipi SA, Ghugre PS (eds) Food irradiation. Agrotech Publishing Academy, Udaipur, pp 40–71

Uragami T (2017) Comparison of pressure-driven membrane separation processes. Sci Technol Sep Membr 13:379–383

Urošević T, Povrenović D, Vukosavljević P, Urošević I, Stevanović S (2017) Recent developments in microfiltration and ultrafiltration of fruit juices. Food Bioprod Process 106:147–161

Van der Bruggen B (2018) Microfiltration, ultrafiltration, nanofiltration, reverse osmosis, and forward osmosis. In: Fundamental modelling of membrane systems. Elsevier, pp 25–70

Velasco RM, Uribe FJ, Pérez-Chavela E (2008) Stratospheric ozone dynamics according to the Chapman mechanism. J Math Chem 44(2):529–539. https://doi.org/10.1007/s10910-007-9326-7

Verma J, Gautam S (2015) Food irradiation and its role in shelf life extension of horticulture produce: a comprehensive evaluation of studies carried out in India and abroad. In: Proceedings of the DAE-BRNS Life Sciences Symposium on Advances in Microbiology of Food, Agriculture, Health and Environment

Wang S, Wang J, Wang T, Li C, Wu Z (2019) Effects of ozone treatment on pesticide residues in food: a review. Int J Food Sci Technol 54(2):301–312. https://doi.org/10.1111/ijfs.13938

Wang S-Q, Huang G-Q, Li Y-P, Xiao J-X, Zhang Y, Jiang W-L (2015) Degradation of aflatoxin B1 by low-temperature radio frequency plasma and degradation product elucidation. Eur Food Res Technol 241:103–113

Wenten IG (2016) Reverse osmosis applications: prospect and challenges. Desalination 391:112–125

Whangchai K, Uthaibutra J, Phiyanalinmat S, Pengphol S, Nomura N (2011) Effect of ozone treatment on the reduction of chlorpyrifos residues in fresh lychee fruits. Ozone Sci Eng 33(3):232–235. https://doi.org/10.1080/01919512.2011.554313

White SD, Murphy PT, Bern CJ, van Leeuwen JH (2010) Controlling deterioration of high-moisture maize with ozone treatment. J Stored Prod Res 46(1):7–12. https://doi.org/10.1016/j.jspr.2009.07.002

World Health Organization (1994) Safety and nutritional adequacy of irradiated food. Available from: https://apps.who.int/iris/bitstream/handle/10665/39463/9241561629-eng.pdf?sequence=4&isAllowed=y. Accessed 04 April 2021

Worobo RW (1999) Assistant professor of food microbiology. Cornell University. Personal Communication

Xuan W, He Y, Huang L, Huang YY, Bhayana B, Xi L, Gelfand JA, Hamblin MR (2018) Antimicrobial photodynamic inactivation mediated by tetracyclines in vitro and in vivo: photochemical mechanisms and potentiation by potassium iodide. Sci Rep 8(1):1–14

Yucel Sengun I, Kendirci P (2018) Potential of ozonated water at different temperatures to improve safety and shelf-life of fresh cut lettuce. Ozone Sci Eng 40(3):216–227. https://doi.org/10.1080/01919512.2017.1416284

Zohair A (2001) Behaviour of some organophosphorus and organochlorine pesticides in potatoes during soaking in different solutions. Food Chem Toxicol 39(7):751–755. https://doi.org/10.1016/S0278-6915(01)00016-3

Part III
Implementation of Novel Nonthermal Technologies in Agri-food-bio Sciences

Part III
Implementation of Novel Nonthermal Technologies in Agri-food-bio Sciences

Chapter 9
Nonthermal Processing Technologies: Synergies and New Applications in Food Engineering

Maria Tsevdou, George Dimopoulos, Eleni Gogou, Efimia Dermesonlouoglou, and Petros Taoukis

1 Introduction

Conventional thermal processes (e.g. pasteurization, sterilization) have been widely used in the food industry. These processes lead to destruction of pathogenic and spoilage microorganisms, and inactivation of enzymes. However, high temperatures applied during the process can cause damage to food components (e.g. bioactive components) affecting negatively the food product quality and acceptability (Fellows 2017). At the same time, the consumers' preference for non or minimally processed food products with "fresh-like" food characteristics has been rapidly increasing. Nonthermal (no use of intense heat during the process) technologies can serve the purpose of producing safe, high quality, minimally processed foods (Knorr et al. 2011). These technologies can extend the shelf life of food products, while retaining or even increasing their nutritional value, and their sensorial characteristics (e.g. colour, flavour). The development of novel technologies can provide a balance between minimal processing and food safety, limited economic resources and increased food quality. It can also lead to less water consumption and energy reduction, and decreased carbon footprint of food processing (Jambrak 2019; Tokusoglu 2015). It has also been recognized that nonthermal technologies play a significant role in food security and sustainability (Knoerzer 2016). Nonthermal technologies are also characterized by mechanical effects able to increase the extraction of bioactive compounds by means of cell rupture (Chan et al. 2016). The above are mainly attributed to the cavitation phenomenon or electroporation of the

M. Tsevdou · G. Dimopoulos · E. Gogou · E. Dermesonlouoglou · P. Taoukis (✉)
National Technical University of Athens, School of Chemical Engineering,
Laboratory of Food Chemistry and Technology, Athens, Greece
e-mail: mtsevdou@chemeng.ntua.gr; gdimop@chemeng.ntua.gr; egogou@chemeng.ntua.gr; efider@chemeng.ntua.gr; taoukis@chemeng.ntua.gr

A. Režek Jambrak (ed.), *Nonthermal Processing in Agri-Food-Bio Sciences*, Food Engineering Series, https://doi.org/10.1007/978-3-030-92415-7_9

cell membranes and organelles, in order to prevent microbial growth and prolong the shelf life of the food product (Li and Farid 2016).

Nonthermal food processes include high pressure, pulsed electric fields, cold atmospheric plasma, continuous or pulsed UV light processing, sonication and ozonation. Extensive research in the field of these nonthermal processes has led to innovative industrial applications. In some cases, nonthermal technologies have replaced conventional preservation processes in the development of high quality food products with extended shelf life. The aim is to have sustainable and optimal food processing (e.g. fast, affordable) for product quality preservation or even modification (novel food design) (Jambrak 2019).

Novel, nonthermal technologies have drawn worldwide attention (scientific community and food manufacturers) (Brennan and Grandison 2012). The increase in the number of publications evidences the interest in these technologies. Hernández-Hernández et al. (2019) reported a significant increase in the number of publications in U.S.A., Latin America and Europe, where emerging nonthermal technologies have been developed, based on the Scopus database results (from 2000 to 2018). Most publications are related to pulsed electric fields, followed by high pressure, ultrasound and ionizing radiation. More recently, there has been an increased interest in the use of cold atmospheric plasma. Apart from the scientific research, the level of commercialization is important. Koutchma et al. (2016) reported the increased trend in commercialization of nonthermal technologies. They reported that, by the year 2025, high pressure technology will remain at the top places, cold plasma and pulsed electric fields will take second and third places, respectively, while ionizing radiation will appear in place ten. According to the results of a recent survey by Khouryieh (2021) for the novel technologies used by the U.S.A food processing industry, high pressure is the most commonly used nonthermal food processing technology (35.6%), followed by pulsed electric field (20%). Rapidly increasing novel technologies are cold atmospheric plasma (14.1%) and oscillating magnetic fields (14.1%). Some interesting conclusions of the survey include: (i) more than 70% of the respondents indicated that the main factor for choosing nonthermal food processing technology is better food quality (nutrition and sensorial properties), and (ii) high investment (41%) was the major limitation for implementing nonthermal food processing technologies. The results also indicated that the main drivers for innovation were equipment manufacturers (43.8%) and government research (42.3%).

In recent years, food engineering research has focused on new applications through the synergy of different nonthermal technologies or the combination of these nonthermal technologies with conventional food processes such as extraction, enzymatic processes, drying and osmotic dehydration. The hurdle concept (combined methods or processes, combination preservation or techniques, barrier technology) has become a working approach that aims to enhance the total quality of foods (e.g. reduce the loss of nutrients, retain or improve the sensorial characteristics) and reduce the high treatment intensities (Leistner 1985; Rahman 2015). In addition, hurdle technology showed synergistic effects while using various mechanisms for the inhibition or inactivation of targeted microorganisms (Rahman 2015). These synergies (High Pressure, Pulsed Electric Fields, Cold Atmospheric Plasma,

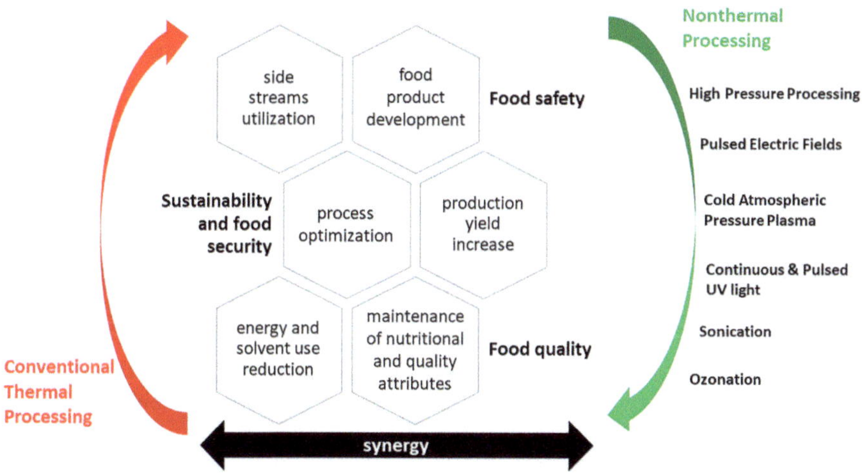

Fig. 9.1 Application of nonthermal processing technologies, and their synergies (Chapter overview)

continuous or pulsed UV light processing, sonication and ozonation) are reviewed in this chapter with regards to their effect on nutritional and quality properties of foods, energy and solvent use reduction, production yield increase, side stream valorization and development of innovative food products (Fig. 9.1).

2 High Pressure Processing

2.1 General Aspects

High Pressure Processing (HPP) has probably gained the most attention among the nonthermal technologies, thus it has successfully advanced through all technology readiness levels (TRLs) reaching TRL9. Through HP technology, food products with equal or even more superior properties than thermally treated foods can be produced in terms of both safety, quality and biofunctionality, as the detrimental effects due to temperature increase are minimized. From the beginning, the application of HP technology was focused in the inactivation of pathogens and spoilage microorganisms, and enzymes that cause quality degradation of foods. The extensive research on this fields over the last 40 years has resulted in the production of foods that fulfill both consumers' demands and industry requirements for "fresh-like" food products and shelf life extension, respectively. Furthermore, the potential environmental benefits compared to the conventional thermal processes in combination with current trend for minimizing food losses play a significant role in the industrial adaptation of this technology.

The first implementation of HPP on food production was related to the inactivation of food pathogenic and/or spoilage microflora. Among vegetative cells, yeasts and molds are the most sensitive microorganisms, for which complete inactivation is achieved at pressures above 400 MPa, while bacteria present higher resistance to pressures and are inactivated at pressures above 500 MPa, often combined with mild thermal treatment at 40–50 °C (Daryaei et al. 2016). With respect to bacterial spores, it is recognized that pressures up to 1200 MPa combined with higher temperatures (>70 °C) are required for commercial sterilization. Alternatively HPP combined with moderately elevated temperatures results in the germination of bacterial spores when moderate pressures are applied (<400 MPa) and then, as the bacterial spores have lost their inherent resistance, they can be inactivated when higher pressures (>500 MPa) are applied (Modugno et al. 2020). In addition, the inactivation of enzymes that lead to quality degradation of foods still present the second most common reason for target of HP application in foods. Food enzyme inactivation involves a number of mechanisms mainly related to the changes induced in their native structure via unfolding and folding, and alterations induced in the substrate-enzyme interactions (Hendrickx and Knorr 2002). In the following paragraphs, the interactions and synergies of HPP and other nonthermal and/or conventional food processes will be discussed mostly with emphasis on other applications, e.g. bioactive compounds extraction, enzymatic treatments and hydrolysis, dehydration, and so on.

It is generally accepted, at least in industrial practices, that the application of HPP is uniform and independent from the food geometry, size, and uniformity. However, one of the major concerns during the application of HPP is the deviations induced in the pressure vessels and are associated with the temperature increase due to adiabatic heating during pressure build-up and the heat transfer phenomena during pressure holding time (Karwe et al. 2015). Therefore, during industrial processes, both pressure and temperature monitoring and control have to be considered and validated to ensure the efficacy of the process, that can be achieved by selecting an appropriate process index which may be either a resistant enzyme and/or microorganism (Gogou and Taoukis 2015).

Apart from the fact that the application of HP cold-pasteurization ensures minimization of post-contamination while minimally affecting the nutritional and sensorial characteristics of HP-treated food products, it is also considered as an environmental friendly process, since it requires significant lesser energy consumption than the corresponding thermal pasteurization processes. Additionally, the fact that HP-treated foods are processed in pack leads to even less water consumption and chemical reagents usage related to the sanitation of the industrial equipment and the packaging materials used, thus leading to lower environmental impact of the process (Wang et al. 2016).

2.2 *Commercialization of HPP*

The ongoing evolution on HPP equipment technology combined with the current consumers' demand for "fresh" products have led several manufacturers in U.S.A, Spain, U.K, Japan, and China to advance the development of HPP equipment.

Current industrial HP equipment is designed to operate at pressures of up to 700 MPa at maximum volumes of 525 L, at ambient or lower temperatures, providing inactivation of foodborne pathogens, spoilage microflora and deteriorative enzymes. Regarding laboratory scale HPP equipment, with capacities ranging from 0.5 to 10 L, the main representatives include Avure, Resato International, Stansted Fluid Power, Baotou, Kobelco, and Toyo Koatsu, while Avure (now JBT) and Hiperbaric, are the main developers and suppliers of industrial scale HPP equipment, representing an annual production capacity of up to 60 million tn (Tsevdou et al. 2019). The main food categories that are HP processed comprise meat (26%), juices and beverages (14%), fruits and vegetables (29%), toll processing (13%), seafood (13%), ready meals (4%), and dairy products (1%).

The industrial scale equipment is mostly of horizontal type due to the ability of loading and unloading containers in the production line, yet there is also available equipment of vertical type (Fig. 9.2a, b). In 2018, the Spanish company Hiperbaric launched the new In-Bulk technology (Fig. 9.2c) as well, represented by two new industrial machines, H525 Bulk and H1050 Bulk, in which liquid foodstuffs are processed in-bulk followed by aseptic packaging. This new industrial equipment can deliver a production of up to 8000 L/h, while achieving 90% of vessel filling. The main advantages of these machines are based on reduced processing costs and energy requirements, while at the same time can be merged into existing production lines, allowing the use of any kind of aseptic packaging after HP processing.

2.3 HP-Assisted Extraction Technology

The key idea of the application of HP for assisting the extraction of intracellular compounds was firstly introduced by Shouqin et al. (2004), who reported the applicability of high pressure for the extraction of essential components from herbs, based on results from preliminary pilot trials. Since then and up to now, there is an ongoing and increasing research interest on the HP assisted extraction (HPE) of proteins, polysaccharides and bioactive compounds, and other components from plant and animal tissues, herbs and spices, algae, as well as from food by-products and side streams. The process of HPE includes mixing the raw material with the

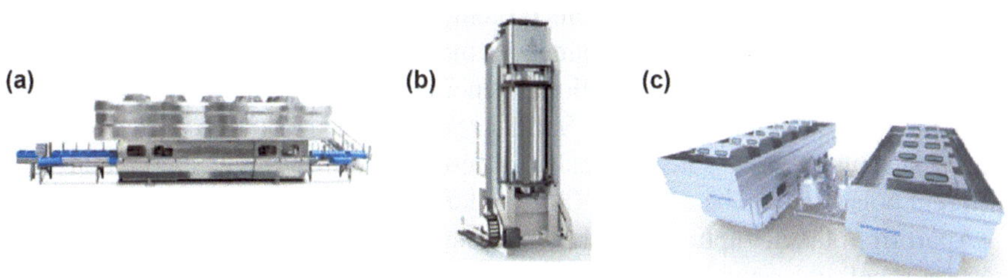

Fig. 9.2 Horizontal (**a**), Vertical (**b**) and In-Bulk (**c**) type of HPP industral equipment. (*Source: Hiperbaric and Avure/JBT Technologies websites*)

appropriate solvent, treating the mixture with HP, and then recovering the target compound from the solvent mixture. The mixture of raw material/solvent can be further concentrated, dried, or purified in order to obtain a single, clear compound, which can be used either as an ingredient in a food or for the preparation of films and coatings.

The application of HPE has been investigated by numerous authors for the extraction of high value compounds, including polyphenols and caffeine from tea leaves (Jun 2009; Jun et al. 2011; Xi et al. 2009), carotenoids from tomato waste (Strati et al. 2015), pectins and polyphenols from orange (Guo et al. 2012) and tomato peel (Ninčević Grassino et al. 2020), phenolic compounds from watercress (Pinela et al. 2018), grape pomace (Cascaes Teles et al. 2021), olive pomace (Andreou et al. 2020b), açai (de Jesus et al. 2020), cape gooseberry (Torres-Ossandón et al. 2018), pomegranate peel (Trigo et al. 2020), stinging nettle leaves (Moreira et al. 2020), palm dates (Sedraoui et al. 2020), and fermented fig by-product (Alexandre et al. 2017), flavonoids and lycopene from tomato pulp (Briones-Labarca et al. 2019), B-phycoerythrin, polyphenols and carotenoids from algae (Tran et al. 2019; Suwal et al. 2019; Bueno et al. 2020; Gallego et al. 2021), tocopherols from pepper seed oil (Ma et al. 2019), chitosan form squid pen waste (Huang and Tsai 2020), cadmium decontamination of long rice grain (Luo et al. 2021), xyloglucan from tamarind (Limsangouan et al. 2020), gelatin from fish skin (Zhang et al. 2011), polysaccharides from longan pulp (Bai et al. 2016) and Umbilicaria (Sun and Jiang, 2020), high molecular weight melanoidins from black garlic (Zhao et al. 2019), and procyanidins from lychee pericarp (Zhang et al. 2017).

The main mechanism of HPE involves the increase in the cell permeability and enhanced diffusion of metabolites from the inside to outside of the cells, according to the phase behaviour theory, i.e. that the solubility of the compound is greater when the pressure increases. Consequently, the main function of pressure holding time is to achieve the equilibrium between intra and extracellular solutions and complete contact between the target compounds and the solvent. The large difference of pressure between inner and outer cell membrane leads to "instant" permeation (Jun 2009). As the applied pressure is found to promote the disruption of tissues, cell walls and organelles, and increases the mass transfer of the solvent into the sample and of compounds to the solvent, it is suggested that the higher the applied pressure is, the more solvent can enter into the cells, and hence the more compounds can permeate out to the solvent (Prasad et al. 2009). Moreover, since HP leads to denaturation of cell membrane protein, they become less selective and other compounds such as phenolic compounds are more accessible for extraction (Cascaes Teles et al. 2021), offering the additional benefit of selective extraction while keeping the extraction of impurities low. Similar results have been also reported depending on the solvent used where by using different solvent, extracts compound with different polarities can be obtained (Alexandre et al. 2017).

Apart from the disruption of cell membrane, Chen et al. (2009) reported that the release of cell compounds is improved by the mechanical effects of high pressure that enhance the penetration of solvent into the cell membranes. As it was observed through scanning electron microscopy, ginseng root treated with HP at 200 MPa for

5 min exhibited a big destruction of cell tissues with hollow breaks and smaller particles compared to untreated or ginseng root treated with heat reflux extraction, confirming that HPP induced considerable changes in the surface tension of the cells that facilitate the diffusion and osmotic process and significantly improve extraction of the target compounds (Zhang et al. 2012). This observation further confirmed that through HPE the elimination of impurities can be achieved and that finally the quality of the extract is improved (Wang et al. 2018).

It is evident that the time needed for the extraction is the same to the HP processing time, which lasts up to few minutes, and the temperature is usually low while no additional energy than that for increasing pressure is needed, suggesting that HPE can be applied in cases where thermosensitive compounds are extracted and it requires less energy consumption than that of thermal extraction processes. In addition to the aforementioned advantages, during HPE, high extraction yields and, in some cases, extraction selectivity are achieved (Tsevdou et al. 2019). Regarding the energy requirements, it is reported that the required energy in HPE is <50 kJ/kg of raw material (Andreou et al. 2020b), and that in order to pressurize water up to 800 MPa the required energy does not exceed 55 kJ/kg (Toepfl et al. 2006) while the heating of water requires approximately 42 kJ/kg for every 10 °C of temperature increase. In addition to short processing time and low temperatures used, HPE is considered, and since 2014 is recognized by the FDA, as an environmental friendly extraction method owing to its lower energy consumption and decreased amounts of solvent that are volatilized.

Table 9.1 summarizes the most recent findings on the extraction of proteins, polysaccharides and bioactive compounds through HPE from plant and animal tissues, and algae.

2.4 HPP Combined with Other Nonthermal Technologies

Over the last decade, several studieshave been published acknowledging the combined effect of HP application with other nonthermal technologies, including Pulsed Electric Fields (PEF), Ultrasonication (US), Ultrafiltration (UF), Microwave (MW) and Ohmic Heating (OM), mainly targeted to the microbial inactivation of pathogens and spoilage microflora of foods (Bermúdez-Aguirre et al. 2016), whereas few studies have addressed their effect on the quality attributes of foods.

Regarding the inactivation of food pathogens and spoilage microflora, HP technology has been combined with either sonication, PEF or microwave, and inactivation is attributed to the induced collapse of the vegetative cells. Abid et al. (2014) showed that the sequential treatments of US (25 kHz/70% amplitude/20 °C/60 min) and HP (450 MPa/10 min) led to inactivation of total aerobic bacteria by 4 log cycles, and of yeasts and molds by 3.7 log cycles in apple juice, compared to those obtained from the single treatment of HP or US, where only 2 or 1 log reduction, respectively, was achieved. A single treatment of US (2 min) or HP (200 MPa/20 min or 400 MPa/5 min) resulted in 1 log reduction of Listeria innocua, while the

Table 9.1 Recent findings on the extraction of proteins, polysaccharides and bioactive compounds through HPE

Raw material	Target compound	HPP conditions/Extraction method	Key findings	Reference
Tomato by-products	Carotenoids	HPP at pressure range of 100–800 MPa for 1–30 min, and ambient temperature / enzymatic extraction using ethanol, hexane, acetone, hexane/ethanol 50/50, hexane/acetone 50/50, ethyl acetate, and ethyl lactate, with solvent to solid ratios of 10:1, 6:1 and 4:1 mL/g	HPE was successfully performed by using significantly ($P < 0.05$) lower ratios of solvent:solid (6:1 and 4:1 mL/g) and reduced processing time (10 min), compared to solvent extraction performed at ambient pressure, where solvent:solid ratio 10:1 mL/g and 30 min extraction time were used. Maximum total carotenoid (165 mg/kg d.w.) and lycopene (83 mg/kg d.w.) extraction yields were obtained in samples extracted with ethyl lactate (solvent:solid= 10:1 mL/g).	Strati et al. (2015)
Orange peel	Pectins	HPP at pressure range 100–600 MPa, extraction time 5–30 min and extraction temperatures ranging from 10 to 30 °C/liquid-solid extraction using 95% v/v ethanol as a solvent and kept overnight at 4 °C	The optimal HPE conditions were determined as pressure 500 MPa, temperature 55 °C, and pressure-holding time 10 min. Under these conditions, the yield of pectin (20.44% ± 0.64) was significantly higher than those extracted by traditional heating (15.47% ± 0.26). The intrinsic viscosity and viscosity-average molecular weight of pectin extracted by HPE (0.7604 L/g and 3.063×10^5 Da) were much higher than those extracted by traditional heating (0.4276 L/g and 1.521×10^5 Da), and the commercial pectin (0.2160 L/g and 0.663×10^5 Da).	Guo et al. (2012)

Raw material	Target compound	HPP conditions/Extraction method	Key findings	Reference
Tomato peel	Pectins, polyphenols, and fatty acids	HPP at constant pressure of 300 MPa and extraction times of 10, 20, 30 and 45 min / *Pectin extraction*: nitric acid (c = 0.1 Mol/L) at 80 °C, sample to solvent ratio was 1:20 Polyphenols extraction: Ultrasounds at maximal power of 400 W, frequency of 30 kHz, working amplitude of 95% with continuous cycle mode, ethanol (70 and 96%), sample to solvent ratio was 1:50 *Fatty acids extraction*: Chloroform and methanol (50:50, V/V) by Soxhlet	HPE enhanced pectin recovery from 14% to 15% after 30 and 45 min of extraction, in comparison with the conventional extraction for 180 min. The extraction of polyphenols was significantly influenced by solvent polarity, and 70% ethanol assured better efficiency than 96% ethanol.	Ninčević Grassino et al. (2020)
Watercress	Phenolic acids, flavonoids	HPP at pressure range 0.1–600 MPa for 1.5–33.5 min / ethanol as solvent; S = 0–100% v/v	The recovery of phenolic compounds was maximized when high pressures, high ethanol concentrations and short extraction times were applied. Optimal HHP conditions were 600 MPa for 3.1 min, and solvent concentration 100%.	Pinela et al. (2018)
Grape pomace	Unsaturated fatty acids and γ-tocopherol	HPP at pressure range 50–200 MPa for 0–30 min / enzyme -assisted extraction, applied individually, under HPP or pretreated with HPP	HHP increased by up to 16 times the activity of the enzymes used in the extraction. The combination of both technologies, as well as the pre-treatment of enzymes by HHP before extraction, showed the best results in bioactive compounds recovery.	Cascaes Teles et al. (2021)

(continued)

Table 9.1 (continued)

Raw material	Target compound	HPP conditions/Extraction method	Key findings	Reference
Açai pulp	Total phenolic compounds, anthocyanins	HPP at 400, 500, and 600 MPa, treatment times of 5 and 15 min, and temperatures of 25 °C and 65 °C / aqueous extraction with the addition of formic acid (1%) to stabilize the anthocyanins	The açaí pulps processed by HHP at optimal conditions (600 MPa/5 min/25 °C) had anthocyanin and phenolic compounds extraction increased by 37% and 10.3%, respectively, while conventional thermal treatment reduced anthocyanin content by 16.3%.	de Jesus et al. (2020)
Cape gooseberry	Total phenols, β-carotene	HPP at 300, 400 and 500 MPa for holding times of 1, 3 and 5 min / supercritical fluid extraction with 50% ethanol in water solvent, 25 MPa and 313.15 K with a CO flow rate of 1 mL/min at constant pressure and an extraction time of 120 min	Treatments at 300 MPa/1 min and 400 MPa/3 min showed an increase of antioxidant capacity post-processing, while treatments at 500 MPa presented the highest antioxidant capacity (12388.3 μmol TE/100 g d.m.) after storage compared to untreated samples. The highest β-carotene content was observed at 300 MPa/3 min (5.51 mg β-carotene/100 g d.m.).	Torres-Ossandón et al. (2018)
Olive pomace	Phenolic compounds, proteins	HPP at pressure range 200–600 MPa for 0–40 min / solid–liquid extraction with solid–liquid ratio 1:10 in different extraction temperatures (25, 40, and 60 °C) and ethanol concentrations (0–70%)	More intense HP conditions resulted in significant increase in the phenolic concentration up to 71.8%. HP-pretreated extracts reached 88.1% higher protein extraction yield than untreated for pressures up to 200 MPa. HP decreased the extraction completion time t_{98} (time needed to recover the equal amount of phenolics and proteins of untreated after 60 min of conventional thermal extraction) to 12 min.	Andreou et al. (2020b)
Pomegranate peel	Phenolic compounds	HPP at 300 or 600 MPa for 30 min at room temperature /enzymatic extraction using 4% (vol.) pectinase and 4% (vol.) cellulase	Maximum level of total phenolic content and antioxidant capacity at 300 MPa. All the extracts exhibited antimicrobial activity against a wide range of food pathogens or spoilage microorganisms, but not against lactic acid bacteria.	Trigo et al. (2020)

Raw material	Target compound	HPP conditions/Extraction method	Key findings	Reference
Fermented fig by-product	Phenolic compounds	HPP performed at 0.1, 300 or 600 MPa for 15, 17.5 or 30 min at ambient temperature / solid–liquid extraction with ethanol concentration of 0%, 40% or 80%	The optimal conditions included 600 MPa, extraction time between 18 and 29 min, and low ethanol concentration (<15%) except for flavonoids (48%). HPE led to an increase of 8–13% of antioxidant activity and an increase of 8–11% of total phenolics, flavonoids and tannins content when compared to extracts obtained at 0.1 MPa. HPE resulted in decreasing extraction time by 30 min.	Alexandre et al. (2017)
Stinging nettle leaves	Phenolic compounds	HPP at pressure range 200–500 MPa for 0–20 min / solid–liquid extraction in different extraction temperatures (25, 40, and 60 °C) and ethanol concentrations (0–70%)	The optimal conditions for the overall maximization of extraction yield, total phenolic content (TPC) and antioxidant activity were 200 MPa, 10.2–15.6 min, and 0% ethanol. HPE allowed increasing the extraction yield about 50.5%; TPC about 84.4%; and antioxidant activity about 77.7% compared to extraction at atmospheric pressure.	Moreira et al. (2020)
Palm dates	Total phenols, tannins, and flavonoids	HPP at pressures 100–600 MPa / ethanol concentration 0–100%, liquid to solid ratio 10–70 mL/g), temperature 25–65 °C)	Ethanol concentration, solvent to sample ratio, and pressure significantly affected extraction yield, while temperature did not. The optimum HPE conditions include ethanol concentration of 60.5%, solvent to sample ratio 70 mL/g, extraction temperature at 65 °C, and pressurizing at 600 MPa.	Sedraoui et al. (2020)

(continued)

Table 9.1 (continued)

Raw material	Target compound	HPP conditions/Extraction method	Key findings	Reference
Tomato pulp	Flavonoids and lycopene	HPP at pressures of 250, 350 and 450 MPa in room temperature for 10 min / extraction with hexane: ethanol solvent in the following ratios, 40:60, 50:50, and 60:40 (v/v), using a pulp/solvent ratio of 1:2, and performed for 24 h	HPE performed at 450 MPa and 60% hexane concentration was the optimal process condition where the maximum extraction yield (8.71%), flavonoid (21.52 ± 0.09 mg QE/g FW) and lycopene content (2.01 ± 0.09 mg QE/100 g FW) were achieved. However, changes in variables such as processing time, which was not included in the study, could interfere with the results and the optimization of the process.	Briones-Labarca et al. (2019)
Red macroalgae *Palmaria palmata* and *Solieria chordalis*	Proteins, polyphenols and polysaccharides	HPP at 400 MPa for 20 min / enzymatic extraction with cellulase (pH 5.0 and 37 °C) and hemicellulase (pH 4.5 and 40 °C) for 28 min	HHP-assisted enzymatic treatment improved the extraction of specific molecules such as proteins, polyphenols and polysaccharides, but their effects are highly dependent on the macroalgae species, probably due to differences in chemical composition and cell wall structure The antioxidant activity of extracted fractions was improved by over 2.8 times for the treatment with HHP with hemicellulose. Antioxidant activity was highly correlated with polysaccharide (89%) and protein (83%) contents for *S. chordalis*, and with polyphenol (65%) for *P. palmata*.	Suwal et al. (2019)
Haematococcus pluvialis microalgae	Carotenoids	HPP at 100–600 MPa and number of cycles ranging between 1 and 3 cycles, for total extraction time of 20 min / ethanol as a solvent at 50 °C Conventional carotenoid extraction: acetone containing 0.1% (w/v) BHT, 24 h in a thermostatic shaker at 500 rpm and 20 °C	HPE improved the selectivity of carotenoid extraction (109.74–119.34 mg per g extract) from *H. pluvialis* in comparison with the conventional extraction. When one extraction cycle was used, a more carotenoid-enriched extract was obtained, reaching values up to 119 mg of total carotenoids per g of extract.	Bueno et al. (2020)

Raw material	Target compound	HPP conditions/Extraction method	Key findings	Reference
Porphyridium cruentum microalgae	B-phycoerythrin, carotenoids, and polyunsaturated fatty acids (PUFAs)	HPP at 100–600 MPa for 20 min / water as a solvent at 25 °C **or** HPP at 100–600 MPa for 20 min / ethanol, ethyl acetate and a mixture of ethanol:D-limonene 1:1 (v/v) as a solvent at 70 °C *Conventional B-phycoerythrin extraction:* Milli-Q water as a solvent, 1 h under agitation at 20 °C, away from direct light exposure *Conventional lipid extraction:* Chloroform/methanol 2:1 (v/v) as a solvent under agitation – Addition of 0.73% NaCl (w/v) strongly agitated – Solvent evaporation under nitrogen stream	A first extraction step using water at 300 MPa provided extracts significantly enriched in B-phycoerythrin with a satisfactory purity index of 2.37 (up to 144.43 mg per g extract). Residual biomass in the second step provided extracts enriched in carotenoids (up to 65.05 mg per g extract) when ethyl acetate was used, and enriched in PUFAs (mainly eicosapentanoic acid and linoleic acid) when a mixture of ethanol and limonene (1:1 v/v) was used.	Bueno et al. (2020)
Pepper seed oil	Tocopherols	HPP at pressures 300, 400 and 500 MPa and temperatures 30, 40 and 50 °C for 3, 5 and 7 min / n-hexane at 50 °C for 6 h	The optimal process parameters were 370 MPa, 50 °C and 5.7 min, where the highest oil extraction efficiency (83.0%) was achieved. This extract exhibited the highest level of unsaturated fatty acids and γ-tocopherol and lowest level of phospholipids, and superior antioxidant capacity.	Ma et al. 2019

(continued)

Table 9.1 (continued)

Raw material	Target compound	HPP conditions/Extraction method	Key findings	Reference
Squid pen waste	Chitosan	HPP at pressures 100, 300 and 500 MPa, for extraction time of 5, 10 and 15 min / acetate concentration of 0, 0.5 and 1% w/w	The optimum HPE condition was achieved at pressure of 500 MPa, extraction time of 10 min, and 1% (w/w) acetate concentration. Maximum yield of chitosan sample from the chitin of squid pens treated by HHP reached 81.9%. In vitro antioxidant assay suggested that samples from HPE had significantly higher DPPH radical scavenging activity, greater reducing power, and a stronger ferrous ion chelating effect than untreated samples did.	Huang and Tsai (2020)
Long rice grain	Cadmium decontamination	HPP treatment at 300, 450 or 600 MPa for 5, 10 or 15 min and pH = 5.5 (single or multiple cycles) / traditional acid washing method at acidic environment (pH: 3.0–5.5) and for processing time of 0–20 h	HPE process at 600 MPa for 10 min at pH 5.5 yielded an extraction efficiency of 43% for rice grain and 82% for rice flour, while no changes in rice quality were observed. HPE reduced significantly the decontamination time (40 and 20 min. respectively for rice gain and flour, compared to traditional method which last *ca.* 16 h) The Cd removal efficiency was further enhanced by multiple HPE cycles, almost reaching the 100% level.	Luo et al. (2021)
Tamarind	Xyloglucan	HPP at pressures 0.1–500 MPa for 5–15 min in ambient temperature / thermal pretreatment of the solution for 30 min at 95 °C – incubation with protease at 45 °C for 3 h – dissolve with 95% ethanol and left for 30 min to form a gel – drying at 60 °C for 8 h	The extraction yields of the xyloglucan obtained using the HPE method at 250–500 MPa were significantly higher compared to the conventional method. However, the viscosity, water absorption index, and average molecular weight of the samples extracted by HPP were significantly lower than those of the conventional extraction method.	Limsangouan et al. (2020)

Raw material	Target compound	HPP conditions/Extraction method	Key findings	Reference
Longan pulp	Polysaccharides	HPP at 300, 400 or 500 MPa, for 2, 6 or 10 min / enzyme to pretreated material ratio 0.4, 0.8 or 1.2% w/w, enzymolysis time 1.0, 1.5 or 2.0 h and water to pretreated material ratio 30, 40 or 50 mL/g	Maximum polysaccharides yield of 8.55% was obtained under the optimal conditions of 407 MPa for 6 min with an enzyme to pretreated material ratio of 1:100, enzymolysis time of 1.7 h and a water to pretreated material ratio of 42 mL/g.	Bai et al. (2016)
Umbilicaria plant	Polysaccharides	HPP at 303 MPa for 14.4 min / ratio of solid to liquid of 1:54	Theoretical HPE yield of polysaccharides was 12.6 ± 1.32%, 1.18 times higher than hot water extraction.	Sun and Jiang (2020)
Black garlic	High molecular weight melanoidins	HPP at 25 °C and 200 MPa for 5, 15, or 25 min and for 5 min at 300, 400, or 500 MPa / the sample was defatted with dichloromethane twice – the aqueous phase was dialyzed with distilled water exchanged every 3 h for 24 h – high-molecular-weight melanoidins obtained were freeze-dried	Total phenolics, flavonoids, and sugar yields were all increased when HPE was applied and the antioxidant and overall reducing power was maximized. The bioactive properties of protein tyrosine phosphatase 1B, angiotensin-converting enzyme, and trypsin inhibitory activities were enhanced compared with the untreated samples. The optimum processing conditions for melanoidin extraction using were considered as 300 MPa for 5 min.	Zhao et al. (2019)
Lychee pericarp	Procyanidins	HPP at pressures 100–500 MPa for 5–20 min / solid–liquid extraction at 50 °C for 2 h with liquid-to-solid ratio 10, 15 and 20 and ethanol concentrations of 60, 70 and 80%	The optimum HPE conditions were 295 MPa for 13 min holding time, 16.0 mL/g liquid- to-solid ratio, and 70% ethanol concentration, with extraction yield of procyanidins being 2.40 ± 0.12%. The yield of procyanidins, including epicatechin, procyanidin A2, and procyanidin B2, was significantly ($p < 0.05$) increased by the HPE process compared to ethanol extraction .	Zhang et al. (2017)

sequential HP-US treatment showed a synergistic effect achieving more than 2.5 log reduction, and the sequential US-HP treatment exhibited an additional reduction by 1.5–2 log cycles (Pyatkovskyy et al. 2018). In the same study, a single HP (200 MPa/20 min) or PEF (20 kV/cm/1 ms) treatment reduced microbial load of L. innocua by 1 log cycle, however more than 3 log reduction was achieved by simultaneous application of HP and PEF, indicating the existence of synergy, and sequential HP-PEF or PEF-HP exhibited an additional microbial reduction by 2 log cycles. The combined effect of HP and UF treatment in fruit and vegetable juices has been also investigated, as an alternative to thermal pasteurization in order to achieve desirable reduction of spoilage microbial populations. The total aerobic bacteria and yeasts and molds of pear juice were reduced by 3.0 and 3.4 log cycles, respectively, after UF treatment, and an additional reduction below the detection limit by 1.7 and 0.68 log cycles, respectively, was achieved after sequential HP treatment (500 MPa/25 °C/10 min) (Zhao et al. 2016). Likewise, in the case of cucumber juice, under similar UF and HP conditions, a reduction by 1.3 and 1.9 log cycles, respectively, for total aerobic bacteria and yeasts and molds after UF treatment was observed, and an additional reduction below the detection limit by 3.2 and 2.5 log cycles, respectively, was achieved after sequential HP treatment (Liu et al. 2016).

Even though nonthermal technologies are recognized to minimally affect the nutritional and quality properties of foods, an additional consideration in their application is the enhancement of biofunctionality to satisfy the consumers' demand for health promoting foods. In this context, the effect of nonthermal technologies on several bioactive compounds of foods (e.g. probiotics, prebiotics, antioxidants, bioactive peptides, etc.) has to be considered, since it is recognized that their functionality may be negatively affected mostly due to the instant increase in the temperature of the food matrix induced during the application of the above mentioned technologies.

On the effect of combined HP and US treatment on the quality and biofunctional attributes of prebiotic cranberry juice, both the applied technologies were reported not to alter the bioactive compounds and antioxidant capacity of the product, with a slight decrease yet good preservation of fructooligosaccharides. It was observed that the sequence of the applied technology impacted differently, with US prior to HP treatment leading to significant decrease of antioxidant capacity of the juice (Gomes et al. 2017).

In a recent study on the combined effect of ultrasound-assisted extraction and HP of Gac leaves results showed that the combined application of US (150 W, 20 min) and HP (300 MPa, 3 min) was in general more efficient compared to either conventional extraction or the single application of each emerging technology, and positive synergistic effects were observed in terms of total phenolic compounds, antioxidant capacity and anti-diabetic activity of the extract (Nguyen et al. 2021).

2.5 HPP Combined with Osmotic Dehydration (OD)

Osmotic dehydration (OD) is an alternative process primary used for the partial dehydration of plant-based foods by immersing them into a hypertonic solution. During OD two main phenomena are observed, i.e. diffusion of water from the plant tissue to osmotic solution and diffusion of solutes from the hypertonic solution into the plant tissue, reaching to an equilibrium after certain time. These phenomena are controlled by several parameters, including the temperature and the concentration of the osmotic solution, the size and geometry of the plant tissue, the osmotic solution to food mass ratio, the permeability of the plant cells, and the level of agitation of the system (Rastogi et al. 2002). Compared to other drying methods, the main advantages of OD are the low energy consumption and temperature requirements, as well as the prevention of loss of volatile compounds, the elimination of appearance and texture damages of the tissue occurring due to heat treatment, and the ability to recycle the osmotic solution in the production lines. However, inherently OD is a slow process, and additional ways to accelerate the mass transfer without affecting the quality of the food need to be sought. In this view, various nonthermal technologies have been suggested for enhancing mass transfer including High Pressure Processing, Pulsed Electric Fields, Ultrasounds, etc. (Rastogi et al. 2002). The effect of HP technology on osmotically dehydrated food products has been studied as a permeability inducing treatment prior to osmodehydration (Rastogi and Niranjan 1998; Rastogi et al. 2000a, b; Verma et al. 2014), as a post OD cold pasteurization technology (Dermesonlouoglou et al. 2017a, b; Andreou et al. 2018) and, as a process simultaneously applied to OD (Nuñez-Mancilla et al. 2011, 2013a, b; Dash et al. 2019).

The application of high pressures is well established that lead to cell structure damages, and thus to more permeable cells, and this is related to increased mass transfer rates during OD as compared to untreated samples. As a result the diffusivity values, as determined by the Fick model, increased by four-fold for water and two-fold for sugar, at applied pressures of 100–800 MPa (Rastogi et al. 2002). In fact, during HP pre-treatment of plant tissues, both compression and decompression that occur, effect the cells permeability and enhance both water removal and solid gain from and into the tissue during a subsequent osmotic dehydration. The enhancement of water removal during osmotic dehydration of HP pre-treated samples has been studied in several plant tissues, including pineapple (Rastogi and Niranjan 1998; Rastogi et al. 2000a, b) and banana slices (Verma et al. 2014). It is worth noticing that under the same mechanism of action during HP pre-treatment, Belmiro et al. (2018) found that the hydration process of common beans was enhanced after applying pressures in the range of 50–600 MPa for 1 min. Moreover, drying was accelerated significantly exhibiting 27% higher water diffusivity for samples treated at 600 MPa compared to HP untreated samples. HP treatment has also been applied to infuse bioactive compounds in foods without altering their matrices. Recent studies have shown that there is higher infusion of bioactive compounds as compared to infusion at atmospheric pressure, as in the case of anthocyanins infusion into apple

(George et al. 2015), and this is directly related to the concentration of the osmotic solution used.

The application of HP subsequently to OD has been studied on plant tissues such as grapefruit jam (Igual et al. 2013), fresh-cut strawberry cubes (Dermesonlouoglou et al. 2017a) and fresh-cut tomatoes (Dermesonlouoglou et al. 2017b), and also on animal tissues such as chicken breast fillets (Andreou et al. 2018), mainly targeting to the improvement of microbial and enzymatic stability, the retention of the quality properties and the extension of the shelf life of the final product. Results showed that HP was not capable to completely inactivate pectin methylesterase (PME) and peroxidase (POD) in osmodehydrated plant tissues, yet it was able to improve the enzymatic stability of grapefruit jam obtained by OD without affecting the bioactive content as expressed by the antioxidant capacity in %DPPH and mM Trolox (Igual et al. 2013). Moreover, it has been reported that HP treatment of both osmotically dehydrated plant and animal tissues at 600 MPa (5 min – ambient temperature) led to an increased shelf life of up to 7–10 months at 5 °C in the case of plant tissues (Dermesonlouoglou et al. 2017a, b), and to 25 days at the same storage temperature for animal tissues (Andreou et al. 2018), while at the same time the sensorial properties of the food product in terms of appearance and texture were enhanced in the cases where HPP was applied.

As mentioned before, besides the inactivation of microorganisms and enzymes and the retention of quality and nutritional properties of foods, HP technology has been reported to enhance mass transfer rates by increasing cell permeability and secondary metabolites diffusion (Corrales et al. 2008). HP-assisted osmotic dehydration has been studied in several plant-based products, e.g. potato (Sopanangkul et al. 2002), strawberries (Nuñez-Mancilla et al. 2011, 2013a, b), and ginger slices (Dash et al. 2019), indicating that the application of HP-assisted osmotic dehydration in the pressure range of 200–400 MPa led to improved diffusion coefficients of water and soluble solids compared to those obtained after treatment at atmospheric pressure, and that the kinetics of water loss and solute uptake can be efficiently described by Weibull model (Nuñez-Mancilla et al. 2011; Dash et al. 2019). Moreover, it was shown that the application of high pressures up to 600 MPa can enhance diffusion and accelerate the mass transfer rates of ingredients into food matrices due to the increase in plant cells permeability in the temperature range of 20–40 °C, however the diffusion may be hindered to some extent in pressures above 400 MPa as a result of the occurrence of other phenomena under HP conditions such as starch gelatinization (Sopanangkul et al. 2002). The same pressure range is suggested for HP-assisted osmotic dehydration of strawberries, since a high level of nutrients and antioxidant capacity is retained after treatment, and no reduction of vitamin C is observed (Nuñez-Mancilla et al. 2013a, 2013b).

2.6 HP-Assisted Enzymatic Treatments

A pre-treatment with microbial transglutaminase (mTGase, EC 2.3.2.13) is been extensively used in order to achieve improved textural and sensorial properties in a wide range of food products, e.g. meat, poultry, fish, pasta, and dairy products. This transferase catalyzes the reaction between the γ-carboxyamide groups of peptide bound glutamyl residues (acyl donor) and several primary amines including ε-amino group of lysine, resulting in protein cross-linking through the formation of both inter- and intramolecular isopeptide linkages (Motoki and Seguro 1998). The pre-treatment with mTGase could be combined with HP technology, as both processes have an excessive effect on the modification and functionality of food proteins. Therefore, their combination may lead to products with enhanced techno-functional properties due to the compact protein networks that are formed. Most research on the combined effect of mTGase and HP on food quality characteristics is focused on animal origin foods, where it is indicated that the application of HP results in increase in mTGase affinity to proteins, and thus in products with enhanced physicochemical characteristics and firmer gel structures (Herranz et al. 2013; Zhu et al. 2014; Cando et al. 2016). Similar results have been observed in the case of HP-assisted mTGase treatment in poultry meat (Trespalacios and Pla 2007a, b), dry-cured ham (Fulladosa et al. 2009) and porcine plasma (Fort et al. 2009). Regarding dairy products, several authors have reported that both the TGase treatment under HP conditions (Anema et al. 2005) and/or the subsequent to HP processing TGase treatment (Tsevdou et al. 2013a, b, 2020) of milk cause extensive denaturation of whey proteins and dissociation of casein micelles, reforming new intra- and inter-molecular cross-links, resulting in strengthened networks with contiguous protein molecules. Consequently, yoghurt- and cheese-making properties are improved and negative effects, such as whey separation phenomena, post-acidification and sensorial disaffects, are minimized, while other characteristics, such as the textural and rheological attributes, flavour evolution and release, and biofunctional profile of the product, are improved.

Apart from the combined effect of HP and enzymes on animal origin proteins in order to achieve products with enhanced mechanical properties, HP-assisted hydrolysis have gained significant interest by several researchers. Since the 90s it is well recognized that the hydrolysis of whey proteins by pepsin, trypsin and chymotrypsin is enhanced under HP conditions (Dufour et al. 1995; Maynard et al. 1998; Peñas et al. 2004). Later studies have indicated that HP pre-treatment of chickpea protein isolate in the pressure range of 100–600 MPa led to increased hydrolysis rate in pressures above 300 MPa, while a pre-treated above 400 MPa led to significantly reduced hydrolysis time accompanied with the release of antioxidant peptides (Zhang et al. 2012). At the same pressure level of 400 MPa, it is observed that the highest value of whey protein unfolding is induced, and that the degree of hydrolysis is increased not only with pressure level applied but also with processing time. As a result, HP-assisted hydrolysis of whey proteins could be a useful tool for the modification of whey proteins antigenic sequences, and thus their antigenic activity

(Ambrosi et al. 2016). Likewise, HP-assisted enzymatic hydrolysis of proteins have reported to be an innovative approach for the production of low-allergen food ingredients with improved functional properties. In particular, the application of HP prior or during enzymatic hydrolysis of soy protein isolate results in completely inhibited or significantly impaired up to 99% of residual immunoreactivity of β-conglycinin (Gly m5) in the pressure range of 400–500 MPa for 15 min processing time. At the same time, the resulting hydrolysates possess improved functional properties, with protein solubility, foaming activity and oil-binding capacity to be increased by 45%, 66% and 210%, respectively (Meinlschmidt et al. 2017). Application of HP-assisted hydrolysis of kidney beans protein isolate at 300 MPa resulted in higher degree of hydrolysis compared to hydrolysis in atmospheric pressure, while at the same time the produced hydrolysates exhibited better rheological and biofunctional (in terms of antioxidant activity and bioactive peptides) properties (Al-Ruwaih et al. 2019).

3 Pulsed Electric Fields Processing

3.1 Introduction

Along with High Pressure Processing, Pulsed Electric Fields (PEF) constitutes one of the most widespread nonthermal processes applied in the food industry. Although the process has a much more recent history compared to other nonthermal processes, it still remains in the spotlight of both research and industrial practice. Apart from its development as a nonthermal pasteurization method for liquid foods, recent applications have highlighted its potential as a pretreatment to conventional mass transfer processes such as extraction, juice expression and dehydration. In this sense, the process lends itself to the valorization of both plant-based and microbial waste streams.

3.2 Mechanism of Action, Equipment and Traditional Applications

PEF processing is based on the exposure of a cellular material (plant, animal or microbial) to pulses of a high-strength (1–80 kV/cm), external electric field which is typically delivered in short duration pulses (time scale ranging from ns to ms). The exposure of the cells to this electric field leads to the formation of a transmembrane potential across the cell membrane. When the transmembrane potential exceeds a certain critical value, pores are formed on the cell membrane. This phenomenon is known as electroporation or electropermeabilization and leads to loss of intracellular contents and death of the cell. The efficiency of electroporation depends on a variety of parameters, mainly electric field strength, pulse characteristics such as pulse width, frequency, shape and polarity, treatment temperature and electrical characteristics of the treated medium. Effective electroporation also depends on the size, shape and physiological characteristics of the treated cells. The effects of electroporation can be harnessed both to inactivate microbial cells and to improve mass transfer of intracellular materials. Typically, because plant and animal cells are larger than microbial cells, their electroporation and cell leakage can occur at field strengths as low as 0.7–3.0 kV/cm (energy input 1.0–20 kJ/kg), as opposed to the inactivation of microbial cells which requires electric field strengths upwards of 15 kV/cm (energy input 40–1000 kJ/kg) (Toepfl et al. 2007). The most common application of PEF is the nonthermal pasteurization of liquid foods. Although microbial inactivation requires intense processing conditions, the benefits of PEF treatment are reflected both in cost savings and preservation of quality characteristics of the pasteurized products. PEF pasteurization of liquid foods has been reported for a wide variety of products including fruit juices such as apple, orange, strawberry, tomato, watermelon (Sampedro et al. 2013; Charles-Rodríguez et al. 2007; Aguilar-Rosas et al. 2007; Yildiz et al. 2019; Aganovic et al. 2017) and other products such as beer (Milani et al. 2015) and milk (Bendicho et al. 2002; Sepulveda et al. 2005). Apart from the benefits of PEF processing as a nonthermal pasteurization technique, the process has exhibited several useful synergies as a pretreatment to other conventional processes such as pressing, extraction and dehydration. Owing to the particular effects of the imposed electric field to cellular materials through electroporation, PEF has the potential to enhance mass transfer phenomena to and from cells, allowing for the improvement of these conventional processes.

3.3 Pulsed Electric Field Assisted Extraction

Compared to other pretreatment methods, PEF treatment has the potential to enhance mass transfer with minimal energy consumption while allowing the preservation of the quality characteristics of the extracted materials. The low treatment temperatures associated with PEF processing enable the preservation of thermolabile compounds during extraction. The process has proven its effectiveness as a cell disintegration pretreatment for microbial, plant and animal cells. Since PEF treatment does not cause cell fragmentation, it also allows for selective extraction of target compounds and facilitates the downstream processing for their separation (Puértolas and Barba 2016).

3.3.1 PEF Assisted Juice and Oil Extraction

Treatment of plant cells with PEF leads to loss of intracellular water due to electroporation, accompanied by a loss in turgor pressure and tissue softening. These effects are especially relevant to plant materials from which juice is expressed such as fruits, vegetables and tubers. Enhancing the pressing efficiency of sugar beet is particularly important, as the sugar beet industry is one of the most energy intensive sectors (Toepfl et al. 2007). Due to the ligneous structure of sugar beet, juice expression is usually preceded by a thermal treatment which, although increases subsequent yield, also leads to the extraction of undesirable compounds such as pectins (Loginova et al. 2011). As shown in Table 9.2, PEF treatment has been demonstrated to increase pressing yields of various plant tissues, including sugar beets. In most reported cases juice yields are significantly increased after PEF treatment, while the electric field strength seldom exceeds 3 kV/cm. Compared to thermal pretreatments, PEF carries the advantage of a very short treatment duration and preservation of the quality of resulting juice (Eshtiaghi and Knorr 2002; Praporscic et al. 2005). Gachovska et al. (2006) also reported that alfalfa juice resulting from PEF treated stems exhibited increased sedimentation rates, which may contribute to an improved separation of solids from the juice. A reduction in juice turbidity was also reported by Praporscic et al. (2007) and was attributed to improved filter cake properties of PEF treated grapes.

Similar to the effectiveness of PEF for juice expression, the process has been demonstrated to positively affect the extraction of oil from oleiferous seeds and fruits (Table 9.2). Oil extraction is usually achieved through pressing, centrifugation or solvent extraction. In the case of olive oil, the ground olive paste is subjected to mechanical agitation at elevated temperatures before centrifugation, known as malaxation. Although malaxation improves the yield of olive oil, it increases oxidation due to the elevated temperatures involved. It has been demonstrated that through the use of PEF as a pretreatment, the process can be performed at temperatures as low as 15 °C, leading to improved oil quality and energy savings (Abenoza et al. 2013; Andreou et al. 2017). Oftentimes, oleiferous plant materials contain high amounts of bioactive compounds such as polyphenols, tocopherols, sterols and chlorophylls. PEF pretreatment leads to the extraction of these compounds from the cake to the oil, improving its nutritional value and overall quality attributes such as sensorial characteristics or oxidative stability (Guderjan et al. 2007; Andreou et al. 2017; Puértolas and de Marañón 2015; Shorstkii et al. 2020; Zeng et al. 2010). At the same time, the electrical pretreatment has not been found to negatively affect the produced oils.

3.3.2 PEF Assisted Extraction of Bioactive Compounds

Extraction of bioactive compounds from plant-based materials involves the use of solvents such as water, ethanol, methanol or hexane which remain in contact with the solid materials for prolonged amounts of time and often at elevated temperatures (exceeding 40 °C). During solid-liquid extraction, bioactive compounds are extracted into the solvent and further recovered by its removal. However, in many cases the target compounds are thermolabile and the elevated temperatures also damage the plant tissue structure, releasing contaminants into the solvent and thus making further separation difficult (Puértolas and Barba 2016). Reduction in the extraction time and volume of solvent used as well as an increase in the extraction

Table 9.2 Application of PEF pretreatment for juice and oil extraction from plant tissues

Plant tissue	PEF treatment conditions	Extraction method	Key conclusions	Reference
Juice expression				
Sugar beet	20–100 μs pulses 100–1000 Hz 1 kV maximum voltage	Multi-plate and frame pressing equipment (5–15 bar) with simultaneous pulse delivery	Juice yield increase from 29% up to 80% PEF treatment increased juice purity	Jemai and Vorobiev (2006)
Sugar beet	1.2–2.5 kV/cm field strength 1–200 pulses at 1–6 Hz	Uniaxial press at 2 or 5 MPa	Twofold increase in sugar yield Reduction in overall processing time	Eshtiaghi and Knorr (2002)
Sugar beet	1.5 kV/cm field strength 400 monopolar pulses with a duration of 100 μs combination with ohmic heating (30–70 °C, 10–30 min)	Ohmic heating -PEF treatment cell with compression capability (25 bar)	Juice yield increase from 40% to 65% combined with ohmic heating (10 min, 60 °C)	Praporscic et al. (2005)
Apple	0.1–0.52 kV/cm field strength 50 pulses of 100 μs	Pressing cell fitted with mobile electrode Expression at 3 bar	Synergistic effect of pressure and electric field to increase juice yield	Bazhal et al. (2001)
Apple	0.65 kV/cm field strength 23.2 ms treatment time	Continuous belt filter press	Juice yield increase from 71.1% to 76.3% due to PEF Increased polyphenol content in final juice Decreased juice turbidity	Turk et al. (2012)
Citrus fruits (orange, lemon, pomelo)	3 kV/cm field strength 10–250 pulses of 70 μs	Laboratory compression chamber at 4 bar	Juice yield increase up to 25% (orange), 37% (pomelo) and 59% (lemon) for PEF treated samples	El Kantar et al. (2018)
White grapes	0.25–1.0 kV/cm field strength	Laboratory filter press system with mobile electrode at 5 bar	Juice yield increase from 49–54% up to 76–78% Juice of superior quality in terms of turbidity and absorbance due to improved filter cake properties	Praporscic et al. (2007)

(continued)

Table 9.2 (continued)

Plant tissue	PEF treatment conditions	Extraction method	Key conclusions	Reference
Alfalfa	1.5 kV/cm field strength 200 pulses delivered at 1 Hz	Cylindrical PEF chamber with built in pressing plate (20 and 40 bar)	Inreased juice yield from 27.3% to 37.8% Increased sedimentation in PEF treated samples	Gachovska et al. (2006)
Tomato	0.5–2.5 kV/cm electric field strength, 0–4000 pulses of 15 μs width at 20 Hz	Lab scale paddle type extractor	Up to 20% juice yield increase Up to 90.2% overall juice yield (compared to untreated 82.2%)	Andreou et al. (2020a)
Oil extraction				
Olive	0–2 kV/cm field strength, 50 pulses of 3 μs at 125 Hz	Malaxation (0,15 30 min at 15 and 26 °C)/ centrifugation (1370 g, 2 min)	Oil yield increase by 54% without malaxation and by 14.1% with malaxation at 15 °C No detrimental effect of PEF on oil quality parameters	Abenoza et al. (2013)
Olive	1.8 kV/cm field strength with pulses of 15 μs width at 300 Hz	Malaxation (30 min, 30 °C)/ Centrifugation (3000 g, 4 min)	Oil yield increase up to 18% after PEF treatment Produced oils with increased polyphenol content and oxidative stability	Andreou et al. (2017)
Sunflower seeds	1–7 kV/cm electric field strength with 10–120 s of treatment time with pulses of 10–50 μs width at 0.5–10 Hz	Solvent extraction with hexane (3 h at room temperature)	Oil yield increase from 39.14% to 48.04% at 7.0 kV/cm	Shorstkii et al. (2017)
Rapeseed	5–7 kV/cm electric field strength with 60–120 pulses of 30 μs pulse width	Mechanical screw press with a capacity of 5 kg/h at 100 °C Soxhlet extraction with hexane	Oil yield increase from 48% to 51% for hulled seeds after treatment at 7 kV/cm Oils with increased chlorophyll, tocopherol and polyphenol content	Guderjan et al. (2007)
Sesame	20 kV/cm electric field strength, 0–2000 exponential pulses at 0.33 Hz	Pressing for 1 h at 100 bar with a texture analyzer	Oil yield increase from 65% to 72% No significant effects of PEF on physicochemical parameters No oil loss in the treatment water	Sarkis et al. (2015)

yield are key requirements during extraction and can reduce processing costs as well as overall solvent use. Pulsed Electric Fields have demonstrated significant benefits in the pretreatment of plant materials for the extraction of bioactive compounds. The effects of the process on the cellular structures can increase the extractability of intracellular compounds. The food sector has been recognized as a major producer of plant-based food wastes and by-products (Arshad et al. 2021). These wastes are discarded when their valorization could result in the efficient extraction of residual bioactive compounds.

Several authors have reported significant yield increases in bioactive compounds extracted from plant materials when PEF was used as a pretreatment (Table 9.3). In certain cases, this yield increase enables the substitution of organic solvents such as methanol and ethanol with greener alternatives such as water and the reduction of the extraction temperatures, due to the yield increase brought about by the pretreatment (Puértolas et al. 2013; Loginova et al. 2011). Extraction can also be significantly accelerated (Pataro et al. 2020; Andreou et al. 2020b). PEF treatment of microbial suspensions has also been demonstrated to effectively increase the extractability of intracellular compounds. As shown in Table 9.3, compared to plant cells, the electric field strength required for electroporation of microbial cells is markedly higher compared to that for plant cells. Nevertheless, the benefits of PEF treatment include increased extraction yields of bioactive compounds while no deleterious effects on the extracted compounds have been reported. On the contrary, compared to other cell disintegration processes, PEF has the advantage of allowing for the electroporated cell envelope to act as a sieve, thereby selectively extracting the intracellular contents (Martínez et al. 2017; Corrales et al. 2008). The selectivity of PEF for the extraction of compounds can be modulated by altering the post-pulsing extraction regimes in terms of pH, temperature or solvent composition (Ganeva et al. 2020; Parniakov et al. 2015a, b; Postma et al. 2016).

3.4 PEF-Assisted Dehydration Processes

The removal of water from foods has always been, and still remains, one of the most critical methods of preservation. The removal of moisture from foods is commonly achieved by air drying, where foods remain in an air stream at elevated temperatures for prolonged periods until the required moisture reduction is achieved. In such processes food materials, especially plant tissues, undergo quality deterioration in terms of colour, shrinkage, sensorial attributes and nutritional value. This fact, along with the high energy demands of drying processes, have led to the emergence of suitable pretreatments that can potentially accelerate water transfer rates thereby reducing the food's exposure to high temperatures. Osmotic dehydration (OD) removes water from food materials by immersion in suitable hypertonic solutions and has emerged as a gentle dehydration method suitable for processing fruits and vegetables (Ade-Omowaye et al. 2001). However, even though OD involves the exposure of foods to lower temperatures compared to air drying, the process is slow

Table 9.3 PEF-assisted extraction from plant and microbial cells

Target system	Target compounds	PEF treatment conditions	Key conclusions	Reference
Plant and animal cells				
Grape and plum peels	Anthocyanins, flavonoids, polyphenols	9.6–25 kV/cm field strength 6 μs pulse width delivered at 10 Hz at a flowrate of 290 L/h	2–4 times increased anthocyanin extraction yield Up to 70% increased flavonoid extraction yield Up to 200% yield increase in extracted polyphenols	Medina-Meza and Barbosa-Cánovas (2015)
Orange peels	Polyphenols	1–7 kV/cm field strength, 20 pulses of 3 μs width	Polyphenol pressing yield increase up to 159% Antioxidant capacity increased up to 192%	Luengo et al. (2013)
Purple-fleshed potatoes	Anthocyanins	1–5 kV/cm field strength, 5–35 pulses of 3 μs width	Possible to replace ethanol with water for anthocyanin extraction (yield increase from 42% to 55% at 25 °C)	Puértolas et al. (2013)_
Grape byproducts	Anthocyanins, polyphenols	3 kV/cm field strength, 30 exponential decay pulses at 2 Hz	Up to four-fold increase in antioxidant capacity Up to 17% anthocyanin yield increase Selective extraction of anthocyanins dependent on sugar moieties	Corrales et al. (2008)
Red beet	Pigments (betalains)	0.375–1.500 kV/cm field strength, trains of 20 pulses of 100 μs width with a total treatment time of 0.1 s	Increased colorant diffusion at lower temperatures Possible to extract colorants avoiding thermal degradation	Loginova et al. (2011)
Flaxseed hulls	Polyphenols	10–20 kV/cm field strength, 1000 exponential decay pulses at 0.33 Hz	Up to 37% polyphenol yield increase Up to 80% of total polyphenols extracted	Boussetta et al. (2014)
Tomato wastes	Phenolic compounds, carotenoids	0.5–2.5 kV/cm electric field strength, 0–4000 pulses of 15 μs width at 20 Hz	Up to 56.4% increased carotenoid extraction Up to 43.6% increased phenolic compound extraction	Andreou et al. (2020a)

(continued)

Table 9.3 (continued)

Target system	Target compounds	PEF treatment conditions	Key conclusions	Reference
Tomato processing byproducts	Lycopene	1–5 kV/cm field strength, pulses of 20 μs width at 10 Hz	Up to 18% increased lycopene extraction yield and 37% increased extraction rate No lycopene degradation due to treatment	Pataro et al. (2020)
Olive pomace	Polyphenols, proteins	1.0–6.5 kV/cm field strength, 1–6000 pulses of 15 μs width at 20 Hz	Up to 91.6% increase in extracted phenolic compound yield and twofold increase in protein extraction yield Extraction time reduced to 12 min and 1 min compared to 60 min for untreated	Andreou et al. (2020b)
Microbial cells				
Saccharomyces cerevisiae	Yeast extract	5–20 kV/cm field strength, 1–2000 pulses of 15 μs width	Increase of amino acid and total solids in extract by 37% and 20%, respectively Up to 78% acceleration of yeast extract production via autolysis	Dimopoulos et al. (2018)
Saccharomyces cerevisiae	Protein, phenolic compounds, glutathione	2.5–5.5 kV/cm field strength, 15 pulses of 0.5–0.8 ms width at 7.5–82.3 Hz	Release of 90% of free amino acids, 80% of intracellular glutathione and 40% of total phenolic compounds extracted at 2 h after treatment	Ganeva et al. (2020)
Saccharomyces cerevisiae	Yeast extract/ amino acids	3–7 kV/cm field strength, 40 pulses of 100 μs width at 1 Hz	Increased aspartic and glutamic acid content of extracts by 232.55% and 209.40%, respectively	Yang et al. (2021)
Rhodotorula glutinis	Carotenoids (torularhodin)	15 kV/cm field strength, 400 exponential decay pulses of 6.8 μs width at 3 Hz	Up to 80% of total carotenoid recovery in subsequent ethanol extraction	Martínez et al. (2020)

(continued)

Table 9.3 (continued)

Target system	Target compounds	PEF treatment conditions	Key conclusions	Reference
Chlorella vulgaris	Carbohydrates, proteins	10–30 kV/cm field strength, pulses of 5 μs width	Selective extraction of 36% of total carbohydrates and 5.2% of total proteins	Carullo et al. (2018)
Auxenochlorella protothecoides	Lipids	40 kV/cm field strength, pulses of 1 μs width at 3 Hz and a flowrate of 0.1 mL/s (2.1 mL chamber volume)	Lipid extraction yields up to 97% with ethanol/hexane mixtures No effects of PEF on fatty acid composition	Silve et al. (2018)
Chlorella vulgaris	Carotenoids, chlorophylls	10–25 kV/cm field strength, 0–50 pulses of 3 μs width at 0.5 Hz	Up to twofold increase of pigment extraction after PEF treatment	Luengo et al. (2014)
Arthrospira platensis	C-phycocyanin	15–25 kV/cm field strength, 15–50 pulses of 3 μs width at 0.5 Hz	Selective extraction of phycocyanin up to 25 times higher Higher phycocyanin purity compared to bead milling	Martínez et al. (2017)

as it relies on the passive diffusion of water towards the osmotic solution. Osmotic dehydration can also benefit from pretreatments that accelerate water diffusion rates. The mechanism that drives the improved extraction of intracellular compounds due to PEF treatment can be extended to include intracellular water. Electroporation-mediated permeabilization of cells causes the outflux of intracellular water and solutes. This phenomenon is harnessed for the enhancement of dehydration processes such as air-drying and osmotic dehydration.

Osmotic dehydration of PEF-pretreated plant tissue samples is enhanced both in terms of water loss as well as uptake of solids from the osmotic solution. A direct relationship between the cell disintegration index and the moisture effective diffusion coefficient has been established. The benefit of solids uptake from the osmotic solution can be seen as a major benefit, as auxiliary compounds of the osmotic formulation contribute to the prevention of quality deterioration or improvement. Several reports exist on the enhanced uptake of calcium chloride of PEF-treated OD samples (Table 9.4), which helps to counteract the loss of firmness caused by electroporation (Tylewicz et al. 2017; Dermesonlouoglou et al. 2016). Uptake of other compounds such as ascorbic acid have also been reported (Dermesonlouoglou et al. 2016, 2018).

Freeze drying is also influenced by PEF pretreatments (Table 9.4). Apart from the reduction of drying time, prevention of tissue shrinkage also appears to be an important benefit of PEF prior to freeze drying. Fauster et al. (2020) reported the

Table 9.4 Application of PEF as a pretreatment to dehydration processes

Target system	PEF treatment conditions	Dehydration method	Key conclusions	Reference
Strawberry	0.1–0.4 kV/cm field strength, 100 pulses of 100 μs width at 100 Hz	Osmotic dehydration (sucrose, 40% trehalose)	12% water loss reduction Preservation of cell viability at 10 kV/cm Slight loss of color due to peroxidase activity Recovery of lost firmness at long OD times due to the presence of calcium	Tylewicz et al. (2017)
Red bell pepper	1–2 kV/cm field strength, 10–80 pulses of 400 μs width	Osmotic dehydration (sucrose/NaCl 21%/2%) Fluidized bed drying (60 °C, 6 h, 1 m/s air velocity)	Significant reduction of drying rates up to 3 h of drying	Ade-Omowaye et al. (2003)
Goji berry	0.9–2.8 kV/cm field strength, 0–32,000 pulses of 15 μs width at 20 Hz	Osmotic dehydration (glycerol 60%/ maltodextrin 20%) 55 °C, 0–180 min) Air drying (60 °C up to 400 min)	Up to 85% increased moisture diffusion coefficients for PEF + OD treatment No significant color deterioration Increased ascorbic acid uptake Up to fivefold increase in drying rate of PEF + OD + air dried samples	Dermesonlouoglou et al. (2018)
Apple	0.8 kV/cm field strength, 10 pulses of 1 ms delivered every 100 ms	Vacuum freeze drying (10 mbar, −10 °C up to 6 h)	PEF treatment causes acceleration of cooling and water evaporation during freeze drying PEF leads to shrinkage avoidance in the dried samples	Parniakov et al. (2016)
Carrot	5 kV/cm field strength, exponential pulses at 0.5 Hz	Convective air drying at 70 °C and an air velocity of 2 m/s	PEF treatment increases both thermal and electrical conductivity Reduction of drying time up to 8.2%	Wiktor et al. (2016)

(continued)

Table 9.4 (continued)

Target system	PEF treatment conditions	Dehydration method	Key conclusions	Reference
Apple	At 10 ms pulse width	Freeze drying at 1 mbar, −55 °C up to 7 h	Reduction of freeze drying time by 57% after PEF treatment Improved rehydration characteristics	Lammerskitten et al. (2019)
Strawberry, red bell pepper	1.0 kV/cm field strength, exponential decay pulses at 100 μs width and 1 Hz	Freeze drying at −40 °C, 0.5 mbar for 72 h	Significant drying acceleration due to mass transfer enhancement Reduced shrinkage due to drying Enhanced water uptake during rehydration	Fauster et al. (2020)
Kiwi	0.7–1.8 kV/cm field strength, 50 pulses of 15 μs width at 20 Hz	Osmotic Dehydration (glycerol, trehalose, maltodextrin) at 1:5 food to solution ratio, 25–45 °C	Significant acceleration of water loss and solids gain Increase in ascorbic acid uptake from the osmotic solution Increased firmness due to calcium uptake	Dermesonlouoglou et al. (2016)
Sea bass fillets	1.6 kV/cm field strength, 50–1500 pulses of 15 μs width at 20 Hz	Osmotic dehydration (40–60% glycerol, 5% NaCl, 15 °C)	Up to 66% increase in the moisture effective diffusion coefficient Abrupt increase of D_{eff} above 5 kJ/kg energy input	Semenoglou et al. (2020)

implementation of PEF for the improved freeze drying of strawberries and red bell peppers. At specific energy inputs as low as 0.3 kJ/kg significantly improved drying rates. Moreover, the authors reported significantly reduced plant tissue shrinkage for PEF treated samples and improved rehydration characteristics of the final products. Similar results were reported for freeze dried apple tissue by Lammerskitten et al. (2019). Shrinkage reduction at high values of cell disintegration is mainly attributed to the facilitated water removal from the individual cells during dehydration. The increased cell permeability caused by PEF is a result that persists even after the food is dried. As such, apart from improved characteristics during drying, PEF treated food materials also exhibit accelerated rehydration kinetics. This observation is particularly important when dried products are produced with the aim of being rehydrated before use.

3.5 PEF as a Pretreatment to Other Food Processes

In recent years, PEF has proven its suitability as a pretreatment to other food related processes such as frying and peeling. The application of the process in the potato processing industry is one of the most successful industrial implementations of the process. PEF pretreated potatoes, even at low intensity processing conditions (1 kV/cm field strength, 0.2 kJ/kg energy input) exhibit significant reductions in required energy for cutting. Also, cuts exhibit a smoother surface and significantly reduced feathering and cutting losses (Fauster et al. 2018). Subsequent frying of potatoes also benefits from PEF pretreatment, as fries resist browning and absorb decreased amounts of fat, effects attributed to loss of reducing sugars and increased water vapor barrier properties, respectively (Fauster et al. 2018; Ignat et al. 2015; Liu et al. 2020; Ostermeier et al. 2020). Genovese et al. (2019) have also reported the loss of free asparagine from potato cuts due to PEF treatment, which contributes to the significant reduction of acrylamide formation during frying. Meat processing has also been found to benefit from a PEF pretreatment. Traditional processes such as curing, salting, tenderization and aging of meat cuts heavily rely on endogenous enzymatic activity, cell structure and are often carried out under refrigeration for prolonged storage periods. PEF treatment of meat has been reported to accelerate tenderization due to the instantaneous release of intracellular proteases and assist processes such as brining and drying (Bhat et al. 2019).

4 High Pressure Homogenization

4.1 Introduction

Homogenization has become an integral part of the food industry, as it has allowed the production of dairy products with improved stability and textural properties. The requirements for increasing the shelf life of food in the context of nonthermal processes have led to significant developments in homogenization. Unlike earlier equipment that could achieve maximum pressures of around 30 MPa, technical developments in the 1990s enabled the manufacturing of equipment capable of achieving significantly higher treatment pressures upwards of 100 MPa, thus paving the way for innovative process applications. These modern systems have led to the process being termed High Pressure Homogenization (HPH). When the processing pressures are in the range of 300–400 MPa, the process is often referred to as Ultra High Pressure Homogenization (UHPH) (Diels and Michiels 2006). Although homogenization is often associated with dairy processing, it is now an integral part of a range of industrial processes such as the formation and stabilization of emulsions and suspensions, microbial inactivation and cell rupture in order to obtain intracellular products from cell suspensions. The high operating pressures of modern systems have also provided new perspectives for the implementation of the process for sterilization and inactivation of bacterial spores (Patrignani and Lanciotti 2016). In recent years, the process has found its place as a pretreatment in order to improve further processing in applications such as the enhancement of mass transfer, enzymatic treatment and the encapsulation of bioactive compounds.

4.2 Mechanisms of Action and Traditional Applications

The most widespread application of HPH is perhaps that of milk homogenization. The underlying physical mechanisms lead to the size reduction of milk fat globules, rendering the processed milk stable against cream separation, a phenomenon mostly undesirable by the modern consumer. High Pressure Homogenization equipment, both on industrial and laboratory scales, consists of a positive displacement pump which forces the fluid to be processed to pass through a valve which is usually made of high abrasion-resistant materials (e.g. ceramics, tungsten carbide, zirconium or synthetic gemstones) (Berk 2018). The valves used in HPH usually consist of a needle-seat or needle- ball-seat combination. The pressure exerted on the fluid passing through the valve is controlled by the force applied to the needle. Some homogenizers feature a nozzle instead of a needle and a seat, with the pressure controlled by the reduction of the cross section along the nozzle and the capacity of the positive displacement pump (Harte 2016). Industrial HPH systems are fitted with cooling or preheating circuits for the treated liquid (Levy et al. 2020). A major challenge in HPH treatments is to achieve a constant pressure during processing, especially in smaller scale systems where the positive displacement pump has a single piston. The reciprocating motion of the piston causes pressure fluctuations with a variation of around 10–15%. This problem is usually solved by using pumps with two or more pistons.

Despite the widespread application of the process, its precise mechanisms of action are still under investigation (Diels and Michiels 2006). The exact mechanism by which HPH causes cell rupture is a combination of natural phenomena with overlapping effects (Brookman 1975; Doulah et al. 1975; Engler and Robinson 1981; Kumar and Pandit 1999). At processing pressures greater than 100 MPa a variety of physical phenomena affect the microorganisms and biomolecules contained in the processed fluid. These include high hydrostatic pressure buildup before the valve, shear stress that develops between the needle and the seat, the intense turbulent flow immediately after the valve, and finally the impingement of the stream following its exit from the valve. The temperature rise during processing falls in the range of 2–2.5 °C per 10 MPa processing pressure (Engler 1990). Commercially available homogenizers achieve pressures up to 400 MPa however with low volumetric flow usually less than 100 L/h. The efficiency of HPH processing depends on processing parameters such as treatment pressure, treatment temperature and number of passes through the valve and valve geometry but also on the characteristics of the treatment medium such as fluid viscosity. In the case of cell treatment, disruption efficiency greatly relies on factors such as the morphology and physiology of the target microbes (size, shape, cell wall presence and structure), their growth state and any pretreatments that might affect the cell's susceptibility to disruptive phenomena.

4.3 HPH Assisted Extraction

The potential of using HPH as a cell disruption method was recognized relatively early in the history of the process. In the following years the process became a staple in laboratory equipment, mainly for the disruption of microbial cells with the aim of obtaining intracellular products. The extraction of intracellular compounds from microbes that typically do no secrete them into the extracellular medium dictates the use of a cell disruption pretreatment in order to increase the yield of obtained compounds. Such microbial cells often include yeasts and bacteria (*Saccharomyces cerevisiae, Escherichia coli*) (Middelberg 2000). HPH has therefore evolved into a pretreatment process prior to extraction of microbial products. More recent studies have undertaken the task to demonstrate the applicability of the process for the pretreatment of plant cells.

HPH has found a widespread application in the context of biorefineries with the aim of extracting lipids from microalgae for the production of biofuels. As microalgae have been recognized as important sources of bioactive compounds related to food applications, more recent works explore the application of the process as a pretreatment for their efficient extraction. The structure of most microalgae is characterized by a thick, external cell wall comprising mainly of polysaccharides (Alhattab et al. 2019). The extraction of intracellular compounds, mainly pigments, proteins, fatty acids and antioxidants requires the cell wall be rendered permeable before extraction. Compared to other extraction procedures, HPH has significant advantages in energy savings and allows for the reduction of solvent use. Yeasts have also been a target organism for the application of HPH. Similarly to microalgae, yeasts feature a thick cell wall which is notoriously resistant to disruption. Products such as intracellular proteins, amino acids and enzymes are commonly extracted by first disrupting the cells by HPH (Table 9.5). More recent works have explored the potential of HPH disruption for the enhancement of yeast extract production by autolysis, as well as the improved extraction of β-glucans from the cell walls (Dimopoulos et al. 2020; Thammakiti et al. 2004; Verduyn et al. 1999; Liu et al. 2008). Both these applications benefit from the intense mechanical disruption of the cells.

Apart from the treatment of microalgae, HPH has recently become a valuable process for the valorization of plant-based waste materials. Food processing byproducts are often rich in compounds which remain unvalorized. HPH has shown the potential to assist with the valorization of such waste streams by disrupting plant cells and allowing for increased extraction yields. These byproducts are usually treated in the form of aqueous suspensions, where the disruption of the plant cells causes profound leakage of water soluble compounds such as polyphenols and

Table 9.5 Application of HPH pretreatment for extraction of intracellular compounds

Target system	Target compounds	HPH conditions	Extraction method	Key results	References
Microbial cells					
Saccharomyces cerevisiae	Cell wall β-glucans	200–800 bar, 1–3 passes	Cellular autolysis (52 °C, pH = 5.5) Extraction with 1 N NaOH and acetic acid (80 °C, 1–2 h)	Increased β-glucan yields in combination with autolysis Increased cellular protein loss due to mechanical disruption	Dimopoulos et al. (2020)
S. cerevisiae (spent brewer's yeast)	Cell wall β-glucans	600 bar, 6 passes	Alkaline extraction (1 N NaOH, 80 °C, 2 h) Acidic extraction (acetic acid 0.5 M, 70 °C, 1 h)	Increased viscosity, emulsion stabilizing capacity and water holding capacity of homogenized glucans	Thammakiti et al. (2004)
S. cerevisiae	Yeast extract	600–1000 bar, 1–6 passes	Cellular autolysis	Up to 80% and 85% total solids and nitrogen release after treatment Development of intense turbidity of extracts	Verduyn et al. (1999)
S. cerevisiae	Yeast extract, β-glucans	200–800 bar, 1–3 passes	Cellular autolysis (52 °C, pH = 5.5)	Significant acceleration in yeast extract production	Dimopoulos et al. (2020)
S. cerevisiae	Cell wall β-glucans	400–800 bar, 1–4 passes	Hot water extraction (121 °C, 4 h)	Hot water extraction improves the extraction of β-glucan due to mannoprotein removal HPH improved the purity of β-glucan from 91% up to 95%	Liu et al. (2008)

(continued)

Table 9.5 (continued)

Target system	Target compounds	HPH conditions	Extraction method	Key results	References
Arthrospira platensis	Polysaccharides, proteins	500–1500 bar, 1–7 passes	Determination of polysaccharides and proteins immediately after treatment	Increased yields in protein (70%) and polysaccharides (12%) compared to hot water treatments Concomitant release of compounds unselectively	Elain et al. (2020)
Chlorella vulgaris	Ionic substances, proteins, carbohydrates	1500 bar, 1–10 passes	Aqueous extraction (shaking at 160 rpm, 25 °C)	41.9% and 54.1 of total carbohydrate and protein, respectively after 5 passes	Carullo et al. (2018)
Yarrowia lipolytica	Total oil	500–1500 bar, 1–10 passes	Extraction with n-hexane (dry and wet cells)	Up to 100% of total oil recovery after HPH pretreatment	Drévillon et al. (2018)
Desmodesmus sp.	Carotenoids	2760 bar, 4 passes	Extraction with diethyl ether	No effect of cell suspension density (up to 90 g/L) Up to eight-fold increase in carotenoid concentration compared to untreated samples	Xie et al. (2016)
Plant cells					
Tomato peels	Proteins, polyphenols, lycopene	1000 bar, up to 10 passes	Determination in aqueous extracts Extraction with acetone or ethyl lactate (lycopene)	Significant increase in extraction of protein and polyphenols Twofold increase in lycopene concentration extracted in the aqueous phase	Jurić et al. (2019)

(continued)

Table 9.5 (continued)

Target system	Target compounds	HPH conditions	Extraction method	Key results	References
Fringed rue (*Ruta chalepensis*)	Flavonoids, alkaloids	1000 bar, up to 10 passes	Determination in the aqueous extract	Rutin and quercetin concentrations increased by 452.7% and 29.8%, respectively in the aqueous extract Improved extraction of water insoluble alkaloids	Gali et al. (2020)
Potato peels	Phenolic acids	1585.8 bar, 2 passes	Alkaline extraction (0–0.4 M NaOH, 40 °C, 0–24 prior to HPH)	Significant increase in extracted phenolic acids with HPH treatment	Zhu et al. (2016)
Broccoli seeds	Sulphoraphane	200–1600 bar, 1–5 passes	Determination of sulphoraphane in aqueous extract	Threefold increase in sulphoraphane content of HPH treated samples	Xing et al. (2019)

antioxidants into the medium. Some authors have also reported increased extraction yields in less water soluble compounds such as lycopene, rutin, quercetin and water insoluble alkaloids (Jurić et al. 2019; Gali et al. 2020). This can significantly reduce the use of organic solvents for the extraction or even eliminate it altogether.

4.4 *HPH Prior to Conventional Dairy Processes*

The effect of HPH on particle size reduction has a profound effect on milk structure and functional properties. During HPH treatment, the size distribution of fat globules is shifted towards lower sizes, which contributes to the stability of milk against creaming without significantly affecting the free fatty acid profile of treated milk (Rodríguez-Alcalá et al. 2009). Evidence also exists that homogenization has an effect on the structure of both caseins and whey proteins. The partial unfolding of micellar caseins has been proposed to alter their hydrophobicity, leading to increased solubility (Han et al. 2020; Sandra and Dalgleish 2005; Guerzoni et al. 1999). The beneficial effects of HPH on caseins has been reported for treatment pressures below 2000 bar, while at higher pressures aggregation of the micelles occurs leading to an increase in their size (Roach and Harte 2008; Escobar et al. 2011). The effect of HPH on whey proteins has been mainly reported for lactalbumins, lactoglobulins and immunoglobulins (Massoud et al. 2016; Serra et al. 2008, 2009). Partial denaturation of these proteins leads to their retention during cheese curd formation thus preventing their loss in the cheese whey (Vannini et al. 2008; Zamora et al. 2007). This increases the cheese yield of HPH treated milk. Other effects that have been reported are the increased binding capacity of whey proteins to caseins and milk fat globules (Massoud et al. 2016). Serra et al. (2009) also reported decreased syneresis, increased water holding capacity and firmness in set-type yoghurt produced from HPH treated milk (2000 and 4000 bar) but concluded that the process is not suitable for stirred-type yoghurt production, due to the disruption of the protein network. The effects of HPH on the structure of all milk proteins can also lead to an increased moisture retention in the cheese curd, contributing to the increased cheese yield (Vannini et al. 2008; López-Pedemonte et al. 2006; Guerzoni et al. 1999). Zamora et al. (2007) also reported improved rennet coagulation behaviour of milk treated at 2000 and 3000 bar, attributed to both protein denaturation and particle size reduction. The effect of HPH on the structure of milk proteins is mainly responsible for the formation of a denser protein network which in turn leads to the production of firmer gels (Guerzoni et al. 1999; Massoud et al. 2016; Serra et al. 2009; Hernández and Harte 2008). Generally, HPH improves the yield, increases the moisture content and the firmness of the produced cheese.

Apart from the effect of HPH in curd formation immediately after treatment, the process has also been shown to improve the sensorial characteristics of cheese during ripening. This has been attributed mainly to two distinct mechanisms: the effect of HPH on lipolysis and proteolysis and the effect of HPH on the microbial population of treated milk (Massoud et al. 2016; Lanciotti et al. 2007; Vannini et al. 2008; Juan et al. 2015; Guerzoni et al. 1999). Processes crucial to ripening such as lipolysis and proteolysis are enhanced in part due to milk fat size reduction (and therefore increased accessibility to microbial lipases) and due to protein denaturation which increases their susceptibility to proteolysis. Enhanced lipase activity promotes the production of free fatty acids and contributes to accelerated ripening and improved sensorial characteristics. Several authors have reported the shift in the microbial makeup of the milk before cheese ripening. The varying sensitivity of microbes during HPH treatment contributes to the shift in the microbial makeup that contributes to ripening. It has also been suggested that the increased production of free fatty acids and amino acids contributes towards the sustenance of probiotic microflora.

4.5 Enzymatic Treatment Assisted by HPH

The effect of HPH treatment on protein and polysaccharide structure enables the process to be used as a pretreatment prior to enzymatic hydrolysis. Protein and polysaccharide unfolding can enhance the accessibility of macromolecular substrates to hydrolytic enzymes and promote the kinetics of hydrolysis. This also has

Table 9.6 Applications of HPH combined with enzymatic treatment

Target system	Enzymatic treatment	HPH conditions	Key findings	References
Bovine serum albumin (BSA) Whey Protein Isolate (WPI)	Hydrolysis by trypsin and α-chymotrypsin	1000–2000 bar, 1–5 passes	Changes in BSA conformation led to more efficient hydrolysis Changes in WPI structure impaired hydrolysis due to formation of disulfide bridges upwards of 200 MPa	Carullo et al. (2020)
Soybean protein isolate (SPI)	Hydrolysis with pepsin	300 bar, 1 pass (post treatment)	Increased protein solubility and foaming capacity for combined treatments Formation of flexible aggregates	Yuan et al. (2012)
Soybean protein isolate (SPI)	Hydrolysis with bromelain	100–900 bar 3 passes	Increased degree of hydrolysis for combined treatment Increased antioxidant activity of resulting hydrolysates Reaggregation upwards of 5 passes at 300 bar	Zhao et al. (2018)
Crayfish shell wastes	Hydrolysis with chitinase	200–800 bar, 1–9 passes	Dispersion of chitin into a netted structure in suspension Increased degree of hydrolysis due to treatment	Wei et al. (2017)
Brown rice bran	Hydrolysis with cellulase	1200 bar, 3 passes (post-treatment)	Improved functional properties of rice bran dietary fiber as a result of combined enzymatic and HPH treatment	Xie et al. (2019)
Sugarcane bagasse	Hydrolysis with xylanase	1034 bar, 30 passes (post-treatment)	Reduction of bleaching chemicals for the production of cellulose nanofibrils	Saelee et al. (2016)

a profound impact on the functional properties of the produced systems. Table 9.6 summarizes reported synergies between HPH and enzymatic treatment.

In most reported cases, an HPH treatment prior to enzymatic hydrolysis of proteins was able to increase the process efficiency. This is mainly attributed to the partial unfolding of proteins and changes in secondary and tertiary structures. However, for certain proteins and treatment conditions studied, reaggregation of proteins was reported, leading to a decrease in the efficiency of hydrolysis. Carullo et al. (2020) reported increased hydrolysis (trypsin and chymotrypsin) rates of bovine serum albumin treated at 2000 bar for 1 pass but the opposite for whey protein isolate. This was attributed to the formation of disulfide bridges in the latter case, which caused structural compaction and impaired the access of the enzymes to the proteins. A similar trend was reported for the hydrolysis of soy protein isolate by bromelain following an HPH treatment at 300 bar for 5 passes (Zhao et al. 2018).

The effect of HPH on the unfolding of polysaccharides can also assist with their enzymatic treatment. Wei et al. (2017) reported significantly increased hydrolysis of crayfish shell chitin to acetylglucosamine by chitinase. HPH treatments at 200–800 bar caused significant conformational changes in the typically compact structure of chitin, leading to a dispersed, netted structure. This increased dispersion of the polysaccharide also caused the improved access of the enzyme in the polymeric matrix.

The combination of HPH with enzymatic treatment has also been reported to be effective when used in reverse i.e. when HPH follows enzymatic treatment. Xie et al. (2019) reported improved functional properties of rice bran dietary fiber by treating it with HPH after the application of cellulase. Similarly, Saelee et al. (2016) combined HPH with a xylanase pretreatment to reduce the amount bleaching chemicals required for the production of cellulose nanofibrils from sugarcane bagasse. Panyasiri et al. (2018) also applied HPH for the formation of cellulose nanofibrils from cassava bagasse pretreated with amylase for the removal of starch.

5 Cold Plasma Technology

5.1 General Aspects and Equipment

Cold Plasma (CP) is an innovative technology which has gained significant scientific interest over the last decade in the agro-food sector as a promising processing technology for food preservation, maintaining food safety and minimally affecting the quality attributes of foods. The main advantages of this technology can be summarized in its effectiveness in microbial inactivation, including pathogens, spoilage fungi and bacterial spores, its simple design and ease of use, cost-effective operation, short processing times of up to few minutes, lack of toxic effects, and significant reduction of water consumption (López et al. 2019).

Generally, it is recognized that matter exists in three natural forms, i.e. solid, liquid, and gas, though a fourth form of matter also exists, namely plasma. Plasma is referred to a mixture of different species comprising of ions – atoms or molecules which have one or more orbital electrons stripped (or, rarely, an extra electron attached), and free electrons, carrying a net neutral charge (Thirumdas et al. 2015). In order plasma to be produced, it is necessary that energy should be supplied to a gas to cause its ionization. Based on the thermodynamic equilibrium, plasma can be classified into two groups, thermal and nonthermal/cold plasma. Among these two groups, scientific interest is focused on cold plasma for food applications since the kinetic energy acquired by electrons from the external electric field is not passed onto the massive species, thus they prevail in a non-equilibrium state (Shaghaleh et al. 2019). In the food processing sector, plasma applications are related to the extension of the shelf life of food products, denaturation of proteins and inactivation of food enzymes that causes quality degradation, structural modification of food and packaging material, as well as degradation of pesticide residues in the horticultural fresh produces (Gavahian et al. 2020).

Nonthermal plasma can be generated by radio frequency power sources such as cold atmospheric pressure plasma (CAPP), radio frequency-discharge atmospheric pressure glow plasma, and one atmosphere uniform glow discharges plasma (Kim et al. 2011). Nonthermal plasma at low pressure is typically generated by an electric discharge in a gas or by microwaves. Generation of CAPP includes corona discharge, dielectric barrier discharges, radio-frequency plasma, and the gliding arc discharge (Guo et al. 2015). However, the most preferred treatments in food processing include dielectric barrier discharge (DBD) and jet plasma (Fig. 9.3), due to their simple designs, the availability of configurations, and ease in handling (Pankaj et al. 2015).

Plasma can be generated by various types of gases, simple such as air or nitrogen, or a mixture of noble gases, i.e. helium, argon, or neon. However, it is showed that the efficacy of operating gas is improved when oxygen is added to the mixture used (Scholtz et al. 2015; Han et al. 2016). As a result of gas generation, several reactive species, e.g. O, O3, OH, NO, NO2, and OH are produced, which subsequently can react with cell components, leading to either food modification and/or microbial inactivation and decontamination, via a mechanism that includes the ionization of neutral, ground-state atoms, molecules, or radicals by electron impact (Mir et al. 2016).

In the following sections, the synergies of CAPP application with conventional food processes and quality preservation procedures in food products will be discussed, mostly focusing on extraction and packaging technologies, that represent the commonest research and industrial applications of CAPP.

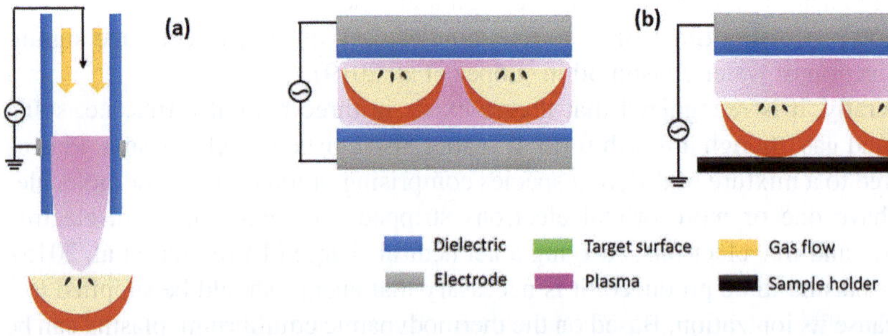

Fig. 9.3 Schematic presentation of (**a**) jet plasma, and (**b**) dielectric barrier discharge (DBD) equipment

5.2 In-Package Cold Plasma

Cold plasma has been recognized as an antimicrobial agent comprising a variety of reactive chemical species. These reactive species are effective in inactivating a wide range of microorganisms, including bacteria, fungi, spores and viruses, as well as pesticides and mycotoxins. In-package cold plasma, that can be considered as a combined technology of CAPP and active packaging, allows to localize and extend the action time of reactive species against microorganisms, while eliminating any post-process contamination.

The concept of in-package plasma is that plasma generates localized reactive species into the package where they can directly act with the food matrix to be disinfected. For this purpose, food products are packed usually in plastic/polymer and rarely in glass containers where the headspace is occupied either with air or a modified atmosphere, and then the package is sealed. When the package is exposed to an electric field, as in the case of plasma, then the gas inside the package is ionized, thus reactive species are generated and consequently act as antimicrobial agents, while the packaging is not affected (Park et al. 2018). The reactive species have a certain long-life up to few hours, and present high diffusivity coefficients that promote the uniformity of the treatment. Then, as there are unreacted plasma species inside the package, they contribute to the transformation of the gas into its original concentration, depending on several parameter, e.g. the energy of the species, temperature and pressure, package characteristics (type, size, volume), the food matrix, etc. As a result, the food product is stored under the same atmosphere (air or modified) as the one initially been packed. Among the different in-package plasma been configurated, the most commonly used include (a) volumetric DBD plasma, (b) surface DBD (SDBD) plasma inside the package, and (c) on package SDBD plasma (Misra et al. 2019), as shown in Fig. 9.4.

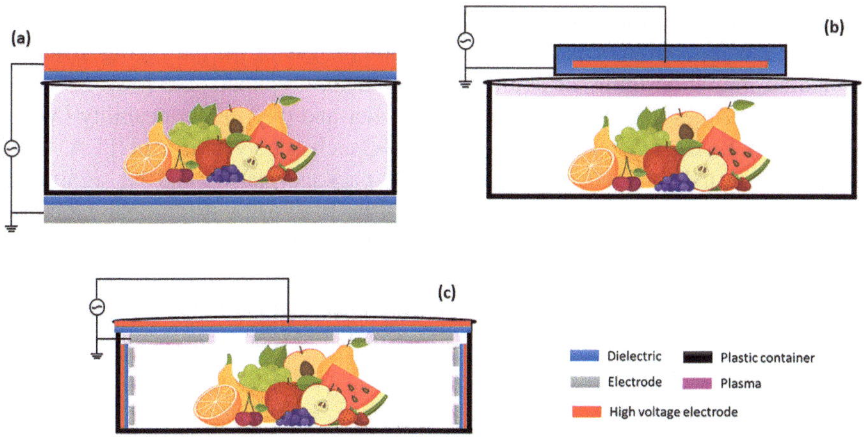

Fig. 9.4 Schematic presentation of (**a**) in-package DBD plasma, (**b**) SDBD plasma inside the package, and (**c**) on package SDBD plasma configurations

As expected most of the studies conducted in in-package plasma concern the inactivation of food pathogens and spoilage microorganisms. The mechanism of the inactivation is related to cell damage due to plasma ions and cell interactions and is associated with the direct oxidative effects on the outer surface of microbial cells (Thirumdas et al. 2015). Min et al. (2016) applied DBD plasma at 35 kV for 5 min in romaine lettuce packed in nylon/PE packaging with 30 mm gap and either air or mixture of N_2-O_2 and found that the counts of E.coli, Salmonella spp., L. monocytogenes and TV-virus exhibited a reduction of 4.5–5.5 log cycles. Similar results were obtained under the same processing conditions when romaine lettuce was packed in PET (Hertrich et al. 2017). When mackerel or herring was packed in PET/film and DBD plasma at 80 kV for 5 min was applied, a reduction of 4.0–4.5 log cycles was observed for both total psychrotrophs, lactic acid bacteria, Pseudomonas spp. and Enterobacteriaceae (Albertos et al. 2017). SDBD inside package plasma has been applied at 250 W for 10 and 15 min in whole milk (Kim et al. 2015) and cheese (Yong et al. 2015), respectively, packed in plastic containers with air and 53 mm, where a reduction by 3.5–4.0 log was observed for E. coli, S. Typhimurium, and L. monocytogenes. Diver and Potts (2015) investigated the effect of in package SDBD plasma applied at 8 kV for half a minute on the inactivation of microorganisms in chicken packed in PE with either air or a mixture of CO_2-O_2 and even though Enterobacteriaceae, E. coli and Campylobacter were found below the detection limit (reduction by 4.0, 3.0, and 2.5 log cycles, respectively) after processing, total aerobes exhibited only a slight reduction by 1.7 log cycles from their initial level of 6.0 log cycles.

It is evident that the inactivation of microorganisms under in-pack conditions have several advantages, including the reduction of post-process contamination, reduction of the spoilage microflora of the packaging material thus less need for disinfection of the packaging materials, use of several types of packaging materials and modified gas atmospheres inside the package, as well as enhancement of the release of antimicrobials and bioactive compounds incorporated to the packaging materials. However, for sake of safety and quality, it is essential to be investigated whether the generation of plasma may lead to changes of the packaging material, including migration limits of additives, monomers, oligomers and low molecular weight substances into the food, as well as water and oxygen permeability (Misra et al. 2019). Biodegradable packaging materials, such as polylactic acid (PLA) films have been tested for their applicability in in-package plasma and no significant changes has been observed (Pankaj et al. 2014a, b, c, d, e). Additionally, natural antimicrobials have been successfully incorporated into biofilms and plasma has been reported to effectively lead to release of the bioactive compound (Pankaj et al. 2017). Therefore, it seems that if appropriate industrial configurations are developed, in-package plasma may be efficiently adapted in food manufacturing.

5.3 CAPP-Assisted Extraction Technology

Apart from the inactivation of food pathogenic and spoilage microflora, in scientific level, CAPP technology is commonly used for polymers' modification and functionalization since it has been recognized that modulates surface wettability and roughness. These characteristics are known to be directly related to the migration tendency of their containing substances from the interior to the surface during extraction process (Van Deynse et al. 2016).

Scientific findings have demonstrated that the application of CAPP lead to modification of both availability and antioxidant capacity of bioactive compounds. Bursać-Kovačević et al. (2016) studied the effect of CAPP on anthocyanins and colour in pomegranate juice, in relation with CAPP treatment time (3, 5, 7 min), sample treated volume (3, 4, 5 cm3), and gas flow (0.75, 1, 1.25 dm3/min), and found that the greatest anthocyanin stability was observed after 3 min, when 5 cm3 sample volume was processed in 0.75 dm3/min gas flow, yielding in higher anthocyanin content from 21% to 35% compared to untreated samples, followed by increase in the total colour of the CAPP-treated sample. Won et al. (2017) also indicated that CAPP treatments increased the bioactive profile of whole mandarin fruits or peels when treated at 0.7 kPa using a microwave CPT system, in a study where different plasma-forming gases (N_2, He, N_2-O_2 mixture of 4:1), variable plasma generation power (400, 650, and 900 W), and different process time (2, 5, and 10 min) were used. They concluded that nitrogen (N_2)-CPT at 900 W for 10 min significantly increased the total phenolic content and antioxidant activity of mandarin peel after the treatment and during storage at 4 and 25 °C.

In a recent study, Rashid et al. (2020) applied high voltage atmospheric cold plasma as a pretreatment during extraction of galactomannans from dry Fenugreek seeds and soaked in NaCl solution Fenugreek seeds, and investigated the physicochemical properties of the extracted polysaccharide. It was observed that plasma pretreatment with air at 80 kV for 30 min caused apparent structural disruption on seeds surface and decreased the pH of extracting solution, resulting in increased galactomannan extraction yields, by 67 and 122% from dry and soaked seeds, respectively. The polysaccharide treated with cold plasma exhibited improved attributes of water-binding capacity, swelling index and viscosity, and lower melting enthalpy. Cold plasma pretreatment also led to modifications on the surface morphology of galactomannan due to plasma etching, yet no significant changes in the molecular and crystalline structures of the polysaccharide were observed.

In addition, Bao et al. (2020a) studied the effect of high voltage atmospheric cold plasma with different working gases (air, Ar, He and N_2) on the modification of the surface of tomato pomace, aiming to facilitate the migration of phenolic compounds during extraction, and consequently increase their extraction efficiency. In addition to rupturing the epidermal cells of tomato pomace, cold plasma pretreatments found to decrease the water contact angle of tomato peels and accelerate the drying of tomato fruit, resulting in the enhancement of the hydrophilic character of the fruit surface. Tomato pomace treated by He- and N_2-generated plasma exhibited slight, yet significant, increased extraction yields of phenolic compounds by approximately 10%. Cold plasma assisted extraction led to increased antioxidant capacity of tomato pomace extracts, while slightly altered the concentration profile of their phenolic compounds. In another study, they found that the application of high voltage (60 kV) atmospheric cold plasma as a pretreatment on grape pomace for different processing times (5, 10 and 15 min) disrupt the epidermal cell structures of grape pomace, reduce the water contact angle of grape peels, and accelerate grape drying, and the effects became more intent as cold plasma treatment extended (Bao et al. 2020b). The application of cold plasma also increased the extraction yield of phenolic compounds, by 10.9–22.8%, and the extract contained a higher concentration of anthocyanins exhibiting an enhanced antioxidant capacity by 16.7–34.7%, indicating an improvement of the bioactive profile of the extract.

A method combining CAPP treatment and ultrasound to extract phenolic compounds from sea asparagus Salicornia neei for application in Italian salami has been reported by Faria et al. (2020). The authors investigated the effect of the glow discharge cold plasma pretreatment on the ultrasound-assisted extraction (UAE) of phenolic compounds from sea asparagus and they found that CAPP pretreatment (discharge power of 14 W for 5 min) prior to UAE increased the antioxidant activity by 22% and 19% measured by the DPPH and ABTS assays, respectively. Furthermore, when the extract was added to salami formulations, apart from a slight colour change, lipid oxidation and texture parameters were similar to the control samples at the end of the ripening period.

5.4 CAPP-Assisted Biomass Hydrolysis

Bioethanol is one of the renewable energies and is produced by fermentation of biomass such as corn, sugarcane, and sugar beets. It can be classified into different generations according to different kinds of raw materials, for example (i) first generation biomass which is mainly from food crops, and (ii) second-generation bioethanol using agricultural byproducts such as sugarcane bagasse, wheat straw, and wood chops as substrates (Robak and Balcerek 2018). Those byproducts are mainly lignocellulosic materials that are composed of cellulose, hemicellulose, and lignin, and a pretreatment process in order to loosen the structure and release cellulose prior to hydrolysis and fermentation is usually needed (Jin et al. 2020). Nevertheless, during these pretreatments, several byproducts that may inhibit subsequent fermentation efficiencies are generated, such as furfural and hydroxymethylfurfural (HMF), formic acid and acetic acid, and phenolic compounds, all of which exhibiting toxicity due to either penetration of cell membranes and cell damage (Klinke et al. 2004), disruption of the ion balance within cells (Liu et al. 2019), and interaction with enzymes to reduce the process efficacy (Qin et al. 2016). As a result, further detoxification process should be conducted to enhance ethanol productivity in order to remove or degrade toxic compounds. CAPP-assisted enzymatic hydrolysis for the production of bioethanol has gained considerably research interest as an alternative to degrade toxic compounds of the used biomass.

Shaghaleh et al. (2019) investigated the effect of CAPP pretreatment on the enzymatic hydrolysis of wheat straw, applying different plasma gases (air, N_2 and CO_2) onto wet substrate environment for varying process times. The obtained results indicated that the composition, structure, and surface of the straw were considerably changed through CAPP pretreatment, promoting enzymatic hydrolysis. In the optimal conditions (90 s, air plasma), high biomass/sugars recoveries as well as high conversion yield of glucan and xylan were achieved, with the overall production accounting 358.4 and 135.8 mg·g-1, respectively, following enzymatic hydrolysis for 24 h. Considering the feasibility of CAPP at the optimal conditions with subsequent enzymatic hydrolysis to glucose and xylose production process of wheat straw, CAPP pretreatment showed a positive impact on the energy balance of the process and low processing cost, demonstrating the significant potential of this technology to be scaled to industrial-level in the biorefineries.

In a recent study, CAPP was investigated as an alternative to degrade the toxic compounds within sulfuric acid-hydrolyzed sugarcane bagasse. After CAPP treatment, significant reduction in toxic compounds (31% of the formic acid, 45% of the acetic acid, 80% of the hydroxymethylfurfural, and 100% of the furfural) were observed (Lin et al. 2020). After adopting optimal CAPP conditions (200 W power for 25 min), the bioethanol productivity improved from 0.25 to 0.65 g/L/h, indicating that CAPP-assisted enzymatic hydrolysis could effectively degrade toxic compounds within the hydrolysate, thus enhancing bioethanol production.

6 Sonication

6.1 General Aspects

In the food industry, ultrasound is applied in food processing by either direct exposure or using an instrument such as ultrasonic water bath. Ultrasound technology uses sound waves with a frequency range greater than 20 kHz. Sound waves applied through a medium i.e. liquid or solid food matrices generate compressions and decompressions in the particles producing high amounts of energy resulting in increased mass transfer phenomena. Ultrasound is among the emerging food technologies studied the last decade with regards the enhancement and efficacy of several food process operations (Leong et al. 2017). Ultrasound applications (low and high-intensity ultrasound combined with temperature and pressure have a synergistic effect.

Ultrasound processing spans over a wide range of acoustic frequencies, starting as low as 20 kHz, up to several MHz (Knoerzer 2016). High-intensity ultrasound uses lower frequencies between 20 and 100 kHz and has applications in emulsification, cleaning, and extraction, improved drying, and defoaming. Low-intensity ultrasound (megasonics) in higher frequencies (>400 kHz) are used as a nondestructive tool to evaluate the composition, structure or flowrate of foods.

The application of low-frequency high intensity ultrasound waves (20–100 kHz) impose significant changes in the mechanical and biochemical properties of foodstuff. Basic food processes including drying, freezing, concentration, filtration, enzyme/microbial inactivation, extraction coupled with ultrasound (Bhargava et al. 2021) can have beneficial effects with regards process efficiency and product quality. Thus US could be applied in almost all food industries covering categories of animal and plant origin products (Table 9.7).

Recently, Khouryieh (2021) published the results of a survey conducted to U.S. food experts to study the major reasons for using novel technologies, the limitations for not implementing specific technologies, and the main drivers for innovation of nonthermal food processing technologies. When the participants (N = 205) were asked to select the nonthermal food processing technologies that they had already been implementing in their organizations, ultrasound technology ranked 5th (3.4%) while high pressure (35.6%) and pulsed electric fields (20.0%) were the top two widely used novel nonthermal technologies.

It has been evidenced that the innovative technologies impact on the physicochemical and functional properties of food components depending on processing parameters and conditions. It is shown that low frequency/high energy ultrasound from 20 to 100 kHz, especially 20, 25, and 40 kHz, are preferred for modification of food components, according to cavitation effect and settings of the instrument. Moreover, it is indicated that ultrasound with frequency of 20–25 kHz is capable to

Table 9.7 Ultrasound applications in food

Category	Products	Effect	Reference
Meat	Chicken, Beef, Pork	Texture modifications	Caraveo et al. (2015), Wang et al. (2018)
Fruits & vegetables	Juices, Purees	Enzyme and microbial inactivation	Chen et al. (2019), Azam et al. (2020)
Cereal	Flour, Batters	Textural and rheological properties	Fox et al. (2004). Zhu and Li (2019)
Dairy	Milk, Yoghurt, Cheese, Ice cream	Milk fat crystal size reduction, cheese yield increase, yoghurt/ice cream viscosity increase	Guimarães et al. (2019), Akdeniz and Akalin (2019), Gregersen et al. (2019), Carrillo-Lopez et al. (2020)
Emulsions	Mayonnaise, Dressings	Emulsion stability enhancement	Carpenter and Saharan (2017); Albano and Nicoletti (2018)
Beverages	Beverages	Degassing	Knoerzer (2016)
Herbs & neutraceuticals	Tea infusions, herbal extracts, bioactive compounds	Increased extraction yields, extraction time reduction	Vinatoru (2001), Albu et al. (2004), Xia et al. (2006)

regulate the particle size and structural conformation of proteins and polysaccharides, further optimizing their functional characteristics in food systems.

Published studies on ultrasound applications are focused on exploring the efficacy of ultrasound in the enhancement and improvement of a wide range of food processing techniques in the food industry. However, broad commercial adaptation of ultrasound is yet to come with the main obstacle to be the required high amounts of energy. Therefore, commercialization and industrialization of ultrasound should be specifically focused on the implementation on an industrial scale. The synergic effect of ultrasound with treatments such as osmotic dehydration, blanching, PEF, extraction, freezing is being extensively studied the last years (Table 9.8).

6.2 *Sonication Synergies with Other Food Processing Technologies*

Recently studies have been published focusing on the potential of ultrasound applications to produce food products with high nutritional value and better quality and physicochemical properties (Gallo et al. 2018). The key principle of ultrasound is that it favors mass and energy transfer processes. Thus dehydration and drying

Table 9.8 Benefits of ultrasound assisted food process operations

Process	Benefit	Reference
Drying	Increased drying rates, shorted processing time	Zhang et al. (2020)
Osmotic dehydration	Mass transfer acceleration, functional compounds retention, increase dehydration rates	Nowacka et al. (2021)
Freezing	Increased freezing rates	Tian et al. (2020), Qiu et al. (2020)
Extraction	Reduced extraction time, increased extraction yields	Dzah et al. (2020), Yang et al. (2017), Carrillo-Hormaza et al. (2020)
Filtration	Increased permeation	Naji et al. (2020), Aliasghari et al. (2015)
Thawing	Process time reduction	Li et al. (2020)
Foaming	Enhanced foaming capacity	Morales et al. (2015), Sheng et al. (2018)
Fermentation	Reduced fermentation time	Carrillo-Lopez et al. (2021), Munir et al. (2019), Barukcic et al. (2015), Zhang et al. (2020)
Emulsification	Improved emulsion stability, improved the emulsifying capacity	Carpenter and Saharan (2017), Sarheed et al. (2020), Li et al. (2020)
Pasteurization	Reduced pasteurization time enhanced temperature homogeneity	Baboli et al. (2020), Scudino et al. (2020), Vercet et al. (2002)
Chemical contaminants removal	Better decontamination effect	Yuan et al. (2021)

processes, the most commonly used techniques for food preservation, are processes in which ultrasound can reduce the loss of bioactive compounds and improve the colour of dehydrated products (Llavata et al. 2020). In the case of freezing, in addition to reducing processing times, US favors the formation of small ice crystals that, when thawed, reduce the loss of water, resulting in a product with high textural properties (Qiu et al. 2020; Cheng et al. 2015).

Numerous studies have focused on the study of US-assisted drying of fruits and vegetables and its effect on the physical (water activity, shrinkage, rehydration, colour, porosity, among others) and chemical (nutrients, antioxidants, vitamins) quality of the dried product. The combination of ultrasound especially with osmotic dehydration has been studied by many research groups for a wide range of plant tissues where ultrasound is used either as a pretreatment processing step, prior to osmotic dehydration, or used to assist during the osmotic dehydration. Ultrasound cavitation phenomenon and compression/ decompression of plant tissues lead to microchannels creation further facilitating the mass transfer between the product and osmotic solution. Ultrasound assisted osmotic dehydration have been proved to accelerate dehydration rates leading to increased water loss and solids gain while improving physicochemical properties and quality attributes of the final dehydrated products. The positive effect on physical properties of osmo-dehydrated plant products with the assistance of ultrasound has been demonstrated in a wide range of fruits and vegetables including blueberries (Spinei and Oroian 2021), cranberries (Nowacka et al. 2017), carrots (Alizehi et al. 2020), apples (Amanor-Atiemoh et al. 2020) and bananas (Farhaninejad et al. 2017), among others. Apart from enhancing mass transfer phenomena and increasing dehydration rates the combination of US with osmotic dehydration has been found to favor functional compounds (polyphenols, vitamins, minerals and fibres) retention.

Shape and distribution of ice crystals is of outmost importance with regards quality and post processing stability of frozen food products. Immersion freezing in ultrasonic baths can allow rapid freezing and the formation of small and numerous ice crystals within the food matrix. Many studies have shown that ultrasound assisted freezing promotes the initiation of nucleation, thus leading to an acceleration of mass and heat transfer, improving key quality attributes of frozen food, i.e. weight loss, texture and microstructure. Ultrasound has been applicable in the crystallization of food products such as ice cream and milk fat.

Ultrasound application enhances the foaming properties of proteins due to decreased particle size, better volume, and increased stability (Higuera-Barraza et al. 2016). Egg white's foaming capacity improvement has been reported (Sheng et al. 2018) while foaming properties were mainly influenced protein structure variations. Increased foaming capacity and stability has been also reported in soy protein. Recent studies have also shown that ultrasound affect serum proteins functional properties, foaming capacity and solubility. Ultrasound combined with temperature in the case of soy protein isolate led to a modification in protein isolate particle size that was linked with improved foaming capacity but with no significant effect on foam stability (Morales et al. 2015).

Ultrasound assisted thermal pasteurization has been found to be less detrimental to nutrients, protein content and functional properties of foods when compared to conventional thermal processing while being effective against vegetative cells. Ultrasound in combination with temperature and pressure has been found to be effective in inactivating thermotolerant enzymes like polyphenoloxidases, peroxidases, lipooxiginase and pectin methylesterase. Recent studies have shown that ultrasound combined with thermal treatment can inactivate the Clostridium perfringens spores in the slurry of beef. Ultrasound can also reduce the required time of heat treatment as it accelerates the pasteurization/sterilization rate of food. Most microorganisms show sensitivity when ultrasound is combined with temperatures greater than 50 °C. Since, the combination of US with lower treatment temperature is effective at low temperatures this leads to improvement of the quality of the heat treated product. US assisted pasteurization at 50 °C can preserve food sensorial attributes such as colour and flavour compounds in comparison to conventional pasteurization techniques that involve higher temperatures. US assisted thermal treatment effects improve with increasing intensity and decrease with increasing frequency.

Ultrasound-assisted extraction is an effective alternative that overcomes the drawbacks of conventional extraction techniques. Ultrasonic implosion and cavitation rupture the cell walls, enhancing the mass transfer from solid to liquid phase leading to higher extraction yields. Also, within the tissues, microchannels are created on an ultrasound application which enhances the solvent penetration into the solid matrix which increases mass transfer. Thus, Ultrasound aided extraction contributes to efficient recovery of compounds in shorter extraction time, requiring less solvent volumes. Ultrasonic assisted extraction is usually performed under continuous wave mode and pulse mode technique is employed over long term extraction. It is generally preferred for the extraction of bioactive compounds due to its versatility, easy operation, the scope of industrial application, with an ability to use less solvent and retain the biological activity. However, the extraction efficiency is greatly affected by ultrasonic power, frequency, solvent, and the matrix to solvent employed.

Ultrasound has been found to enhance the processes when used in conjunction with various unit operations in the food industry (Table 9.8). Filtration has long been used as an important process in the food industry for efficient separation such as solids from their mother liquor or in the production of solid free liquid. The problems of fouling or concentration polarization caused by the deposition of the filtrate or filter cake on the membrane surface is a major issue in this process. These problems cause a reduction in filtration efficiency. However, ultrasound energy is effective against this issue. On the application of ultrasound during the process of filtration, the retentate layers accumulated on the membrane surface causing concentration polarization, are disrupted while the membrane's intrinsic permeability remains unaffected. This results in an increase in flux and a decrease in flow resistance. Acoustic filtration has been effectively used for the enhancement of industrial wastewater filtration. Ultrasound, when combined with filters, enhances the life of the filter, by preventing the caking and clogging of the membrane, enabled by continuous cavitation at the surface of the filter.

Ultrasound has proven its abilities in the food industry into preservation, extraction, and processing. Ultrasound is being increasingly used to enhance various processes in the food industry and has become an extremely promising technology on the processing front. However, the combination of ultrasound with other techniques generates better results on the overall quality of the final product and could be the focus of further research. The further development for the application of ultrasound on an industrial level requires the optimization of parameters and ground-level research to analyze the effect of acoustic treatment on the bulk production of food (Bhargava et al. 2021). Ultrasound with its abilities to increase efficiency and reduce the time required for various processing operations has promised a progressive future with a positive environmental impact. The food industry is in a phase where the focus has shifted to the development of novel processing and preservation technologies while maintaining the economic feasibility. However, need for upscaling at an industrial level would require proper modelling and analysis for improvement in design variables required to offer novel and additional opportunities in the field of liquid food processing. Ultrasound techniques in combination with other nonthermal processes has shown a tremendous potential in terms of efficiency and effectiveness of the process operation. It could be summarized Ultrasound is a processing technology of great significance especially for liquid foods, however there is a need to carefully develop it and carry process optimization for each specific process.

7 Continuous and Pulsed UV-Light Processing

7.1 Introduction

Pulsed light (PL) technology is among the emerging nonthermal technologies explored to destroy both pathogens and spoilage microorganisms (e.g. bacteria, yeasts, molds and viruses) in several food products. Short duration and high power light pulses used in PL technology trigger photochemical, photothermal and photophysical mechanisms that lead to microbial inactivation (Wang et al. 2005). While, PL is effective in killing bacteria it minimally affects food quality attributes making it a good candidate minimal processing technology in the food industry. PL treated food products have extended shelf life with superior nutritional and quality properties when compared to conventional thermal treatments. In most studies PL is proposed to be used in combination with other technologies in the framework of hurdle technology.

Recently, Khouryieh (2021) published the results of a survey conducted to U.S. food experts to study the major reasons for using novel technologies, the limitations for not implementing specific technologies, and the main drivers for innovation of non-thermal food processing technologies. When the participants (N = 205) were asked to select the non-thermal food processing technologies that they had already been implementing in their organizations, pulsed light technology ranked 3rd (13.2%) following high pressure (35.6%) and pulsed electric fields (20.0%) which were the top two widely used novel nonthermal technologies. This ranking demonstrates a great potential of pulsed light in market penetration. The results also showed that pulsed light processing technology was primarily used to provide better nutritional/sensorial qualities, followed by solution of safety problems, shelf life extension and cost savings. Industrial PL equipment for the food industry are already available by commercial companies (Cacace and Palmieri 2014). However, PL is a surface decontamination technology for pathogens highly dependent on environmental factors such as type of food and quality of surface of food as PL treatment is more effective for the sterilization of surfaces rather than liquid media. Microbial inactivation using PL is limited to the surface of opaque products due to the decreasing light intensity as absorption and scattering are reduced when the light travels through the treated food matrix (Abida et al. 2014). A very interesting aspect of PL is the in-pack application that enables the treatment in already packaged food products, assuring no further cross-contamination.

PL technology uses light in the form of a single pulse, a burst of pulses (timed mode) (1–20 Hz) and a pulse width of 300 ns to 1 ms, or a continuous array of pulses or in random sequences in continuous mode (John and Ramaswamy 2018). Irrespective of the number and duration of the pulses, the power provided by the pulses is estimated to be at least 20,000 times more intense than that provided by a continuous light radiation of equivalent total energy (Keener and Krishnamurthy 2014).

7.2 PL Synergies with Other Food Preservation Technologies

PL technology efficacy in microbial inactivation can be enhanced with the combination of other mild preservation technologies such as the use of washing in the case of fruits and vegetables, the use of edible coatings in fresh cut fruit. Recently the combination of PL with ultrasounds has been proposed for the treatment of liquid food products. In the case of apple juice (Ferrario and Guerrero 2017), the combination of PL was identified as an alternative for inactivating S. cerevisiae KE 162 while it resulted in an up to 6.4 log reduction in yeasts.

PL has potential applications for the treatment of foods that require a rapid disinfection. Foods with smooth surfaces such as fresh whole fruit and vegetable, hard cheeses, or smooth surface meat slices are suitable for treatment with PL where surface contamination is a concern for microbial contamination.

The significant microbial reductions in very short treatment times, the limited energy cost of PL, the lack of residual compounds, and its great flexibility are some of the major benefits of the technique (Oms-Oliu et al. 2010).

8 Ozonation Technology

8.1 Introduction

Ozone processing has been widely used in the food industry due to the following advantages of ozone: (i) spontaneous decomposition to oxygen without forming any hazardous residues on food surface, (ii) high oxidation potential, and (iii) wide anti-microbial spectrum (against gram-positive and gram-negative bacteria, fungi and yeasts, viruses, protozoa, bacterial and fungal spores (Brodowska et al. 2018; Cullen et al. 2009; Guzel-Seydima et al. 2004; Manousaridis et al. 2005; Rodoni et al. 2010). In 1997, ozone was recognized as GRAS (Generally Recognized as Safe), and in 2001, it was approved as a direct additive to food (U.S. FDA). Ozone has been regarded as an additive-free, safe, and cost-effective processing in the food industry. In gaseous or liquid form (as ozonated water), it has plenty of applications in the food industry such as disinfection, preservation or storage of food products, sanitation of plant equipment, sanitation of food packaging materials and other applications (air treatment, water treatment, industrial processes to reduce the use of chemical agents or pesticides on food). It is used for processing of food products such as raw fruit and vegetables, dried fruit and vegetables, fruit and vegetable juices, dairy products, grain products, meat and meat products, fish and seafood, and their products. It reduces the microbial load of the food products and extends their shelf life (Li et al. 2015).

8.2 Mode of Action, Mechanisms and Applications

Ozone, in gaseous or liquid form (as ozonated water), has been widely used for fruit preservation. Typically, it has been shown to inhibit the growth of spoilage bacteria and yeasts at concentrations of 0.15–5.0 ppm (Jay et al. 2005). Degradation of ozone produces free radicals (e.g. hydroxyl) leading to enzyme inactivation, nucleic acid alteration, and lipid peroxidation of the microbial membrane (Patil et al. 2011). Despite the known suitability of ozone for microbial inactivation, the effect of ozone on fruit quality is still scarce. Oxidation capacity of ozone can lead to quality loss of the food product depending on the dosage applied (Alwi and Ali 2014; Rodoni et al. 2010). While reports have confirmed the positive effect of ozone on bacterial growth of fruit juices, the impact on bioactive compounds should be investigated (Gibson et al. 2019; Torlak et al. 2013). Studies demonstrate that high ozone concentrations affect negatively the juice quality characteristics (the sensory characteristics and consequently the acceptability of the food product); on the other hand, low concentrations may put at risk the safety of juices (Akbas and Ozdemir 2006; Tiwari et al. 2010). Ozone has also used for washing of fruit and vegetables (e.g. leafy vegetables), as an alternative to chlorine (Fan and Sokorai 2015; Gibson et al. 2019; Gómez-López et al. 2013; Papachristodoulou et al. 2018). The resulting destruction or lysis associated with ozone processing is a faster inactivation mechanism than other disinfectants, which require the disinfectant to permeate through the cell membrane to be effective (Cullen et al. 2009). As it was expected, disinfectants including ozone, can cause some detrimental effects on sensory characteristics and nutritional value of the processed food product (Kenny and O'Beirne 2009) (Table 9.9).

Processes that combine disinfection treatments can be used to control microbial growth in fruit and vegetables and to retain their quality characteristics. Ozone processing has been combined with modified atmosphere packaging (MAP) to reduce the respiration rate and stimulate the antioxidant defense system of green peppers (Chen et al. 2016), and to control microbial decay and the accumulation of acetaldehyde in organic table grapes (Admane et al. 2018). Combined effect of a gaseous ozone pre-treatment and MAP (13 mg·m − 3 of gaseous ozone – about 6 ppm – for 16 h at 1 °C, and storage for 15 days under MAP – 10 kPa O_2 and 40 kPa CO_2- or in air at 4 °C) has shown to cause a significant reduction of the yeast and mold counts (compared to samples stored in air and MAP) while the main quality parameters of the three types of fruits (blueberries, raspberries, strawberries) were not affected by combined ozone pre-treatment and MAP storage. The combined effects of ozone and MAP led to a better preservation of the total and individual anthocyanins in the blueberries during cold storage. The exposure to gaseous ozone before packaging followed by storage under MAP proved to be a useful technological approach to extend the postharvest storage (Pinto et al. 2020).

The ozone application depending on the dose and post-drying of fruit by-products can make a significant contribution to the knowledge of alternatives for agricultural waste processing. Depending on the dose of ozone used, ozone can enhance the drying process, increase the diffusivity and improve the antioxidant capacity of the product, activating its protection system against oxidative stress. After ozone and convective drying treatments, a slight increase in oil yield of citrus by-products, and a reduction in microbiological contaminants present on the surface of the citrus peel was found (Bechlin et al. 2020). The use of ozone as a postharvest processing step before the partial dehydration of grapes for the production of dry and sweet wines in Italy was also reported. Ozone did not accelerate grape water loss but induced the antioxidant system and increased polyphenol content (processing conditions: ozone gas at 10 °C, in 70% relative humidity and air flow, for 12 h; kept at 20 °C for 12 h.; partial dehydration at 10 °C until a 30% weight loss was reached) (Modesti et al. 2018).

Ozone has widely applied in food grain processing (Chittrakorn et al. 2014; Ding et al. 2015; Gozé et al. 2016; Mei et al. 2016; Obadi et al. 2018; Tiwari et al. 2010). It largely eliminates insects, mycotoxins and pathogens in grain products (Tiwari et al. 2010). However, some functional properties of the grain products can be modified by ozone. For example, it was reported that ozone application on wheat kernels reduced the required milling energy, while decreasing the germination rate (Tiwari et al. 2010). More recently, the effects of ozone on the composition and properties of grain components (e.g. protein and starch properties) have been investigated (Gozé et al. 2016; Obadi et al. 2018). The functional properties of food products

Table 9.9 Representative recent research studies on the effect of ozone processing on the quality characteristics of fruit and vegetables or juices

Food product	Ozone processing conditions	Quality parameter	Conclusions	References
Banana, guava and pineapple	Flow rate: 8 ± 0.2 ml/s, Time: 0, 10, 20, and 30 min	Total phenol, flavonoid contents (FRAP, DPPH), Vitamin C content	Total phenol and flavonoid contents increased significantly when exposed to ozone for up to 20 min, with a concomitant increase in FRAP and DPPH values. The opposite was observed for guava. Ozone processing significantly decreased the vitamin C content	Alothman et al. (2019)
Blueberries, raspberries	Discontinuous daily gaseous ozone treatment (10 or 15 ppm)	Flavonoids, anthocyanins and ascorbic acid	Ozone processing reduced the loss of flavonoids, anthocyanins and ascorbic acid in blueberries and raspberries stored at low and room temperatures.	Piechowiak et al. (2019)
Kiwi fruit	Gaseous ozone concentration: 4 mg/h in the chamber at a temperature of 0 °C and a humidity of 90–95% (Storage conditions)	Vitamin C (ascorbic acid) content	No significant change in ascorbic acid content of kiwi fruit over a 7 month storage period.	Barboni et al. (2010)
Raspberries		Phenolic compounds profile and glutathione status	Ozone processing inhibited the loss of polyphenols and glutathione in fruit.	Piechowiak et al. (2020)
Strawberries	Gaseous ozone 18 mg L − 1	Fruit weight loss	Ozone processing reduced the yeast and mould loads and total mesophilic bacteria by 1–3.5 log cfu g − 1 after 6 days of cold storage, and the fruit weight loss by 60%.	Alves et al. (2019)

(continued)

Table 9.9 (continued)

Food product	Ozone processing conditions	Quality parameter	Conclusions	References
Strawberries	Processing conditions: 0.1 ppm aqueous ozone, Time intervals: 1–4 min). Storage conditions: 25 ± 2 °C – 45–50% RH and 2 ± 1 °C – 90% RH	Weight loss, firmness, colour	Ozone processing resulted in 21% lower weight loss, 16% higher firmness and 15% lesser change in fruit colour during 2 days in ambient storage. Under low temperature storage, 2 min ozone processed fruits exhibited ~21% lower PLW, 19% higher firmness and 46% lesser colour change as compared to control fruits during 14 days of storage.	Nayak et al. (2020)
Apple juice	10–48 mg O3 L – 1, Time: 10 min	Colour, rheological properties, phenolic content	Negative effects colour, rheological properties and total phenolics.	Torres et al. (2011)
Blackberry juice	Ozone concentration: 7.8% w/w, Time: 10 min	Cyanidin-3-glucoside content	The cyanidin-3-glucoside content was reduced (>90%).	Tiwari et al. (2008)
Strawberry juice	Ozone concentration: 7.8% w/w, Time: 10 min	Pelargonidin-3-glucoside content	The contents of anthocyanins (up to 98.2%) and ascorbic acid (up to 85.8%) were reduced.	Tiwari et al. (2009)

made from ozone processed grains have been evaluated (Bai et al. 2017; Chittrakorn et al. 2014; Mei et al. 2016; Li et al. 2015). Zhu (2018) reported that ozone treatment under suitable conditions strengthens dough, increases bread volume and enhances shelf life of grain products (e.g. noodles). Though, sensorial characteristics of the food products made from ozone processed flour should be studied. In any case, ozone processing of grain products can be combined with other processes (e.g. food additives).

The need for the combination of ozone with other thermal or nonthermal (mild) processes is evident to overcome quality issues mentioned above, optimize the process or even create novel food products. Only few studies on the combined effect of ozone and other processes on nutritional and quality properties of foods, energy and solvent use reduction, production yield increase, side stream valorization and development of innovative food products are available.

Sung et al. (2014) investigated the combined effects of ozone and (followed by) heat treatment on the inactivation of foodborne pathogens, E. coli O157:H7,

L. monocytogenes, and S. typhimurium in apple juice. The results revealed that the combined treatment synergistically reduced the microorganism load, maintained the food quality and reduced the concentration of residual ozone (<0.4 mg/L). Treating eggs containing Salmonella with gaseous ozone followed by heating significantly inactivated the pathogens (Perry and Yousef 2013). Increased inactivation of Cl. perfringens spores and vegetative cells was achieved using a combination of ozone gas followed by mild heat applied on the surface of beef (Novak and Yuan 2004). Proctor et al. (2004) used a combination of heat and gaseous ozone to treat peanut kernels and peanut flour against aflatoxin. The results demonstrated effective reduction of aflatoxin contamination in peanut kernels and peanut flour.

Nonthermal processes, combined with mild heating, cannot completely inactivate all the spores within a food product. Markland et al. (2013) demonstrated that a combination of ozone and High Pressure (HP) may be able to effectively reduce the number of superdormant spores (e.g. Bacillus cereus and Bacillus weihenstephanensis). They studied the ability of a combined ozone (aqueous form) and HP process, to inactivate superdormant spore populations of mesophilic and psychrotolerant isolates of B. cereus. According to results, these spores were approximately 20% more resistant to ozone treatment than heterogeneous spore populations and psychrotolerant species were approximately 31.9% more resistant than mesophilic species. The combined ozone-HP process achieved a maximum 2.67 log CFU/mL reduction of superdormant spores (Markland et al. 2013).

Garcia-Mateos et al. (2019) studied a possible synergistic effect of an ozone (low ozone dosage 24 mg·L − 1 of juice min − 1) and HP (low pressures, 179–321 MPa, 5 min) process on the quality parameters of pitaya juice. The antimicrobial effectiveness of HP in pitaya juice was previously achieved at high pressure levels for long processing times (550–600 MPa, 12–16 min) (Quiroz-González et al. 2018). Application of 7 min of ozone (24 mg·L − 1·min − 1) plus 316 MPa (5 min) was the optimal, synergistic treatment to inactivate >5 log10 (CFU·mL − 1) the population of L. innocua in pitaya juice. The combined application retained pitaya juice microbiologically safe for 30 d (5 ± 2 °C). Although the treatment affected negatively the sensorial characteristics of the juice (decreased lightness and chroma), a higher preference was observed compared with to the respective non processed juices. Processed juices had lower content of bioactive compounds and ascorbic acid.

Oliviera et al. (2018) studied the combined use of ozone and ultrasound processes. Two ultrasound energy densities (19 kHz 350 and 700 J·mL − 1, 5 min; Processing temperature: 32 ± 1.2 °C) and two ozone processing times (direct immersion of ozone gas for 5, 10 min; ozone concentration: 1.50 ppm and processing temperature: 25 °C) were analyzed for pure açai juice. Authors concluded that sonication increased the cloud value and stability of açai juice, ozone was more effective in inactivating the enzymes, especially PPO and ultrasound was more effective in reducing microbial load. The ultrasound-assisted ozone processing methods reduced the viscosity of the juices but impaired the values of bioactive compounds such as antioxidant activity, phenolic compounds and anthocyanin content.

A synergetic method integrating both pulsed electric fields (PEF) and ozone process was developed as a novel approach to investigate the degradation of high molecular weight chitosan (Mw = 4.5·105 Da) (Yue et al. 2008a, b, 2009; Dan et al. 2014). Results showed that the highest degradation percentage of chitosan was achieved with the combination of PEF and ozone generated at experimental conditions of 1.2 L/min of ozone flow rate, 100 mL/min of 0.6% (w/v) chitosan solution flow rate, and 26.7 kV/cm of PEF intensity. The chitosan samples obtained from the combined PEF and ozone treatment have smaller molecular weight (Mw < 2300 Da) than the sole PEF and ozone processes. Moreover, the degradation product obtained could be completely dissolved in water (Dan et al. 2014).

9 Concluding Remarks and Future Prospects

In this chapter, the most commonly explored and applied nonthermal technologies including High Pressure, Pulsed Electric Fields, Cold Atmospheric Pressure Plasma, Sonication, UV-continuous and pulsed light, and Ozonation were presented and analyzed in terms of their main effects on food safety and quality, focusing on the interactions and synergies between them, as well as their synergies with conven tional food processes and preservation technologies such as extraction, drying and osmotic dehydration, enzymatic hydrolysis, packaging technologies, etc. All the above technologies are to a great extent considered as "green technologies" since they usually involve less energy requirements compared to conventional thermal processes, less water consumption, reduced sanitazing agents due to in-package application, and decreased carbon footprint. The evolution of these technologies, both in terms of scientific interest and industrial implementation, was initially based on their effect on food pathogens and spoilage microorganisms elimination. However, it is now evident that these novel food processes present a variety of other advantages beyond their antimicrobial effects, and may be used in order to improve the techno-functional, quality, sensorial and biofunctional properties of food products. Still there is much research for many of these technologies to be conducted in order to fully understand and determine both the exact mechanisms of their synergistic action and their effect on foods components and properties. Apart from High Pressure which is a widely industrially implemented food process even if there are still issues to be resolved, most other discussed technologies, aside important specific successful niche applications, are still under investigation and development in order to reach the highest TRL and be more extensively commercialized. This strive is even more important and promising when the aforementioned synergies of these technologies, either between them or with other food processes, are factored in. In this view, taking into consideration the purpose of implementation and the target food product, for each of the discussed technologies there are different parameters to be taken into consideration and be further investigated. Yet, the successful commercialization of these technologies is straightly related to the design of an appropriate equipment that which can be easily adopted in a processing line, and at the same time will be compact, environmental friendly, and cost-affordable. Taking into account all the discussed synergies and the potential arising for industrial and commercial applications, expectedly in the frame of "green technologies", the exciting requirement and opportunity for validation, optimization and effective and cost efficient scaling up remains a cutting edge prospect. This is reflected in the growing research highlighted in this chapter and clearly points to the increasing need for the continuation and intensification of another essential synergy, that of the experts in different relevant fields of food science and engineering.

References

Abenoza M, Benito M, Saldaña G, Álvarez I, Raso J, Sánchez-Gimeno AC (2013) Effects of pulsed electric field on yield extraction and quality of olive oil. Food Bioproc Technol 6:1367–1373

Abid M, Jabbar S, Hu B, Hashim MM, Hu T, Lei S et al (2014) Thermosonication as a potential quality enhancement technique of apple juice. Ultraso Sonochem 21:984–990

Abida J, Rayees B, Masoodi FA (2014) Pulsed light technology: A novel method for food preservation. Int Food Res J 21:839–848

Ade-Omowaye BIO, Angersbach A, Taiwo KA, Knorr D (2001) Use of pulsed electric field pretreatment to improve dehydration characteristics of plant based foods. Trends Food Sci Technol 12:285–295

Ade-Omowaye BIO, Rastogi NK, Angersbach A, Knorr D (2003) Combined effects of pulsed electric field pre-treatment and partial osmotic dehydration on air drying behaviour of red bell pepper. J Food Eng 60:89–98

Admane N, Genovese F, Altieri G, Tauriello A, Trani A, Gambacorta G et al (2018) Effect of ozone or carbon dioxide pre-treatment during long-term storage of organic table grapes with modified atmosphere packaging. LWT-Food Sci Technol 98:170–178

Aganovic K, Smetana S, Grauwet T, Toepfl S, Mathys A, Van Loey A, & Heinz V (2017. Pilot scale thermal and alternative pasteurization of tomato and watermelon juice: An energy comparison and life cycle assessment. J cleaner Prod 141:514–525

Aguilar-Rosas SF, Ballinas-Casarrubias ML, Nevarez-Moorillon GV, Martin-Belloso O, Ortega-Rivas E (2007) Thermal and pulsed electric fields pasteurization of apple juice: effects on physicochemical properties and flavour compounds. J Food Eng 83:41–46

Akbas MY, Ozdemir M (2006) Effectiveness of ozone for inactivation of Escherichia coli and Bacillus cereus in pistachios. Int J Food Sci Technol 41:513–519

Akdeniz V, Akalın AS (2019) New approach for yoghurt and ice cream production: high-intensity ultrasound, Trends Food Sci Technol 86:392–398

Albano KM, Nicoletti VR (2018) Ultrasound impact on whey protein concentrate pectin complexes and in the O/W emulsions with low oil soybean content stabilization, Ultrason. Sonochem. 41:562–571

Albertos I, Martin-Diana A, Cullen PJ, Tiwari B, Ojha S, Bourke P, Álvarez C, Rico D (2017) Effects of dielectric barrier discharge (DBD) generated plasma on microbial reduction and quality parameters of fresh mackerel (*Scomber scombrus*) fillets. Innovative Food Sci Emerg Technol 44:117–122

Albu S, Joyce E, Paniwnyk L, Lorimer J, Mason T (2004) Potential for the use of ultrasound in the extraction of antioxidants from Rosmarinus officinalis for the food and pharmaceutical industry. Ultrason Sonochem 11:261–265

Alexandre EMC, Araújo P, Duarte MF, de Freitas V, Pintado M, Saraiva JA (2017) High-pressure assisted extraction of bioactive compounds from industrial fermented fig by-product. J Food Sci Technol 54:2519–2531

Alhattab M, Kermanshahi-Pour A, Brooks MSL (2019) Microalgae disruption techniques for product recovery: influence of cell wall composition. J Appl Phycol 31:61–88

Aliasghari M, Aghdama, Mirsaeedghazib H, Aboonajmia M, Kianmehra MH (2015) Innovative Food Science & Emerging Technologies 30:127–131

Alizehi MH, Niakousari M, Fazaeli M, Iraji M (2020) Modeling of vacuum- and ultrasound-assisted osmodehydration of carrot cubes followed by combined infrared and spouted bed drying using artificial neural network and regression models. J Food Process Eng 43:e13563

Alothman M, Kaur B, Fazilah A, Bhat R, Karim AA (2019) Ozone-induced changes of antioxidant capacity of fresh-cut tropical fruits. Innovative Food Sci Emerg Technol 11:666–671

Al-Ruwaih N, Ahmed J, Mulla MF, Arfat YA (2019) High-pressure assisted enzymatic proteolysis of kidney beans protein isolates and characterization of hydrolysates by functional, structural, rheological and antioxidant properties. LWT – Food Sci Technol 100:231–236

Alves H, Alencar ER, Ferreira WFS, Silva CR, Ribeiro JL (2019) Microbiological and physical-chemical aspects of strawberry exposed to ozone gas at different concentrations during storage. Braz J Food Technol 22:e201800

Alwi NA, Ali A (2014) Dose-dependent effect of ozone fumigation on physiological characteristics, ascorbic acid content and disease development on bell pepper (*Capsicum annuum L.*) during storage. Food Bioproc Technol 8:558–566

Amanor-Atiemoh R, Zhou C, Wahia H, Mustapha AT, Rashid MT, Sampson G, Amoa-Owusu A, Ma H, Zhou R (2020) Acoustically-aided osmo-dehydration pretreatments under pulsed vacuum dryer for apple slices: drying kinetics, thermodynamics, and quality attributes. J Food Sci 85:3909–3919

Ambrosi V, Polenta G, Gonzalez C, Ferrari G, Maresca P (2016) High hydrostatic pressure assisted enzymatic hydrolysis of whey proteins. Innovative Food Sci Emerg Technol 38:294–301

Andreou V, Dimopoulos G, Alexandrakis Z, Katsaros G, Oikonomou D, Toepfl S et al (2017) Shelf-life evaluation of virgin olive oil extracted from olives subjected to nonthermal pretreatments for yield increase. Innovative Food Sci Emerg Technol 40:52–57

Andreou V, Dimopoulos G, Dermesonlouoglou E, Taoukis P (2020a) Application of pulsed electric fields to improve product yield and waste valorization in industrial tomato processing. J Food Eng 270:109778

Andreou V, Psarianos M, Dimopoulos G, Tsimogiannis D, Taoukis P (2020b) Effect of pulsed electric fields and high pressure on improved recovery of high-added-value compounds from olive pomace. J Food Sci 85:1500–1512

Andreou V, Tsironi T, Dermesonlouoglou E, Katsaros G, Taoukis P (2018) Combinatory effect of osmotic and high pressure processing on shelf life extension of animal origin products-application to chilled chicken breast fillets. Food Packag Shelf Life 15:43–51

Anema SG, Lauber S, Lee SK, Henle T, Klostermeyer H (2005) Rheological properties of acid gels prepared from pressure- and transglutaminase-treated skim milk. J Food Hydrocoll 19:879–887

Arshad RN, Abdul-Malek Z, Roobab U, Qureshi MI, Khan N, Ahmad MH et al (2021) Effective valorization of food wastes and by-products through pulsed electric field: a systematic review. J Food Process Eng 44:e13629

Azam SMR, Ma H, Xu B, Devi S, Siddique MAB, Stanley SL, Bhandari B, Zhu J (2020) Efficacy of ultrasound treatment in the removal of pesticide residues from fresh vegetables: a review, Trends Food Sci Technol 97:417–432

Baboli ZM, Williams L, Chen G (2020) Design of a batch ultrasonic reactor for rapid pasteurization of juices, J Food Eng 268:109736

Bai Y, Liu L, Zhang R, Huang F, Deng Y, Zhang M (2016) Ultrahigh pressure-assisted enzymatic extraction maximizes the yield of longan pulp polysaccharides and their acetylcholinesterase inhibitory activity in vitro. Int J Biol Macromol 96:214–222

Bai YP, Guo XN, Zhu KX, Zhou HM (2017) Shelf-life extension of semi-dried buckwheat noodles by the combination of aqueous ozone treatment and modified atmosphere packaging. Food Chem 237:553–557

Barukcic I, Lisak Jakopovi´c K, Herceg Z, Karlovi´c S, Boˇzani´c RM (2015) Influence of high intensity ultrasound on microbial reduction, physico-chemical characteristics and fermentation of sweet whey, Innov. Food Sci Emerg Technol 27:94–101

Bao Y, Reddivari L, Huang JY (2020b) Enhancement of phenolic compounds extraction from grape pomace by high voltage atmospheric cold plasma. LWT – Food Sci Technol 133:109970

Bao Y, Reddivari L, Huang J-Y (2020a) Development of cold plasma pretreatment for improving phenolics extractability from tomato pomace. Innovative Food Sci Emerg Technol 65:102445

Barboni T, Cannac M, Chiaramonti N (2010) Effect of cold storage and ozone treatment on physicochemical parameters, soluble sugars and organic acids in *Actinidia deliciosa*. Food Chem 121:946–951

Bazhal MI, Lebovka NI, Vorobiev E (2001) Pulsed electric field treatment of apple tissue during compression for juice extraction. J Food Eng 50:129–139

Bechlin TR, Granella SJ, Christ D, Coelho SRM, Paz CHO (2020) Effects of ozone application and hot-air drying on orange peel: moisture diffusion, oil yield, and antioxidant activity. Food Bioprod Proc 123:80–89

Belmiro RH, Tribst AAL, Cristianini M (2018) Impact of high pressure processing in hydration and drying curves of common beans (*Phaseolus vulgaris* L.). Innovative Food Sci Emerg Technol 47:279–285

Bendicho S, Barbosa-Cánovas GV, & Martıń O (2002) Milk processing by high intensity pulsed electric fields. Trends in Food Sci Technol 13(6–7):195–204

Berk Z (2018) Food process engineering and technology. Academic, Oakville

Bermúdez-Aguirre D, Corradini MG, Candŏgan K, Barbosa-Cánovas GV (2016) High pressure processing in combination with high temperature and other preservation factors. In: Balasubramaniam VM, Barbosa-Cánovas GV, Lelieveld HLM (eds) High pressure processing of foods: principles, technology and applications, 1st edn. Springer, New York, pp 193–215

Bhargava B, Mor RS, Kumar K, Sharanagat VS (2021) Advances in application of ultrasound in food processing: a review. Ultrason Sonochem 70:105293

Bhat ZF, Morton JD, Mason SL, Bekhit AEDA (2019) Current and future prospects for the use of pulsed electric field in the meat industry. Crit Rev Food Sci Nutr 59:1660–1674

Boussetta N, Soichi E, Lanoisellé JL, Vorobiev E (2014) Valorization of oilseed residues: extraction of polyphenols from flaxseed hulls by pulsed electric fields. Ind Crops Prod 52:347–353

Brennan J, Grandison A (2012) C.h.H. Wiley-VCH, Weinheim

Briones-Labarca V, Giovagnoli-Vicuña C, Cañas-Sarazúa R (2019) Optimization of extraction yield, flavonoids and lycopene from tomato pulp by high hydrostatic pressure-assisted extraction. Food Chem 278:751–759

Brodowska AJ, Nowak A, Smigielski K (2018) Ozone in the food industry: principles of ozone treatment, mechanisms of action, and applications: an overview. Crit Rev Food Sci Nutr 58:2176–2201

Brookman JS (1975) Further studies on the mechanism of cell disruption by extreme pressure extrusion. Biotechnol Bioeng 17:465–479

Bueno M, Gallego R, Chourio AM, Ibáñez E, Herrero M, Saldãna MDA (2020) Green ultra-high pressure extraction of bioactive compounds from *Haematococcus pluvialis* and *Porphyridium cruentum* microalgae. Innovative Food Sci Emerg Technol 66:102532

Bursać-Kovačević D, Putnik P, Dragović-Uzelac V, Pedisić S, Režek Jambrak A, Herceg Z (2016) Effects of cold atmospheric gas phase plasma on anthocyanins and color in pomegranate juice. Food Chem 190:317–323

Cacace D, Palmieri L (2014) High intensity pulsed light technology. In: Sun DW (ed) Emerging technologies for food processing. Academic, Amsterdam, pp 239–258

Cando D, Borderías AJ, Moreno HM (2016) Combined effect of amino acids and microbial transglutaminase on gelation of low salt surimi content under high pressure processing. Innovative Food Sci Emerg Technol 36:10–17

Caraveo O, Alarcon-Rojo AD, Renteria A, Santellano E, Paniwnyk L (2015) Physicochemical and microbiological characteristics of beef treated with highintensity ultrasound and stored at 4° C, J. Sci. Food Agric 95:2487–2493

Carpenter J, Saharan VK (2017) Ultrasonic assisted formation and stability of mustard oil in water nanoemulsion: effect of process parameters and their optimization, Ultrason. Sonochem. 35:422–430

Carrillo-Hormaza L, Duque L, López-Parra S, Osorio E (2020) High-intensity ultrasound-assisted extraction of Garcinia madruno biflavonoids: Mechanism, kinetics, and productivity. Biochem Eng J 161:107676

Carrillo-Lopez LM, Juarez-Morales MGMG, Garcia-Galicia IA, Alarcon-Rojo ADAD, Huerta-Jimenez M (2020) The effect of high-intensity ultrasound on the physicochemical and microbiological properties of Mexican panela cheese, Foods 9:1–14

Carullo D, Abera BD, Casazza AA, Donsì F, Perego P, Ferrari G, Pataro G (2018) Effect of pulsed electric fields and high pressure homogenization on the aqueous extraction of intracellular compounds from the microalgae Chlorella vulgaris. Algal Res 31:60–69

Carullo D, Donsì F, Ferrari G (2020) Influence of high-pressure homogenization on structural properties and enzymatic hydrolysis of milk proteins. LWT-Food Sci Technol 130:109657

Cascaes Teles AS, Chávez DWH, Coelho MAZ, Rosenthal A, Gottschalk LMF, Tonon RV (2021) Combination of enzyme-assisted extraction and high hydrostatic pressure for phenolic compounds recovery from grape pomace. J Food Eng 288:110128

Chan CHH, Yeoh HK, Yusoff R, Ngoh GC (2016) A first-principles model for plant cell rupture in microwave-assisted extraction of bioactive compounds. J Food Eng 188:98–107

Charles-Rodríguez AV, Nevárez-Moorillón GV, Zhang QH, Ortega-Rivas E (2007) Comparison of thermal processing and pulsed electric fields treatment in pasteurization of apple juice. Food Bioprod Proc 85:93–97

Chen R, Meng F, Zhang S, Liu Z (2009) Effects of ultrahigh pressure extraction conditions on yields and antioxidant activity of ginsenoside from ginseng. Sep Purif Technol 66:340–346

Chen J, Hu Y, Wang J, Hu H, Cui H (2016) Combined effect of ozone treatment and modified atmosphere packaging on antioxidant defense system of fresh-cut green peppers. J Food Process Preserv 40:1145–1150

Cheng X, Zhang M, Xu B, Adhikari B, Sun J (2015) The principles of ultrasound and its application in freezing related processes of food materials: a review. Ultrason Sonochem 27:576–585

Chen F, Zhang M, Yang CH (2019) Application of ultrasound technology in processing of ready-to-eat fresh food: a review, Ultrason. Sonochem. 104953

Chittrakorn S, Earls D, MacRitchie F (2014) Ozonation as an alternative to chlorination for soft wheat flours. J Cereal Sci 60:217–221

Corrales M, Toepfl S, Butz P, Knorr D, Tauscher B (2008) Extraction of anthocyanins from grape by-products assisted by ultrasonics, high hydrostatic pressure or pulsed electric fields: A comparison. Innovative Food Sci Emerg Technol 9:85–91

Cullen PJ, Tiwari BK, O'Donnell CP, Muthukumarappan K (2009) Modelling approaches to ozone processing of liquid foods. Trends Food Sci Technol 20:125–136

Dan G, Zhang ZH, Zeng XA, Han Z, Luo WB, Tang C, Quek SY (2014) Synergetic effects of pulsed electric field and ozone treatments on the degradation of high molecular weight chitosan. Int J Food Eng 10:775–784

Daryaei H, Yousef AE, Balasubramaniam VM (2016) Microbiological aspects of high-pressure processing of food: inactivation of microbial vegetative cells and spores. In: Balasubramaniam VM, Barbosa-Cánovas GV, Lelieveld HLM (eds) High pressure processing of foods: principles, technology and applications, 1st edn. Springer, New York, pp 271–294

Dash KK, Balasubramaniam VM, Kamat S (2019) High pressure assisted osmotic dehydrated ginger slices. J Food Eng 247:19–29

de Jesus ALT, Cristianini M, dos Santos NM, Maróstica MR Jr (2020) Effects of high hydrostatic pressure on the microbial inactivation and extraction of bioactive compounds from açaí (*Euterpe oleracea* Martius) pulp. Food Res Int 130:108856

Dermesonlouoglou E, Chalkia A, Dimopoulos G, Taoukis P (2018) Combined effect of pulsed electric field and osmotic dehydration pre-treatments on mass transfer and quality of air dried goji berry. Innovative Food Sci Emerg Technol 49:106–115

Dermesonlouoglou EK, Andreou V, Alexandrakis Z, Katsaros GJ, Giannakourou MC, Taoukis PS (2017b) The hurdle effect of osmotic pretreatment and high pressure cold pasteurisation on the shelf-life extension of fresh-cut tomatoes. Int J Food Sci Technol 52:916–926
Dermesonlouoglou EK, Bimpilas A, Andreou V, Katsaros GJ, Giannakourou MC, Taoukis PS (2017a) Process optimization and kinetic modeling of quality of fresh-cut strawberry cubes pretreated by high pressure and osmosis. J Food Process Preserv 41:e13137
Dermesonlouoglou EK, Giannakourou M, Taoukis PS (2016) Kinetic study of the effect of the osmotic dehydration pre-treatment with alternative osmotic solutes to the shelf life of frozen strawberry. Food Bioprod Proc 99:212–221
Diels AM, Michiels CW (2006) High-pressure homogenization as a non-thermal technique for the inactivation of microorganisms. Crit Rev Microbiol 32(4):201–216
Dimopoulos G, Stefanou N, Andreou V, Taoukis P (2018) Effect of pulsed electric fields on the production of yeast extract by autolysis. Innovative Food Sci Emerg Technol 48:287–295
Dimopoulos G, Tsantes M, Taoukis P (2020) Effect of high pressure homogenization on the production of yeast extract via autolysis and beta-glucan recovery. Innovative Food Sci Emerg Technol 62:102340
Ding W, Wang Y, Zhang W, Shi DYC (2015) Effect of ozone treatment on physicochemical properties of waxy rice flour and waxy rice starch. Int J Food Sci Technol 50:744–749
Diver DA, Potts H (2015) Plasma generation and use of plasma generation apparatus, Patent # 8,980,190
Doulah MS, Hammond TH, Brookman JSG (1975) A hydrodynamic mechanism for the disintegration of Saccharomyces cerevisiae in an industrial homogenizer. Biotechnol Bioeng 17:845–858
Drévillon L, Koubaa M, Vorobiev E (2018) Lipid extraction from *Yarrowia lipolytica* biomass using high-pressure homogenization. Biomass Bioenergy 115:143–150
Dufour E, Herve G, Cedex Y, Haertle T (1995) Hydrolysis of b-lactoglobulin by thermolysin and pepsin under high hydrostatic pressure. Biopolymers 35:475–483
Dzah CS, Duan Y, Zhang H, Wen C, Zhang J, Chen G, Ma H (2020) The effects of ultrasound assisted extraction on yield, antioxidant, anticancer and antimicrobial activity of polyphenol extracts: a review, Food Biosci 100547
El Kantar S, Boussetta N, Lebovka N, Foucart F, Rajha HN, Maroun RG et al (2018) Pulsed electric field treatment of citrus fruits: improvement of juice and polyphenols extraction. Innovative Food Sci Emerg Technol 46:153–161
Elain A, Nkounkou C, Le Fellic M, Donnart K (2020) Green extraction of polysaccharides from Arthrospira platensis using high pressure homogenization. J Appl Phycol 32:1719–1727
Engler CR (1990) Cell disruption by homogenizer. Bioprocess Technol 9:95–105
Engler CR, Robinson CW (1981) Disruption of *Candida utilis* cells in high pressure flow devices. Biotechnol Bioeng 23:765–780
Escobar D, Clark S, Ganesan V, Repiso L, Waller J, Harte F (2011) High-pressure homogenization of raw and pasteurized milk modifies the yield, composition, and texture of queso fresco cheese. J Dairy Sci 94:1201–1210
Eshtiaghi MN, Knorr D (2002) High electric field pulse pretreatment: potential for sugar beet processing. J Food Eng 52:265–272
Fan X, Sokorai KJ (2015) Formation of trichloromethane in chlorinated water and fresh-cut produce and as a result of reaction with citric acid. Postharv Biol Technol 109:65–72
Farhaninejad Z, Fathi M, Shahedi M, Sadeghi M (2017) Osmotic dehydration of banana slices using direct and indirect sonication: optimization and microstructure analysis. J Food Process Eng 40:e12336
Faria GYY, Souza MM, Oliveira JRM, Costa CSB, Collares MP, Prentice C (2020) Effect of ultrasound-assisted cold plasma pretreatment to obtain sea asparagus extract and its application in Italian salami. Food Res Int 137:109435
Fauster T, Giancaterino M, Pittia P, Jaeger H (2020) Effect of pulsed electric field pretreatment on shrinkage, rehydration capacity and texture of freeze-dried plant materials. LWT-Food Sci Technol 121:108937

Fauster T, Schlossnikl D, Rath F, Ostermeier R, Teufel F, Toepfl S, Jaeger H (2018) Impact of pulsed electric field (PEF) pretreatment on process performance of industrial French fries production. J Food Eng 235:16–22

Fellows PJ (2017) Food processing technologies: principles and practices, 4th edn. Woodhead Publishing, Elsevier, Cambridge, MA

Ferrario M, Guerrero S (2017) Impact of a combined processing technology involving ultrasound and pulsed light on structural and physiological changes of *Saccharomyces cerevisiae* KE 162 in apple juice. Food Microbiol 65:83–94

Fort N, Lanier T, Amato PM, Carretero C, Saguer E (2009) Simultaneous application of microbial transglutaminase and high hydrostatic pressure to improve heat induced gelation of pork plasma. Meat Sci 80:939–943

Fulladosa E, Serra X, Gou P, Arnau J (2009) Effects of potassium lactate and high pressure on transglutaminase restructured dry-cured hams with reduced salt content. Meat Sci 82:213–218

Gachovska TK, Ngadi MO, Raghavan GSV (2006) Pulsed electric field assisted juice extraction from alfalfa. Can Biosyst Eng 48:3

Gali L, Bedjou F, Velikov KP, Ferrari G, Donsì F (2020) High-pressure homogenization-assisted extraction of bioactive compounds from *Ruta chalepensis*. J Food Measur Charact 14:2800–2809

Gallego R, Bueno M, Chourio AM, Ibáñez E, MDA S, Herrero M (2021) Use of high and ultra-high pressure based-processes for the effective recovery of bioactive compounds from *Nannochloropsis oceanica* microalgae. J Supercrit Fluids 167:105039

Gallo M, Ferrara L, Naviglio D (2018) Application of ultrasound in food science and technology: a perspective. Foods 7:164

Ganeva V, Angelova B, Galutzov B, Goltsev V, Zhiponova M (2020) Extraction of proteins and other intracellular bioactive compounds from Baker's yeasts by pulsed electric field treatment. Front Bioeng Biotechnol 8:1433

García-Mateos MR, Quiroz-González B, Corrales-García J, Ybarra-Moncada MC, Leyva-Ruelas G (2019) Ozone-high hydrostatic pressure synergy for the stabilization of refrigerated pitaya (*Stenocereus pruinosus*) juice. Innovative Food Sci Emerg Technol 56:102187

Gavahian M, Meng-Jen T, Khaneghah AM (2020) Emerging techniques in food science: the resistance of chlorpyrifos pesticide pollution against arc and dielectric barrier discharge plasma. Qual Assur Saf Crops Foods 12(SP1):12–19

Genovese J, Tappi S, Luo W, Tylewicz U, Marzocchi S, Marziali S et al (2019) Important factors to consider for acrylamide mitigation in potato crisps using pulsed electric fields. Innovative Food Sci Emerg Technol 55:18–26

George JM, Selvan TS, Rastogi NK (2015) High-pressure-assisted infusion of bioactive compounds in apple slices. Innovative Food Sci Emerg Technol 33:100–107

Gibson KE, Almeida G, Jones SL, Wright K, Ae Lee J (2019) Inactivation of bacteria on fresh produce by batch wash ozone sanitation. Food Control 106:106747

Gogou E, Taoukis P (2015) High-pressure process design and evaluation, chapter 11. In: Varzakas T, Tzia C (eds) Handbook of food processing Vol.1: food preservation. CRC Press, Taylor & Francis Group, Boca Raton, pp 415–437

Gomes WF, Tiwari BK, Rodriguez O, de Brito ES, Fernandes FAN, Rodrigues S (2017) Effect of ultrasound followed by high pressure processing on prebiotic cranberry juice. Food Chem 218:261–268

Gómez-López VM, Marín A, Medina-Martínez MS, Gil MI, Allende A (2013) Generation of trihalomethanes with chlorine-based sanitizers and impact on microbial, nutritional and sensory quality of baby spinach. Postharv Biol Technol 85:210–217

Gozé P, Rhazi L, Pauss A, Aussenac T (2016) Starch characterization after ozone treatment of wheat grains. J Cereal Sci 70:207–213

Gregersen SB, Frydenberg RP, Hammershøj M, Dalsgaard TK, Andersen U, Wiking L (2019) Application of high intensity ultrasound to accelerate crystallization of anhydrous milk fat and rapeseed oil blends, Eur. J. Lipid Sci Technol 121

Guderjan M, Elez-Martínez P, Knorr D (2007) Application of pulsed electric fields at oil yield and content of functional food ingredients at the production of rapeseed oil. Innovative Food Sci Emerg Technol 8:55–62

Guerzoni ME, Vannini L, Lopez CC, Lanciotti R, Suzzi G, Gianotti A (1999) Effect of high pressure homogenization on microbial and chemico-physical characteristics of goat cheeses. J Dairy Sci 82:851–862

Guimarães JT, Balthazar CF, Scudino H, Pimentel TC, Esmerino EA, Ashok kumar M (2019) High-intensity ultrasound: A novel technology for the development of probiotic and prebiotic dairy products, Ultrason. Sonochem. 57:12–21

Guo J, Huang K, Wang J (2015) Bactericidal effect of various non-thermal plasma agents and the influence of experimental conditions in microbial inactivation: a review. Food Control 50:482–490

Guo X, Han D, Xi H, Rao L, Liao X, Hu X, Wu J (2012) Extraction of pectin from navel orange peel assisted by ultra-high pressure, microwave or traditional heating: a comparison. Carbohydr Polym 88:441–448

Guzel-Seydima ZB, Greene AK, Seydim AC (2004) Use of ozone in the food industry. LWT-Food Sci Technol 37:453–460

Han L, Patil S, Boehm D, Milosavljevic V, Cullen PJ, Bourke P (2016) Mechanism of inactivation by high voltage atmospheric cold plasma differs between *Escherichia coli* and *Staphylococcus aureus*. Appl Environ Microbiol 82:450–458

Han T, Wang M, Wang Y, Tang L (2020) Effects of high-pressure homogenization and ultrasonic treatment on the structure and characteristics of casein. LWT-Food Sci Technol 130:109560

Harte F (2016) Food processing by high-pressure homogenization. High pressure processing of food, p 123–141. Springer, USA.

Hendrickx MEG, Knorr D (2002) Ultra-high pressure treatment of foods, Aspen food engineering series. Kluwer Academic, New York

Hernández A, Harte FM (2008) Manufacture of acid gels from skim milk using high-pressure homogenization. J Dairy Sci 91:3761–3767

Hernández-Hernández HM, Moreno-Vilet L, Villanueva-Rodríguez SJ (2019) Current status of emerging food processing technologies in Latin America: novel non-thermal processing. Innovative Food Sci Emerg Technol 58:102233

Herranz B, Tovar CA, Borderias AJ, Moreno HM (2013) Effect of high-pressure and/or microbial transglutaminase on physicochemical, rheological, and microstructural properties of flying fish surimi. Innovative Food Sci Emerg Technol 20:24–33

Hertrich SM, Boyd G, Sites J, Niemira BA (2017) Cold plasma inactivation of salmonella in prepackaged, mixed salads is influenced by cross-contamination sequence. J Food Prot 80:2132–2136

Higuera-Barraza OA, Del Toro-Sanchez CL, Ruiz-Cruz S, Márquez-Ríos E (2016) Effects of high-energy ultrasound on the functional properties of proteins. Ultrason Sonochem 31:558–562

Huang Y-L, Tsai Y-H (2020) Extraction of chitosan from squid pen waste by high hydrostatic pressure: effects on physicochemical properties and antioxidant activities of chitosan. Int J Biol Macromol 160:677–687

Ignat A, Manzocco L, Brunton NP, Nicoli MC, Lyng JG (2015) The effect of pulsed electric field pre-treatments prior to deep-fat frying on quality aspects of potato fries. Innovative Food Sci Emerg Technol 29:65–69

Igual M, Sampedro F, Martínez-Navarrete F, X. (2013) Combined osmodehydration and high pressure processing on the enzyme stability and antioxidant capacity of a grapefruit jam. J Food Eng 114:514–521

Jambrak AN (2019) Non-thermal and innovative processing technologies. Ref Mod Food Sci Encycl Food Sec Sustain 1:477–483

Jay J, Loessner M, Golden D (2005) Modern food microbiology, 7th edn. Springer Science + Business Media, New York

Jemai AB, Vorobiev E (2006) Pulsed electric field assisted pressing of sugar beet slices: towards a novel process of cold juice extraction. Biosyst Eng 93:57–68
Jin Y, Shi Z, Xu G, Yang H, Yang J (2020) A stepwise pretreatment of sugarcane bagasse by alkaline and hydroxymethyl reagent for bioethanol production. Ind Crops Prod 145:112136
John D, Ramaswamy HS (2018) Pulsed light technology to enhance food safety and quality: a mini-review. Curr Opin Food Sci 23:70–79
Juan B, Quevedo JM, Zamora A, Guamis B, Trujillo AJ (2015) Lipolysis of cheeses made from goat milk treated by ultra-high pressure homogenization. LWT-Food Sci Technol 60:1034–1038
Jun X (2009) Caffeine extraction from green tea leaves assisted by high pressure processing. J Food Eng 94:105–109
Jun X, Deji S, Ye L, Rui Z (2011) Micromechanism of ultrahigh pressure extraction of active ingredients from green tea leaves. Food Control 22:1473–1476
Jurić S, Ferrari G, Velikov KP, Donsì F (2019) High-pressure homogenization treatment to recover bioactive compounds from tomato peels. J Food Eng 262:170–180
Karwe MV, Maldonado J, Mahadevan S (2015) High pressure processing: current status. In: Bhattacharya S (ed) Conventional and advanced food processing technologies, 1st edn. Wiley, Chichester, pp 595–616
Keener L, Krishnamurthy K (2014) Shedding light on food safety: application of pulsed light technology. Food Safety Magazine, 2014
Kenny O, O'Beirne D (2009) The effects of washing treatment on antioxidant retention in ready-to-use iceberg lettuce. Int J Food Sci Technol 44:1146–1156
Khouryieh HA (2021) Novel and emerging technologies used by the U.S. food processing industry. Innov Food Sci Emerg Technol 67:102559
Kim B, Yun H, Jung S, Jung Y, Jung H, Choe W et al (2011) Effect of atmospheric pressure plasma on inactivation of pathogens inoculated onto bacon using two different gas compositions. Food Microbiol 28:9–13
Kim H-J, Yong HI, Park S, Kim K, Choe W, Jo C (2015) Microbial safety and quality attributes of milk following treatment with atmospheric pressure encapsulated dielectric barrier discharge plasma. Food Control 47:451–456
Klinke HB, Thomsen A, Ahring BK (2004) Inhibition of ethanol-producing yeast and bacteria by degradation products produced during pre-treatment of biomass. Appl Microbiol Biotechnol 66:10–26
Knoerzer K (2016) Nonthermal and innovative food processing technologies. Ref Mod Food Sci:767–771
Knorr D, Froehling A, Jaeger H, Reineke K, Schlueter O, Schoessler K (2011) Emerging technologies in food processing. Ann Rev Food Sci Technol 2:203–235
Koutchma T, Popović V, Ros-Polski V, Popielarz A (2016) Effects of ultraviolet light and high-pressure processing on quality and health-related constituents of fresh juice products. Compr Rev Food Sci Food Saf 15:844–867
Kumar PS, Pandit AB (1999) Modeling hydrodynamic cavitation. Chem Eng Technol 22:1017–1027
Lammerskitten A, Mykhailyk V, Wiktor A, Toepfl S, Nowacka M, Bialik M et al (2019) Impact of pulsed electric fields on physical properties of freeze-dried apple tissue. Innovative Food Sci Emerg Technol 57:102211
Lanciotti R, Patrignani F, Iucci L, Saracino P, & Guerzoni ME (2007) Potential of high pressure homogenization in the control and enhancement of proteolytic and fermentative activities of some Lactobacillus species. Food Chem 102(2):542–550
Leistner L (1985) Hurdle technology applied to meat products of the shelf stable product and intermediate moisture food types. In: Properties of water in foods. Springer, pp 309–329
Leong T, Juliano P, Knoerzer K (2017) Advances in ultrasonic and megasonic processing of foods. Food Eng Rev 9:237–256
Levy R, Okun Z, Shpigelman A (2020) High-pressure homogenization: principles and applications beyond microbial inactivation. Food Eng Rev 13:490–508

Li X, Farid M (2016) A review on recent development in non-conventional food sterilization technologies. J Food Eng 182:33–45
Li D, Zhao H, Muhammad AI, Song L, Guo M, Liu D (2020) The comparison of ultrasound-assisted thawing, air thawing and water immersion thawing on the quality of slow/fast freezing bighead carp (Aristichthys nobilis) fillets, Food Chem:126614
Li MM, Guan EQ, Bian K (2015) Effect of ozone treatment on deoxynivalenol and quality evaluation of ozonised wheat. Food Addit Contamin Part A 32:544–553
Limsangouan N, Charunuch C, Sastry SK, Srichamnong W, Jittanit W (2020) High pressure processing of tamarind (*Tamarindus indica*) seed for xyloglucan extraction. LWT – Food Sci Technol 134:110112
Lin S-P, Kuo T-C, Wang H-T, Ting Y, Hsieh C-W, Chen Y-K, Hsu H-Y, Cheng K-C (2020) Enhanced bioethanol production using atmospheric cold plasma-assisted detoxification of sugarcane bagasse hydrolysate. Bioresour Technol 313:123704
Liu C, Grimi N, Lebovka N, Vorobiev E (2020) Impacts of preliminary vacuum drying and pulsed electric field treatment on characteristics of fried potatoes. J Food Eng 276:109898
Liu F, Zhang X, Zhao L, Wang Y, Liao X (2016) Potential of high-pressure processing and high-temperature/short-time thermal processing on microbial, physicochemical and sensory assurance of clear cucumber juice. Innovative Food Sci Emerg Technol 34:51–58
Liu H, Zhang J, Yuan J, Jiang X, Jiang L, Zhao G, Huang D, Liu B (2019) Omicsbased analyses revealed metabolic responses of Clostridium acetobutylicum to lignocellulose-derived inhibitors furfural, formic acid and phenol stress for butanol fermentation. Biotechnol Biofuels 12:101
Liu XY, Wang Q, Cui SW, Liu HZ (2008) A new isolation method of β-D-glucans from spent yeast Saccharomyces *cerevisiae*. Food Hydrocoll 22:239–247
Llavata B, Garcia-Perez JV, Simal S, Carcel JA (2020) Innovative pre-treatments to enhance food drying: a current review. Curr Opin Food Sci 35:20–26
Loginova KV, Lebovka NI, Vorobiev E (2011) Pulsed electric field assisted aqueous extraction of colorants from red beet. J Food Eng 106:127–133
López M, Calvo T, Prieto M, Múgica-Vidal R, Muro-Fraguas I, Alba-Elías F, Alvarez-Ordóñez A (2019) A review on non-thermal atmospheric plasma for food preservation: mode of action, determinants of effectiveness, and applications. Front Microbiol 10:622
López-Pedemonte T, Brinẽz WJ, Roig-Sagues AX, Guamis B (2006) Fate of *Staphylococcus aureus* in cheese treated by ultrahigh pressure homogenization and high hydrostatic pressure. J Dairy Sci 89:4536–4544
Luengo E, Álvarez I, Raso J (2013) Improving the pressing extraction of polyphenols of orange peel by pulsed electric fields. Innovative Food Sci Emerg Technol 17:79–84
Luengo E, Condón-Abanto S, Álvarez I, Raso J (2014) Effect of pulsed electric field treatments on permeabilization and extraction of pigments from *Chlorella vulgaris*. J Membr Biol 247:1269–1277
Luo Z, Duan H, Yang Y, Zhang W, Ramaswamy HS, Bai W, Wang C (2021) High pressure assisted extraction for cadmium decontamination of long rice grain. Food Control 125:107987
Ma Y, Wu X, Zhao L, Wang Y, Liao X (2019) Comparison of the compounds and characteristics of pepper seed oil by pressure-assisted, ultrasound-assisted and conventional solvent extraction. Innovative Food Sci Emerg Technol 54:78–86
Manousaridis G, Nerantzaki A, Paleologos EK, Tsiotsias A, Savvaidis IN, Kontominas MG (2005) Effect of ozone on microbial, chemical and sensory attributes of shucked mussels. Food Microbiol 22:1–9
Markland SM, Kniel KE, Setlow P, Hoover DG (2013) Nonthermal inactivation of heterogeneous and superdormant spore populations of Bacillus cereus using ozone and high pressure processing. Innov Food Sci Emerg Technol 19:44–49
Martínez JM, Delso C, Aguilar DE, Álvarez I, Raso J (2020) Organic-solvent-free extraction of carotenoids from yeast *Rhodotorula glutinis* by application of ultrasound under pressure. Ultrason Sonochem 61:104833
Martínez JM, Luengo E, Saldaña G, Álvarez I, Raso J (2017) C-phycocyanin extraction assisted by pulsed electric field from *Artrosphira platensis*. Food Res Int 99:1042–1047

Masooma, Munir Muhammad, Nadeem Tahir Mahmood, Qureshi Thomas SH, Leong Charitha J, Gamlath Gregory JO, Martin Muthupandian, Ashokkumar (2019) Effects of high pressure microwave and ultrasound processing on proteins and enzyme activity in dairy systems — A review. Innovative Food Science & Emerging Technologies 57102192. https://doi.org/10.1016/j.ifset.2019.102192

Massoud R, Belgheisi S, Massoud A (2016) Effect of high pressure homogenization on improving the quality of milk and sensory properties of yogurt: a review. Int J Chem Eng Appl 7:66

Maynard F, Weingand A, Hau J, Jost R (1998) Effect of high pressure treatment on the tryptic hydrolysis of bovine β-lactoglobulin AB. Int Dairy J 8:125–133

Medina-Meza IG, Barbosa-Cánovas GV (2015) Assisted extraction of bioactive compounds from plum and grape peels by ultrasonics and pulsed electric fields. J Food Eng 166:268–275

Mei J, Liu G, Huang X, Ding W (2016) Effects of ozone treatment on medium hard wheat (*Triticum aestivum L.*) flour quality and performance in steamed bread making. CyTA-J Food 14:449–456

Meinlschmidt P, Brode V, Sevenich R, Ueberhame E, Schweiggert-Weisz J, Lehmanne J, Rauh C, Knorr D, Eisner P (2017) High pressure processing assisted enzymatic hydrolysis – an innovative approach for the reduction of soy immunoreactivity. Innovative Food Sci Emerg Technol 40:58–67

Middelberg AP (2000) Microbial cell disruption by high-pressure homogenization. In: Downstream processing of proteins. Humana Press, Totowa, pp 11–21

Milani EA, Alkhafaji S, & Silva FV (2015) Pulsed electric field continuous pasteurization of different types of beers. Food Control, 50:223–229

Min SC, Roh SH, Niemira BA, Sites JE, Boyd G, Lacombe A (2016) Dielectric barrier discharge atmospheric cold plasma inhibits *Escherichia coli* O157:H7, *Salmonella*, *Listeria monocyto genes*, and Tulane virus in Romaine lettuce. Int J Food Microbiol 237:114–120

Mir SA, Shah MA, Mir MM (2016) Understanding the role of plasma technology in food industry. Food Bioproc Technol 9:734–750

Misra NN, Yepez X, Xu L, Keener K (2019) In-package cold plasma technologies. J Food Eng 244:21–31

Modesti M, Petriccione M, Forniti R, Zampella L, Scortichini M, Mencarelli F (2018) Methyl jasmonate and ozone affect the antioxidant system and the quality of wine grape during post-harvest partial dehydration. Food Res Int 112:369–377

Modugno C, Peltier C, Simonin H, Dujourdy F, Sandt C, Perrier-Cornet J-M (2020) Understanding the effects of high pressure on bacterial spores using synchrotron infrared spectroscopy. Front Microbiol 10:3122

Morales R, Martínez KD, Ruiz-Henestrosa VMP, Pilosof AM (2015) Modification of foaming properties of soy protein isolate by high ultrasound intensity: particle size effect. Ultrason Sonochem 26:48–55

Moreira SA, Silva S, Costa EM, Saraiva JA, Pintado M (2020) Effect of high hydrostatic pressure on biological activities of stinging nettle extracts. Food Funct 11:921–931

Motoki M, Seguro K (1998) Transglutaminase and its use for food processing: a review. Trends Food Sci Technol 9:204–210

Naji O, Al-juboori RA, Bowtell L, Alpatova A, Ghaffour N (2020) Direct contact ultrasound for fouling control and flux enhancement in air-gap membrane distillation, Ultrason. Sonochem. 61:104816

Nayak SL, Sethi S, Sharma RR, Sharma RM, Singh S, Singh D (2020) Aqueous ozone controls decay and maintains quality attributes of strawberry (*Fragaria* × *ananassa Duch.*). J Food Sci Technol 57:319–326

Nguyen TMC, Gavahian M, Tsai P-J (2021) Effects of ultrasound-assisted extraction (UAE), high voltage electric field (HVEF), high pressure processing (HPP), and combined methods (HVEF+UAE and HPP+UAE) on Gac leaves extraction. LWT – Food Sci Technol 143:111131

Ninčević Grassino A, Ostojić J, Miletić V, Djaković S, Bosiljkov T, Zorić Z, Ježek D, Rimac Brnčić S, Brnčić M (2020) Application of high hydrostatic pressure and ultrasound-assisted extractions as a novel approach for pectin and polyphenols recovery from tomato peel waste. Innovative Food Sci Emerg Technol 64:102424

Novak JS, Yuan JTC (2004) Increased inactivation of ozone-treated Clostridium perfringens vegetative cells and spores on fabricated beef surfaces using mild heat. J Food Prot 67:342–346

Nowacka M, Fijalkowska A, Wiktor A, Rybak K, Dadan M, Witrowa-Rajchert D (2017) Changes of mechanical and thermal properties of cranberries subjected to ultrasound treatment. Int J Food Eng 13(6):20160306

Nowacka M, Dadan M, Tylewicz U (2021) Current Applications of Ultrasound in Fruit and Vegetables Osmotic Dehydration Processes. Appl Sci 11:1269

Nuñez-Mancilla Y, Vega-Gálvez A, Pérez-Won M, Zura L, García-Segovia P, Di Scala K (2013a) Effect of osmotic dehydration under high hydrostatic pressure on microstructure, functional properties and bioactive compounds of strawberry (*Fragaria vesca*). Food Bioproc Technol 7:516–524

Nuñez-Mancilla Y, Pérez-Won M, Uribe E, Vega-Gálvez A, Di Scala K (2013b) Osmotic dehydration under high hydrostatic pressure: effects on antioxidant activity, total phenolics compounds, vitamin C and colour of strawberry (*Fragaria vesca*). LWT Food Sci Technol 52:151–156

Nuñez-Mancilla Y, Perez-Won M, Vega-Gálvez A, Arias V, Tabilo-Munizaga G, Briones-Labarca V, Lemus-Mondaca R, Di Scala K (2011) Modeling mass transfer during osmotic dehydration of strawberries under high hydrostatic pressure conditions. Innovative Food Sci Emerg Technol 12:338–343

Obadi M, Zhu K-X, Peng W, Sulieman AA, Mohammed K, Zhou H-M (2018) Effects of ozone treatment on the physicochemical and functional properties of whole grain flour. J Cereal Sci 81:127–132

Oliveira AFA, Mar JM, Santos SF, da Silva Júnior JL, Kluczkovski AM, Bakry AM et al (2018) Non-thermal combined treatments in the processing of açai (*Euterpe oleracea*) juice. Food Chem 265:57–63

Oms-Oliu G, Martín-Belloso O, Soliva-Fortuny R (2010) Pulsed light treatments for food preservation: a review. Food Bioproc Technol 3:13–23

Ostermeier R, Hill K, Toepfl S, Jäger H (2020) Pulsed electric field as a sustainable tool for the production of healthy snacks. In: In Pulsed electric fields to obtain healthier and sustainable food for tomorrow. Academic, London, pp 103–128

Pankaj SK, Wan Z, Colonna W, Keener KM (2017) Effect of high voltage atmospheric cold plasma on white grape juice quality. J Sci Food Agric 97:4016–4021

Pankaj SK, Bueno-Ferrer C, Misra NN, Bourke P, Cullen PJ (2014a) Zein film: effects of dielectric barrier discharge atmospheric cold plasma. J Appl Polym Sci 131:40801–40806

Pankaj SK, Bueno-Ferrer C, Misra NN, Milosavljević V, O'Donnell CP, Bourke P, Keener KM, Cullen PJ (2014b) Applications of cold plasma technology in food packaging. Trends Food Sci Technol 35:5–17

Pankaj SK, Bueno-Ferrer C, Misra NN, O'Neill L, Jiménez A, Bourke P, Cullen PJ (2014c) Characterization of polylactic acid films for food packaging as affected by dielectric barrier discharge atmospheric plasma. Innovative Food Sci Emerg Technol 21:107–113

Pankaj SK, Bueno-Ferrer C, Misra NN, O'Neill L, Jiménez A, Bourke P, Cullen PJ (2014d) Surface, thermal and antimicrobial release properties of plasma-treated zein films. J Renew Mater 2:77–84

Pankaj SK, Bueno-Ferrer C, Misra NN, O'Neill L, Tiwari BK, Bourke P, Cullen PJ (2014e) Physicochemical characterization of plasma-treated sodium caseinate film. Food Res Int 66:438–444

Pankaj SK, Bueno-Ferrer C, Misra NN, O'Neill L, Tiwari BK, Bourke P, Cullen PJ (2015) Dielectric barrier discharge atmospheric air plasma treatment of high amylose corn starch films. LWT-Food Sci Technol 63:1076–1082

Panyasiri P, Yingkamhaeng N, Lam NT, Sukyai P (2018) Extraction of cellulose nanofibrils from amylase-treated cassava bagasse using high-pressure homogenization. Cellulose 25:1757–1768

Papachristodoulou M, Koukounaras A, Siomos AS, Liakou A, Gerasopoulos D (2018) The effects of ozonated water on the microbial counts and the shelf life attributes of fresh-cut spinach. J Food Process Preserv 42:e13404
Park S, Choe W, Jo C (2018) Interplay among ozone and nitrogen oxides in air plasmas: rapid transition of plasma chemistry. Chem Eng J 352:1014–1021
Parniakov O, Bals O, Lebovka N, Vorobiev E (2016) Pulsed electric field assisted vacuum freeze-drying of apple tissue. Innovative Food Sci Emerg Technol 35:52–57
Parniakov O, Barba FJ, Grimi N, Marchal L, Jubeau S, Lebovka N, Vorobiev E (2015a) Pulsed electric field and pH assisted selective extraction of intracellular components from microalgae Nannochloropsis. Algal Res 8:128–134
Parniakov O, Barba FJ, Grimi N, Marchal L, Jubeau S, Lebovka N, Vorobiev E (2015b) Pulsed electric field assisted extraction of nutritionally valuable compounds from microalgae Nannochloropsis spp. using the binary mixture of organic solvents and water. Innovative Food Sci Emerg Technol 27:79–85
Pataro G, Carullo D, Falcone M, Ferrari G (2020) Recovery of lycopene from industrially derived tomato processing by-products by pulsed electric fields-assisted extraction. Innovative Food Sci Emerg Technol 63:102369
Patil S, Valdramidis V, Karatzas K, Cullen P, Bourke P (2011) Technological University Dublin. Assessing the mechanism of microbial inactivation during ozone processing. ICEF11 international conference on engineering and food, Athens, Greece, May 2011. https://doi.org/10.21427/D7N62X
Patrignani F, Lanciotti R (2016) Applications of high and ultra high pressure homogenization for food safety. Front Microbiol 7:1132
Peñas E, Préstamo G, Gómez R (2004) High pressure and the enzymatic hydrolysis of soybean whey proteins. Food Chem 85:641–648
Perry JJ, Yousef AE (2013) Factors affecting thermal resistance of Salmonella enterica serovar enteritidis ODA 99-30581-13 in shell egg contents and use of heat-ozone combinations for egg pasteurization. J Food Prot 76:213–219
Piechowiak T, Grzelak-Błaszczyk K, Sójka M, Balawejder M (2020) Changes in phenolic compounds profile and glutathione status in raspberry fruit during storage in ozone-enriched atmosphere. Postharvest Biol Technol 168:111277
Piechowiak T, Antos P, Józefczyk R, Kosowski P, Skrobacz K, Balawejder M (2019) Impact of ozonation process on the microbiological contamination and antioxidant capacity of highbush blueberry (*Vaccinum corymbosum* L.) fruit during cold storage. Ozone Sci Eng 41:376–385
Pinela J, Prieto MA, Barros L, Carvalho AM, Oliveira BPP, Saraiva JA, Ferreira ICFR (2018) Cold extraction of phenolic compounds from watercress by high hydrostatic pressure: process modelling and optimization. Sep Purif Technol 192:501–512
Pinto L, Palma A, Cefola M, Pace B, D'Aquino S, Carboni C, Baruzzi F (2020) Effect of modified atmosphere packaging (MAP) and gaseous ozone pre-packaging treatment on the physico-chemical, microbiological and sensory quality of small berry fruit. Food Packag Shelf Life 26:100573
Postma PR, Pataro G, Capitoli M, Barbosa MJ, Wijffels RH, Eppink MHM et al (2016) Selective extraction of intracellular components from the microalga Chlorella vulgaris by combined pulsed electric field–temperature treatment. Bioresour Technol 203:80–88
Praporscic I, Ghnimi S, Vorobiev E (2005) Enhancement of pressing of sugar beet cuts by combined ohmic heating and pulsed electric field treatment. J Food Process Preserv 29:378–389
Praporscic I, Lebovka N, Vorobiev E, Mietton-Peuchot M (2007) Pulsed electric field enhanced expression and juice quality of white grapes. Sep Purif Technol 52:520–526
Prasad KN, Yang E, Yi C, Zhao M, Jiang Y (2009) Effects of high pressure extraction on the extraction yield, total phenolic content and antioxidant activity of longan fruit pericarp. Innovative Food Sci Emerg Technol 10:155–159
Proctor A, Ahmedna M, Kumar JV, Goktepe I (2004) Degradation of aflatoxins in peanut kernels/flour by gaseous ozonation and mild treatment. Food Addit Contam 21:786–973

Puértolas E, Barba FJ (2016) Electrotechnologies applied to valorization of by-products from food industry: Main findings, energy and economic cost of their industrialization. Food Bioprod Process 100:172–184

Puértolas E, Cregenzán O, Luengo E, Álvarez I, Raso J (2013) Pulsed-electric-field-assisted extraction of anthocyanins from purple-fleshed potato. Food Chem 136:1330–1336

Puértolas E, de Marañón IM (2015) Olive oil pilot-production assisted by pulsed electric field: impact on extraction yield, chemical parameters and sensory properties. Food Chem 167:497–502

Pyatkovskyy TI, Shynkaryk MV, Mohamed HM, Yousef AE, Sastry SK (2018) Effects of combined high pressure (HPP), pulsed electric field (PEF) and sonication treatments on inactivation of *Listeria innocua*. J Food Eng 233:49–56

Qin L, Li W-C, Liu L, Zhu J-Q, Li X, Li B-Z, Yuan Y-J (2016) Inhibition of lignin derived phenolic compounds to cellulase. Biotechnol Biofuels 9:70

Qiu L, Zhang M, Chitrakar B, Bhandari B (2020) Application of power ultrasound in freezing and thawing processes: effect on process efficiency and product quality. Ultrason Sonochem 68:105230

Quiroz-González B, Rodríguez-Martínez V, del Rosario García-Mateos M, Torres JA, Welti-Chanes J (2018) High hydrostatic pressure inactivation and recovery study of Listeria innocua and Saccharomyces cerevisiae in pitaya (Stenocereus pruinosus) juice Innovative Food Sci Emerg Technol 50:169–173

Rahman MS (2015) Hurdle technology in food preservation. Springer, Minimally processed foods, pp 17–33

Rashid F, Bao Y, Ahmed Z, Huang J-Y (2020) Effect of high voltage atmospheric cold plasma on extraction of fenugreek galactomannan and its physicochemical properties. Food Res Int 138:109776

Rastogi NK, Angersbach A, Knorr D (2000a) Synergistic effect of high hydrostatic pressure pretreatment and osmotic stress on mass transfer during osmotic dehydration. J Food Eng 45:25–31

Rastogi NK, Angersbach A, Knorr D (2000b) Evaluation of mass transfer mechanisms during osmotic treatment of plant materials. J Food Sci 65:1016–1021

Rastogi NK, Niranjan K (1998) Enhanced mass transfer during osmotic dehydration of high pressure treated pineapple. J Food Sci 63:508–511

Rastogi NK, Raghavarao KSMS, Niranjan K, Knorr D (2002) Recent developments in osmotic dehydration: methods to enhance mass transfer. Trends Food Sci Technol 13:48–59

Roach A, Harte F (2008) Disruption and sedimentation of casein micelles and casein micelle isolates under high-pressure homogenization. Innovative Food Sci Emerg Technol 9:1–8

Robak K, Balcerek M (2018) Review of second generation bioethanol production from residual biomass. Food Technol Biotechnol 56:174

Rodoni L, Casadei N, Concellón A, Chaves Alicia AR, Vicente AR (2010) Effect of short-term ozone treatments on tomato (*Solanum lycopersicum* L.) fruit quality and cell wall degradation. J Agric Food Chem 58:594–599

Rodríguez-Alcalá LM, Harte F, Fontecha J (2009) Fatty acid profile and CLA isomers content of cow, ewe and goat milks processed by high pressure homogenization. Innovative Food Sci Emerg Technol 10:32–36

Saelee K, Yingkamhaeng N, Nimchua T, Sukyai P (2016) An environmentally friendly xylanase-assisted pretreatment for cellulose nanofibrils isolation from sugarcane bagasse by high-pressure homogenization. Ind Crops Prod 82:149–160

Sampedro F, McAloon A, Yee W, Fan X, Zhang HQ, Geveke DJ (2013) Cost analysis of commercial pasteurization of orange juice by pulsed electric fields. Innovative Food Sci Emerg Technol 17:72–78

Sandra S, Dalgleish DG (2005) Effects of ultra-high-pressure homogenization and heating on structural properties of casein micelles in reconstituted skim milk powder. Int Dairy J 15:1095–1104

Sarheed O, Shouqair D, Ramesh KVRNS, Khaleel T, Amin M, Boateng J, Drechsler M (2020) Formation of stable nanoemulsions by ultrasound-assisted two-step emulsification process for

topical drug delivery: effect of oil phase composition and surfactant concentration and loratadine as ripening inhibitor, Int J Pharm 576:118952
Sarkis JR, Boussetta N, Tessaro IC, Marczak LDF, Vorobiev E (2015) Application of pulsed electric fields and high voltage electrical discharges for oil extraction from sesame seeds. J Food Eng 153:20–27
Scholtz V, Pazlarova P, Souskova H, Khun J, Julak J (2015) Nonthermal plasma – a tool for decontamination and disinfection. Biotechnol Adv 33:1108–1119
Scudino H, Eric Keven Silva, Andresa Gomes, Jonas T. Guimarães, Rosiane L. Cunha, Anderson S. Sant'Ana, M. Angela A. Meireles, Adriano G. Cruz (2020) Ultrasound stabilization of raw milk: Microbial and enzymatic inactivation, physicochemical properties and kinetic stability, Ultrasonics Sonochemistry 67:105185
Sedraoui S, Badr A, Barba MGM, Doyen A, Tabka Z, Desjardins Y (2020) Optimization of the ultrahigh-pressure–assisted extraction of phenolic compounds and antioxidant activity from palm dates (*Phoenix dactylifera L.*). Food Anal Methods 13:1556–1569
Semenoglou I, Dimopoulos G, Tsironi T, Taoukis P (2020) Mathematical modelling of the effect of solution concentration and the combined application of pulsed electric fields on mass transfer during osmotic dehydration of sea bass fillets. Food Bioprod Proc 121:186–192
Sepulveda DR, Góngora-Nieto MM, Guerrero JA, Barbosa-Cánovas GV (2005) Production of extended-shelf life milk by processing pasteurized milk with pulsed electric fields. J Food Eng 67:81–86
Serra M, Trujillo AJ, Jaramillo PD, Guamis B, & Ferragut V (2008) Ultra-high pressure homogenization-induced changes in skim milk: Impact on acid coagulation properties. J Dairy Res, 75(1):69–75
Serra M, Trujillo AJ, Guamis B, Ferragut V (2009) Evaluation of physical properties during storage of set and stirred yogurts made from ultra-high pressure homogenization-treated milk. Food Hydrocoll 23:82–91
Shaghaleh H, Xu X, Liu H, Shifa Wang S, Hamoud YA, Dong F, Luo J, Sharifian et al (2019) The effect of atmospheric pressure plasma pretreat-ment with various gases on the structural characteristics and chemical com-position of wheat straw and applications to enzymatic hydrolysis. Energy 176:195–210
Shaofeng, Yuan Changjian, Li Yichi, Zhang Hang, Yu Yunfei, Xie Yahui, Guo Weirong, Yao (2021) Ultrasound as an emerging technology for the elimination of chemical contaminants in food: A review Trends in Food Sci Technol 109:374–385. https://doi.org/10.1016/j.tifs.2021.01.048
Sheng L, Wang Y, Chen J, Zou J, Wang Q, Ma M (2018) Influence of high-intensity ultrasound on foaming and structural properties of egg white. Food Res Int 108:604–610
Shorstkii I, Khudyakov D, Mirshekarloo MS (2020) Pulsed electric field assisted sunflower oil pilot production: impact on oil yield, extraction kinetics and chemical parameters. Innovative Food Sci Emerg Technol 60:102309
Shorstkii I, Mirshekarloo MS, Koshevoi E (2017) Application of pulsed electric field for oil extraction from sunflower seeds: electrical parameter effects on oil yield. J Food Process Eng 40:e12281
Shouqin Z, Junjie Z, Changzhen W (2004) Novel high pressure extraction technology. Int J Pharm 278:471–474
Silve A, Kian CB, Papachristou I, Kubisch C, Nazarova N, Wüstner R et al (2018) Incubation time after pulsed electric field treatment of microalgae enhances the efficiency of extraction processes and enables the reduction of specific treatment energy. Bioresour Technol 269:179–187
Sopanangkul A, Ledward DA, Niranian K (2002) Mass transfer during sucrose infusion into potatoes under high pressure. Food Eng Phys Prop 67:6
Spinei M, Oroian M (2021) The influence of osmotic treatment assisted by ultrasound on the physico-chemical characteristics of blueberries (*Vaccinium myrtillus* L.). Ultrasonics 110:106298
Strati IF, Gogou E, Oreopoulou V (2015) Enzyme and high pressure assisted extraction of carotenoids from tomato waste. Food Bioprod Proc 94:668–674

Sun, k., Jiang, W. (2020) Study on ultra – high pressure assisted extraction of polysaccharides from Umbilicaria in Yellow Mountain. Mater Sci Forum 980:187–196

Sung H-J, Song W-J, Kim K-P, Ryu S, Kang D-H (2014) Combination effect of ozone and heat treatments for the inactivation of *Escherichia coli* O157:H7, *Salmonella Typhimurium*, and *Listeria monocytogenes* in apple juice. Int J Food Microbiol 171:147–153

Suwal S, Perreault V, Marciniak A, Tamigneaux E, Deslandes E, Bazinet L, Jacques L, Doyen A (2019) Effects of high hydrostatic pressure and polysaccharidases on the extraction of antioxidant compounds from red macroalgae, *Palmaria palmata* and *Solieria chordalis*. J Food Eng 252:53–59

Thammakiti S, Suphantharika M, Phaesuwan T, Verduyn C (2004) Preparation of spent brewer's yeast β-glucans for potential applications in the food industry. Int J Food Sci Technol 39:21–29

Thirumdas R, Sarangapani C, Annapure US (2015) Cold plasma: a novel non-thermal technology for food processing. Food Biophys 10:1–11

Tiwari BK, O'Donnell CP, Muthukumarappan K, Cullen PJ (2008) Anthocyanin and colour degradation in ozone treated blackberry juice. Innovative Food Sci Emerg Technol 10:70–75

Tiwari BK, O'Donnell CP, Patras A, Brunton N, Cullen PJ (2009) Effect of ozone processing on anthocyanins and ascorbic acid degradation of strawberry juice. Food Chem 113:1119–1126

Tiwari BK, Brennan CS, Curran T, Gallagher E, Cullen PJ et al (2010) Application of ozone in grain processing. J Cereal Sci 51:248–255

Toepfl S, Heinz V, & Knorr D (2007) High intensity pulsed electric fields applied for food preservation. Chemical engineering and processing. Process intensification, 46(6):537–546

Toepfl S, Mathys A, Heinz V, Knorr D (2006) Review: potential of high hydrostatic pressure and pulsed electric fields for energy efficient and environmentally friendly food processing. Food Rev Intl 22:405–423

Tokusoglu O (2015) Introduction to innovative food processing. Nat Sci Dis 1:85–87

Torlak E, Sert D, Ulca P (2013) Efficacy of gaseous ozone against Salmonella and microbial population on dried oregano. Int J Food Microbiol 165:276–280

Torres B, Tiwari BK, Patras A, Wijngaard HH, Brunton N, Cullen PJ, O' Donnell, C.P. (2011) Effect of ozone processing on the colour, rheological properties and phenolic content of apple juice. Food Chem 124:721–726

Torres-Ossandón MJ, Antonio Vega-Gálvez A, Jéssica López J, Karina Stucken K, Julio Romero J, Karina Di Scala K (2018) Effects of high hydrostatic pressure processing and supercritical fluid extraction on bioactive compounds and antioxidant capacity of Cape gooseberry pulp (*Physalis peruviana L.*). J Supercrit Fluids 138:215–220

Tran T, Denimal E, Lafarge C, Journaux L, Lee JA, Winckler P, Perrier-Cornet J-M, Pradelles R, Loupiac C (2019) Effect of high hydrostatic pressure on extraction of B-phycoerythrin from *Porphyridium cruentum*: use of confocal microscopy and image processing. Argal Res 38:101394

Trespalacios P, Pla R (2017a) Simultaneous application of transglutaminase and high pressure to improve functional properties of chicken meat gels. Food Chem 100:264–272

Trespalacios P, Pla R (2017b) Synergistic action of transglutaminase and high pressure on chicken meat and egg gels in absence of phosphates. Food Chem 104:1718–1727

Trigo JP, Alexandre EMC, Silva S, Costa E, Saraiva JA, Pintado M (2020) Study of viability of high pressure extract from pomegranate peel to improve carrot juice characteristics. Food Funct 11:3410

Tsevdou M, Eleftheriou E, Taoukis P (2013a) Transglutaminase treatment of thermally and high pressure processed milk: effects on the properties and storage stability of set yoghurt. Innovative Food Sci Emerg Technol 17:144–152

Tsevdou M, Gogou E, Taoukis P (2019) High hydrostatic pressure processing of foods, chapter 4. In: Vorobiev E, Chemat F (eds) Green food processing techniques: preservation, transformation and extraction. Academic, Elsevier, London, pp 87–137

Tsevdou M, Soukoulis C, Cappellin L, Gasperi F, Taoukis P, Biasioli F (2013b) Monitoring and modeling of endogenous flavour compounds evolution during fermentation of thermally, high hydrostatic pressure or transglutaminase treated milk using PTR-TOF-MS. Food Chem 138:2159–2167

Tsevdou M, Theodorou G, Pantelaiou S, Chatzigeorgiou A, Politis I, Taoukis P (2020) Impact of type and enzymatic/high pressure treatment of milk on the quality and bio-functional profile of yoghurt. Foods 9:49

Turk MF, Vorobiev E, Baron A (2012) Improving apple juice expression and quality by pulsed electric field on an industrial scale. LWT-Food Sci Technol 49:245–250

Tylewicz U, Tappi S, Mannozzi C, Romani S, Dellarosa N, Laghi L et al (2017) Effect of pulsed electric field (PEF) pre-treatment coupled with osmotic dehydration on physico-chemical characteristics of organic strawberries. J Food Eng 213:2–9

Van Deynse A, Morent R, De Geyter N (2016) Surface modification of polymers using atmospheric pressure cold plasma technology. In: Méndez-Vilas A, Solano A (eds) Polymer science: research advances, practical applications and educational aspects. Formatex Research Center S.L, Badajoz, pp 506–516

Vannini L, Patrignani F, Iucci L, Ndagijimana M, Vallicelli M, Lanciotti R, Guerzoni ME (2008) Effect of a pre-treatment of milk with high pressure homogenization on yield as well as on microbiological, lipolytic and proteolytic patterns of "Pecorino" cheese. Int J Food Microbiol 128:329–335

Vercet A, Sánchez C, Burgos J, Montañés L, Buesa PL (2002) The effects of manothermosonication on tomato pectic enzymes and tomato paste rheological properties, J Food Eng 53(3):273–278

Verduyn C, Suksomcheep A, Suphantharika M (1999) Effect of high pressure homogenization and papain on the preparation of autolysed yeast extract. World J Microbiol Biotechnol 15:57–63

Verma D, Kaushik N, Rao PS (2014) Application of high hydrostatic pressure as a pretreatment for osmotic dehydration of Banana slices (*Musa cavendishii*) finish dried by dehumidified air drying. Food Bioproc Technol 7:1281–1297

Vinatoru M (2001) An overview of the ultrasonically assisted extraction of bioactive principles from herbs. Ultrason Sonochem 8:303–313

Wang T, MacGregor SJ, Anderson JG, Woolsey GA (2005) Pulsed ultra-violet inactivation spectrum of *Escherichia coli*. Water Res 39:2921–2925

Wang C-Y, Huang H-W, Hsu C-P, Yang BB (2016) Recent advances in food processing using high hydrostatic pressure technology. Crit Rev Food Sci Nutr 56:527–540

Wang H, Ma X, Cheng Q, Wang L, Zhang L (2018) Deep eutectic solvent-based ultrahigh pressure extraction of Baicalin from *Scutellaria Baicalensis* Georgi. Molecules 23:3233

Wei G, Zhang A, Chen K, Ouyang P (2017) Enzymatic production of N-acetyl-D-glucosamine from crayfish shell wastes pretreated via high pressure homogenization. Carbohydr Polym 171:236–241

Wiktor A, Nowacka M, Dadan M, Rybak K, Lojkowski W, Chudoba T, Witrowa-Rajchert D (2016) The effect of pulsed electric field on drying kinetics, color, and microstructure of carrot. Dry Technol 34:1286–1296

Won MY, Lee SJ, Min SC (2017) Mandarin preservation by microwave-powered cold plasma treatment. Innovative Food Sci Emerg Technol 39:25–32

Xi J, Shen D, Zhao S, Lu B, Li Y, Zhang R (2009) Characterization of polyphenols from green tea leaves using a high hydrostatic pressure extraction. Int J Pharm 382:139–143

Xia T, Shi S, Wan X (2006) Impact of ultrasonic-assisted extraction on the chemical and sensory quality of tea infusion. J Food Eng 74:557–560

Xie F, Zhao T, Wan H, Li M, Sun L, Wang Z, Zhang S (2019) Structural and physicochemical characteristics of rice bran dietary fiber by cellulase and high-pressure homogenization. Appl Sci 9:1270

Xie Y, Ho SH, Chen CNN, Chen CY, Jing K, Ng IS et al (2016) Disruption of thermo-tolerant *Desmodesmus* sp. F51 in high pressure homogenization as a prelude to carotenoids extraction. Biochem Eng J 109:243–251

Xing JJ, Cheng YL, Chen P, Shan L, Ruan R, Li D, Wang LJ (2019) Effect of high-pressure homogenization on the extraction of sulforaphane from broccoli (*Brassica oleracea*) seeds. Powder Technol 358:103–109

Yang RF, Geng LL, Lu HQ, Fan XD (2017) Ultrasound-synergized electrostatic field extraction of total flavonoids from Hemerocallis citrina baroni, Ultrason. Sonochem. 34:571–579
Yang G, Wang R, Gao JR, Niu D, Li J, Wen QH, Zeng XA (2021) The effect of moderate pulsed electric fields on autolysis of *Saccharomyces cerevisiae* and the amino acid content in autolysates. Int J Food Sci Technol 56:441–451
Yildiz S, Pokhrel PR, Unluturk S, Barbosa-Cánovas GV (2019) Identification of equivalent processing conditions for pasteurization of strawberry juice by high pressure, ultrasound, and pulsed electric fields processing. Innovative Food Sci Emerg Technol 57:102195
Yong HI, Kim H-J, Park S, Alahakoon AU, Kim K, Choe W, Jo C (2015) Evaluation of pathogen inactivation on sliced cheese induced by encapsulated atmospheric pressure dielectric barrier discharge plasma. Food Microbiol 46:46–50
Yuan B, Ren J, Zhao M, Luo D, Gu L (2012) Effects of limited enzymatic hydrolysis with pepsin and high-pressure homogenization on the functional properties of soybean protein isolate. LWT-Food Sci Technol 46:453–459
Yue W, Yao PJ, Wei YN (2009) Ultraviolet radiation-induced accelerated degradation of chitosan by ozone treatment. Carbohydr Polym 77:639–642
Yue W, Yao PJ, Wei YN, Li SQ, Lai F, Liu XM (2008a) An innovative method for preparation of acid-free-water-soluble low-molecular-weight chitosan (AFWSLMWC). Food Chem 108:1082–1087
Yue W, Yao PJ, Wei YN, Mo HT (2008b) Synergetic effect of ozone and ultrasonic radiation on degradation of chitosan. Polym Degrad Stab 93:1814–1821
Zamora A, Ferragut V, Jaramillo PD, Guamis B, Trujillo AJ (2007) Effects of ultra-high pressure homogenization on the cheese-making properties of milk. J Dairy Sci 90:13–23
Zeng XA, Han Z, Zi ZH (2010) Effects of pulsed electric field treatments on quality of peanut oil. Food Control 21:611–614
Zhang Y, Abatzoglou N (2020) Fundamentals, applications and potentials of ultrasoundassisted drying, Chem. Eng Res Des 154:21–4.
Zhang R, Su D, Hou F, Liu L, Huang F, Dong L, Deng Y, Zhang Y, Wei Z, Zhang M (2017) Optimized ultra-high-pressure-assisted extraction of procyanidins from lychee pericarp improves the antioxidant activity of extracts. Biosci Biotechnol Biochem 81:1576–1585
Zhang T, Jiang B, Miao M, Mu W, Li Y (2012) Combined effects of high-pressure and enzymatic treatments on the hydrolysis of chickpea protein isolates and antioxidant activity of the hydrolysates. Food Chem 135:904–912
Zhang Y, Ma L, Shi X (2011) Ultra-high pressure assisted extraction of gelatin from fish skin. Food Sci 32:99–103
Zhao F, Zhang D, Li X, Dong H (2018) High-pressure homogenization pretreatment before enzymolysis of soy protein isolate: the effect of pressure level on aggregation and structural conformations of the protein. Molecules 23:1775
Zhao L, Wang Y, Hu X, Sun Z, Liao X (2016) Korla pear juice treated by ultrafiltration followed by high pressure processing or high temperature short time. LWT-Food Sci Technol 65:283–289
Zhao Y, Jiand Y, Ding Y, Wang D, Deng Y (2019) High hydrostatic pressure-assisted extraction of high-molecular-weight Melanoidins from black garlic: composition, structure, and bioactive properties. J Food Qual 2019:1682749
Zhu F (2018) Effect of ozone treatment on the quality of grain products. Food Chem 264:358–366
Zhu X, Cheng Y, Chen P, Peng P, Liu S, Li D, Ruan R (2016) Effect of alkaline and high-pressure homogenization on the extraction of phenolic acids from potato peels. Innovative Food Sci Emerg Technol 37:91–97
Zhu Z, Lanier TC, Farkas BE, Li B-S (2014) Transglutaminase and high pressure effects on heat-induced gelation of Alaska pollock (*Theragra chalcogramma*) surimi. J Food Eng 131:154–160

Chapter 10
Implementation of a Novel Nonthermal Plasma Air Cleaner in a Plant Factory

Masaaki Okubo

1 Introduction

Lettuce and iceplant are increasingly being cultivated and harvested in plant factories. However, the cultivated items and their uses, such as their consumption as general raw foods, processed foods, and medical products, are diversifying, and the plant factory specifications for each of these uses are very different. For example, when we produce high-value-added plants that suppress the number of viable bacteria and can be eaten without being washed or used as raw materials for cosmetics and medical treatments, the plant cultivation environment should be optimized and cleaned. Therefore, artificially-lit plant factories are often used, as they have closed environments. Such plant factories form the focus of this study. In order to clean the environment of the plant factory, high-efficiency particulate air (HEPA) filters are generally used. However, higher capital and running costs are required because of the large pressure drop and the higher cost of the filter.

We focus on an artificially-lit plant factory and review the application of an ion emission-type nonthermal plasma air cleaner as a low-cost technology for realizing sterilization in the factory (Kanayama 2014; Maruyama 2016). Furthermore, we develop a numerical prediction technology for ion concentration and experimentally examine the sterilization ability of the cleaner. Finally, we report on the validity of the nonthermal plasma air cleaner.

M. Okubo (✉)
Department of Mechanical Engineering, Osaka Prefecture University, Sakai, Japan
e-mail: mokubo@me.osakafu-u.ac.jp

A. Režek Jambrak (ed.), *Nonthermal Processing in Agri-Food-Bio Sciences*,
Food Engineering Series, https://doi.org/10.1007/978-3-030-92415-7_10

2 Clean-up Technologies for the Environment of an Artificially-Lit Plant Factory

2.1 Conventional Technology with HEPA Filter Air Cleaner

In an artificially-lit factory (Kanayama 2014; Maruyama 2016), plants can be produced via pesticide-free cultivation. These plants can be eaten without being washed, have a low viable cell count, can be stored for a long duration, and appeal to consumers as they can be safely consumed. When cultivated in this manner, the plants produced do not have to be cleaned at the time of processing, which is an advantage. In order to realize pesticide-free cultivation, a clean environment is required, and cleanliness achieved through hygiene management, such as in food factories, is important. Plant factories using artificial light have a lower risk of dust, germs, and pest invasion than solar-powered plant factories.

In general, dust, germs, and pests enter and exit artificially-lit plant factories through the entry and exit points for workers and items; further, there is a high possibility that the dust, germs, and pests are brought in by the workers and items. Therefore, installing an air curtain at the entrance and an air cleaning system in a cleanroom can effectively prevent the entry of dust, germs, and pests. After minimizing the number of entrances and exits, a front room, air shower, airlock room equipment, and handwashing station provide effective protection when workers enter and leave the room. Furthermore, the wearing of a clean suit, sterilization of carried-in items, and possible further measures are also important. A certain degree of cleanliness should be ensured in the cultivation room; however, no clear standards that should be followed exist, and cleanliness is managed according to the cultivated item and its use.

The cleanliness of a cleanroom in a plant factory varies from ISO class 6 to 8 depending on the facility, but in several cases, it is ISO class 8. Cleanliness in these facilities is obtained through general hygiene control, such as the number of airborne particles, the concentration of CO_2, and temperature control. Since the generation of suspended particles that will affect the cleanliness is expected to differ from case to case, criteria for determining whether a cultivation room has been efficiently cleaned may be obtained by understanding the relationship between actual cleanliness and the number of viable bacteria. An air cleaning system using a HEPA filter is generally used to clean a cultivation room. A HEPA filter is used to remove dust and fumes from the air to create clean air. This is the primary filter used in a cleanroom for manufacturing electronic components. Based on the Japan Industrial Standard (JIS), JIS Z 8122 stipulates that a HEPA air filter should have a minimum particle collection rate of 99.97% and a maximum initial pressure loss of 245 Pa for 0.3 μm particles with a rated air volume. The higher the performance of the filter, the larger the ventilation pressure loss, and the larger the fan power required.

2.2 New Technology with Nonthermal Plasma Air Cleaner

In this study, we aim to replace a HEPA-filtered air cleaner with a low pressure-loss nonthermal plasma air cleaner. An ion emission-type plasma air cleaner is a device that generates positive and negative ions via plasma discharge and releases them into the air. Figure 10.1 shows the sterilization process (Kanayama 2014; Sharp Corporation website 2020). After adhering to the surface of an airborne bacterium, the released positive and negative ions are transformed into OH radicals with extremely high oxidizing power. Then, by the OH radicals' instant extraction of hydrogen atoms from the protein on the cell wall of the floating bacterium, the protein is decomposed, and the floating bacterium whose cell wall has been destroyed ceases to pose any hazard. The OH radical binds to the hydrogen atom extracted from the protein and returns to the air as water. Plasma air cleaners can suppress the action of airborne bacteria, eradicate airborne mold bacteria, and suppress the growth of adherent mold bacteria. The decomposition and removal of adherent odors are also realized by such a mechanism.

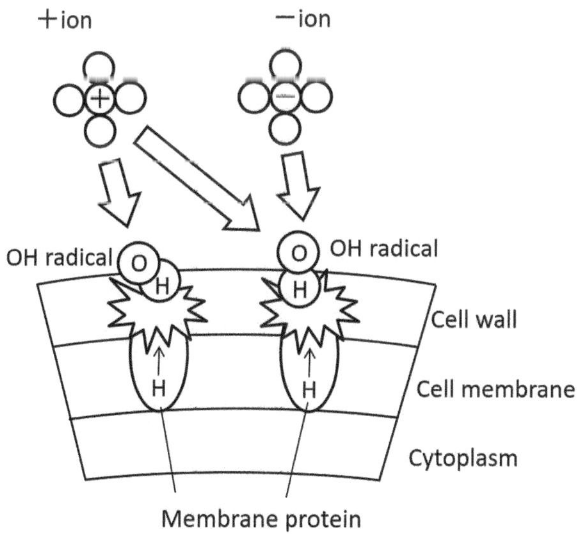

Fig. 10.1 Sterilization process of airborne bacteria with ions (Sharp Corporation website 2020)

3 Design and Simulation of Nonthermal Plasma Air Cleaner

3.1 Generation of Positive and Negative Ions in Nonthermal Plasma Air Cleaner

We focus on the plasma air cleaner design, which was difficult to realize (Okubo et al. 2020). We intend to build an analysis model for the numerical simulation of the plasma reactor part (ion generation device), which generates plasma in the plasma air cleaner. To develop a plasma reactor part similar to the actual ion generation device, we aim to build an analysis model by adopting an axisymmetric two-dimensional numerical simulation. To this end, among the various plasma parameters, the following are numerically predicted: electron temperature, electron number density, positive ion ($H_3O^+(H_2O)_n$ [n = 0–5]), molecular number density, and negative ion ($O_2^-(H_2O)_n$ [n = 0–5]). Numerical simulations are conducted on the ion cluster formation of such ion generation devices. The behavior of streamer evolution and cluster formations are simulated, and we demonstrate that the clustering progresses toward $H_3O^+(H_2O)_4$ and $O_2^-(H_2O)_4$ when the chemical species of $H_3O^+(H_2O)_n$ and $O_2^-(H_2O)_n$ (n = 0–5) are considered, as shown in Fig. 10.2. The ion number densities measured in the study are smaller than those calculated considering the known differences between the numerical model and the experimental apparatus (in particular, the difference due to the loss of ions in the path of the chamber). The evaluation of the lifetime and the diffusion coefficients of the ion clusters can be predicted.

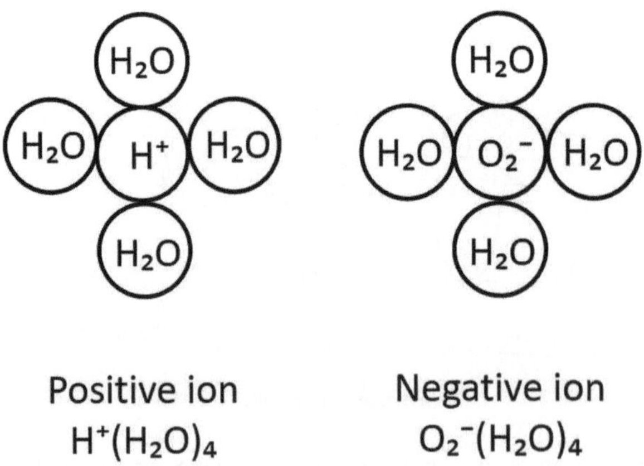

Fig. 10.2 Dominant ion clusters generated in the plasma air cleaner

3.2 Numerical Prediction of Formation of Ion Clusters

3.2.1 Nonthermal Plasma Reactor

In the present study, the nonthermal plasma (NTP) reactor or ion generation device inside a plasma electric air cleaner that uses a pulsed corona discharge is analyzed. When a pulsed high voltage is applied to a pair of electrodes at atmospheric pressure, both negative and positive ions are induced as well as high-speed electrons and radicals. These ions are useful for cleaning out hazardous gases, e.g., odorous gases such as ammonia, from the atmosphere. A numerical simulation of the induced atmospheric plasma is carried out for such air cleaners or ion generation devices. It is noted that the induced negative/positive ions may improve human health and decompose allergens.

3.2.2 Numerical Model

As shown in Fig. 10.3, the analytical model consists of discharge and ground electrodes based on a real corona discharge device. Figure 10.3a displays an r-z plane view. The centerline is at the bottom of the analytical model, along the z axis, which is perpendicular to the r and θ axes. An axisymmetric quasi-two-dimensional analysis is carried out based on cylindrical (r, θ, z) coordinates. The model considered in the numerical method is based on a coaxial needle-to-ring electrode atmospheric NTP reactor. The reactor consists of a needle discharge and a ring-shaped grounded electrode. The needle discharge electrode, which has a negative polarity, is 1 mm in diameter and 4 mm in length. The tip of the needle is spherical. On the ring-shaped grounded electrode, the distance from the origin to the surface of the ring is 5 mm, and the diameter of the cross-section is 1 mm. In the radial direction, the calculation region ranges from $r = 0$ to $r = 9.5$ mm. The sharp tip of the high-voltage needle electrode is positioned at $r = 0$ mm and $z = 0$ mm, while the surface of the grounded ring electrode is placed at $r = 5$ mm. It is noted that both the needle and the ring are metal electrodes, and no dielectric barrier exists in the configuration. In the axial z-direction, the calculation region ranges from $z = -4$ mm to 4 mm. Except for the electrode region, the other boundaries in the model are provided as the outlet or free boundary conditions. Electrons, ions, and radicals, which are induced by the discharge, flow out of the calculation region. Figure 10.3b shows a perspective view of the coaxial needle-to-ring electrodes.

Figure 10.4 shows the nonuniform computational grid and cells. A three-dimensional (r, θ, z) numerical model is employed, and the gradients of the radial r-direction and the axial z-direction are spatially considered with the assumption that $\partial/\partial\theta = 0$, implying that the model is axisymmetric. The nonstructural analytical mesh used in this study is shown in the figure. The total number of cells is 14,952. The cell used in the study (shown in the figure) is almost uniform. Some trials using

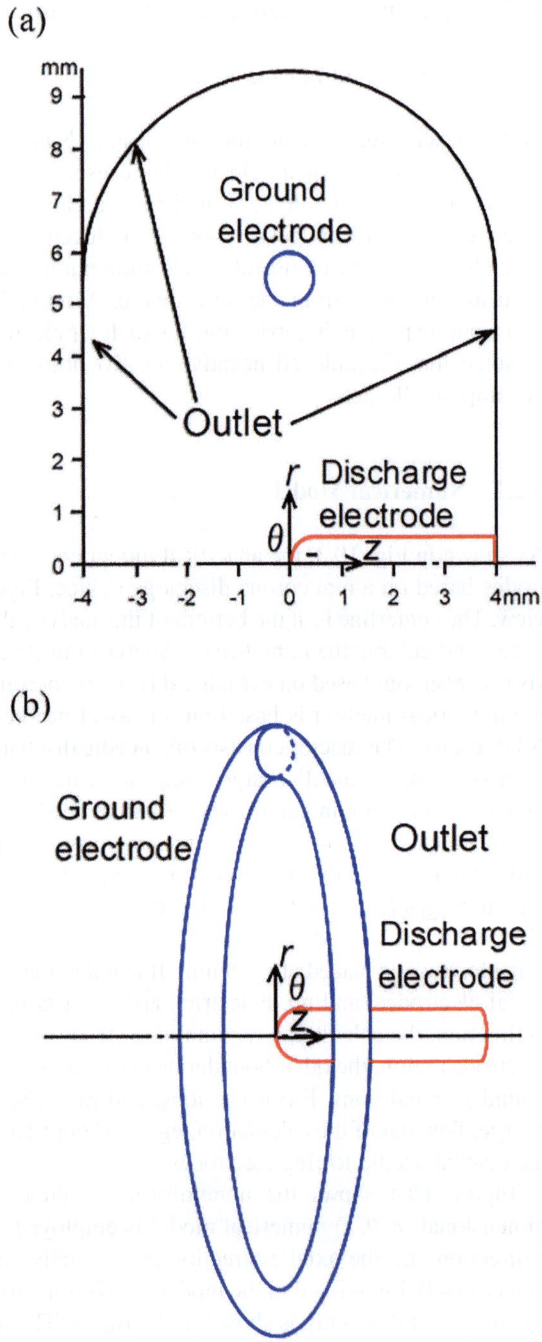

Fig. 10.3 Analytical model composed of discharge and ground electrodes based on the real corona discharge device. (**a**) *r-z* plane view, and (**b**) perspective view (Okubo et al. 2020)

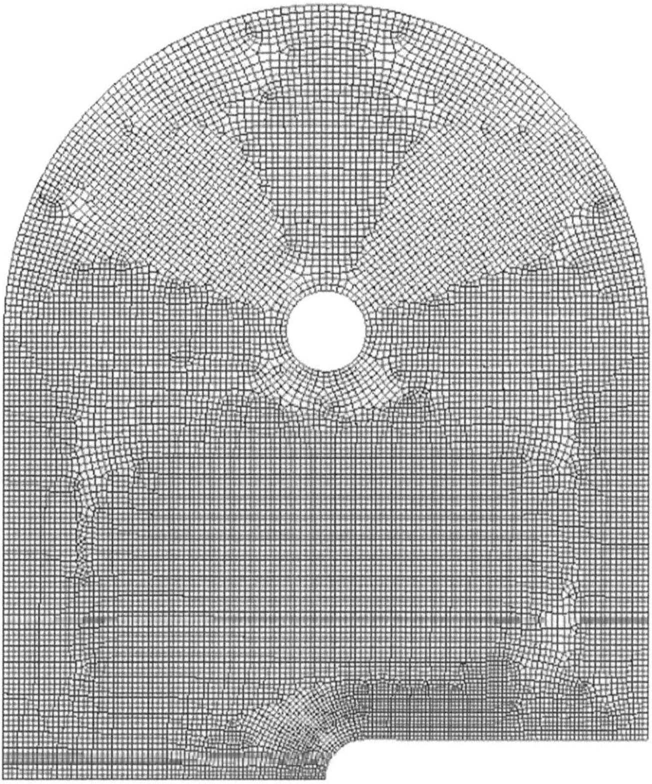

Fig. 10.4 Computational grids and cells for the numerical calculation (14,952 cells) (Okubo et al. 2020)

fewer cells and a nonuniform mesh model were made before employing the mesh system shown in Fig. 10.2. A satisfactory result could not be obtained in the system with fewer cells.

3.2.3 Analysis Procedure

The following continuum fluid equations (Okubo 2015; Okubo et al. 2008; Kuroki et al. 2007) for a two-temperature nonequilibrium NTP are used as the governing equations.

Continuity equation:

$$\frac{\partial \rho}{\partial t} + \nabla \cdot (\rho \mathbf{u}) = 0 \tag{10.1}$$

Momentum equation:

$$\frac{\partial(\rho\mathbf{u})}{\partial t}+\nabla\cdot(\rho\mathbf{u}\mathbf{u})=-\nabla p+\nabla\cdot(\mu\nabla\mathbf{u}) \tag{10.2}$$

Equation of state:

$$p=\rho RT \tag{10.3}$$

Energy equation:

$$\frac{\partial(\rho h)}{\partial t}+\nabla\cdot(\rho\mathbf{u}h)=\nabla\cdot(\lambda\nabla T)+\psi_{\mathrm{D}}+S_{\mathrm{c}}+n_{\mathrm{e}}\sum_{\mathrm{k}}\varepsilon_{\mathrm{k}}n_{\mathrm{k}}K_{\mathrm{k}} \tag{10.4}$$

Transport equations of chemical species:

$$\frac{\partial(\rho Y_{\mathrm{i}})}{\partial t}+\nabla\cdot(\rho\mathbf{u}Y_{\mathrm{i}})=\nabla\cdot\mathbf{J}_{\mathrm{i}}+M_{\mathrm{i}}\omega_{\mathrm{i}} \tag{10.5}$$

$$\mathbf{J}_{\mathrm{i}}=-\rho D_{\mathrm{i}}\nabla Y_{\mathrm{i}}+\rho Y_{\mathrm{i}}\mathbf{u}_{\mathrm{di}}+\mathbf{J}_{\mathrm{ci}} \tag{10.6}$$

$$\omega_{\mathrm{i}}=\sum_{\mathrm{j}=1}^{\mathrm{N_R}}\left(\nu''_{\mathrm{ij}}-\nu'_{\mathrm{ij}}\right)q_{\mathrm{j}} \tag{10.7}$$

$$q_{\mathrm{j}}=k_{\mathrm{j}}\left(\prod_{\mathrm{m}=1}^{\mathrm{N_S}}c_{\mathrm{m}}^{\nu'_{\mathrm{mj}}}-K_{\mathrm{cj}}\prod_{\mathrm{m}=1}^{\mathrm{Ns}}c_{\mathrm{m}}^{\nu''_{\mathrm{mj}}}\right) \tag{10.8}$$

$$k_{\mathrm{j}}=A_{\mathrm{j}}T_{\mathrm{j}}^{n}\exp\left(-E_{\mathrm{j}}/RT_{\mathrm{j}}\right) \tag{10.9}$$

$$k_{\mathrm{j}}=\int\sqrt{T_{\mathrm{e}}}\,\sigma(T_{\mathrm{e}})\,\mathrm{f}(T_{\mathrm{e}})\,\mathrm{d}T_{\mathrm{e}} \tag{10.10}$$

Electron transport equation:

$$\frac{\partial n_{\mathrm{e}}}{\partial t}+\nabla\cdot\Gamma_{\mathrm{e}}=S_{\mathrm{e}} \tag{10.11}$$

$$\Gamma_{\mathrm{e}}=\mu_{\mathrm{e}}n_{\mathrm{e}}\nabla\varphi-D_{\mathrm{e}}\nabla n_{\mathrm{e}} \tag{10.12}$$

Electron energy equation:

$$\frac{3}{2}\frac{\partial(n_{\mathrm{e}}T_{\mathrm{e}})}{\partial t}+\nabla\cdot\left(\frac{5}{2}T_{\mathrm{e}}\Gamma_{\mathrm{e}}-\chi\nabla T_{\mathrm{e}}\right)=P_{\mathrm{elec}}-n_{\mathrm{e}}\sum_{\mathrm{k}}n_{\mathrm{k}}K_{\mathrm{k}} \tag{10.13}$$

Poisson's equation:

$$\nabla \cdot \left(\varepsilon_{\mathrm{r}} \varepsilon_0 \nabla \varphi \right) = -\sigma_{\mathrm{c}}, \tag{10.14}$$

where T_{j} (= T_{e} or T) is the absolute temperature of species j. The quantities A_{j}, n, and E_{j} are the constants corresponding to species j, σ is the collision cross-section as a function of T_{e}, and f is the electron energy distribution function (a Maxwell-Boltzmann distribution is assumed). The definitions of the symbols for physical quantities are the same as those stated in our prior publications (Okubo 2015; Okubo et al. 2008).

In our previous studies on atmospheric air plasma (Okubo 2015; Okubo et al. 2008), we considered 197 gas-phase reactions and 21 surface reactions for 25 chemical species, concerning nitrogen and oxygen ions and radicals as well as their excited states. In a previous paper (Okubo et al. 2008), the electron temperature and the abundances of other chemical species calculated with the present plasma model were compared with the experimental data obtained using optical emission spectroscopy. The results showed the validity of the present plasma model. In the present study, an atmospheric air NTP, including moisture, is tested. The chain of chemical reactions of the positive H_3O^+ ions and the negative O_2^- ions that combine with the moisture in the air to form ion clusters of $H_3O^+(H_2O)_n$ and $O_2^-(H_2O)_n$, respectively, are studied, taking into account the new chemical species. These chain reactions correspond with the reactions based on the chemical reactions reported in the literature (Nagato et al. 2006; Huertas et al. 1978; Ikebe et al. 1989; Dorai 2000; Person et al. 1988; Parts and Luts 2004; Huertas and Fontan 1975). Note that n is either a positive integer or zero. For $n = 0$, although no water molecules are connected to the ions, they are still termed ion clusters in the study. Because of the lack of information on certain details of the chemical reactions, especially the rate constants, the targets are negative and positive ion clusters with $n = 0$–5. Consequently, 225 gas-phase reactions for 40 chemical species (N, N^+, N_2, N_2^+, $N_2(a^1\Sigma_u^-)$, $N_2(A^3\Sigma_u^+)$, $N_2(B^3\Pi_g)$, $N_2(C^3\Pi_u)$, N_3^+, N_4^+, O, $O(^1D)$, $O(^1S)$, O^+, O^-, O_2, O_2^{**}, $O_2(a^1\Delta)$, $O_2(b^1\Sigma)$, O_2^+, O_2^-, $O_2(v)$, O_3, O_3^-, O_4^+, H_3O^+, H_2O, $H_3O^+(H_2O)$, $H_3O^+(H_2O)_2$, $H_3O^+(H_2O)_3$, $H_3O^+(H_2O)_4$, $H_3O^+(H_2O)_5$, $O_2^-(H_2O)$, $O_2^-(H_2O)_2$, $O_2^-(H_2O)_3$, $O_2^-(H_2O)_4$, $O_2^-(H_2O)_5$, H, OH, and H_2), and 34 surface reactions on the electrodes are considered in a moisture air plasma under atmospheric pressure, based on Refs. (Nagato et al. 2006; Huertas et al. 1978; Ikebe et al. 1989; Dorai 2000; Person et al. 1988; Parts and Luts 2004; Huertas and Fontan 1975).

A partial set of the reactions of the ion clusters is as follows:

$$\begin{aligned} & n = 0-4 \\ & O_2^- \left(H_2O \right)_n + H_2O + X \rightarrow O_2^- \left(H_2O \right)_{n+1} + X, X = O_2, k_n \end{aligned} \tag{10.15}$$

$$\begin{aligned} & m = 1-4 \\ & O_2^- \left(H_2O \right)_{m+1} + X \rightarrow O_2^- \left(H_2O \right)_m + X + H_2O, X = O_2, k_m \end{aligned} \tag{10.16}$$

$$n = 0-4$$
$$H_3O^+(H_2O)_n + H_2O + X \rightarrow H_3O^+(H_2O)_{n+1} + x, x = O_2, X = N_2, k'_n \quad (10.17)$$

$$m = 0-4$$
$$H_3O^+(H_2O)_{m+1} + X \rightarrow H_3O^+(H_2O)_m + X + H_2O, X = O_2, X = N_2, k'_m \quad (10.18)$$

$$e + H_2O \rightarrow e + H + OH(E/N = 100\text{Td}), k = 2.0 \times 10^{-9}\text{cm}^3/\text{s} \quad (10.19)$$

$$H + H_2O \rightarrow H_3O^+ + e, (\text{rapid}) \quad (10.20)$$

where k_n, k_m, k'_n, and k'_m are the rate coefficients of cluster formations and are supplied in Table 10.1. It is important to note that the unit of the reaction rate coefficient becomes cm^3/s for two-body reactions and cm^6/s for three-body reactions.

Figure 10.5 shows the applied pulse voltage waveform of the reactor. The waveform is approximated as a sinusoidal function with a negative peak voltage of −4 kV and a duration time T of 600 ns. These characteristics are the same as those reported in previous studies (Okubo 2015; Sato et al. 2005). The use of a pulsed voltage with a very short rise-time (~100 ns) and short duration (1 μs) is considered appropriate for efficient gas cleaning (Okubo 2015; Matsumoto et al. 2010; Wang et al. 2010).

The key initial conditions are as follows: gas temperature $T = 300$ K, $p = 100$ kPa, electron temperature $T_e = 0.2$ eV under the quasi-neutrality condition, uniform electron number density $n_e = 8 \times 10^8\ m^{-3}$, electric potential $\phi = 0$, and the mass ratios of atmospheric air including water moisture are assumed to be N_2: 76.75%, O_2: 23.15%, and H_2O: 0.1%. It should be noted that calculations with higher moisture could not be successfully executed in the present scheme and should be studied in future studies. The viscosity is calculated using Sutherland's law. The Schmidt and Prandtl numbers are set to 0.7 and 0.707, respectively. The thermal conductivity and specific heat of the stainless steel are set to 16 W/(m K) and 0.5 kJ/(kg K), respectively.

Table 10.1 Rate coefficients of formation reactions of ion clusters (Okubo et al. 2020)

unit: cm^3/s or cm^6/s	O_2^- $X = O_2$		H_3O^+ $X = O_2$ (n = 0–3, m = 1–3) X = Air (n = 4, m = 4)		$X = N_2$	
m or n	k_n	k_m	k'_n	k'_m	k'_n	k'_m
0	1.6×10^{-28}	–	3.4×10^{-27}	–	3.7×10^{-27}	7.0×10^{-26}
1	5.4×10^{-28}	1.1×10^{-14}	1.85×10^{-27}	1.4×10^{-17}	2.3×10^{-27}	7.0×10^{-18}
2	2.1×10^{-28}	3.0×10^{-13}	1.5×10^{-27}	4.0×10^{-14}	2.4×10^{-27}	3.0×10^{-14}
3	1.0×10^{-28}	1.2×10^{-12}	9.0×10^{-28}	6.0×10^{-12}	–	–
4	5.0×10^{-29}	4.5×10^{-12}	5.0×10^{-31}	2.0×10^{-15}	–	–

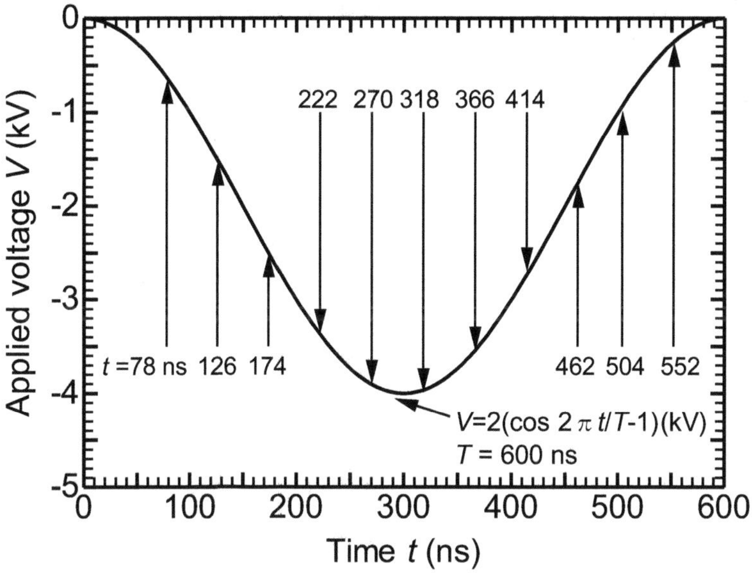

Fig. 10.5 Waveform of the nanosecond pulse voltage applied at the needle electrode (applied voltage peak = −4 kV, duration time = 600 ns) (Okubo et al. 2020)

The system of these governing equations is solved simultaneously using the CFD-ACE+ (ESI Group, France) solver (Okubo 2015; Okubo et al. 2008; Kuroki et al. 2007; Sato et al. 2005). In this solver, the time-implicit semi-implicit method for the pressure-linked equations-consistent (SIMPLEC) method is used to complete the flow, heat transfer, and chemical analysis using Eqs. 10.11, 10.12, and 10.13. An implicit Poisson solver is used for the electric potential analysis with Eq. (10.14) under the given boundary conditions. For the unsteady (transient) calculation of the plasma flow, the first-order Euler implicit scheme is adopted. In general, a plasma flow calculation requires a very small time step because the discharge process progresses rapidly. In the current calculation, $\Delta t = 6 \times 10^{-12}$ s = 6 ps is adopted. When Δt exceeds this value, the calculation diverges. The total number of time steps and the full timespan are 100,000 and 600 ns, respectively. The flow and energy fields must be solved with the same time step used in the CFD-ACE+ solver, although the flow and heat parameters are scarcely changed after the 600 ns calculation. In the current simulation, the plasma is only simulated during the first period of the applied voltage because the computation time is fairly long (approximately one month), even with a high-performance PC. It is also noted that grid convergence studies have been carried out. A more divergent grid yields unresolved results. In the future, parallel computing will be attempted to reduce the computing time.

3.2.4 Predicted Results for Electron Number Density

Figure 10.6 shows the time-dependent electron number density (n_e) distribution obtained from the calculation. A color bar, which indicates the logarithmic value of

Fig. 10.6 Result obtained for the time-dependent electron number density (n_e) distribution. The color bar indicates the logarithmic value of n_e (Okubo et al. 2020)

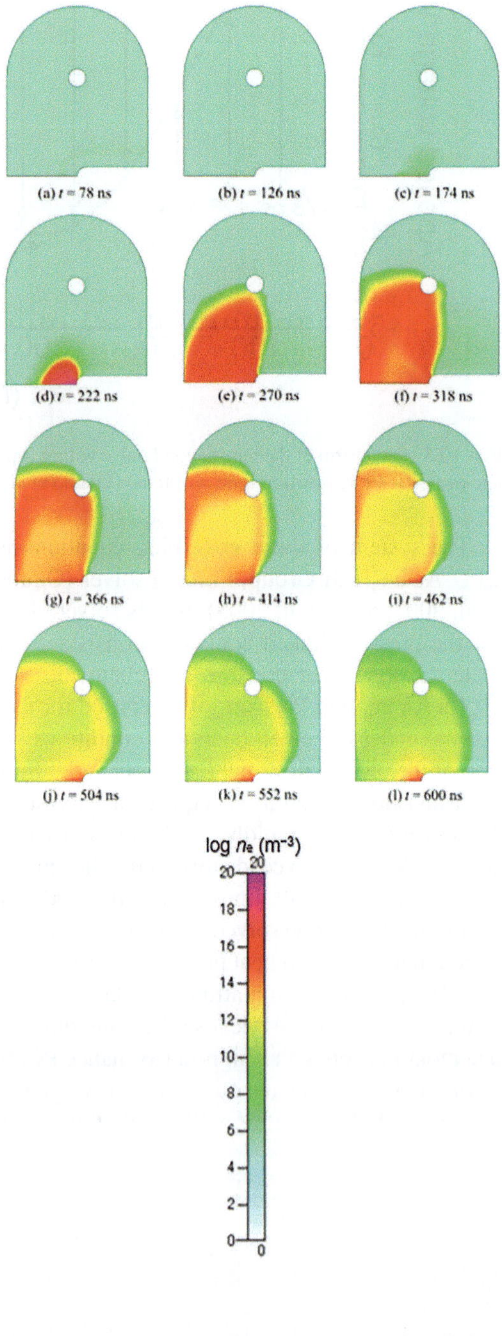

n_e (in units of m^{-3}), is provided in the figure. Almost no change is observed until a time t = 150 ns has elapsed from the moment when the plasma is turned on. The primary streamer discharge is observed at t = 174 ns, after which the number density gradually increases, starting at the tip of the discharge electrode and extending toward the grounded electrode. The primary streamer proceeds and almost completely crosses the gap between the electrodes. The peak voltage is attained at t = 300 ns, and after t = 318 ns, n_e gradually decreases over the entire region. The value of n_e remains higher near the discharge electrode than in the other regions. It is thought that this is a secondary streamer, followed by the first streamer propagation. The maximum n_e is 4.79×10^{19} m^{-3} at the peak voltage. No obvious streamer head is observed, although it was clearly simulated in our earlier study (Okubo 2015).

3.2.5 Predicted Results for Ion Cluster Number Density

Figure 10.7 shows the calculated number-density distribution for the time-dependent positive ion cluster $H_3O^+(H_2O)_n$ for n = 4. Although it is shown as an example of a calculated result, similar results can also be calculated for n = 0, 1, 2, 3, and 5. Almost no change is observed before t = 150 ns from the time the plasma is turned on. However, the number density does increase with an increase in voltage. After t = 174 ns, the density increases gradually, starting at the tip of the discharge electrode and extending toward the grounded electrode. After t = 300 ns, which coincides with the peak voltage, the density remains higher.

Although not all the calculated results are displayed in the figure, it is found that the number densities increase between the discharge gap from the tip to the grounded electrode for all positive ions with n = 0–5. After the peak voltage is attained at t = 300 ns, the number densities of $O_2^-(H_2O)_n$ (n = 3–5) increase slightly. On the other hand, those of $O_2^-(H_2O)_u$ (n = 0–2) decrease. The maximum values of the ion cluster densities for n = 0–5 are as follows: n = 0: 1.10×10^{18} m^{-3}, n = 1: 1.86×10^{18} m^{-3}, n = 2: 1.91×10^{18} m^{-3}, n = 3: 1.70×10^{19} m^{-3}, n = 4: 5.25×10^{19} m^{-3}, n = 5: 7.08×10^{18} m^{-3}.

Figure 10.8 shows the calculated number-density distribution for the time-dependent negative ion cluster ($O_2^-(H_2O)_n$) for n = 4. Although the densities of the negative ion clusters are slightly lower than those of the positive ion clusters, the qualitative behaviors for negative n = 0–5 ion clusters are almost the same as those for positive n = 0–5 ion clusters. The maximum values of the ion cluster densities for n = 0–5 are as follows: n = 0: 4.27×10^{18} m^{-3}, n = 1: 1.10×10^{18} m^{-3}, n = 2: 2.04×10^{18} $m^{-3,}$ n = 3: 2.27×10^{18} m^{-3}, n = 4: 4.27×10^{18} m^{-3}, n = 5: 1.74×10^{18} m^{-3}.

3.2.6 Predicted Results for the Total Generation of Ion Clusters

Figure 10.9a shows the time-dependent total number of positive ion clusters generated by a pair of needle-to-ring electrodes, for six types of clusters. In this figure, the horizontal axis is assumed to represent elapsed time, and the vertical axis represents

Fig. 10.7 Result obtained for the time-dependent positive ion cluster $H_3O^+(H_2O)_n$ number-density distribution for $n = 4$. The color bar indicates the logarithmic value of the ion number density n_{pi} (Okubo et al. 2020)

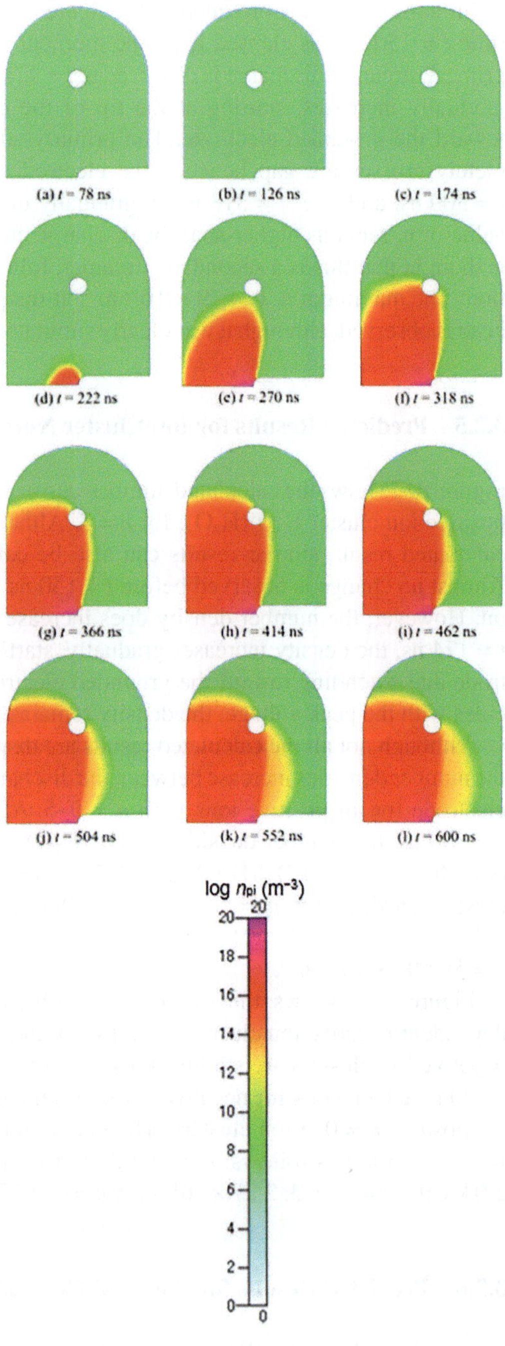

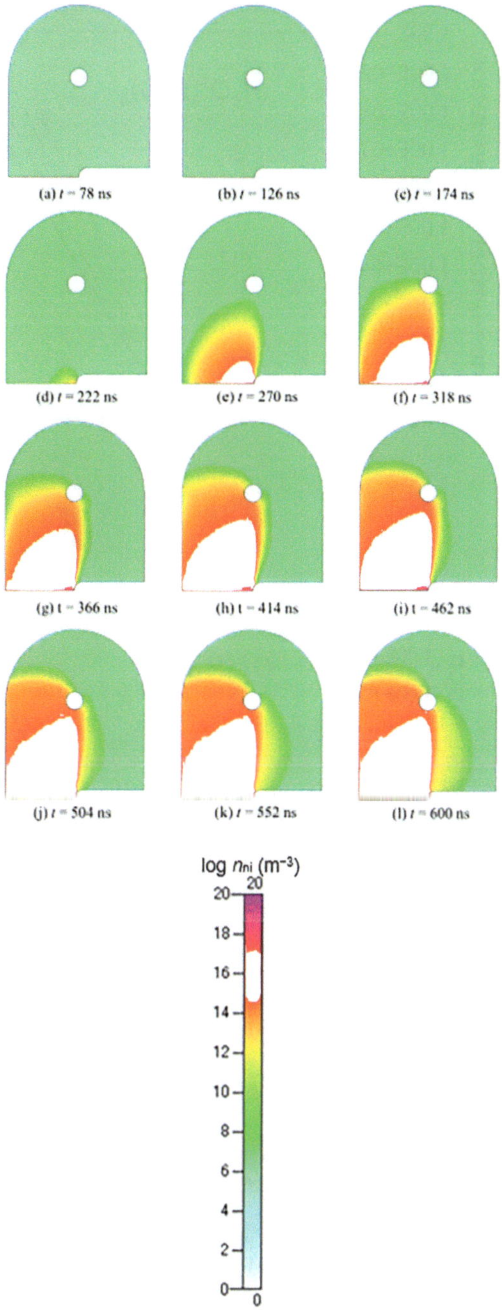

Fig. 10.8 Result obtained for the time-dependent negative ion cluster ($O_2^-(H_2O)_n$) number-density distribution for $n = 4$. The color bar indicates the logarithmic value of the ion number density n_{ni} (Okubo et al. 2020)

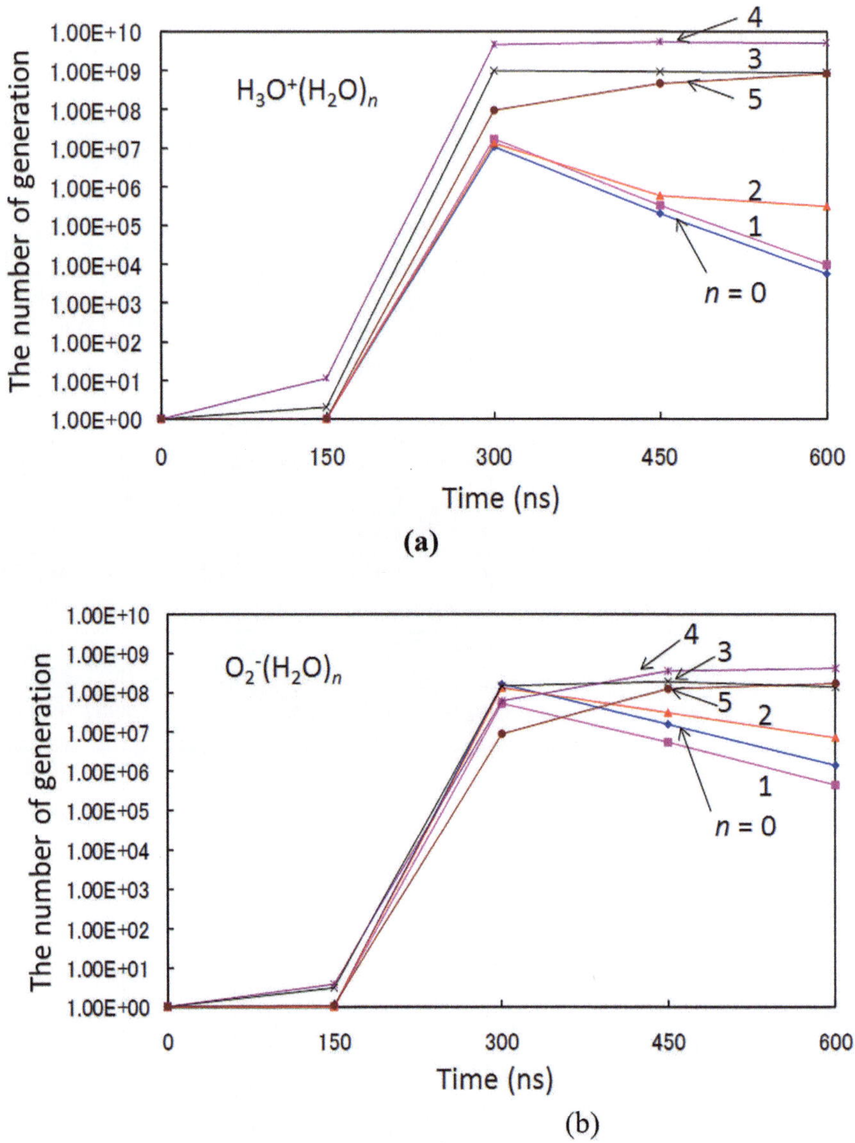

Fig. 10.9 Results obtained for the time-dependent total number of ion clusters generated by a pair of needle-to-ring electrodes. The horizontal axis represents elapsed time and the vertical axis represents the number of ions generated, the unit of which is the molecule. (**a**) Positive ion clusters $H_3O^+(H_2O)_n$ (n = 0–5) and (**b**) negative ion clusters $O_2^-(H_2O)_n$ (n = 0–5) (Okubo et al. 2020)

the number of ions generated, the unit of which is the molecule. As seen in the graph, all the types of ion clusters begin to be generated at approximately 150 ns. At the peak voltage coinciding with t = 300 ns, the number of clusters with n = 0–5 has increased significantly. After this peak, clusters with $n = 0$–2 decrease in number, clusters with $n = 3$ stay constant, and clusters with $n = 4$ and 5 increase in number. It is concluded that positive ion clustering progresses over time. At the end of the pulse, the generation of $H_3O^+(H_2O)_4$ ($n = 4$) is at its highest: 5.17×10^9 #, where # means the number. The differing generation numbers for the respective types of ion clusters are due to the differences in the rate coefficients of the chemical reactions forming the clusters, as shown in Table 10.1.

Figure 10.9b shows the time-dependent total number of negative ion clusters generated by a pair of needle-to-ring electrodes for six types of clusters. It is evident that the graph exhibits a similar qualitative tendency to that in Fig. 10.9a. The clustering of negative ions progresses over time. At the end of the pulse, all the counts of the generated negative ion clusters tend to be constant. The number of $O_2^-(H_2O)_4$ ($n = 4$) ions generated is the highest at 4.25×10^8 #.

We conclude that the clustering processes for both the negative and positive ions have been successfully simulated in the current numerical simulation. Approximately a total of 10^5 cm^{-3} of negative and positive ions are generated with a single ion-generated device. In the next section, the results of the application of the NTP air cleaner to an actual plant factory are described. It is noted that the positive and negative ion clusters ($n > 0$) targeted in the current study have a relatively long lifetime compared to the O_2^- and H_3^+ radicals themselves. The cleansing effect of air decomposes viruses, hazardous materials, and air pollutants suspended in the atmosphere when ion clusters are discharged into the air (Sharp Corporation website 2020; Okubo et al. 2020). The positive and negative ion clusters bond to the surfaces of airborne viruses and other substances and change into OH radicals (Sharp Corporation website 2020; Okubo et al. 2020), which have cleansing effects. Therefore, it is important to find the plasma condition that yields the optimum number of ion clusters to achieve the optimal design goals of the electrode systems.

4 Application of a Nonthermal Plasma Air Cleaner

4.1 *Introduction of the Air Cleaner to Cleanroom for Plant Factory*

In addition to the numerical simulation, we test a plasma air cleaner to determine its effect in a cleanroom in a plant factory. Originally, an air cleaner with a HEPA filter was installed in the cleanroom. However, HEPA filters are expensive and induce a large pressure loss. In addition to experimentally validating the numerical simulation, we propose a low-cost plasma air cleaner as an alternative to expensive HEPA filters. Therefore, in this study, the plasma air cleaner is operated in a cleanroom with and without the HEPA filter in operation to investigate the sterilization effect, particularly for airborne bacteria.

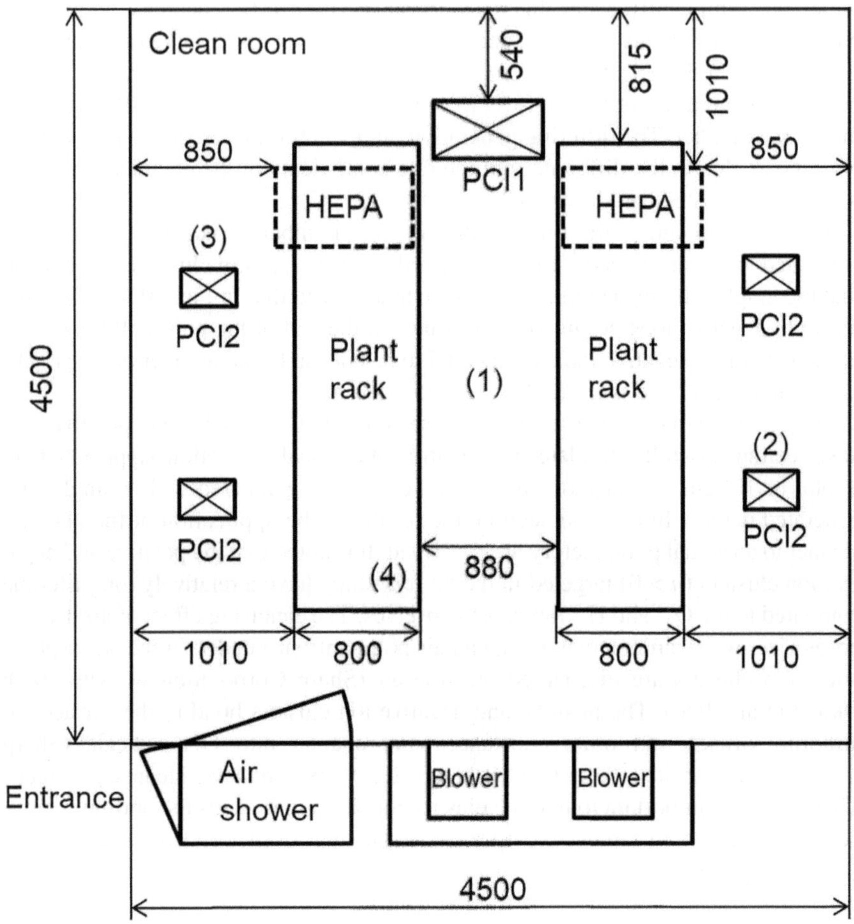

Fig. 10.10 Overview of the cleanroom for the plant factory

The top view of the cleanroom at Osaka Prefecture University, C21 building for measuring airborne bacteria is shown in Fig. 10.10. The photograph is shown in Fig. 10.11. The positions of the two HEPA filters are indicated in Fig. 10.10. PCI is used as an acronym for "plasma cluster ion" devices; the positions of these devices are also indicated in Fig. 10.10. PCI1 is a floor-standing-type plasma air cleaner, and the four PCI2 units are ceiling-mounted plasma air cleaners. The external dimensions of PCI1 are 392 mm in width × 182 mm in depth × 500 mm in height, and the flow rate is 4.5 $m^3 min^{-1}$. The dimensions of the exit of the PCI2 units at the ceiling are 281 mm × 170 mm. A flow rate of 0.58 $m^3 min^{-1}$ is used.

In addition, there are two plant racks (total height 222 mm) shown on the left and right of Fig. 10.10 in the cleanroom, and iceplants are planted on both of them, as shown in Fig. 10.11. Each rack is divided into three shelves: lower, middle, and upper. The plants are planted in the lower and middle tiers. The positions where the

(a)

(b)

Fig. 10.11 Photograph of the cleanroom for the plant factory

airborne bacteria and ion concentration are measured are shown by (1)–(4) in Fig. 10.10. Position (1) indicates the central part of the cleanroom, (2) is near the PCI2 in the lower right of the figure, (3) is near the PCI2 in the upper left of the figure, and (4) is the position in the left plant rack. In addition, positions (1) to (3) are located at a height of 1 m from the floor surface, and position (4) is determined by placing a measuring instrument in the middle shelf of the left plant rack. For the ion concentration, the spatial concentration of positive and negative ions is measured using a spatial ion concentration meter.

4.2 Measurement Condition and Apparatus for Suspended Bacteria and True Fungi

Four experiments were conducted; two experiments each for two types of airborne bacteria. The types of bacteria are general bacteria such as *Staphylococcus aureus*, and true fungi. Each type was measured with the HEPA filters on and off, respectively. Sampling was performed with two devices, two times each day for total six days. At each day, a period of one week was taken. The plasma air cleaner was switched on for the first, third, and fifth days, and it was switched off for the second, fourth, and sixth days.

In these experiments, two devices were used to measure airborne bacteria. The first was the MBS-1000D (Midori Anzen Co., ltd.) airborne bacterium measuring instrument. Air was sucked into the measuring instrument from its upper part, and the sucked air was blown onto the surface of the agar medium in the measuring instrument to collect airborne bacteria. The sampling flow rate was 100 L min^{-1}. Measurements was made for 5 min, and the sampling volume was 500 L. The second device used to measure airborne bacteria was the RCS air sampler (Merck KGaA). The wing of the device rotated and blowed air onto the surface of the medium to collect airborne bacteria. The sampling flow rate was 40 L min^{-1}. The measurement was performed for 8 min, and the sampling volume was 320 L. The measurement result was converted to a sampling volume of 500 L. The culture medium for each sample is shown below.

For bacteria (*Staphylococcus aureus*, etc.), standard agar medium (Nihon Pharmaceutical Co., Ltd.), and Agar Strip TC (total count) (Biotest Co., Ltd.) was used. For true fungi, potato dextrose agar with chloramphenicol (Kohjin Bio Co., Ltd.) and Agar Strip YM (yeasts and molds) (Biotest Co., Ltd.) was used. Bacteria and fungi were measured twice each, with two types of devices at each of the locations (1)–(4), as shown in Fig. 10.10. The measurements were performed for six days. The medium was cultured at a constant temperature for a certain period of time. The method is as follows: Bacterial medium was incubated for 72 h at a temperature of 35 °C. Similarly, the fungal medium was cultured for 120 h at 25 °C.

4.3 Results and Discussion

4.3.1 Results for Bacteria

Figure 10.12 shows the measurement results of the relationship between the average ion concentration and the average number of bacteria with the HEPA filters switched on. The horizontal axis represents the average concentration of negative and positive ions at each measurement point, the vertical axis represents the average number of microorganisms in 500 L of air detected with the two types of measuring devices, and the error bar indicates the standard deviation $\pm\sigma$. The same descriptors apply to Figs. 10.13, 10.14, and 10.15, except that the measurements shown in Figs. 10.13 and 10.15 are taken with the HEPA filters off. The measurements indicate that the average number of bacteria is small and that the number of bacteria does not depend on the ion concentration. No effect due to the plasma air cleaner is observed.

Figure 10.13 shows the measurement results of the relationship between the average ion concentration and the average number of bacteria with the HEPA filters off. The measurements indicate that the average number of bacteria is relatively large (because the HEPA filters are not used), and the measurement data fluctuate greatly. However, the number of bacteria could show no dependence on the ion concentration, and no effect of the plasma air cleaner is observed.

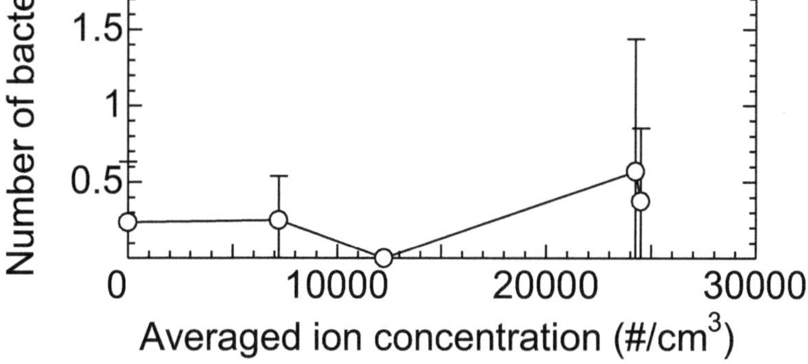

Fig. 10.12 The measured relationship between the average number of ions and the number of bacteria (HEPA: on)

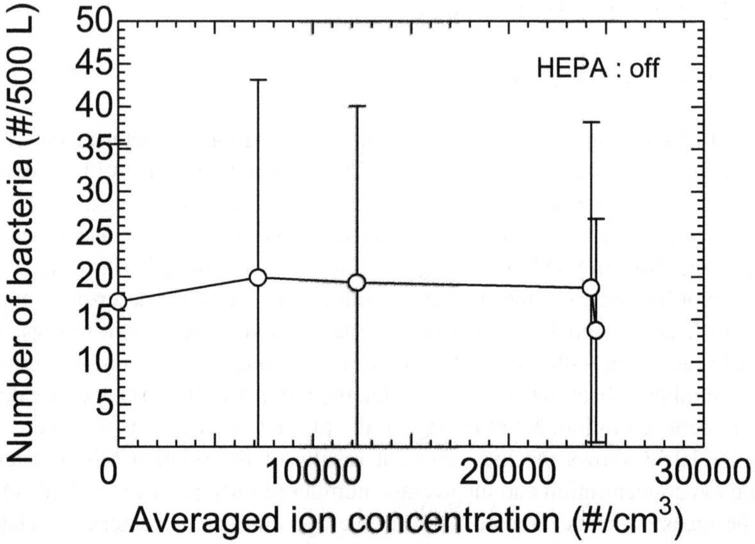

Fig. 10.13 The measured relationship between the average number of ions and the number of bacteria (HEPA: off)

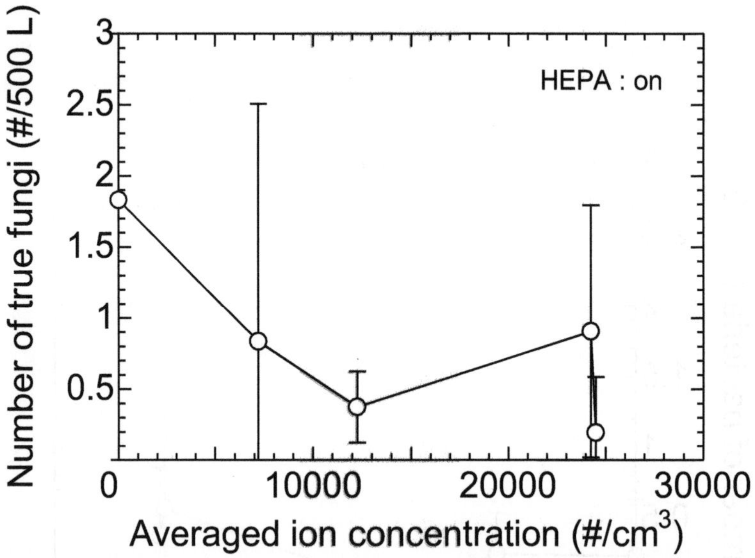

Fig. 10.14 The measured relationship between the average number of ions and the number of true fungi with HEPA filters on

4.3.2 Results for True Fungi

Figure 10.14 shows the measurement results of the relationship between the average ion concentration and the average number of fungi with the HEPA filters switched on. The measurements indicate that the average number of fungi is small (because

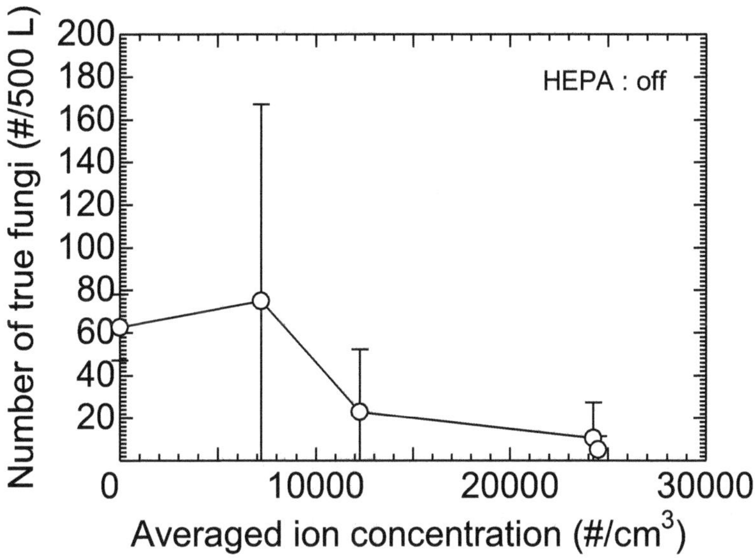

Fig. 10.15 The measured relationship between the average number of ions and the number of true fungi with HEPA filters off

the HEPA filters are used), and the measurement data fluctuate greatly. However, the number of fungi decreases with an increase in ion concentration. When an ion air cleaner is used in combination with HEPA filters, the maximum removal efficiency of true fungi is 99.7%.

Figure 10.15 shows the measurement results of the relationship between the average ion concentration and the average number of true fungi with the HEPA filters off. As shown in the figure, the average number of fungi is relatively large (because the HEPA filters are not used), and the measurement data fluctuate greatly. However, the number of true fungi decreases with an increase in ion concentration. When only the ion air cleaner is used, the maximum removal efficiency of true fungi is 91.8%.

5 Conclusions

This study demonstrates that the low pressure-loss plasma air cleaner could replace a HEPA air cleaner. Numerical and experimental studies of pulsed, high-voltage-induced ion cluster formations are carried out. The findings of the current study are useful for the design of ion generator type of indoor air cleaners. The ion generation parts of the air cleaner can be designed with a numerical simulation. Furthermore, the measurements of the average number of airborne bacteria in the cleanroom of the plant factory do not demonstrate the effect of the plasma cluster on bacteria. The measurements of the average number of true fungi in the cleanroom of the plant factory demonstrate that the plasma cluster is effective for true fungi. When an ion air cleaner is used in combination with HEPA filters, the maximum removal efficiency of true fungi is 99.7%. When the ion air cleaner is used solely, the maximum removal efficiency of true fungi is approximately 91.8%.

Acknowledgments The author would like to thank Mr. K. Kanayama and Mr. Y. Hiroyasu (former graduate students at Osaka Prefecture University) for conducting the numerical simulations and experiments. The present text is a review of their authored thesis (Kanayama 2014) and papers (Okubo et al. 2020).

References

Dorai R (2000) Modeling of plasma remdediation of NO_x using global kinetic models accounting for hydrocarbons, Ph.D. Thesis, Univ. Illinois at Urbana-Champaign

Huertas ML, Fontan J (1975) Evolution times of tropospheric positive ions. Atmos Environ 12:1018–1026

Huertas ML, Fontan J, Gonzalez J (1978) Evolution times of tropospheric negative ions. Atmos Environ 12:2351–2362

Ikebe K, Nakanishi K, Arai S (1989) Evolution process of microdischarge products in ozonizer. Trans IEE Japan 109-A(11):474–480. (in Japanese)

Kanayama D (2014) 2D Numerical simulation considering ion cluster chemical reactions, Master's Thesis, Department of Mechanical Engineering at Osaka Prefecture University (in Japanese)

Kuroki T, Tanaka S, Okubo M, Yamamoto T (2007) Numerical investigation for CF_4 decomposition using RF low pressure plasma. IEEE Trans Ind Appl 43(4):1075–1083

Maruyama S (2016) Control of culture environment and cleaning techniques for closed plant factory. Earozoru Kenkyu 31:104–109., (in Japanese). https://doi.org/10.11203/jar.31.104

Matsumoto T, Wang D, Namihira T, Akiyama H (2010) Energy efficiency improvement of nitric oxide treatment using nanosecond pulsed discharge. IEEE Trans Plasma Sci 38:2639

Nagato K, Matsui Y, Miyata T, Yamauchi T (2006) An analysis of the evolution of negative ions produced by a corona ionizer in air. Int J Mass Spectrometry 248:142–147

Okubo M (2015) Evolution of streamer groups in nonthermal plasma. Phys Plasmas 22(12):123515

Okubo M, Yoshida K, Yamamoto T (2008) Numerical and experimental analysis of nanosecond pulse dielectric barrier discharge-induced nonthermal plasma for pollution control. IEEE Trans Ind Appl 44:1410

Okubo M, Hiroyasu Y, Kuroki T (2020) Ion cluster formation by nonthermal plasma induced by pulse Corona discharge toward indoor air cleaning. IEEE Trans Ind Appl 56(5):5480–5488

Parts TE, Luts A (2004) Observed and simulated effects of certain pollutants on small air ion spectra: I. positive ions. Atmos Environ 38(9):1283–1289

Person JC, Personm JC, Ham DO (1988) Removal of SO_2 and NO_x from stack gases by electron beam irradiation. Radiat Phys Chem 31(1–3):1–8

Sato T, Ito D, Nishiyama H (2005) Reactive flow analysis of nonthermal plasma in a cylindrical reactor. IEEE Trans Ind Appl 41(4):900–905

Sharp Corporation website (2020) [Online]. Available: https://global.sharp/pci/en/about_pci/

Wang D, Matsumoto T, Namihira T, Akiyama H (2010) Development of higher yield ozonizer based on nano-seconds pulsed discharge. J Adv Oxid Technol 13:71

Chapter 11
LED-Based Photosensitization – a Prospect for Visible Light-Driven Nonthermal Fresh Produce Sanitation

Zivile Luksiene

1 Introduction

Recently microbial food safety is one of the major issues of global society (Akhtar et al. 2014; CDC 2018a). Despite fantastic technological advances, fast industrial development and broad education of population, in 2010 about 600 million people all over the world suffered from foodborne diseases and 420,000 died due to consumption of contaminated food (WHO 2015). The Center for Disease Control and Prevention (CDC 2017) estimated that foodborne diseases have been diagnosed for about 48 million people in USA per year, and 3.000 of them died. About 1.0 million people per year in USA got salmoneliosis, and as a consequence, there were 19.336 hospitalizations and 378 deaths (Scallan et al. 2011). *Listeria monocytogenes* which induce listeriosis is mostly foodborne (98%) and accounts for $200 million in monetary loss, whereas overall foodborne diseases cost the USA economy $ 15.6 billion in loss of productivity and medical expenses (CDC 2018a).

The modern consumer requires fresh, non-thermally processed, nutritionally healthy food. But consumption of fresh fruits and vegetables lead to increased number of foodborne diseases, and according to (Pui et al. 2011) fresh produce became second leading cause of foodborne illnesses. Most frequent pathogenic bacteria which cause foodborne illnesses are *Salmonella* spp., *Campylobacter* spp., *Escherichia coli* and *Listeria monocytogenes* (CDC 2017). In May 2011 *Escherichia coli* O104:H4 bacteria outbreak due to the consumption of fresh produce began in Germany. After one month, even 3228 cases and 35 deaths had been reported. It spread in Switzerland, Poland, Netherlands Sweden, Denmark, United Kingdom, Canada and the USA (Nillian et al. 2011). To this end, the prevention of foodborne infections and better control of microbial contamination of fresh produce remains a

Z. Luksiene (✉)
Vilnius university, Institute of Photonics and nanotechnology, Vilnius, Lithuania
e-mail: Zivile.Luksiene@ff.vu.lt

A. Režek Jambrak (ed.), *Nonthermal Processing in Agri-Food-Bio Sciences*, Food Engineering Series, https://doi.org/10.1007/978-3-030-92415-7_11

world problem with huge social and economic impact (Newell et al. 2010). Moreover, it must be noticed that this problem is permanently growing due to globalization of food market, frequent international travels and, by no means, increasing consumption of minimally processed or ready-to-eat food. The other important reason of prevalent outbreaks is the increasing resistance of microorganisms to existing chemical antimicrobials. For instance, in 2018, an outbreak in the USA that caused 21 hospitalizations was induced by multidrug-resistant *Salmonella Infantis* (CDC 2018b).

Hence, the development of innovative and more effective technology for microbial control of fresh produce remains an urgent task. In order to replace thermal sterilization or pasteurization, a number of new non-thermal emerging technologies recently are in progress (high power pulsed light, high pressure processing, pulsed electric field, application of organic acids, modified atmosphere packaging, coating, chlorine dioxide, etc.). According to the ideal non-thermal processing technology must meet the following requirements:

(a) it must be effective against broad range of microorganisms, including spores, biofilms, yeasts/fungi;
(b) it must prolong the shelf-life of fresh produce without thermal effects, with no changes in organoleptic and nutritional characteristics, with no residues left on treated food;
(c) it must be cost-effective and easy to maintain.

A technology that promises to answer all these needs and requirements is photosensitization, based on natural photoactive compounds extracted from plants and visible light. Combined action of photoactive compound (photosensitizer), visible light and oxygen is the background of this antimicrobial treatment. This interaction eventually produces plethora of reactive oxygen species (ROS) and singlet oxygen ($^1O_2$) in the target cell what eventually induces multiple lethal damages and selective death of the microorganism (Luksiene 2005; Luksiene 2021). The most important feature of photosensitization is its high efficiency against Gram (+) and Gram (−) bacteria, especially their spores and biofilms which usually are extremely resistant to chemical and physical antimicrobial treatments. Moreover, photosensitization is effective against pathogenic fungi/ yeasts and viruses without development of resistant forms (Hadi et al. 2020a; Kashef et al. 2017; Maisch 2015). This is very important advantage if compare with other emerging food processing technologies, for instance ultrasonication, pulsed electric field or ultraviolet light (Cebrian et al. 2016). Moreover, this strategy has potential to eradicate multidrug-resistant bacteria (Maisch 2015). Eventually, the most important is the fact, that this treatment is non-thermal, environmentally friendly and cost-effective.

This review summarizes the first pioneering attempts to apply photosensitization phenomenon for prevention of infectious foodborne diseases. It includes the description of the mechanism of action of this treatment, selected photosensitizers and light sources which can be used for decontamination of foods. Review shortly presents first experiments on the effective inactivation of foodborne pathogens and first successful results on reduced microbial contamination of fresh produce when photosensitization treatment is applied. In addition, basic evaluation of main organoleptic and nutritional quality of treated fresh produce is discussed. All factors which have impact on antimicrobial efficiency of photosensitization and are important for treatment optimization are analysed. Moreover, the possibility to apply photosensitization for production of photoactive antimicrobial bio-based coatings is presented.

2 History of Photosensitization Phenomenon

About two thousand years ago in ancient India and Egypt patients with skin disease *vitiligo* were treated with plant-based photoactive compound psoralen, and afterwards the damaged skin was exposed to the sunlight. And it worked. So, at that time people empirically without any knowledge about photosensitization applied this treatment to cure skin diseases.

About one hundred years ago, Oscar Raab, being student discovered that the combination of sunlight and acridine orange can kill *Paramecium caudatum*, whereas the sunlight or the dye separately were absolutely nontoxic. Afterwards, von Tappeiner and Jodlbauer found that combined action of photoactive compound and light can produce cytotoxic effect. This phenomenon was termed photodynamic effect or photosensitization. In the 1970's it was found that several photoactive compounds can selectively accumulate in the highly proliferating tumor cells. After exposure of such tumors to visible light it was possible to achieve perfect and selective their destruction, without any damage of surrounding tissue. These first experiments became a background for development of new cancer therapy- photodynamic therapy, which recently is established as a successful modality for malignancies (Hamblin and Hasan 2004; Dai et al. 2012). Since several photosensitizers can accumulate in the highly proliferating microorganisms as well, recently in the era of increased microbial resistance the photosensitization has been proposed as an alternative approach to cure local infections (Kashef et al. 2017).

About twenty years ago innovative approach to use photosensitization for prevention infectious foodborne diseases, controlling the pathogens before they enter human body, i.e. on foods, surfaces, surrounding was published (Luksiene 2004a, b). First time photosensitization was applied for inactivation of food pathogens on food matrix few years later (Luksiene et al. 2007).

3 Photosensitizers

Photoactive compounds or photosensitizers may be present naturally inside the bacterial cell (endogenous photosensitizer) or may be externally supplied (exogenous photosensitizer).Over the years a great deal of work has been carried out to evaluate the photosensitizing efficiency of different compounds. Thus, first of all it was important to find the correlation between the structure of photosensitizer and its antimicrobial efficiency. So, the structure of most identified photosensitizers is based on tetrapyrrole ring. Such molecules have low toxicity, can form long-lived triplet excited state for effective production of ROS and have excellent biocompatibility. Several lines of evidence indicate that physico-chemical properties of any photosensitizer have huge impact on their photosensitizing efficacy. For instance, hidrophilicity, ionization, light-absorption region, photostability and the yield of singlet oxygen production must be included in a putative photoantimicrobial profile (Alves et al. 2014).

Moreover, when we want to apply photosensitization for sanitation of food or food-contact surface the selected photosensitizer must exhibit some special and some additional properties. For example, it must be of highest chemical purity, high photostability, affinity to target microorganism and antimicrobial efficiency. It must be absent of mutagenicity or genotoxicity. In addition, it must be natural compound, food additive or food component, without strong color, taste and flavour, working at minimal concentration. What is very important for acceptability of such food by consumers- it must have no effects on nutritional as well as organoleptic properties of the foods. Eventually it must be low-cost (Luksiene 2005; Luksiene and Zukauskas 2009; Luksiene and Brovko 2013; Luksiene 2014).

3.1 *Endogenous Photosensitizers or Their Precursors*

It is slowly recognized that many bacteria have a sufficiently high concentration of endogenous photosensitizers, such as heterocyclic macromolecules coproprophyrin III, protoporphyrin IX und uroporphyrin III) which after exposure to visible light generate reactive oxygen species (ROS) that damage functionally important cell compartments and lead to cell death. The presence of endogenous photosensitizer means that adding an exogenous one in some cases is not necessary. In study (Kumar et al. 2015) it was attempted to understand the differential susceptibility of various Gram (+) and Gram (−) bacteria to photosensitization through their amount of endogenous coproporphyrins. It was concluded that many different porphyrins are present in bacterial cells, and the bacterial susceptibility mostly depends on the porphyrin composition (Ghate et al. 2019).

The other approach that has been popular in last three decades is the use of PS precursor 5-aminolevulinic acid (5-ALA). This precursor after entering the microbial cell participates in the production of previously described endogenously localized porphyrins (uroporphyrin, coproporphyrin and protoporphyrin) (Luksiene et al. 2007). So, exploiting the endogenous porphyrins or their precursor 5-ALA presents an excellent opportunity to use photosensitization as new generation chemical-free antimicrobial treatment. Recently, when consumers require fresh food free of chemical antimicrobials and food additives, this is a significant advantage. However, one has to be prepared for very long treatment time (5–48 hours), what is not always acceptable by food processing industry, may be more for crop preservation during storage.

3.2 *Plant-Produced or Natural Exogenous Photosensitizers Suitable for Photosensitization-Based Preservation of Fresh Produce*

Exogenous photosensitizers seemed more attractive for food industry, since they enhance the rate of the pathogen inactivation, and treatment time continues 20–30 min. For instance, Riboflavin is widely used as essential micronutrient vitamin B_2 for human health and as colorant (E101) in food industry. Moreover, it is non-toxic, labelled as "generally recognized as safe (GRAS), is included in all living systems since is part of complex flavoproteins (Yin and Hamblin 2015). Besides all these features this compound is water soluble and exhibits excellent photosensitizing properties.

After excitation with UV or visible light Riboflavin produces ROS which render the redox status of bacterial cells into a compromised state what lead to significant membrane damage ultimately causing bacterial death (Fig. 11.1). Its usage for developing of antimicrobial technology has shown significant results (Thakur et al. 2017). The main disadvantage of Riboflavin is low photostability (Sheraz et al. 2014).

It is interesting to note that plant pigment Hypericin can be isolated not just from herbs *Hypericum perforatum* L. (St. John's worth), but also from basidiomycetes (*Dermocybe* spp.) or endophytic fungi (*Thielavia subthermophila*). In recent decades *H. perforatum*-derived products are often used for medical purposes as tinctures. Food industry uses these extracts as a flavour for beverages, juices, oils and teas (Maskovic et al. 2011).

Hypericin activated by 585 nm visible light was identified as advantageous photosensitizer few decades ago. Recently this natural compound has status of "one of the most powerful photosensitizers in nature"(Fig. 11.2). Most important is the fact

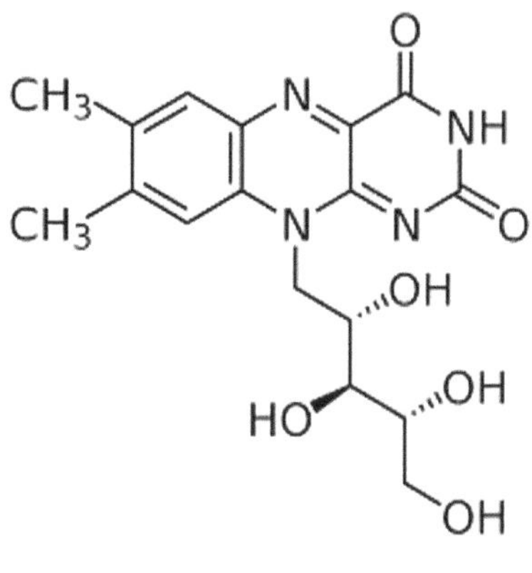

Fig. 11.1 Chemical structure of Riboflavin

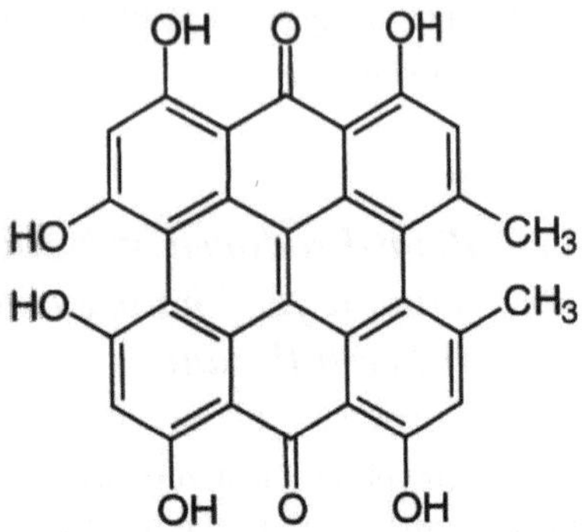

Fig. 11.2 Chemical structure of Hypericin

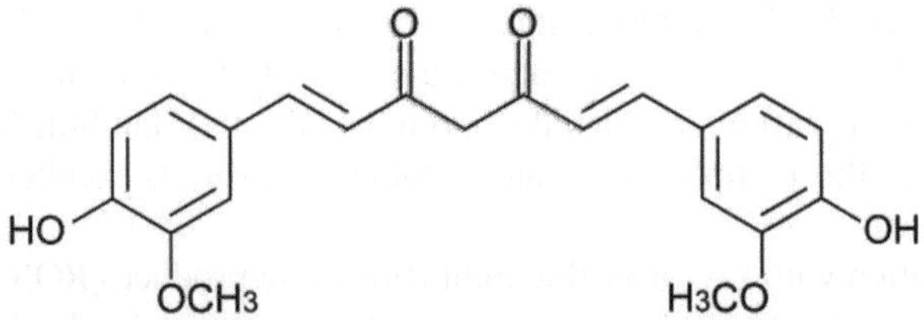

Fig. 11.3 Chemical structure of Curcumin

that plant-derived Hypericin has no toxicity or genotoxicity (Luksiene and de Witte 2003; Bernal et al. 2011). Since this compound is negatively charged and exhibits high lipophilicity, it does not accumulate inside the cell and preferentially binds to the cellular membrane, and after illumination this treatment produces multiple lethal destructions in cell membrane (Kairyte et al. 2012).

In the roots of *Curcuma longa* a yellow poliphenolic pigment Curcumin has been identified. Since ancient times it has been used as a spice and colorant (Rao and Khanum 2016). Recently it was found that Curcumin exhibit a wide spectrum of different biologically active properties. For instance, it is effective antioxidant (Rao and Khanum 2016), antimicrobial (Arutselvi et al. 2012), exhibits pronounced anti-inflammatory and anticancer properties (Sharma et al. 2005). Curcumin absorbs blue light (400–500 nm), i.e. is photoactive, thus can be used as a potential natural photosensitizer (Fig. 11.3). Since Curcumin is food additive E100, its possible application in food industry is controlled by European regulations (directive 94/36/EC) (EFSA 2010, 2014). It is important to note that no toxicity from dietary use of Curcumin has been found. EFSA also confirmed that this compound is safe and not carcinogenic (EFSA 2010, 2014). However, the efficient usage of Curcumin is limited, mostly due to its low water solubility and poor chemical stability (Dias et al. 2020).

Erythrosine B belongs to the class of xanthenes and is really approved for usage in food products by FDA. It absorbs visible light at 500–550 nm, and after photoactivation exhibits favourable inactivating ability to Gram (+) bacteria. Its chemical structure is presented in Fig. 11.4.

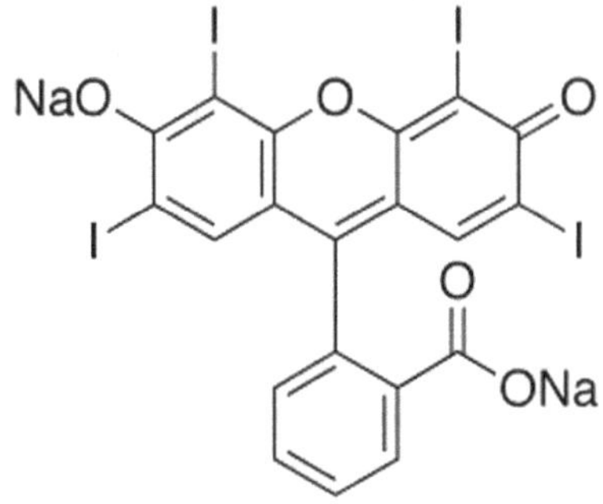

Fig. 11.4 Chemical structure of erythrosine B

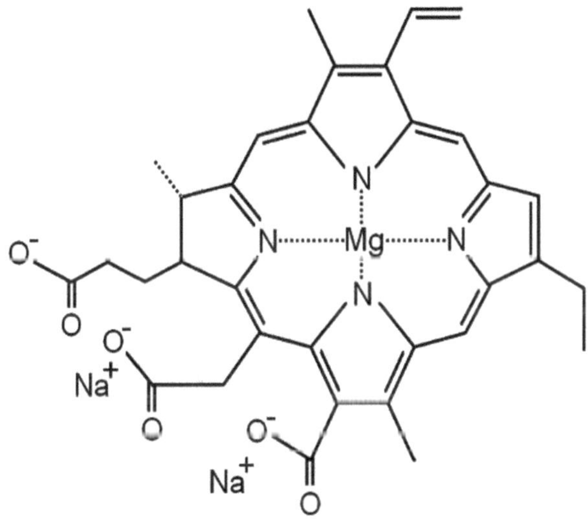

Fig. 11.5 Chemical structure of Chlorophyllin

The main destruction of these bacteria occurs through the lipid peroxidation in cell membranes. One of the drawbacks of this photosensitizer is lower efficiency against Gram (−) bacteria if compare with Gram (+) ones (Wood et al. 2006).

Chlorophyllin belongs to porphyrin-type compounds and is water soluble chlorophyll derivative. One of the greatest its advantages is that Chlorophyllin is authorized for use in the food industry for food coloring (EU license number 473 E141), what confirms that this compound is harmlessness for humans (Fig. 11.5).

Recently it became popular to use Chlorophyllin as food supplement or food additive (Viera-Alcaide et al. 2019). In study authors found that Chlorophyllin exhibited pronounced anti-mutagenic and anti-carcinogenic activities against the mycotoxin aflatoxin B1, heterocyclic amines, and polycyclic aromatic hydrocarbons. These properties of Chlorophyllin may be attributed to its ability to interact with carcinogens forming with them tight complexes and at the same time diminishing their bioavailability. It is important to emphasize that the antioxidant activity of

Chlorophyllin was found to be much higher than that of natural chlorophylls a and b, pheophytins and pheophorbids. Moreover, Chlorophyllin exhibits even radioprotective activity and can be used in radiotherapy (Geric et al. 2019).

It can be seen from Fig. 11.6. that Chlorophyllin is perfect pretender to be a good photosensitizer, since its absorption spectrum consists of the main absorption band (Soret band) which is around 405 nm and a progressively diminishing set of Q bands, with the last one at 630 nm region.

In our laboratory for the first time it was confirmed that Chlorophyllin exhibited pronounced photosensitizing properties (Luksiene and Zukauskas 2009). But, there is one problem. Chlorophyllin tends to photobleach, thus is not enough photostable when exposed to daylight. Thus, prepared Chl solution must be kept in the dark all time before experiments.

It is interesting to note that photosensitizing activity of Chlorophyllin, besides its application for food safety, has been successfully applied for reduction of different fish ectoparasites as well as mosquito larvae in aquaculture (Richter et al. 2014).

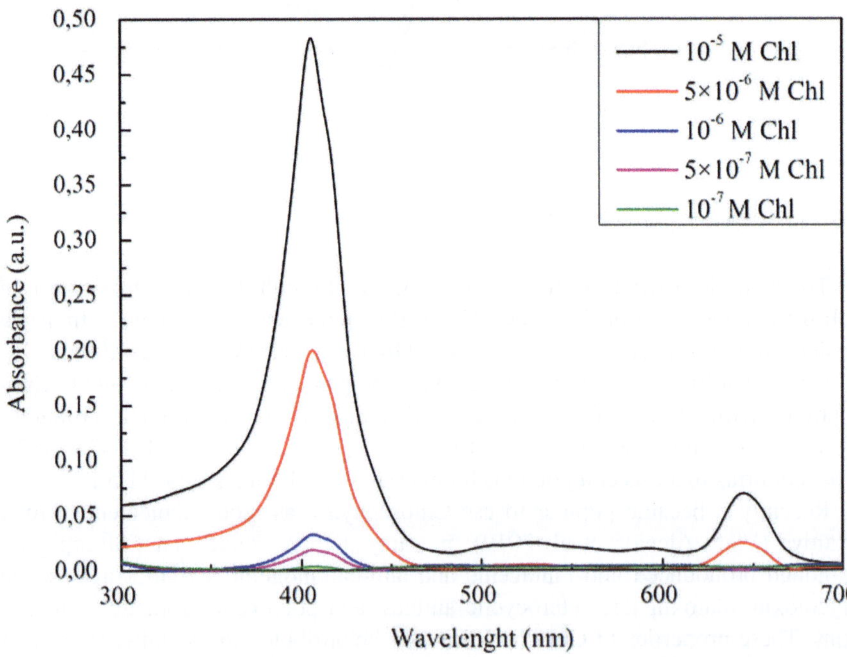

Fig. 11.6 Absorption spectra of Chlorophyllin measured at different concentrations (10^{-7} M–10^{-5} M) in water (Paskeviciute et al. 2018)

4 LED-Based Light Sources Necessary for Photosensitization Treatment

One of the advantages of photosensitization- based antimicrobial technology is the fact that this treatment requires mostly visible light which does not pose the risk of other light-based treatments, for instance harmful ultraviolet light. Actually, the optimized light source for photosensitization should emit the light with the same wavelength what is the maximum absorption wavelength of the photosensitizer. But, it must be emphasized that different photosensitizers have different absorption spectra and absorption peaks even in visible region of spectrum (400–700 nm). Moreover, the applicability of photosensitizer depends on the existing tasks. For instance, for cancer therapy the preferable absorbance of photosensitizers is in the red region of visible spectrum (630–700 nm), since the light of this wavelength penetrates deeper (6 mm) into tumor and better destroys it (Luksiene 2003). On contrary, for the food safety the priority is such photosensitizer which absorbs in blue light region (400–430 nm). First, the penetration of blue light into food matrix is lower (2 mm) and impact on food nutritional and organoleptic quality is less. Second, blue light never produces heating effects. Third, if compare with UV light, visible blue light is of higher safety and enhanced transmissibility (Vaitonis and Luksiene 2010).

In the past, first attempts to find light sources for photosensitization were performed using incandescent lamps equipped with collection of selective glass filters. The continuing rapid development of optoelectronics and progress in solid-state lighting technology creates more cost-effective illumination systems with significant energy savings. Solid state lighting using light-emitting diodes (LEDs) represents a fundamentally different, cheap and advanced technology if compare with previously used lamps. Such properties like narrow bandwidth (10 nm), high photoelectric efficiency, monochromatic light emission, compactness, portability are very attractive to any industry. Moreover the possibility to regulate photon flux, to control spectral composition without radiant heat emission makes this technology potentially one of the most significant advances in lighting. The unique properties of LEDs enable the convenient selection of the spectral characteristics as well as control of temporal settings of the light produced. Most important is the fact, that such lighting systems can present a wide emission variation- from ultraviolet A light to infrared light with light intensities from very high to very low (D'Souza et al. 2014). Due to the fact that lifetime of diodes can reach hundred thousands of hours, the LED-based lighting systems have the potential to be a very cost-effective technology to adopt and integrate. Moreover, visible light LEDs used for photosensitization-based decontamination of fresh produce do not induce thermal damage and do not destroy the food structure (Vaitonis and Luksiene 2010).

Thus, it is possible to conclude that recently LED-based devices achieved maturity, sufficient for their application in different fields, including food industry and agriculture (Luksiene and Zukauskas 2009; Prasad et al. 2020). The scheme of LED-based light source prototype, constructed in our laboratory for

Chlorophyllin-based photosensitization is presented in Fig. 11.7. It is clear that the intensity of light distribution from the LEDs is higher in the center of the chamber in comparison with periphery. Skewed light intensity distribution can also occur when the light source is placed too close to the sample, since it can lead to the higher light energy density in central part. Not equal distribution of light can make additional undesirable effects such as higher susceptibility of fresh produce to physico-chemical changes in the centrum and compromised action at periphery. In order to overcome this challenge only central part of the chamber was used for decontamination of fruits and vegetables.

As mentioned before, for the excitation of Chlorophyllin molecule blue light (405 nm) was used. It is important to note that according to (Hadi et al. 2020b) blue light has minimal impact on the nutritional and organoleptic properties of treated food. Moreover, it is necessary to check the intensity of delivered light every time before treatment. Actually, in all cases it can'nt be higher than 100 mW/cm^{-2}, since higher light intensities may induce undesirable thermal effects which can change the structure, nutritional and organoleptic properties of treated produce. The illumination intensity required for food-processing (220–540 lux) corresponds to the light intensity 0.3–0.54 mW/cm^{-2} and, hence, is sufficient to kill harmful and pathogenic microorganisms on the different surfaces (Brovko 2010).

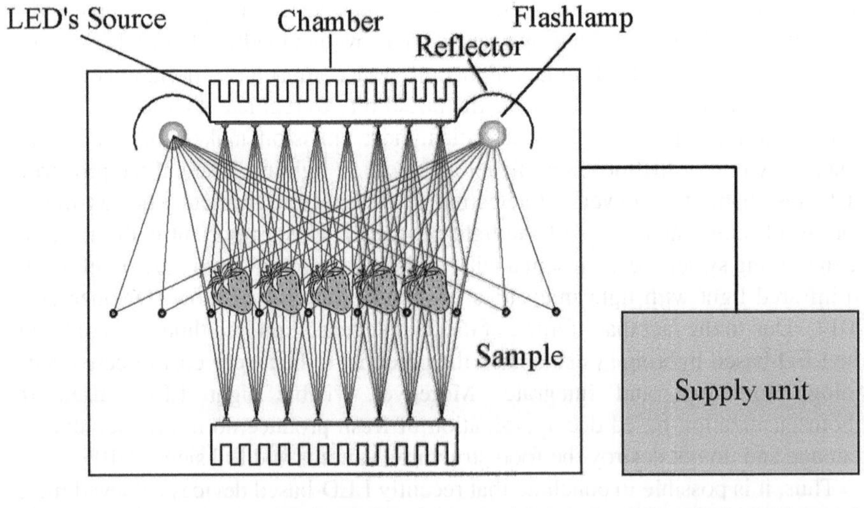

Fig. 11.7 Two- dimentional LED-based light source prototype tailored for microbial decontamination of fruits and vegetables using Chlorophyllin as photosensitizer ($\lambda = 405$ nm, power density 10–20 mW cm^{-2}). Additionally this prototype has 2 flash lamps for combined treatment of photosensitization with high power pulsed light

5 Mechanism of Action of Photosensitization: From First Photochemical Reactions to Total Destruction of Microorganisms

There is an expanding body of literature regarding photosensitization mechanism. Thus, recently it is well accepted that principle of photosensitization is based on the combination of three indispensable factors, i.e. non-toxic photosensitizer, non-toxic visible light and molecular oxygen (Alves et al. 2014; Wainwright et al. 2017) This interaction is described well in other studies (Luksiene and Brovko 2013; Maisch 2015; Wainwright and Crossley 2004; Luksiene 2021). Shortly, as presented in Fig. 11.8. the photosensitizer (PS^0) after absorption of light moves from ground singlet state (S_0) to the higher energy excited singlet state ($^1S^*$). After nanoseconds the excited photosensitizer ($^1PS^*$) has probability to return to the ground state (PS^0) in two ways:

(a) it can release the excitation energy in the form of fluorescence;
(b) it can undergo an intersystem crossing to the longer-lived (ms) excited triplet state($^3PS^*$). From this state it has two ways of relaxation and return to the ground state (PS^0): first, it can release off the excitation energy as phosphorescence, second- it can transfer excitation energy to a surrounding acceptor biomolecules/triplet oxygen producing reactive oxygen species (ROS: H_2O_2; $HO^{\cdot}$; $O2^{\cdot-}$) (Type I) or singlet oxygen $^1O_2$ respectively (Type II) (Luksiene and Zukauskas 2009) (Fig. 11.8).

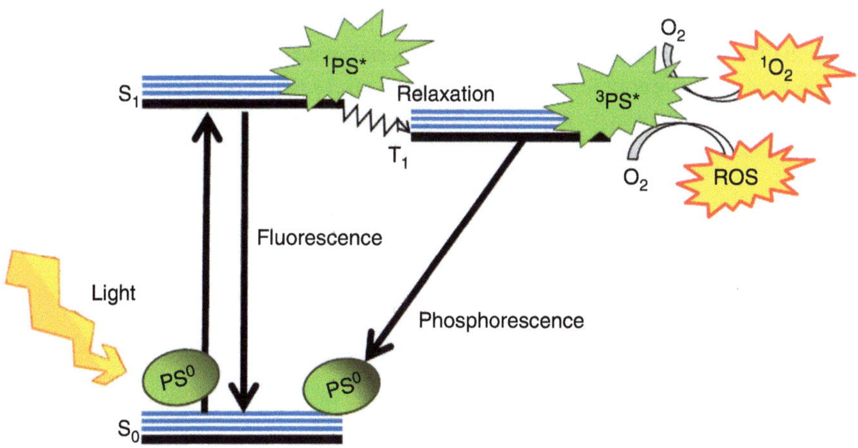

Fig. 11.8 First photophysical - photochemical stages of photosensitization. They involve the interaction of photosensitizer (PS) with light and transfer to excited S_1, relaxation to triplet T_1 and production of ROS: PS^0 – ground state, $^1PS^*$ – singlet excited state, $^3PS^*$ – triplet excited state. (Adapted from Dai et al. 2012)

Both ways of relaxation of photosensitizer to the ground state (PS^0) can occur separately or simultaneously. Which of the two pathways is dominant depends on the intracellular levels of molecular oxygen, the structure of photosensitizer and nature of microorganism. When photosensitizer is accumulated inside the bacterium or attached to it, both external components of cell wall and intracellular organelles are potential targets for the destructions, induced by ROS and $^1O_2$. It is important to note, that in both cases this treatment has multiple targets, so the development of antimicrobial resistance is not supposed to occur (Silva et al. 2018).

The photosensitization-induced damage can result in considerable morphological and functional changes in the microbial cells (Fig. 11.9). One of the main targets for produced ROS is cell membrane. Time-dependent irreversible destruction of membrane proteins and lipids had clear correlation with inactivation efficiency. Damage of it induces the leakage of cellular contents and following disintegration of the membrane-associated transport system. Moreover, such damage as loss of enzymatic activities, increased protein oxidation and unfolding, formation of protein-protein cross-links, lipid peroxidation and inhibition of different metabolic processes (e.g. DNA synthesis, glucose transport) have been observed (Catala et al. 2000; Sperandio et al. 2013). Morphological alterations include destruction of the mesosome, ribosome, chromosome structure, blebbing of outer membrane, swelling of mitochondria and cells (Kashef et al. 2017). Not direct cleavage of genomic DNA, plasmid and RNA has been observed after photosensitization in Gram (+) and Gram (−) bacteria (Alves et al. 2014). Thus, the probability to develop bacterial resistance to photosensitization is around zero.

Nucleic acids and proteins have been detected in the extracellular matrix after Chlorophyllin-based photosensitization of *S. enterica* Typhimurium. This indicated that the cell membrane was disintegrated after photosensitization treatment. The addition of $^1O_2$ quencher sodium azide to the suspension significantly reduced the bactericidal effect, revealing that $^1O_2$ plays the main role in the inactivation (Zudyte et al. 2020). Results obtained analyzing bacterial genes indicated that photosensitization treatment upregulated the expression of stress response genes related to oxidative stress (oxyR, grxA, ahpC), extracytoplasmic stress (STM0225), acid stress (atpC), thermal stress (groES), and SOS (sulA) (Luksiene 2021).

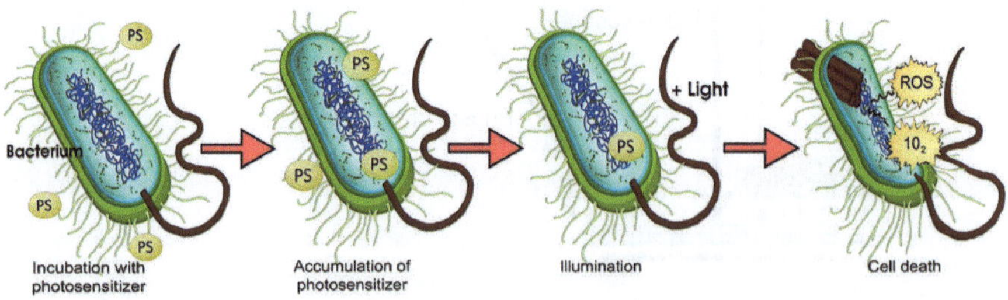

Fig. 11.9 Photosensitization-induced destruction of microorganisms: incubation with photosensitizer (PS), it's accumulation into bacteria, illumination with visible light and cell destructions by induced ROS or $^1O_2$

6 Main Factors Affecting the Antimicrobial Potential of Photosensitization in Suspension

6.1 Concentration of Photosensitizer Interacting with Bacteria

Concentration of photosensitizer plays one of the key roles in the inactivation efficiency of the treatment, since the inactivation of microorganism depends on obtained light dose (photons) and number of photosensitizer molecules which are in the binding with target bacteria and can be photoactivated by these photons. This interaction of photons with molecules of photosensitizer triggers plethora of ROS, photocytotoxic reactions and death of microorganism. But, by no means, there are limits. Too high concentration of photosensitizer leads to the decrease in its phototoxic ability, due to higher optical shadowing effect unabling the light to reach all photosensitizer molecules.

6.1.1 Structure of Bacteria and Fungi

Different microorganisms at the molecular level have very wide variation in the cellular structure and organization. These variations, by no means, have impact on the interaction of photosensitizer with different cellular components, what eventually affect the effectiveness of killing of different pathogens. For instance, the differences in the cell wall of Gram (+) and Gram (−) bacteria play a key role in the susceptibility of bacteria to photosensitization. The cell wall of Gram (+) pathogens is composed of lipoteichoic and teichoic acids organized in multiple layers of peptidoglycan, which confers easy penetration of photosensitizer into the cell. In Gram (−) bacteria the presence of an intricate outer membrane (composed from phospholipid bilayer and peptidoglycan) in the cell wall makes it less permeable (Alves et al. 2014). For effective antimicrobial treatment, the photosensitizer needs to penetrate (or at least to attach to) the cell wall of the bacteria. However, the membrane of Gram (−) bacteria is physical barrier for the diffusion of photosensitizer into the bacterial cytosol and, by no means, the inactivation of Gram (−) bacteria is lower.

It must be noticed that, Gram (+) and Gram (−) bacteria carry either cationic or anionic electric charge. These charges are essential for photosensitizer uptake and localization in the bacterial cell membrane. It is more or less accepted that the neutral, anionic, or cationic photosensitizers were more effective against Gram (+) bacteria, whereas just cationic photosensitizers were able to significantly inactivate Gram (−) species (Luksiene and Brovko 2013).

The structure of cell envelope of yeasts and fungi is variable, but mostly it consists of moderately porous layer of β-glucan and mannan polysaccharides (Kashief et al. 2017). Such structure makes them inherently more permeable to external molecules than in the case of Gram (−) bacteria. May be, in addition to the protective impermeable barrier, the susceptibility of the investigated fungi to the photosensitization treatment could be attributed to the various number of catalase genes as defensive mechanism against ROS. For instance microfungi, filamentous fungi and yeasts containing very small pores are more resistant to Chlorophyllin-based photosensitization than bacteria and require higher doses of photosensitizer (10^{-3}–10^{-4} M whereas for bacteria we use just 10^{-5} M) (Fig. 11.10).

Meanwhile, the produced ROS damage cell wall and membrane, including inactivation of enzymes or other proteins, peroxidation of lipids, destruction of lysosomes and mitochondria (Dai et al. 2012).

Eventually fungal spores due to low permeability to chemicals and photosensitizers have rather high resistance to various antimicrobial techniques, including photosensitization. To inactivate them the higher concentrations of photosensitizer and higher light dose necessary to use.

6.1.2 Incubation Time

As a rule, the microorganisms in the first stage of experiment must be incubated with photosensitizer for better interaction with them. Incubation time must be elaborated for every other object. For instance, in the case when interaction of photosensitizer with microbe is superficial (just attachment to the cell wall without accumulation inside), it is not necessary to apply long incubation time (enough

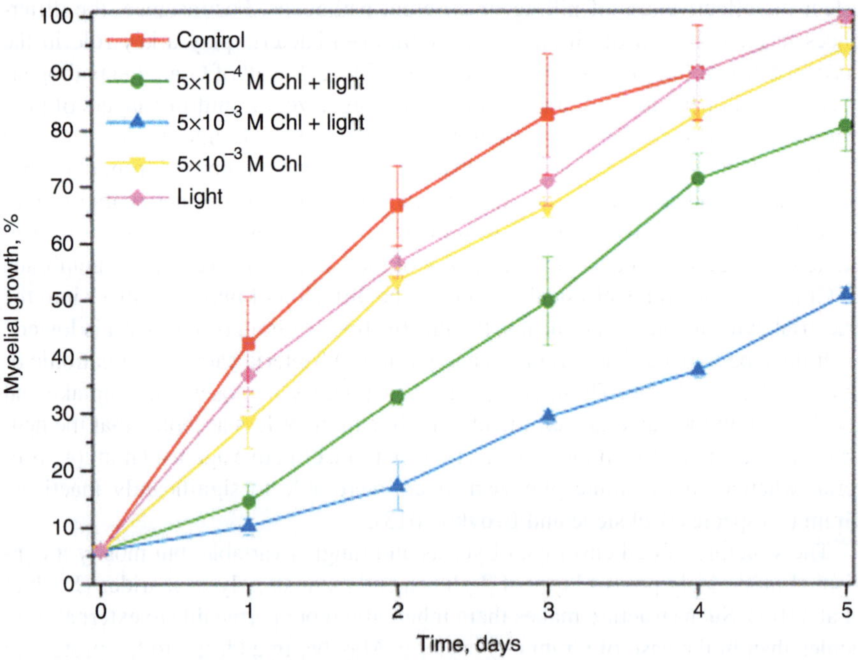

Fig. 11.10 *Fusarium oxysporum* mycelium growth delay after Chl-based photosensitization as function of Chlorophyllin concentration (light dose- 18 J/cm^2) (Luksevicuite and Luksiene 2020)

2–10 min). In the case the photosensitizer tends to penetrate inside microbe it is necessary to apply longer incubation time (4 hours and more). For instance, *E. coli* or *S. enterica*, being Gram (−) bacteria interact weakly with Chlorophyllin, mostly just from outside. In this case the extension of incubation time from 2 min to 120 min did not change the inactivation level (Luksiene and Paskeviciute 2011a). Experience with yeasts/moulds indicated that these microorganisms always need longer incubation time than bacteria (Lukseviciute and Luksiene 2020).

6.2 *Light Intensity and Light Dose*

Theoretically, the photoactivation of photosensitizer molecule, the consequent production of ROS, eventually the inactivation efficiency of microorganisms are strictly correlated to the absorbed photons. So, light dose delivered to the object can be expressed by following formula:

$$E = It,$$

Were E- is light dose, I- is light intensity, t- is exposure time.

It is obvious, that light intensity (*I*) of our LED-based light source is very important parameter for effective inactivation of microorganisms. When light intensity is low, we need to apply longer illumination time to obtain the same light dose. When light intensity is enough high, illumination time *t* can be shorter. As a rule, food industry usually needs technologies which are as short as possible. Meanwhile, when the light intensity is higher than 100 mW/cm^2, the undesirable thermal effects always take place (Luksiene and Brovko 2013).

In study (Le et al. 2020) authors evaluated the experimental inactivation curves of *S. aureus* and *E. coli* when Curcumin-based photosensitization was applied. Data indicated that inactivation kinetic in both cases was strictly depended on the number of absorbed photons. In other study the inactivation kinetic of bacteria treated with Chlorophyllin-based photosensitization was analyzed. It was confirmed that the inactivation as function of light dose mostly was biphasic, entailing either an initial lag phase (shoulder), or a tapering off (tailing) towards the end. A Weibull model was the most suitable to describe this light dose dependent inactivation of bacteria (Vaitonis and Luksiene 2010).

Concluding, the efficiency of antimicrobial inactivation of photosensitization strictly depends on the object, photosensitizer concentration, light intensity and light dose. Moreover, the interrelationship between the concentration of bound photosensitizer, light dose and degree of inactivation could be used to optimize the treatment protocol.

7 Efficiency of Photosensitization Against Food-Borne Pathogens and Spoilage Microorganisms *In Vitro*

7.1 Antibacterial Efficiency of 5-Aminolevulinic Acid-Based Photosensitization

First results obtained on using 5-aminolevulinic acid (5-ALA), a non-photoactive natural metabolite of hem synthesis pathway and precursor of endogenous photosensitizers in the bacterial cell indicated that most bacteria are able to synthesize endogenous photosensitizers from the precursor 5-ALA and after light exposure, significant inactivation of food pathogens such as *B. cereus* (6.5 log), *L. monocytogenes* (4 log), *S. typhimurium* (6 log) *in vitro* was observed (Vaitonis and Luksiene 2010). Application of 5-ALA-based photosensitization in food industry is really possible, since ALA is tastless, yet its use is effective against a wide range of food-borne pathogens, their spores and biofilms (Luksiene and Brovko 2013). The main disadvantage of ALA-based photosensitization as treatment is its high cost due to high cost of 5-ALA.

7.2 Antibacterial Efficiency of Curcumin-Based Photosensitization

In study (Penha et al. 2017) the efficiency of Curcumin-based photosensitization against several food pathogens was investigated. Obtained data revealed that Curcumin in the presence of blue LED-based light (420 nm) significantly inactivated *S. aureus*, *A. hydrophila*, *P. aeruginosa E. coli.* Just *S. enterica* Typhimurium was most resistant to Curcumin-based photosensitization and just partial inactivation of this pathogen was observed.

Other authors (Bhavya and Hebbar 2019) investigated the efficiency of curcumin-based photosensitization against *S. aureus* and *E. coli* and found inactivation of both pathogens by 5.9 log. Membrane injuries were the main mechanism of bacterial cell damage after this treatment.

Obviously, such features of Curcumin as low solubility, instability and scarce bioavailability hinder its more wide application (Dias et al. 2020). But it is possible to overcome this challenge synthesizing Curcumin incorporated in polymeric nanoparticles (Gutierrez et al. 2017).

7.3 Antibacterial Efficiency of Hypericin-Based Photosensitization

Results, obtained in our previous study (Kairyte et al. 2012) revealed that Hypericin-based photosensitization can effectively destroy food pathogens. Analysis of fluorescence emission spectra of bacteria-associated Hypericin indicated that this compound tended to interact with Gram-(+) bacteria more intensively than with Gram (−) ones. As the result, the inactivation of Gram (+) *L. monocytogenes* was more effective than inactivation of Gram (−) *S. enterica* (Dementavicius et al. 2016). It must be noted that interaction of Hypericin with both bacteria needed just few minutes, but accumulation of Hypericin in *Listeria* was higher than in *Salmonella.*

On contrary, Zhang and coauthors (2018) found that *E. coli*, despite it is Gram (−) bacterium interacted strongly with Hypericin and after exposure to light plethora of intracellular ROS has been detected. The inactivation of bacteria reached 4.1 log.

7.4 Antibacterial Efficiency of Eosin Y- and Erythrosine Based Photosensitization

Bonin et al. (2018) investigated the possibility to apply Eosin Y-mediated photosensitization (light absorption 490–570 nm) for inactivation of main food pathogens. Obtained data indicate that the selective susceptibility of different pathogens to this treatment was observed. For instance, *S. enterica* Typhimurium and *B. cereus* exhibited moderate susceptibility to the treatment, *E. coli* was slightly reduced, *P. aeruginosa* and *S. aureus* were completely inactivated. Thus, the main drawback for this treatment is variation of susceptibility of different pathogens to Eosin Y-mediated photosensitization treatment.

The efficiency of erythrosine-based photosensitization against foodborne pathogens and spoilage bacteria (*A. hydrophila*, *S. enterica*, *E. coli*, *S. aureus* and *P. aeruginosa*) was evaluated in (Yassunaka et al. 2015). Obtained results confirmed the data from previous study (Bonin et al. 2018) that *S. aureus* was more photosensitive to this treatment than Gram (−) bacteria.

7.5 Antibacterial Efficiency of Chlorophyllin-Based Photosensitization

For the first time it was found in studies (Luksiene and Zukauskas 2009; Luksiene and Paskeviciute 2011a) that Chlorophyllin (Chl), well known food additive (E140) after exposure to visible light ($\lambda = 405$ nm) exhibited pronounced photosensitizing properties. Gram (+) *B. cereus, L. monocytogenes and E. faecalis, and* Gram (−) ones such as *S. enterica, E. coli, P. aeruginosa,* have been inactivated by this treatment (). Even thermoresistant strains of *L. monocytogenes* and *B. cereus* were susceptible to this treatment and were inactivated 7 log. What is more, the spores and biofilms, which are extremely resistant to all chemical antimicrobials, were inactivated by this treatment using higher photosensitizer concentration and light dose (Luksiene and Paskeviciute 2011a; Luksiene 2021). But, it must be noted that inactivation of Gram (−) bacteria in all cases was lower that Gram (+) ones (Luksiene and Paskeviciute 2011a).

7.6 Photosensitization Against Spores and Biofilms

Particular concern for the agro-food industry pose microbial spores - they can survive even conventional sterilization. The point is, that spores usually formed by bacteria and fungi have very high resistance to chemical and physical biocides (Wainwright and Crossley 2004). In study (Pirttijarvi et al. 1996) it is evaluated that more than 90% of food packaging contamination is composed of spore-forming bacteria. In this context it is import to note that Chlorophyllin-based photosensitization treatment reduced the *Bacillus* spores population by 3.1–4 log. It is possible to achieve higher level of spore inactivation (5 log) when higher Chl concentration is used for experiments (Luksiene and Paskeviciute 2011a).

Sometimes fresh produce or food-related surfaces can be contaminated with sessile cells which can form mature biofilms Actually, most of the microorganisms (99%) exist in the environment not in planktonic form, but as biofilms. Biofilms cover the bacterial cells from outside with polymeric substance which include polysaccharides, proteins, lipids, and extracellular DNA. Bacteria living in biofilm community have some advantages, for instance higher structural stability, better adherence to different surfaces and increased virulence. Moreover, the protected environment of the biofilm together with significantly different phenotypic properties of the cells have been implicated in giving rise as much as 1000-fold resistance to antimicrobials if compare with planktonic cells (Dai et al. 2012; Silva et al. 2018). In study (Bonifacio et al. 2018) authors compared the efficacy of a Curcumin-enriched ´ extract of *Curcuma longa* with that of commercial Curcumin. Obtained data indicated that pure Curcumin was more effective photosensitizing agent than Curcumin-enriched extract, since inactivated *L. innocua* biofilms by 4.9 log. The ALA-based photoinactivation of *L. monocytogenes* biofilms attached to polyolefine by 3.1 log indicated that this treatment also has potential to combat biofilms on different food-contact surfaces. Moreover, Chlorophyllin-based photosensitization is also effective tool against *Listeria* biofilms and can reduce them by 4 log (Luksiene et al. 2011a). Hence, it seems that photosensitization can represent an innovative and effective approach in sanitation of different surfaces contaminated by bacterial biofilms (Silva et al. 2018).

7.7 Photosensitization Against Harmful Microfungi

Uncontrolled mold growth on fresh produce as well as food-contact surfaces induces fast spoilage of fresh produce and dangerous contamination with harmful mycotoxins. Several studies have been published on the inactivation of yeasts and microfungi by photosensitization. Data indicated that the yeast *Saccharomyces cerevisiae* as well as micromycetes *Ulocladium oudemansii*, *Trichotecium roseum* and *Aspergillus flavus* are susceptible to hematoporphyrin dimethyl ether (HPde)-based photosensitization. Fluorescence micrographs revealed that HPde tended to accumulate inside the fungal conidia (Luksiene et al. 1989, 2004a, b). Moreover, the used treatment induced remarkable delay of spore germination in several investigated fungi, i.e. *Aureobasidium* sp., *Rhodotorula* sp., *Penicillium stoloniferum*, *Aspergillus fumigatus*, *Aureobasidium pullulans*, *Ulocladium chartarum*, *Alternaria alternata*, *Rhizopus oryzea*, *Fusarium avenaceum* and *Acremonium strictum* (Luksiene et al. 2004a, b; Luksiene 2005; Luksiene et al. 2005). The amount of photosensitizer accumulated inside the fungal cell was found to be 30 mol/μM protein (Luksiene et al. 2005).

Usually the fungal cells have an overall negative/ neutral charge. So, obviously, cationic photosensitizer interacts better with cell-wall components than anionic photosensitizer. Meanwhile, after exposure to the light, the generated ROS cause oxidative stress and disintegration of the fungal cell wall what eventually leads to the leakage of intracellular material. Probability to damage fungal DNA mostly is very low.

Filamentous moulds can survive under stressful environmental conditions and remain dormant until better conditions. Such increased resistance of spores can be explained by their special envelope (60–80 nm) which is made from chitin, α-β glucans and mannoproteins (Al-Asmari et al. 2017). Data presented in (Al-Asmari et al. 2017) revealed that Curcumin-based photosensitization effectively inactivated spores of *A. niger*, *A. flavus*, *P. griseofulvum*, *P. chrysogenum*, *F. oxysporum*, *C. albicans* and *Z. bailii*. An increased light dose (96 J/cm^2) reduced the fungal populations by 75–100%. Obtained data are promising and indicate that a plethora of harmful and pathogenic micromycetes residing on packaging, food processing surfaces or fruits and vegetables might be inactivated by photosensitization treatment in completely safe and sustainable way.

8 Photosensitization for Preservation of Fresh Produce

8.1 Exploiting Endogenous Photosensitizers for Decontamination of Freshly-Cut Fruits

The possibility to control foodborne pathogens on fresh-cut papaya by photosensitization which is based on naturally existing endogenous porphyrins is evaluated in the study (Kim et al. 2017a). The population of *Salmonella* on cut fruits was reduced by 0.3–1.3 log at chilling temperatures. It is important to note that LED-based illumination had no impact on the physicochemical and nutritional properties of cut papaya. These results indicate that a food chiller equipped with 405 ± 5 nm LEDs can preserve fresh-cut papayas in retail stores, minimizing the risk of spoilage and salmonellosis. Meanwhile, the main drawback of this treatment is long illumination time (36–48 h; 1.3–1.7 kJ/cm^2).

In the other study (Kim et al. 2017a, b) authors evaluated the possibility to decontaminate other fruit- fresh-cut mango using photosensitization based on endogenous porphyrins. Results revealed that *E. coli* O157:H7, *L. monocytogenes*, and *Salmonella* spp. on the surface of fresh-cut mango were inactivated by 1.0–1.6 log CFU/cm^2, but the treatment again continued 36–48 h (total dose, 2.6–3.5 kJ/cm^2). There were no significant changes in color, no depletion of antioxidant capacity, ascorbic acid, *β*-carotene, and flavonoid in treated fruits.

Josewin et al. (2018) evaluated the decontamination efficiency of sodium Chlorophyllin and 405 nm light on *Salmonella spp.* and *Listeria monocytogenes* inoculated on cantaloupe rinds. Obtained data indicate that this treatment reduced by 3 log both pathogens and can be effective tool to minimize the risk of listeriosis and salmoneliosis.

8.2 Hypericin-Based Photosensitization for Decontamination of Fruits and Vegetables

Optimistic results have been obtained using Hypericin-based photosensitization for decontamination of several selected fruits and vegetables. Apricots (*Prunus armeniaca*), plumes (*Prunus domestica*), and cauliflowers (*Brassica oleracea*) inoculated with *B. cereus* were decontaminated using Hypericin-based photosensitization by 0.8–1.3 log. Moreover, the population of naturally distributed mesophiles on the surface of fruits and vegetables was reduced by 0.3–0.72 log. But it must be emphasized that the treatment strongly depended on the fruit shape and surface (Aponiene et al. 2015; Dementavicius et al. 2016). It must be noted that the higher inactivation levels can be easily achieved using higher illumination intensity or longer illumination time.

8.3 *Curcumin-Based Photosensitization for Decontamination of Fresh Produce*

Several studies using Curcumin-based photosensitization for microbial control of fresh produce look promising. For instance, the effect of photoactivated Curcumin in decontamination of fresh date fruit (*Phoenix dactylifera*) when sprayed on the fruit surface was assessed. Following illumination with visible blue light (420 nm, 180 J/cm^2) extended the shelf life to 98 days whereas in control fruit it reached just 28 days (Al-Asmari et al. 2017; Al- Asmari et al. 2018).

In other study acidified Curcumin was sprayed on lettuce, spinach, and tomato which have been inoculated with *E. coli* O157:H7 and *L. innocua* followed by illumination. It led to 3 log reduction with no changes of color of the treated produce. In addition, lower pH increased the antimicrobial efficacy of Curcumin-based photosensitization and reduced the time of illumination from 10 min to 2 min. The obtained inactivation of *E. coli* O157:H7 reached 5 log (Oliveira et al. 2018).

Inactivation of *E. coli* inoculated on fresh-cut Fuji apple slices increased up to 1 log with an increase in Curcumin concentration (0.5–50 μM), incubation time (5–30 min), and illumination time (150–510 s). Reduction in browning, weight loss, enzyme activity (peroxidase by 48% and polyphenol oxidase by 51%), with few negative impact on bioactive components was observed after storage for 8 days (Tao et al. 2019).

Curcumin- based photosensitization reduced total aerobic microorganisms on fresh-cut Hami melons by 1.8 log without reduction in soluble solid content, water content and firmness. Moreover, this treatment delayed the browning rate and maintained luminosity as well as sensory quality (Lin et al. 2019).

Some fungi that reside on fruits and vegetables can produce toxic secondary metabolites known as mycotoxins. It is well documented that mycotoxins are harmful to human health, have huge impact on crop production and economics. Curcumin-based photosensitization reduced *A. flavus* spores on maize kernels about 2 log (Temba et al. 2016).

These data suggest that Curcumin-based photosensitization treatment could be a promising weapon against foodborne pathogens and harmful microfungi, residing on fresh produce.

8.4 *Chlorophylin-Based Photosensitization for Decontamination of Fruits and Vegetables with Irregular Surfaces*

Theoretically, the antimicrobial efficiency of photosensitization as light-based technology can strongly depend on the structure and regularity of fruit surface. Hence, Chlorophyllin-based inactivation of naturaly distributed mesophilic bacteria on plums (*Prunus domestica*), tomatoes (*Solanum lycopersicum*), cauliflowers

(*Brassica oleracea*) and red peppers (*Capsicum annum*) was rather effective. The highest decontamination level was observed for red pepper (1.3 log), tomatoes (1.2 log), plums (1.5 log) The lowest microbial decontamination was obtained in the case of cauliflowers, which have very irregular surface and shape (1.1 log) (Luksiene and Zukauskas 2009; Luksiene 2013a, b).

FAO and WHO in 2007 concluded that leafy vegetables currently presented the greatest concern in terms of microbiological contamination (FAO-WHO 2008). In this context the highly contaminated fresh produce also can be a target of bioterrorism (Wallin et al. 2007). Data presented in the study (Paskeviciute et al. 2019) clearly indicated that Chlorophyllin-based photosensitization can reduce significantly the microbial load (1.3 log) on the surface of basil as well as inoculated thermoresistant *L. monocytogenes* (1.6 log).

8.5 Chlorophyllin-Based Photosensitization for Preservation of Perishable Fruits

Strawberry (*Fragaria* × *ananassa* Duch.) is a major crop of production worldwide. Meanwhile, this fruit has short ripening period and is highly susceptibile to mechanical injuries. All these circumstances reduce significantly strawberry shelf life. Sometimes, especially in developing countries, strawberries spoilage losses can reach 40–80% (Feliziani and Romanazzi 2013; Lafarga et al. 2019).

Consumption of fresh strawberries induced several outbreaks in Europe. According to Food and Drug Administration (FDA) data in 1 out of 143 imported strawberries *Salmonella* was found (FDA 2015). Moreover, since chemical fungicides are used to preserve the strawberry crop from spoilage inducing microfungi, their residues have been identified in more than 60% of strawberries (Jianglian and Shaoying 2013).

Thus, during the last two decades postharvest fruit protection shifted from using chemical technologies to physical treatments (Lafarga et al. 2019). However, it was found that, for instance, irradiation of strawberries by ionizing radiation (2–3 kGy) was rather effective against fungi and prolonged significantly the shelf life of berries, but besides it, this treatment induced changes of strawberry texture and color (Yu et al. 1995).

After application of Chlorophyllin-based photosensitization for microbial control of strawberries, the population of mesophilic bacteria naturally distributed on the surface of fruits reduced by 1.7 log. Moreover, this treatment reduced the population of inoculated *L. monocytogenes* by 1.8 log. It is important to note that the spoilage-related yeasts and microfungi were less susceptibile to photosensitization (the reduction reached just 0.9 log). Meanwhile, the overall microbial inactivation of bacteria and fungi eventually prolonged the shelf-life of fruits by 2 days what is very important for agricultures and industry (Luksiene and Paskeviciute 2011b).

9 Chlorophyllin-Based Photosensitization for Microbial Control of Ready-to-Eat Meals

The consumption of ready-to-eat fresh produce seems to be a nowadays choice of fashion (Vivek et al. 2019). Especially sprouts as healthy and unprocessed food are gaining popularity across the world (Galieni et al. 2020). In parallel, sprouted seeds are among the most contaminated fresh produce and have been recognized as source of such food pathogen as *E. coli* O157:H7, *B. cereus*, *S. enterica*. Moreover, in one of our previous studies (Luksiene et al. 2007) more than 10 different microfungi families on the wheat grains have been identified.

In order to increase the microbial safety of sprouts, U.S. Food and Drug Administration in 1999 suggested chlorine-based sanitizers (FDA 1999). However, recently there are a lot of published experimental evidences, confirming that widely applicable calcium hypochlorite is a concern, since interacting with biomatrix can produce highly toxic trihalomethanes, haloketones, chloropicrins, and haloacetic acids. Thus, due to public health issues chlorine-based sanitizers are prohibited in some European countries including Belgium, Denmark, Germany, and the Netherlands (Meireles et al. 2016). Moreover, rather long treatment time using hypochlorite reduced the seed germination rate, so the sprout industry did not fully welcome this technique (Ding et al. 2013).

According to the results published in (Zudyte and Luksiene 2019) the Chlorophyllin-based photosensitization treatment remarkably reduced the viability of surface-attached mesophilic bacteria (2.5 log), inoculated *E. coli* (1.5 log) and naturally distributed yeasts/fungi (1.5 log). Moreover, SEM images confirmed that this treatment did not damage grain surface microstructure. Most important is the fact that photosensitization did not reduce such important parameters as seed germination rate and seedling growth or visual quality of sprouts.

10 Effects of Photosensitization on Nutritional and Organoleptic Properties of Treated Fresh Produce

It is obvious, that the main criterion of a prospective preservation technique is the maintaining of quality attributes of treated fresh produce, main organoleptic and nutritional properties. To this end, the probability exists, that photosensitization being an effective antimicrobial treatment can reduce the quality of treated fruits. For instance, photosensitization being light-based technology can make impact on phenolics and anthocyanins, which tend to accumulate in epidermal and cortex tissues of fruits and vegetables. So far, studies that have used the photosensitization for decontamination of fresh produce indicated that no undesirable changes in the physicochemical and nutritional profiles of the treated fruits, vegetables and berries have been found (Kim et al. 2017a, b; Luksiene and Paskeviciute 2011b, Aponiene et al. 2015). Obtained data indicated that Chlorophyllin-based photosensitization did not change the amount of main antioxidants, i.e. soluble phenolics and anthocyanins in strawberries (Luksiene, Paskeviciute 2011b). Eventually, the treatment did not reduce the firmness of the treated strawberries (Luksiene and Paskeviciute 2011b; Rasiukeviciute et al. 2015).

Moreover, it has also been reported that illumination of fresh produce may result in enhancing the formation and accumulation of different secondary metabolites, which enhance the growing process in germinated sprouts (Lukseviciute and Luksiene 2020).

Browning of the cut surface of fresh produce is important and undesirable problem. It depends on the presence and activity of enzymes in the plant cells called polyphenoloxidases. In this context the activity of this enzyme after Chlorophyllin-based treatment of freshly cut basil was tested. Data indicated that this treatment had no impact on the activity of polyphenoloxidases and browning of freshly cut basil (Paskeviciute et al. 2019).

One of the important quality indicators in appearance and acceptance of fresh produce is color. Moreover, it has been considered to have a main role in food choice and preference. Some risk exists that the color of treated fresh produce can be compromised by photosensitization treatment. As a rule, every light technology can change the color of treated produce. For instance, using longer illumination time it is possible to achieve bleaching of the treated object. Moreover, since most photosensitizers absorb in the visible range and have intensive color, for instance Curcumin, Chlorophyll and Riboflavin, it is a risk to change the color of produce (Ormond and Freeman 2013). One of the possible ways to minimize effects on food color is optimization of the process, wherein color change should be incorporated as one of the constraints. In such a scenario, minimal effective, but not coloring concentration for every photosensitizer must be found and applied. Thus, to determine whether Chlorophyllin-based photosensitization has any negative effects on the color of the treated fruits and vegetables, fruits and vegetables were investigated immediately after treatment (Luksiene and Paskeviciute 2011b). Obtained data confirmed, that photosensitization had no effects on the color of cucumber, carrot, paprika, apricots, plums, cauliflower, strawberry, tomato, when Chlorophyllin concentrations did not exceed mM. Using higher Chlorophyllin concentration will make better antimicrobial effect, but also can change the color of the treated fruit. Thus, decontamination of sprouts by Chlorophyllin-based photosensitization using the optimized light dose and optimized Chlorophyllin concentration had no impact on visual quality and color of treated sprouts (Zudyte and Luksiene 2019).

11 Comparison of Antimicrobial Efficiency of Chl-Based Photosensitization with Some Conventional and Emerging Treatments

Thus, obviously, it is interesting and important to evaluate the antimicrobial efficiency of photosensitization in comparison with conventional treatments. For this purpose cherry tomatoes (*Solanum lycopersicum var. cerasiforme*) - one of the most important vegetable crops have been used. Thus, in the study (Paskeviciute et al. 2018) the obtained antimicrobial efficiency of Chlorophyllin-based photosensitization was compared with that of high power pulsed light (HPPL) (Luksiene et al. 2007; Luksiene et al. 2011). Results indicated that Chlorophyllin-based photosensitization reduced not just mesophilic bacteria (1.6 log), but also inoculated on tomatoes the thermoresistant strains of *B. cereus* and *L. monocytogenes* by 1.5 log and 1.6 log respectively. As described previously the application of Chlorophyllin-based photosensitization for microbial control of strawberries reduced mesophilic bacteria by 1.7 log, inoculated *L. monocytogenes* -by 1.8 log, fungi- by 0.9 log.

In comparison, HPPL reduced mesophilic bacteria on the surface of strawberry by 2.2 log, inoculated *B. cereus* and *L. monocytogenes* by 1.5 log and 1.1 log respectively. Yeasts/microfungi distributed on the surface of strawberries were inactivated by 1 log, and as a result of all these microbial inactivations, the shelf-life of treated strawberries was extended by 2 days (Luksiene et al. 2013a, b). No effects on visual or nutritional properties of strawberries have been observed (Cao et al. 2019). After treatment of other fruits and vegetables (plums, tomatoes, cauliflower) by HPPL very similar inactivation of mesophilic bacteria (1.0–1.3 log) and inoculated *B. cereus* (1.3–2 log) depending on surface characteristics have been found (Luksiene et al. 2012). Higher antimicrobial effect of HPPL is tailored by higher temperature on the surface of fresh produce (Gudelis and Luksiene 2010). Thus, antimicrobial efficiency of photosensitization is comparable with FDA approved technology of high power pulsed light.

In addition, the observed decontamination efficiency of Chlorophyllin-based photosensitization was significantly higher in comparison with conventional washing with water (0.6–0.8 log) or hypochlorite (1.4–1.6 log). It must be noted that NaOCl issued widely in fresh produce industry, since is cost- effective and exhibits antimicrobial activity. Results presented in (Allende et al. 2009) revealed that traditional washing with sodium chlorite of cut cilantro reduced the population of mesophilic bacteria and *E. coli* O157:H7 about 1 log. In other study it was shown that washing lettuce, arugula, parsley with water containing 200 ppm chlorine for 10 min reduced the populations of mesophilic bacteria by 1.18–1.49 log (Temiz et al. 2011). But treatment of strawberries with hypochlorite (NaOCl) (200 μg mL^{-1}) reduced their microbial contamination just by 0.45 log. Moreover, it is well known that chemical treatment with hypochlorous acid releasing chlorine, is very dangerous and gives rise to toxicity concerns, since can form highly mutagenic compounds (Wright et al. 2017).

12 Main Advantages of Chlorophyllin-Based Photosensitization

Photosensitization as antimicrobial treatment has several advances (Fig. 11.11). Meanwhile, one of the most important advantages of this treatment is its non-thermal nature. It was confirmed by extensive experimental data that during the photosensitization treatment using LED-based light there was no significant temperature increase on the surface of fresh produce that would be harmful and change the organoleptic properties or reduce the nutritional quality of fruit (Luksiene and Paskeviciute 2011b; Paskeviciute et al. 2019). Usually, the surface temperature of apricots, plums, tomatoes, strawberies and cauliflowers held at room temperature (20 °C) increased during treatment to a maximum of 25–26 °C (Aponiene et al. 2015; Luksiene and Paskeviciute 2011b). In comparison, high power pulsed light is

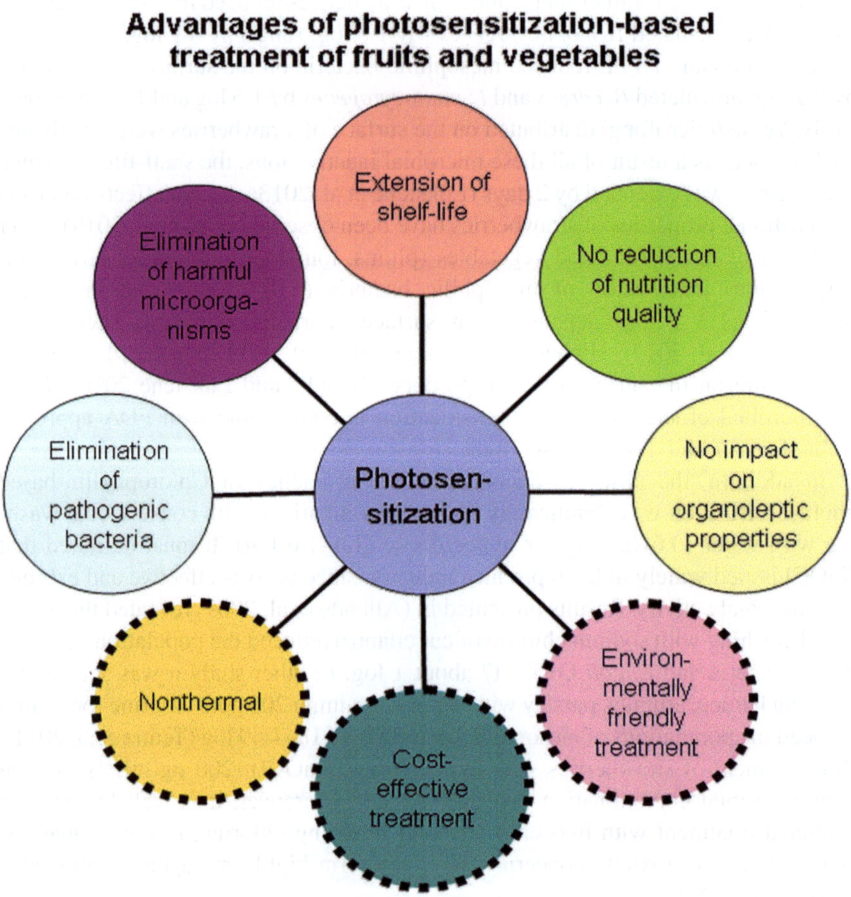

Fig. 11.11 Benefits of photosensitization

an effective antimicrobial treatment and approved by FDA for microbial control of fresh produce, but at longer light exposures induced thermal effects on food matrix up to 80 °C, what is unacceptable (Gudelis and Luksiene 2010).

A major reason why photosensitization is so attractive and promising is that so far, no resistance has been observed among the plethora of microorganisms that have been treated. This is a major advantage over other novel emerging food processing technologies, for instances ultrasonication, pulsed electric field or ultraviolet light, all of which have been known to induce microbial resistance (Cebrian et al. 2016). The lack of resistance to photosensitization treatment can be explained by two reasons. First, the ROS generated during the photosensitization have an extremely short half-life. For example, the half-life of singlet oxygen, produced during photosensitization is about 10 ns (Luksiene 2005). At such circumstances the microorganism has not enough time to respond to the stress. Second, the photosensitization targets multiple cellular compartments and cellular repair systems are unable to do all necessary reparation. By no means, it is reasonable to expect that resistance could develop with time and greater popularization of the technology (Kashef et al. 2017; Maisch 2015).

13 Drawbacks of Chlorophyllin-Based Photosensitization and Ways to Eliminate Them

Inactivation of pathogenic and harmful microorganisms by Chl-based photosensitization is effective, selective and environmentally friendly treatment. However, this treatment has one notable disadvantage: Gram (−) bacteria mostly due to more complicated cell wall structure and lower photosensitizer uptake have lower susceptibility to Chlorophyllin-based photosensitization than Gram (+) bacteria (Kairyte et al. 2012; Alves et al. 2014; Aponiene and Luksiene 2015). To make more evidence Kruger et al. (2019) tested Gram (−) bacteria *E. coli* NR698, which has a defective outer membrane. Obtained results indicate that this strain was effectively inactivated by Chl-based photosensitization and inactivation level reached that of Gram (+) *B. subtilis*. So, it is clear that the outer membrane of Gram-negative bacteria may be a barrier for chlorophyllin uptake, consequently accumulation and efficiency of photoinactivation.

To overcome this permeability problem, Chl-based photosensitization of Gram (−) pathogens might be combined with other antimicrobial or membrane-destabilizing treatments exploiting concept of hurdle antimicrobial technologies (Afrasiabi et al. 2020). It is well documented that successful combination of several antimicrobial agents can significantly increase the antimicrobial efficiency. What is the mechanism of this inactivation? First, such combination enable us to use lower intensities of every treatment. Second, such approach do not cause so great destructions of food matrix or lethal injuries to microorganism as intensive treatment do. Thus, it generates a number of sublethal stresses and sublethal damages in microbial cell that microorganizm of concern should not be able to overcome. Eventualy microorganism becames metabolically exhausted and die.

For instance, in order to increase the uptake of Chlorophyllin by Gram (−) bacteria and herewith enhance the inactivation efficiency it is possible to use sub-lethal concentration of membrane - destabilizing agent colistin with sublethal dose of photosensitization. Data presented in (Richter et al. 2019) revealed that such combined action significantly increased inactivation of *E. coli* and *S. enterica* and gave rise to the reasonable assumption that Chlorophyllin-based photosensitization in combination with membrane-destabilizing agents may be a promising alternative to control pathogens.

Moreover, it is possible to combine Chlorophyllin-based photosensitization with 5-ALA-based photosensitization. Mathematical modeling of bacterial inactivation curves, presented in (Luksiene et al. 2013a) allowed to make a hypothesis that cocktails of Chlorophyllin and ALA can be an effective combination to enhance antimicrobial efficiency against Gram(+) and Gram(−) food pathogens and to inactivate them in a more uniform way. Experimental data confirmed that combining Chlorophyllin- and ALA-based photosensitization increased the inactivation of all investigated Gram(+) *L. monocytogenes* and Gram(−) *S. enterica* bacteria by 7 log.

In order to increase the inactivation of Gram (−) bacteria it seems promising to combine photosensitization with FDA approved high power pulsed light technology (HPPL) (Mandal et al. 2020). It is well documented, that the main cellular target for HPPL ("working" wavelength, λ = 260 nm) is microbial DNA (Gudelis and Luksiene 2010). Thus, combining Chlorophyllin-based photosensitization induced membrane injuries with DNA damages produced by HPPL produced lethal destructions in Gram (−) bacteria. Data presented in study (Kairyte et al. 2012) indicated that such combined treatment reduced viable counts of Gram (−) *Salmonella* and Gram (+) *Listeria* by 6–7 log. Data indicated that, combined treatment of HPPL and photosensitization can be useful strategy to combat both Gram-positive and Gram-negative bacteria on different surfaces in an effective and uniform way.

The innovative approach on the combination of Chlorophyllin-based photosensitization with chitosan in the form of Chlorophyllin-Chitosan conjugate is presented in (Rodriges-Lopez et al. 2020). In this case both treatments have the same target-cell membrane. The overall induced lethal damages in *S. enterica* reduced its viability by 7 log. It is important to note that the destructions in *Salmonella* were so multiple and lethal that no *Salmonella* regrowth after treatment was observed. It looks like such combination potentiates the interaction between negatively- charged Chlorophyllin and negatively charged microorganism, since Chitosan being positively charged with protonated NH^{+3} groups interacts electrostatically with negatively charged lipopolisaccharides of bacterial membrane. From other side chitosan electrostatically interacts with Chlorophyllin.Thus, Chlorophyllin-Chitosan conjugate can work as mediator between bacterial cell wall and photosensitizer and enhances the possibility of cell-photosensitizer interaction. As a result, the increased membrane permeability, internal osmotic disbalance, disintegration of cellular structures induced by photoactivated Chlorophyllin-Chitosan conjugate lead bacteria to total destruction and death (Goy et al. 2009). By no means, the right combination of several sub-lethal treatments that would push the microbial cell from growth to inactivation is really an art. In other study (Li et al. 2020) chitosan with loaded curcumin was used for significant photosensitization –based inactivation of *Staphylococcus aureus* and its biofilms on stainless-steel. Results confirmed that antimicrobial efficiency of this conjugate depended on curcumin concentration, incubation time ans light dose.

Obviously, antimicrobial efficiency of Chlorophyllin-based photosensitization is significantly higher in experiments *in vitro*, when bacteria are in suspension or attached to the flat surface. When the task is to inactivate bacteria on the surface of fresh commodities which are 3-dimentional, not simple geometry (spherical, round, irregular) the antimicrobial efficiency is going down. The point is that the surface of fruits and vegetables very often is not flat and smooth, very often rough and complicated, and microorganisms may reside in crevices or in surface irregularities and injuries, where light cannot reach them. Obviously, this factor slightly decreases the efficiency of the treatment. Moreover, due to the fact that fruits and vegetables have different shapes, a lot of photons are not absorbed due to increased light scattering. This factor also reduces the overall antimicrobial efficiency of the light-based treatment. Nevertheless, in order to increase the inactivation efficiency of treatment it is always possible to use:

(a) more powerful LEDs or to make more frequent arangement of diodes in illumination platform (Fig. 11.12., middle) what by no means will increase the energy density and antimicrobial efficiency;

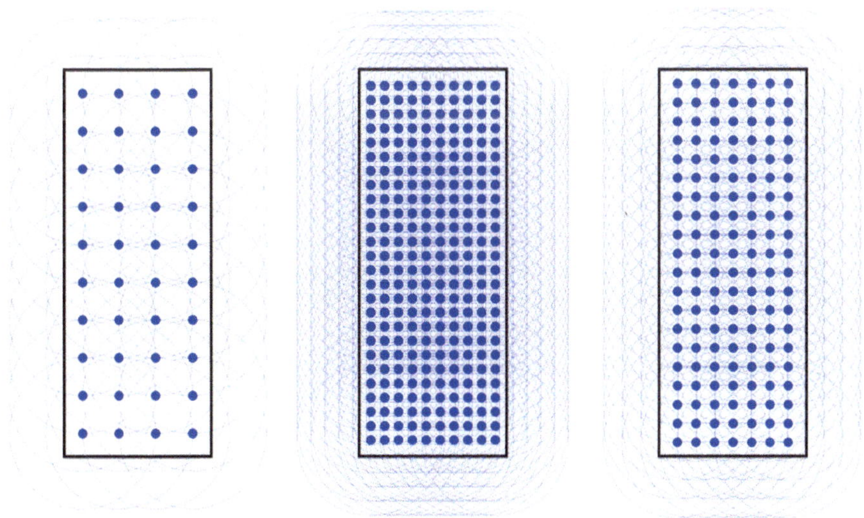

Fig. 11.12 The density of LEDs in the illumination platform can be more or less dense. Higher density of LEDs (middle picture) always produces higher light intensity and better photosensitizing effect

(b) rotating or shaking of the plate with placed fresh produce will enable all-round illumination. Such solution will reduce the shadow effect and will assist the delivered photons to reach the target surface of fruit, since antimicrobial efficiency of photosensitization depends on amount of photons which reached the surface.
(c) obviously, the development of three-dimentional light source with rotating/shaking function would be the best solution.

14 Prospects of Photosensitization-Based Edible Biopolymer Coatings

During last decade the use of different films and coatings for better safety of fresh produce has received special attention (Lopez-Carballo et al. 2018). Such "packaging" can increase the microbial safety of fruit and preserve its nutritional/sensorial properties. Moreover, such edible coating is perfect regulator of water migration both in and out of fruit and significantly extends the shelf-life of products (Radev and Pashova 2020). Recently different edible biopolymers, such as polysaccharide-based starch, cellulose derivatives or chitosan are used for formation of coatings for fresh produce.

Chitosan is well known natural compound, cationic linear polysaccharide and primarily by-product of crustacean, fish and seafood processing (Kumar et al. 2020). It has no specific taste and is approved as nutritional supplement. Most attractive chitosan's features are biodegradability, expressed antimicrobial activity and possibility to form films (Jiang et al. 2020). These properties opened the door for formation of edible active antimicrobial coatings which have inserted Chlorophyllin (Rodríguez López et al. 2020). Data obtained revealed that strawberries coated with Chlorophyllin-Chitosan in the presence of visible light (405 nm) were protected not just from pathogenic Gram (+) and Gram (−) bacteria, but also from spoilage inducing microfungi. Such coating significantly delayed the spoilage of treated berries and prolonged their shelf-live by 3 days (Luksiene 2021). It is important to confirm that photoactivated Chl-CHS coating of strawberries had no effects on their color, texture or overall visual quality. Moreover such treatment did not induce any long-lasting free radicals in the strawberries (Luksiene 2021), as for instance do 2 kGy ionizing radiation (Raffi and Stocker 1996). Moreover, such photoactive coating reduced water evaporation and preserved the strawberry weight if compare with control.

In the other study (Lopez-Carballo et al. 2008) authors investigated gelatin-based edible films and coatings containing inserted photosensitizer sodium magnesium Chlorophyllin and sodium copper Chlorophyllin. Such coating was very effective against *S. aureus* and *L. monocytogenes,* since reduced their counts by 5 log and 4 log respectively. Thus, it is clear that such active gelatin-based films and coatings can be successfully applied for better preservation of fruits and vegetables.

Meanwhile we must take into account that hurdle approach has one serious drawback, i.e. application of several technologies at sub-lethal level can induce microbial resistance in the survived population.

15 Conclusions and Future Outlook

The main priority of all agro-food industry is to extend the shelf life of foods saving their organoleptic and nutritional properties. Recently various conventional and emerging preservation technologies have been introduced. However, most of them have specific drawbacks.

During last two decades photosensitization has gained serious attention due to its high antimicrobial efficiency against wide spectrum of microorganisms, multi-target nature and overcoming multidrug resistance. Moreover, this treatment is non-thermal. Due to the fast development of optoelectronics LED-based photosensitization pretends to become cost- effective antimicrobial technology. Eventually, it is human and environmentally friendly, since no hazards, volatile compounds, mutagens or carcinogens are produced during photosensitization treatment.

First insights into application of photosensitization for microbial control of fruits and vegetables were realistic and successful. After 20 years of hard work it looks like photosensitization treatment is effective non-thermal and not-chemical tool to control microbial contamination of fruits and vegetables. Next to perfect antimicrobial efficiency Chlorophyllin-based photosensitization maintains food quality attributes, such as nutritional quality, color and texture.

It is obvious that photosensitization efficiency is lower when we treat fruits and vegetables with different shapes and irregular surfaces in comparison with experiments *in vitro*. First, the treatment efficiency can be enhanced constructing more powerful LED-based lighting system or simply increasing light exposure time. Second, it can be increased combining photosensitization with other antimicrobials to create a hurdle effect. High power pulsed light, chitosan, nanoparticles or other food-grade photosensitizers are perfect antimicrobial tools for hurdle treatment.

It must be highlighted that photosensitization can be used for development of natural, edible biopolymer-based photoactive coatings. Such coatings are able to prolong significantly the spoilage-free period of treated fresh produce, saving water evaporation, nutritional and organoleptic properties.

In our opinion, photosensitization opened a new window for the development of innovative, non-thermal, sustainable, effective and completely safe antimicrobial treatment. Its proper and gentle application for the microbial control of fresh produce suggests the potential of this treatment as an alternative for traditional thermal or chemical treatments and might be really powerful tool to prevent the foodborne diseases in the future.

Acknowledgements Author is thankful to co-worker Saule Girdzijauskaite for her technical support.

References

Afrasiabi S, Pourhajibagher M, Chiniforush N et al (2020) Propolis nanoparticle enhances the potency of antimicrobial photodynamic therapy against Streptococcus mutants in a synergistic manner. Sci Rep 10:15560. https://doi.org/10.1038/s41598-020-72119-y

Akhtar S, Sarker MR, Hossain A (2014) Microbiological food safety: a dilemma of developing societies. Crit Rev Microbiol 40:348–359

Al-Asmari F, Mereddy R, Sultanbawa Y (2017) A novel photosensitization treatment for the inactivation of fungal spores and cells mediated by Curcumin. J Photochem Photobiol 173:301–306

Al-Asmari F, Mereddy R, Sultanbawa Y (2018) The effect of photosensitization mediated by Curcumin on storage life of fresh date (*Phoenix dactylifera* L.) fruit. Food Control. https://doi.org/10.1016/j.foodcont.2018.06.005

Allende A, McEvoy J, Tao Y et al (2009) Antimicrobial effect of acidified sodium chlorite, sodium chlorite, sodium hypochlorite, and citric acid on *Escherichia coli* O157:H7 and natural microflora of fresh-cut cilantro. Food Control 20:230–234

Alves E, Faustino M, Neves M et al (2014) An insight on bacterial cellular targets of photodynamic inactivation. Future Med Chem 6(2):1–24

Aponiene K, Paskeviciute E, Reklaitis I et al (2015) Reduction of microbial contamination of fruits and vegetables by Hypericin-based photosensitization: comparison with other emerging antimicrobial treatments. J Food Eng 144:29–35

Aponiene K, Luksiene Z (2015) Effective combination of LED-based visible light, photosensitizer and photocatalyst to combat Gram (−) bacteria. J Photochem Photobiol 142:257–263

Arutselvi R, Balasaravanan T, Ponmurugan P et al (2012) Phytochemical screening and comprative study of anti microbial activity of leaves and rhizomes of turmeric varieties. Asian J Plant Sci Res 2:212–219

Bhavya ML, Hebbar HU (2019) Efficacy of blue LED in microbial inactivation: effect of photosensitization and process parameters. Int J Food Microbiol 290:296–304

Bernal C, Rodrigues JAO, Guimaraes APP et al (2011) Selective photoinactivation of *C. albicans* and *C. dubliniensis* with Hypericin. Laser Phys 21(1):245–249

Bonifacio D, Martins C, David B et al (2018) Photodynamic inactivation of *Listeria innocua* biofilms with food-grade photosensitizers: a Curcumin-rich extract of *Curcuma longa* vs commercial Curcumin. J Appl Microbiol 125(1):282–294

Bonin E, Santos A, Fiori di Silva A et al (2018) Photodynamic inactivation of foodborne bacteria by Eosin Y. J Appl Microbiol 124:1317–1328

Brovko L (2010) Photodynamic treatment a new efficient alternative for surface sanitation. Adv Food Nutr Res 61:119–147

Cao XN, Huang RZ, Chen HQ (2019) Evaluation of food safety and quality parameters for shelf-life extension of pulsed light treated strawberries. J Food Sci 84(6):1494–1500

Català EC, Sumalla JT, Ros Salvador J (2000) Oxidative stress in bacteria and protein damage by reactive oxygen species. Int Microbiol 3(1):3–8

CDC, Centers for Disease Control and Prevention (2017) Foodborne Germs and illnesses, Food safety, CDC. https://www.cdc.gov/foodsafety//foodborne-germs.html. Accessed 2017 Nov 26

CDC, Center for Disease Control and Prevention (2018a). Available at: https://www.cdc.gov/VitalSigns/foodsafety/. Accessed on 16 December 2018

CDC, Centers for Disease Control and Prevention. (2018b) Outbreak of multidrug-resistant Salmonella infections linked to raw chicken products - multistate outbreak of Salmonella infections linked to raw chicken products. https://www.cdc.gov/salmonella/infantis-10-18/index.html. Retrieved 7 November 2018

Cebrian G, Manas P, Condon S (2016) Comparative resistance of bacterial foodborne pathogens to non-thermal technologies for food preservation. Frontiers Microbiol 7:1–17

Dai T, Fuchs BB, Coleman JJ et al (2012) Concepts and principles of photodynamic therapy s an alternative antifungal discovery platform. Front Microbiol 3:120. https://doi.org/10.3389/fmicb.2012.00120

Dementavicius D, Lukseviciute V, Gómez-López VM et al (2016) Application of mathematical models for bacterial inactivation curves using Hypericin based photosensitization. J Appl Microbiol 120:1492–1500

Dias LD, Blanco KC, Mfouo-Tynga I et al (2020) Curcumin as a photosensitizer: from molecular structure to recent advances in antimicrobial photodynamic therapy. J Photochem Photobiol C 45:100384

Ding HT, Fu J, Smith MA (2013) Microbial contamination in sprouts: how effective is seed disinfection treatment? J Food Sci 78(R):495–R501

D'Souza C, Yuk H-G, Khoo GH et al (2014) Application of light-emitting diodes in food production, postharvest preservation, and microbiological food safety. Compr Rev Food Sci Food Saf 14(6):719–740

EFSA, European Food Safety Authority (2010) Scientific opinion on the re-evaluation of Curcumin (E 100) as a food additive. Eur Food Safety Author J 8(9):1–46

EFSA, European Food Safety Authority (2014) Refined exposure assessment for Curcumin (E 100). Eur Food Safety Author J 12(10):1–43

Feliziani E, Romanazzi G (2013) Preharvest application of synthetic fungicides and alternative treatments to control postharvest decay of fruit. Stewart Postharvest Rev 9. https://doi.org/10.2212/spr.2013.3.4

Food and Drug Administration (FDA) (1999) Guidance for Industry: Reducing Microbial Food Safety Hazards For Sprouted Seeds. Available at: http://www.fda.gov/Food/GuidanceRegulation/GuidanceDocumentsRegulatoryInformation/ucm120244.htm. Accessed 15.12.11

Food and Agriculture Organization of the United Nations (FAO), World Health Organization (WHO) (2008) Microbiological Hazards in Fresh Fruits and Vegetables Food and Agriculture Organization of the United Nations World Health Organization 1–28. https://www.who.int/foodsafety/publications/micro/MRA_FruitVeges.pdf

Food and Drug Administration (FDA) (2015) FDA Survey of Imported Fresh Produce FY 1999 Field Assignment . http://www.fda.gov/Food/GuidanceRegulation/GuidanceDocumentsRegulatoryInformation/ProducePlantProducts/ucm118891.htm. Accessed February 2015

Galieni A, Falcinelli B, Stagnari F (2020) Sprouts and microgreens: trends, opportunities and horizonts for novel research. Agronomy Basel 10:1424. https://doi.org/10.3390/agronomy10091424

Geric M, Gajski G, Mihaljevic B et al (2019) Radioprotective properties of food colorant sodium copper chlorophyllin on human peripheral blood cells in vitro. Mutat Res Genet Toxicol Environ Mutagen 845:403027. https://doi.org/10.1016/j.mrgentox.2019.02.008

Ghate VS, Zhou W, Yuk HG (2019) Perspectives and trends in the application of photodynamic inactivation for microbiological food safety. Compr Rev Food Sci Food Saf 18:402–424

Rodrigez-Lopez MI, Gomez-Lopez VM, Lukseviciute V et al (2020) Modeling the inactivation and possible regrowth of Salmonella enterica treated with Chlorophyllin-chitosan complex and visible light. Food Technol Biotechnol 58(1):64–70

Goy RC, de Britto D, Assis OBG (2009) A review of the antimicrobial activity of chitosan. Polímeros 19:241–247

Gudelis V, Luksiene Z (2010) High power pulsed light for decontamination of chicken from food pathogens. Lith J Phys 50:147–150

Gutierrez JKT, Zanatta GC, Ortega ALM et al (2017) Encpsulation of curcumin in polymeric nanoparticles for antimicrobial photodynamic therapy. PLoS One 6. https://doi.org/10.1371/journal.pone.0187418

Hadi J, Dunowska M, Wu SY et al (2020a) Control measures of SARS-CoV-2: a review on light-based inactivation of single-stranded RNA viruses. Pathogens 9:737. https://doi.org/10.3390/pathogens9090737

Hadi J, Wu SY, Brightwell G (2020b) Antimicrobial blue light versus pathogenic bacteria: mechanism, application in the food industry, hurdle technologies and potential resistance. Foods 9(12):1895. https://doi.org/10.3390/foods9121895

Jiang YL, Yu L, Hu YW (2020) The preservation performance of chitosan coating with different molecular weight on strawberry using electrostatic spraying technique. Int J Bio Macromol 151:278–285

Jianglian D, Shaoying Z (2013) Application of chitosan based coating in fruit and vegetable preservation: a review. J Food Process Technol 4:227

Yin R, Hamblin M (2015) Antimicrobial photosensitizers: drug discovery under spotlight. Curr Med Chem 22:2159–2185. https://doi.org/10.2174/0929867322666150319120134

Josewin SW, Kim MJ, Yuk HG (2018) Inactivation of Listeria monocytogenes and salmonella spp on cantaloupe rinds by blue light emitting diodes (LEDS). Food Microbiol 76:219–225

Kairyte K, Lapinskas S, Gudelis V et al (2012) Effective inactivation of food pathogens *Listeria monocytogenes* and *Salmonella enterica* by combined treatment of Hypericin-based photosensitization and high power pulsed light. J Appl Microbiol 112(6):1144–1151

Kashef N, Huang YY, Hamblin MR (2017) Advances in antimicrobial photodynamic inactivation at the nanoscale. Nano 6(5):853–879

Kim MJ, Bang WS, Yuk HG (2017a) 405 ± 5 nm light emitting diode illumination causes photodynamic inactivation of Salmonella spp. on fresh-cut papaya without deterioration. Food Microbiol 62:124–132

Kim MJ, Tang CH, Bang WS et al (2017b) Antibacterial effect of 405 ± 5 nm light emitting diode illumination against *Escherichia coli* O157:H7, *Listeria monocytogenes*, and Salmonella on the surface of fresh-cut mango and its influence on fruit quality. Int J Food Microbiol 244:82–89

Kumar A, Ghate V, Kim MJ et al (2015) Kinetics of bacterial inactivation by 405 nm and 520 nm light emitting diodes and the role of endogenous coproporphyrin on bacterial susceptibility. J Photochem Photobiol B 149:37–44

Kumar S, Mukherjee A, Dutta J (2020) Chitosan based nanocomposite films and coatings: emerging antimicrobial food packaging alternatives. Trends Food Sci Technol 97:196–209

Krüger M, Richter P, Strauch SM et al (2019) What an E. coli mutant can teach us about an antibacterial effect of Chorophyllin. Microorganisms 7(2):59. https://doi.org/10.3390/microorganisms7020059

Lafarga T, Colas-Meda P, Abadias M et al (2019) Strategies to reduce microbial risk and improve the quality of fresh and processed strawberries: a review. Innov Food Sci Emerg Technol 52:197–212

Le TD, Phasupan P, Nguyen LT et al (2020) Antimicrobial photodynamic efficacy of selected natural photosensitizers against food pathogens: impacts and interrelationship of process parameters. Photodiagn Photodyn Ther 32:102024. https://doi.org/10.1016/j.pdpdt.2020.102024

Li TM, Zhao YL, Metthews K et al (2020) Antibacterial activity of Staphylococcus aureus of curcumin-loaded chitosan spray coupled with photodynamic treatment. LWT-Food Sci Technol 134:110073. https://doi.org/10.1016/J.lwt.2020.110073

Lin Y, Hu J, Li S et al (2019) Curcumin-based photodynamic sterilization for preservation of fresh-cut hami melon. Molecules 24(13):2374

Lopez-Carballo G, Hernandez-Munoz R, Gavara R et al (2008) Photoactivated Chlorophyllin-based gelatin films and coatings to prevent microbial contamination of food products. Int J Food Microbiol 126(1–2):65–70

Lopez-Carballo G, Hernandez-Munoz P, Gavara R (2018) Photoactivated self-sanitizing chlorophyllin-containing coatings to prevent microbial contamination in packaged food. Coatings 8:328. https://doi.org/10.3390/coatings8090328

Lukseviciute V, Luksiene Z (2020) Inactivation of molds on the surface of wheat sprouts by Chlorophyllin-chitosan coating in the presence of visible LED-based light. J Photochem Photobiol 202:111721

Luksiene Z (2003) Photodynamic therapy: mechanism of action and ways to improve the efficiency of treatment. Medicina 12:1137–1150

Luksiene Z, Bagdoniene I, Smilgevicius V et al (1989) Action of sensitizers and light on biomolecules and cells: inactivation of *S. cerevissiae* growth after laser and porphyrin treatment. Biology 27:161–169

Luksiene Z, de Witte PAM (2003) Hypericin as novel and promising photodynamic therapy tool: studies on intracellular accumulation capacity and growth inhibition efficiency. Medicina 39(7):677–682

Luksiene Z, Peciulyte D, Lugauskas A (2004a) Inactivation of fungi in vitro by photosensitization: preliminary results. Ann Agric Environ Med 11(2):215–220

Luksiene Z, Peciulyte D, Lugauskas A (2004b) Photodynamic inactivation of harmful and pathogenic microorganisms. Vet Med Zoot 26(48):58–60

Luksiene Z, Peciulyte D, Jurkoniene S et al (2005) Inactivation of possible fungal food contaminants by photosensitization. Food Technol Biotechnol 43(4):335–341

Luksiene Z (2005) New approach to inactivation of harmful and pathogenic microorganisms by photosensitization. Food Technol Biotechnol 43(4):411–418

Luksiene Z, Danilcenko H, Taraseviciene Z et al (2007) New approach to the decontamination of germinated wheat from microfungi: effects of aminolevulinic acid and light. Int J Food Microbiol 116:153–158

Luksiene Z, Zukauskas A (2009) Prospects of photosensitization in control of pathogenic and harmful micro-organisms. J Appl Microbiol 107(5):1415–1424

Luksiene Z, Paskeviciute E (2011a) Novel approach to decontaminate food-packaging from pathogens in non-thermal and not chemical way: Chlorophyllin-based photosensitization. J Food Eng 106(2):152–158

Luksiene Z, Paskeviciute E (2011b) Novel approach to the microbial decontamination of strawberries: Chlorophyllin-based photosensitization. J Appl Microbiol 110(5):1274–1283

Luksiene Z, Buchovec I, Kairyte K et al (2012) High- power pulsed light for microbial decontamination of some fruits and vegetables with different surfaces. J Food Agric Environ 10:162–167

Luksiene Z, Brovko L (2013) Antibacterial photosensitization-based treatment for food safety. Food Engin Rev 5(4):185–189

Luksiene Z, Kokstaite R, Katauskis P et al (2013a) Novel approach to effective and uniform inactivation of gram-positive listeria monocytogenes and gram-negative salmonella enterica by photosensitization. Food Technol Biotechnol 51(3):338–344

Luksiene Z, Buchovec I, Viskelis P (2013b) Impact of high power pulsed light on microbial contamination, health promoting components and shelf-life of strawberries. Food Technol Biotechnol 51:284–292

Luksiene Z (2014) Novel approach to control pathogenic and harmful microorganisms in nonthermal way: photosensitization. In: Boziaris J (ed) Novel food preservation and microbial assessment techniques. Taylor Francis group, Boca Raton, pp 184–217

Luksiene Z (2021) Photosensitization: principles and applications in food processing. In: Knoerzer K, Muthukumarappan K (eds) Innovative food processing technologies: a comprehensive review, vol 2. Elsevier, pp 368–384

Maisch T (2015) Resistance in antimicrobial photodynamic inactivation of bacteria. Photochem Photobiol Sci 14(8):1518–1526

Mandal R, Mohammadi X, Viktor A et al (2020) Applications of pulsed light decontamination technology in food processing: an overview. Appl Sci-Basel 10(10). https://doi.org/10.3390/app10103606

Maskovic PZ, Mladenovic JD, Cvijovic MS et al (2011) Phenolic content, antioxidant and antifungal activities of acetonic, ethanolic and petroleum ether extracts of *Hypericum perforatum* L. Hemijska Industrija 65(2):159–164

Meireles A, Giaouris E, Simoes M (2016) Alternative disinfection methods to chlorine for use in the fresh-cut industry. Food Res Int 82:71–85

Newell DG, Koopmans M, Verhoef L et al (2010) Food-borne diseases – the challenges of 20 years ago still persist while new ones continue to emerge. Int J Food Microbiol 139:3–15

Nillian E, Ching C, Fung P et al (2011) Simultaneous detection of *Salmonella* spp., *Salmonella* Enteritidis and *Salmonella* Typhimurium in raw salad vegetables and vegetarian burger patties. Food Nutr Sc 2:1077–1081

Oliveira EF, Tikekar R, Nitin N (2018) Combination of aerosolized Curcumin and UV—A light for the inactivation of bacteria on fresh produce surfaces. Food Res Int 114:133–139

Ormond AB, Freeman HS (2013) Dye sensitizers for photodynamic therapy. Materials 6(3): 817–840

Paskeviciute E, Zudyte B, Luksiene Z (2018) Towards better microbial safety of fresh produce: Chlorophyllin-based photosensitization for microbial control of foodborne pathogens on cherry tomatoes. J Photochem Photobiol B 182:130–136

Paskeviciute E, Zudyte B, Luksiene Z (2019) Innovative Nonthermal Technologies: Chlorophyllin and visible light significantly reduce microbial load on basil. Food Technol Biotechnol 57(1):126–132

Penha CB, Bonin E, da Silva AF et al (2017) Photodynamic inactivation of foodborne and food spoilage bacteria by Curcumin. Food Sci Technol 76:198–202

Pirttijarvi TS, Graeffe TH, Salkinoja Salonen MS (1996) Bacterial contaminants in liquid packaging boards: assessment of potential for food spoilage. J Appl Bacteriol 81(4):445–458

Prasad A, Du LH, Zubar M (2020) Applications of light-emitting diodes (LEDs) in food processing and water treatment. Food Engin Rev 12(3):268–289

Pui CF, Wong WC, Chai LC et al (2011) Simultaneous detection of *Salmonella* spp., *Salmonella* Typhi and *Salmonella* Typhimurium in sliced fruits using multiplex PCR. Food Control 22(2):337–342

Raffi J, Stocker P (1996) Electron paramagnetic resonance detection of irradiated foodstuffs. Appl Magnet Resonanse 10:357–573

Rasiukeviciute N, Bastienė L, Valiuskaitė A et al (2015) New nonchemical postharvest technologies reducing berry contamination. Zemdirbyste-Agriculture 102:411–416

Rao PJ, Khanum H (2016) A green chemistry approach for nanoencapsulation of bioactive compound – Curcumin. LWT – Food Sci Technol 65:695–702

Radev R, Pashova S (2020) Application of edible films and coatings for fresh fruit and vegetables. Quality-Access Success 21(177):108–112

Richter P, Kruger M, Prasad B et al (2019) Using Colistin as a Trojan horse:inactivation of Gram (−) bacteria with Chlorophyllin. Antibiotics-Basel 8:158. https://doi.org/10.3390/antibiotics8040158

Richter PR, Strauch SM, Azizulah A (2014) Chlorophyllin as a possible measure against vectors of human parasites and fish parasites. Front Environ Sci 2:2–4. https://doi.org/10.3389/fenvs.2014.00018

RodríguezLópez MI, Gómez-LópezV M, Lukseviciute V et al (2020) Modelling the inactivation and possible regrowth of Salmonella enterica treated with Chlorophyllin-Chitosan complex and visible light. Food Technol Biotechnol 58(1):64–70

Scallan E, Hoekstra RM, Angulo FJ et al (2011) Foodborne illness acquired in the United States – major pathogens. Emerg Infect Dis 17:7–15

Sharma RA, Gescher AJ, Steward WP (2005) Curcumin: the story so far. Eur J Cancer 41:1955–1968

Silva AF, Borges A, Giaouris E et al (2018) Simoes M, Photodynamic inactivation as an emerging strategy against foodborne pathogenic bacteria in planktonic and sessile states. Crit Rev Microbiol 44:1–51

Sheraz MA, Kazi SH, Ahmed S (2014) Photo, thermal and chemical degradation of riboflavin. Beilstein J Org Chem 10:1999–2012. https://doi.org/10.3762/bjoc.10.208

Sperandio F, Huang YY, Hamblin MR (2013) Antimicrobial photodynamic therapy to kill Gram-negative bacteria. Recent Pat Antiinfect Drug Discov 8(2):108–120

Tao R, Zhang F, Tang QJ et al (2019) Effects of Curcumin based photodynamic treatment on the storage quality of freshcut apples. Food Chem 274:415–421

Temba BA, Fletcher T, Fox GP et al (2016) Inactivation of *Aspergillus flavus* spores by Curcumin-mediated photosensitization. Food Control 59:708–713

Temiz A, Bağcı U, Toğay SÖ (2011) Efficacy of different decontamination treatments on microbial population of leafy vegetables. GIDA 36(1):9–15

Thakur K, Tomar SK, Singh AK et al (2017) Riboflavin and health: a review of recent human research. Crit Rev Food Sci Nutr 57:3650–3660
Yassunaka NN, Freitas CF, Rabello RB et al (2015) Photodynamic inactivation mediated by Erythrosine and its derivatives on foodborne pathogens and spoilage bacteria. Curr Microbiol 72:243–251
Yu L, Reitmeier CA, Gleason ML, Nonnecke GR et al (1995) Qulity of electron-beam irradiated strawberries. J Food Sci 60(5):1084–1087
Vaitonis Z, Luksiene Z (2010) LED-based sources for decontamination of food: modeling photosensitization-based inactivation of pathogenic bacteria. Lith J Phys 50:141–145
Viera-Alcaide I, Perez-Galvez A, Roca M (2019) Green natural colorants. Molecules 24:154. https://doi.org/10.3390/molecules24010154
Vivek K, Singh SS, Riesh W et al (2019) A review on postharvest management and advances in minimal prpcessing of fresh-cut fruits and vegetables. J Microbiol Biotechnol Food Sci 8(5):1178–1187
Wallin A, Luksiene Z, Zagminas K et al (2007) Public health and bioterrorism: renewed threat of anthrax and smallpox. Medicina (Kaunas) 43(4):278–284
Wainwright M, Crossley KB (2004) Photosensitising agents—circumventing resistance and breaking down biofilms: a review. Int Biodeterior Biodegradation 53(2):119–126
Wainwright MT, Maisch K, Nonell S et al (2017) Photoantimicrobials - are we afraid of the light? Lancet Infect Dis 17:49–55
WHO (2015). http://www.who.int/mediacentre/factsheets/fs399/en/. Retrieved 25 Nov 2015
Wood S, Metcalf D, Devine D et al (2006) Erythrosine is a potential photosensitizer for the photodynamic therapy of oral plaque biofilms. J Antimicrob Chemother 57(4):680–684
Wright P, Bill P, Walsh L (2017) Alkaline sodium hypochlorite irrigant and its chemical interactions. Materials (Basel) 10(10):1147
Zudyte B, Luksiene Z (2019) Towards innovative nonthermal technologies: chlorophyllin in the presence of visible light significantly reduces microbial load on basil. Food Technol Biotechnol 57(1):126–132
Zudytė B, Velicka M, Sablinskas V et al (2020) Understanding *Escherichia coli* damages after Chlorophyllin-based photosensitization. Biophotonics 2. https://doi.org/10.1002/jbio.202000144
Zhang JN, Zhang F, Tang QJ (2018) Effect of photodynamic inactivation of *Escherichia coli* by hypericin. World J Microbiol Biotechnol 34(7):100. https://doi.org/10.1007/sl1274-018-2464-1

Chapter 12
Electrospinning Technology: Its Process Conditions and Food Packaging Applications

Gulden Goksen, P. Nisha, and H. Ibrahim Ekiz

The electrospinning technology has been emerged as a promising nonthermal processing technology that has begun to attract the attention of the food industry, as much as in the scientific community, for developing encapsulated delivery systems for food applications. This technology is effective in the manufacturing of nano and micro scale fibres for encapsulation of nutritionally important and bioactive compounds and packaging applications. It is a simple, cost-effective, efficient, versatile, and scalable technology to manufacture dry nano- and microencapsulation structures in the form of electrospun mats by subjecting a polymeric solution jet to an external high-voltage electrostatic field. Electrospinning nano/microstructures possess large surface to volume ratio, amendable size and morphology, nano porous features, intertwined fibrous structure, easy encapsulation and its efficiency, and sustained release property. The most important advantage of electrospun against the conventional techniques is processing tat low temperatures that prevent the degradation of bioactive compounds. Electrospinning process have open up new platform for developing food related applications such as encapsulation of bioactive compounds with improved stability and bioavailability, and development of active and

G. Goksen (✉)
Department of Food Technology, Vocational School of Technical Sciences at Mersin Tarsus Organized Industrial Zone, Tarsus University, Mersin, Turkey
e-mail: guldengoksen@tarsus.edu.tr

P. Nisha
Agro Processing and Technology Division, CSIR – National Institute for Interdisciplinary Science and Technology (NIIST), Thiruvananthapuram, Kerala, India

Academy of Scientific and Innovative Research (AcSIR), CSIR-NIIST Campus, Ghaziabad, Uttar Pradesh, India

H. Ibrahim Ekiz
Department of Food Engineering, Faculty of Engineering, Mersin University, Ciftlikkoy, Mersin, Turkey

A. Režek Jambrak (ed.), *Nonthermal Processing in Agri-Food-Bio Sciences*, Food Engineering Series, https://doi.org/10.1007/978-3-030-92415-7_12

intelligent packaging materials with improved shelf life, biodegradability, and sustainability. The chapter summarises recent development in the electrospun technology for food and packaging application.

1 Introduction

Electrospinning is a non-thermal technology with functions on an external electric field to manufacture continuous sub-micron or nano-scale (10–1000 nm) fibres (Ghorani and Tucker 2015). It really is a highly versatile and simple technique capable of generating high porosity and high surface to volume ratio electrospun fibrous structure. The process is influenced by polymer concentration, molecular weight and solution viscosity, applied voltage, tip-to-collector distance and flow rate affect (Echegoyen et al. 2017; Göksen et al. 2020). The method is based on applying electrical field onto a droplet of polymer solution to convert to nanofiber assisted by collection onto a grounded plate (Göksen et al. 2021; Neo et al. 2013). The polymer solution is extruded from the needle tip at a constant rate by a syringe pump, forming a droplet at the tip. When a droplet of polymer solution is exposed to high voltage, it is induced with free charges. The hemispherical surface of the droplet is distorted into a conical shape known as the Taylor cone at the tip of the needle, due to two main electrostatic forces, namely the electrostatic repulsion of identical charges and the Coulombic force of the external electric field. A charged polymer jet is expelled from the tip of the Taylor cone as soon as the electrostatic force counteracts the surface tension. The unevenly distributed charges cause the jet's whipping or bending motion as the jet is guided towards the collector. As a result, jet elongation and rapid solvent evaporation occur and firm, fine fiber is collected as an arbitrarily directed on the grounded collector (Anu Bhushani and Anandharamakrishnan 2014; Moghe and Gupta 2008). A basic electrospinning setup is schematically illustrated in Fig. 12.1. It consists of a high-voltage power supply, is applied on the needle containing the spinning solution (1–30 kV), a syringe pump for feeding the spinning solution, and a grounded collector, present in different geometries (flat or roll). (Rostamabadi et al. 2020; Wen et al. 2017).

Several electrospinning methods have been developed and one of the key critical points for providing an effective bioactive targeted and/or control delivery is to select an optimal electrospinning technique in this context. The electrospinning methods can be categorized based on the configuration or number of the nozzles/needle such as: (1) single nozzle/needle or uniaxial electrospinning; the most popular and simplest technique of manufacturing electrospun fibers from single polymer solutions; (2) Coaxial electrospinning; the method through coaxial capillaries to generate core-shell structured fibers. It allows the achievement of electrospun nanofibers embedded in porous, hollow, multi-channel, and even thin-wall. Selectable internal structure, high surface area, as well as high porosity of coaxially produced structures give distinctive features for innovative delivery applications (Rostamabadi et al. 2020; Wen et al. 2017; C. Zhang et al. 2020).

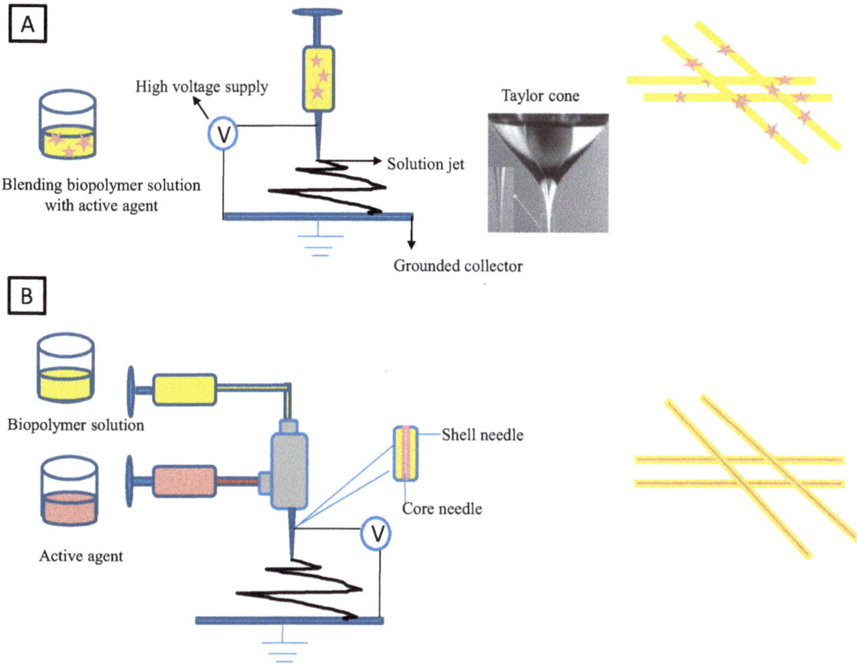

Fig. 12.1 Electrospinning methods

2 Parameters Affecting the Electrospinning System

The major parameters that influence successful electrospinning process is the nature of solution includes concentration, viscosity, surface tension, molecular weight, electric conductivity etc. Apart from solution properties, the process conditions such as flow rate, applied voltage, tip to collector distance, the diameter of the needle have significant impact on the process. The third factor is the environmental conditions i.e. temperature, humidity, and airflow. To get the desired fiber morphology, it is very critical that these parameters are provided with the best consideration (Senthil Muthu Kumar et al. 2019; C. Zhang et al. 2020).

Nonetheless, it is a technology influenced by certain processing parameters, because changing the conditions of the system can lead to the acquisition of various morphologies, such as porous fibers, flattened, branched, elliptical and hollow. Likewise, the characteristics of the polymer matrix, primarily natural and solubility influence the process resulting in fibers of various morphologies and sizes are produced, due to differences in viscosity, electrical conductivity, and surface tension (Rodríguez-Sánchez et al. 2020).

2.1 Polymer Solution and Properties

Electrospinning utilizes polymeric solutions as wall structures that are biocompatible, biodegradable and are of food grade. The use of food-grade polymers and biopolymers which are GRAS (generally regarded as safe), is a favourable approach for developing electrospinning systems for food related applications such as encapsulation and food packaging (Wen et al. 2017). Whey protein isolate, whey protein concentrate, plant protein isolate and concentrate, egg albumen, collagen, gelatin, zein and casein are popular protein based encapsulating solutions used in electrospinning. The significance of solution properties for the morphology of protein-based electrospun fibers has been reported by many researchers (Anu Bhushani and Anandharamakrishnan 2014; Göksen et al. 2020; Neo et al. 2013; Okutan et al. 2014).

Pullulan, alginate β-glucan, cellulose, modified starch, chitosan, cyclodextrin, or dextran are among the most utilized polysaccharides to manufacture nanofibers. Carbohydrate-based delivery systems can interact with several bioactive compounds, rendering them effective carriers for encapsulating a variety of bioactive agents (Lim et al. 2019; Rostami et al. 2019; Topuz and Uyar 2020).

Furthermore, several synthetic polymers that are biodegradable and biocompatible (e.g., PVA, PCL, PEO, PLA, PVP, etc.) can be converted into electrospun matrices (Noruzi 2016; Rostamabadi et al. 2020). Depending on the process, they could be used alone or in blending with other biopolymers (Castro Coelho et al. 2021; Drosou et al. 2017).

The concentration of the prepared solution determines successful formation of efficient fiber and morphology. The increase in concentration increases fiber diameter. If the solution concentration is low, droplet formation is seen, while if it is high, fiber formation becomes difficult. Highly concentrated solutions may accumulate at the syringe tip, which prevents fiber formation or causes it to collect in a smaller area on the collector plate. Therefore, the solution concentration must be adjusted whose morphology is suitable for proper fiber formation (Bhardwaj and Kundu 2010; Okutan et al. 2014). However, at a very high concentration, helix-shaped, curly, and wavy fibers will be produced (Aman Mohammadi et al. 2020).

The viscosity of the polymer solution affects the morphology and diameter of the fiber formed during the electrospinning process. Since a stable fiber production cannot be achieved from very low viscosity polymer solutions, and solution feeding is difficult for very high viscosity solutions, optimum viscosity must be achieved in electrospining process according to the structure of the polymer. The concentration, molecular weight and viscosity of the polymer are related to each other. If there is an increase in the concentration or viscosity of the polymer solution, larger diameter fiber formation occurs (Mendes et al. 2017).

The surface of a liquid exhibits properties similar to a flexible layer is expressed as surface tension. As the surface tension of the polymer solution used in in the electrospinning increases, the applied voltage required for nanofiber formation increases. The surface tension of the polymer solution generally decreases with increasing polymer concentration. The high surface tension causes the formation of jet irregularity and beaded structure and stops the electro-spinning process. When the surface tension is a bit low, thinner and beadless fiber formation is observed. However, all polymer with low surface tension may not be suitable for electrospinning (Bhardwaj and Kundu 2010; Leidy and Maria Ximena 2019).

The ability to draw nanofibers from a solution prepared from synthetic or natural polymers depends on the electrical conductivity. The polymer jet is formed and stretched by the flow of charges on the surface of the jet. The higher the electrical conductivity of the solution, the more it will be carried over the jet, so the higher the charges, the greater the elongation of the solution. Consequently, nanofibers with thinner diameters are obtained with the lengthening of the path taken by the polymer jet (Rezaei et al. 2015).

2.2 Process Conditions of Electrospinning Technology

A voltage must be applied to the solution for electrospinning to take place. Depending on the voltage applied, the fiber diameter changes. At values below critical voltage, the solution does not take the Taylor cone shape at the syringe tip and fiber formation does not occur. Values above the critical voltage value will result in a longer and unstable jet formation as there will be more charge accumulation in the solution. While the increase in the voltage value creates fibers with lower diameters in the first stage, then the fiber diameters become thicker and beads are formed due to the Taylor cone gaining an inverted structure rather than an extrovert (Ghorani and Tucker 2015; Wen et al. 2017).

Flow rate plays a role in determining the yield because the difference in flow rate negatively affects fiber morphology. A low flow rate causes the formation of a truncated Taylor cone by affecting the solution drawn into the collector plate, while if the flow rate is high, the solution dripsleading to the formation of thick diameter fibers. Flow rate should be the same as the amount of solution exiting the syringe tip for continuous fiber formation and regular morphology (Aman Mohammadi et al. 2020; Torres-Giner et al. 2008).

The distance between syringe needle to collector is adjusted to create a jet of the solution and allow the solvent to evaporate away. Since the short distance shortens the evaporation time, it causes instability in the Taylor cone in the collector plate, resulting in wet and beaded fibers, and consequently the fibers lose their cylindrical structure and become flat and ribbon. If the distance is long, the evaporation time will increase, so fiber formation occurs in dry and small diameters (Tseng and Zhao 2013; Zhao et al. 2020).

2.3 Ambient Conditions

Temperature, relative humidity and airflow affect the electrospinning process. Increased temperature decreases the solution viscosity and increases the evaporation rate, resulting in thinning across the fibers. It has been observed that changes in relative humidity affect fiber diameter and drafting performance. Due to the decrease in the electrostatic force on the jet surface with the increase of relative humidity, the porous structure and increased diameter of the fibers on the collector plate is observed. However, the fact that the relative humidity of the environment is so high that the solvent does not evaporate completely causes the solution to be sticky and wet fibers when it reaches the collector plate. At low relative humidity, when the solvent evaporates quickly, it can cause a blockage at the syringe tip, stopping the process (Ghorani and Tucker 2015; Mendes et al. 2017). Therefore, the atmospheric conditions has a major impact on the he process.

3 Advantages of Electrospinning Technology of Versus Thermal Encapsulation Processing-Spray Drying

Spray drying is one of the most commonly used encapsulation techniques in the food industry. The main advantage of spray drying is good quality powder particles that is produce. The active material that is dispersed in a carrier polymer solution is atomized into tiny droplets during the spray drying process. Using a hot air flow with controlled temperature and relative humidity, the solvent is evaporated to produce a powder (Drosou et al. 2017). This process, however, uses a relatively high drying temperature (up to 150 °C), which can degrade ingredients that are thermally unstable. Therefore, spray drying is not recommended for the protection of temperature sensitive bioactive agents such as essential oils, essential fatty acids (PUFA) and amino acids, antioxidants, plant extracts, enzymes and vitamins etc. The selection of non-thermal encapsulation is of particular significance for sensitive agents (Jacobsen et al. 2018). For this reason, the emerging electrospinning process as an alternative technology are gained increasing popularity lately due to their high encapsulation efficiency and ability to successfully protect heat- and oxidation-sensitive ingredients (Castro Coelho et al. 2021; Yang et al. 2017). Because hydrophobic and hydrophilic compounds could be encapsulated without the use of organic solvent or heat and, they could also be included within lipophilic/hydrophobic biopolymer systems, the nanofiber of bioavailability is usually higher than that of the other encapsulation methods counterparts owing to their submicron scale (Jacobsen et al. 2018).

Some of the prominent advantages of electrospinning are as given below

1. Effective integration of bioactive compounds into electrospun matrices
2. Simple and economical process

3. Reduced size, increased surface volume ratio and porous structure that can be incorporated in to food systems without affecting the product's sensory characteristics;
4. Nonthermal process;
5. Provide control release with desired time and where target;
6. Increase the bioavailability of bioactive compounds in vitro digestion (Jiang et al. 2014; Rostami et al. 2019; Vilchez et al. 2020).

The importance of the electrospinning process can be understood by the fact that the increasing researches worldwide are determining multiple perspectives of the electrospinning process and the fiber it makes, as well as by the rise in the number of patents for electrospinning-based applications in recent years (Bhardwaj and Kundu 2010). As a result, electrospinning applications have been optimised in the lab scale for the food related industry for the production of micro and nanosized fibers and its large-scale commercial exploitation have been working on. This process have also been used for the encapsulation of various food and bioactive compounds and suggested utilized different applications of food processing and packaging, according to literature researches (Fabra et al. 2016).

4 Applications in Food Encapsulation

Electrospinning technology has major advantages in the protection and delivery of bioactive ingredients. It is a process of entrapping an active ingredient for instance food bioactive components, probiotics or enzymes within or surface or between nanofiber) that depended on varieties of method (uniaxial or coaxial) (Ghorani and Tucker 2015). In addition, it can develop the organoleptic properties of the product, masking unwanted the flavour or odour without altering the taste, colour, texture, and enhance the release of the compounds (Drosou et al. 2017). It may improve the stability and bioavailability of bioactive components and enhance the viability of the probiotics need for the production of novel functional foods (Wen et al. 2017). Furthermore, the nanofibers acquired are also used as nanocarriers for the targeted delivery and controlled release of bioactive compounds in gastrointestinal system (Leidy and Maria Ximena 2019; Rezaei et al. 2015). Nanoencapsulated bioactive and nutritionally important compounds when incorporated in the food matrix, can enhance the -life and sensory quality (Castro Coelho et al. 2021).

5 Food Packaging Applications

The electrospun mats may consist of synthetic biopolymers, synthesized from bio derivative monomers (Polylactic acid-PLA, Polyvinyl alcohol-PVA etc.) or natural biopolymers, especially protein-based and carbohydrate-based with biodegradable features of the structures or mixtures of them. Therefore, electrospinning promises advantages of simplicity of scale-up, repeatability and reproducibility as well as enabling manufacture of continuous nano/microstructure that has high strength, elasticity and surface area, and also allowing the loading of bioactive molecules and the control release. Natural polymers are commonly used in the food industry because, they display low toxicity, good biocompatible and biodegradable properties when compared to synthetic polymers (Rodríguez-Sánchez et al. 2020).

The food packaging applications are categorized based on the purpose; augmented, intelligent and active packaging.

The demand for the development of food packaging materials with loaded active agents is increasing, because the materials can be fine-tuned to release bioactive compounds in response to unfavourable external conditions of food matrix (Table 12.1). Electrospun mats permit to load high amount of active agents as it possess higher loading capacity and encapsulation efficiency. Moreover, these mats allow release of bioactive compounds at appropriate time during storage to protect food against oxidations, undesirable microorganisms, especially pathogen, moulds, fungi and foodborne virus etc. The small spaces among fibers that are formed as nanofibers which come on top during production, acts as a barrier preventing bacteria penetration (Topuz and Uyar 2020). Plant based natural antioxidants and natural antimicrobial agents that possess General Recognized As Safe-GRAS status have a great potential in electrospun food packaging development (Anu Bhushani and Anandharamakrishnan 2014; Zhao et al. 2020).

In recent years intelligent packaging have gained more attention among researchers. Fast reacting biosensors are perhaps the most prevalent use of nanofibre application allowing fast response, higher sensitivity, and selectivity for smart or intelligent packaging systems (Senthil Muthu Kumar et al. 2019). The electrospun mats have been explored use on applications that are defined as a chemical or biosensor which variations observed in the external or internal environment, monitors food safety and quality (Anu Bhushani and Anandharamakrishnan 2014; Ghaani et al. 2016). Electrospun-based intelligent applications including indicators (for temperature, freshness, integrity, leakage, and pH tracking), data carriers (bar codes) and sensors (gas sensors and biosensors) have been explored in efforts to realize actual tracking of a commodity during the food supply chain (Aman Mohammadi et al. 2020; Müller and Schmid 2019) (Table 12.2). Electrospun structures are employed for loading different compounds and substances such as food indicators. The pH indicators prepared via electrospinning technology are suitable candidates for intelligent food packaging due to their high surface area, porosity, flexibility, absorption capacity, low-cost production, and portable. Furthermore, electrospun mats loaded with indicator reagent show high sensibility and real time response and thus providing ease in terms of quality assessment of food (Aman Mohammadi et al. 2020; Kalpana et al. 2019).

Apart from active and intelligent packaging materials, a multilayered structure design can be applied to enhance the barrier and functional efficiency of biodegradable polymers by incorporating the electrospinning process (Fabra et al. 2013). The principle problem, therefore, is the production of biodegradable interlayer that

Table 12.1 Last 5 years active food packaging studies by electrospinning

Biopolymer	Bioactive compounds	Process conditions	Average diameter size	Results	References
Zein	Laurus nobilis and Rosemary officinalis EOs	0.4 mL/h 11 cm 18 kV	118.63–201.78 nm 73.31–86.06%	Both EOs were incorporated into homogeneous and smooth electrospun fibre mats which showed a significant antimicrobial activity against L. *monocytogenes*, S. *aureus* and aerobic mesophilic bacteria when applied as coatings on cheese slices. When compared to cast films (ZF), the fast release of EO from the ZF rapidly reduced bacterial counts at short storage times, but later on the effect was diminished.	Göksen et al. (2020)
Zein γ-Cyclodextrin (γ-CD)	Thymol (THY)	0.5 mL/h 17 cm 17 kV	155–415 nm	The release of THY was higher from zein-THY/γ-CD-IC-NF (2:1) compared to other nanofibers, due to the higher stability and preservation rate of THY in THY/γ-CD-IC (2:1). Zein-THY/γ-CD-IC-NF nanofibrous webs were most effective at reducing the bacterial count in meat stored up to 5 days at 4 °C.	Aytac et al. (2017)

(continued)

Table 12.1 (continued)

Biopolymer	Bioactive compounds	Process conditions	Average diameter size	Results	References
Zein, Soy protein isolate (SPI), poly(ethylene oxide) (PEO)	Ginger essential oil (GEO)	1 mL/h 15 cm 24 kV	220–625 nm	GEO encapsulated in the ultrafine fibers showed antimicrobial activity against L. *monocytogenes* by the action of volatiles. In situ antimicrobial activity of the fibers containing 12% GEO on fresh Minas cheese significantly reduced the proliferation of L. *monocytogenes* during refrigerated storage of 12 days	F. T. d. Silva et al. (2018)
Zein, β-cyclodextrin (β –CD)	Eucalyptus essential oil (EEO)	1 mL/h 15 cm 18 kV	333–389 nm	The EEO showed antimicrobial activity against Gram-positive and Gram-negative bacteria when evaluated by disk diffusion tests, MIC, MBC and micro-atmosphere. The citronelal was the major constituent of the essential oil. Nanofibers with IC-β-CD/EEO has antimicrobial activity against S. *aureus* and L. *monocytogenes*.	Dias Antunes et al. (2017)
Gelatin, ε-polylysine, β-CD	Thyme essential oil (TEO)	0.4 mL/h 15 cm 20 kV	169–206 nm	The thyme essential oil/β-cyclodextrin ε-polylysine nanoparticles (TCPNs) were included into gelatin nanfibers that exhibited antimicrobial affect against C. jejuni in chicken.	Lin et al. (2018)

(continued)

Table 12.1 (continued)

Biopolymer	Bioactive compounds	Process conditions	Average diameter size	Results	References
Chitosan	Chrysanthemum essential oil (CEO)	0.2 mL/h 15 cm 25 kV	50–250 nm	The chrysanthemum oil loaded to chitosan nanofibers (CHEO/CS/NF) were manufactured. CHEO/CS/NF has antibacterial activity and antioxidant properties on beef. The practical application in beef showed that the CHEO/CS/NF had long-term anti-L. *monocytogenes* effect during 7-day storage.	Lin et al. (2019)
Zein/κ--carrageenan	Rosemary essential oil (REO)	1 mL/h 10 cm 15 kV	495–877 nm	The fabricated nanofiber exhibited the excellent antimicrobial and antioxidant activities. The in-vitro cell cytotoxicity assay approved the biocompatibility property.	Amjadi et al. (2020)
Chitosan, Gelatin	Thyme essential oil (TEO)	5 cm 25 kV	127–343 nm	TEO and encapsulated TEO showed antimicrobial activity against *Clostridium perfringens*. Nitrite of sausage could be replaced by nanofiber containing TEO via nozzle-less electrospinning technique.	Vafania et al. (2019)

(continued)

Table 12.1 (continued)

Biopolymer	Bioactive compounds	Process conditions	Average diameter size	Results	References
Gelatin	Peppermint (PO) and Chamomile (CO) essential oils	0.3 mL/h 10 cm 15 kV	293–462 nm	The antioxidant activity was significantly found for the nanofibers loaded with CO. The antibacterial activity against E. *coli* and S. *aureus* was better for the nanofibers with the addition of PO. The 3-(4,5-dimethylthiazol-2-yl)-2,5-diphenyltetrazolium bromide assay with NIH-3 T3 fibroblasts demonstrated the absence of cytotoxicity of the gelatin/EO nanofibers.	Tang et al. (2019)
Gelatin	Orange essential oil (OEO), Tannic acid	0.3 mL/h 15 cm 13 kV	83–145 nm	Both gelatin and gelatin-cross-linked tannic acid were provided targeted controlled release of OEO. They increased OEO storage stability.	Tavassoli-Kafrani et al. (2018)
Chitosan, flaxseed mucilage	*Ziziphora clinopodioides* essential oil	2 ml/h 25 cm 15 kV	137–285 nm	Active chitosan-flaxseed nanofibers possessed good antimicrobial and antioxidant properties due to sustained release within 96 h. These nanofibers has good tensile strength, puncture force, puncture deformation, water vapor transmission rate, water vapor permeability, and swelling index.	Karami et al. (2021)

(continued)

Table 12.1 (continued)

Biopolymer	Bioactive compounds	Process conditions	Average diameter size	Results	References
Pullulan, γ-CD	Eugenol	0.5 ml/h 15 cm 15 kV	265–525 nm	Pullulan/eugenol-γCD nanofibers proved high thermal stability of eugenol. It exhibited 93% encapsulation efficiency. The antioxidant performance of eugenol was effectively showed even after long-term storage (98.2 ± 1.4%, 3-months) and heat-treatment (92.8 ± 1.0%, 175 °C for 1 h) for pullulan/eugenol-γCD nanofibers.	Celebioglu and Uyar (2021)
Gelatin, chitosan	Moringa oil (MO)	0.2 ml/h 15 cm 23 kV	92–142 nm	The fabricated MO-CNPs embedded gelatin nanofibers demonstrated excellent physicochemical properties, improving stability, and action time of MO. The nanofibers had high antibacterial activity against L. *monocytogenes* and S. *aureus* on cheese.	Lin et al. (2019)
Gelatin, chitosan	Clove oil (CO)	0.5 ml/h 15 cm 25 kV	274–325	The active gelatin nanofibers can eliminate E. *coli* O157:H7 biofilms on cucumber during 4 days of storage. The active nanofibers can maintain the sensory quality of cucumber especially the colour and taste for 4 days.	Cui et al. (2018)

(continued)

Table 12.1 (continued)

Biopolymer	Bioactive compounds	Process conditions	Average diameter size	Results	References
Polyvinyl alcohol (PVA), β-cyclodextrin	Cinnamon essential oil	0.6–0.9 ml/h 15 cm 12–15 kV	308–751 nm	PVA/β-cyclodextrin sustained-release antimicrobial nanofibers was prepared The chemical crosslinking PVA/β-CD/CEO nanofibrous film delay the rapid decay of mushrooms during storage	Pan et al. (2019)
Lentil flour, polyethylene oxide (PEO)	Gallic acid	0.6 ml/h 30 cm 15 kV	184–334 nm	Gallic acid was loaded to lentil flour/PEO nanofibers. Gallic acid encapsulated nanofibers showed no enthalpic peak related to melting of unique crystalline structure of gallic acid. Active packaging electrospun mats enhance oxidative stability of walnuts.	Aydogdu et al. (2019)
Poly(lactide-co-glycolide) (PLGA)	Thymol	PLGA: 0.8–1.2 mL/h, Thymol: 0.1 mL/h. 17 cm 20 kV	Not given	The antibacterial of the nanofiber mat was tested on strawberries. Active elctrospun mat can effectively inhibit the growth of bacteria, fungi, and yeast and extend the shelf life of strawberries.	Y. Zhang et al. (2019)

promote layer adhesion, and this can be accomplished through the use of the electrospinning process. This route creates high oxygen barrier structures, even at high relative humidity, by combining biopolyester layers with protein and polysaccharide-based interlayers (Fabra et al. 2013, 2016). When used as intermediate layers in packaging materials, the electrospun nanofibers improve their structural, optical and functional characteristics and thus supply advantages according to the other multi-layered structures (Echegoyen et al. 2017). Figure 12.2 illustrates the application of electrospinning in food packaging.

Food packaging materials with heat management or oxygen barrier properties are especially important for frozen food transport (Anu Bhushani and Anandharamakrishnan 2014). Protein-based electrospun nanofibers can also be

Table 12.2 The intelligent packaging applications used electrospun mats

Biopolymer	Bioactive compounds	Processing conditions	Morphology	Advantages and Applications	References
Polyvinyl alcohol (PVA)	Red cabbage extract	15 kV, 15 cm, 1 mL/h	255–749 nm Ribbon like fibers	The pH sensor was designed sensitivity to different pH. The pH sensor improved stability and reversibility and could monitor pH changes in packaged fresh dates (Rutab) during storage.	Maftoonazad and Ramaswamy (2019)
Zein	Red cabbage extract	16 kV, 16 cm, 1 ml/h	444–510 nm Ultrafine fibers with smooth surfaces	The nanofibers with anthocyanins exhibited vivid color changes depending on pH, from pink in an acidic condition to green in a basic medium.	Prietto et al. (2018)
poly (lactic acid) (PLA), poly (ethylene oxide) (PEO)	Phycocyanin	15 kV, 14 cm, 0.6 ml/h	921–1357 nm Reduction in the mean diameter of up to 43%	PLA/PEO nanofibers loaded with 3% of the phycocyanin presented the best responses regarding the perception of color change (ΔE). The nanofibers thickness of 68.7 μm showed 76% of ΔE values above 12.	Moreira et al. (2018)

(continued)

Table 12.2 (continued)

Biopolymer	Bioactive compounds	Processing conditions	Morphology	Advantages and Applications	References
Polycaprolactone (PCL), polyethylene oxide (PEO)	açaí (*Euterpe oleracea*) extract	20 kV, 12.5 cm, 0.6 ml/h	885–1635 nm uniform fibers	The pH sensor was developed with an açaí extract and possessed a melting point of 58 °C and a maximum degradation temperature of 408 °C. The total color difference (ΔE) of the colorimetric response was greater than 5, corresponding to the human ability for color differentiation.	C. K. D. Silva et al. (2019)
Pullulan, Zein, Glycerol	Purple sweet potato extract (PSPE), Carvacrol	9 kV, 12 cm, 0.7 ml/h	281–745 nm	PSPE can change color over the pH range of 2–3 to 7.2 used as a halochromic dye. The change in ΔE values of double-layer electrospun mats was 7.56 and that provided faster and more sensitive detection, and had also reversible property. Effective monitoring pork freshness and extending the shelf life by 24 h at 25 °C.	Guo et al. (2020)

(continued)

Table 12.2 (continued)

Biopolymer	Bioactive compounds	Processing conditions	Morphology	Advantages and Applications	References
Zein	Alizarin	20 kV, 15 cm, 1 ml/h	79–619 nm Fibers free of beads	An electrospun nanosensor successfully monitored trout fish freshness. Colorimetric results agreed with the microbial and physicochemical changes in the fish. The nanosensor provides high sensitivity real-time alerts for trout at 4 °C.	Aghaei et al. (2020)
Cellulose acetate	Alizarin	25 kV, 15 cm, 0.5 ml/h	210–304 nm	A halochromic sensor of cellulose acetate nanofibres and alizarin as a fish spoilage indicator in real-time is designed. The colour of sensor tended towards violet on the 12th day; the colorimetric result proved the expected visual colour change in the electrospun nanosensor.	Aghaei et al. (2018)

(continued)

Table 12.2 (continued)

Biopolymer	Bioactive compounds	Processing conditions	Morphology	Advantages and Applications	References
Polyethylene oxide (PEO), Chitosan	Curcumin	12 kV, 30 cm, 0.6 ml/h	283–338 nm	The electrospun nanofiber halochromic pH sensor mat using curcumin, chitosan (CS) and polyethylene oxide (PEO) to monitor chicken freshness was developed. Electrospun mat showed drastic color changes with respect to chicken deterioration which was demonstrated by pH and TVB-N alteration.	Yildiz et al. (2021)

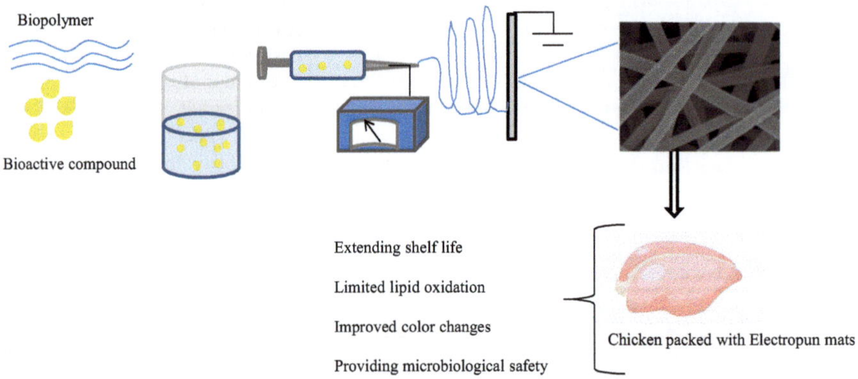

Fig. 12.2 Electro spun mats in packaging application

used as strengthening agents in multilayer structure not only reduced oxygen and/or aroma permeability and enhanced barrier properties but also development of transparency of the composite (Fabra et al. 2013, 2014).

6 Conclusion

The main advantages of bioactive compounds encapsulated with electrospun nanofibers are sustained and controlled release, process at room temperature, efficient encapsulation capacity, enhanced stability achieved by food grade polymers and biopolymers. Furthermore, by changing processing parameters, the diameter of the encapsulant can be adjusted. Over the past several years, electrospinning has firmly established its advantages as an encapsulation methodology in food application for developing innovative products and packaging materials and therefore, the demand for electrospinning equipment, both for lab and industry scale up, is expected to widen substantially. There exist many barriers the implementation of electrospinning in the food industry. The low efficiency, which limits its large-scale industrial utilization, is one key obstacle of electrospinning. Therefore, by changing the structural aspects of the electrospinning setup apparatus, resolving this limitation provides the potential use as commercialization. The more Academic-Industry cooperation should be done because electrospinning technology is need lower cost and developed performance for converting to from prototype to commercialization.

References

Aghaei Z, Emadzadeh B, Ghorani B, Kadkhodaee R (2018) Cellulose acetate nanofibres containing alizarin as a halochromic sensor for the qualitative assessment of rainbow trout fish spoilage. Food Bioprocess Technol 11(5):1087–1095. https://doi.org/10.1007/s11947-017-2046-5

Aghaei Z, Ghorani B, Emadzadeh B, Kadkhodaee R, Tucker N (2020) Protein-based halochromic electrospun nanosensor for monitoring trout fish freshness. Food Control 111:107065. https://doi.org/10.1016/j.foodcont.2019.107065

Aman Mohammadi M, Hosseini SM, Yousefi M (2020) Application of electrospinning technique in development of intelligent food packaging: a short review of recent trends. Food Sci Nutr 8(9):1656–1665. https://doi.org/10.1002/fsn3.1781

Amjadi S, Almasi H, Ghorbani M, Ramazani S (2020) Reinforced ZnONPs/rosemary essential oil-incorporated zein electrospun nanofibers by κ-carrageenan. Carbohydr Polym 232:115800. https://doi.org/10.1016/j.carbpol.2019.115800

Anu Bhushani J, Anandharamakrishnan C (2014) Electrospinning and electrospraying techniques: potential food based applications. Trends Food Sci Technol 38(1):21–33. https://doi.org/10.1016/j.tifs.2014.03.004

Aydogdu A, Yildiz E, Aydogdu Y, Sumnu G, Sahin S, Ayhan Z (2019) Enhancing oxidative stability of walnuts by using gallic acid loaded lentil flour based electrospun nanofibers as active packaging material. Food Hydrocoll 95:245–255. https://doi.org/10.1016/j.foodhyd.2019.04.020

Aytac Z, Ipek S, Durgun E, Tekinay T, Uyar T (2017) Antibacterial electrospun zein nanofibrous web encapsulating thymol/cyclodextrin-inclusion complex for food packaging. Food Chem 233:117–124. https://doi.org/10.1016/j.foodchem.2017.04.095

Bhardwaj N, Kundu SC (2010) Electrospinning: a fascinating fiber fabrication technique. Biotechnol Adv 28(3):325–347

Castro Coelho S, Nogueiro Estevinho B, Rocha F (2021) Encapsulation in food industry with emerging electrohydrodynamic techniques: electrospinning and electrospraying – a review. Food Chem 339:127850. https://doi.org/10.1016/j.foodchem.2020.127850

Celebioglu A, Uyar T (2021) Electrohydrodynamic encapsulation of eugenol-cyclodextrin complexes in pullulan nanofibers. Food Hydrocoll 111:106264. https://doi.org/10.1016/j.foodhyd.2020.106264

Cui H, Bai M, Rashed MMA, Lin L (2018) The antibacterial activity of clove oil/chitosan nanoparticles embedded gelatin nanofibers against Escherichia coli O157:H7 biofilms on cucumber. Int J Food Microbiol 266:69–78. https://doi.org/10.1016/j.ijfoodmicro.2017.11.019

Dias Antunes M, da Silva Dannenberg G, Fiorentini ÂM, Pinto VZ, Lim L-T, da Rosa Zavareze E, Dias ARG (2017) Antimicrobial electrospun ultrafine fibers from zein containing eucalyptus essential oil/cyclodextrin inclusion complex. Int J Biol Macromol 104:874–882. https://doi.org/10.1016/j.ijbiomac.2017.06.095

Drosou CG, Krokida MK, Biliaderis CG (2017) Encapsulation of bioactive compounds through electrospinning/electrospraying and spray drying: a comparative assessment of food-related applications. Dry Technol 35(2):139–162. https://doi.org/10.1080/07373937.2016.1162797

Echegoyen Y, Fabra MJ, Castro-Mayorga JL, Cherpinski A, Lagaron JM (2017) High throughput electro-hydrodynamic processing in food encapsulation and food packaging applications: viewpoint. Trends Food Sci Technol 60:71–79. https://doi.org/10.1016/j.tifs.2016.10.019

Fabra MJ, Busolo MA, Lopez-Rubio A, Lagaron JM (2013) Nanostructured biolayers in food packaging. Trends Food Sci Technol 31(1):79–87

Fabra MJ, Lopez-Rubio A, Lagaron JM (2013) High barrier polyhydroxyalcanoate food packaging film by means of nanostructured electrospun interlayers of zein. Food Hydrocoll 32(1):106–114

Fabra MJ, López-Rubio A, Lagaron JM (2014) On the use of different hydrocolloids as electrospun adhesive interlayers to enhance the barrier properties of polyhydroxyalkanoates of interest in fully renewable food packaging concepts. Food Hydrocoll 39:77–84

Fabra MJ, López-Rubio A, Lagaron JM (2016) Use of the electrohydrodynamic process to develop active/bioactive bilayer films for food packaging applications. Food Hydrocoll 55:11–18

Ghaani M, Cozzolino CA, Castelli G, Farris S (2016) An overview of the intelligent packaging technologies in the food sector. Trends Food Sci Technol 51:1–11

Ghorani B, Tucker N (2015) Fundamentals of electrospinning as a novel delivery vehicle for bioactive compounds in food nanotechnology. Food Hydrocoll 51:227–240. https://doi.org/10.1016/j.foodhyd.2015.05.024

Göksen G, Fabra MJ, Ekiz HI, López-Rubio A (2020) Phytochemical-loaded electrospun nanofibers as novel active edible films: characterization and antibacterial efficiency in cheese slices. Food Control 112:107133. https://doi.org/10.1016/j.foodcont.2020.107133

Göksen G, Fabra MJ, Pérez-Cataluña A, Ekiz HI, Sanchez G, López-Rubio A (2021) Biodegradable active food packaging structures based on hybrid cross-linked electrospun polyvinyl alcohol fibers containing essential oils and their application in the preservation of chicken breast fillets. Food Packag Shelf Life 27:100613

Guo M, Wang H, Wang Q, Chen M, Li L, Li X, Jiang S (2020) Intelligent double-layer fiber mats with high colorimetric response sensitivity for food freshness monitoring and preservation. Food Hydrocoll 101:105468. https://doi.org/10.1016/j.foodhyd.2019.105468

Jacobsen C, García-Moreno PJ, Mendes AC, Mateiu RV, Chronakis IS (2018) Use of electrohydrodynamic processing for encapsulation of sensitive bioactive compounds and applications in food. Annu Rev Food Sci Technol 9(1):525–549. https://doi.org/10.1146/annurev-food-030117-012348

Jiang H, Wang L, Zhu K (2014) Coaxial electrospinning for encapsulation and controlled release of fragile water-soluble bioactive agents. J Control Release 193:296–303

Kalpana S, Priyadarshini S, Leena MM, Moses J, Anandharamakrishnan C (2019) Intelligent packaging: trends and applications in food systems. Trends Food Sci Technol 93:145–157

Karami N, Kamkar A, Shahbazi Y, Misaghi A (2021) Electrospinning of double-layer chitosan-flaxseed mucilage nanofibers for sustained release of Ziziphora clinopodioides essential oil and sesame oil. LWT 140:110812. https://doi.org/10.1016/j.lwt.2020.110812

Leidy R, Maria Ximena Q-C (2019) Use of electrospinning technique to produce nanofibres for food industries: a perspective from regulations to characterisations. Trends Food Sci Technol 85:92–106. https://doi.org/10.1016/j.tifs.2019.01.006

Lim L-T, Mendes AC, Chronakis IS (2019) Chapter Five – Electrospinning and electrospraying technologies for food applications. In: Lim L-T, Rogers M (eds) Advances in food and nutrition research, vol 88. Academic Press, pp 167–234

Lin L, Gu Y, Cui H (2019) Moringa oil/chitosan nanoparticles embedded gelatin nanofibers for food packaging against Listeria monocytogenes and Staphylococcus aureus on cheese. Food Packag Shelf Life 19:86–93. https://doi.org/10.1016/j.fpsl.2018.12.005

Lin L, Mao X, Sun Y, Rajivgandhi G, Cui H (2019) Antibacterial properties of nanofibers containing chrysanthemum essential oil and their application as beef packaging. Int J Food Microbiol 292:21–30. https://doi.org/10.1016/j.ijfoodmicro.2018.12.007

Lin L, Zhu Y, Cui H (2018) Electrospun thyme essential oil/gelatin nanofibers for active packaging against Campylobacter jejuni in chicken. LWT 97:711–718. https://doi.org/10.1016/j.lwt.2018.08.015

Maftoonazad N, Ramaswamy H (2019) Design and testing of an electrospun nanofiber mat as a pH biosensor and monitor the pH associated quality in fresh date fruit (Rutab). Polym Test 75:76–84. https://doi.org/10.1016/j.polymertesting.2019.01.011

Mendes AC, Stephansen K, Chronakis IS (2017) Electrospinning of food proteins and polysaccharides. Food Hydrocoll 68:53–68

Moghe AK, Gupta BS (2008) Co-axial electrospinning for nanofiber structures: preparation and applications. Polym Rev 48(2):353–377. https://doi.org/10.1080/15583720802022257

Moreira JB, Terra ALM, Costa JAV, Morais, M. G. d. (2018) Development of pH indicator from PLA/PEO ultrafine fibers containing pigment of microalgae origin. Int J Biol Macromol 118:1855–1862. https://doi.org/10.1016/j.ijbiomac.2018.07.028

Müller P, Schmid M (2019) Intelligent packaging in the food sector: a brief overview. Foods 8(1):16

Neo YP, Ray S, Jin J, Gizdavic-Nikolaidis M, Nieuwoudt MK, Liu D, Quek SY (2013) Encapsulation of food grade antioxidant in natural biopolymer by electrospinning technique: a physicochemical study based on zein–gallic acid system. Food Chem 136(2):1013–1021

Noruzi M (2016) Electrospun nanofibres in agriculture and the food industry: a review. J Sci Food Agric 96(14):4663–4678. https://doi.org/10.1002/jsfa.7737

Okutan N, Terzi P, Altay F (2014) Affecting parameters on electrospinning process and characterization of electrospun gelatin nanofibers. Food Hydrocoll 39:19–26

Pan J, Ai F, Shao P, Chen H, Gao H (2019) Development of polyvinyl alcohol/β-cyclodextrin antimicrobial nanofibers for fresh mushroom packaging. Food Chem 300:125249. https://doi.org/10.1016/j.foodchem.2019.125249

Prietto L, Pinto VZ, El Halal SLM, de Morais MG, Costa JAV, Lim LT et al (2018) Ultrafine fibers of zein and anthocyanins as natural pH indicator. J Sci Food Agric 98(7):2735–2741

Rezaei A, Nasirpour A, Fathi M (2015) Application of cellulosic nanofibers in food science using electrospinning and its potential risk. Compr Rev Food Sci Food Saf 14(3):269–284

Rodríguez-Sánchez IJ, Fuenmayor CA, Clavijo-Grimaldo D, Zuluaga-Domínguez CM (2020) Electrospinning of ultra-thin membranes with incorporation of antimicrobial agents for applications in active packaging: a review. Int J Polym Mater Polym Biomater:1–24

Rostamabadi H, Assadpour E, Tabarestani HS, Falsafi SR, Jafari SM (2020) Electrospinning approach for nanoencapsulation of bioactive compounds; recent advances and innovations. Trends Food Sci Technol

Rostami M, Yousefi M, Khezerlou A, Mohammadi MA, Jafari SM (2019) Application of different biopolymers for nanoencapsulation of antioxidants via electrohydrodynamic processes. Food Hydrocoll 97:105170

Senthil Muthu Kumar T, Senthil Kumar K, Rajini N, Siengchin S, Ayrilmis N, Varada Rajulu A (2019) A comprehensive review of electrospun nanofibers: food and packaging perspective. Compos Part B 175:107074. https://doi.org/10.1016/j.compositesb.2019.107074

Silva CKD, Mastrantonio DJDS, Costa JAV, Morais MGD (2019) Innovative pH sensors developed from ultrafine fibers containing açaí (Euterpe oleracea) extract. Food Chem 294:397–404. https://doi.org/10.1016/j.foodchem.2019.05.059

Silva FTD, Cunha KFD, Fonseca LM, Antunes MD, Halal SLME, Fiorentini ÂM et al (2018) Action of ginger essential oil (Zingiber officinale) encapsulated in proteins ultrafine fibers on the antimicrobial control in situ. Int J Biol Macromol 118:107–115. https://doi.org/10.1016/j.ijbiomac.2018.06.079

Tang Y, Zhou Y, Lan X, Huang D, Luo T, Ji J et al (2019) Electrospun gelatin nanofibers encapsulated with peppermint and chamomile essential oils as potential edible packaging. J Agric Food Chem 67(8):2227–2234. https://doi.org/10.1021/acs.jafc.8b06226

Tavassoli-Kafrani E, Goli SAH, Fathi M (2018) Encapsulation of orange essential oil using cross-linked electrospun gelatin nanofibers. Food Bioprocess Technol 11(2):427–434. https://doi.org/10.1007/s11947-017-2026-9

Topuz F, Uyar T (2020) Antioxidant, antibacterial and antifungal electrospun nanofibers for food packaging applications. Food Res Int 130:108927. https://doi.org/10.1016/j.foodres.2019.108927

Torres-Giner S, Gimenez E, Lagaron JM (2008) Characterization of the morphology and thermal properties of Zein Prolamine nanostructures obtained by electrospinning. Food Hydrocoll 22(4):601–614. https://doi.org/10.1016/j.foodhyd.2007.02.005

Tseng A, Zhao Y (2013) Wine grape pomace as antioxidant dietary fibre for enhancing nutritional value and improving storability of yogurt and salad dressing. Food Chem 138(1):356–365. https://doi.org/10.1016/j.foodchem.2012.09.148

Vafania B, Fathi M, Soleimanian-Zad S (2019) Nanoencapsulation of thyme essential oil in chitosan-gelatin nanofibers by nozzle-less electrospinning and their application to reduce nitrite in sausages. Food Bioprod Process 116:240–248. https://doi.org/10.1016/j.fbp.2019.06.001

Vilchez A, Acevedo F, Cea M, Seeger M, Navia R (2020) Applications of electrospun nanofibers with antioxidant properties: a review. Nano 10(1):175

Wen P, Wen Y, Zong M-H, Linhardt RJ, Wu H (2017) Encapsulation of bioactive compound in electrospun fibers and its potential application. J Agric Food Chem 65(42):9161–9179. https://doi.org/10.1021/acs.jafc.7b02956

Wen P, Zong M-H, Linhardt RJ, Feng K, Wu H (2017) Electrospinning: a novel nano-encapsulation approach for bioactive compounds. Trends Food Sci Technol 70:56–68. https://doi.org/10.1016/j.tifs.2017.10.009

Yang H, Wen P, Feng K, Zong MH, Lou WY, Wu H (2017) Encapsulation of fish oil in a coaxial electrospun nanofibrous mat and its properties. RSC Adv 7(24):14939–14946

Yildiz E, Sumnu G, Kahyaoglu LN (2021) Monitoring freshness of chicken breast by using natural halochromic curcumin loaded chitosan/PEO nanofibers as an intelligent package. Int J Biol Macromol 170:437–446. https://doi.org/10.1016/j.ijbiomac.2020.12.160

Zhang C, Li Y, Wang P, Zhang H (2020) Electrospinning of nanofibers: potentials and perspectives for active food packaging. Compr Rev Food Sci Food Saf 19(2):479–502. https://doi.org/10.1111/1541-4337.12536

Zhang Y, Zhang Y, Zhu Z, Jiao X, Shang Y, Wen Y (2019) Encapsulation of thymol in biodegradable nanofiber via coaxial eletrospinning and applications in fruit preservation. J Agric Food Chem 67(6):1736–1741. https://doi.org/10.1021/acs.jafc.8b06362

Zhao L, Duan G, Zhang G, Yang H, He S, Jiang S (2020) Electrospun functional materials toward food packaging applications: a review. Nano 10(1):150. Retrieved from https://www.mdpi.com/2079-4991/10/1/150

Chapter 13
Application of Encapsulation Technology in the Agri-Food Sector

Marko Vinceković and Slaven Jurić

1 Introduction Into the Encapsulation Technology to the Agri-Food-Bio Sciences

Encapsulation might be defined as a process of substance (*internal phase*, *payload*, or *payload phase*) insertion into another substance (*membrane*, *shell*, *capsule*, *carrier material*, *external phase*, *wall*, or *matrix*) (Vinceković et al. 2021; Nedović et al. 2011). Throughout the encapsulation process, various sizes of particles can be produced, from a few nm (nanoparticles) to a few mm (microparticles) (Lengyel et al. 2019). The encapsulation technology was first introduced to the area of biotechnology to increase the efficiency of products. The developed technology became huge interest in the other areas like pharmaceuticals and cosmetics industry, as well as agricultural and food industry.

There are several advantages towards using the encapsulation process in agri-food-bio sciences: (i) easier handling (*e.g.* converting liquid ingredients into a powder form, which can be completely free of certain impurities with better rheological and sensory (smell/odor) properties, (ii) immobilization of encapsulated material for various production processes, (iii) better stability of encapsulated material during technological preparation and in the final product (*i.e.* significant reduction of volatiles evaporation, reduced degradation/decomposition and reduction of reaction with other ingredients in the complex matrix of the product), (iv) increase the safety and security (*e.g.* reduced flammability and explosive behavior of volatile compounds and easier handling), (v) improving visible and textural effects (visual signs) in the final product (cosmetic, food and agricultural industry), (vi) tunning the properties of encapsulated material (particle size distribution, structure, solubility in organic and inorganic solvents, color), (vii) time adjustment of the release of

M. Vinceković (✉) · S. Jurić
Department of Chemistry, University of Zagreb Faculty of Agriculture, Zagreb, Croatia
e-mail: mvincekovic@agr.hr

A. Režek Jambrak (ed.), *Nonthermal Processing in Agri-Food-Bio Sciences*,
Food Engineering Series, https://doi.org/10.1007/978-3-030-92415-7_13

encapsulated material which may be activated under certain conditions or with the ingredients of a complex product system or by the action of external factors (Jurić et al. 2020a).

In addition to the above advantages, encapsulation processes have certain disadvantages, as (i) increasing the costs of upscaling the encapsulation process, (ii) the industrial production and/or supply chain process are complex, (iii) the final appearance of the product is not in line with customer expectations (visually or sensory does not meet customer criteria) products, and this is especially problematic in the food production sector, (iv) stability problems of prepared nano- and microparticle formulations during storage, transport and application in certain complex system products (Vinceković et al. 2021). Despite these shortcomings mentioned, encapsulation technology is increasingly being advanced and developed and continuously used as a process in the preparation of new products in the fields of agriculture, food technology, cosmetics, and nutraceuticals.

The encapsulation technology applies to the food industry as a useful tool to improve the delivery of bioactive compounds (*e.g.* antioxidants, minerals, vitamins, phytosterols, lutein, fatty acids, lycopene, esters, aromas, colors) and living cells (*e.g.* probiotics, yeast) in real food products (Jurić et al. 2021; Mrkonjić Fuka et al. 2021; Belščak-Cvitanović et al. 2017; Vos et al. 2010). Furthermore, there is an increasing trend towards using encapsulation technology for agricultural purposes to increase the viability and to control the delivery of living microorganisms into the field. These methods proved efficient and superior to the other formulations in terms of living organisms' protection against the harsh environment (Jurić et al. 2020c; Vinceković et al. 2016). The encapsulation process can be applied for the production of particles loaded with biological and chemical agents as an advanced tool for ecological and sustainable plant production. Encapsulation in biopolymer matrices has been recognized as an effective method for the controlled release of agents used for plant protection and nutrition (Jurić et al. 2019, 2020b). In the cosmetic industry encapsulation process have been proposed to increase the stability of the material, to protect it against degradation, and also to direct and control the release of encapsulated material used in cosmetic products (Casanova and Santos 2016).

The stability of encapsulated compounds mainly depends on a combination of environmental and chemical factors (*i.e.* pH, metal ions, light, high temperatures, enzymes, and oxygen) (Mahdavee Khazaei et al. 2014). In Table 13.1 we have outlined some of the recently used stabilization techniques and carriers for encapsulation of natural pigments. Usually, with regards to the encapsulation of various ingredients, research papers deal with the fabrication and development of new production methods but worryingly the research on the inclusion of encapsulated material into *e.g.* real food products is still scarce (Jurić et al. 2020a). Due to the stability issues under environmental conditions during product manufacturing and later storage, the incorporation of particles loaded with active ingredients into final products is still extremely challenging. Even though encapsulation is always advancing and represents an effective way to protect encapsulated material, incorporation of nano- and microparticles into products is still not investigated enough (Jurić et al. 2020a).

Table 13.1 Stabilization techniques and materials used for the protection of water and lipid-soluble natural pigments (Adapted from Jurić et al. 2020a, 2020b, 2020c)

Pigment	Co-pigment/Wall materials	Stabilization technique	Reference
Water soluble			
Anthocyanins	Dairy proteins	Complexation	Chung et al. (2015); He et al. (2016)
	Pectins	Complexation	Lin et al. (2016)
	Whey proteins and pectins	Complexation/physical entrapment	Arroyo-Maya et al. (2016)
	Gum arabic	Complexation	Chung et al. (2016a); Guan and Zhong (2015)
	β-cyclodextrins	Molecular inclusion	Howard et al. (2013); Fernandes et al. (2013)
	Green tea extracts	Complexation	Chung et al. (2016b)
	Ferric ion	Chelation	Tachibana et al. (2014)
	Stearic acid	Lyophilization	Cruz et al. (2015)
	Oleic acids	Lyophilization	Cruz et al. (2016)
	Different fatty acids	Lyophilization	Cruz et al. (2017, 2018); Luo et al. (2017); Yang et al. (2019)
	Montmorillonite	Hybridization	Kohno et al. (2009)
	Methoxyl pectin	Ionic gelation	de Moura et al. (2018)
	Sodium alginate	Ionic gelation	da Silva Carvalho et al. (2019)
	Polyethylene glycol (PEG)	Ionic gelation	Santos et al. (2013)
	Alginate	Ionic gelation	Belščak-Cvitanović et al. (2016)
	Pectin amide	Ionic gelation	Oidtmann et al. (2012)
	Whey protein isolate	Microemulsions	Oidtmann et al. (2012)
	Maltodextrin, pectin amide	Spray drying	Oidtmann et al. (2012)
	Glycerol mono-oleate, soy Lecithin, maltodextrin, poloxamer 338	Spray drying	Ravanfar et al. (2018)
	Supercritical carbon dioxide	Liposomes	Zhao et al. (2017)
Betalains	Sucrose	Co-crystallization	Karangutkar and Ananthanarayan (2020)
	Sodium alginate, sodium alginate-bovine serum	Ionic gelation	Otálora et al. (2016)
	Rapeseed oil, guar gum, xanthan gum	Double emulsions	Kaimainen et al. (2015)
Lipid Soluble			
Carotenoids	Whey protein concentrate, gum arabic	Spray drying	Chuyen et al. (2019)
β-carotene	Wheat gluten nanoparticles, wheat gluten nanoparticle-xanthan gum	Pickering emulsion	Fu et al. (2019)

(continued)

Table 13.1 (continued)

Pigment	Co-pigment/Wall materials	Stabilization technique	Reference
	Maltodextrin, gum arabic, chitosan, gelatin	Spray drying	Bonilla-Ahumada et al. (2018)
	Native and hydrolyzed Pinhao starches	Freeze-drying	da Silva Carvalho et al. (2019)
Lycopene	Gelatin, sucrose	Spray drying	Shu et al. (2006)
	Lecithin, α-tocopherol	Supercritical antisolvent co-precipitation (SAS)	Cheng et al. (2017)
Lutein	Gelatin, gum arabic	Coacervation	Qv et al. (2011)
Chlorophylls	Polycaprolactone	Microfluidic emulsification	Hsiao et al. (2020)
	Gum arabic, maltodextrin	Spray drying	Kang et al. (2019)
	Whey protein isolate	Spray drying	Zhang et al. (2020)

2 Classification of Next-Generation Biopolymer-Based Carriers as Sustainable Materials

Research is nowadays more focused on the investigation of alternative carriers such as biopolymers. Biodegradable polymers are suitable materials for the production of NPs because of their abundance, relative stability, and durability throughout the encapsulation processes. One of the most important advantages of encapsulation in biopolymeric particles is also high food compatibility and safety which is connected with the availability of polysaccharides, proteins, and lipids (Fathi et al. 2021). Biopolymeric carriers are generally easily prepared from natural biodegradable polymers. These types of materials are usually used because they are generally regarded as safe for the consumer and environment. Also, prepared particles have superior properties especially considering controlled and targeted release (Jana et al. 2020). Biopolymeric particles are usually spherical with some deviations. It is possible to distinguish two types of particles, the *reservoir* type, and the *matrix* type. The reservoir type (*capsule*) has a shell around the bioactive component (*filler*). These types can further be divided into a single-core/mono-core or a core-shell type. The release of the payload from reservoirs can be achieved via the application of force (pressure) or under specific conditions which are generally resulting in capsule breakage. Poly- or multiple-core type particles have several reservoir chambers loaded with encapsulated material in a single particle (Vinceković et al. 2021). In the matrix type of particles, the payload is usually dispersed over the biopolymer matrix carrier and it can be in the form of homogeneously dispersed small droplets or it can be adsorbed on the surface.

Currently, various materials of different origins and properties are being used for the encapsulation process either in solid, liquid, or gaseous forms. Materials used in the production of carrier systems can be divided into three groups as proteins, polysaccharides, and lipids (Fig. 13.1). Biopolymer carriers can be prepared in various

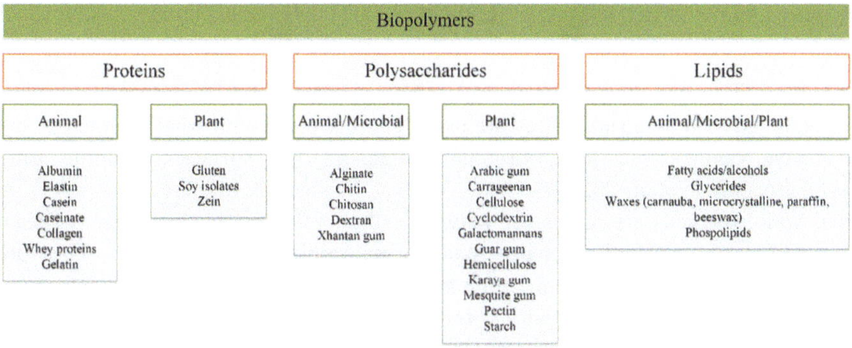

Fig. 13.1 Major natural biomolecules groups which are most often used in the preparation of nanosized carriers

shapes (films, spheres, irregular particles), or structures (compact or porous, amorphous or crystalline, rubbery or glassy) which depends mainly on the type of payload and application (Madene et al. 2006). Encapsulation in biopolymeric carrier systems is continuously developed and mostly advances to improve physicochemical, functional, and release properties while keeping in mind cost-effectiveness and the use of environmentally friendly material throughout the process (Lozano-Vazquez et al. 2015). The chosen material for the production depends on the purpose of encapsulation and final application. Generally, there are a couple of aims that are to be considered when choosing the right material for the encapsulation process: (i) the improvement of shelf life, (ii) type of encapsulation method, (iii) masking of taste or aroma, (iv) easier handling, (v) controlled and/or targeted release, (vi) improvement of appearance. Chosen materials for biopolymeric particle production are required to have several of the following requirements (Desai and Park 2005a). Furthermore, carriers should not react with a component which is to be encapsulated and should have good rheological properties and behavior even at very high concentration. Also, it should not release the encapsulated component during the storage or transport, provide maximal protection against environmental conditions, and should be completely solvent-free or from any other material used during encapsulation under drying or other desolvating conditions. Concerning the economical aspect, encapsulation material should be inexpensive, available in large quantities, and of constant quality (Desai and Park 2005b).

3 Modern Nanocarrier Systems

When considering the use of desired nanocarrier system, few basic points need to be considered. The main point is to take into the account type of used payload, consider its physicochemical stability, consider the overall sustainability of the production process, and possible health risks (Fathi et al. 2021). There are carrier systems that can be considered to be used in agri-food sectors and these include:

(i) *Nanofibers* are used due to their desirable properties for encapsulation of various payload materials. They are usually lightweight with small diameters and have controllable pore structures with a high surface-to-volume ratio (Vinceković et al. 2021).
(ii) *Nanohydrogels* are nanosized networks of chemically/physically cross-linked polymers consisted of chains that are hydrophilic or amphiphilic. They are three-dimensional biocompatible materials with a large amount of water content. For delivery applications, few key properties are necessary: high water content/swellability, biocompatibility, and adjustable chemical/mechanical properties. Hydrogels can retain a large quantity of water or biological fluid without disturbing their basic polymeric chain structure. Hydrogels prepared from natural polymers have drawn huge attention due to their applications in pharmacy, agriculture, medicine, tissue engineering, cancer therapy, and drug delivery (Akram and Hussain 2017; Khoee and Asadi 2016).
(iii) *Nanoemulsions* are kinetically stable liquid-in-liquid dispersions with droplet sizes in the range of 100 nm. They are characterized by high surface area per volume, robust stability, optically transparent appearance, and tunable rheology. Nanoemulsions are applicable in different areas from drug delivery, food production, agriculture, cosmetics, pharmaceuticals, to material synthesis (Gupta et al. 2016).
(iv) *Nanostructured lipid carriers* are delivery systems composed of both solid and liquid lipids as core matrices. These types of carriers have advantages for drug therapy over conventional carriers. These include higher solubility, increased storage stability, better permeability, and bioavailability, decreased adverse effects, prolonged half-life, and tissue-targeted delivery (Nie et al. 2020).
(v) *Bionanocrystals* are especially interesting due to their unique properties and have received considerable attention for the delivery of bioactive compounds. They are biocompatible, rigid, biodegradable, easy to modify, and are renewable (produced from food and agriculture waste) (Koshani and Madadlou 2018). For the food industry, especially are interesting starch, chitin, and cellulose nanocrystalline particles. They are considered promising contenders for the fabrication of reinforced, biodegradable carrier systems (Kasiri and Fathi 2018; Hao et al. 2018).

Despite many advantages to the nanocarrier systems, some problems are in the future to be overcome. These include (i) aggregation and adhesion of particles, (ii) special storage conditions and limited stability time of prepared formulations, (iii) difficulties in encapsulating some payloads of different hydrophilic properties in the same matrix, (iv) difficulties in regulating the polydispersity of particles (Vinceković et al. 2021).

To choose the optimal encapsulation method and the process of production of particles several things are needed to be considered. Mainly the type of material and encapsulated component because this is determinant when regarding pore size, payload size, molecular weight, solubility in the carrier, the volumetric size of the carrier, and complex interactions between the payload and the carrier since this will govern the release mechanisms. Knowledge of this can significantly increase efficiency, loading capacity, and release properties which are the most important parameters (Panyam et al. 2004).

4 Encapsulation Technology in Agriculture – Present and Future

The encapsulation technology is widely used in different sectors from medicine, agriculture, food processing through the cosmetics and pharma industries. The scientific investigation in agricultural science technology and development in the last several years have concluded the huge necessity to set up a new type of microparticles (microspheres/microcapsules) as a delivery system of biological agents (fungi, bacteria, microalgae) and chemical agents (micro-and macronutrients, esters, peptides, amino acids, hormones, pesticides, etc.) (Vinceković et al. 2016; Jurić et al. 2020a, 2020b, 2020c; Slattery et al. 2019; Pereira et al. 2019; Rodríguez Nogales et al. 2020; Tsuji 2001).

One of the most important properties of the prepared microparticle formulations (microspheres/microcapsules) is the protection of active ingredients from external conditions and their decomposition and loss of activity in a particular environment. It can also improve their bioavailability and regulate the time release of ingredients over a longer period. Also, it brings the possibility of longer storage without loss of their activity. All setup properties depend on several important factors of encapsulation technology in agricultural application: (i) various types of wall materials and their concentration, (ii) encapsulation method/encapsulation process – microparticle production, (iii) pH and temperature, (iv) particle size (especially important for the method of application in agriculture (plant protection/nutrition)), (v) type and amount of encapsulated ingredients/additives and their interaction, as well as interaction with the carrier material. All these factors have a significant influence on the microparticle loading capacity, encapsulation efficiency, swelling degree, the strength of the membrane, and the type of release mechanism of bioactive components from microparticles (Li et al. 2019).

Present scientific investigation of microparticle formulations is focused on the preparation of a complex biopolymer-based network containing several bioactive components (synergistic effect). With this intensive research, new insights were gained connected to the complex processes and mechanisms of inter-and intramolecular interactions in biopolymer-based microparticles (microspheres/microcapsules). Inter- and intramolecular interactions are influencing the structural properties of microparticles loaded with active agents (biological and chemical) which in turn have an impact on their overall properties, especially on tunning their release mechanism from microparticle formulation in a specific environment. With this knowledge, it is possible to prepare a new generation of microparticle formulations with the desired properties for different types of application in the agricultural production in the open field, greenhouse, hydroponics, or foliar approach of agroecological plant nutrition or protection and functional food production (higher level of plant metabolites) (Jurić et al. 2020a, 2020b, 2020c; Vinceković et al. 2019).

Besides the use of microparticles in agriculture, nanoparticles are also extensively utilized. Nanocapsules are vesicular systems consisted of a polymeric porous membrane that encapsulates an inner liquid core at the nanoscale. Some of the preferentially used nanoparticles are:

- Polymeric nanoparticles have superior biocompatibility and a minimal impact on non-targeted organisms. Polymeric types of nanomaterials are widely used in agriculture are polyethylene glycol, poly(epsilon-caprolactone), poly(lactide-co-glycolides), and poly (γ-glutamic acid) (Chand Mali et al. 2020; Clemente et al. 2014; Grillo et al. 2013; Ranganathan et al. 2019; Xu et al. 2013; Govender 1999).
- Silver nanoparticles are very effective against different phytopathogens (pesticide activity) with low toxicity and also in some cases, they are showing plant growth promotor properties. They are efficiently used for site-targeted delivery of important agrochemical products and diagnosis purpose tools in case of prior detection of plant diseases (Sadak 2019).
- Nano alumino-silicates, used as an effective pesticide (for different insects) in agriculture. They have very good properties: non-toxicity, biocompatibility, low costs, and environment-friendly nature (Singh et al. 2021; Mittal et al. 2020).
- Titanium dioxide nanoparticles (TiO_2) are one of the forms of titanium in the environment. TiO_2 nanoparticles are widely used for plant protection and environmental remediation because of their photoprotective and photocatalytic properties (Lyu et al. 2017).
- Carbon nanomaterials (graphene, graphene oxide, carbon dots, fullerenes, carbon nanotubes, fullerenes, carbon nanoparticles, and carbon nano-horns) have beneficial and stimulatory effects on plants *in vitro* or culture conditions. They are used to improve the seed germination process (Mukherjee et al. 2016; Husen and Siddiqi 2014).

Most nanotechnology products utilized in agriculture are used for plant protection and nutrition (nano herbicides, nano pesticides, nano fertilizers, and nanosystems for disease protection). All these systems explore the possible use of nanotechnology primarily in the process of controlled delivery of active ingredients that could be used as pesticides, herbicides, or fertilizers, but also secondary to improve the safety of the products which are applied in the process of plant protection and nutrition. Because of that, their group name is nanoagroparticles (Baker et al. 2017). In Table 13.2 we have presented examples of the application of colloidal delivery systems for essential oils in agriculture.

Viruses, fungi, and bacteria infections are causing huge economical losses in agricultural production. The preparation of nanomaterials enriched with certain components which are having specific antimicrobial properties against phytopathogenic fungi (*Colletotrichum gloeosporioides, Fusarium oxysporum, Fusarium*

Table 13.2 Examples of application of colloidal delivery systems for essential oils in agriculture (Adapted from Fathi et al. 2021)

Field of application	Essential oil delivery system	Preparation procedure	Claimed advantages
Organic farming	Carvacrol in alginate-whey protein biopolymeric particles	Emulsification and extrusion	Targeted release in chicken jejunum and ileum (Zhang et al. 2014)
Pest control	*Aegeratum conyzoides*, *Achillea fragrantissima*, and *Tagetes minuta* EOs nanoemulsions	High pressure homogenization	Higher toxicity against eggs and adults of beetle *Callosobruchus maculates* than free oils (Nenaah et al. 2015)
	Zanthoxylum rhoifolium EO in biodegradable polycaprolactone nanospheres	Nanoprecipitation of the pre-formed polymer	Significantly higher reduction of *Bemisia tabaci* eggs and nymphs compared with control (Christofoli et al. 2015)
	Carum copticum EO in myristic acid–chitosan nanogels	Self-assembly	4–8-fold higher fumigant toxicity against *Sitophilus granarius* and *Tribolium confusum* than the free oil (Ziaee et al. 2014)
	Geranium and bergamot EOs in poly(ethylene glycol) nanoparticles	Self-assembly	Higher toxicity against *Tribolium castaneum* and *Rhizopertha dominica* than free oils (Werdin González et al. 2014)
	Artemisia arborescens EO in solid lipid nanoparticles	Hot high-pressure homogenization	Reduced volatility with respect to emulsions (Lai et al. 2006)
	Lippia sidoides EO in chitosan/cashew gum nanoparticles	Spray drying of nanoemulsion	Mortality rate of *St. aegypti* larvae correlates with EO loading (Abreu et al. 2012)
Pest luring	Geraniol in chitosan/gum arabic nanoparticles	Ionic gelation method	Improved EO stability and luring effect toward whitefly *Bemisia tabaci* (de Oliveira et al. 2018)
Repellant textile	Citronella EO in chitosan/gelatin microcapsules	Complex coacervation	Higher repellant effect and lasting protection from insects compared to textiles sprayed with EO in ethanol (Specos et al. 2010)

solani, Dematophora necatrix, etc.) can be used in the plant disease protection process. Prepared cobalt and nickel ferrite nanoparticles ($CoFe_2O_4$ and $NiFe_2O_4$) are successfully tested for antimycotic activity against three plant-pathogenic fungi: *Fusarium oxysporum*, *Colletotrichum gloeosporioides,* and *Dematophora necatrix* (Sharma et al. 2017). Copper nanoparticles with chitosan and celluloses showed antifungal and antibacterial properties against *Escherichia coli, Staphylococcus aureus, Alternaria solani, Fusarium oxysporum, Klebsiella pneumoniae, Pseudomonas aeruginosa, Aspergillus niger, Aspergillus flavus, and Candida albicans.* Besides plant protection, copper nanoparticles are also used in plant nutrition (Rai et al. 2018). Copper is a micronutrient that can be found in high concentrations in chloroplasts. Almost 70% of the total Cu is found in chloroplasts. Cu has an important role in the process of synthesis of chlorophyll, other pigments and has a crucial role in the process of protein and carbohydrate metabolism (Mengel et al. 2001).

Despite the significant advantages of encapsulation technologies in the process of nano- and microparticle production, the preparation process still has several significant obstacles that must be addressed in the coming years to be able to achieve production in larger quantities:

(a) insufficient number of methods used in the characterization of micro- and nanoparticle formulations,
(b) the balance between biosafety and compatibility of wall materials,
(c) various types of active ingredients release mechanisms,
(d) stability during long-term storage at variable environmental conditions and temperatures.

Due to the abovementioned, it is necessary to conduct further research that will focus on increasing the stability of nano- and microcapsule formulations, control the uniformity of formulation sizes and release mechanisms of bioactive components in certain time intervals, testing their effectiveness on certain phytopathogenic fungi and bacteria and testing their action as new green formulations with 3 in 1 effects (plant protection, plant nutrition, and time-release mechanism). It can be concluded that from the above scientific research, technologies of encapsulation and production of nano- and microparticle formulations will more effectively promote the development in the agroecological agriculture and functional food production process.

5 Implementation of Encapsulated Material Into Final Food Products

There is a significant gap in the research with regards to the implementation of encapsulated natural pigments in real food products. Usually, with regards to this topic, research papers deal with bioactive compounds encapsulation procedures but the research on their inclusion in real food products is worryingly scarce and a couple of available examples are listed in Table 13.3.

It is important to observe the behavior of encapsulated bioactive compounds in food matrices and their influence on the sensory characteristics of food products. This would significantly advance the knowledge of ingredient behavior when considering implementation during food production.

Table 13.3 Examples of application of stabilized natural pigments in real food products (Adapted from Jurić et al. 2020a, 2020b, 2020c)

Encapsulated compounds	Compound donor	Stabilization method	Material	Functional food	References
Anthocyanin	Barberry (*Berberis vulgaris* L.)	Spray drying	Gum arabic, maltodextrin, gelatin	Jelly	Mahdavi et al. (2016)
	Black bean (*Phaseolus vulgaris* L.) coat	Molecular inclusion	β-cyclodextrin	Sport beverage	Aguilera et al. (2016)
	Grape (*Vitis vinifera* L.) skin	Spray drying	Maltodextrin	Apple puree	Lavelli et al. (2016)
	Sour cherry (*Prunus cerasus* L.) pomace extract	Freeze-drying	Whey and soy proteins	Cookies	Tumbas Šaponjac et al. (2016)
Betalains	Barbary fig (*Opuntia ficus-indica* L.)	Ionic gelation	Calcium alginate/Gelatin	Gummy candy	Otálora et al. (2019)
	Barbary fig (*Opuntia ficus-indica* L.)	Spray drying	Soluble fiber [(1–3)(1–4)-*β*-D-glucan	Extruded cereal	Ruiz-Gutiérrez et al. (2017)
	Beetroot (*Beta vulgaris* L.)	Freeze-drying	Maltodextrin	Chewing gum	Chranioti et al. (2015)
	Red beet (*Beta vulgaris* L.) extract diluted with dextrin	Thin-film hydration-sonication technique	Lecithin liposome	Gummy candy	Amjadi et al. (2018)
Carotenoids	Yellow bell pepper (*Capsicum annuum* L.)	Ultrasonic homogenization, kneading	β-cyclodextrin	Isotonic beverage	Lobo et al. (2018)
	Saffron (*Crocus sativus* L.)	Freeze-drying	Maltodextrin	Chewing gum	Chranioti et al. (2015)
Chlorophylls	Alfalfa (*Medicago sativa* L.)	Emulsification + Freeze-drying	Canola oil, glycerol monostearate, gelatin, agar	Gummy candy	Raei et al. (2017)

When considering the application of encapsulated ingredients into final food products the matrix can significantly affect its release behavior and particles physicochemical properties. Diffusion of ingredients may be affected by the presence of proteins, carbohydrates, fatty acids, pH, water activity, packaging material, and trace metals. Furthermore, physicochemical stress factors like food processing, preservation, storage, food ingredients, etc. may start the degradation and collapse of the encapsulation system making it inefficient.

There are a couple of examples (Table 13.3) that have proven to have problems when the implementation of encapsulated bioactive compounds like anthocyanins into food products have negative effects on the shape, size, and uniformity of products (*e.g.* cookies, and other dough-based products). Furthermore, protein-based coatings (encapsulation systems) might induce significant changes in the structure of dough-based products (proteins absorb water resulting in increased hardness) while the presence of an additional particulate phase can increase the fragility of dough-based products (hindering the formation of a continuous starch network).

The main applications of encapsulated ingredients can be classified into four groups (Table 13.4). (1) Direct mixing with liquid foods or mixing with the food ingredients before food preparation; (2) washing the product surface with carrier systems in an aqueous dispersion; (3) infusion in porous food matrices; (4) coating with a biopolymeric layer incorporating the active ingredients delivery systems. Details about the strategies for the utilization of different colloidal systems for active ingredients like essential oils (EO) alongside the examples of application in food products are listed in Table 13.4 (Fathi et al. 2021).

Published research mainly deals with the fabrication procedures and the work on implementation into food products is scarce (even often contradicts the *in vitro* results). Thus it is important to understand the issues related to the application of encapsulated bioactive compounds (or other ingredients like microorganisms) into various food matrices. It is also necessary to investigate the behavior of carrier systems (*i.e.* protein-based) in complex food matrices alongside the influence on the sensory characteristics of final food products (Jurić et al. 2021).

6 Future Remarks

Encapsulation is becoming essential for the sustainable and economic development of various products, from agri-food to nutraceuticals. Respectively, it can resolve some problems regarding the stability of payload during industrial processing and storage. Even though still there is a significant gap when considering using this technology for a large scale production due to the limitations. Advancing the encapsulation methodology and technology these limitations could be minimized significantly. Advanced nanocarrier systems are becoming popular due to the low-cost materials necessary for their manufacturing especially when introduced to industrial-scale production.

Table 13.4 Strategies of the utilization of different colloidal systems for essential oils (EO) delivery, together with examples of application in food products (Adapted from Fathi et al. 2021)

	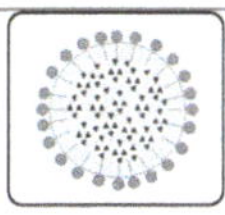 **Nanoemulsions**	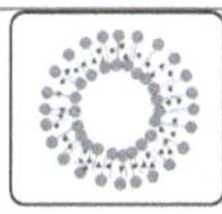 **Liposomes**	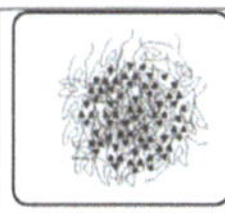 **Biopolymeric nanoparticles**
Advantages	*Active barrier with controlled release of antimicrobial, low amount needed (surface treatment), low impact on organoleptic properties*	*High efficiency of delivery through the biological membranes, additional loading of hydrophilic molecules*	*High food compatibility, several natural polysaccharides and proteins available*
Disadvantages	*Need for surfactant in formulation, high costs of nanoemulsion production*	*Limited loading of bioactives, high costs of phospholipids and of fabrication*	*Formation of particles in aqueous systems requires chemical/physical modification of hydrophilic polymers, or use of solvents for hydrophobic polymers*
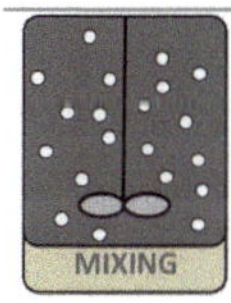	Shelf life extension of milk and quality preservation by encapsulation of thyme EO (Ben Jemaa et al. 2017) Microbial stabilization of orange and pear juices by encapsulation of carvacrol (Donsì et al. 2011) Microbiological stabilization of chicken pâté by encapsulation of oregano EO (Moraes-Lovison et al. 2017)	Microbial stabilization of tofu by clove EO encapsulation (Cui et al. 2015)	Shelf life extension in bakery products by encapsulation of thyme EO by complex coacervation (Gonçalves et al. 2017)
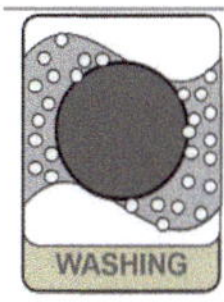	Microbial stabilization of fresh lettuce by washing with oregano EO nanoemulsions (Bhargava et al. 2015) or of spinach leaves by carvacrol or eugenol nanoemulsions (Ruengvisesh et al. 2015)	–	–
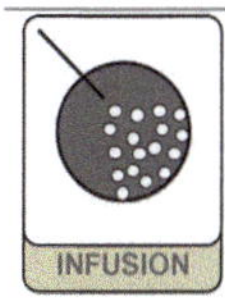	Enhancement of organoleptic properties and extension of the shelf life of trout fillets by infusion of nanoemulsions of rosemary, laurel and thyme EOs (Ozogul et al. 2017)	Preservation of minced beef by *Zataria multiflora* EO liposomes (Khosravi-Darani et al. 2016)	Reduction of lipid oxidation and microbial growth by infusion in meat patties of chitosan nanoparticles containing cinnamon EO (Ghaderi-Ghahfarokhi et al. 2017)

(continued)

Table 13.4 (continued)

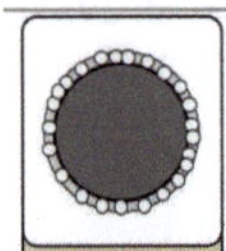 COATING	Shelf life extension of rucola leaves coated with chitosan containing lemon EO nanoemulsions (Sessa et al. 2015) Microbial stabilization of green beans (Severino et al. 2014a, 2015) or broccoli florets (Severino et al. 2014b) by modified chitosan coatings containing citrus EO nanoemulsions Preservation of fresh-cut cheese by encapsulation of Oregano EO (Artiga-Artigas et al. 2017) Microbial stabilization of bread slice by an edible coating containing clove bud or oregano EO nanoemulsions (Otoni et al. 2014)	Shelf life extension and quality improvement of banana slices by a mucilage coating containing rosemary EO liposomes (Alikhani-Koupaei 2015)	Microbial stabilization of beef cutlet by spraying with a chitosan-myristic acid nanogels containing clove EO (Rajaei et al. 2017) Enhanced antimicrobial and antioxidant activity of cinnamon EO on pork by encapsulation in chitosan nanoparticles (Hu et al. 2015)

References

Abreu FOMS, Oliveira EF, Paula HCB, de Paula RCM (2012) Chitosan/cashew gum nanogels for essential oil encapsulation. Carbohydr Polym 89(4):1277–1282. https://doi.org/10.1016/j.carbpol.2012.04.048

Aguilera Y, Mojica L, Rebollo-Hernanz M, Berhow M, De Mejía EG, Martín-Cabrejas MA (2016) Black bean coats: new source of anthocyanins stabilized by β-cyclodextrin copigmentation in a sport beverage. Food Chem 212(1):561–570. https://doi.org/10.1016/j.foodchem.2016.06.022

Akram M, Hussain R (2017) Nanohydrogels: history, development, and applications in drug delivery. In: Jawaid M, Mohammad F (eds) Nanocellulose and nanohydrogel matrices: biotechnological and biomedical applications, Weinheim, Germany, pp 297–330. https://doi.org/10.1002/9783527803835.ch11

Alikhani-Koupaei M (2015) Liposomal and edible coating as control release delivery systems for essential oils: comparison of application on storage life of fresh-cut banana. Qual Assur Saf Crop Foods 7(2):175–185. https://doi.org/10.3920/QAS2013.0297

Amjadi S, Ghorbani M, Hamishehkar H, Roufegarinejad L (2018) Improvement in the stability of betanin by liposomal nanocarriers: its application in gummy candy as a food model. Food Chem 256(1):156–162. https://doi.org/10.1016/j.foodchem.2018.02.114

Arroyo-Maya IJ, Campos-Terán J, Hernández-Arana A, McClements DJ (2016) Characterization of flavonoid-protein interactions using fluorescence spectroscopy: binding of pelargonidin to dairy proteins. Food Chem 213(1):431–439. https://doi.org/10.1016/j.foodchem.2016.06.105

Artiga-Artigas M, Acevedo-Fani A, Martín-Belloso O (2017) Improving the shelf life of low-fat cut cheese using nanoemulsion-based edible coatings containing oregano essential oil and mandarin fiber. Food Control 76(1):1–12. https://doi.org/10.1016/j.foodcont.2017.01.001

Baker S, Volova T, Prudnikova SV, Satish S, Prasad NMN (2017) Nanoagroparticles emerging trends and future prospect in modern agriculture system. Environ Toxicol Pharmacol 53(1):10–17. https://doi.org/10.1016/j.etap.2017.04.012
Belščak-Cvitanović A, Bušić A, Barišić L, Vrsaljko D, Karlović S, Špoljarić I, Vojvodić A, Mršić G, Komes D (2016) Emulsion templated microencapsulation of dandelion (*Taraxacum officinale* L.) polyphenols and β-carotene by ionotropic gelation of alginate and pectin. Food Hydrocoll 57:139–152. https://doi.org/10.1016/j.foodhyd.2016.01.020
Belščak-Cvitanović A, Jurić S, Đorđević V, Barišić L, Komes D, Ježek D, Bugarski B, Nedović V (2017) Chemometric evaluation of binary mixtures of alginate and polysaccharide biopolymers as carriers for microencapsulation of green tea polyphenols. Int J Food Prop 20(9):1971–1986. https://doi.org/10.1080/10942912.2016.1225762
Ben Jemaa M, Falleh H, Neves MA, Isoda H, Nakajima M, Ksouri R (2017) Quality preservation of deliberately contaminated milk using thyme free and nanoemulsified essential oils. Food Chem 217(1):726–734. https://doi.org/10.1016/j.foodchem.2016.09.030
Bhargava K, Conti DS, da Rocha SRP, Zhang Y (2015) Application of an oregano oil nanoemulsion to the control of foodborne bacteria on fresh lettuce. Food Microbiol 47(1):69–73. https://doi.org/10.1016/j.fm.2014.11.007
Bonilla-Ahumada FDJ, Khandual S, Lugo-Cervantes EDC (2018) Microencapsulation of algal biomass (*Tetraselmis chuii*) by spray-drying using different encapsulation materials for better preservation of beta-carotene and antioxidant compounds. Algal Res 36(1):229–238. https://doi.org/10.1016/j.algal.2018.10.006
Casanova F, Santos L (2016) Encapsulation of cosmetic active ingredients for topical application – a review. J Microencapsul 33(1):1–17. https://doi.org/10.3109/02652048.2015.1115900
Chand Mali S, Raj S, Trivedi R (2020) Nanotechnology a novel approach to enhance crop productivity. Biochem Biophys Rep 24(1):e100821. https://doi.org/10.1016/j.bbrep.2020.100821
Cheng YS, Lu PM, Huang CY, Wu JJ (2017) Encapsulation of lycopene with lecithin and α-tocopherol by supercritical antisolvent process for stability enhancement. J Supercrit Fluids 130(1):246–252. https://doi.org/10.1016/j.supflu.2016.12.021
Chranioti C, Nikoloudaki A, Tzia C (2015) Saffron and beetroot extracts encapsulated in maltodextrin, gum Arabic, modified starch and chitosan: incorporation in a chewing gum system. Carbohydr Polym 127(1):252–263. https://doi.org/10.1016/j.carbpol.2015.03.049
Christofoli M, Costa ECC, Bicalho KU, Domingues VC, Peixoto M, F., Alves, C. C. F., Araújo, W. L., Cazal, C. M. (2015) Insecticidal effect of nanoencapsulated essential oils from *Zanthoxylum rhoifolium* (Rutaceae) in *Bemisia tabaci* populations. Ind Crop Prod 70(1):301–308. https://doi.org/10.1016/j.indcrop.2015.03.025
Chung C, Rojanasasithara T, Mutilangi W, McClements DJ (2015) Enhanced stability of anthocyanin-based color in model beverage systems through whey protein isolate complexation. Food Res Int 76(3):761–768. https://doi.org/10.1016/j.foodres.2015.07.003
Chung C, Rojanasasithara T, Mutilangi W, McClements DJ (2016a) Enhancement of colour stability of anthocyanins in model beverages by gum arabic addition. Food Chem 201(1):14–22. https://doi.org/10.1016/j.foodchem.2016.01.051
Chung C, Rojanasasithara T, Mutilangi W, McClements DJ (2016b) Stabilization of natural colors and nutraceuticals: inhibition of anthocyanin degradation in model beverages using polyphenols. Food Chem 212(1):596–603. https://doi.org/10.1016/j.foodchem.2016.06.025
Chuyen HV, Roach PD, Golding JB, Parks SE, Nguyen MH (2019) Encapsulation of carotenoid-rich oil from Gac peel: optimisation of the encapsulating process using a spray drier and the storage stability of encapsulated powder. Powder Technol 344:373–379. https://doi.org/10.1016/j.powtec.2018.12.012
Clemente Z, Grillo R, Jonsson M, Santos NZ, Feitosa LO, Lima R, Fraceto LF (2014) Ecotoxicological evaluation of poly(epsilon-caprolactone) nanocapsules containing triazine herbicides. J Nanosci Nanotechnol 14(7):4911–4917. https://doi.org/10.1166/jnn.2014.8681
Cruz L, Benohoud M, Rayner CM, Mateus N, de Freitas V, Blackburn RS (2018) Selective enzymatic lipophilization of anthocyanin glucosides from blackcurrant (*Ribes nigrum* L.) skin

extract and characterization of esterified anthocyanins. Food Chem 266(1):415–419. https://doi.org/10.1016/j.foodchem.2018.06.024

Cruz L, Fernandes I, Guimarães M, De Freitas V, Mateus N (2016) Enzymatic synthesis, structural characterization and antioxidant capacity assessment of a new lipophilic malvidin-3-glucoside-oleic acid conjugate. Food Funct 7(6):2754–2762. https://doi.org/10.1039/c6fo00466k

Cruz L, Fernandes VC, Araújo P, Mateus N, de Freitas V (2015) Synthesis, characterisation and antioxidant features of procyanidin B4 and malvidin-3-glucoside stearic acid derivatives. Food Chem 174:480–486. https://doi.org/10.1016/j.foodchem.2014.11.062

Cruz L, Guimarães M, Araújo P, Évora A, De Freitas V, Mateus N (2017) Malvidin 3-glucoside-fatty acid conjugates: from hydrophilic toward novel lipophilic derivatives. J Agric Food Chem 65(31):6513–6518. https://doi.org/10.1021/acs.jafc.6b05461

Cui H, Zhao C, Lin L (2015) The specific antibacterial activity of liposome-encapsulated Clove oil and its application in tofu. Food Control 56(1):128–134. https://doi.org/10.1016/j.foodcont.2015.03.026

da Silva Carvalho AG, da Costa Machado MT, de Freitas Queiroz Barros HD, Cazarin CBB, Maróstica Junior MR, Hubinger MD (2019) Anthocyanins from jussara (*Euterpe edulis* Martius) extract carried by calcium alginate beads pre-prepared using ionic gelation. Powder Technol 345(1):283–291. https://doi.org/10.1016/j.powtec.2019.01.016

de Moura SCSR, Berling CL, Germer SPM, Alvim ID, Hubinger MD (2018) Encapsulating anthocyanins from *Hibiscus sabdariffa* L. calyces by ionic gelation: pigment stability during storage of microparticles. Food Chem 241(1):317–327. https://doi.org/10.1016/j.foodchem.2017.08.095

de Oliveira JL, Campos EVR, Pereira AES, Nunes LES, da Silva CCL, Pasquoto T, Lima R, Smaniotto G, Polanczyk RA, Fraceto LF (2018) Geraniol encapsulated in chitosan/gum Arabic nanoparticles: a promising system for pest management in sustainable agriculture. J Agric Food Chem 66(21):5325–5334. https://doi.org/10.1021/acs.jafc.8b00331

Desai KGH, Park HJ (2005a) Recent developments in microencapsulation of food ingredients. Dry Technol 23(7):1361–1394. https://doi.org/10.1081/DRT-200063478

Desai KGH, Park HJ (2005b) Encapsulation of Vitamin C in triphosphate cross-linked chitosan microspheres by spray drying. J Microencapsul 22(2):179–192. https://doi.org/10.1080/02652040400026533

Donsì F, Annunziata M, Sessa M, Ferrari G (2011) Nanoencapsulation of essential oils to enhance their antimicrobial activity in foods. LWT Food Sci Technol 44(9):1908–1914. https://doi.org/10.1016/j.lwt.2011.03.003

Fathi M, Vinceković M, Jurić S, Viskić M, Režek Jambrak A, Donsì F (2021) Food-grade colloidal systems for the delivery of essential oils. Food Rev Int 37(1):1–45. https://doi.org/10.1080/87559129.2019.1687514

Fernandes A, Sousa A, Azevedo J, Mateus N, De Freitas V (2013) Effect of cyclodextrins on the thermodynamic and kinetic properties of cyanidin-3-O-glucoside. Food Res Int 51(2):748–755. https://doi.org/10.1016/j.foodres.2013.01.037

Fu D, Deng S, McClements DJ, Zhou L, Zou L, Yi J, Liu C, Liu W (2019) Encapsulation of β-carotene in wheat gluten nanoparticle-xanthan gum-stabilized Pickering emulsions: enhancement of carotenoid stability and bioaccessibility. Food Hydrocoll 89(1):80–89. https://doi.org/10.1016/j.foodhyd.2018.10.032

Ghaderi-Ghahfarokhi M, Barzegar M, Sahari MA, Ahmadi Gavlighi H, Gardini F (2017) Chitosan-cinnamon essential oil nano-formulation: application as a novel additive for controlled release and shelf life extension of beef patties. Int J Biol Macromol 102(1):19–28. https://doi.org/10.1016/j.ijbiomac.2017.04.002

Gonçalves ND, Pena F d L, Sartoratto A, Derlamelina C, Duarte MCT, Antunes AEC, Prata AP (2017) Encapsulated thyme (*Thymus vulgaris*) essential oil used as a natural preservative in bakery product. Food Res Int 96(1):154–160. https://doi.org/10.1016/j.foodres.2017.03.006

Govender T, Stolnik S, Garnett MC, Illum L, Davis SS (1999) PLGA nanoparticles prepared by nanoprecipitation: drug loading and release studies of a water soluble drug. J Control Release 57(2):171–185. https://doi.org/10.1016/s0168-3659(98)00116-3

Grillo R, Rosa AH, Fraceto LF (2013) Poly(ε-caprolactone) nanocapsules carrying the herbicide atrazine: effect of chitosan-coating agent on physico-chemical stability and herbicide release profile. Int J Environ Sci Technol 11(6):1691–1700. https://doi.org/10.1007/s13762-013-0358-1

Guan Y, Zhong Q (2015) The improved thermal stability of anthocyanins at pH 5.0 by gum arabic. LWT Food Sci Technol 64(2):706–712. https://doi.org/10.1016/j.lwt.2015.06.018

Gupta A, Eral HB, Hatton TA, Doyle PS (2016) Nanoemulsions: formation, properties and applications. Soft Matter 12(11):2826–2841. https://doi.org/10.1039/c5sm02958a

Hao Y, Chen Y, Li Q, Gao Q (2018) Preparation of starch nanocrystals through enzymatic pretreatment from waxy potato starch. Carbohydr Polym 184(1):171–177. https://doi.org/10.1016/j.carbpol.2017.12.042

He Z, Xu M, Zeng M, Qin F, Chen J (2016) Interactions of milk α- and β-casein with malvidin-3-O-glucoside and their effects on the stability of grape skin anthocyanin extracts. Food Chem 199(1):314–322. https://doi.org/10.1016/j.foodchem.2015.12.035

Howard LR, Brownmiller C, Prior RL, Mauromoustakos A (2013) Improved stability of chokeberry juice anthocyanins by β-cyclodextrin addition and refrigeration. J Agric Food Chem 61(3):693–699. https://doi.org/10.1021/jf3038314

Hsiao C-J, Lin J-F, Wen H-Y, Lin Y-M, Yang C-H, Huang K-S, Shaw J-F (2020) Enhancement of the stability of chlorophyll using chlorophyll-encapsulated polycaprolactone microparticles based on droplet microfluidics. Food Chem 306(1):e125300. https://doi.org/10.1016/j.foodchem.2019.125300

Hu J, Wang X, Xiao Z, Bi W (2015) Effect of chitosan nanoparticles loaded with cinnamon essential oil on the quality of chilled pork. LWT Food Sci Technol 63(1):519–526. https://doi.org/10.1016/j.lwt.2015.03.049

Husen A, Siddiqi KS (2014) Carbon and fullerene nanomaterials in plant system. J Nanobiotechnol 12(1):e16. https://doi.org/10.1186/1477-3155-12-16

Jana P, Shyam M, Singh S, Jayaprakash V, Dev A (2020) Biodegradable polymers in drug delivery and oral vaccination. Eur Polym J 142(1):e110155. https://doi.org/10.1016/j.eurpolymj.2020.110155

Jurić S, Đermić E, Topolovec-Pintarić S, Bedek M, Vinceković M (2019) Physicochemical properties and release characteristics of calcium alginate microspheres loaded with *Trichoderma viride* spores. J Integr Agric 18(11):2534–2548. https://doi.org/10.1016/S2095-3119(19)62634-1

Jurić S, Jurić M, Król-Kilińska Ż, Vlahoviček-Kahlina K, Vinceković M, Dragović-Uzelac V, Donsì F (2020a) Sources, stability, encapsulation and application of natural pigments in foods. Food Rev Int, in press. https://doi.org/10.1080/87559129.2020.1837862

Jurić S, Jurić M, Siddique MAB, Fathi M (2020b) Vegetable oils rich in polyunsaturated fatty acids: nanoencapsulation methods and stability enhancement. Food Rev Int, in press. https://doi.org/10.1080/87559129.2020.1717524

Jurić S, Sopko K, Król-Kilinska Ż, Žutić I, Fabek Uher S, Đermić E, Topolovec-Pintarić S, Vinceković M (2020c) The enhancement of plant secondary metabolites contents in *Lactuca sativa* L. by encapsulated bioactive agents. Sci Rep 10(1):e3737. https://doi.org/10.1038/s41598-020-60690-3

Jurić S, Tanuwidjaja I, Mrkonjić Fuka M, Vlahoviček-Kahlina K, Marijan M, Boras A, Udiković Kolić N, Vinceković M (2021) Encapsulation of two fermentation agents, *Lactobacillus sakei* and calcium ions in microspheres. Colloids Surf B Biointerfaces 197(1):e111387. https://doi.org/10.1016/j.colsurfb.2020.111387

Kaimainen M, Marze S, Järvenpää E, Anton M, Huopalahti R (2015) Encapsulation of betalain into w/o/w double emulsion and release during invitro intestinal lipid digestion. LWT Food Sci Technol 60(2):899–904. https://doi.org/10.1016/j.lwt.2014.10.016

Kang YR, Lee YK, Kim YJ, Chang YH (2019) Characterization and storage stability of chlorophylls microencapsulated in different combination of gum Arabic and maltodextrin. Food Chem 272(1):337–346. https://doi.org/10.1016/j.foodchem.2018.08.063

Karangutkar AV, Ananthanarayan L (2020) Co-crystallization of *Basella rubra* extract with sucrose: characterization of co-crystals and evaluating the storage stability of betacyanin pigments. J Food Eng 271(1):e 109776. https://doi.org/10.1016/j.jfoodeng.2019.109776

Kasiri N, Fathi M (2018) Production of cellulose nanocrystals from pistachio shells and their application for stabilizing pickering emulsions. Int J Biol Macromol 106(1):1023–1031. https://doi.org/10.1016/j.ijbiomac.2017.08.112

Khoee S, Asadi H (2016) Nanogels: chemical approaches to preparation. In: Mishra MK (ed) Encyclopedia of biomedical polymers and polymeric biomaterials. Taylor and Francis, New York, NY, pp 5266–5293. https://doi.org/10.1201/b19038-60

Khosravi-Darani K, Khoosfi ME, Hosseini H (2016) Encapsulation of *Zataria multiflora* Boiss. Essential oil in liposome: antibacterial activity against *E. Coli* O157:H7 in broth media and minced beef. J Food Saf 36(4):515–523. https://doi.org/10.1111/jfs.12271

Kohno Y, Kinoshita R, Ikoma S, Yoda K, Shibata M, Matsushima R, Tomita Y, Maeda Y, Kobayashi K (2009) Stabilization of natural anthocyanin by intercalation into montmorillonite. Appl Clay Sci 42(3–4):519–523. https://doi.org/10.1016/j.clay.2008.06.012

Koshani R, Madadlou AA (2018) A viewpoint on the gastrointestinal fate of cellulose nanocrystals. Trends Food Sci Technol 71(1):268–273. https://doi.org/10.1016/j.tifs.2017.10.023

Lai F, Wissing SA, Müller RH, Fadda AM (2006) *Artemisia arborescens* L essential oil-loaded solid lipid nanoparticles for potential agricultural application: preparation and characterization. AAPS PharmSciTech 7(1):10–18. https://doi.org/10.1208/pt070102

Lavelli V, Sri Harsha PSC, Spigno G (2016) Modelling the stability of maltodextrin-encapsulated grape skin phenolics used as a new ingredient in apple puree. Food Chem 209(1):323–331. https://doi.org/10.1016/j.foodchem.2016.04.055

Lengyel M, Kállai-Szabó N, Antal V, Laki AJ, Antal I (2019) Microparticles, microspheres, and microcapsules for advanced drug delivery. Sci Pharm 87(3):e20. https://doi.org/10.3390/scipharm87030020

Li T, Teng D, Mao R, Hao Y, Wang X, Wang J (2019) Recent progress in preparation and agricultural application of microcapsules. J Biomed Mater Res 107(10):2371–2385. https://doi.org/10.1002/jbm.a.36739

Lin Z, Fischer J, Wicker L (2016) Intermolecular binding of blueberry pectin-rich fractions and anthocyanin. Food Chem 194(1):986–993. https://doi.org/10.1016/j.foodchem.2015.08.113

Lobo FATF, Silva V, Domingues J, Rodrigues S, Costa V, Falcão D, de Lima Araújo KG (2018) Inclusion complexes of yellow bell pepper pigments with β-cyclodextrin: preparation, characterisation and application as food natural colorant. J Sci Food Agric 98(1):2665–2671. https://doi.org/10.1002/jsfa.8760

Lozano-Vazquez G, Lobato-Calleros C, Escalona-Buendia H, Chavez G, Alvarez-Ramirez J, Vernon-Carter EJ (2015) Effect of the weight ratio of alginate-modified tapioca starch on the physicochemical properties and release kinetics of chlorogenic acid containing beads. Food Hydrocoll 48(1):301–311. https://doi.org/10.1016/j.foodhyd.2015.02.032

Luo S-Z, Chen S-S, Pan L-H, Qin X-S, Zheng Z, Zhao Y-Y, Pang M, Jiang S-T (2017) Antioxidative capacity of crude camellia seed oil: impact of lipophilization products of blueberry anthocyanin. Int J Food Prop 20(2):1627–1636. https://doi.org/10.1080/10942912.2017.1350974

Lyu S, Wei X, Chen J, Wang C, Wang X, Pan D (2017) Titanium as a beneficial element for crop production. Front Plant Sci 8(1):e597. https://doi.org/10.3389/fpls.2017.00597

Madene A, Jacquot M, Scher J, Desobry S (2006) Flavour encapsulation and controlled release – a review. Int J Food Sci Technol 41(1):1–21. https://doi.org/10.1111/j.1365-2621.2005.00980.x

Mahdavee Khazaei K, Jafari SM, Ghorbani M, Hemmati Kakhki A (2014) Application of maltodextrin and gum Arabic in microencapsulation of saffron Petal's anthocyanins and evaluating their storage stability and color. Carbohydr Polym 105(1):57–62. https://doi.org/10.1016/j.carbpol.2014.01.042

Mahdavi SA, Jafari SM, Assadpour E, Ghorbani M (2016) Storage stability of encapsulated Barberry's anthocyanin and its application in jelly formulation. J Food Eng 181(1):59–66. https://doi.org/10.1016/j.jfoodeng.2016.03.003

Mengel K, Kirkby EA, Kosegarten H, Appel T (2001) Soil copper. In: Principles of plant nutrition. Springer, Dordrecht, pp 599–611

Mittal D, Kaur G, Singh P, Yadav K, Ali SA (2020) Nanoparticle-based sustainable agriculture and food science: recent advances and future outlook. Front Nanotechnol 2(1):e579954. https://doi.org/10.3389/fnano.2020.579954

Moraes-Lovison M, Marostegan LFP, Peres MS, Menezes IF, Ghiraldi M, Rodrigues RAF, Fernandes AM, Pinho SC (2017) Nanoemulsions encapsulating oregano essential oil: production, stability, antibacterial activity and incorporation in chicken pâté. LWT Food Sci Technol 77(1):233–240. https://doi.org/10.1016/j.lwt.2016.11.061

Mrkonjić Fuka M, Žgomba Maksimović A, Hulak N, Kos I, Marušić Radovčić N, Jurić S, Tanuwidjaja I, Karolyi D, Vinceković M (2021) The survival rate and efficiency of nonencapsulated and encapsulated native starter cultures to improve the quality of artisanal game meat sausages. J Food Sci Technol 58(1):710–719. https://doi.org/10.1007/s13197-020-04587-z

Mukherjee A, Majumdar S, Servin AD, Pagano L, Dhankher OP, White JC (2016) Carbon nanomaterials in agriculture: a critical review. Front Plant Sci 7(1):e172. https://doi.org/10.3389/fpls.2016.00172

Nedović V, Kalusević A, Manojlović V, Lević S, Bugarski B (2011) An overview of encapsulation technologies for food applications. Proced Food Sci 1(1):1806–1815. https://doi.org/10.1016/j.profoo.2011.09.265

Nenaah GE, Ibrahim SIA, Al-Assiuty BA (2015) Chemical composition, insecticidal activity and persistence of three Asteraceae essential oils and their nanoemulsions against *Callosobruchus maculatus* (F.). J Stored Prod Res 61(1):9–16. https://doi.org/10.1016/j.jspr.2014.12.007

Nie X, Chen Z, Pang L, Wang L, Jiang H, Chen Y et al (2020) Oral nano drug delivery systems for the treatment of type 2 diabetes mellitus: an available administration strategy for anti diabetic phytocompounds. Int J Nanomedicine 15(1):10215–10240. https://doi.org/10.2147/ijn.s285134

Oidtmann J, Schantz M, Mäder K, Baum M, Berg S, Betz M, Kulozik U, Leick S, Rehage H, Schwarz K, Richling E (2012) Preparation and comparative release characteristics of three anthocyanin encapsulation systems. J Agric Food Chem 60(3):844–851. https://doi.org/10.1021/jf2047515

Otálora MC, Carriazo JG, Iturriaga L, Osorio C, Nazareno MA (2016) Encapsulating betalains from *Opuntia ficus-indica* fruits by ionic gelation: pigment chemical stability during storage of beads. Food Chem 202(1):373–382. https://doi.org/10.1016/j.foodchem.2016.01.115

Otálora MC, de Jesús Barbosa H, Perilla JE, Osorio C, Nazareno MA (2019) Encapsulated Betalains (*Opuntia Ficus-indica*) as natural colorants. Case study: gummy candies. LWT Food Sci Technol 103(1):222–227. https://doi.org/10.1016/j.lwt.2018.12.074

Otoni CG, Pontes SFO, Medeiros EAA, Soares NDFF (2014) Edible films from methylcellulose and nanoemulsions of clove bud (*Syzygium aromaticum*) and oregano (*Origanum vulgare*) essential oils as shelf life extenders for sliced bread. J Agric Food Chem 62(22):5214–5219. https://doi.org/10.1021/jf501055f

Ozogul Y, Yuvka İ, Ucar Y, Durmus M, Kösker AR, Öz M, Ozogul F (2017) Evaluation of effects of nanoemulsion based on herb essential oils (rosemary, laurel, thyme and sage) on sensory, chemical and microbiological quality of rainbow trout (*Oncorhynchus mykiss*) fillets during ice storage. LWT Food Sci Technol 75(1):677–684. https://doi.org/10.1016/j.lwt.2016.10.009

Panyam J, Williams D, Dash A, Leslie-Pelecky D, Labhasetwar V (2004) Solid-state solubility influences encapsulation and release of hydrophobic drugs from PLGA/PLA nanoparticles. J Pharm Sci 93(7):1804–1814. https://doi.org/10.1002/jps.20094

Pereira ADES, Oliveira HC, Fraceto LF (2019) Polymeric nanoparticles as an alternative for application of gibberellic acid in sustainable agriculture: a field study. Sci Rep 9(1):e7135. https://doi.org/10.1038/s41598-019-43494-y

Qv X-Y, Zeng Z-P, Jiang J-G (2011) Preparation of lutein microencapsulation by complex coacervation method and its physicochemical properties and stability. Food Hydrocoll 25(6):1596–1603. https://doi.org/10.1016/j.foodhyd.2011.01.006

Raei A, Yasini Ardakani SA, Daneshi M (2017) Microencapsulation of the green pigment of alfalfa and its applications on heated food. J Food Process Eng 40(5):e12529. https://doi.org/10.1111/jfpe.12529

Rai M, Ingle AP, Pandit R, Paralikar P, Shende S, Gupta I, Biswas JK, da Silva SS (2018) Copper and copper nanoparticles: role in management of insect-pests and pathogenic microbes. Nanotechnol Rev 7(4):303–315. https://doi.org/10.1515/ntrev-2018-0031

Rajaei A, Hadian M, Mohsenifar A, Rahmani-Cherati T, Tabatabaei M (2017) A coating based on clove essential oils encapsulated by chitosan-myristic acid nanogel efficiently enhanced the shelf-life of beef cutlet. Food Packag Shelf Life 14(1):137–145. https://doi.org/10.1016/j.fpsl.2017.10.005

Ranganathan A, Manabe Y, Sugawara T, Hirata T, Shivanna N, Baskaran V (2019) Poly (D, L-lactide-co-glycolide)-phospholipid nanocarrier for efficient delivery of macular pigment lutein: absorption pharmacokinetics in mice and antiproliferative effect in Hep G2 cells. Drug Deliv Transl Res 9(1):178–191. https://doi.org/10.1007/s13346-018-0590-9

Ravanfar R, Comunian TA, Abbaspourrad A (2018) Thermoresponsive, water-dispersible microcapsules with a lipid-polysaccharide shell to protect heat-sensitive colorants. Food Hydrocoll 81(1):419–428. https://doi.org/10.1016/j.foodhyd.2018.03.030

Rodríguez Nogales JM, Simo G, Pérez Magarino S, Cano Mozo E, Fernández Fernández E, Ruiperez V, Vila Crespo J (2020) Evaluating the influence of simultaneous inoculation of SiO_2-alginate encapsulated bacteria and yeasts on volatiles, amino acids, biogenic amines and sensory profile of red wine with lysozyme addition. Food Chem 327(1):e126920. https://doi.org/10.1016/j.foodchem.2020.126920

Ruengvisesh S, Loquercio A, Castell-Perez E, Taylor TM (2015) Inhibition of bacterial pathogens in medium and on spinach leaf surfaces using plant-derived antimicrobials loaded in surfactant micelles. J Food Sci 80(11):2522–2529. https://doi.org/10.1111/1750-3841.13085

Ruiz-Gutiérrez MG, Amaya-Guerra CA, Quintero-Ramos A, Pérez-Carrillo E, Meléndez-Pizarro CO (2017) Use of red cactus pear (*Opuntia Ficus-indica*) encapsulated powder to pigment extruded cereal. J Food Qual 1:1–12. https://doi.org/10.1155/2017/7262464

Sadak MS (2019) Impact of silver nanoparticles on plant growth, some biochemical aspects, and yield of fenugreek plant (*Trigonella foenum-graecum*). Bull Natl Res Cent 43(1):e38. https://doi.org/10.1186/s42269-019-0077-y

Santos DT, Albarelli JQ, Beppu MM, Meireles MAA (2013) Stabilization of anthocyanin extract from jabuticaba skins by encapsulation using supercritical CO_2 as solvent. Food Res Int 50(2):617–624. https://doi.org/10.1016/j.foodres.2011.04.019

Sessa M, Ferrari G, Donsì F (2015) Novel edible coating containing essential oil nanoemulsions to prolong the shelf life of vegetable products. Chem Eng Trans 43(1):55–60. https://doi.org/10.3303/CET1543010

Severino R, Ferrari G, Vu KD, Donsì F, Salmieri S, Lacroix M (2015) Antimicrobial effects of modified chitosan based coating containing nanoemulsion of essential oils, modified atmosphere packaging and gamma irradiation against *Escherichia coli* O157:H7 and *Salmonella Typhimurium* on green beans. Food Control 50(1):215–222. https://doi.org/10.1016/j.foodcont.2014.08.029

Severino R, Vu KD, Donsì F, Salmieri S, Ferrari G, Lacroix M (2014a) Antibacterial and physical effects of modified chitosan based-coating containing nanoemulsion of mandarin essential oil and three non-thermal treatments against *Listeria innocua* in green beans. Int J Food Microbiol 191(1):82–88. https://doi.org/10.1016/j.ijfoodmicro.2014.09.007

Severino R, Vu KD, Donsì F, Salmieri S, Ferrari G, Lacroix M (2014b) Antimicrobial effects of different combined non-thermal treatments against *Listeria monocytogenes* in broccoli florets. J Food Eng 124(1):1–10. https://doi.org/10.1016/j.jfoodeng.2013.09.026

Sharma P, Sharma A, Sharma M, Bhalla N, Estrela P, Jain A, Thakur A (2017) Nanomaterial fungicides: *In Vitro* and *In Vivo* antimycotic activity of cobalt and nickel nanoferrites on phytopathogenic fungi. Glob Chall 1(9):e1700041. https://doi.org/10.1002/gch2.201700041

Shu B, Yu W, Zhao Y, Liu X (2006) Study on microencapsulation of lycopene by spray-drying. J Food Eng 76(4):664–669. https://doi.org/10.1016/j.jfoodeng.2005.05.062
Singh H, Sharma A, Bhardwaj SK, Arya SK, Bhardwaj N, Khatri M (2021) Recent advances in the applications of nano-agrochemicals for sustainable agricultural development. Environ Sci Process Impacts 23(2):213–239. https://doi.org/10.1039/D0EM00404A
Slattery M, Harper B, Harper S (2019) Pesticide encapsulation at the nanoscale drives changes to the hydrophobic partitioning and toxicity of an active ingredient. Nano 9(1):e81. https://doi.org/10.3390/nano9010081
Specos MM, García JJ, Tornesello J, Marino P, Vecchia MD, Tesoriero MV, Hermida LG (2010) Microencapsulated citronella oil for mosquito repellent finishing of cotton textiles. Trans R Soc Trop Med Hyg 104(10):653–658. https://doi.org/10.1016/j.trstmh.2010.06.004
Tachibana N, Kimura Y, Ohno T (2014) Examination of molecular mechanism for the enhanced thermal stability of anthocyanins by metal cations and polysaccharides. Food Chem 143(1):452–458. https://doi.org/10.1016/j.foodchem.2013.08.017
Tsuji K (2001) Microencapsulation of pesticides and their improved handling safety. J Microencapsul 18(2):137–147. https://doi.org/10.1080/026520401750063856
Tumbas Šaponjac V, Ćetković G, Čanadanović-Brunet J, Pajin B, Djilas S, Petrović J, Lončarević I, Stajčić S, Vulić J (2016) Sour cherry pomace extract encapsulated in whey and soy proteins: incorporation in cookies. Food Chem 207(1):27–33. https://doi.org/10.1016/j.foodchem.2016.03.082
Vinceković M, Jalšenjak N, Topolovec-Pintarić S, Đermić E, Bujan M, Jurić S (2016) Encapsulation of biological and chemical agents for plant nutrition and protection: chitosan/alginate microcapsules loaded with copper cations and *Trichoderma viride*. J Agric Food Chem 64(43):8073–8083. https://doi.org/10.1021/acs.jafc.6b02879
Vinceković M, Jurić S, Marijan M, Viskić M, Vlahoviček-Kahlina K, Maslov Bandić L (2021) Encapsulation of herb extracts (aromatic and medicinal herbs). In: Galanakis CM (ed) Aromatic herbs in food: bioactive compounds, processing, and applications. Academic Press (Elsevier), Cambridge, MA, pp 263–322. https://doi.org/10.1016/B978-0-12-822716-9.00008-1
Vinceković M, Maslov Bandić L, Jurić S, Jalšenjak N, Ćaić A, Živičnjak I, Đermić E, Karoglan M, Osrečak M, Topolovec-Pintarić S (2019) The Enhancement of bioactive potential in *Vitis vinifera* leaves by application of microspheres loaded with biological and chemical agents. J Plant Nutr 42(6):543–558. https://doi.org/10.1080/01904167.2019.1568467
Vos P, Faas MM, Spasojevic M, Sikkema J (2010) Review: encapsulation for preservation of functionality and targeted delivery of bioactive food components. Int Dairy J 20(4):292–302. https://doi.org/10.1016/j.idairyj.2009.11.008
Werdin González JO, Gutiérrez MM, Ferrero AA, Fernández Band B (2014) Essential oils nanoformulations for stored-product pest control – characterization and biological properties. Chemosphere 100(1):130–138. https://doi.org/10.1016/j.chemosphere.2013.11.056
Xu Z, Wan C, Xu X, Feng X, Xu H (2013) Effect of poly (γ-glutamic acid) on wheat productivity, nitrogen use efficiency and soil microbes. J Soil Sci Plant Nutr 13(3):744–755. https://doi.org/10.4067/S0718-95162013005000059
Yang W, Kortesniemi M, Ma X, Zheng J, Yang B (2019) Enzymatic acylation of blackcurrant (*Ribes nigrum*) anthocyanins and evaluation of lipophilic properties and antioxidant capacity of derivatives. Food Chem 281(1):189–196. https://doi.org/10.1016/j.foodchem.2018.12.111
Zhang Y, Gong J, Yu H, Guo Q, Defelice C, Hernandez M, Yin Y, Wang Q (2014) Alginate-whey protein dry powder optimized for target delivery of essential oils to the intestine of chickens. Poult Sci 93(10):2514–2525. https://doi.org/10.3382/ps.2013-03843
Zhang Z-H, Peng H, Woo MW, Zeng X-A, Brennan M, Brennan CS (2020) Preparation and characterization of whey protein isolate-chlorophyll microcapsules by spray drying: effect of WPI ratios on the physicochemical and antioxidant properties. J Food Eng 267(1):e 109729. https://doi.org/10.1016/j.jfoodeng.2019.109729

Zhao L, Temelli F, Chen L (2017) Encapsulation of anthocyanin in liposomes using supercritical carbon dioxide: effects of anthocyanin and sterol concentrations. J Funct Foods 34(1):159–167. https://doi.org/10.1016/j.jff.2017.04.021

Ziaee M, Moharramipour S, Mohsenifar A (2014) Toxicity of Carum copticum essential oil-loaded nanogel against *Sitophilus granarius* and *Tribolium confusum*. J Appl Entomol 138(10):763–771. https://doi.org/10.1111/jen.12133

Part IV
Nonthermal Processing Legislation

Chapter 14
Overview of Legislation Across the Globe, Diagnostics and Standards Which Provide a Legal and Regulatory Framework in Which NTP Is Used Worldwide

Nada Smigic and Ilija Djekic

1 Introduction

Food-borne diseases include a number of illnesses ranging from mild gastrointestinal issues to life threatening conditions such as botulism, hemolytic–uremic syndrome, Guillain-Barre syndrome, etc. (Bari and Yeasmin 2018). Despite numerous projects and studies performed in the field of food safety, and various preventive and control measures defined and implemented in the food industry, the number of reported food-borne cases remains at an unacceptable high level (Havelaar et al. 2010; Newell et al. 2010; Nyachuba 2010). World Health Organization (WHO) has estimated the burden of food-borne diseases, with more than 420,000 millions of people that fall ill, and 230,000 die every year from diarrheal diseases, caused by consumption of contaminated food and/or water (Kirk et al. 2015). Although the majority of cases occur in developing countries, a great number of people in developed countries still experience some food-borne threats (Painter et al. 2013; EFSA 2019). In the European Union (EU), more than 48,000 cases of food-borne illnesses, with 4588 hospitalizations and 40 deaths occurred only in 2018 (EFSA 2019).

The reasons for this are numerous. On one side, food-borne pathogens are triggered to survive and grow in different environments, and often occur in food not commonly associated with the specific pathogen, e.g. *Escherichia coli* O157:H7 was found in fresh produce (Gelting et al. 2011), *E. coli* O104:H4 in sprouts (Buchholz et al. 2011), *Listeria monocytogenes* in celery (Gaul et al. 2013), melons (Martinez et al. 2016), or caramelized apples (Angelo et al. 2017). Also, the

N. Smigic (✉) · I. Djekic
Department of Food Safety and Quality Management, Faculty of Agriculture, University of Belgrade, Belgrade, Republic of Serbia
e-mail: nadasmigic@agrif.bg.ac.rs; idjekic@agrif.bg.ac.rs; idjekic@mts.rs; http://www.agrif.bg.ac.rs/; http://www.agrif.bg.ac.rs/

A. Režek Jambrak (ed.), *Nonthermal Processing in Agri-Food-Bio Sciences*, Food Engineering Series, https://doi.org/10.1007/978-3-030-92415-7_14

developments in the field of food microbiology and food analytics resulted in more precise and sensitive methods for the detection and quantification of microbial hazards (Ankireddy and Kim 2018). On the other side, consumers' eating habits are continuously changing (Zink 1997). Compared with the food markets in the past, today's marketplaces offer more fresh products and more innovative packaging. Nowadays consumers try to avoid traditionally prepared and preserved food, they prefer less-processed, fresher-tasting, nutritive foods without the use of heat and/or chemical preservatives and safe food products with extended shelf-life. Consumers' food choices are also influenced by other factors, such as food origin, price, animal welfare, environmental impact, technology used, etc. (Nielsen et al. 2009). Therefore, food producers are facing new challenges and constantly search for new and enhanced preservation treatments to improve microbial safety of food products.

Food preservation treatments are used to slow down or inhibit chemical deterioration and microbial multiplication in food, including both pathogenic and spoilage microorganisms. For the aim of reducing or completely inactivating present microflora in food, thermal treatments, such as pasteurisation and sterilization have been most commonly used (Tucker and Featherstone 2010). Although these treatments usually ensure safety of food products, often they unfavourably affects flavour, chemical composition and nutritional quality of the treated food. In the last few decades, together with the application and development of thermal treatments, food industries and research institutions have been developing and exploring novel non-thermal processing (NTP) technologies. The term NTP is used to entitle and describe technologies that are effective at ambient or sub-lethal temperatures. They include treatments such as irradiation, high pressure processing (HPP), pulsed electric fields (PEF), pulsed light (PL), ultrasound (US), ultraviolet (UV) radiation, etc. (Smigic and Rajkovic 2014; Zhang et al. 2019). The aim of NTP application is twofold, firstly to inactivate food-borne pathogens and spoilage microorganisms present in food, and secondly to maintain nutritional and sensorial characteristics of food products. In order to minimise limitations and drawbacks of NTP technologies, they are combined with low temperature, mild heat, modified atmosphere packaging (MAP), antimicrobial agents or other preservation treatments, within so called hurdle technology (Rajkovic et al. 2010; Smigic and Rajkovic 2014).

However, the commercialization and industrial application of NTP is still quite limited, mainly due to high investment costs, lack and often complicated procedure for the regulatory approval, and the lack of the consumer acceptance (Luckose et al. 2019; Nath et al. 2019). The aim of this chapter was to give an overview of legislation and standards adopted in different countries towards commercial application of NTP technologies.

2 NTP Treated Foods as Novel Foods

In their regulatory boundaries, many developed countries of the world perceive NTP treated foods as novel foods, as it is in EU, Canada, Australia/New Zealand, China, etc. Novel foods approval in these countries is based on a risk or safety evaluation model, with most also requiring final notification and approval. In general, foods that result from a production process that has not been previously used for food production should be initially considered as novel foods.

In EU, the assessment and authorization of novel foods was regulated by the old Regulation (EC) 258/97 (EU 1997) until first of January 2018, when new Regulation (EC) 2015/2283 (EU 2015) entered into force. According to new Regulation (EU 2015), novel foods are defined by two key elements. Firstly, novel foods should not have been consumed to a significant degree within the EU before 15 May 1997, and secondly, novel foods have to fall within, at least, one of ten defined food categories. The most relevant food category for NTP treated foodstuffs could be the category (vii), which is defined as: "Food resulting from a production process not used for food production within the Union before 15 May 1997, which gives rise to significant changes in the composition or structure of a food, affecting its nutritional value, metabolism or level of undesirable substances." This definition corresponds to the novel foods definition given in the old Regulation (EC) 258/97 (EU 1997). Despite the different wording used in old and new EU Regulations, the concept remains the same. It is the aim to cover and assess novel foods produced by novel production processes. Hence, the first question that arises is which treatments could be considered as production processes not used before in EU. Neither old nor new Regulations define this. Only one legal document, EU Regulation (EC) 1334/2008 (EU 2008), gives a list of traditional food preparation processes (Table 14.1). Although this is a list of processes permitted in the manufacturing of natural

Table 14.1 List of traditional food preparation processes (EU 2008)

Chopping	Coating
Heating, cooking, baking, frying (up to 240 °C at atmospheric pressure) and pressure cooking (up to 120 °C)	Cooling
Cutting	Distillation/ rectification
Drying	Emulsification
Evaporation	Extraction
Fermentation	Filtration
Grinding	Stepping
Infusion	Maceration
Microbiological preparation	Mixing
Peeling	Percolation
Pressing	Refrigeration/ freezing
Roasting/grilling	Squeezing

flavouring preparations, it is still a useful indication for discussing novelty of production processes. In addition, the development of novel foods through the usage of NTP treatments could result in unintended changes in the composition and nutritional value of final foods. Therefore, these changes have to be evaluated and currently available literature data should indicate if significant changes in structure, composition or nutritional value occurred in foods, as a consequence of NTP treatments. The term "significant changes" is giving the most confusion, and even some authors have perceived the regulation of novel foods as a grey area (Sprong 2014). Nowadays, European Commission (EC) conducts an initial evaluation of all applications and decides whether further assessment is needed or not. If no additional assessment is necessary, and if there are no objections from the EC or other member states, an authorization decision and approval is made. A risk assessment for novel foods is based on the data provided by the applicant and is in a charge of European Food Safety Authority (EFSA) (Ververis et al. 2020). All authorized novel foods are available on a positive list and can be marketed in the EU. Currently, this positive list includes only few products that have been obtained using NTP treatments, namely UV-treated mushrooms (radiation process in ultraviolet light within the wavelength of 200–800 nm), UV-treated baker's yeast, UV-treated bread (radiation process in ultraviolet light within the wavelength of 240–315 nm for maximum of 5 seconds with energy input of 10–50 mJ/cm^2) and UV-treated milk (radiation process in ultraviolet light within the wavelength of 200–310 nm with energy input of 1045 J/l) (EU 2017). In addition to UV-treated products, EC also authorized the placing on the market HPP fruit-based preparations (EU 2001). It is of note that some other HPP treated foods have been marketed in the EU, without EC approval, due to the fact that these products had not been classified and assessed as novel foods. This was based on the evidence that no significant changes occurred in HPP treated foods (Kurowska et al. 2016). The reason for this can be found in different interpretations and approaches to novel food assessments that national competent authorities had in the past, based on the requirements set out in the old Regulation. Regarding EU novel food legislation, three main conclusions arise: (i) shift from old to new regulation widens the gap in (ambiguous) interpretation of novel foods; (ii) only technologies with a technology readiness level (TRL) above 5 (industrial application) may be considered for novel food applications; (iii) not all NTP technologies (regardless of their TRL level) have the potential of creating novel food.

The Food Standards of Australia and New Zealand, FSANZ, 1.5.1 – Novel Foods was adopted to regulate the area of novel foods production and distribution (Australia 2015). Within this document non-traditional foods have been defined as a food which does not have a history of significant human consumption by the broad community, while novel foods as a non-traditional food that requires an assessment of the public health and safety. A commonly consumed food can be considered non-traditional if prepared by a process not previously applied to the food, therefore NTP treated foods fall under this definition. Nevertheless, the Advisory Committee on Novel Food (ANCF) assessed and adopted only one application of HPP for the production of yoghurt, and afterward in 2013, they have published its opinion that HPP-treated food products should not be treated anymore as novel foods (FSANZ

2018). There is no available data for the assessment and approval of other NTP treated foods.

Regarding the regulatory aspects of novel foods in Canada, Ministry of Justice had adopted the Food and Drug Regulations, in which novel foods is covered by part B, division 28 (Canada 2021c). Health Canada is responsible for the safety assessment of novel foods proposed for sale in Canada. In the period from 2004-2016, this body has concluded that no reasoned safety objections could be posted for several NTP treated foods, including UV and HPP treated foods. Consequently, in 2016 they have published a document in which they have concluded that HPP-treated products does not pose an increased risk to the safety compared to their untreated counter-parts (Canada 2016).

3 Other Regulatory Aspects for NTP Treated Foods

On contrary to countries in which NTP treated foods have to be accessed as novel foods, the United States do not recognise novel foods as such. The US Food and Drug Administration (US FDA) has adopted "broader" definition of food additives. According to 21 CFR 170.3 "Food additives includes all substances not exempted by section 201(s) of the act, the intended use of which results or may reasonably be expected to result, directly or indirectly, either in their becoming a component of food or otherwise affecting the characteristics of food" (US FDA 2016). EU, Australian or Canadian food regulatory regime define food additive as substances added to food only to achieve a technological function. This further means that under the US regulatory approach, food additive could be also the source of radiation, including irradiation, UV radiation and PL (US FDA 2021). According to 21 CFR 179.39, UV radiation can be used for the processing and treatment of food under following conditions:

- Food and food products, with the aim to control surface microorganism, without ozone production; high fat-content food irradiated in vacuum or in an inert atmosphere; intensity of radiation, 1 W (of 2537 A. radiation) per 5 to 10 ft^2.
- Potable water, with the aim to sterilize potable water in food processing production, without ozone production; coefficient of absorption, 0.19 per cm or less; flow rate, 100 gal/h per watt of 2537 A. radiation; water depth, 1 cm or less; lamp-operating temperature, 36 to 46 °C.
- Juice products, with the aim to reduce foodborne pathogens, Turbulent flow through tubes with a minimum Reynolds number of 2200.
- US FDA approved UV treatment to be used for sterilization of water for dairy industry within Pasteurized Milk Ordinance (PMO) (USPHS/FDA 2017).

Beside UV radiation, US FDA has also approved PL treatment for the purpose of controlling surface microorganisms in food, with the total cumulative effect up to 12 J/cm^2 (21 CFR 179.41) and irradiation of food (21 CFR 179.26) (Table 14.2).

Table 14.2 Overview of the food approved to be irradiated in different countries

Food	Purpose	Legally approved irradiation dose			
		USA (US FDA 2021)	Canada (Canada 2021b)	Australia/New Zealand (Australia 2017)	European Union (EU 2016) [a]
Astronaut foods	Sterilization	Min. 44 kGy	N/A	N/A	N/A
Dry or dehydrated aromatic vegetable substances	Control of foodborne pathogens	Max. 30 kGy	Max. 10 kGy	N/A	Max. 10 kGy
Seeds for sprouting	Control of foodborne pathogens	Max. 8.0 kGy	N/A	N/A	N/A
Chilled or frozen raw, cooked or dried crustaceans	Control of foodborne pathogens	Max. 6.0 kGy	N/A	N/A	N/A
Fresh/frozen, uncooked poultry products	Control of foodborne pathogens	Max. 4.5 kGy non-frozen Max. 7.0 kGy frozen	N/A	N/A	N/A
Fresh/frozen, uncooked red meat products	Control of foodborne pathogens	Max. 4.5 kGy non-frozen Max. 7.0 kGy frozen	Max. 4.5 kGy fresh Max. 7.0 kGy frozen	N/A	N/A
Fresh/frozen molluscan shellfish	Control of foodborne pathogens such as *vibrio* spp.	Max. 5.5 kGy	N/A	N/A	N/A
Fresh lettuce and spinach	Control of foodborne pathogens	Max. 4.0 kGy	N/A	N/A	N/A
Fresh shell eggs	Control of *salmonella*	Max. 3.0 kGy	N/A	N/A	N/A
Fresh foods	Growth and maturation inhibition	Max. 1 kGy	N/A	0.15–1 kGy [b]	N/A
Pork carcasses	Control of *Trichinella spiralis*	0.3–1 kGy	N/A	N/A	N/A
Wheat	Insect control	Max. 0.5 kGy	Max. 0.75 kGy	N/A	N/A
Potatoes	Sprout inhibition	Max. 0.15 kGy	Max. 0.15 kGy	0.15–1 kGy	N/A

(continued)

Table 14.2 (continued)

Food	Purpose	Legally approved irradiation dose USA (US FDA 2021)	Canada (Canada 2021b)	Australia/New Zealand (Australia 2017)	European Union (EU 2016) [a]
Onions	Sprout inhibition	N/A	Max. 0.15 kGy	0.15–1 kGy	N/A

[a] Only spices have approved for the irradiation within EU, while each country may authorize additional products
[b] Selected fresh fruits

One important aspect of commercialization of NTP treated foods in the US is laid down in the broader definition of pasteurisation. Accordingly, in 2004 the US Department of Agriculture (USDA) National Advisory Committee on Microbiological Criteria for Foods (NACMCF) has redefined definition on pasteurization as "any process, treatment, or combination thereof that is applied to food to reduce the most resistant microorganism(s) of public health significance to a level that is not likely to present a public health risk under normal conditions of distribution and storage" (NACMCF 2004). This was done as a consequence of novel, emerging new physical methods that appeared worldwide and up scaling of TRL level from proof-of-concept and lab-scale to industrial application. This definition allowed easier application of NTP technologies in food industry (HPP, UV, irradiation, PEF, ultrasound, high voltage arc discharge, etc.). Nevertheless, several aspects have to be taken into account when developing pasteurization processes (either thermal or non-thermal), being:

- Conducting hazard analysis for the food;
- Determining the most resistant microorganisms of public health concern that is likely to survive the pasteurization process;
- Assessing and validating the required level of inactivation of the target microorganism, to make sure it is "not likely to present a public health risk";
- Evaluating appropriate distribution and storage temperature and shelf life;
- Considering the impact of the food matrix on pathogen survival;
- Defining the critical limits that need to be met during processing;
- Defining the specific equipment and operating parameters for the proposed pasteurization process (NACMCF 2004).

Opposed to EU legislation focusing on food product and its physical/chemical/microbial characteristics as a result of any type of food processing (conventional and/or NTP), the US legislation is "process oriented" aimed at promoting technology development, with treated food as a consequence. To present the application of current legislation associated with NTP technologies, the authors have analysed two case studies.

4 A Case Study of Irradiation

Food irradiation is a processing treatment that exposes food to radiation in the form of electron beams, X-rays or gamma rays. This treatment may be successfully used in destroying pathogenic microorganisms, reducing spoilage of foodstuffs, reducing the loss of foodstuffs by premature ripening, germination or sprouting, etc. The first international recognition of the possibility to use irradiation in food processing appeared in 1981, when World Health Organisation (WHO) issued a document titled "Wholesomeness of Irradiated Foods". Following this, Codex Alimentarius Commission adopted The Codex General Standard for Irradiated Foods (Codex Alimentarius 1983), where it was underlined that the irradiation of foods up to a dose of 10 kGy does not introduce any special nutritional or microbiological problems. This publication was very important for defining and adopting legal documents related to food irradiation in many countries of the world (Carreño and Vergano 2012).

The first EU Directives on food irradiation were adopted in 1999, and they are still in force. First one is Directive 1999/2/EC, which covers general and technical aspects for carrying out the process, labelling of irradiated foods and conditions for authorising food irradiation (EU 1999a), and the second one is Directive 1999/3/EC which establishes an EU list of food and food ingredients authorised for treatment with ionising radiation (EU 1999b). Nevertheless, this positive list currently contains only one product dried aromatic herbs, spices and vegetable seasonings (EU 2016). In addition to herbs and spices, all foodstuffs that have been subject to a favourable opinion of the EU Scientific Committee on Food (SCF) may be authorised for irradiation by national committees of Member States, such as fruits, vegetables, cereals, starchy tubers, fish and shellfish, fresh meats, poultry, camembert cheeses manufacture from raw milk, frog legs, shrimp, gum Arabic, casein/caseinates, egg white, cereal flakes, rice flour and blood products (SCF 1998). It is of note that despite this opinion, only several EU countries including Belgium, Czech Republic, France, Italy, the Netherlands, Poland and the UK authorised additional foodstuffs to be irradiated (EU 2009). The other Member States may still restrict or ban these irradiated foodstuffs because they are not on the positive EU list.

Compared to EU, the list of authorized irradiated food in the US involves variety of foodstuffs, from astronauts' foods, fresh and frozen poultry, beef, eggs and seeds, till fresh foods, iceberg salads and spinach (US FDA 2021) (Table 14.2). In Canada, foods as potatoes, onions and wheat, whole or ground spices and dehydrated seasoning preparations, have been complemented with fresh and frozen raw beef in 2017 (Canada 2021b). However, in Australia and New Zealand selected fruits and vegetables can be irradiated for the purpose of pest disinfestation for a phytosanitary objective, with the dose 0.15–1 kGy (Australia 2017). Currently, FSANZ is analysing the possibility to give the permission for the irradiation to all fruit and vegetables that are not produced and consumed within the same quarantine region.

Despite the advantages that irradiation may provide for the treated foods and long regulatory history in many countries of the world, the commercial success is still limited due to negative public perception of irradiated foods (Nielsen et al. 2009).

5 A Case Study of High Pressure Processing

The drawbacks and public aversion towards commercialization of ionizing radiation for food preservation highlighted the challenges that had to be overcome with other NTP technologies to be widely accepted. All interested parties, consumers, food processors, equipment manufactures, and regulatory agencies had to accept novel technology. The great potential for broad acceptance is seen in HPP treatment. This technology might be an alternative to thermal food preservation, in which a food is processed under very high pressure, leading to the inactivation of most microorganisms and enzymes in the treated food, without significant alteration of the organoleptic characteristics of food. The inactivation mechanism of HPP proceeds through low energy and does not promote formation of unwanted chemical compounds, or free radicals (Müjica-Paz et al. 2011). Currently, HPP is the most successfully commercialized NTP technology worldwide. In most cases, HPP has been applied as a final step after packaging, to extend the shelf-life of the product or to enhance microbial safety. As mentioned in previous paragraphs, the application of HPP has been regulated by the novel food Regulations in EU, Australia/New Zealand and Canada. In the last few decades, several scientific committees have analysed and evaluated the safety of HPP treated foods. They all agree that there is no reason for concern regarding industrial application of HPP technology (Canada 2016; FSANZ 2018). In EU, all HPP foodstuffs have to pass premarket evaluation of novel foods (EU 1997, 2015), and until now only a HPP treated fruit-based preparation had been approved on the EU level. There is no agreement by all Member States, on whether HPP treated foods are considered to be novel. For example, the Spanish Food Safety Agency and the UK Food Standards Agency permitted the sale of HPP treated food products without EU approval because they consider that HPP does not produce novel foods. The detailed evaluation of its fate in last few decades has been given in the document written by Kurowska et al. (2016).

Health Canada has assessed a great number of HPP treated foods, including Ready-To-Eat (RTE) meats, raw meats, fruit and vegetable-based juices/smoothies, egg products and other spreads, etc. Based on these assessments and available scientific literature on HHP, they concluded that HPP treated foods should no longer be considered as novel foods, starting from December 2016 (Canada 2016). Recently, Canada Health has also issued Guidance on food products treated with HPP (Canada 2020). Although their conclusion was that HHP treatment of no more than 87,000 psi (600 MPa) for no more than 27 minutes can be applied to a food without compromising nutritional quality or chemical safety relative to its untreated counterpart, it has been indicated that the food processors have to validate treatment parameters in the case of food safety implication (Table 14.3).

Table 14.3 Validation requirements for specific HPP applications (Canada 2020)

The target of HHP treatment	Validation requirements
Shelf life extension	Validation is not required unless a food safety hazard(s) may be introduced by the extended shelf life
Quality improvement	Validation is not required for non-food safety purposes
Lethality treatment	Validation is required demonstrating the intended log reduction is achieved
Post-lethality treatment of ready-to-eat (RTE) food to support applying a reduced listeria monitoring frequency based on a lowered risk	Validation is required demonstrating at least a 3-log reduction of *listeria monocytogenes* is achieved

As the US regulatory system does not recognize novel foods, they have accepted HPP as an antimicrobial treatment and as a treatment to improve microbial safety of RTE products. USDA Food Safety and Inspection Service (FSIS) adopted Directive 6120.2 (FSIS 2012), with a request to verify that the hazard analysis supports the use of HPP treatment in controlling pathogens in the product. The critical parameters for this evaluation include operating parameters such as target pressure, holding time, pre-compression and pressurised temperatures, compression and decompression times and the absence or presence of added CO_2 within packaging, but also food specific factors such as pH, water activity, composition and added preservatives.

HPP could be also used as post-lethality treatments (PLT) in order to control *Listeria monocytogenes* in RTE such as sliced ham, chicken and cured meats. As such it has been incorporated in the US legislation (9 CFR 430.4) (FSIS 2003). PLT is an additional lethality treatment that is applied or is effective after post-lethality exposure of the product. It is applied to the final product or sealed package of product in order to reduce or eliminate *L. monocytogenes* should contamination occur during post-lethality exposure. Some examples of post-lethality treatments include steam pasteurization, hot water pasteurization, radiant heating, and HPP. The possible pressure resistance of foodborne pathogens has been also taken into account. The FSIS requires HPP treatment to obtain at least 5-log reduction of *E. coli* O157:H7 as the indicator microorganism for reprocessing, to ensure microbial safety, due to its great pressure resistance (FSIS 2012).

6 Labeling Requirements for NTP Foods

Food labelling presents an important regulatory issue for all foodstuffs. It is the question whether and how the NTP treated should be declared. When it comes to declaring irradiated food, legal requirements have been very explicitly stated (Maherani et al. 2016). According to Codex Alimentarius standard, when the food or an ingredient product is treated with ionizing radiation a written statement shall be placed in proximity to the food to indicate that the treatment was done. The form of labelling requirement might slightly differ from country to country. In the US, the Code of Federal Regulation (179.26 (c) of 21 CFR) requires that the label of retail packages of foods irradiated shall bear the "Radura" logo along with either the statement "Treated with radiation" or the statement "Treated by irradiation". In Canada, when the whole product was irradiated or more than 10% of the ingredients that compose the final product, than the written statements such as "irradiated," "treated with radiation," or "treated by irradiation" and usage of Radura logo is required (Canada 2021a). In EU, any irradiated food, or food containing an irradiated ingredient must carry the word "irradiated", while the usage of "Radura" log is optional (EU 1999a). However, Australia and New Zealand demands labelling for even minor ingredients (Australia 2017).

Nevertheless, for other NTP treatments, the legal requirements related to labelling have not been very clearly defined. EU Regulation (EC) No 1169/2011 (EU 2011) requires that the name of food shall include or be accompanied by particulars as to the physical condition of the food or the specific treatment which it has undergone (for example, powdered, refrozen, freeze-dried, quick-frozen, concentrated, smoked) in all cases where omission of such information could mislead the purchaser. Now the question is whether the information related to NTP treatments, other than irradiation are essential for the consumers and whether without this information consumers would be misled. The major purpose for the application of majority of NTP treatments is to prolong shelf life and to obtain microbial safety of treated foods. In this sense, some authors have concluded that the omission of information on the NTP treatments, such as HPP would not mislead the consumers and it seems that mentioning production process with the food label is not a mandatory requirement (Kurowska et al. 2016). Nevertheless, within the Article 2 of Commission Decision authorizing the placing on the market of high-pressure pasteurized fruit-based preparations, it has been required to use following statement "pasteurised by high-pressure treatment" on the label of fruit preparations (EU 2001). In addition, the Food Safety Authority of Ireland (FSAI) recommended that the food label should have clearly stated that the food has been treated with HPP in a non-abbreviated form, as its absence is potentially misleading to the consumer (FSAI 2015). It is also important to note that food treated with HPP should not be marketed in such a way as to mislead the consumer, e.g. fruit juice treated with HPP should not be marketed as "fresh" or "raw". Anyhow, the uniformed and clear guidance on the labelling requirements would be helpful for all novel NTP treatments that might be expected to be approved and marketed in the future.

7 Hygiene Aspects for NTP Foods

NTP treated foods have to fulfil the requirements of the food legislation in terms of hygiene, microbiological limits, and chemical contaminants, as it is applicable for all other foods. According to EU Regulation (EC) 178/2002 (EU 2002), the main responsibility for the production of safe foodstuffs is set on food producer. At the same time, EU machinery directive requires from the manufacturer of equipment to lay out cleaning rules (EU 2006). Therefore, the producer of NTP treated foodstuffs has to take the responsibility for the production of NTP treated foodstuffs. Food producer has to comply with the general hygiene requirements, as requested by Regulation (EC) 852/2004 (EU 2004a), which cover Good Hygiene Practices (GHP), Good Manufacturing Practices (GMP), so-called prerequisite programs (PRPs). They include infrastructure, cleaning and disinfection, technical maintenance and calibration, physical and chemical contaminations from production environment, allergens, waste management, water and air control, personnel, raw materials, temperature control, etc.

When one is aiming to produce NTP treated products of animal origin, then additional requirements given in Regulation (EC) No 853/2004 (EU 2004b) should be also followed. Following implementation of PRPs, food producer using NTP treatment have to put in place, implement and maintain a permanent procedure based on Hazard Analysis and Critical Control Point principles (HACCP) as requested by legislation. In addition, it is of great importance for FBO to apply effective traceability and recall system in place (EU 2002). As for all other foods, NTP treated foods should be in compliance with legal microbiological criteria given in Regulation (EC) 2073/2005 (EU 2005).

8 Standards Associated with NTP Food Technologies

When it comes to standards that are directly or indirectly linked with NTP technologies, three main perspectives may be recognized (Fig. 14.1): (i) standards associated with food characteristics; (ii) standards related to the technology and (iii) standards connected with the equipment.

The first dimension covers standards from the "food safety" perspective. International food safety management system standards have not directly included NTP technologies, but have left the door open for their possible application. The International Featured Standards (IFS) recognizes irradiation of food, HPP and

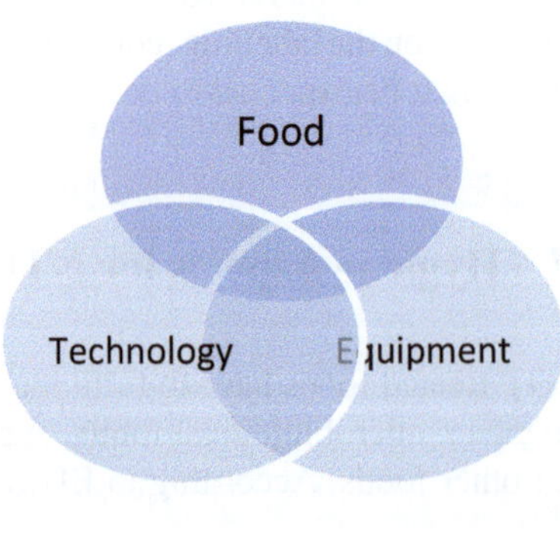

Fig. 14.1 Perspectives of standardization of NTP food technologies

microwave as certifiable technological scopes (IFS 2017), while the BRC Global Food standard allows that alternative thermal processes may be accepted or required if they meet national guidelines and/or are supported with validated scientific data (BRC 2018). Both standards recognize and approve use of X-ray systems for detecting foreign bodies in food. As both technological processes and control measures need to be trustful, ISO 22000:2018 standard requires that all food companies validate their control measures prior to their implementation either as a part of their operational prerequisite plans or through Hazard Analysis Critical Control Points (HACCP) plans to justify that these measures are effective and capable of achieving the intended level of control (ISO 2018).

When it comes to validation, this activity includes several stakeholders occurring in the food supply chain continuum. Food companies plays the most significant role in validating its control mechanisms aiming to achieve the intended level of food safety hazard control, as the most responsible player for placing food on the market. However, national competent authorities, such as inspection services and/or governments should support companies in their food safety risk management decisions by performing validation studies (Codex Alimentarius 2008). Also, the Food and Agriculture Organisation and the WHO need to develop standards, guidelines and codes of practice contributing to food safety having the foundation of these standards supported by scientific evidence (FAO/WHO 2020). Currently, the technological dimension is represented by ISO 14470:2011 covering control of food irradiation process (ISO 2011). It specifies requirements for the development, validation and routine control of the irradiation process applied for the treatment of food by using the radionuclides ^{60}Co or ^{137}Cs, electron beams or X-ray generators. This standard addressed only aspects of the food production related to the irradiation process that affects the safety or quality of food.

Finally, when it comes to the equipment, hygienic design standards are the common denominator for most of these standards, especially since they occur and are in use at different TRLs. The most popular hygienic design standard (EHEDG 2018) outlining principles in designing equipment from a hygienic perspective to prevent food contamination, joint with standards for materials used in constructing equipment (EHEDG 2005) and cleaning validation guidelines (EHEDG 2016). It is worth mentioning that there are similar standards in the US prescribing minimum sanitation requirements for materials and finishes used in manufacturing food equipment (NSF 2019) or ISO 14159 concerned with the associated hygiene risks of the machinery to the consumer of the product being processed (ISO 2002).

9 Conclusion

Nowadays, two regulatory approaches associated with approval of NTP treatments at the global level prevail. The first consider NTP treated foods as novel foods like in EU, Canada, Australia and New Zealand. Second, being in force in the US, where the NTP treatments have been seen as either food additives or pasteurisation processes. In both cases, the successful industrial application of NTP treatments is dependent on the ability to prove compliance with the relevant legislation. However, although regulatory aspects are important, they are not the only limitation factor for the commercialization of NTP treatments since TRL level, marketing of products and labelling have specific impact on NTP development and industrialization. In addition to legal acts, numerous international standards governing the field of food safety give the opportunity for the application of NTP treatments, with mandatory compliance with national legislation.

This chapter reveals that in spite of a large number of different NTPs, only two merits to be considered as currently applied in accordance with legal framework – irradiation and HPP. The first has a long legislative history in most countries in the world. Although numerous advisory bodies have concluded that irradiated food products are safe for human usage, its wider application is limited due to the negative attitude that consumers have towards this treatment. On the contrary, HPP has a much shorter history in legislation acts but its great efficiency, high TRL levels and general consumer acceptance, creates an environment for a wider application. Legal committees in Canada, Australia and some EU countries have even concluded that HPP can be safely used for the food production and processing, if applied in line with hygiene principles and should not be considered as a novel treatment anymore.

References

Angelo KM et al (2017) Multistate outbreak of listeria monocytogenes infections linked to whole apples used in commercially produced, prepackaged caramel apples: United States, 2014–2015. Epidemiol Infect 145(5):848–856

Ankireddy SR, Kim J (2018) Status and recent developments in analytical methods for the detection of foodborne microorganisms. In: Recent developments in applied microbiology and biochemistry. Elsevier, Oxford, United Kingdom, pp 323–334

Australia (2015) Australia New Zealand Food Standards Code – Standard 1.5.1 – Novel foods

Australia (2017) Australia New Zealand Food Standards Code – Standard 1.5.3 – Irradiation of food

Bari ML, Yeasmin S (2018) Foodborne diseases and responsible agents. In: Food safety and preservation. Elsevier, pp 195–229

BRC (2018) BRC global standard for food safety, issue 8. BRC Trading Ltd, London

Buchholz U et al (2011) German outbreak of Escherichia coli O104:H4 associated with sprouts. N Engl J Med 365(19):1763–1770

Canada (2016) High Pressure Processing (HPP) Is No Longer a Novel Process, Health Canada

Canada (2020) Foods treated with high pressure processing (HPP), Canadian Food Inspection Agency

Canada (2021a) Food and Drug Regulations, Part B, division 1 – General, section 0.35

Canada (2021b) Food and Drug Regulations, part B, divison 26 – Irradiation

Canada (2021c) Food and Drug Regulations, part B, divison 28 – Novel foods

Carreño I, Vergano PR (2012) Food irradiation: the EU regulatory framework, risk assessment and international trade considerations. Eur J Risk Regulat 3(3):373–392

Codex Alimentarius (1983) Codex Standard 106-1983, Rev. 1-2003, General standard for irradiated foods. World Health Organisation & Food and Agriculture Organization of the United Nation., Rome, Italy

Codex Alimentarius (2008) CAC/GL 69-2008 Guidelines for the validation of food safety control measures. World Health Organisation & Food and Agriculture Organization of Rome, Italy the United Nations
EFSA (2019) The European Union one health 2018 Zoonoses report. EFSA J 17(12):1–109
EHEDG (2005) Materials of construction for equipment in contact with food. EHEDG guidelines. European Hygienic Engineering and Design Group, Frankfurt
EHEDG (2016) Cleaning validation in the food industry – general principles. European Hygienic Engineering & Design Group, Brussels
EHEDG (2018) Hygienic design principles. EHEDG Guidelines. European Hygienic Engineering & Design Group, Brussels
EU (1997) Regulation (EC) No 258/97 of the European Parliament and of the Council of 27 January 1997 concerning novel foods and novel food ingredients
EU (1999a) Directive 1999/2/EC of the European Parliament and of the Council of 22 February 1999 on the approximation of the laws of the Member States concerning foods and food ingredients treated with ionising radiation
EU (1999b) Directive 1999/3/EC of the European Parliament and of the Council of 22 February 1999 on the establishment of a Community list of foods and food ingredients treated with ionising radiation
EU (2001) Commission Decision of 23 May 2001 authorising the placing on the market of pasteurised fruit-based preparations produced using high-pressure pasteurisation under Regulation (EC) No 258/97 of the European Parliament and of the Council
EU (2002) Regulation (EC) No 178/2002 of the European Parliament and of Council of 28 January 2002 laying down the general principles and requirements of food law, establishing the European Food SafetyAuthorityand laying down procedures in matters of food safety
EU (2004a) Regulation (EC) No 852/2004 of the European Parliament and of the Council of 29 April 2004 on the hygiene of foodstuffs
EU (2004b) Regulation (EC) No 853/2004 of the European Parliament and of the Council of 29 April 2004 laying down specific hygiene rules for food of animal origin
EU (2005) Commission Regulation (EC) No 2073/2005 of 15 November 2005 on microbiological criteria for foodstuffs
EU (2006) Directive 2006/42/EC of the European Parliament and of the Council of 17 May 2006 on machinery, and amending Directive 95/16/EC
EU (2008) Regulation (EC) No 1334/2008 of the European Parliament and of the Council of 16 December 2008 on flavourings and certain food ingredients with flavouring properties for use in and on foods
EU (2009) 'List of Member States' authorisations of food and food ingredients which may be treated with ionising radiation
EU (2011) Regulation (EU) No 1169/2011 of the European Parliament and of the Council of 25 October 2011 on the provision of food information to consumers
EU (2015) Regulation (EU) 2015/2283 of the European Parliament and of the Council of 25 November 2015 on novel foods
EU (2016) List of approved facilities for the treatment of foods and food ingredients with ionising radiation in the Member States
EU (2017) Commission implementing Regulation (EU) 2017/2470 of 20 December 2017 establishing the Union list of novel foods in accordance with Regulation (EU) 2015/2283 of the European Parliament and of the Council on novel foods
FAO/WHO (2020) About Codex Alimentarius. FAO HQ. http://www.fao.org/fao-who-codexalimentarius/about-codex/en/#c453328. Accessed December, 2020
FSAI (2015) High pressure processing of food, microbial fact sheets. Food Safety Authority of Ireland
FSANZ (2018) Records in views formed in response to inquiries, Advisory Committee on Novel Foods, Food Standards Australia New Zealand

FSIS (2003) Code of federal Register, CFR title 9, Part 430—Requirements for specific classes of product. Food Safety Inspection Services
FSIS (2012) Directive 6120.2 - High pressure processing (HPP) and inspection program personnel (IPP) verification responsibilities. Food Safety and Inspection Service
Gaul LK et al (2013) Hospital-acquired listeriosis outbreak caused by contaminated diced celery-Texas, 2010. Clin Infect Dis 56(1):20–26
Gelting RJ et al (2011) Irrigation water issues potentially related to the 2006 multistate E. coli O157:H7 outbreak associated with spinach. Agric Water Manag 98(9):1395–1402
Havelaar AH et al (2010) Future challenges to microbial food safety. Int J Food Microbiol 139(SUPPL. 1):S79–S94
IFS (2017) IFS food, version 6.1. International Featers Standard Management GmbH, Berlin
ISO (2002) ISO 14159:2002 safety of machinery — hygiene requirements for the design of machinery. International Organization for Standardization, Geneva
ISO (2011) ISO 14470:2011 food irradiation — requirements for the development, validation and routine control of the process of irradiation using ionizing radiation for the treatment of food. International Organization for Standardization, Geneva
ISO (2018) ISO 22000:2018 food safety management systems – requirements for any organization in the food chain. International Organization for Standardization, Geneva
Kirk MD et al (2015) World Health Organization estimates of the global and regional disease burden of 22 foodborne bacterial, protozoal, and viral diseases, 2010: a data synthesis. PLoS Med 12(12):1–21
Kurowska A, Szajkowska A, van der Meulen B (2016) EU regulatory approach to high-pressure processing. In: High pressure processing of food, principles, technology and applications. Springer, New York, USA, pp 717–732
Luckose F, Mamatha BS, Chauhan OP (2019) Commercialization and regulatory issues of non-thermal processed foods. In: Non-thermal processing of foods. CRC Press, Boca Raton, USA, pp 417–431
Maherani B et al (2016) World market development and consumer acceptance of irradiation technology. Foods 5(4):79
Martinez MR et al (2016) Capacity of listeria monocytogenes strains from the 2011 cantaloupe outbreak to adhere, survive, and grow on cantaloupe. J Food Prot 79(5):757–763
Mújica-Paz H et al (2011) High-pressure processing technologies for the pasteurization and sterilization of foods. Food Bioprocess Technol 4:969–985
NACMCF (2004) National Advisory Committee on Microbiological Criteria for Foods Scientific parameters for establishing the equivalence of alternative methods of pasteurization
Nath P, Kale SJ, Bhushan B (2019) Consumer acceptance and future trends of non-thermal-processed foods. In: Non-thermal processing of foods. CRC Press, Boca Raton, USA, pp 433–454
Newell DG et al (2010) Food-borne diseases – the challenges of 20 years ago still persist while new ones continue to emerge. Int J Food Microbiol 139(SUPPL. 1):S3–S15
Nielsen HB et al (2009) Consumer perception of the use of high-pressure processing and pulsed electric field technologies in food production. Appetite 52(1):115–126. https://doi.org/10.1016/j.appet.2008.09.010
NSF (2019) NSF/ANSI 51-2019: food equipment materials. NSF International Standard / American National Standard, Ann Arbor
Nyachuba DG (2010) Foodborne illness: is it on the rise? Nutr Rev 68(5):257–269
Painter JA et al (2013) Attribution of foodborne illnesses, hospitalizations, and deaths to food commodities by using outbreak data, United States, 1998-2008. Emerg Infect Dis 19(3):407–415
Rajkovic A, Smigic N, Devlieghere F (2010) Contemporary strategies in combating microbial contamination in food chain. Int J Food Microbiol 141(SUPPL):S29–S42
SCF (1998) Opinion of the Scientific Committee on Food on the irradiation of eight foodstuffs. Scientific Committee on Food

Smigic N, Rajkovic A (2014) Hurdle technology. In: Microbial food safety and preservation techniques. CRC Press, Boca Raton, USA, pp 383–399

Sprong C (2014) Grey area novel foods: an investigation into criteria with clear boundaries. Eur J Nutrit Food Saf 4(4):342–363

Tucker, G. and Featherstone, S. (2010) Essentials of thermal processing, Blackwell Publishing Ltd., Iowa, USA

US FDA (2016) Code of federal Register, CFR title 21, Part 170 – Food Additives

US FDA (2021) Code of federal Register, CFR title 21, Part 179.26 – Irradiation in the production, processing and handling of food

USPHS/FDA (2017) Grade "A" Pasteurized Milk Ordinance (PMO), pp. 1–426

Ververis E et al (2020) Novel foods in the European Union: scientific requirements and challenges of the risk assessment process by the European Food Safety Authority. Food Res Int 137:109515

Zhang Z-H et al (2019) Non-thermal technologies and its current and future application in the food industry: a review. Int J Food Sci Technol 54(1):1–13

Zink DL (1997) The impact of consumer demands and trends on food processing. Emerg Infect Dis 3(4):467–469

Chapter 15
Current Technology Readiness Levels (TRL) of Nonthermal Technologies and Research Gaps for Improved Process Control and Integration into Existing Production Lines

Edward Ebert and Kemal Aganovic

1 Introduction

It is within the human being to judge and estimate characteristics observed in nature, and then, often using numerical values, assign them to different scales and rank them to different levels. Technological development has been a powerful driving force for human progress and the search for solutions to social problems, and it is an inevitable process and a natural course of things. Although technological innovations and changes from "conventional-to-novel" represent a long-term perspective for competitive and sustainable production, development often comes with economic and success risks. Looking at the food sector, as one of the largest manufacturing sectors in terms of jobs and value-added in the European Union, it has enormous innovation potential, but compared to other industries is rather slow and conservative. The majority of food production still relies on traditional products and processes with low R&D expenditures. Development and introduction of new technologies typically depends on the quality of previous research and level of investigation as a starting point, inputs and efforts put in the development, and risk ready to take. In the food research community, significant amount of resources has been spent on R&D activities on the investigation of new technologies and production concepts, their suitability for different applications and products, all with the goal to improve food quality, process efficiency and sustainability, along with increasing their readiness level. In general, many technologies undergo similar development paths and are usually first investigated at the lab-scale level, following validation on a pilot-scale, and eventual test applications in an industrial environment. The application at the industrial level still does not guarantee the success of the technology,

E. Ebert · K. Aganovic (✉)
German Institute of Food Technologies DIL e.V., Quakenbrück, Germany
e-mail: K.Aganovic@dil-ev.de

A. Režek Jambrak (ed.), *Nonthermal Processing in Agri-Food-Bio Sciences*,
Food Engineering Series, https://doi.org/10.1007/978-3-030-92415-7_15

where numerous examples are available in the food industry. Exceptions in these development paths are technologies already applied in industries other than the food sector, with already high maturity level, which can be translated in the food sector with marginal efforts. The development and maturing of a technology and related activities often face three major challenges: performance, schedule and budget (Mankins 2009), and include a great deal of risk and uncertainties for machine manufacturers and technology users. The risk and uncertainty are reflected in technology performances, possible integration into an existing system, and reliability (Olechowski et al. 2020). Thus, the task of developing engineers is not only the development of the technology itself, but at the same time, it includes planning and consideration for the technology to be easily integrated into the existing production lines and operational reliability. In contrast, premature attempts to integrate the technology can lead to increased costs and delays, and in certain cases failures (Meier 2008; Dubos et al. 2008).

At the same time there is a significant amount of uncertainties and challenges present related to the point of knowing how to measure the progress of the technology development; if the technology is ready for the transition from research level to pilot scale and to industrial environment; and how to ensure that a process or a technology is reliable, and meets the standards and expectations of the technology users (Olechowski et al. 2020). For the purpose of this book chapter, the focus will be on food sector stakeholders. In order to try to overcome these issues, engineers of the national aeronautics and space administration (NASA) introduced the technology readiness level (TRL) scale in the 1970s for easier understanding using a standard language that can be used across different disciplines, organizations, and functions. (Banke 2019). The TRL is a tool for assessing the maturity and level of development of a technology, from the early beginning, through its development, until its integration into the production system. Nowadays the TRL scale is commonly accepted and used for consistent assessment of technology maturity, monitoring its progress and managing risks associated with its development. At the same time, TRL is a scale for comparison of the technologies and plays an important role in decision making, not only in aerospace but also in projects in other industries such as energy, transportation, food production and similar.

The TRL assessment is in general quite a challenging task, and the description for the technologies selected in this book chapter is based on the current status and available literature, supported by information from the machine manufacturers. TRL is a rather subjective assessment, where different levels are not perfectly distinct from one another, and such assessment would typically require a consensus among several stakeholders, that may hold conflicting perspectives when it comes to how mature the technology really is, and in which fields it has been validated. Further, when TRL is used for decision making, it is recommended to have criteria, not only machine- or technology-specific, but also product-line or product-type specific (Olechowski et al. 2015).

1.1 Terminology and Focus

We use the term technology to describe a processing step based on a phenomenon of the certain processing technology. It is difficult to consider different applications of one technology to describe its readiness level, thus we will rather focus on the technology as a system. In this context, parameters such as level of technological development (size and production scale), design maturity, risk assessment, reliability, and certification are considered. It is important to notice, that the TRL is only a measure of the maturity for an individual technology, and not necessarily a system readiness. Moreover, TRLs offer only limited information on the integration of the technology into a system, and how these changes would affect the system or other technologies in general.

2 Technology Readiness Level as a Tool for Development Maturation

Soon after the implementation of the NASA TRL system, it was widely adopted and developed into a nine-level measurement system for assessing the maturity of a technology. It allows consistent comparison of maturity between different types of technologies (Sauser et al. 2006). In the following, a brief description of each of nine TRL is given, based on the description by Moorhouse (2002) and Mankins (2009), and an overview is depicted in Fig. 15.1.

TRL 1 – fundamental principle observed – is the lowest level of technology maturation, and includes fundamental elements of basic theoretical research, where scientific questions are translated into applied research i.e., basic working principles of a technology or a working concept. On the example of food science, these can be fundamental questions e.g., if a phenomenon or a technology is suitable for the treatment of food and framing the working parameters with a particular goal. Very often the results are supported by analytical or theoretical predictions and test data.

TRL 2 – technology concept and/or application – represents the next step in the maturity of a technology, after the physical principle was observed. At this level, the practical applications are investigated; on previous examples, if the technology can be applied for inactivation of microorganisms in selected matrices, surface, or in other fields like biotechnology or medicine. Usually, TRL1 and TRL2 are driven by scientists, and costs to achieve this stage and risk related to it remain usually low, except for the risk of application feasibility due to unknown effects, and costs in certain other research disciplines (e.g. nuclear research). The application remains still rather speculative, as usually no supportive experimental protocols exist.

At **TRL 3 – experimental proof of function** – more intensive research and development activities are performed, and the first proof of concepts are designed. From the food science perspective, at this stage scientist are mostly looking into

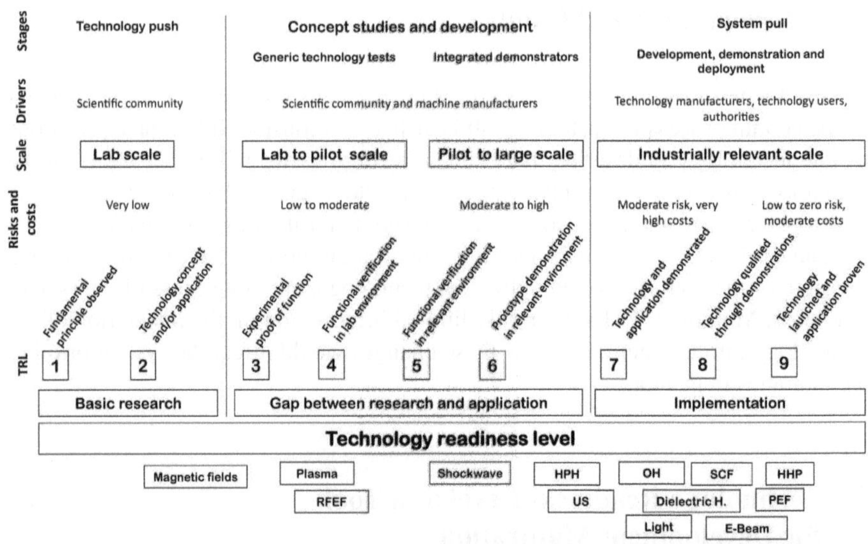

Fig. 15.1 Technology readiness level and risk and cost assessment of selected non-thermal processing technologies. (Adapted from (ESA 2019))

analytical and experimental setups for validation of previous observations, as well as general applicability of a technology, along with the impact on quality and safety aspects in terms of any untypical changes, chemical reactions or residues. Most of the research can be still assigned to the lab scale, thus the costs to achieve this level remain relatively low and in certain cases moderate, compared to the costs of the eventual system with proven functions.

TRL 4 – functional verification in lab environment – follows activities beyond the proof of concept and moves from investigating core elements of the innovation to the integration of these elements into a single system or a unit. This usually aims at the development of first prototypes, which can support the concept proven earlier, but also considers requirements for the final "future" system. The validation at this level can still be considered as of low accuracy if compared to the eventual system applications. At this stage, the costs are rather moderate, as the first investments are usually required for setting up test systems and prototypes.

TRL 5 – functional verification in relevant environment – applies for functional investigation of the technology in a relevant environment, larger than a lab scale. At this stage, processing parameters must be realistic, comparable, and relevant to those applied in the final "future" system. The system of sensors or monitoring controls does not have to be at the required level of development, but the supportive elements should depict a realistic and relevant working environment. This level can be allocated to prototype testing on a pilot scale, preferably integrated into a small production line (e.g. transporting, pre-heating or cooling etc.). The costs at this stage can be considered as moderate to high and are typically significantly higher compared to TRL4 in the same area. This is related to necessary investments by scientists, but sometimes in collaboration with machine manufacturers and equipment providers.

TRL 6 – model system or prototype demonstration in relevant environment – represents a sort of breaking point in the development maturity. At this stage, technology is demonstrated in a relevant environment and the maturation is driven by validation of R&D outcomes. At this level mostly assuring of upper management and investors in the technology takes place. Besides the innovation technology itself, other system relevant peripheral technologies can be integrated into the demonstrator. The technology is characterized by a moderate risk, but costs can be moderate to high, depending on the requirements of the prototype and relevant working environment; however, still significantly lower compared to costs for reaching TRL 7 for the same technology.

TRL 7 – technology and application demonstrated – represents a significant advancement of the prototype, which requires its demonstration in the expected operational environment. In the case of food science, the prototype demonstration requires operation in the food production environment, at least close to the full production scale. It serves also for operational managers and assuring technology users, and as such requires a prototype level of an actual application. Accordingly, the TRL7 usually requires very high costs and typically represents a large portion of the costs for the development of the final operational system. The risk to achieve this level is considered moderate, having already proven concept and small scale prototype functionality.

TRL 8 – technology qualified through demonstrations – is achieved through completed demonstration and qualified tests in the industrial environment, and the final system is ready for operation. In most cases, this stage represents the end of the final system development. The costs to achieve TRL8 can be considered moderate, as TRL 8 represents the end of the development, and typically only minor changes and adjustments of the systems are required. The risk can be recognized as low. However, both costs and risk might vary for different applications and different technologies.

TRL 9 – technology launched and application proven – According to Mankins 2009, all technologies that succeed in application actually go through the TRL 8 and TRL 9, while sometimes some levels (e.g. TRL7) do not have to be necessarily implemented in the development and can be skipped. Bug-fixing and final tuning that takes place at the very end is a good indication of TRL 9, and can only occur at that level. The major difference between the TRL8 and TRL9 is the operation – building an industrially qualified machine and testing it in the industrial environment is TRL 8; launching and operation during the regular production is TRL9. The costs for TRL9 can be considered as marginal, at least for the food sector or more precisely for machine manufacturers.

3 TRL of Non-thermal Food Processing Technologies

After getting familiar with the 9 TRL scales, we can take a look in the brief overview of the TRL for selected non-thermal technologies. For certain technologies, information was obtained directly from machine manufacturers using questionnaires containing questions related to the TRL assessment. It is worth mentioning that it is challenging to group certain technologies with different characteristics into a single group. However, in an attempt to group them according to their working phenomenon as a major driving force, still, there are certain overlappings between the technologies, as clear distinction is not always possible. Beside TRL assessment, limitations for their implementation and increasing TRL, as well as research gaps are briefly discussed.

3.1 Technologies Based on Pressure

Within this chapter technologies that use pressure as a physical phenomenon as a major working principle are considered, and include high hydrostatic pressure (HHP), ultra-high pressure homogenization (UHPH) and application of supercritical fluids (in particular water and carbon dioxide).

3.1.1 High Hydrostatic Pressure

Thank its industrial use since the 1990s, continuous machine development and broadening of applications for different food products, the TRL for high hydrostatic pressure (HHP) technology is estimated at the level of 8 and 9. It can be considered as one of the highest among the non-thermal processing technologies. By looking at different applications of HHP, the TRL might vary, but for the machine and the technology, typically used for major applications like batch-preservation of different food and beverages at the pasteurization equivalent, or seafood opening, can be considered TRL 9. The bulk treatment, where pumpable products are directly pumped in and out of the pressure vessel and aseptically filled after the treatment, is recently introduced into the industry by a French company Ateliers Hermes Boissons, and as such can be considered slightly lower at TRL 8. In 2019, over 550 industrial machines were installed in operation worldwide, with most of them in North America, followed by Europe and Asia (González-Angulo et al. 2021; Tonello-Samson et al. 2020).

Although the capital investment for the pressure machines has decreased around 15% over the last five years, investment and operation costs remain one of the greatest challenges for the technology implementation, especially for the small-scale producers. On the other hand, for large scale producers, a higher level of automation, a higher number of intensifiers or even more vessels, are required to meet the

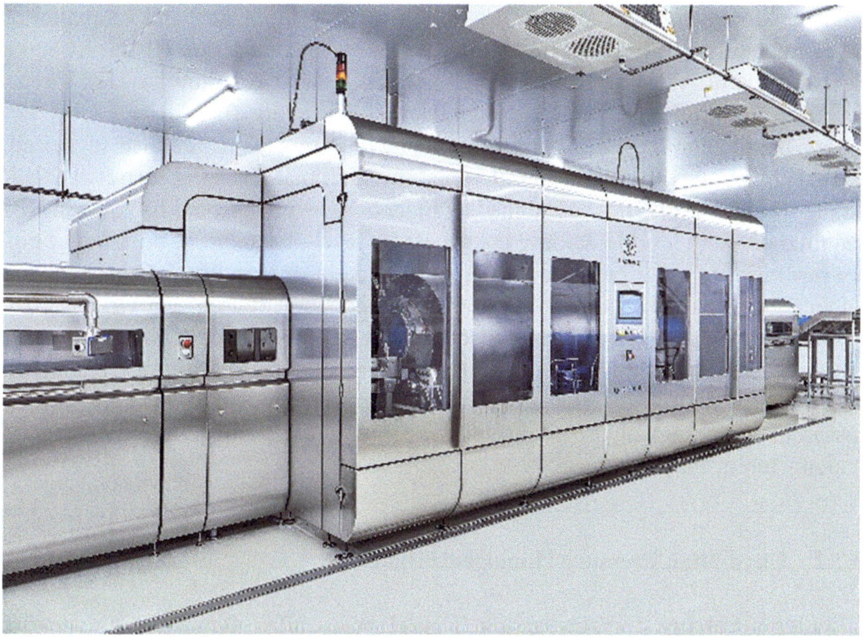

Fig. 15.2 HHP Uhde 350–60. (Source: Uhde High Pressure Technologies GmbH, HPP Center Quakenbrück, Germany)

capacity demands, resulting in higher investment costs. Products suitable for HHP treatment have to be contained in a flexible packaging able to withstand high pressures of up to 6000 bar, thus, sometimes adjustments of an existing product packaging are required, resulting in additional investment costs. Every change of the packaging material needs to be evaluated concerning food safety and possible interaction between the food and the packaging material, especially if the packed food is treated by high pressure. As an alternative to the conventional fossil-based packaging materials, more eco- and environmentally friendly bioplastic materials are available, for which a lack of knowledge exists with respect to the impact of the HHP-treatment on the packaging materials exists (Aganovic et al. 2021b). Nevertheless, unlike other technologies, HHP technology can be found in toll processing centres of machine owners, offering processing facilities and HHP services to other food companies. This is especially attractive for young start-ups and companies with small production batches (Fig. 15.2).

On the other hand, pressure combined with thermal treatments (high-pressure thermal sterilization - HPTS) with the purpose of inactivation of all microorganisms including bacterial spores (sterilization equivalent), has a lower TRL between 6 and 7. In particular, for this application, the cost and risk related to achieving a higher TRL level seem to be higher than moderate. In order to improve the technology and achieve increased process control, a better understanding of temperature distribution over the processing cycle at various pressure and temperatures should be studied. Moreover, development of wireless sensors, modelling and validation of temperature distribution is of major importance when considering HPTS. Also, the development of systems and packaging materials, that would allow for HPTS treatment at a large industrial scale is required (Aganovic et al. 2021b).

For broadening the HHP application, combined treatments are also suggested, e.g. combination with natural antimicrobials or microbial phages (Chien et al. 2017; Komora et al. 2018). These combined treatments have the potential for lowering the required pressure level to achieve the necessary level of safety and are suitable e.g. for raw meat, or other pressure-sensitive and protein-rich products. Other pressure applications, like structure modification, salt reduction (Olsen and Orlien 2016; Bolumar et al. 2020), or development of new functional foods, are still under research and have usually lower TRL than existing applications for preservation. In a conclusion, despite the high TRL of the HHP technology, there are still certain research topics that would help this technology and increase its usefulness for food manufacturers and consumers.

3.1.2 Ultra-High Pressure Homogenization

Ultra-high pressure homogenization (UHPH) also called dynamic high-pressure processing, is well developed and established for large scale applications in the biotechnology or pharmaceutical industry, where typically production capacities are lower compared to traditional food scale production. In this case also the margins and added value are higher, compared to food. In this case, the UHPH technology has a high level of TRL between 8 and 9. However, with relevance for this chapter and the food industry, the TRL of the UHPH can be considered at level 6. Depending on the application, manufacturers are offering machines with production capacities from a few L/h up to a few thousand L/h (Table 15.1). Increasing production capacity typically results in a decrease of the maximum achievable pressure. Still, compared to conventional milk homogenization and production lines reaching

Table 15.1 UHPH machine manufacturers with maximum operating pressure and production capacity[a]

Manufacturer	Maximum operating pressure	Maximum throughput
Stansted fluid power (UK)	3500 bar	100 L/h
PSI (Italy)	2000 bar	84 L/h
IKA (Germany)	2000 bar	100 L/h
Microfluidics MPT (USA)	2068 bar	900 L/h
GEA Niro Soavi (Germany)	4000 bar 1500 bar	250 L/h 5000 L/h
SPX flow (USA)	1500 bar	800 L/h
Avestin (Canada)	2000 bar	1000 L/h
Bee international (USA)	3100 bar 1370 bar	60 L/h 120 L/h

[a]The data is based on web search

capacities of up to 50.000 L/h, the actually achievable capacities are still low to match the needs and requirements of the food industry. When considered as a tool for sterilization of pumpable products (Georget et al. 2014), the UHPH TRL can be considered even lower. Nevertheless, for small scale production of a high-value product, the TRL can be considered high as 8.

To reach the next TRL in applications for food processing, the UHPH still needs certain research supported developments. At first, from material science and mechanical engineering perspective, there is a need for the developments of new concepts and materials, that would allow large scale UHPH treatment at a technical scale. This can be achieved by increasing the capacity and endurance of the core components of the machine (valves, needle, seats, fittings etc.). The occurring physical phenomena responsible for the treatment effects, shear effects, turbulence and cavitation, are often discussed in the literature, but it is not fully clear which of these effects is the major driver and which factors can be optimally combined. More research is required for modelling the flow behaviour and energy dissipation of different valve types and configurations, and related different effects on food (cell disruption, droplet size reduction, hydrocolloid modification etc.) (Martínez-Monteagudo et al. 2017). Modification of hydrocolloids and obtaining their improved functionality, gain attention from many researcher groups, but are not completely elucidated (Levy et al. 2020). With a better understanding, the industry would gain different possibilities to develop new foods with unique properties, e.g. fat-reduced products without artificial emulsifiers and thickeners (Aganovic et al. 2018), dairy-free yoghurt-like fermented product (Levy et al. 2021), inactivation of spores (Dong et al. 2016) or stabilization of plant-based milk alternatives (Poliseli-Scopel et al. 2013).

The UHPH has a great potential for the production of new structures and healthy food products, while at the same time can be used for microbial inactivation (e.g. simultaneous homogenization and preservation) and cell disruption. Therefore, a significant amount of research is needed with respect to up-scaling, valve design and understanding the interactions and phenomenon occurring during the UHPH treatment.

3.1.3 Supercritical Fluids (CO_2 and Water)

Supercritical fluids (SCF) is a well-established technology mostly used for mass-transfer processes, phase-transition processes and reactive system processes, among others (Brunner 2010). Particularly relevant processes for food production include removal of certain substances like fat or alcohol, or enrichment of food with certain functional substances (Brunner 2005). It is important to note that SC-CO_2 has a higher TRL compared to SC water, and as such fulfils the criteria for the TRL 9, whereas SC water is estimated lower (at least in the food and agricultural sector). Despite the high TRL and several plants operating worldwide, the technology still did not find a wider application in food production systems. Similar to other non-thermal technologies, increasing the production capacity, reduction of investment

Fig. 15.3 2×200L Supercritical Extraction Plant, USA, with permission of Uhde GmbH

cost due to new material for the high-pressure equipment or due to new design concepts for pressure build-up, are desired. Also, machine manufacturers of this technology have a huge interest in the development of new methods and equipment for the characterization of the product properties (concentration, thermal, chemical) under high-pressure extraction conditions (Zhou et al. 2021) (Fig. 15.3). Similar to the HHP, in some countries there are toll processing centres offering SC-CO_2 production of small batches.

A better understanding of kinetics during the extraction processes is of high interest, which could lead to an increase in extraction yield or development of more selective extraction processes (e.g. for pesticides), which in turn would increase the number of applications and allow for further equipment optimization. So far extraction of molecules with a relatively large molecular size is challenging, thus requiring more research effort to close the knowledge gap and create possibilities for an industrial-scale extraction of valuable macromolecules, in an efficient and energy-saving way (Zhou et al. 2021; Ziero et al. 2020). Moreover, increase of the solubilization of molecules with high-molecular-weight (Steiner et al. 2018) and possible utilization of side streams and agricultural goods for fermentation purposes (Yin et al. 2014) or for energy production (Saqib et al. 2019) are recognised as promising future applications for SC-H_2O.

In recent years, promising results have been achieved for the development of SCF-impregnated active packaging or encapsulated bioactive substances with the aim of slowing down and controlling the release of these substances into the food matrix during the shelf-life. However, this type of applications is still in the early stage with relatively low TRL. Further research is needed to develop large-scale and cost-effective production of active packaging materials or encapsulations of biosubstances to be of greater use, and thus increase the TRL level for this application (Bastante et al. 2017; Franco et al. 2019).

3.2 Technologies Based on Pressure Waves

In this section, the TRL of different technologies based on effects of wave energy are discussed: ultrasound and shockwaves, as a form of hydrodynamic pressure treatment.

3.2.1 Ultrasound

Ultrasound (US) technology is typically used for emulsification, defoaming, viscosity alteration, encapsulation, supporting the drying, tenderization of meat, freezing and thawing processes, and inactivation of microorganisms or modification of enzymes (Gallo et al. 2018; Singla and Sit 2021). Typically the applications can be found in the food industry, but also in the chemical, biotechnology and pharmaceutical industry (Jiang et al. 2020; Taha et al. 2020). Several companies offer machines for industrial processing or cleaning applications (Table 15.2). However, mostly the US technology can be found still in a prototype and testing phase, resulting in TRL classification between 7 and 8.

Although some industrial applications exist and the TRL for these applications can be considered high, there is still potential for other applications where fundamental principles are proven, to be improved and transferred on a large industrial scale. These include preservation, emulsification or US-assisted extraction (Chemat

Table 15.2 US machine manufacturers with different reactor systems and maximum production capacity[a]

Manufacturer	Reactor system	Maximum throughput
Industrial Sonomechanics (USA)	Batch/bath	250 L/h depending on the application
Bandelin Electronic GmbH & Co. KGB (Germany)	Batch/bath	600 L/h
Hielscher Ultrasonics GmbH (Germany)	Batch and continuous	1000 L/h or more depending on the application
Sonics & Materials, Inc. (USA)	Batch and continuous	30 L/h
Advanced Sonic Processing Systems (USA)	Batch and continuous	n.a.
Synetude Sas (France)	Batch and continuous	500 L/h

[a]The data is based on web search

et al. 2020). During the development of large-scale applications, it is necessary to investigate potential negative effects of transmitted soundwaves on workers, as well as the maximum exposure to these soundwaves, and these findings could be framed into safety regulations (Harvey et al. 2014).

Besides developments in machine and engineering aspects to increase the technical TRL, potential further research could be identified in promising US treatment on (anti-)nutritional properties, e.g. the impact of high-intensity US on allergens or speed of degradation of certain chemicals. So far there is a lack of knowledge on the formation of potentially harmful substances, as a result of the US treatment, and certainly more studies would help to assess possible risks due to radical formation and chemical alteration of ingredients. This is of particular importance for decision-makers in order to give the technology push for legal authorities, as well as for machine manufactures, and at the end more certainty to food producers (Bhargava et al. 2020). A better understanding of the complex physiochemical reactions, in particular process-product interactions and resulting impact on technological and functional properties of foods, would support wider commercialization of US in the food industry (Soria and Villamiel 2010).

Finally, US is a proven technology for emulsification and defoaming and some other applications. However, it is expected that the scientific community would take an effort for broadening the range of application and allow for the scaling up of the technology for these applications.

3.2.2 Hydrodynamic Shockwaves

Shockwave (SW) technology has been introduced in the late 1950s and 60s with the first patents reporting applications of hydrodynamic pressure for modification of food structure by the detonation of explosives in the vicinity of the product placed in water (Simjian 1958; Godfrey 1970). The technology is recognized as promising in terms of improving the tenderness of low-value meat cuts, but also for cracking of shells of seafood, like oyster, lobsters and mussel (Bolumar and Toepfl 2016). Since then significant scientific effort has been taken to increase the TRL of the technology which today is still at the TRL 6. Within the EU FP7-SME project "ShockMeat" a continuous shock wave plant prototype was built by the German Institute of Food Technologies (DIL e.V.) and tests in relevant industrial conditions have been performed. In 2018 second improved pilot-scale machine was built by the DIL and installed at CSIRO in Australia. To the best of our knowledge, these two pilot-scale prototypes are the only ones currently operating at the pilot scale for research purposes, feasibility tests and small-scale production (Aganovic et al. 2021a).

Despite improvements in meat quality and tenderness (McDonnell et al. 2021; Bolumar et al. 2014; Chian et al. 2021), to further increase the TRL of this technology, a significant amount of research is required, especially from the machine manufacturing perspective. At first, the generation of shockwave can be improved, and operator safety and quality of the operational environment has to be considered. In that sense, an improved way of conversion of electrical energy into mechanical energy is required, optimally resulting in the generation of more less-intensive single shockwaves, at a comparable level of mechanical energy like one high-intensive shockwave. With more less-intensive shockwaves, the peak stress to the packaging and the machine components would decrease, which is beneficial for the life span of the machine components and would most likely cause less damage to the packaging material. Although certain packaging solution exists (McDonnell et al. 2021), the packaging material needs to be improved further, to be able to withstand damages caused by the pressure shock or eroded material from the wire or electrodes.

Another possibility could be the development of a machine with a membrane, e.g. made from silicon, to shield the product against the shockwave transmitting medium (ShockMeat 2016). Further, from the mechanical engineering and material sciences perspective, the durability of the electrodes and insulators, but also electrical and mechanical components in general, could be improved to allow for lower costs and faster uptake by the industry. To the best of our knowledge, so far, the impact of long-term operation on durability and erosion of the electrodes and insulators has not been fully evaluated. Finally, the production scale should be increased to meet the industry requirements, by either enlargement of the treatment zone or more powerful shockwave generators should be developed (Aganovic et al. 2021a; Bolumar and Toepfl 2016). The cost-benefit of this application is also not completely assessed so far and would be crucial for bringing the TRL higher.

In order to better assess the applicability of SW in food processing, a better understanding of the effects of different shockwave intensities, pulse counts and time between shockwaves for different muscle types (e.g. striploin, brisket), as well as the impact on bones is recommended (McDonnell et al. 2021). The acceptability of SW treated meat cuts so far was mostly investigated with trained or untrained sensory panels, but mostly the panel was already familiar with the technology to a certain extent. However, the question remains if the acceptance of these small test groups could be translated into general consumer acceptability. Besides tenderization of meat, the focus could be also on other applications, where perhaps the potential benefits are higher, e.g. cracking seafood shells. Another application is softening of plant tissues and increasing the extraction of valuable compounds of interest, thus reducing extraction time and extraction energy. Although these types of experiments are conducted in a lab-scale environment (Kuraya et al. 2017; Yasuda et al. 2017), and thus have a lower TLR, with available pilot-scale equipment it can be assumed that a higher level could be achieved with minor costs and risks.

Finally, the SW technology is a promising tool for improvement of meat tenderness, acceleration of meat ageing, or for shucking of oysters or disintegration of grains or similar plant products. However, the limited availability of the processing equipment and related challenges, limit the applications so far to pilot-scale testing at TRL 6. In order to reach a higher level, these technical challenges have to be overcome.

3.3 Technologies Based on Electromagnetic Phenomenon

In this subchapter, the TRL of technologies whose effects are based on electromagnetic phenomenon are discussed. These include pulsed electric fields, radiofrequency and magnetic fields.

3.3.1 Pulsed Electric Fields

Along with the HHP, pulsed electric fields (PEF) is another technology already used worldwide which can be assessed with a high TRL. The technology is used primarily for pre-treatment of plant materials for softening the tissue prior to cutting and facilitating further processing, for example in potato processing. Further, less numbered applications include increasing the extraction yield, or gentle preservation of liquid products, such as juices and fruit preparations. Currently, there are over 250 PEF machines in total worldwide, operating in different fields: around 110 machines operate in the production of French fries, 50 in the processing of potato and vegetable chips, 15 machines for preservation and shelf-life extension of juices, around 25 for enhancing the extraction processes. Besides these machines, another 50 are owned by different universities and research facilities for tests and feasibility studies. Based on the development and number of installations, this technology can be classified in TRL 8 or 9.

Even if this technology is well developed and researched, it is still considered emerging and its users are pioneers. Thus, certain research topics need to be addressed in order to increase the number of applications. At first, further development is seen in the improvement of treatment electrodes, in order to reduce erosion during the processing. Therefore, the experimental evaluation of different materials will be of great interest to the scientific community for the next years. Also, the process parameter and product properties could affect the erosion, and more research is needed to evaluate the connection between the process parameter and the electron eroding (Arshad et al. 2020; Pataro and Ferrari 2020).

As for the most other emerging technologies, several theories on pore creation, PEF-induced cellular interactions, mechanism of microbial inactivation, or application in (bio-)medicine, can be found in the literature, however, it seems not fully understood so far (Garner 2019; Saulis 2010; Perrier et al. 2017). Evaluation of the underlying mechanism of action at a cellular level could allow for even more targeted treatment and could help increase the number of new applications, also outside of the food industry. Especially the impact of PEF on animal-based cells, proteins or products thereof should be further elucidated. In recent years, several studies with promising results have been published, which demonstrated that PEF treatment can be used as a tool for improving the drying process of several goods, by decreasing drying time or drying intensity (Ostermeier et al. 2018; Lebovka et al. 2007; Shorstkii et al. 2020; Telfser and Galindo 2019; Wiktor et al. 2016). Also so far there are no applications in the dairy sector, for which some studies reported promising results (Buckow et al. 2014; Bendicho et al. 2002; Sharma et al. 2014), or in the processing of insects (Alles et al. 2020; Shorstkii et al. 2020), or algae (Smetana et al. 2020).

Needless to mention that during food production, sometimes significant quantities of side products are generated. Disposal of these products is sometimes associated with costs and environmental issues. Nevertheless, in certain cases, these side streams still contain valuable ingredients, which can be obtained and further utilized. The PEF was suggested as one of the technologies that can help in the extraction of valuable compounds and improved the utilization of side streams (Rocha et al. 2018).

3.3.2 Radiofrequency

The radiofrequency (RF) technology is studied for more than 60 years, where initial studies were mostly focusing on the assessment of thermal effects during the RF treatment. In the last 20 years, the research in this field intensified, focusing on the inactivation of microorganisms in culture media and different foods, as well as applications of RF for enhanced heating (Altemimi et al. 2019; Awuah et al. 2014).

In general, RF is a technology comparable to microwaves (MW), with the difference in wavelengths, where RF operates at much longer wavelengths than the MW. In the US, only frequencies of 13.56, 27.12 and 40.68 MHz are permitted for use in industrial applications (Wang et al. 2011), for these frequencies the wavelengths are 22.1, 11.1 and 7.4 m respectively. The wavelengths at the allowed working frequencies of MW (915; 2450 MHz) are 0.33 and 0.12 m (Jiao et al. 2018). The dielectric field generated at these frequencies has a higher penetration depth than the MW or conventional heating, resulting in a faster temperature increase in a more homogenous way (Altemimi et al. 2019). During the treatment, the dielectric molecules in the foods (e.g. water) rotate, while the electromagnetic field alternates. This movement generates frictions in the food, increasing the temperature in a rapid and volumetric way. Next to the faster heating, another benefit is, that construction complexity, as well as the construction costs of RF, are lower in comparison to MW (Jiao et al. 2018).

Radiofrequency electric fields (RFEF) is a technology comparable to PEF, with the difference that it applies sinusoidal waves, with a high-frequency range from a few kHz to a few hundred MHz. The most promising results were obtained in the frequency range of 15–70 kHz (Trujillo and Geveke 2014). The sinusoidal waveform is beneficial, considering that the potentially occurring electrochemical reaction could be reduced or avoided compared to monopolar pulse generation, which potentially can reduce erosion of the electrodes and lower maintenance costs.

So far, the RFEF was mostly studied as a potential technology for the inactivation of microorganisms, with less focus on quality aspects of treated food. Performed research and development in recent years led to the development of first prototypes and trials at a larger scale (Geveke and Brunkhorst 2008). Nevertheless, the RFEF technology can be considered still in an early stage and classified at a TRL 4. For an increase in TRL, it is necessary to broaden the existing knowledge on the RFEF inactivation fundamentals, and impact on food quality, especially when compared to traditionally processed and well-established foods and technologies (Masood et al. 2017).

The development of prototypes on a technical scale requires in general a significant amount of research, resources and time. Important factors for a successful upscaling include the design of the treatment chamber, design of the electrodes, mapping the flow profile, one or more treatment zones (with/without cooling between the treatment zones) and how temperature profile changes during the treatment (Masood et al. 2017; Trujillo and Geveke 2014).

Similar to other non-thermal, but also in some cases thermal technologies, not necessarily all possible side effects of the treatments have been investigated and understood so far. It is important to consider whether chemical reactions take place during the treatment, in which potentially harmful substances are created.

Another promising application is the RF supported drying or radiofrequency hot air drying (RF-HAD). The product is placed between two electrodes, which generate an alternating electromagnetic field, while hot air is flowing over the product. The term hot air is somewhat misleading in this application because usually, the air has a temperature between 40 and 60 °C (Chian et al. 2021; Xie et al. 2020; Wang et al. 2014; Zhou et al. 2018).

Worldwide only few companies offer industrial-scale RF-HAD equipment, mostly for post-backing drying to adjust the final moisture. Compared to conventional drying, the RF-HAD has the potential for reduced drying time, drying at a lower temperature, resulting in improved quality and sensory properties (McHugh 2016; Zhou et al. 2018). Until now, the RF-HAD industrial applications are very limited, mostly due to high investment costs compared to conventional drying techniques. The overall TRL for this RF application can be classified between 6 and 7. To increase the TRL and the industrial penetration for RF-HAD more research is needed with respect to broader applications, improved benefits of the technology and for reduction of the investment costs, as well as equipment development to avoid dielectric breakdowns (McHugh 2016).

Furthermore, the type of products suitable for treatment with a combination of hot air and RF should be investigated, as the results of some existing studies are sometimes contradictory (Wang et al. 2014; Naidu et al. 2016; Zhou et al. 2018). As food is a very complex and usually inhomogeneous system, the drying kinetics need to be evaluated for more different types of foods, to assess whether RF-HAD is suitable for applications other than post-baking drying. It may be beneficial to explore the combination of different non-thermal processing technologies with RF-HAD to further reduce drying time and/or heat generation, to improve product quality and increase treatment homogeneity (Zhang et al. 2017).

3.3.3 Microwave

Microwave (MW) is an established technology for heating, finding increasing application in a wide variety of domestic, commercial and industrial food processing operations, mostly applied in industrial and commercial food applications for tempering of frozen meat and poultry products, precooking of bacon for food service, to the cooking of sausages (Ahmed and Ramaswamy 2007). As this book focuses on non-thermal technologies, in the following focus will not be given to heating, but rather the increasing interest microwave-assisted freezing (MAF) and microwave-assisted drying (MAD). MAF is an interesting technology, with however low TRL, in the range between 2 and 3. In the literature, it is discussed how the MW affect the nucleation and the growth of ice crystals during the freezing process. Xanthakis et al. 2014 proposed that the water molecules rotate with the applied alternating electromagnetic field which has two main effects. First, the rotation generates a marginal heat in the product, which causes melting of the ice crystals and could induce a second nucleation which is associated with generation of more small crystals. The second effect of the rotation is related to disrupted H-bond network between the water molecules during both the nucleation and growing phase (Sadot et al. 2017). MAD is new approach for MW vacuum drying, where high energy water molecules diffuse to the surface and evaporate due to low pressure. As such, it results in improved dried product quality, by preventing oxidation due to absence of air and by reduced drying temperature (Zhang et al. 2006). TRL for this application can be considered between 7 to 8.

Few studies showed beneficial effects of applying MW during the freezing in sense of a reduction of the size of ice crystals associated with a reduction in damage to the tissue and improved product quality (Xanthakis et al. 2014; Jha et al. 2020; Sadot et al. 2020).

However, still there is a limited number of studies and lack of fundamental understanding of the responsible phenomena and more research is needed, supported by real-time temperature measurements and modelling of the wave behaviour, as well as the heat and energy distribution during the treatment. The reported sample size used in these studies were rather small (Xanthakis et al. (2014), Sadot et al. (2020)), indicating the need for further evaluation on a larger scale, in particular for freezing of whole foods. To the best of authors knowledge, so far, there are no indications of the technical and large scale systems in operation.

3.3.4 Magnetic Fields

The magnetic fields (MF) is a technology in its early development phase, and as such can be assessed at TRL 2. Only a few studies are reporting promising results for the inactivation of microorganisms, partly in food systems (Lipiec et al. 2004; Lipiec et al. 2005; Grigelmo-Miguel et al. 2011). To the best of our knowledge, the inactivation effects and kinetics are not completely elucidated so far, and there is still a need for further improvements and better understanding of the process. With good explanatory models, prototypes could be built for proof of concept, which could lead to increased maturity. According to Barbosa-Canovas et al. (2000) the following aspects should be further studied and elaborated:

- identification of the inactivation kinetics and models,
- determination and validation of critical process parameters and
- isolate major specific resistant microorganisms.

At the beginning of the twenty-first century, certain patents for MF-assisted freezers were filed (Hirasawa et al. 2001; Owada and Saito 2010). Since that time, there is a lack of evidence on the benefits of MF-assisted freezing, and in some cases the results are contradictory, indicating that more research is needed (Otero et al. 2016).

Another proposed application of MF was the solid-solid separation, without any use of chemicals, and improved separation compared for example to air separation (Maki and Hirota 2014). However, to the best of our knowledge, this is the only study reporting this application, thus more extensive research and development of prototypes is required to confirm this topic.

In conclusion, the MF could have beneficial effects for either food processing, purification, or preservation; however, the mechanisms of action and scalability are still considered under-investigated. Until then the technology remains at TRL of 2.

3.4 Radiation-Based Technologies

In this chapter, the TRL of technologies whose basic working principles are on the radiation of energy in form of electrons or light, or a combination, are discussed. These include irradiation by electrons and gamma irradiation, light-based technologies and applications of cold plasma.

3.4.1 Ionizing Radiation

Radiation is called ionizing radiation (irradiation) when it has a sufficiently high frequency so that it results in the production of charged particles or ions in the material that it comes in contact with. Irradiation energy can be generated in different ways, as gamma rays from radioisotopes of cobalt-60 and cesium-137, x-rays (using electricity) and electron beam (EB) (using electricity). Irradiation technology is investigated over a long period of time, in particular for inactivation of microorganisms, shelf-life extension, inhibiting sprouting, ripening delay or insect disinfestation (Farkas et al. 2014). Several studies have shown that irradiation has a positive effect in terms of decontamination of food, degradation of mycotoxins and reduction of allergenic potential (Khaneghah et al. 2020), although for degradation of mycotoxins could not always be confirmed (Woldemariam et al. 2021).

Besides the classification based on energy generation, it can be also divided into two groups, depending on the intensity of the treatment: low energy EB (LEEB,

≤300 keV) and high energy EB (HEEB,>300 keV), and from that resulting penetration depth. While HEEB has the necessary energy to penetrate up to few centimetres and throughout the product, the LEEB penetrates only up to a few hundred micrometres from the food surface. Comparing EB to gamma rays, the advantage of EB, is in the generation of the electron beams, where for EB no radioactive material is required. Although it has a high TRL level of 8 to 9 in many cases, current technical limitations for radiation technology (HEEB) is generally recognised in a relatively low throughput and, the need for special radiation-shielded facilities. In the case of LEEB, the equipment has a much smaller footprint and machines could be self-shielded, but this increases the weight and the production costs. Based on the technical developments and industrial tests and applications so far, the LEEB can be estimated at the TRL of 8.

The main limitation of wider commercialisation of this technology is seen in no harmonised regulation in different regions of the world, which makes a global implementation challenging. In most cases, all products treated with EB must be labelled either with the term "irradiated" or "treated with ionizing radiation" and/or accompanied with a certain logo. This label faces a problem with the retailer and consumer acceptance and is mostly related to regulation, understanding the process and with it related product benefits. Usually, consumer acceptance increases if the consumers are better informed about the benefits of the treatment, related to product safety and quality (Balatsas-Lekkas et al. 2020).

Besides the market barriers related to acceptance, EB treatment of food is limited to few product groups (e.g. spices, herbs and vegetable seasonings in the EU, Directive 1999/3/EC). In the EU an evaluation process of food irradiation directives, that may lead to a new legislative proposal, is ongoing. More extensive studies on retailer and consumer perspectives on the communication of different preservation technologies and possibly their labelling would be useful. Also, an evaluation study that would compare the effects of chemicals and EB treated foods in terms of acceptance, safety, wholesomeness, and impact on the environment etc., would be beneficial for the decision-makers.

Also, more research is needed regarding the building of radiolytic substances for different foods at different doses and whether these substances have toxicological, genotoxic, cytotoxic concentration and/or any other adverse effects. For the cytotoxic substances, it is known that irradiation could build 2-alkylcyclobutanones (2-ACBs) during the treatment of fat-rich products, but more studies are required to determine whether the concentration of these products is high enough to cause a cytotoxic effect. This could have a positive impact on the breakdown of consumer preconceptions and on the regulatory approval of new fields of applications for EB, especially in the EU (Ravindran and Jaiswal 2019).

Further improvements in the machine-building sector are required in order to allow for scaling up, which include a competitive manufacturing, as well as the establishment of stable supply chains, to be able to decrease the price of the equipment and so far high investment costs. In combination with broadening the fields of application and a better-informed consumer could help higher implementation of

EB on the market. In the course of development, the improvement of measuring devices and instruments should be improved as well. In particular, dosimetry is an important and critical aspect of this technology, should be further improved. Despite appropriate tools for ensuring traceability and different dosimeters are available, there are described issues that should be solved in the future (Kuntz and Strasser 2016). Moreover, research on novel dosimetry methods such as product coating dosimeter or in-product dosimetry should be continued.

In conclusion, EB-irradiation is a useful technology for providing food safety in a fast and energy-saving way, with minimal quality impact for selected foods. The main limitation so far is strict regulation (especially in the EU) and consumer and retail acceptance. For the new approach, LEEB, acting only on the surface of the product, scalability, operational safety, and affordability should be improved.

3.4.2 Light Technologies

Pulsed light (PL) is a so far well-developed technology that applies light pulses of very high intensity at wavelengths of the whole light spectrum, where most of the inactivation effects are based on the UV-C spectrum (Mandal et al. 2020). The FDA approved PL for decontamination of food and food packaging surface treatments (Food and Drug Administration 1996). Nevertheless, the technology is not yet widely used and most of the applications so far found are based on the approved ones - systems for decontamination of surfaces of food packaging, like bottle preforms or trays, and few for foods. Based on these facts, the PL can be estimated at a TRL in the range between 7 and 8.

The FDA also defined treatment parameters and a processing window, as a cumulative total fluence that must not exceed 12 J/cm^2 and a pulse duration of <2 ms (FDA 1996). A harmonization and uniform guidelines would be beneficial, especially having in mind studies reporting different parameters, which makes sometimes the comparison of results difficult.

Several studies reported a successful reduction of microorganisms by PL, either in culture media, liquid products or on the surface of solids (Gomez-Lopez et al. 2007; Mandal et al. 2020; Heinrich et al. 2015). However, more research is needed for a better understanding of the inactivation mechanisms with respect to different morphologies, classification (Gram + vs. Gram -) and growth phases of different microorganisms. Regarding to the impacts of the PL treatment on organoleptic properties of food, varying results can be found. In general, more research is needed for a better understanding of PL suitability for specific food types, and how specific food groups, depending on their composition (fat content, fat composition, or moisture) are affected by the treatment (Mandal et al. 2020).

Finally, to reach a higher TRL for this technology, further up-scaling and modifications of equipment are required, in particular modifications of the treatment zones for various range of products, solids, liquids or particulate food. One limitation of the PL technology is the generation of heat by the lamps during continuous production, thus, an improved cooling system for equipment cooling is necessary to avoid thermal changes. Furthermore, a longer life span of the lamps is desired and should be in the focus of development, in order to reduce the operating costs (Gomez-Lopez et al. 2007).

Ultraviolet light (UV) is a technology for decontamination of surfaces of solid foods or shelf-life extension of liquid foods. In this technology, a light source (mostly low or medium pressure mercury lamps) emits the UV light continuously at wavelengths between 140–400 nm, where the highest germicidal effect is recognised in the range between 250 and 280 nm (UV-C). UV treatment is recognised as a successful approach for shelf-life extension of various liquid products, decontamination of microorganisms on fruits, meat and seafood (Keklik et al. 2012; Koutchma 2008; Müller et al. 2011). The FDA approved UV-C treatment for the reduction of human pathogens and other microorganisms in juices. Considering that the majority of the results are generated at a lab scale, and based on few machines that exist in the industry, the current TRL for UV can be estimated in the range between 6 and 7.

In order to reach higher TRL for UV technology, the development of large-scale industrial equipment, with improved treatment homogeneity is required. Most of the reported studies are performed on custom-made and lab-scale equipment, in either batch or continuous systems, making the comparison of the results challenging. The treatment homogeneity is especially important for the treatment of particulate foods, powders, liquids, as the most challenging products for the UV light treatment. Another challenge represents the design of a continuous UV system, and characterization of the dose, as well as mixing efficiency in order to be able to assess process performance and delivery of UV homogeneously (Koutchma et al. 2016).

Another field of further development is the development of new lamps, as an alternative for the existing and mostly used mercury lamps. It was recommended that this type of alternative lamp emits poly-wavelengths within the UV-spectrum (200–280 μm), therefore different microbiological species can be inactivated. LED is a promising alternative for mercury lamps, however, they have higher costs, they are weak light emitters (especially in the needed wavelength) and they have a lower life span. By overcoming these drawbacks, the UV treatment could be established as an environmentally friendly and energy-saving solution, to provide safe food, without using chemicals and without contact with food (Koutchma et al. 2019).

The UV treatment is recognised also for helping in the degradation of undesirable substances, like mycotoxins, pesticides and process-induced contaminants (Zhu et al. 2014). Still, this application is not extensively investigated, and it is necessary to generate more knowledge on the effects of UV light on the degradation of a wide range of food contaminants. In most of the studies, very few targeted substance typically in high concentration were in focus, so that the effects of the UV treatment on multi-target contamination or the effect on lower concentrated compounds remain unclear. Next to the focus on the degradation of undesirable substances, further studies are recommendable on the toxicity of potentially produced degradation products or synthesis of new compounds. As the effect of UV treatment is limited only to the area exposed to the light, it was recommended to use it as part of a hurdle concept and in combination with other technologies

(Hernández-Carranza et al. 2021) or approaches, such as photo-active or light diffraction substance.

Besides the applications of UV for preservation, this technology is used for vitamin D enrichment in mushrooms and bread (Kalaras et al. 2012; Urbain et al. 2016). UV-treated mushrooms and bread are authorised to be placed as novel food on the EU market, with the insinuation that the products are labelled to inform consumers of the UV treatment (EC 2018). Based on the tested and validation applications and approval, the TRL of this UV application can be considered in the range between 8 and 9.

3.4.3 Cold Plasma

The cold plasma (CP) technology is based on treatment with partially or fully ionized gas composed of highly reactive species like electrons, negative ions, positive ions, free radicals, excited or non-excited atoms and photons. It usually takes place at around ambient temperature and can be applied for the inactivation of microorganisms (Whitehead 2016; Misra et al. 2011; Mandal et al. 2018). Most of the research activities so far were performed on custom-made lab-scale equipment, as a proof of concept for different applications. To the best of our knowledge, so far, no technical scale prototypes exist, classifying the CP at a TRL 4.

Several studies focused on the inactivation of different microorganisms using CP and its effects on surfaces of solid foods, but also for decontamination of liquid foods (Ziuzina et al. 2013, 2014; Lacombe et al. 2015; Hertwig et al. 2015). It was mostly described that the impact of CP on food quality strongly depends on the composition of food, along with selected processing parameters and intensity of the treatment. Therefore, optimisation of product-specific processing conditions, in particular for products challenging for other preservation approaches (e.g. nuts, herbs or spices), along with mapping the treatment intensity and homogeneity would be of great importance. The microbial inactivation data is up to this moment scarce, as well as scientific data on CP impact on food quality attributes, compositional changes, especially related to potentially harmful substances (Schlüter et al. 2013).

Since CP is a relatively new technology, there are no standardized approaches for the characterization of systems, methods and treatments, making it very difficult to compare results from different studies. Only one study demonstrated low-pressure microwave plasma as a potential technology to be upscaled for disinfection of food packaging surface (Schneider et al. 2005). Like several other non-thermal technologies, the exact mechanisms of inactivation of microorganisms are still not completely elucidated. So far there are three models discussed, namely: (1) permeabilization of the cell membrane, (2) intercellular damage of the proteins from oxygen or nitrosative species, and (3) damage of the DNA (Mai-Prochnow et al. 2014; Niemira 2012; Mandal et al. 2018). A more general understanding of the process-product interactions, in particular, any plasma-induced changes in physical, chemical, biochemical or microbiological aspects, including more precise studies on penetration depths are necessary (Schlüter et al. 2013).

Besides the inactivation of microorganisms using CP, the technology delivered promising results in the immobilisation of bioactive compounds on the surface of the material, making it a potential technology for further studies with the aim of developing new active surfaces (Mastromatteo et al. 2011; Popelka et al. 2012). CP could potentially contribute to the degradation of food undesired compounds and contaminants, like pesticides and allergens (Wu et al. 2014; Ng et al. 2021; Venkataratnam et al. 2019, 2020), however, this should be also further studied and evaluated as the current number of studies is very limited.

In summary, CP is a very promising technology for the reduction of microbial contamination on different surfaces, but there is still a limited amount of scientific data to accelerate its technological development and increase its TLR. At first, the number of studies on microbial inactivation should be improved, and quality changes of food and risk assessment should be carried out, in order to encourage the construction of larger prototypes and further technological development.

4 Conclusion

For most of the discussed technologies, there are proven concepts and defined major applications; in certain cases, there are machine manufacturers and disruptive start-ups, along with large companies, successfully using them. Even where selected non-thermal technologies exhibit high TRLs, they are still largely considered emerging and are facing implementation and commercialisation barriers. These are mostly related to throughput and cost-related barriers, legal constraints, and consumer acceptance. For the technologies with high TRL, further improvements through research and development might not be urgently necessary but are certainly highly desired, in particular for smaller and medium-sized equipment building enterprises, where their own R&D resources are rather limited. For the technologies with low TRL, most of the activities are related to improved mapping of the treatment, standardization of treatment conditions, and risk assessment for the formation of undesired substances, which would allow for the elaboration of scale-up strategies and validation of concepts on the technical scale.

In terms the legal barriers, different rules may apply for different areas. In the EU, most of the food that is produced by described technologies may be considered as a Novel food and as such subject to authorization. Whether a product is subject to Novel Food Regulation ((EC) No 258/97), strongly depends on the technology and place where it is applied, as well as on the final product itself. Thus, before placing products produced using some of these technologies on the market, the responsible local food authority must be informed, and it must be demonstrated that these products are safe for consumers. This can be a time-consuming and expensive process, especially for start-ups and small and medium-sized enterprises. Therefore, one of the tasks of the scientific community in the coming years should be to prove the safety aspects of the technologies and products produced by them in order to support the decision-makers and to eliminate any concerns regarding their usage.

Nevertheless, most of the considered technologies are rapidly maturing and continually going through improvements and optimization in their design, allowing for custom made and less expensive systems. The number of industrial machines for some of the technologies was increasing over the past few years, so it is to be expected that also the number of product applications and industrial systems will be increased in the future.

Acknowledgement The authors would like to thank the following persons and companies for their support in gathering the information and assessing the TRL of different technologies: Jasna Ivanovic and Boris Brockhaus, Uhde High Pressure Technologies (Germany), Mario González-Angulo and Carole Tonello, Hiperbaric (Spain), Stefan Toepfl, Claudia Siemer and Oleksii Parniakov, Elea (Germany), Ciara McDonnell und Aarti Tobin CSIRO (Australia); Mark Freeman Homogenising Systems Ltd. (United Kingdom); Heidi-Maria Kotilainen Bühler (Switzerland). We appreciate your help very much.

References

Aganovic K, Bindrich U, Heinz V (2018) Ultra-high pressure homogenisation process for production of reduced fat mayonnaise with similar rheological characteristics as its full fat counterpart. Innovative Food Sci Emerg Technol 45:208–214

Aganovic K, Bolumar T, Toepfl S, Heinz V (2021a) Fundamentals of shockwave processing for food. In: Knoerzer K, Muthukumarappan K (eds) Innovative food processing technologies: a comprehensive review, Elsevier, pp 395–411. ISBN 9780128157824. https://doi.org/10.1016/B978-0-12-815781-7.00023-8. (https://www.sciencedirect.com/science/article/pii/B9780128157817000238)

Aganovic K, Hertel C, Vogel RF, Johne R, Schlüter O, Schwarzenbolz U, Jäger H, Holzhauser T, Bergmair J, Roth A, Robert S, Bandick N, Kulling SE, Knorr D, Engel K-H, Heinz V (2021b) Aspects of high hydrostatic pressure food processing: perspectives on technology and food safety. Compr Rev Food Sci Food Saf 20(4):3225–3266

Ahmed J, & Ramaswamy HS (2007) Microwave pasteurization and sterilization of foods. In Handbook of food preservation (pp. 709–730). CRC Press

Alles MC, Smetana S, Parniakov O, Shorstkii I, Toepfl S, Aganovic K, Heinz V (2020) Biorefinery of insects with pulsed electric field pre-treatment. Innovative Food Sci Emerg Technol 64:102403

Altemimi A, Aziz SN, Al-Hiiphy ARS, Lakhssassi N, Watson DG, Ibrahim SA (2019) Critical review of radio-frequency (RF) heating applications in food processing. Food Qual Saf 3:81–91

Arshad RN, Abdul-Malek Z, Munir A, Buntat Z, Ahmad MH, Jusoh YM, Bekhit AE-D, Roobab U, Manzoor MF, Aadil RM (2020) Electrical systems for pulsed electric field applications in the food industry: an engineering perspective. Trends Food Sci Technol, pp 1–13

Awuah GB, Ramaswamy HS, & Tang J (Eds.). (2014) Radio-Frequency heating in food processing: Principles and applications. CRC Press

Balatsas-Lekkas A, Arvola A, Kotilainen H, Meneses N, Pennanen K (2020) Effect of labelling fresh cultivated blueberry products with information about irradiation technologies and related benefits on Finnish, German, and Spanish consumers' product acceptance. Food Control 118:107387

Banke J (2019) Technology readiness levels demystified. 2010 [Online]. https://www.nasa.gov/topics/aeronautics/features/trl_demystified.html. [Accessed [Last accessed: 08.04.2021]]

Barbosa-Canovas GV, Schaffner DW, Pierson MD, Zhang QH (2000) Oscillating magnetic fields. J Food Sci 65:86–89

Bastante CC, Cardoso LC, Serrano CM, de la Ossa EM (2017) Supercritical impregnation of food packaging films to provide antioxidant properties. J Supercrit Fluids 128:200–207

Bendicho SL, Barbosa-Cánovas GV, Martín O (2002) Milk processing by high intensity pulsed electric fields. Trends Food Sci Technol 13:195–204

Bhargava N, Mor RS, Kumar K, Sharanagat VS (2020) Advances in application of ultrasound in food processing: a review. Ultrason Sonochem 70:105293

Bolumar T, & Toepfl S (2016) Application of shockwaves for meat tenderization. In Innovative food processing technologies (pp. 231–258). Woodhead Publishing

Bolumar T, Bindrich U, Toepfl S, Toldrá F, Heinz V (2014) Effect of electrohydraulic shockwave treatment on tenderness, muscle cathepsin and peptidase activities and microstructure of beef loin steaks from Holstein young bulls. Meat Sci 98:759–765

Bolumar T, Orlien V, Sikes A, Aganovic K, Bak KH, Guyon C, Stübler AS, De Lamballerie M, Hertel C, Brüggemann DA (2020) High-pressure processing of meat: molecular impacts and industrial applications. Compr Rev Food Sci Food Saf 20(2):332–368. https://doi.org/10.1111/1541-4337.12670

Brunner G (2005) Supercritical fluids: technology and application to food processing. J Food Eng 67:21–33

Brunner G (2010) Applications of supercritical fluids. Ann Rev Chem Biomol Eng 1:321–342

Buckow R, Chandry PS, Ng SY, McAuley CM, Swanson BG (2014) Opportunities and challenges in pulsed electric field processing of dairy products. Int Dairy J 34:199–212

Chemat F, Vian MA, Fabiano-Tixier A-S, Nutrizio M, Jambrak AR, Munekata PE, Lorenzo JM, Barba FJ, Binello A, Cravotto G (2020) A review of sustainable and intensified techniques for extraction of food and natural products. Green Chem 22:2325–2353

Chian FM, Kaur L, Astruc T, Vénien A, Stübler A-S, Aganovic K, Loison O, Hodgkinson S, Boland M (2021) Shockwave processing of beef brisket in conjunction with sous vide cooking: effects on protein structural characteristics and muscle microstructure. Food Chem 343:128500

Chien S-Y, Sheen S, Sommers C, Sheen L-Y (2017) Modeling the inactivation of Escherichia coli O157: H7 and Uropathogenic E. coli in ground beef by high pressure processing and citral. Food Control 73:672–680

Dong P, Georget E, Aganovic K, Heinz V, Mathys A (2016) Inactivation of Bacillus amyloliquefaciens spores by continuous high-pressure-assisted thermal sterilization in an oil-in-water (o/w) emulsion with 10% soybean oil. Eur Food Res Technol 242:935–942

Dubos GF, Saleh JH, Braun R (2008) Technology readiness level, schedule risk, and slippage in spacecraft design. J Spacecr Rocket 45:836–842

EC (2018) Commission Implementing Regulation (EU) 2018/1011 of 17 July 2018 authorising an extension of use levels of UV-treated mushrooms as a novel food under Regulation (EU) 2015/2283 of the European Parliament and of the Council, and amending Commission Impleme. In: Commission E (ed.) Official Journal of the European Union

ESA (2019) Future launchers preparatory programme – Technology Readiness Levels [Online]. https://www.esa.int/ESA_Multimedia/Images/2019/06/Technology_Readiness_Levels: The European Space Agency (ESA) [Accessed [Last accessed: 28.04.2021]]

Farkas J, Ehlermann DAE, Mohácsi-Farkas C (2014) Food technologies: Food irradiation. In: Motarjemi Y (ed) Encyclopedia of food safety, Academic Press, pp 178–186, ISBN 9780123786135. https://doi.org/10.1016/B978-0-12-378612-8.00259-6. (https://www.sciencedirect.com/science/article/pii/B9780123786128002596)

FDA, FADA (1996) Code of Federal Regulations, 21CFR179.41. In: (FDA), FADA (ed) Title 21, Volume 3

Franco P, Incarnato L, De Marco I (2019) Supercritical CO_2 impregnation of α-tocopherol into PET/PP films for active packaging applications. J CO_2 Utilizat 34:266–273

Gallo M, Ferrara L, Naviglio D (2018) Application of ultrasound in food science and technology: a perspective. Foods 7:164

Garner AL (2019) Pulsed electric field inactivation of microorganisms: from fundamental biophysics to synergistic treatments. Appl Microbiol Biotechnol 103:7917–7929

Georget E, Miller B, Aganovic K, Callanan M, Heinz V, Mathys A (2014) Bacterial spore inactivation by ultra-high pressure homogenization. Innovative Food Sci Emerg Technol 26:116–123

Geveke DJ, Brunkhorst C (2008) Radio frequency electric fields inactivation of Escherichia coli in apple cider. J Food Eng 85:215–221

Godfrey CS (1970) Apparatus for tenderizing food. Google Patents

Gomez-Lopez VM, Ragaert P, Debevere J, Devlieghere F (2007) Pulsed light for food decontamination: a review. Trends Food Sci Technol 18:464–473

González-Angulo M, Serment-Moreno V, Queirós RP, Tonello-Samson C (2021) 1.03 – food and beverage commercial applications of high pressure processing. In: Muthukumarappan KKK (ed) Innovative food processing technologies. Elsevier, Oxford

Grigelmo-Miguel N, Soliva-Fortuny R, Barbosa-Cánovas GV, Martín-Belloso O (2011) Use of oscillating magnetic fields in food preservation. Nonthermal processing technologies for food, 45:222–243

Harvey G, Gachagan A, Mutasa T (2014) Review of high-power ultrasound-industrial applications and measurement methods. IEEE Trans Ultrason Ferroelectr Freq Control 61:481–495

Heinrich V, Zunabovic M, Bergmair J, Kneifel W, Jäger H (2015) Post-packaging application of pulsed light for microbial decontamination of solid foods: a review. Innovative Food Sci Emerg Technol 30:145–156

Hernández-Carranza P, Peralta-Pérez A, Avila-Sosa R, Ruiz-López II, Benitez-Rojas AC, Ochoa-Velasco CE (2021) Effect of ultraviolet-C light and mild thermal treatment on the storage life of orange juice. Czech J Food Sci 39:106

Hertwig C, Reineke K, Ehlbeck J, Knorr D, Schlüter O (2015) Decontamination of whole black pepper using different cold atmospheric pressure plasma applications. Food Control 55:221–229

Hirasawa K, Shu R, Goto H, Okamoto M (2001) Method for freezing and freezer using variance of magnetic field or electric field. Japan Patent, 86967

Jha PK, Chevallier S, Xanthakis E, Jury V, Le-Bail A (2020) Effect of innovative microwave assisted freezing (MAF) on the quality attributes of apples and potatoes. Food Chem 309:1–12

Jiang Q, Zhang M, Xu B (2020) Application of ultrasonic technology in postharvested fruits and vegetables storage: a review. Ultrason Sonochem 69:105261

Jiao Y, Tang J, Wang Y, Koral TL (2018) Radio-frequency applications for food processing and safety. Annu Rev Food Sci Technol 9:105–127

Kalaras MD, Beelman RB, Elias RJ (2012) Effects of postharvest pulsed UV light treatment of white button mushrooms (Agaricus bisporus) on vitamin D2 content and quality attributes. J Agric Food Chem 60:220–225

Keklik NM, Krishnamurthy K, & Demirci A (2012) Microbial decontamination of food by ultraviolet (UV) and pulsed UV light. In Microbial decontamination in the food industry (pp. 344–369). Woodhead Publishing

Khaneghah AM, Moosavi MH, Oliveira CA, Vanin F, Sant'Ana AS (2020) Electron beam irradiation to reduce the mycotoxin and microbial contaminations of cereal-based products: an overview. Food Chem Toxicol 143:111557

Komora N, Bruschi C, Ferreira V, Maciel C, Brandão TRS, Fernandes R, Saraiva JA, Castro SM, Teixeira P (2018) The protective effect of food matrices on listeria lytic bacteriophage P100 application towards high pressure processing. Food Microbiol 76:416–425

Koutchma T (2008) UV light for processing foods. Ozone Sci Eng 30:93–98

Koutchma T, Popović V, Ros-Polski V, Popielarz A (2016) Effects of ultraviolet light and high-pressure processing on quality and health-related constituents of fresh juice products. Compr Rev Food Sci Food Saf 15:844–867

Koutchma T, Popović V, & Green A (2019) Overview of Ultraviolet (UV) LEDs technology for applications in food production. In Ultraviolet LED technology for food applications (pp. 1–23). Academic Press

Kuntz F, Strasser A (2016) The specifics of dosimetry for food irradiation applications. Radiat Phys Chem 129:46–49

Kuraya E, Touyama A, Nakada S, Higa O, Itoh S (2017) Underwater shockwave pretreatment process to improve the scent of extracted citrus junos Tanaka (Yuzu) juice. Int J Food Sci 2017:2375181
Lacombe A, Niemira BA, Gurtler JB, Fan X, Sites J, Boyd G, Chen H (2015) Atmospheric cold plasma inactivation of aerobic microorganisms on blueberries and effects on quality attributes. Food Microbiol 46:479–484
Lebovka NI, Shynkaryk NV, Vorobiev E (2007) Pulsed electric field enhanced drying of potato tissue. J Food Eng 78:606–613
Levy R, Okun Z, Shpigelman A (2020) High-pressure homogenization: principles and applications beyond microbial inactivation. Food Eng Rev 13(3):1–19
Levy R, Okun Z, Davidovich-Pinhas M, Shpigelman A (2021) Utilization of high-pressure homogenization of potato protein isolate for the production of dairy-free yogurt-like fermented product. Food Hydrocoll 113:106442
Lipiec J, Janas P, Barabasz W (2004) Effect of oscillating magnetic field pulses on the survival of selected microorganisms. International Agrophysics, 18(4)
Lipiec J, Janas P, Barabasz W, Pysz M, Pisulewski P (2005) Effects of oscillating magnetic field pulses on selected oat sprouts used for food purposes. Acta Agrophysica 5:357–365
Mai-Prochnow A, Murphy AB, McLean KM, Kong MG, Ostrikov KK (2014) Atmospheric pressure plasmas: infection control and bacterial responses. Int J Antimicrob Agents 43:508–517
Maki S, Hirota N (2014) Magnetic separation technique on binary mixtures of sorbitol and sucrose. J Food Eng 120:31–36
Mandal R, Singh A, Singh AP (2018) Recent developments in cold plasma decontamination technology in the food industry. Trends Food Sci Technol 80:93–103
Mandal R, Mohammadi X, Wiktor A, Singh A, Pratap Singh A (2020) Applications of pulsed light decontamination technology in food processing: an overview. Appl Sci 10:3606
Mankins JC (2009) Technology readiness assessments: A retrospective. Acta Astronaut 65: 1216–1223
Martínez-Monteagudo SI, Yan B, Balasubramaniam VM (2017) Engineering process characterization of high-pressure homogenization—from laboratory to industrial scale. Food Eng Rev 9(3):143–169
Masood H, Razaeimotlagh A, Cullen PJ, Trujillo FJ (2017) Numerical and experimental studies on a novel Steinmetz treatment chamber for inactivation of Escherichia coli by radio frequency electric fields. Innovative Food Sci Emerg Technol 41:337–347
Mastromatteo M, Lecce L, De Vietro N, Favia P, Del Nobile MA (2011) Plasma deposition processes from acrylic/methane on natural fibres to control the kinetic release of lysozyme from PVOH monolayer film. J Food Eng 104:373–379
McDonnell CK, Fitzgerald AG, Burt P, Hughes J, Mellor GE, Barlow RS, Sikes AL, Li Y, Tobin AB (2021) The effect of electro-hydrodynamic shockwaves on the quality of striploin and brisket beef muscles during long-term storage. Innovative Food Sci Emerg Technol 68:102627
McHugh T (2016) Radio frequency processing of food. INST Food Technologists 525 West Van Buren, STE 1000, Chicago, IL 60607-3814 USA
Meier SR (2008) Best project management and systems engineering practices in the preacquisition phase for federal intelligence and defense agencies. Proj Manag J 39:59–71
Misra N, Tiwari B, Raghavarao K, Cullen P (2011) Nonthermal plasma inactivation of food-borne pathogens. Food Eng Rev 3:159–170
Moorhouse DJ (2002) Detailed definitions and guidance for application of technology readiness levels. J Aircr 39:190–192
Müller A, Stahl MR, Graef V, Franz CMAP, Huch M (2011) UV-C treatment of juices to inactivate microorganisms using dean vortex technology. J Food Eng 107:268–275
Naidu MM, Vedashree M, Satapathy P, Khanum H, Ramsamy R, Hebbar HU (2016) Effect of drying methods on the quality characteristics of dill (Anethum graveolens) greens. Food Chem 192:849–856

Ng SW, Lu P, Rulikowska A, Boehm D, O'Neill G, Bourke P (2021) The effect of atmospheric cold plasma treatment on the antigenic properties of bovine milk casein and whey proteins. Food Chem 342:128283

Niemira BA (2012) Cold plasma decontamination of foods. Annu Rev Food Sci Technol 3:125–142

Olechowski A, Eppinger SD, Joglekar N (2015) Technology readiness levels at 40: a study of state-of-the-art use, challenges, and opportunities. 2015 Portland international conference on management of engineering and technology (PICMET), 2015. IEEE, pp 2084–2094

Olechowski AL, Eppinger SD, Joglekar N, Tomaschek K (2020) Technology readiness levels: shortcomings and improvement opportunities. Syst Eng 23:395–408

Olsen K, & Orlien V (2016) High-pressure processing for modification of food biopolymers. In Innovative food processing technologies (pp. 291–313). Woodhead Publishing

Ostermeier R, Giersemehl P, Siemer C, Töpfl S, Jäger H (2018) Influence of pulsed electric field (PEF) pre-treatment on the convective drying kinetics of onions. J Food Eng 237:110–117

Otero L, Rodríguez AC, Pérez-Mateos M, Sanz PD (2016) Effects of magnetic fields on freezing: application to biological products. Compr Rev Food Sci Food Saf 15:646–667

Owada N, Saito S (2010) Quick freezing apparatus and quick freezing method. Google Patents

Pataro G, & Ferrari G (2020) Limitations of pulsed electric field utilization in food industry. In Pulsed electric fields to obtain healthier and sustainable food for tomorrow (pp. 283–310). Academic Press

Perrier DL, Rems L, Boukany PE (2017) Lipid vesicles in pulsed electric fields: fundamental principles of the membrane response and its biomedical applications. Adv Colloid Interf Sci 249:248–271

Poliseli-Scopel FH, Hernández-Herrero M, Guamis B, Ferragut V (2013) Characteristics of soymilk pasteurized by ultra high pressure homogenization (UHPH). Innovative Food Sci Emerg Technol 20:73–80

Popelka A, Novák I, Lehocký M, Chodák I, Sedliačik J, Gajtanska M, Sedliačiková M, Vesel A, Junkar I, Kleinová A (2012) Anti-bacterial treatment of polyethylene by cold plasma for medical purposes. Molecules 17:762–785

Ravindran R, Jaiswal AK (2019) Wholesomeness and safety aspects of irradiated foods. Food Chem 285:363–368

Rocha CMR, Genisheva Z, Ferreira-Santos P, Rodrigues R, Vicente AA, Teixeira JA, Pereira RN (2018) Electric field-based technologies for valorization of bioresources. Bioresour Technol 254:325–339

Sadot M, Curet S, Rouaud O, Le-Bail A, Havet M (2017) Modélisation numérique d'un processus innovant de congélation assistée par micro-ondes. Int J Refrig 80:66–76

Sadot M, Curet S, Chevallier S, Le-Bail A, Ruaud O, Havet M (2020) Microwave assisted freezing part 2: impact of microwave energy and duty cycle on ice crystal size distribution. Innov Food Sci Emerg Technol 62:102359

Saqib NU, Sharma HB, Baroutian S, Dubey B, Sarmah AK (2019) Valorisation of food waste via hydrothermal carbonisation and techno-economic feasibility assessment. Sci Total Environ 690:261–276

Saulis G (2010) Electroporation of cell membranes: the fundamental effects of pulsed electric fields in food processing. Food Eng Rev 2:52–73

Sauser B, Verma D, Ramirez-Marquez J, Gove R (2006) From TRL to SRL: the concept of systems readiness levels. Conference on Systems Engineering Research, Los Angeles, CA, pp 1–10

Schlüter O, Ehlbeck J, Hertel C, Habermeyer M, Roth A, Engel KH, Holzhauser T, Knorr D, Eisenbrand G (2013) Opinion on the use of plasma processes for treatment of foods. Mol Nutr Food Res 57:920–927

Schneider J, Baumgärtner KM, Feichtinger J, Krüger J, Muranyi P, Schulz A, Walker M, Wunderlich J, Schumacher U (2005) Investigation of the practicability of low-pressure microwave plasmas in the sterilisation of food packaging materials at industrial level. Surf Coat Technol 200:962–966

Sharma P, Oey I, Everett DW (2014) Effect of pulsed electric field processing on the functional properties of bovine milk. Trends Food Sci Technol 35:87–101
ShockMeat (2016) Development of shock wave technology for packed meat. Grant agreement ID: 287034. https://cordis.europa.eu/project/id/287034/reporting. Last accessed 11.05.2020: Coordinated by: DIL DEUTSCHES INSTITUT FUR LEBENSMITTELTECHNIK EV
Shorstkii I, Comiotto Alles M, Parniakov O, Smetana S, Aganovic K, Sosnin M, Toepfl S, Heinz V (2020) Optimization of pulsed electric field assisted drying process of black soldier fly (Hermetia illucens) larvae. Dry Technol:1–9
Simjian LG (1958) Method of tenderizing food. Google Patents
Singla M, Sit N (2021) Application of ultrasound in combination with other technologies in food processing: a review. Ultrason Sonochem 73:105506
Smetana S, Mhemdi H, Mezdour S, & Heinz V (2020) Pulsed electric field–treated insects and algae as future food ingredients. In Pulsed Electric Fields to Obtain Healthier and Sustainable Food for Tomorrow (pp. 247–266). Academic Press
Soria AC, Villamiel M (2010) Effect of ultrasound on the technological properties and bioactivity of food: a review. Trends Food Sci Technol 21:323–331
Steiner J, Franke K, Kießling M, Fischer S, Töpfl S, Heinz V, Becker T (2018) Influence of hydrothermal treatment on the structural modification of spent grain specific carbohydrates and the formation of degradation products using model compounds. Carbohydr Polym 184:315–322
Taha A, Ahmed E, Ismaiel A, Ashokkumar M, Xu X, Pan S, Hu H (2020) Ultrasonic emulsification: an overview on the preparation of different emulsifiers-stabilized emulsions. Trends Food Sci Technol 105. https://doi.org/10.1016/j.tifs.2020.09.024
Telfser A, Galindo FG (2019) Effect of reversible permeabilization in combination with different drying methods on the structure and sensorial quality of dried basil (Ocimum basilicum L.) leaves. LWT 99:148–155
Tonello-Samson C, Queirós RP, & González-Angulo M (2020) Advances in high-pressure processing in-pack and in-bulk commercial equipment. In Present and future of high pressure processing (pp. 297–316). Elsevier
Trujillo FJ, Geveke DJ (2014) Chapter 14 – nonthermal processing by radio frequency electric fields. In: Sun D-W (ed) Emerging technologies for food processing, 2nd edn. Academic, San Diego
Urbain P, Valverde J, Jakobsen J (2016) Impact on vitamin D 2, vitamin D 4 and agaritine in Agaricus bisporus mushrooms after artificial and natural solar UV light exposure. Plant Foods Hum Nutr 71:314–321
Venkataratnam H, Sarangapani C, Cahill O, Ryan CB (2019) Effect of cold plasma treatment on the antigenicity of peanut allergen Ara h 1. Innovative Food Sci Emerg Technol 52:368–375
Venkataratnam H, Cahill O, Sarangapani C, Cullen PJ, Barry-Ryan C (2020) Impact of cold plasma processing on major peanut allergens. Sci Rep 10:17038
Wang Y, Li Y, Wang S, Zhang L, Gao M, Tang J (2011) Review of dielectric drying of foods and agricultural products. Int J Agric Biol Eng 4:1–19
Wang Y, Zhang L, Johnson J, Gao M, Tang J, Powers JR, Wang S (2014) Developing hot air-assisted radio frequency drying for in-shell macadamia nuts. Food Bioprocess Technol 7:278–288
Whitehead JC (2016) Chapter 3 – the chemistry of cold plasma. In: Misra NN, Schlüter O, Cullen PJ (eds) Cold plasma in food and agriculture. Academic, San Diego
Wiktor A, Nowacka M, Dadan M, Rybak K, Lojkowski W, Chudoba T, Witrowa-Rajchert D (2016) The effect of pulsed electric field on drying kinetics, color, and microstructure of carrot. Dry Technol 34:1286–1296
Woldemariam HW, Kießling M, Emire SA, Teshome PG, Töpfl S, Aganovic K (2021) Influence of electron beam treatment on naturally contaminated red pepper (Capsicum annuum L.) powder: kinetics of microbial inactivation and physicochemical quality changes. Innovative Food Sci Emerg Technol 67:102588
Wu Y, Liang Y, Wei K, Li W, Yao M, Zhang J (2014) Rapid allergen inactivation using atmospheric pressure cold plasma. Environ Sci Technol 48:2901–2909

Xanthakis E, Le-Bail A, Ramaswamy H (2014) Development of an innovative microwave assisted food freezing process. Innov Food Sci Emerg Technol 26:176–181
Xie Y, Zhang Y, Xie Y, Li X, Liu Y, Gao Z (2020) Radio frequency treatment accelerates drying rates and improves vigor of corn seeds. Food Chem 319:126597
Yasuda A, Kuraya E, Touyama A, Higa O, Hokamoto K, Itoh S (2017) Underwater shockwave pretreatment process for improving carotenoid content and yield of extracted carrot (Daucus carota L.) juice. J Food Eng 211:15–21
Yin J, Wang K, Yang Y, Shen D, Wang M, Mo H (2014) Improving production of volatile fatty acids from food waste fermentation by hydrothermal pretreatment. Bioresour Technol 171:323–329
Zhang M, Chen H, Mujumdar AS, Tang J, Miao S, Wang Y (2017) Recent developments in high-quality drying of vegetables, fruits, and aquatic products. Crit Rev Food Sci Nutr 57:1239–1255
Zhou X, Xu R, Zhang B, Pei S, Liu Q, Ramaswamy HS, Wang S (2018) Radio frequency-vacuum drying of kiwifruits: kinetics, uniformity, and product quality. Food Bioprocess Technol 11:2094–2109
Zhou J, Gullón B, Wang M, Gullón P, Lorenzo JM, Barba FJ (2021) The application of supercritical fluids technology to recover healthy valuable compounds from marine and agricultural food processing by-products: a review. PRO 9:357
Zhu Y, Koutchma T, Warriner K, Zhou T (2014) Reduction of patulin in apple juice products by UV light of different wavelengths in the UVC range. J Food Prot 77:963–971
Ziero HDD, Buller LS, Mudhoo A, Ampese LC, Mussatto SI, Carneiro TF (2020) An overview of subcritical and supercritical water treatment of different biomasses for protein and amino acids production and recovery. J Environ Chem Eng 8:104406
Ziuzina D, Patil S, Cullen P, Keener K, Bourke P (2013) Atmospheric cold plasma inactivation of E scherichia coli in liquid media inside a sealed package. J Appl Microbiol 114:778–787
Ziuzina D, Patil S, Cullen PJ, Keener K, Bourke P (2014) Atmospheric cold plasma inactivation of Escherichia coli, salmonella enterica serovar typhimurium and listeria monocytogenes inoculated on fresh produce. Food Microbiol 42:109–116

Chapter 16
Industry Implementation (Scale-Up): Clients' Experience Towards Understanding of How Regulations Are Affecting Novel Product Development

Yuthana Phimolsiripol, Noppol Leksawasdi, and Choncharoen Sawangrat

1 Regulation on Novel Food

The current "Thailand 4.0" policy focuses on driving the economy through innovation, transforming the production of everyday mundane products to innovative products, changing the conventional industrial sector to be led by novel and creative technology. In addition, it is aimed to switch from service-oriented towards product manufacturing with food industry as one of the five target industries. Food and Drug Administration (FDA), the main food safety regulator under the Food Act, has revised its missions and strategies in responding to this policy. The emphasis is made on adjusting the rules and regulations to satisfy demands by the business sector according to the risks involved with quick response time and competitiveness. Along with the establishment of a reliable monitoring system which can respond quickly to identify unsafe products and thus building consumers' confidence (FDA 2020).

However, the risks of food consumption can be altered according to consumer behavior. Most consumers have now become more aware of health care which influencing their healthy food choices. Industries have also aimed to develop new/novel food ingredients and food products. Several new food ingredients and novel food processing technologies have been validated and applied in the attempt to satisfy insatiable marketing demands. These include providing relatively high nutritional value food products in a convenient and easy-to-buy form as adaptation to the changing lifestyles of society such as dietary supplements. Several food companies have taken further steps by taking the mixtures or compounds without prior

Y. Phimolsiripol (✉) · N. Leksawasdi
Faculty of Agro-Industry, Chiang Mai University, Chiang Mai, Thailand
e-mail: yuthana.p@cmu.ac.th

C. Sawangrat
Faculty of Engineering, Chiang Mai University, Chiang Mai, Thailand

A. Režek Jambrak (ed.), *Nonthermal Processing in Agri-Food-Bio Sciences*, Food Engineering Series, https://doi.org/10.1007/978-3-030-92415-7_16

historical record of being consumed as food. These compounds may be herbs used for therapeutic purposes or extracts from plants that are previously restrict for animal feed. The addition of such ingredients to food or selling them as food accompanying the development of new processes or production technologies, may result in unsafe food products for consumers. The FDA has therefore issued a notification of the Ministry of Public Health (No. 376) B.E. 2559 (A.D. 2016) regarding the new food (novel food), effective from July 16, 2016 to protect the health of consumers. Many oversea governing bodies have also already established novel food ingredients, such as Regulation (EC) No. 258/97 of the European Parliament and of the Council of 27 January 1997 concerning novel foods and novel food ingredients of the European Union (European Commission, 1997). In addition, Food Standards Code 1.5.1 Novel Foods of Australia, New Zealand (Food Standards Australia New Zealand, 2021) and the Generally Recognized As Safe (GRAS) Substance Regulations under Code of Federal Regulations Title 21 of the United States have been announced (U.S. Food and Drug Administration, 2020).

The requirements of these governing bodies including Thailand based on a consistent principle that categorized the novel food into the defined definition. It is necessary to pass a safety assessment and obtain permission from the responsible government bodies first. Then, the permits for production or import for sale can be obtained. The safety assessment is based on reliable academic evidence and followed internationally recognized auditing processes. Therefore, in order to bridge understanding and awareness of all involved parties including manufacturers, importers and consumers, the Ministry of Public Health Announcement of "novel food" definition has thus been announced to the public in response to the innovative economic development of Thailand according to the Thailand 4.0 policy.

2 What Is New/Novel Food?

According to the announcement of the Ministry of Public Health, the definition of new/novel food are as following:

1. Food substances or ingredients that have been shown by academic evidence that, historically, they have been consumed for less than 15 years,
2. Materials used as food or as an ingredient of food obtained from the production process that is not typical of that food. In addition, food components, food structure or the pattern of that food may have been changed significantly resulting in alteration of the nutritional value, chemical processes within the body of living organisms or their metabolism or level of undesirable substances, or
3. Food products containing substance from (1) or (2). This does not include food additives and food derived from genetic modification techniques.

From the above definition, the "new/novel food" may stem from two important parts, namely, dietary history and uncommon production process. Dietary history referred to the way that particular food has been consumed based on accepted historical academic data. For example, the root portion of ginseng is normally consumed as food. If the other parts such as ginseng leaves are consumed with the historical academic data for less than 15 years, the leaves of ginseng are considered as new/novel food. The second part is production process that is considered unusual. This implies to any manufacturing process that may result in composition change of the food, food structure, or food patterns that may affect the nutritional value or chemical processes in the body after consumption. The process may also influence the level of unwanted substances such as environmental contaminants, mycotoxin, allergens, natural toxins, nutrient inhibitors, and harmful microorganisms, etc. in comparison with the conventional processes. For instance, manufacturing process based on nanotechnology that generates smaller particles of food components than the traditional production methods or non-thermal food pasteurization process.

3 Clients' Experience Towards Understanding of How Regulations Are Affecting Novel Product Development

Manufacturers or importers who are going to produce or import novel food for sales must be concerned with the aforementioned regulations. The food producers must pass a safety assessment first then all documents and evidence relating to the new food must be submitted as specified in the attached list of the announcement such as quality or standard (specification) information on historical consumption of that food, results of production process analysis, recommended intake method or dietary advice and safety information. This also includes toxicology evidence in laboratory animals or humans, nutrition information, report of safety review from a relevant international agency or granted permit by safety assessment unit recognized by the FDA.

The food operator must deliver safety assessment results and related evidence to the FDA for approval and guidance for the use of new, safe and appropriate food. There are currently three agencies that can provide novel food safety assessment services which are recognized by the FDA, namely, (1) the Food Quality and Safety Bureau, Department of Medical Sciences, Ministry of Public Health, (2) Food Institute, Ministry of Industry, and (3) Thailand Risk Assessment Center Institute of Nutrition, Mahidol University. After passing the safety assessment, the results of the assessment can be submitted along with the official permission request to produce or import new food. The legal compliance of relevant standards of quality control must be followed and other requirements must also be met such as

1. Good production process control (GMP)
2. Contaminant control and pathogenic microorganisms
3. Use of food additives and containers, labeling, in compliance with the notification of the Ministry of Public Health Re: Labeling of Food in Containers
4. The expiration date must be printed in the format of date, month, year in this specific order and labeled "manufactured" or "expired"

The critical part is the labeling display on the product which indicates (1) the name of the substance (if any), for example. "Cactus/*Caralluma fimbriata* extract 100% contains important substances *Pregnane glycosides* and *Saponin glycosides*" and (2) method of consumption, methods of use, or conditions of use according to the safety assessment results, for example, "Oligonol extracted from lychee fruit and green tea leaves can be used in dietary supplements in the amount not exceeding 200 mg per day". The novel food regulatory system is presented in Fig. 16.1.

However, this announcement of the Ministry of Public Health does not apply to (1) new food produced for export, or (2) new food that manufacturers or importers have been allowed to produce or import prior this announcement comes into effect. Novel food can be used as food in general if it passes a safety assessment by supporting academic documents confirming that innovation or technology applied in food production or components of that food still remain safe for consumers. One case study of novel food products or ingredients is longan syrup. The production process of this product includes simple and conventional processing such as grinding, precipitation, filtration and evaporation. However, the longan syrup can be considered as novel food due to the active ingredients being released from seed and peel of longan fruit during production process (Leksawasdi et al. 2021). Therefore, further perusal of longan syrup safety for consumption by Chiranthanut et al. (2020) on acute and chronic oral toxicity assessment of longan sugar extracts derived from whole fruit and from fruit pulp in rats were carried out. This study revealed that whole fruit longan did not cause any toxic effect in rats for consideration of the FDA approval.

For non-thermal processing in Thailand, the high-pressure processing (HPP) technique has already been approved for industrial commercialization and implementation. Other processes such as plasma and pulsed electric field are still in progress of consideration. Radio-frequency process is allowed to be applied in rice industry processing for removing insects. Therefore, the industrial manufacturers who utilize HPP have to be concerned with FDA guidelines (FDA 2020). General criteria and product scope for HPP are as following:

1. General criteria

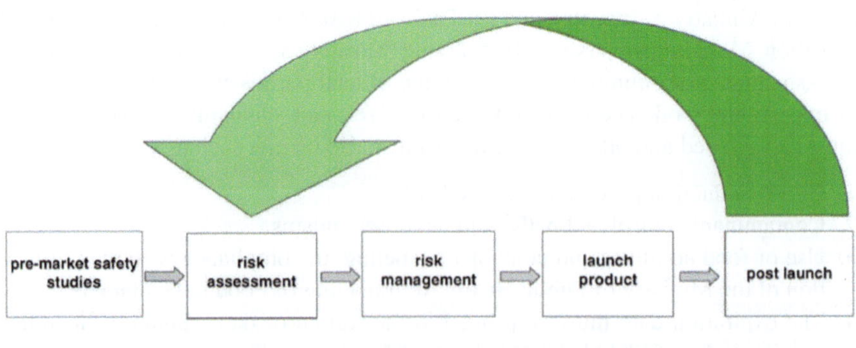

Fig. 16.1 Novel food regulatory system. (Source: Adapted from Hepburn et al. (2008))

Implementation of HPP technology on pasteurization process instead of heat or used in conjunction with heat can be divided into two pH criterion subgroups including the acidic group (pH ≤ 4.6) and the low acidic group (pH > 4.6) as shown in Table 16.1.

2. Scope of food products processed by HPP pasteurization in accordance with established general criteria

This describes the product group relevant to HPP for ready-to-eat liquid products and concentrates including (1) Drink in a sealed container according to the announcement of the Ministry of Public Health: Drinks in a sealed container, (2) Tea according to the Notification of the Ministry of Public Health: Tea, (3) Coffee according to the Notification of the Ministry of Public Health: Coffee, (4) Soy milk

Table 16.1 General criteria for HPP categorized by pH criterion subgroups

Product group	Parameter	Conditions
Ready-to-eat liquid products with pH ≤ 4.6	Storage conditions before pasteurization	Initial microbial load does not exceed 10^6 CFU/g and must be stored at <10 °C
	Pasteurization conditions[a] Pressure for pasteurization Holding time	≥400 MPa 1–20 min and must have scientific evidence indicating mitigation of the pathogenic microorganisms[b] by more than 5 log CFU
	Storage conditions after pasteurization	Lower than 5 °C
	Shelf life	Results of product's shelf life study
Ready-to-eat liquid products with pH > 4.6	Storage conditions before pasteurization	Initial microbial load does not exceed 10^6 CFU/g and must be stored at <10 °C
	Pasteurization conditions[a] Pressure for pasteurization Holding time	≥600 MPa 3–20 min and must have scientific evidence indicating mitigation of the pathogenic microorganisms[b] by more than 5 log CFU
	Storage conditions after pasteurization	≤5 °C
	Shelf life	Results of product's shelf life study (must include detection of *Clostridium botulinum*)

Source: FDA (2020)

Initial microbial load at storage conditions before pasteurization is the number of Total plate count (TPC) of raw materials which have been prepared before the pasteurization process. The unit of pressure 1 MPa is approximated to 10 Bar

[a]The pasteurization condition is a guideline. Inoculated pack / challenge study results are required study, showing the reduction of the reference pathogenic microorganism by at least 5 log CFU/mL (5 log reduction)

[b]Reference pathogens are (1) *Listeria monocytogenes* and (2) *Escherichia coli* O157: H7. In this way, microbial agents (surrogate organism) can be used in the test which has the ability to resist pressurized conditions equal to or greater than those of the underlying pathogenic microorganisms, such as *L. innocua* for *L. monocytogenes*, non-pathogenic microorganisms *E. coli* (e.g. *E. coli* K12) for *E. coli* O157: H7

according to the announcement of the Ministry of Public Health: Soy milk. For inoculated pack or challenge study, FDA by the food board committee provides the guidelines for implementing HPP on food industry as follows:

1. Types of studies: for assessing the capability of the HPP pasteurization process, a pathogen inactivation study in conjunction with a shelf life study or combined growth and inactivation study must be used.
    - Pathogen Inactivation study demonstrates the ability of HPP pasteurization process at a predetermined pressure and duration which can destroy pathogenic microorganisms in the food product by more than 5 log CFU.
    - Combined growth and inactivation studies examine both the capability of the pasteurization process with HPP to destroy the underlying pathogenic microorganisms by more than 5 log CFU and assess the change in the number of microorganisms after the pasteurization process by HPP throughout the storage period at the desired temperature and time of study. An educational institution or specialized service organization must perform the said study to ensure the precision and accuracy of the results.
2. Experimental design
    - Type and number of strains: studies should consider strains diversity for each pathogen. In growth and survival studies, more than one reference pathogen for each type (strain) must be used. Strains of reference pathogens used in the study must be a pure culture with evidence confirming the strain and purity. They must be compliance to the requirements of the Act Pathogens and Animal Toxins and then cultured in the culture medium to increase the microbial load at the stationary phase before testing.
    - Sample preparation: The test samples must be representatives the product being sold to the consumers. The standard quality measurements (pH, a_w, % total solids) of the test samples must be within the acceptable ranges. In the case of multiple products or formulas, manufacturers can choose any formula as sample for testing. However, the formula with the highest pH must be selected for the test. The specimen must be prepared for at least 2 repetitions per test condition (pressure and time) against the underlying pathogenic microorganisms. Sampling at least 5 samples for each iteration must be used to support product variability.
    - Inoculum level: Inoculation of each reference pathogen must have relatively high levels of vegetative cells of approximately 6–7 log CFU/mL.
    - Inactivation level: for both inactivation study as well as combined growth and inactivation studies, the amount of pathogenic microorganisms or agent microorganisms included in the tested food samples must be mitigated by at least 5 log CFU/mL to ensure that the pasteurization conditions (pressure and duration) are appropriate and sufficient to ensure safety of the end product. After the pasteurization process for at least 24 h, the samples will be collected at conditions of 7 ± 2 °C for analysis of reference pathogens. The control group must be done in conjunction with the test group. It is a food sample of

the same type or formula as the experimental group with added reference pathogens to have a high initial pathogenic microorganism (6–7 log CFU/mL) without passing through pasteurization process.

- Shelf life study: this can be performed by a specialist from an educational institution or staff with expertise in the factory. The test sample must represent the product being sold or representatives of product groups that are likely to pose risks related to microbial growth without the need to add any underlying pathogenic microorganisms. The sample will be processed by HPP at the same conditions as the inactivation study test, it is then stored in a cool state not higher than 5 °C. The random sample will be sampling for at least 5 periods, covering 60 days from the date of production or the expiration date. In order to ensure the safety of storage, it is recommended to test the retention period at 25% longer than the specified expiration date, for example, if the shelf life is 100 days, the test period shall be about 125 days.

3. Inoculated pack/challenge study results

Reports of studies for pathogen inactivation study, shelf life study as well as combined growth and inactivation studies must be elaborated. The quality and standard of the products used for testing protocol, study design, study results and conclusions are then submitted to the FDA for approval in extending the permit.

In conclusion, the importers and food producers must ensure that their new / novel food products followed the regulations stipulated by Thai FDA in Food Act. Novel food with short dietary history (less than 15 years) and uncommon production process are products of particular concern. HPP is one non-thermal processing technology that has been approved by Thai FDA for industrial commercialization and implementation. The parameters classified by pH subgroups to be used as safety criteria include storage and pasteurization conditions as well as shelf life.

Acknowledgement Authors would like to thank to all participated of the interview from the Thai FDA.

Conflict of Interest Authors declare that they have no conflict of interest.

References

Chiranthanut N, Teekachunhatean S, Panthong A, Lertprasertsuke N (2020) Acute and chronic oral toxicity assessment of longan sugar extracts derived from whole fruit and from fruit pulp in rats. J Ethnopharmacol 263:113184

European Commission (1997) Regulation (EC) No. 258/97 of the European Parliament and of the Council of 27 January 1997 concerning novel foods and novel food ingredients. Off J Eur Communities 43(40):1–7

Food and Drug Administration, Thailand (2020) Law and regulations. [Internet]. Retrieved 5 Mar 2021, from: https://www.fda.moph.go.th/sites/FDA_EN/SitePages/Drug.aspx?IDitem=LawsAndRegulations

Food Standards Australia New Zealand (2021) Regulation of novel foods [Internet]. Retrieved 5 Mar 2021, from: https://www.foodstandards.gov.au/industry/novel/Pages/default.aspx

Hepburn P, Howlett J, Boeing H, Cockburn A, Constable A, Davi A, de Jong N, Moseley B, Oberdörfer R, Robertson C, Wal JM, Samuels F (2008) The application of post-market monitoring to novel foods. Food Chem Toxicol 46:9–33

Leksawasdi N, Porninta K, Khemacheewakul J, Techapun C, Phimolsiripol Y, Nunta R, Trinh NTN, Reungsang A (2021) Longan syrup and related products: processing technology and new product developments. In: Xiao G, Xu Y, Yu Y (eds) Asian berries: health benefits. CRC Press, London, pp 123–148

U.S. Food and Drug Administration (2020) CFR – Code of Federal Regulations Title 21 [Internet]. Retrieved 5 Mar 2021, from: https://www.accessdata.fda.gov/scripts/cdrh/cfdocs/cfcfr/CFRSearch.cfm?CFRPart=182&showFR=1

Chapter 17
Supercritical Fluids as a Tool for Sustainable Manufacturing of Added Value Products

Maša Knez Marevci, Darija Cör, and Željko Knez

1 Introduction

1.1 Development of High Pressure Processes

Application of high pressure in "food sector" was "digester" from Denis Papin in 1680. He made experiments in high pressure vessel to prove that the boiling point of water could be increased by elevation of pressure. The practical application was cooking meals for King Charles II at elevated pressure (Reverchon 2002).

Several hundred years later high-pressure steam engines were invented by James Watt round 1785 (Reverchon 2002). In 1822 baron Cagniard de la Tour published the first manuscript on the observation of the critical point (Segura-Campos et al. 2014). The thermodynamic fundamentals on P-V-T behavior of substances reported in year 1869 by Thomas Andrews (Jessop and Leitner 2008). Hannay and Hogarth did in years 1879 and 1880 several studies on the solubility of substances in supercritical fluids (Jessop and Leitner 2008; Knez Hrnčič et al. 2018).

One of the most important invention for the application of high pressure in industry was synthesis of ammonia by Haber and Bosch. For this invention they received Nobel prize in 1918. This application of high pressure is an important milestone for several processes which results in industry today. One of the most sophisticated is process for synthesis of polyethylene where pressures are between 2500 and 3000 bar and temperatures round 300 °C.

M. Knez Marevci · D. Cör
Faculty of Chemistry and Chemical Engineering, University of Maribor, Maribor, Slovenia

Ž. Knez (✉)
Faculty of Chemistry and Chemical Engineering, University of Maribor, Maribor, Slovenia

Faculty of Medicine, University of Maribor, Maribor, Slovenia
e-mail: zeljko.knez@um.si

A. Režek Jambrak (ed.), *Nonthermal Processing in Agri-Food-Bio Sciences*, Food Engineering Series, https://doi.org/10.1007/978-3-030-92415-7_17

In high pressure pasteurization of food stuff static pressures up to 8000 bars are used. Today high pressure is applied in several industrial sectors. The highest pressures are used for production of diamonds from graphite at pressures of 120.000 bar at 3000 °C.

Terminology of high pressure of industrial processes is presented in Table 17.1.

The definition of supercritical fluids is defined by IUPAC. In literature (Jessop and Leitner 2008) more precise definition is given where "SCF is defined as a state of a compound, mixture or element above its critical pressure (*P*c) and critical temperature (*T*c) but below the pressure required to condensate into solid" (Fig. 17.1). Considering the fact that supercritical fluid (SCF) solvents are intermediates between liquid and gases, by increasing the density of fluid often an increased solubility is achieved. Viscosity, which is similar to viscosity of gases, enables better transport properties. The main property of SCFs is the possibility to change the characteristics of solvent near the critical point dramatically. Also, selectivity of a solvent is an important characteristic. It varies by changing pressure and/or temperature.

The advantages on use of high-pressure processes especially the use of supercritical fluids have numerous advantages, but the main are environmental, health and safety, process and chemical benefits. Nowadays, there is a trend to develop alternative technologies with minimal environmental impact for products with special custom designed properties. Reduced energy consumption, less toxic residues, efficient conversion of reactants to products, less byproducts and higher quality and safety of final products are crucial requirements for the future processes. High pressure technologies are a relatively new tool to satisfy the mentioned demands. Technologies using high pressures developed several processes which resulted in completely new products with special characteristics (Aymonier et al. 2011; Reverchon 2002).

Using supercritical fluids (SCF) as solvents in chemical processes means an advantage in many points of view. It has health, safety, environmental and also chemical benefits. Commonly SCFs are termed as green solvents. Using high pressure as a processing tool surpasses legal limitations for solvent residues and restrictions on use of conventional solvents in chemical processes. Additionally, particulate products can be also achieved by means of SCF.

Supercritical fluids are already applied in several processes developed to the commercial scale; – from pharmacy, food sciences to the textile industry. Most of

Table 17.1 Terminology of high pressure of industrial processes

Pressure (bar)		
From	To	Term
1	20	Low pressure
20	100	Medium pressure
100	1000	High pressure
>1000	–	Ultra-high pressure

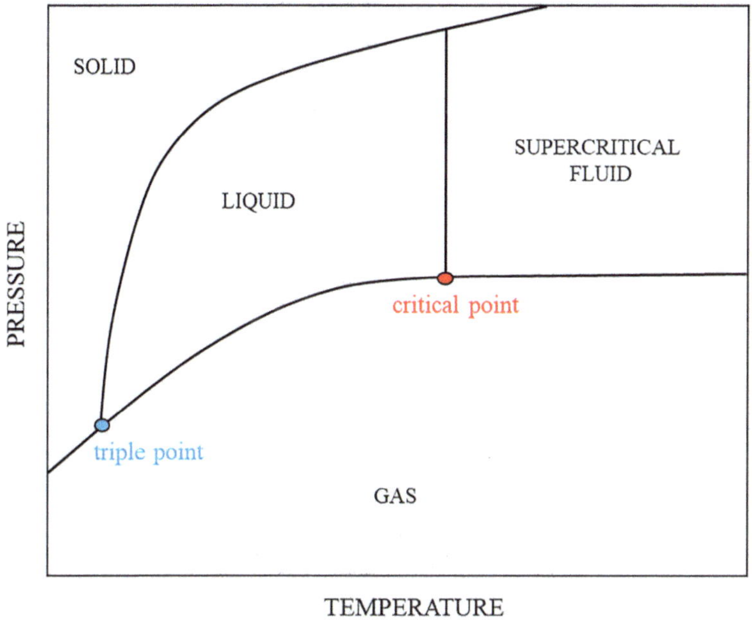

Fig. 17.1 Phase diagram of CO_2

them are available in a relatively pure grade at reasonable costs as compared with the industrial grade liquid solvents. Material processing using sub- and supercritical fluids including particle formation techniques and formulation of materials with special characteristics is today an important field of research. The highest capacities are installed for coffee and tea decaffeination. The second largest application is for the extraction of hop (Knez Hrnčič et al. 2018). The extraction of spices for production of oleoresins and the extraction of bioactive material from plants are two very diverse applications of SC for extraction. Dense gases can also be used for impregnation of solid particles, formation of solid powderous emulsions, particle coating, etc

One of the most noticeable applications is the extraction of oil from degumming residue to obtain highly concentrated and very pure lecithin.

1.2 Selective Extraction of Components Using Dense Gases

Selective extraction of components or fractionation of total extracts is possible by use of different gases for the isolation/fractionation of components and/or changing the process parameters. The limitation on further application of extracts obtained by high pressure technology lies in the price of the product, which in comparison with conventionally obtained products, is relatively high. The legal limitations on solvent residues and solvents (for products meant to be used in human applications) and isolation/fractionation of special components from total extracts, in combination with various formulation and sterilisation processes (controlled release, for example) will increase the use of dense gases for extraction applications.

As is known from thermodynamics, the solubility of compounds in SCF/dense gases is based on the density of SCF/dense gas, which is dependent on the pressure and temperature of SCF. Another very important parameter that influences the solubility of compounds in SCF is the dielectric constant of SCF, which is influenced by the temperature and/or pressure of SCF. A general flow sheet of the extraction process is presented in Fig. 17.2.

In the extraction step, the solubility of the compound or mixture of the compound has to be the highest, while in the separation step, the solubility of the compound in SCF has to be the lowest.

Therefore, the phase equilibrium data are the most important data for the design of operating pressures and temperatures for SCF at an extraction plant. Based on the phase equilibrium data, the theoretical mass of SCF necessary for separation of the compound from the solid or liquid mixture can be calculated.

Design process parameters have a very important influence on the investment cost for a high-pressure plant and consequently on the economy of the process.

Besides, as mentioned above, the solubility data for a solute in SCF mass transfer also exert an enormous influence on the economy of the extraction process. Mass transfer models usually describe extraction yield vs. extraction time, but a better presentation for the design of extraction apparatuses is yield vs. S/F mass of SC

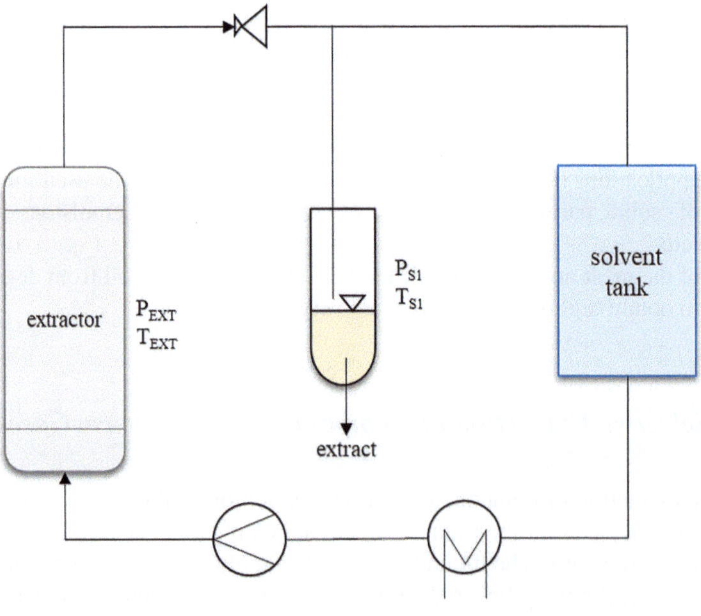

Fig. 17.2 Basic flow sheet of an SCF extraction plant

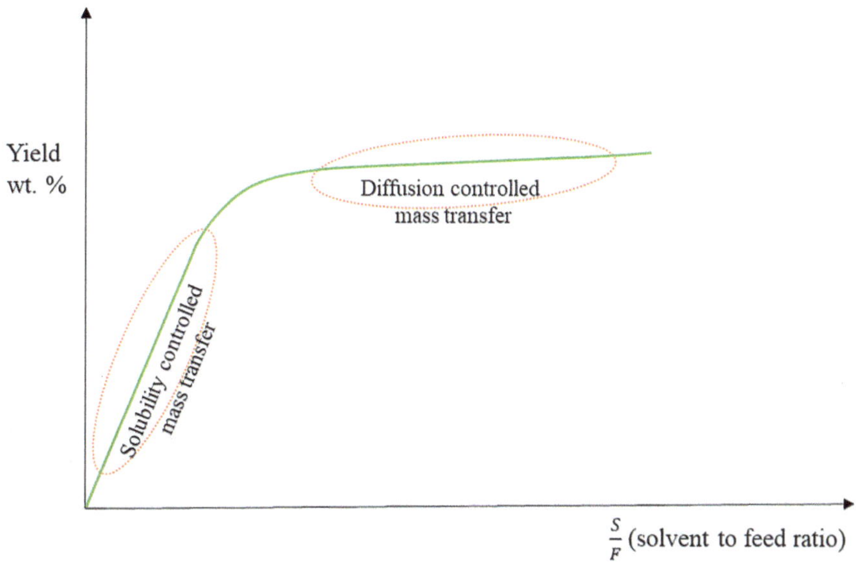

Fig. 17.3 Typical extraction curves for isolation of a substance from solids

solvent (to solid material). In Fig. 17.3, typical extraction curves for isolation of a substance from solids is presented.

There are fewer industrial units for separation of components from liquid mixtures using supercritical fluids. Extraction of liquid mixtures with supercritical fluids is comparable to liquid-liquid extraction, where compressed gas is used instead of an organic solvent. In liquid- SCF extraction processes, the pressure plays an important role. In changing pressure and/or temperature, the physico-chemical properties of the SCF, like density, viscosity, surface tension, dielectric constant etc., are changed. Selective extraction of components or fractionation of total extracts is possible by using different gases for isolation/fractionation of components and/or changing the process parameters. Another advantage is that, depending on the feed material, the density difference between the two counter-current flowing phases can be adjusted.

In several years, several extraction methods, techniques and solvents, were used for producing the different oils from seeds. However, detailed studies to characterize the oil and investigate the influence of different extraction methods and conditions on the composition are still limited. In addition, the extraction solvents must be compatible with requirements of the food industry. It is known that usage of different extraction methods are causing variation in the extraction yields, quality, and content of fatty acids, the content of dietary fibers, antioxidant content, etc.

Segura-Campos et al. (2014) and Knez Hrnčič et al. (2018) report port extraction of Chia oil conventionally by Soxhlet extraction. Non-polar conventional organic solvents such as n-hexane or ether are used. Advantages of using conventional solvent (CS) extraction are mainly the simplicity of the method, relatively high extraction yield, suitable functional characteristics of the oil (such as water holding, absorption capacity, organic molecule absorption, molecule stability), meanwhile the disadvantages are decreased antioxidant activity, due to the decomposition of thermolabile antioxidants, environmental and health concerns involved by using n-hexane. As most suitable extraction method has been recently used is supercritical fluid extraction (SC) where carbon dioxide (CO_2) is the most used solvent for SC (Coelho and Salas-Mellado 2018). Advantages of using SFE comparing to other techniques are usage of a solvent with low density, viscosity, surface tension, mild conditions of temperature and pressure, which leads to no degradation of the compounds.

Silva et al. (2016) and Knez Hrnčič et al. (2018) demonstrates that extraction yield and composition of extracts from Chia seeds are slightly affected by the different process parameters studied (pressure, temperature). Additionally, it has been demonstrated that the composition of both, Black and White chia seed extract is similar. The use of different extraction conditions resulted in differing extraction yields, but did not significantly affect the composition of extract. Higher operating pressures contributed to higher extraction yields when operating at higher temperature. It has been demonstrated, that the composition of oils, obtained from both seed varieties is similar (Table 17.2).

GC analyses were performed for palmitic, palmitoleic, stearic, oleic, linoleic and linolenic acids, but presence of palmitoleic acid has not been demonstrated. From the results can be seen that Chia seeds are generally rich with ω-3 and ω-6 acids. The differences in the content in each extract are minimal, however, the content of palmitic, stearic, oleic and linoleic acids is higher in extracts, attained from White Chia seeds, whilst the content of linolenic acid was higher in black seed extracts. The majority compound identified in oils was linolenic acid (almost 60%), followed by linoleic acid (around 20%), oleic (10%), the content of stearic and palmitic acid

Table 17.2 GC analysis of free fatty acids from Chia seeds

		FFA (%)				
	Conditions	Palmitic	Stearic	Oleic	Linoleic (ω-6)	Linolenic (ω-3)
White Chia	100 bar – 40 °C	8.20 ± 0.09	7.89 ± 0.07	10.17 ± 0.09	20.04 ± 0.05	53.67 ± 0.08
	100 bar – 60 °C	8.03 ± 0.09	6.39 ± 0.03	9.18 ± 0.07	19.80 ± 0.12	56.60 ± 0.09
	200 bar – 40 °C	8.05 ± 0.08	6.61 ± 0.05	9.26 ± 0.06	19.97 ± 0.25	56.11 ± 0.11
	200 bar – 60 °C	7.83 ± 0.07	6.19 ± 0.06	9.00 ± 0.07	19.55 ± 0.04	57.42 ± 0.08
	300 bar – 40 °C	7.69 ± 0.07	6.50 ± 0.02	9.42 ± 0.05	19.81 ± 0.06	56.58 ± 0.05
	300 bar – 60 °C	7.90 ± 0.05	6.10 ± 0.04	8.97 ± 0.04	20.11 ± 0.09	56.92 ± 0.18
Black Chia	100 bar – 40 °C	7.54 ± 0.07	6.00 ± 0.04	8.03 ± 0.06	19.23 ± 0.06	59.20 ± 0.09
	100 bar – 60 °C	7.48 ± 0.06	5.93 ± 0.07	7.98 ± 0.07	19.46 ± 0.32	59.15 ± 0.12
	200 bar – 40 °C	7.50 ± 0.03	5.76 ± 0.03	7.75 ± 0.07	19.49 ± 0.28	59.50 ± 0.29
	200 bar – 60 °C	7.69 ± 0.08	5.67 ± 0.05	7.77 ± 0.09	19.80 ± 0.06	59.06 ± 0.07
	300 bar – 40 °C	7.46 ± 0.06	5.60 ± 0.05	7.74 ± 0.07	19.47 ± 0.08	59.73 ± 0.11
	300 bar – 60 °C	7.50 ± 0.06	5.83 ± 0.04	7.91 ± 0.07	19.47 ± 0.05	59.29 ± 0.06

is lower than 10% each. Obviously, the higher solvent power, characteristic for the higher operating pressures, did not contribute to a higher solubility of the four free fatty acids: palmitic, stearic, oleic and linoleic. These free fatty acids occur in a similar percentage, independently on operating pressures and temperatures. On the contrary, the content of linolenic acid, which was also higher in black seed extracts, increased with increasing operating pressure. The influence of temperature is almost imperceptible, except for White chia seed extracts, attained at lower operating pressures, where the elevation of temperature increased the content of the linolenic acid for almost 3%. The content of linolenic acid was generally by far the highest and also most affected by changing operating conditions, almost 57% in white seed extracts attained at 300 bar and 60 °C and 59.73% in extracts, obtained from black seeds 300 bar and 40 °C. Despite this fact, at equal pressures, generally the percentage of the volatile compounds within the extract was higher at higher temperatures, the most probable explanation is that they were more easily dissolved by the solvent.

A great deal of interest has been devoted to the extraction of active components from natural sources, aiming at satisfying the increasing request of natural products not only for therapeutic use but also as preventing and protecting agents. Among the large number of active substances in the focus, polyphenols have received particular attention. The identification and development of phenolic compounds or extracts from different plants has become a major area of food, health- and medical-related research. Divided into two major groups (nonflavonoid and flavonoids), phenolic compounds show antioxidant and radical scavenging activities possibly responsible for many health benefit effects and for the yellow, orange and red pigments in a large variety of plants and animal kingdoms.

An investigation by Knez and co-workers (2019) has been oriented towards utilization of fruits and their specific parts with a high bioactive compounds content. The main aims of the present study were to maximize the recovery of phenolic compounds in the extracts by application of different extraction methods and variation of experimental parameters. Materials that were investigated are available in the phytogeographical regions of Central Europe (eg Rosa canina L.) or even constitute waste in processing (grape skin) and have been relatively poorly studied so far.

Due to the low polarity of CO_2, EtOH-modified SC CO_2 extraction has been performed to obtain extracts with a high phenolic content. Alteration of operating pressure has been assumed to influence the extraction rate of phytochemicals.

According to the analyses, the yield of phenolic compounds in extracts attained by SC extraction was lower than in the extracts attained by the conventional solvent extraction. This may be explained by the fact that CO_2 usually yields good recoveries for nonpolar compounds, but polar compounds may remain partially unextracted because of their lower solubility in this fluid. For this reason, EtOH as organic modifier has been added as cosolvent to the primary fluid to boost the extraction effectiveness.

The extraction output considering mass yield was higher in case of conventional extraction methods, but the amount of anthocyanins and phenols successfully extracted with supercritical solvent is considerable. The results show, that generally SC CO_2 in combination with a polar entrainer, represents a good extraction media

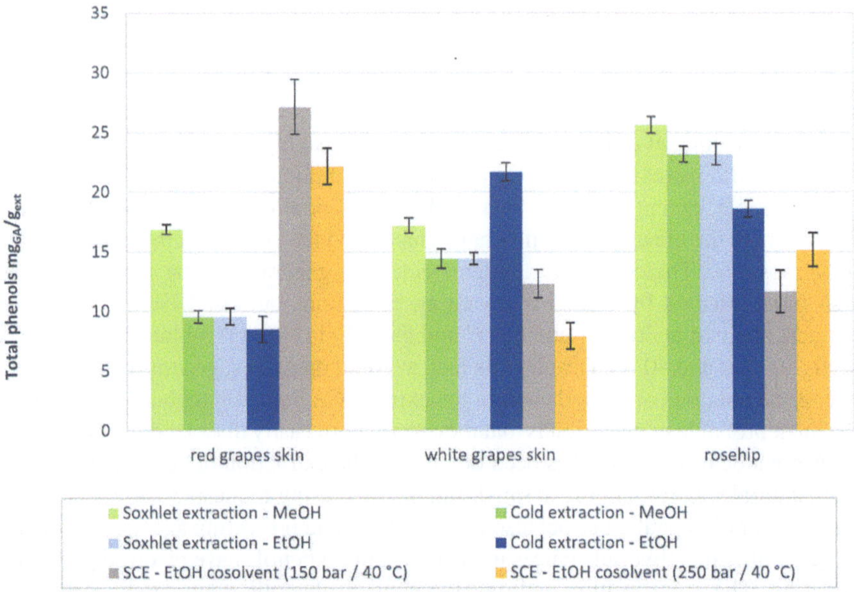

Fig. 17.4 Comparison of total phenolic content, expressed as mg of gallic acid/g of extract (mg_{GA}/g_{EXT}) at different extraction procedures

for isolation of total phenols, while the amount of extracted anthocyanins is low. LC-MS/MS analyses show that gallic, ellagic acid and resveratrol were identified in SC extracts. Therefore, we could consider that SC extraction with CO_2 therefore provides an alternative method to replace extractions with organic solvents for the recovery of phenolic compounds (Fig. 17.4).

2 Other Applications

Micronisation using sub or supercritical fluids is currently subject of intensive research (Knez et al. 2015). There is certain time gap before research is converted to industrial application, and based on developments in area of high pressure extraction, we could be sure that the number of high pressure micronisation units will increase in the near future. Main advantage of the use of sub or supercritical fluids for production of fine particles is the tunability of solvent properties. Again the unique thermodynamics and fluid dynamic properties of sub or SCFs give the products with unique customer designed properties. Micronisation processes could be easily connected to sub or SFE processes, or to a downstream processing after chemical and biochemical reactions. Use of CO_2 also prevents oxidation of products during micronisation steps. Processing of substances using PGSS process could be performed even below their melting point. The other reasons for the fast development in the field of application of sub and SCFs are restrictions regarding the use of conventional organic solvents in food, feed, and pharmaceutical industries.

Most of the research on bio-catalysis in sub and SCFs was made by use of CO_2 as solvent. There are quite several limitations of using CO_2 as reaction media like low solubility of reactants in dense CO_2, change of enzyme activity or deactivation of enzymes due to carbamate formation or acidification of reaction media, deactivation of enzymes or enzyme preparation due to pressure and/or temperature and due to cyclic pressure changes in batch processes. However, CO_2 has extremely favorable attributes for it application as reaction media in biocatalysed reactions such as inflammability, nontoxicity and low costs. The limiting factors of CO_2 as media for chemical and biochemical reactions could be exceeded by use of co-solvents or other dense gasses which increase the solubility of substrates.

CO_2 at high pressures has antimicrobial properties and therefore could be used for high pressure pasterization- sterilization (Perrut 2012). It is known for long that supercritical fluid extraction processes do protect the processed materials from oxidation and contamination with organic solvents and prevent bio-burden increase.

Table 17.3 Range of materials encapsulated using PGSS™ in the past 7 years

Material	Solvent	Ref.
Acizol® pharmaceutical substance	CO_2	Bogorodski et al. (2015)
Alpha lipoic acid/hydrogenated colza oil	CO_2	Mishima et al. (2015)
Avocado oil	CO_2	Aredo et al. (2021)
C. aurantifolia essential oil	CO_2	Akolade et al. (2020)
Citrus oil	CO_2	Ndayishimiye and Chun (2018)
Curcumin	CO_2	Pedro et al. (2016)
Elderberry juice	CO_2	Bánvölgyi et al. (2016)
Epigallocatechin gallate (EGCG) solid formulations	CO_2	Gonçalves et al. (2016) and Shi et al. (2018)
Eucalyptol	CO_2	Akolade et al. (2019)
Fenofibrate	CO_2	Pestieau et al. (2015)
Fish oil	CO_2	Yang and Ciftci (2017)
Fucoxanthin	CO_2	Vo et al. (2018)
Hydrogenated canola oil	CO_2	Ciftci and Temelli (2016)
Lactoferrin	CO_2	Ono et al. (2020)
Limonene in modified starch	CO_2	Machado et al. (2016)
Lipid microparticles	CO_2	López-Iglesias et al. (2020)
Liposomal microencapsulation	CO_2	Tsai and Rizvi (2016)
Omega-3 polyunsaturated fatty acids	CO_2	Melgosa et al. (2019); Haq and Chun (2018)
Pea protein	CO_2	Saldanha do Carmo et al. (2016)
PEGylated biodegradable polyesters	CO_2	Perinelli et al. (2016)
Quercetin	CO_2	Lévai et al. (2017)
Resveratrol on lecithin and β-glucans	CO_2	Salgado et al. (2015)
Rice bran oil	CO_2	Benito-Román et al. (2020)
Vitamin B2	CO_2	Couto et al. (2017)

Table 17.4 Some applications of SFE for separation of flavonoids and other phenolic compounds from plants, with CO_2 as a solvent

Compound(s)	Plant	Ref.
Anthocyanin	Blueberry Cranberry Juçara Haskap berry	Paes et al. (2014) Tamkutė et al. (2020) del et al. (2017) Jiao and Kermanshahi Pour (2018)
Apigenin	Korean perilla	Mishima et al. (2014)
Catechin	Gren tea Peanut Areca nut Guarana	Sökmen et al. (2018) Putra et al. (2018) Ruslan et al. (2015) Santana et al. (2019)
Catechin	Gren tea Peanut Areca nut Guarana	Sökmen et al. (2018) Putra et al. (2018) Ruslan et al. (2015) Santana et al. (2019)
Coumarins	Perennial Lavender	Torres et al. (2017), Medeiros-Neves et al. (2020) Jerković et al. (2017)
Gallic acid	Grape	Da Porto et al. (2014)
Kampferol	Jamun	Prakash Maran et al. (2014)
Lignan	Magnolia berry	Lin et al. (2015)
Myricetin	Bilberry Myrtle Leptocarpha	Babova et al. (2016) Pereira et al. (2016) Marillán and Uquiche (2020)
Orientin	Yarrow	Villalva et al. (2019)
Polyphenols	Propolis	De Zordi et al. (2014)
Quercetin	Taxus chinensis Myrtle	Ruan et al. (2014) Pereira et al. (2016)
Resveratrol	Peanut Sorrel	Jitrangsri et al. (2020) Santos et al. (2017)
Rutin	Morus Asparagus	Radojković et al. (2016) Solana et al. (2015)
Tannin	Stinkingtoe	Veggi et al. (2014)
Vitexin	Snake grass Moringa	Mustapa et al. (2015) Rodríguez-Pérez et al. (2016)
Wogonin	Barbed skullcap	Yang et al. (2019)

Moreover, supercritical fluids were also shown to have the ability to kill most microorganisms and to "inactivate viruses", including human pathogenic strains. Supercritical fluid sterilization/pasteurization and virus inactivation as an alternative "green" method to classical processes that cannot be used in a growing number of cases: thermolabile products degrading by heat sterilization, or compounds reacting with sterilizing chemicals (hydrogen peroxide, ethylene oxide, peracetic acid, etc.), or radiolysis of bio-molecules during irradiation Effectiveness of this application has been shown for various products, but the mechanisms of inactivation have not been fully understood although they have been investigated for more than 60 years.

From an economic point of view, technologies involving elevated pressures require high investment costs for high–pressure equipment. Because of this, it is reasonable to apply supercritical fluid extraction (SFE) for the separation of components with high added value, such as nutraceuticals, pharmaceuticals, food additives, or components with a high feed-to-solvent (F/S) extraction ratio.

A "green" revolution as one part of necessary sustainable development, can also use high pressure as a tool. The main impetus for this conversion driven, on the one hand, by concern for the environment in reducing the usage of solvents and energy. On the other hand, increasing consumer demand for new and natural products sees high pressure as a tool to design and produce natural products with completely new characteristics.

In recent times, there has been greater emphasis on the recovery of high value-added products by using sustainable technologies. One of the ways to achieve this is the application of sub- and supercritical fluids (SCF). Some applications of PGSS™ process on different types of material are presented in the Table 17.3, whilst Table 17.4 introduces some applications of SFE for separation of flavonoids and other phenolic com-pounds from plants, with CO_2 as a solvent.

3 Conclusion

Supercritical fluid based technologies offer important advantages over organic solvent technology, such as ecological friendliness and ease of product fractionation.

With regard to contemporary issues, future research will be focused towards utilization of the many valuable compounds present in natural materials. These can be of economic importance in the food industry and are known for their beneficial health effects. Technologists will be encouraged to develop new processes and soft technologies for preserving the beneficial characteristics of these compounds. Hence, future research efforts should be oriented to developing methods for isolation and identifying new compounds and preserving them after minimal processing and storage of various nutritional products, pharmaceuticals, cosmetics and other materials. Recent demand has tended towards implementation of extraction and formulation processes that enable the transition to "green" technologies, without further use of environmentally and health-hazardous organic solvents. Furthermore, according to a basic concept of bio refineries, the supercritical fluid extraction (SFE) process allows extraction of very pure, high-value product from materials which otherwise would be considered by-products or waste and sold cheaply, or simply disposed of. Such processing concepts promote reuse of residues from the food industry. A possible solution to the low bioavailability of the relevant compounds could be represented by nano formulation. However, in the field of material processing, comprising particle size reduction and foam formation, in-depth research is still needed to obtain the necessary data for designing and optimizing the technologies.

Acknowledgments Special thanks to the Slovenian Research Agency (ARRS) for financial support of the research programme group P2 – 0046: Separation processes and production design, contract No. 1000-15-0552 and project J2-1725: Smart materials for bioaplications.

References

Akolade JO, Balogun M, Swanepoel A, Ibrahim RB, Yusuf AA, Labuschagne P (2019) Microencapsulation of eucalyptol in polyethylene glycol and polycaprolactone using particles from gas-saturated solutions. RSC Adv 9:34039–34049. https://doi.org/10.1039/C9RA06419B

Akolade JO, Nasir-Naeem KO, Swanepoel A, Yusuf AA, Balogun M, Labuschagne P (2020) CO_2-assisted production of polyethylene glycol / lauric acid microparticles for extended release of Citrus aurantifolia essential oil. J CO_2 Util 38:375–384. https://doi.org/10.1016/j.jcou.2020.02.014

Aredo V, Bittencourt GM, de Jesus Agnolon Pallone EM, Guimarães FEC, de Oliveira AL (2021) Formation of edible oil-loaded beeswax microparticles using PGSS – particles from gas-saturated solutions. J Supercrit Fluids 169:105106. https://doi.org/10.1016/j.supflu.2020.105106

Aymonier C, Le Meur AC, Heroguez V (eds) (2011) Nanocomposite particles for bio-applications: materials and bio-interfaces. Pan Stanford, Singapore

Babova O, Occhipinti A, Capuzzo A, Maffei ME (2016) Extraction of bilberry (Vaccinium myrtillus) antioxidants using supercritical/subcritical CO_2 and ethanol as co-solvent. J Supercrit Fluids 107:358–363. https://doi.org/10.1016/j.supflu.2015.09.029

Bánvölgyi S, Vatai T, Molnár Z, Kiss I, Knez Ž, Vatai G, Škerget M (2016) Integrated process to obtain anthocyanin enriched palm-fat particles from elderberry juice. Acta Aliment 45:206–214. https://doi.org/10.1556/AAlim.2015.0009

Benito-Román Ó, Sanz T, Beltrán S (2020) Microencapsulation of rice bran oil using pea protein and maltodextrin mixtures as wall material. Heliyon 6:e03615. https://doi.org/10.1016/j.heliyon.2020.e03615

Bogorodski SE, Krotova LI, Kursakov SV, Minaeva SA, Popov VK, Sevast'yanov VI (2015) Supercritical fluid encapsulation of acizol into aliphatic polyether microparticles. Russ J Phys Chem B 9:1011–1017. https://doi.org/10.1134/S1990793115070052

Ciftci ON, Temelli F (2016) Formation of solid lipid microparticles from fully hydrogenated canola oil using supercritical carbon dioxide. J Food Eng 178:137–144. https://doi.org/10.1016/j.jfoodeng.2016.01.014

Coelho MS, Salas-Mellado M d l M (2018) How extraction method affects the physicochemical and functional properties of chia proteins. LWT 96:26–33. https://doi.org/10.1016/j.lwt.2018.05.010

Couto R, Alvarez V, Temelli F (2017) Encapsulation of vitamin B2 in solid lipid nanoparticles using supercritical CO 2. J Supercrit Fluids 120:432–442. https://doi.org/10.1016/j.supflu.2016.05.036

Da Porto C, Natolino A, Decorti D (2014) Extraction of proanthocyanidins from grape marc by supercritical fluid extraction using CO_2 as solvent and ethanol–water mixture as co-solvent. J Supercrit Fluids 87:59–64. https://doi.org/10.1016/j.supflu.2013.12.013

De Zordi N, Cortesi A, Kikic I, Moneghini M, Solinas D, Innocenti G, Portolan A, Baratto G, Dall'Acqua S (2014) The supercritical carbon dioxide extraction of polyphenols from Propolis: a central composite design approach. J Supercrit Fluids 95:491–498. https://doi.org/10.1016/j.supflu.2014.10.006

del M, Garcia-Mendoza P, Espinosa-Pardo FA, Baseggio AM, Barbero GF, Maróstica Junior MR, Rostagno MA, Martínez J (2017) Extraction of phenolic compounds and anthocyanins from juçara (Euterpe edulis Mart.) residues using pressurized liquids and supercritical fluids. J Supercrit Fluids 119:9–16. https://doi.org/10.1016/j.supflu.2016.08.014

Gonçalves VSS, Poejo J, Matias AA, Rodríguez-Rojo S, Cocero MJ, Duarte CMM (2016) Using different natural origin carriers for development of epigallocatechin gallate (EGCG) solid formulations with improved antioxidant activity by PGSS-drying. RSC Adv 6:67599–67609. https://doi.org/10.1039/C6RA13499H

Haq M, Chun B-S (2018) Microencapsulation of omega-3 polyunsaturated fatty acids and astaxanthin-rich salmon oil using particles from gas saturated solutions (PGSS) process. LWT 92:523–530. https://doi.org/10.1016/j.lwt.2018.03.009

Jerković I, Molnar M, Vidović S, Vladić J, Jokić S (2017) Supercritical CO_2 extraction of *Lavandula angustifolia* Mill. Flowers: optimisation of oxygenated monoterpenes, coumarin and herniarin content: optimization of supercritical CO_2 extraction of *Lavandula angustifolia*. Phytochem Anal 28:558–566. https://doi.org/10.1002/pca.2705

Jessop PG, Leitner W (2008) Chemical synthesis using supercritical fluids. https://nbn-resolving.org/urn:nbn:de:101:1-2014081311511. Accessed 25 Mar 2021

Jiao G, Kermanshahi Pour A (2018) Extraction of anthocyanins from haskap berry pulp using supercritical carbon dioxide: influence of co-solvent composition and pretreatment. LWT 98:237–244. https://doi.org/10.1016/j.lwt.2018.08.042

Jitrangsri K, Chaidedgumjorn A, Satiraphan M (2020) Supercritical fluid extraction (SFE) optimization of transresveratrol from peanut kernels (Arachis hypogaea) by experimental design. J Food Sci Technol 57:1486–1494. SpringerLink. https://link.springer.com/article/10.1007/s13197-019-04184-9. Accessed 25 Mar 2021

Knez Hrnčič M, Cör D, Knez Ž (2018) Subcritical extraction of oil from black and white chia seeds with n-propane and comparison with conventional techniques. J Supercrit Fluids 140:182–187. https://doi.org/10.1016/j.supflu.2018.06.017

Knez Hrnčič M, Cör D, Kotnik P, Knez Ž (2019) Extracts of white and red grape skin and rosehip fruit: phenolic compounds and their antioxidative activity. Acta Chim Slov 751–761. https://doi.org/10.17344/acsi.2019.5253

Knez Ž, Knez Hrnčič M, Škerget M (2015) Particle formation and product formulation using supercritical fluids. Annu Rev Chem Biomol Eng 6:379–407. https://doi.org/10.1146/annurev-chembioeng-061114-123317

Lévai G, Martín Á, Moro A, Matias AA, Gonçalves VSS, Bronze MR, Duarte CMM, Rodríguez-Rojo S, Cocero MJ (2017) Production of encapsulated quercetin particles using supercritical fluid technologies. Powder Technol 317:142–153. https://doi.org/10.1016/j.powtec.2017.04.041

Lin C-H, Lin H-W, Wu J-Y, Houng J-Y, Wan H-P, Yang T-Y, Liang M-T (2015) Extraction of lignans from the seed of Schisandra chinensis by supercritical fluid extraction and subsequent separation by supercritical fluid simulated moving bed. J Supercrit Fluids 98:17–24. https://doi.org/10.1016/j.supflu.2014.12.025

López-Iglesias C, López ER, Fernández J, Landin M, García-González CA (2020) Modeling of the production of lipid microparticles using PGSS® technique. Molecules 25:4927. https://doi.org/10.3390/molecules25214927

Machado LC, Pelegati VB, Oliveira AL (2016) Study of simple microparticles formation of limonene in modified starch using PGSS – particles from gas-saturated suspensions. J Supercrit Fluids 107:260–269. https://doi.org/10.1016/j.supflu.2015.09.023

Marillán C, Uquiche E (2020) Inhibition of α-amylase activity by extracts from Leptocarpha rivularis stalks obtained with supercritical CO_2. J Supercrit Fluids 161:104849. https://doi.org/10.1016/j.supflu.2020.104849

Medeiros-Neves B, Diel KAP, Eifler-Lima VL, Teixeira HF, Cassel E, Figueiró Vargas RM, von Poser GL (2020) Influence of the supercritical CO_2 extraction in the stability of the coumarins of Pterocaulon lorentzii (Asteraceae). J CO_2 Util 39:101165. https://doi.org/10.1016/j.jcou.2020.101165

Melgosa R, Benito-Román Ó, Sanz MT, de Paz E, Beltrán S (2019) Omega–3 encapsulation by PGSS-drying and conventional drying methods. Particle characterization and oxidative stability. Food Chem 270:138–148. https://doi.org/10.1016/j.foodchem.2018.07.082

Mishima K, Kawakami R, Yokota H, Harada T, Kato T, Kawamura H, Matsuyama K, Mustofa S, Hasanah F, Siregar YDI, Putri LSE, Salim A (2014) Extraction of luteolin and apigenin from leaves of Perilla frutescens (L.) Britt. with liquid carbon dioxide. Solvent Extr Res Dev Jpn 21:55–63. https://doi.org/10.15261/serdj.21.55

Mishima K, Honjo M, Sharmin T, Ito S, Kawakami R, Kato T, Misumi M, Suetsugu T, Orii H, Kawano H, Irie K, Sano K, Mishima K, Harada T, Ouchi M (2015) Gas-saturated solution process to obtain microcomposite particles of alpha lipoic acid/hydrogenated colza oil in supercritical carbon dioxide. Pharm Dev Technol 1–12. https://doi.org/10.3109/10837450.2015.1049707

Mustapa AN, Martin Á, Mato RB, Cocero MJ (2015) Extraction of phytocompounds from the medicinal plant Clinacanthus nutans Lindau by microwave-assisted extraction and supercritical carbon dioxide extraction. Ind Crop Prod 74:83–94. https://doi.org/10.1016/j.indcrop.2015.04.035

Ndayishimiye J, Chun BS (2018) Formation, characterization and release behavior of citrus oil-polymer microparticles using particles from gas saturated solutions (PGSS) process. J Ind Eng Chem 63:201–207. https://doi.org/10.1016/j.jiec.2018.02.016

Ono K, Sakai H, Tokunaga S, Sharmin T, Aida TM, Mishima K (2020) Encapsulation of lactoferrin for sustained release using particles from gas-saturated solutions. PRO 9:73. https://doi.org/10.3390/pr9010073

Paes J, Dotta R, Barbero GF, Martínez J (2014) Extraction of phenolic compounds and anthocyanins from blueberry (Vaccinium myrtillus L.) residues using supercritical CO_2 and pressurized liquids. J Supercrit Fluids 95:8–16. https://doi.org/10.1016/j.supflu.2014.07.025

Pedro AS, Villa SD, Caliceti P, de Melo SABV, Albuquerque EC, Bertucco A, Salmaso S (2016) Curcumin-loaded solid lipid particles by PGSS technology. J Supercrit Fluids 107:534–541. https://doi.org/10.1016/j.supflu.2015.07.010

Pereira P, Cebola M-J, Oliveira MC, Bernardo-Gil MG (2016) Supercritical fluid extraction vs conventional extraction of myrtle leaves and berries: comparison of antioxidant activity and identification of bioactive compounds. J Supercrit Fluids 113:1–9. https://doi.org/10.1016/j.supflu.2015.09.006

Perinelli DR, Cespi M, Bonacucina G, Naylor A, Whitaker M, Lam JKW, Howdle SM, Casettari L, Palmieri GF (2016) PEGylated biodegradable polyesters for PGSS microparticles formulation: processability, physical and release properties. Curr Drug Deliv 13:673–681. https://doi.org/10.2174/1567201813666151207111034

Perrut M (2012) Sterilization and virus inactivation by supercritical fluids (a review). J Supercrit Fluids 66:359–371. https://doi.org/10.1016/j.supflu.2011.07.007

Pestieau A, Krier F, Lebrun P, Brouwers A, Streel B, Evrard B (2015) Optimization of a PGSS (particles from gas saturated solutions) process for a fenofibrate lipid-based solid dispersion formulation. Int J Pharm 485:295–305. https://doi.org/10.1016/j.ijpharm.2015.03.027

Prakash Maran J, Priya B, Manikandan S (2014) Modeling and optimization of supercritical fluid extraction of anthocyanin and phenolic compounds from Syzygium cumini fruit pulp. J Food Sci Technol 51:1938–1946. SpringerLink. https://link.springer.com/article/10.1007/s13197-013-1237-y. Accessed 25 Mar 2021

Putra NR, Yunus MAC, Ruslan MSH, Idham Z, Idrus FN (2018) Comparison extraction of peanut skin between CO_2 supercritical fluid extraction and soxhlet extraction in term of oil yield and catechin. Pertanika J Sci Technol 26(2):799–810

Radojković M, Zeković Z, Mašković P, Vidović S, Mandić A, Mišan A, Đurović S (2016) Biological activities and chemical composition of Morus leaves extracts obtained by maceration and supercritical fluid extraction. J Supercrit Fluids 117:50–58. https://doi.org/10.1016/j.supflu.2016.05.004

Reverchon E (2002) Micro and nano-particles produced by supercritical fluid assisted techniques: present status and perspectives. Chem Eng Trans 2:1–10

Rodríguez-Pérez C, Mendiola JA, Quirantes-Piné R, Ibáñez E, Segura-Carretero A (2016) Green downstream processing using supercritical carbon dioxide, CO_2-expanded ethanol and pressur-

ized hot water extractions for recovering bioactive compounds from Moringa oleifera leaves. J Supercrit Fluids 116:90–100. https://doi.org/10.1016/j.supflu.2016.05.009
Ruan X, Yan L-Y, Li X-X, Liu B, Zhang H, Wang Q (2014) Optimization of process parameters of extraction of amentoflavone, quercetin and ginkgetin from Taxus chinensis using supercritical CO_2 plus co-solvent. Molecules 19:17682–17696. https://doi.org/10.3390/molecules191117682
Ruslan MSH, Che Yunus MA, Idham Z, Morad NA, Ali A (2015) Parametric evaluation for extraction of catechin from Areca catechu linn seeds using supercritical CO_2 extraction. J Teknol 74. https://doi.org/10.11113/jt.v74.4704
Saldanha do Carmo C, Nunes AN, Silva I, Maia C, Poejo J, Ferreira-Dias S, Nogueira I, Bronze R, Duarte CMM (2016) Formulation of pea protein for increased satiety and improved foaming properties. RSC Adv 6:6048–6057. https://doi.org/10.1039/C5RA22452G
Salgado M, Rodríguez-Rojo S, Alves-Santos FM, Cocero MJ (2015) Encapsulation of resveratrol on lecithin and β-glucans to enhance its action against Botrytis cinerea. J Food Eng 165:13–21. https://doi.org/10.1016/j.jfoodeng.2015.05.002
Santana ÁL, Queirós LD, Martínez J, Macedo GA (2019) Pressurized liquid- and supercritical fluid extraction of crude and waste seeds of guarana (Paullinia cupana): obtaining of bioactive compounds and mathematical modeling. Food Bioprod Process 117:194–202. https://doi.org/10.1016/j.fbp.2019.07.007
Santos ÊRM, Oliveira HNM, Oliveira EJ, Azevedo SHG, Jesus AA, Medeiros AM, Dariva C, Sousa EMBD (2017) Supercritical fluid extraction of Rumex Acetosa L. roots: yield, composition, kinetics, bioactive evaluation and comparison with conventional techniques. J Supercrit Fluids 122:1–9. https://doi.org/10.1016/j.supflu.2016.11.019
Segura Campos MR, Ciau-Solís N, Rosado-Rubio G, Chel-Guerrero L, Betancur-Ancona D (2014) Chemical and functional properties of chia seed (*Salvia hispanica* L.) gum. Int J Food Sci 2014:1–5. https://doi.org/10.1155/2014/241053
Shi M, Shi Y-L, Li X-M, Yang R, Cai Z-Y, Li Q-S, Ma S-C, Ye J-H, Lu J-L, Liang Y-R, Zheng X-Q (2018) Food-grade encapsulation systems for (−)-epigallocatechin gallate. Molecules 23:445. https://doi.org/10.3390/molecules23020445
Silva C, Garcia VAS, Zanette CM (2016) Chia (Salvia hispanica L.) oil extraction using different organic solvents: oil yield, fatty acids profile and technological analysis of defatted meal. Int Food Res J 23:998
Sökmen M, Demir E, Alomar SY (2018) Optimization of sequential supercritical fluid extraction (SFE) of caffeine and catechins from green tea. J Supercrit Fluids 133:171–176. https://doi.org/10.1016/j.supflu.2017.09.027
Solana M, Boschiero I, Dall'Acqua S, Bertucco A (2015) A comparison between supercritical fluid and pressurized liquid extraction methods for obtaining phenolic compounds from Asparagus officinalis L. J Supercrit Fluids 100:201–208. https://doi.org/10.1016/j.supflu.2015.02.014
Tamkutė L, Liepuoniūtė R, Pukalskienė M, Venskutonis PR (2020) Recovery of valuable lipophilic and polyphenolic fractions from cranberry pomace by consecutive supercritical CO_2 and pressurized liquid extraction. J Supercrit Fluids 159:104755. https://doi.org/10.1016/j.supflu.2020.104755
Torres FC, Medeiros-Neves B, Ferreira Teixeira H, Kawanoa D, Eifler-Lima VL, Cassel E, Vargas RMF, von Poser GL (2017) Supercritical CO_2 extraction as a selective method for the obtainment of coumarins from Pterocaulon balansae (Asteraceae). J CO_2 Util 18:303–308. https://doi.org/10.1016/j.jcou.2017.02.008
Tsai W-C, Rizvi SSH (2016) Liposomal microencapsulation using the conventional methods and novel supercritical fluid processes. Trends Food Sci Technol 55:61–71. https://doi.org/10.1016/j.tifs.2016.06.012
Veggi PC, Prado JM, Bataglion GA, Eberlin MN, Meireles MAA (2014) Obtaining phenolic compounds from jatoba (Hymenaea courbaril L.) bark by supercritical fluid extraction. J Supercrit Fluids 89:68–77. https://doi.org/10.1016/j.supflu.2014.02.016

Villalva M, Jaime L, Villanueva-Bermejo D, Lara B, Fornari T, Reglero G, Santoyo S (2019) Supercritical anti-solvent fractionation for improving antioxidant and anti-inflammatory activities of an Achillea millefolium L. extract. Food Res Int 115:128–134. https://doi.org/10.1016/j.foodres.2018.08.027

Vo DT, Saravana PS, Woo H-C, Chun B-S (2018) Fucoxanthin-rich oil encapsulation using biodegradable polyethylene glycol and particles from gas-saturated solutions technique. J CO_2 Util 26:359–369. https://doi.org/10.1016/j.jcou.2018.05.019

Yang J, Ciftci ON (2017) Encapsulation of fish oil into hollow solid lipid micro- and nanoparticles using carbon dioxide. Food Chem 231:105–113. https://doi.org/10.1016/j.foodchem.2017.03.109

Yang Y-C, Wang C-S, Wei M-C (2019) Kinetics and mass transfer considerations for an ultrasound-assisted supercritical CO_2 procedure to produce extracts enriched in flavonoids from Scutellaria barbata. J CO_2 Util 32:219–231. https://doi.org/10.1016/j.jcou.2019.04.008

Part V
Mechanisms of Validation of Nonthermal Processes in Biomaterials and Agri-food Applications

Chapter 18
Current Validation of NTP Technologies and Overview of Their Current and Potential Implementation in the Production Chain Including Agri-food Wastes

Urszula Tylewicz, Juan Manuel Castagnini, Silvia Tappi, Santina Romani, Pietro Rocculi, and Marco Dalla Rosa

1 Introduction

Non-thermal processing (NTP) technologies could be used for different scopes in the food industry. Principally, they have been designed and utilized as a decontamination/pasteurization process. In this sense, the quality management systems have to be flexible enough to allow the same tools to be used to achieve both goals, product safety and product quality. A quality system can achieve its objectives irrespective of the area where it is applied, the processed product and the technologies used in the manufacturing process when it is properly designed and implemented. On the other hand, NTP technologies could be used as a pre treatment for structure modification and enhancing the mass transfer of different kind of processes like extraction, drying, osmotic dehydration, freezing and so on. In these cases, the validation and verification of the process may not be intrinsically related with the safety but with the quality, the reduction of energy consumption and/or the reduction of food waste.

U. Tylewicz (✉) · S. Tappi · S. Romani · P. Rocculi · M. D. Rosa
Department of Agricultural and Food Sciences, Alma Mater Studiorum-University of Bologna, Cesena, Italy

Interdepartmental Centre for Agri-Food Industrial Research, University of Bologna, Cesena, Italy
e-mail: urszula.tylewicz@unibo.it

J. M. Castagnini
Interdepartmental Centre for Agri-Food Industrial Research, University of Bologna, Cesena, Italy

Instituto de Ciencia y Tecnología de los Alimentos de Entre Ríos. Consejo Nacional de Investigaciones Científicas y Técnicas – Universidad Nacional de Entre Ríos, Concepción del Uruguay, Entre Ríos, Argentina

A. Režek Jambrak (ed.), *Nonthermal Processing in Agri-Food-Bio Sciences*, Food Engineering Series, https://doi.org/10.1007/978-3-030-92415-7_18

Different NTP technologies have been validated on the laboratory and pilot scale, showing a good potentiality for the industrial implementation. In fact, some of them are already used at the industrial level for a specific application, while others still require further studies to overcome the limits of these technologies.

Pulsed electric field (PEF), thanks to the electroporation of the plant cell membrane, which occurs upon the application of low and moderate electric field strength (up to 10 kV/cm), has been shown to modify the plant tissue structure and to improve the mass transfer within the tissue. PEF pre-treatment has been already successfully applied at the industrial scale in potato industry, showing high benefits in improved texture and crispness, less oil uptake, lower acrylamide content (Ostermeier et al. 2020). In addition to the structure improvement, PEF-assisted processing promotes higher extraction yields, shorter processing time, a decrease of process intensity (temperature, solvent usage), and better preservation of heat-sensitive compounds as compared to conventional techniques (Barba et al. 2015a; Picart-Palmade et al. 2019). The other important advantage of low and moderate PEF treatment is to modulate the process conditions, giving the possibility of pore resealing (reversible electroporation) after treatment and the formation of pore with different sizes in the electroporated cell membrane. In this way, the different molecular size intracellular compounds may be selectively recovered and more easily purified (Vorobiev and Lebovka 2017). However, there are also some factors limiting the industrial implantation of PEF technology in mass transfer enhancement, such as the capital investment costs, design of the equipment, suitability and characteristics of food products. Therefore, there is need to evaluate the economic and environmental impact of using PEF-assisted processes in comparison to currently available technologies before their scaling up (Golberg et al. 2016).

Power US application is considered a green, eco-friendly and emerging technology, which was validated for the structure modification (French Fries, tomato peeling, meet tenderisation) and as a pre-treatment of treatment assisted process for mass transfer enhancement. It was also validated for the valorisation of agri-food waste and by-products, by the recovery of various molecules and biomaterials, including polysaccharides, essential oils, proteins, peptides, pigments, and bioactive molecules of commercial importance (Tiwari 2015). The advantage of this technology is the reduced use of solvents, short extraction time and its relatively easy to use, versatility, and flexibility. Moreover, UAE presents also a low investment cost compared to other novel extraction techniques (e.g., supercritical fluid extraction, PEF-assisted extraction, pressurized solvent extraction) (Flores et al. 2021). However, still further studies of the economic and environmental impact are necessary to build up the multi-criteria decision tools that would help to select the most suitable technology and associated operating conditions for a given application.

High hydrostatic pressure (HHP) with a pressure of 100–700 MPa is able to damage cell membranes, increasing their permeability and thus allowing extraction of value-added compounds from fruit and vegetable products as well from the agri-food waste and by-products (Barba et al. 2015b). Moreover, HHP is able to modify the protein structure, resulting in meat tenderisation. Anyhow, at the moment, the benefit of meat tenderization and mass-transfer enhancement by HHP is not as evident as in the case of cold pasteurization, which is a standard practice in many meat companies for ensuring food safety, therefore limits its commercial application (Bolumar et al. 2021).

The aim of this book chapter is to show the current validation of the NTP technologies namely PEF, US and HHP for structure modification and mass transfer enhancement with a particular attention agri-food waste valorisation. Moreover, this chapter will show current and potential implementation of these NT technologies in the food production chain and in agri-food waste.

2 NTP for Structure Modification

2.1 Potato Industry

The main processed products from potatoes are French fries and chips, although their characteristics are very different. French fries are characterized by a crispy crust and firm-mealy core, while potato chips (or crisps) are characterized by a crispy structure (Botero-Uribe et al. 2017). Different factors may affect the crispness of French fries and chips, such as moisture, starch content and the entity of the oil uptake during frying (Pedreschi et al. 2007). However, the potato industry is currently facing different challenges. One of them is to develop products with low acrylamide content, according to the Commission Regulation (EU) 2017/2158 of 20 November 2017, which established the rules for the mitigation of acrylamide formation in foods (European Union 2017). The snacks with reduced-fat content are, nowadays, more and more appreciated by the consumers, therefore the attention needs to be focused also on the reduction of oil uptake during frying. It has been shown that the manufacturing process in terms of different pre-treatments before frying and the frying conditions (time/temperature) influence the final crispness, oil uptake and acrylamide content of the products. Traditionally, at the industrial scale blanching is performed as a mitigation strategy; however, the traditional blanching treatment presents several practical drawbacks and leads to undesirable changes in the product quality (Genovese et al. 2019).

Non-thermal technologies such as PEF and US have been validated as alternatives to traditional processing. In general PEF processing on potato causes tissue softening, which results in better cutting performance (Liu et al. 2017; Fauster et al. 2018). As reported by (Botero-Uribe et al. 2017), the cause of the tissue softening is the cell structure modification during PEF treatment. In fact, the electric field promotes irreversible changes to the cell membrane permeability, increasing mass transfer during further drying. Tissue softening during PEF application also allows the longer knife durability of cutting machines in the industry. Moreover, homogeneous starch gelatinization leads to texture improvement by increasing crispness (Ostermeier et al. 2020; Genovese et al. 2019) obtained a reduction of acrylamide content in potato chips treated by PEF by 30% without compromising their crispness if compared to the blanched ones. (Ostermeier et al. 2021) observed that the application of PEF or US alone reduced only slightly the acrylamide content while combining PEF pre-treatment and ultrasound-assisted frying resulted in a 66% reduction of acrylamide, and about 25% lower oil uptake. Moreover, PEF treatment has been reported to reduce water and energy consumptions, for a typical 50,000 kg/h line, 60–70,000,000 l of water and 20 GJ of thermal energy are saved per year.

Since 2010, PEF treatment has also been used in the industrial-scale production of potato chips and French fries worldwide to improve the quality of the final product. The implementation of PEF in the potato industry is quite simple, since the PEF treatment is very short (microseconds), and PEF systems can take place either before or after the peeling plant. The costs for a 6 t/h PEF system, considering an annual production of 24,000 t raw material, would amount to 7€/t raw material, including also expenses such as electricity consumption. Therefore, there is a positive impact on the return of investment for a PEF system installation, considering also the processing benefits and the improved quality of the final product (Lammerskitten et al. 2020; Ostermeier et al. 2020).

2.2 *Tomato Industry*

Tomato (*Lycopersicon eculentum* L.) fruits could be consumed as a raw fresh product; however, the majority of them are processed into a whole, diced or sliced peeled tomato, juices, sauce and ketchup. Therefore, very often tomatoes need to be peeled before the industrial transformation. Traditional peeling is usually performed by using either hot lye solutions or steam blanching, which presents several drawbacks, such as disposal of caustic, high pH waste solution, and excessive water and energy consumption (Barrett 2015). To overcome these disadvantages, in recent years, different innovative solutions have been proposed and investigated.

PEF pre-treatment has been validated as strategy to facilitate the peelability of tomato fruits and also reduce energy consumption during the thermo-physical peeling phase. (Andreou et al. 2020) showed that PEF treatments (0.5–1.5 kV/cm, 0–8000 pulses, 15 μs pulse width) applied to whole tomatoes improved peeling and were able to reduce the work required for peel detachment up to 72.3%. (Pataro et al. 2018) suggested that applying PEF pre-treatment (<1 kV/cm and 1 kJ/kg) prior to steam blanching represents a less energy-intensive peeling method if compared to the conventional one. In addition to the increased peeling performance, the extractability of carotenoids, mainly lycopene with high antioxidant activity, from tomato peels has been also improved (Andreou et al. 2020; Pataro et al. 2020).

The implementation of PEF in the tomato industry was studied by (Arnal et al. 2018). PEF system was successfully integrated into the processing line of the FPD Srl Company (Fisciano, Italy). PEF chamber was installed in-line at the washing phase, which was identified as the best location for PEF pre-treatment of raw tomato fruits before the thermo-physical peeling phase. The exposure of tomatoes to the

PEF pre-treatment (0.45 kV/cm and 0.40 kJ/kg) showed that a steam pressure of 80 kPa caused a similar peeling performance product quality as steam peeling alone (at 100 kPa), demonstrating that PEF was able to reduce by 20% the energy consumption in the peeling phase (Arnal et al. 2018). Moreover, the environmental assessment was performed in terms of LCA study, demonstrating that PEF technology is environmentally friendly since it improved by 17–20% all the considered environmental indicators. However, the authors suggested that in order to fully validate the implementation of the PEF technology at an industrial scale further studies are necessary, taking into consideration different tomato cultivars in semi/real industrial conditions, also as a function of the ripening stage of the raw material.

The validation of the power US application as a "green, eco-friendly and emerging technology" for tomato peeling has been reviewed by (Rock et al. 2012). In a previous work, (Rock et al. 2010) observed that power US treatments (1500 W, 20 kHz) applied to Roma var. tomatoes without the addition of any chemical were just as effective as lye peeling in terms of the ease of peeling. Also, (Gao et al. 2018) showed that 4% (w/v) lye treatment at 97 °C for 30 s with a post-assistance of a 31.97 W/L US treatment at 24 kHz and 70 °C for 50 s achieved a 100% peelability. The benefits of using power US not only include its potential in eliminating the use of lye in tomato peeling thus reducing environmental impacts but also an increase in production yield and lycopene content in the peeled product as compared to the conventional hot lye peeling method. The authors explained in detail also the mechanism of the hot lye and US join action. Briefly, hot lye dissolves the waxy crust and penetrates across the de-waxed cuticle via a pitting model, causing the degradation of the cell wall and the middle lamella which produces fissures at the boundary of the epidermal cell and hypodermal cell layers. Finally, the US application increases the fissures, cracks on the skin and causes the detachment of the epidermal layer from the fruits (Gao et al. 2018). However, further research is recommended to investigate the mechanism affecting the lycopene content, as well as the scaling-up of the present method.

2.3 *Meat Industry*

Tenderness is one of the most relevant features for consumers when purchasing meat products. Among different sensory parameters, the tenderness of meat plays a prominent role and is considered a decisive sensory attribute, which can directly affect its quality. Therefore, the increased tenderness can improve its eating quality and final value (Madhusankha and Thilakarathna 2021). Traditionally, meat tenderness is improved by different methodologies such as mechanical (physical), chemical and enzymatic ones. Recently, many studies have been performed in order to verify the possibility to apply non-thermal technologies for this goal (Buckow et al. 2013; Turantaş et al. 2015; Warner et al. 2017; Gómez et al. 2019).

The most studied and validated technology is HPP since it has been demonstrated that muscle proteins are very sensitive to the pressure application (Buckow et al. 2013; Bolumar et al. 2021). Moreover, meat tenderization depends on the time postmortem when HPP is applied, type of muscle and applied processing condition. It is well established that the application of HPP at 100–225 MPa and 10–35 °C, to pre-rigour meat, improves the tenderness of beef, lamb, poultry and pork (Warner et al. 2017). In particular, (Morton et al. 2018) showed that pre-rigour HPP (175 MPa, 3 min) significantly disrupted the sarcomere structure of longissimus thoracis beef muscle improving its eating quality. These effects were associated with changes in glycon phosphorylase and an increase in ultimate pH. In addition, the same treatment was able to produce a beef meat at 1-day postmortem as tender or even more tender than muscles aged for 28 days (Morton et al. 2017). To obtain tender meat by applying HPP on post-rigour muscle, usually, the application of higher than ambient temperature is required (Warner et al. 2017; Bolumar et al. 2021).

The disadvantage of HPP technique is that high pressure can induce undesirable colour changes in treated meat (Morton et al. 2017) particularly for the red ones, giving to the product the typical visual quality of the cooked one. Therefore, HPP can be considered as a viable option for ready-to-eat food, but not for the 'fresh food' shelves, limiting its commercial application. Concerning the cost of HPP processing, it comprises high initial investment, high operative and maintenance cost (Aganovic et al. 2017; Cacace et al. 2020). However, (Cacace et al. 2020) concluded through LCA study, that the use of HPP for Parma ham was not only less expensive but also had a lower impact in most of the impact categories if compared to modified atmosphere packaging (MAP), since MAP requires a significant amount of packaging materials and food gases.

Anyhow, at the moment, the benefit of meat tenderization by HPP is not as evident as in the case of cold pasteurization, which is a standard practice in many meat companies for ensuring food safety. Moreover, the industrial implementation of HPP for this kind of application requires important modification of process layout. This is a reason why it has not yet been embraced by the meat industry (Bolumar et al. 2021).

Recent studies on the effect of PEF on the tenderization of different beef muscle types have shown promising results (Suwandy et al. 2015a, b, c). This effect is associated with membrane damage, which could promote calcium release from the cellular organelles, the release of cathepsins from lysosomes and/or accelerated glycolysis in pre-rigour meat (Warner et al. 2017). Moreover, the increased proteolysis of myofibrillar proteins in low pH meat samples by the application of PEF (0.58–0.73 kV/cm; 10 Hz; pulse width 20 μs) was observed (Suwandy et al. 2015b) Some studies reported that PEF-induced effect depends on the muscle type (e.g. beef loins and topsides) to be treated (Suwandy et al. 2015a, c; Bekhit et al. 2016). For instance, (Bekhit et al. 2016), showed that PEF treatment (10 kV, 90 Hz, 20 μs) applied once or twice had no effect on the tenderisation of hot-boned beef loins and topsides muscles for all ageing treatment times. The application of 3x PEF treatment reduced the tenderness in the loins, while increasing it in topsides at 3 days post-treatment time. Some studies reported that PEF at 1.5 kV/cm combined with sous vide cooking 60 °C for 20.8–23.7 h promoted an improved tenderness of beef briskets (Alahakoon et al. 2018, 2019) by increasing the in vitro protein digestibility of the meat (Chian et al. 2021).

The implementation of the PEF equipment in the meat industry is still a huge challenge and, at the moment, it is rather limited, although it could be highly favourable due to the low energy consumption and short processing times required in PEF processing. The main limits of its implementation in the meat industry are related to the high capital investment, the necessity of changes in the processing plant, and the need for optimization of PEF conditions for specific product application. Therefore, further investigations on the impact of PEF on quality parameters of meat products are required to favour the transfer of this technology into industrial application (Gómez et al. 2019).

Recent studies showed the potential of power US application on fresh meat, mainly for meat tenderization, processing (brining, cooking, freezing) and microbial inactivation (Peña-González et al. 2017; Alarcon-Rojo et al. 2019; Bhargava et al. 2021). It has been shown by different authors that US was able to promote meat tenderness and shorten the period of ageing, without compromising other quality parameters (Chang et al. 2015; Peña-González et al. 2017, 2019). In particular, (Peña-Gonzalez et al. 2019) showed that US application at 40 kHz, 11 W/cm^2 for 60 min reduced the shear force by 11% and 15% after a storage at 4 °C for 0 and 14 days, respectively. The increased tenderness of meat could be explained by the rupture of the myofibrillar structure of the protein, collagen macromolecules fragmentation, protein migration responsible for the proteolysis acceleration or protein denaturation (Barekat and Soltanizadeh 2018; Alarcon-Rojo et al. 2019).

The implementation of US technology requires further studies, in order to understand more in-depth, the effect of US on meat (microstructure, enzyme activity, final tenderness), as well as on the optimization of the process parameters (frequency, intensity, treatment time) for improved efficiency and penetration into the meat matrix (Warner et al. 2017).

3 NTP for Mass Exchange

3.1 Extraction by Diffusion

Although the wide research related to PEF-assisted extraction by diffusion (Table 18.1) demonstrated that this methodology is promising for the development of modern industrial technology, the conventional equipment for sucrose extraction from sugar beetroot, betalain from red beet, inulin from chicory, beta-carotene from carrot, did not change during the last century (Barba et al. 2015a). The estimated power consumptions, for PEF-treated tissues were found to be rather low and typically lying within 1–15 kJ/kg (Vorobiev and Lebovka 2010). So, from the standpoint of power consumption, PEF is practically the ideal method to favour the damaging of plant tissues as compared to other methods (Barba et al. 2015a).

Table 18.1 Applications of PEF for the extraction of bioactive compounds from plant tissue

Product	Process parameters	Results	References
Sugar beet	E: 600 V/cm Pulse number: 500 Pulse duration: 100 μs Pulse repetition: 200 Hz	High purity sugar beet juices (95.3 ± 0.4%). Also demonstrated that the combination of mild heating (50 °C) and PEF allows shortening the diffusion time.	Loginova et al. (2011a) Loginov et al. (2011) Loginova et al. (2012)
	E: 600 V/cm Pulse number: 100 Pulse duration: 100 μs Pulse repetition: 10 ms	PEF allowed high yield sucrose extraction and purification with less energy consumption.	Zhu and Mhemdi (2016)
Chicory	E: <600 V/cm Pulse number: 4 Pulse duration: 1 ms Pulse repetition: 10 ms Pulse repetition: 1–1000 trains	Enhancement of the soluble matter extraction and increased diffusion even at low temperatures within 20–40 °C.	Loginova et al. (2010)
	E: 600 V/cm Pulse number: 100–500 Pulse duration: 100 μs Pulse repetition: 5 ms Total time: 10–50 ms	PEF facilitates extraction of inulin at conventional temperature; diffusion temperature can even be reduced by 10–15 °C with comparable juice inulin concentration.	Zhu et al. (2012) Zhu et al. (2015)
Red beet	E: 1 kV/cm Temperature: 30–80 °C	PEF accelerates the betalains extraction and yield.	Loginova et al. (2011b)
	E: 4.38 kV/cm, Energy input: 6 kJ/kg Pulse number: 30	The highest increase in the content of betalain compounds in the red beet's extract (betanin by 329%, vulgaxanthin by 244%), compared to the control sample	Nowacka et al. (2019)

(continued)

Table 18.1 (continued)

Product	Process parameters	Results	References
Olive	E: 1.7 kV/cm (17 kJ/kg)	Oil yield increased from 2.3% to 6%, and increased concentration of hydrophilic phenols, from 3.2% to 14.3%	Veneziani et al. (2019)
	E: 3.3 kV/cm Pulse duration: 0.3 ms	Increased extraction yield by 13.3%; total phenolic content, total phytosterols and total tocopherols significantly higher (11.5%, 9.9% and 15.0%, respectively). The use of PEF had no negative effects on general chemical and sensory characteristics	Puértolas and Martínez De Marañón (2015)
	E: 1–2 kV/cm (1.47–5.22 kJ/kg) Pulse number: 50 Pulse duration: 3 μs Pulse repetition: 125 Hz	Yield improved by 54% with PEF (2 kV/cm) without malaxation. With malaxation at 15 °C and PEF, the extraction yield improved by 14.1%	Abenoza et al. (2013)
Saffron	E: 5 kV/cm Pulse duration: 35 μs Pulse number: 100	Enhanced extraction 14.1% in colour, 15.5% in bitterness and 10.2% in taste	Pourzaki et al. (2013)
	E: 2 kV/cm (1.5 kJ/kg) Pulse duration: 20 μs Pulse number: 50	Enhanced concentration of picrocrocin, crocin and safranal in the PEF saffron samples after ageing for 3 months	Neri et al. (2021)

In the conventional sugar production process, counter-current diffusion in hot water (70–75 °C) for more than 1 h is widely applied for satisfactory extraction of juice from sugar beet slices. To reduce energy consumption and to improve juice quality, researchers have investigated the cold extraction of sugar beet juice (Zhu and Mhemdi 2016). The implementation of the PEF treatment of sugar beet cossettes at an industry scale was studied in a counter-current extractor as reported by Loginova et al. (2011a). The authors stated that PEF-assisted "cold" extraction (between 30 and 50 °C) is a promising method for the preparation of high purity sugar beet juices (95.3% ± 0.4%). They also demonstrated that the combination of mild heating at 50 °C and PEF-treatment can be a useful tool for shortening the extraction time. Finally, their results showed that sugar beet juices obtained, had a lower concentration of colloidal impurities (especially pectins), lower colouration and better filterability through polyether sulfone ultrafiltration membrane. The different research articles related to sugar extraction from beet concluded that PEF-assisted processing could allow more environmentally friendly and efficient processes.

Inulin is widely used in food industries mainly due to its ability to substitute fat and sugar ingredients in formulations. The current industrial production from chicory roots is mainly carried out by diffusion in hot water (70–80 °C), followed by a relatively complex purification process, due to the presence of a large number of impurities generated by the application of high temperatures (Zhu et al. 2016). Loginova et al. (2010) studied the effect of PEF application for the enhancement of the soluble matter extraction from chicory. With a field strength between 100 and 600 V/cm they achieved membrane permeabilization and a high level of tissue disintegration. They also reported that the temperature contribution (40–50 °C) to electroporation efficiency is very important and that this synergetic effect increases at small fields (especially at $E \leq 200$ V/cm). Zhu et al. (2012) validated the inulin extraction from chicory roots treated with PEF at industrial conditions, confirming the feasibility of PEF assisted extraction in a pilot counter-current extractor. When treated with PEF, the extraction at conventional temperatures of 70–80 °C produced a juice reacher in inulin and better exhausted cossettes. They also stated that the temperature can be decreased by 10–15 °C as, for example, at 60 °C the juice inulin concentration was 11.65 g/100 mL and the juice purity 87.1%. Finally, Zhu et al. (2012), stated that the benefit obtained by the reduction of the diffusion temperature after the PEF treatment of chicory roots, could cover largely the PEF electrical energy consumption, and the economic profit (related to diffusion step) could be 34.76 € / t inulin. Moreover, if the purification and concentration step are also considered, the total economic profit will be even higher.

Different research groups also investigated the inulin US-assisted extraction (Table 18.2). Lingyun et al. (2007) reported that indirect ultrasound-assisted (sample immersed in an ultrasound bath) extraction from Jerusalem artichoke tubers is more suitable for the inulin extraction because the direct method (probe immersed into the sample) induces its partial degradation and releases more oligosaccharides in the extract; Milani et al. (2011) used US with high intensity, and improved significantly the inulin extraction from Burdock roots, powdered and sieved through a 0.125 mm sieve. The obtained results showed that increasing the amplitude and the extraction time the extraction yield increases, with a minor effect on temperature. Finally, Abbasi and Farzanmehr (2009) reported a significant increase in inulin recovery from Iranian artichokes proportionally to the power input of US.

Betalain pigments are mostly used as food dyes due to their non-precarious, non-toxic, non-carcinogenic, and non-poisonous nature (Chhikara et al. 2019) but are also associated with health benefits like antioxidants, anti-cancer, anti-lipidemic, and can act as antimicrobial agents (Gengatharan et al. 2015; Tanabtabzadeh et al. 2019). Therefore, food industries are increasingly interested in the extraction of this natural food colourant. To extract the pigments from plant material, the disruption of membranes is necessary, and this is usually obtained through the application of detergents, solvents, or thermal treatments (Celli and Brooks 2017). As an alternative, Loginova et al. (2011b) demonstrated that the extractability of betalains could be enhanced by PEF, releasing about 90% of the total red pigment. Another study reported that PEF treatment caused changes in the colour of red beetroot tissue associated with better extraction of pigments (Nowacka et al. 2019).

Table 18.2 Applications of US for the extraction of inulin from plant tissue

Product	Process parameters	Results	References
Chicory	pH: neutral Extraction time: 20 min Temperature: 77 °C Solvent:solid ratio: 11:1 (v:w)	The optimal conditions for maximizing inulin extraction yield (83.6%) were at natural pH for 20 min at 76.65 °C and solvent:solid ratios of 10.56:1 (v/w)	Lingyun et al. (2007)
Artichoke	Treatment duration: 5 min Temperature: 80 °C Solvent:solid ratio: 5:1.	Optimal aqueous extraction of inulin could be achieved at 80 °C, for a duration of 5 min at a solvent:solid ratio of 1:5.	Abbasi and Farzanmehr (2009)
Burdock	Treatment duration: 25 min Temperature: 37 °C Amplitude: 83%	Optimum extraction conditions: sonication time: 25 min; sonication amplitude 83.22% and temperature: 36.76 °C.	Milani et al. (2011)

The PEF potential to improve extraction yield and quality of olive oil was studied by different authors (Abenoza et al. 2013; Puértolas and Martínez De Marañón 2015; Veneziani et al. 2019). The results obtained shows that PEF allows a reduction in the malaxation temperature without detriment of the extraction yield nor the quality of the olive oil. The low energy requirements and the short processing times could be key advantages to implementing the technology in the industry (Abenoza et al. 2013). Puértolas and Martínez De Marañón (2015) studied the implementation of PEF for olive oil extraction at pilot scale. They stated that PEF is an appropriate technology to improve the yield of virgin and extra-virgin olive oil (VOO and EVOO) and that from the chemical and sensory point of view, PEF treatment not only has no negative effects but increases the content of human-health-related compounds, such as polyphenols, phytosterols and tocopherols, maintaining the EU legal standards of the highest quality olive oil.

An increase in the extraction of major components (crocin – colour, safranal – flavour and picrocrocin – taste) of saffron (*Crocus sativus*) stigma and saffron pomace after PEF application were also observed (Pourzaki et al. 2013). Moreover, Neri et al. (2021) observed that ageing of PEF treated saffron stigmas significantly contributed to enhancing the concentration of picrocrocin, crocin and safranal, responsible for colouring and flavouring properties, allowing their ISO-like quality classification (ISO 3632-2:2010). Finally, high hydrostatic pressure (100–600 MPa) in combination with elevated temperatures (30–70 °C) was also studied (Shinwari and Rao 2018). The authors reported a significantly positive effect of pressure on the extract quality with an increase of 52–63%, 54–85%, and 55–62% in crocin, picrocrocin, and safranal content, respectively, as the pressure increased.

3.2 Drying

PEF treatment could also be used as a pre-treatment to enhance drying. Hot air dryers are the most widespread dryers in different industries, not only for food and agriculture applications but also for paper, textile and chemical ones (Mousakhani-Ganjeh et al. 2021). However, these dryers are characterized by a high energy consumption, low efficiency, and a high impact on the quality of the final product. On the other hand, freeze dryers and vacuum dryers are alternative technologies for products that contain thermolabile compounds, but both require a higher initial investment, and they usually operate in batch. Among different innovative technologies, PEF and US drying are receiving great interest because of their higher efficiency and lower operation time (Mousakhani-Ganjeh et al. 2021). The reduction of drying time by PEF pre-treatment depends on the applied parameters, drying method, and raw material. As a consequence, there is no standard PEF application method, and the effectiveness of electroporation depends on many processing-related parameters. Among them, electric field strength and energy input are the most frequently modulated ones and could be used to compare results of different research groups. Besides, the effective water diffusion coefficient provides valuable information and could be used in order to contrast the drying kinetics of different processes (Barba et al. 2015a).

Several researchers studied the effect of PEF treatment before drying (Table 18.3). Alam et al. (2018) reported a reduction of 21% on the drying time of carrot slices at 60 °C, a reduction in lightness values (L*) and an increase in a* values and that PEF did not affect the mechanical properties of the samples when compared to untreated ones; Liu et al. (2018) found that applying PEF before vacuum drying of potatoes resulted in a reduction of 22–27% of drying time at 40–70 °C. They also reported changes in the textural properties of treated and untreated samples (cutting force), as also described by (Castagnini et al. 2020). PEF could also enhance the freeze-drying kinetics of apple samples as demonstrated by Lammerskitten et al. (2019) which determined a 44% increase in the diffusion coefficient and a drying time reduction of 57%. After thermal properties measurements, they also found a higher crystallinity of the PEF samples. The freeze-dried PEF-treated fruits absorbed more water than the untreated ones, while no changes were reported for hygroscopicity and loss of the soluble solids during rehydration.

In the vacuum drying of fresh blueberries, a 33%, 30% and 42% reduction on drying time at 75, 60 and 45 °C, respectively, was reported by Yu et al. (2017). The authors concluded that vacuum drying at 75 °C provided the optimal conditions for minimizing the loss of anthocyanin, total phenolics, vitamin C, and antioxidant activity in dried samples and maximizing the drying rate.

Won et al. (2015) applied PEF before hot air drying of red pepper and reported a 34.7% reduction of drying time at 45 °C. This reduction was also beneficial for the colour quality of dried red pepper because an increase greater than 10% in b* value and in the ASTA (American Spice Trade Association) unit was observed.

Table 18.3 Applications of PEF before drying of different plant tissues

Product	Process parameters	Results	References
Carrot	E: 0.9 kV/cm (65.2 kJ/kg) Pulse duration: 20 μs Pulse repetition: 50 Hz	Reduction by 21% on the drying time at 60 °C, a reduction in lightness values (L*) and an increase in a* values and that PEF did not affect the mechanical properties	Alam et al. (2018)
Potato	E: 600 V/cm Pulse duration: 100 μs Pulse number: 100 Treatment time: = 0.1 s	PEF treatment allowed noticeable decreasing of drying time (by 22–27% at 40–70 °C).	Liu et al. (2018)
Apple	E: 1.07 kV/cm (0.5 and 1 kJ/kg) Pulse duration: 10 and 40 ms Pulse repetition: 2 Hz	PEF before freeze-drying makes possible to obtain high quality freeze-dried products; better rehydration of PEF treated samples	Lammerskitten et al. (2019)
	E: 0.5–1.5 kV/cm (0.121–1.086 kJ/kg) Pulse duration: 10 μs Pulse repetition: 100 Hz	Modification of water holding capacity induced by PEF and drying	Castagnini et al. (2020)
Blueberry	Electric field strength: 2 kV/cm	The PEF significantly increased the drying efficiency, reducing the drying time and did not significantly influence the nutritive quality.	Yu et al. (2017)
Red bell pepper	Electric field strength: 1.0–2.5 kV/ cm Pulse width: 30 μs Pulse frequency: 100 Hz. Treatment time: 1, 2, and 4 s	The 34.7% reduction in drying time was also beneficial to colour quality.	Won et al. (2015)

Lastly, Mannozzi et al. (2020) combined osmotic dehydration (OD) (trehalose at 40%) with PEF (200 V/cm) before air-drying of kiwifruit slices. They observed good colour retention, a higher content of total polyphenols, vitamin C content and antioxidant activity after drying at 60 °C. Besides, they stated that the combination of OD-PEF could have an additional positive effect on the sustainability of the overall process, by saving the energy required for the process, because osmotic dehydration is less energy-intensive than drying.

US was also applied as a pre-treatment prior to or during drying (Table 18.4). Ricce et al. (2016) found that carrot slices pre-treated using an US bath for 30 and 60 min enhanced not only the drying but also the rehydration rate at lower temperatures. They also corroborated that the use of high drying temperatures hides the US pre-treatment effects since the temperature has a greater effect on drying. Therefore, US pre-treatment is a good option to reduce the drying time and/or temperature. Chen et al. (2016) proposed a novel dehydration technique that applies US during vacuum drying (USV-drying). With this technique, they achieved a 44 and 55% reduction in drying time at 65 and 75 °C, respectively, when compared to

Table 18.4 Applications of US before/during drying of different plant tissues

Product	Process parameters	Results	References
Carrot	Power: 700 W Frequency: 25 kHz Volumetric power: 41 W/L Treatment time: 30 and 60 min.	Enhanced drying and rehydration rate at lower temperatures.	Ricce et al. (2016)
	Power: 200 W Frequency: 40 kHz Mode: 10 s on and 5 s off.	Drying time reduced around 50% and quality properties enhanced (rehydration potential, nutritional value, colour, and texture)	Chen et al. (2016)
	Ultrasonic probe: 65, 75 and 85 W Ultrasonic bath: 10, 20 and 30 °C Treatment time: 3, 5 and 10 min	Drying time significantly reduced (up to 20%); β-carotene and rehydration ratio significantly enhanced.	Yilmaz et al. (2019)
Potato	Power: 100 and 200 W Frequency: 25 kHz Temperature: 50 °C	Reduced drying time by 40%; improved dried potato quality (colour and rehydration)	Kroehnke et al. (2014)
Apple	Power: 200 W Frequency: 25 kHz Temperature: 50 °C	Greatly increased the drying kinetics; Reduced the total colour change by about 32%	Mierzwa and Kowalski (2016)
Red/green pepper	Power: 0.5 kW/m³ Frequency: 21.7 kHz Temperature: 30, 50 and 70 °C	Improved drying kinetics. US reduced the loss of antioxidant properties at 50 °C but produced greater degradation at 70 °C.	Cárcel et al. (2019)
	Power: 100 and 200 W Treatment time: 8.6 and 7.9 h Temperature: 54 °C	Reduced the overall drying time by about 35%; decreased discolouration; better rehydration ability and retention of vitamin C.	Szadzińska et al. (2017)
	Power: 590 W Amplitude: 100% Temperature: 45, 55, 65 and 75 °C	Reduced drying time at the same temperature.	Tekin and Baslar (2018)

vacuum drying without US. The rehydration potential, nutritional value, colour, and textural properties of USV-dried carrot slices were also greatly improved. Finally, the source of ultrasonic waves (probe or bath) was studied by Yilmaz et al. (2019). At the investigated operating conditions (probe at 65, 75 and 85 W or ultrasonic bath at 10, 20 and 30 °C for 3, 5 and 10 min before air drying) they reported a 20% reduction of the dying time at 60 °C and a higher β-carotene content for the samples treated in the water bath.

In the case of potato convective drying, Kroehnke et al. (2014) described that the US application during convective drying of potato reduced the total colour change by about 8% and 12% at 100 and 200 W, respectively. They further reported that US application did not influence the water activity (a_w) of the dried potato.

US pre-treatment applied to apple slices reduced the total colour change by about 32% at 200 W in comparison to the convective drying. US application did not influence the water activity of the dried apple and greatly increased the drying kinetics, which resulted in a reduction of the overall drying time (Mierzwa and Kowalski 2016).

Cárcel et al. (2019) studied the US influence on the drying kinetics and antioxidant properties of red pepper. The experiments were carried out in an ultrasonic-assisted dryer with a high-intensity ultrasonic field (up to 154.3 dB) produced inside the drying chamber. They found that temperature significantly influences drying (at 70 °C the time needed was 11% of the experiments carried out at 30 °C). Besides, the application of US significantly increased the drying rate for every tested temperature (30, 50 and 70 °C); the lower the tested temperature, the shorter the drying time. For the drying at 70 °C, the application of ultrasound reduced the drying time by 32%. This reduction increased up to 62% and 54% in experiments carried out at 50 and 30 °C respectively. Finally, the use of high drying temperatures reduced the loss in the antioxidant properties of red pepper. However, the application of US only significantly reduced the degradation at intermediate temperatures but increased them at the highest. Therefore, ultrasonically assisted red pepper drying can only be considered in terms of reducing the drying temperature to save energy. On the other hand, Szadzińska et al. (2017) studied the US-assisted drying of green pepper and demonstrated that applying US at 100 and 200 W during dying at 54 °C reduced drying time by 32 and 37% respectively; decreased the colour changes and allowed retaining most of the vitamin C (up to 70%). Finally, the authors stated that the application of ultrasound in convective drying could significantly reduce the total electric energy consumption. Lastly, another application found in the literature was red pepper US-assisted vacuum drying. In this case, the US was applied in an ultrasonic water bath during the vacuum drying process (Tekin and Baslar 2018). The authors found that US reduced by 26, 12, 25 and 11% the drying time of red pepper at 45, 55, 65 and 75 °C respectively; the rehydration ratio slightly decreased for red peppers USV dried and the total phenolic compounds also decreased as the drying temperature increased. Taking into account the radical scavenging capability, at 45 °C, no significant difference in the antioxidant activity (DPPH) was found but at 75 °C.

4 NTP for Agri-food Waste and By-products Valorisation

Eco-friendly and sustainable extraction of high-value compounds from food waste and by-products is one of the main challenges in the food industry in recent years. The conventional techniques (mechanical, thermal, chemical and enzymatic) used for the extraction of the valuable compounds often require a significant amount of mechanical or thermal energy, long time steps, use of toxic solvents, or high temperatures that can degrade thermolabile compounds, and lead to a non-selective extraction. In the last few decades, emerging non-thermal technologies generally used for food preservation have shown a promising result for more efficient and sustainable extraction/separation processes (Režek Jambrak et al. 2018; Picart-Palmade et al. 2019).

4.1 Pulsed Electric Field

PEF-assisted extraction has been particularly investigated for the extraction of bioactive compounds from different agri-food wastes, among others, orange (Luengo et al. 2013; El Kantar et al. 2018) tomato (Pataro et al. 2018, 2020) and plum (Medina-Meza and Barbosa-Cánovas 2015) peels, grape by-products (Corrales et al. 2008; Medina-Meza and Barbosa-Cánovas 2015), rapeseed by-products (Yu et al. 2015) and papaya seeds (Parniakov et al. 2015). Table 18.5 summarises some of the many studies related to the effect of PEF on the extraction of functional components from agri-food by-products.

Barbosa-Pereira et al. (2018) used PEF to obtain added-value compounds from cocoa and coffee beans by-products, showing increased extraction yield of bioactive compounds (e.g. polyphenols and methylxanthines), which could be used in several applications in the food (as food ingredients and nutraceuticals), pharmaceutical and cosmetic industries. Martín-García et al. (2020) optimised the extraction of the phenolic compounds by using PEF pre-treatment from brewer's spent grain, validating the PEF method also in this field of application.

One of the recent and interesting segments of the scientific investigation is the opportunity to exploit some microorganisms as an alternative source of proteins, pigments, antioxidant compounds, carbohydrates and lipids. In particular, blue coloured and water-soluble phycocyanins could be used as a natural pigment, exerting also a high antioxidant activity, for different food product formulations. Phycocyanin recovery at industrial level is still limited due to extraction methods drawbacks and to the low stability of the final product. Different studies showed that PEF resulted a valid and promising technology for phycocyanin extraction, since the extracts present relatively high phycocyanins concentration and purity (Käferböck et al. 2020; Pez Jaeschke et al. 2021), also if compared to other non-thermal techniques, such as US and HPP (Pez Jaeschke et al. 2021). These results could be useful in the implementation step of PEF processing in the phycocyanins

Table 18.5 Effect of PEF on the agri-food waste and by-products valorisation

Product	Process parameters	Results	References
Orange peel	E: 1–7 kV/cm Energy input: 0.06–3.77 kJ/kg Pulse number: 5–50 Pulse duration: 3 μs Pulse repetition: 1 Hz	Improvement in flavonoids extraction yield and antioxidant activity up to 159 and 192%, respectively	Luengo et al. (2013)
	E: 10 kV/cm Pulse duration: 70 μs	Synergic effect of PEF and ethanol concentration (50%) Improved polyphenols recovery by 83%	El Kantar et al. (2018)
Tomato peels	E = 0.25–0.75 kV/cm Energy input: 1 kJ/kg Pulse duration: 20 μs Pulse repetition:10 Hz	Increased extraction of carotenoids (up to 188%) and antioxidant power (up to 372%)	Pataro et al. (2018)
	E = 5 kV/cm Energy input: 5 kJ/kg Pulse duration: 20 μs Pulse repetition:10 Hz	Increased extraction rate of lycopene yields (12–18%) and the antioxidant power (18.0–18.2%) in either acetone and ethyl lactate extracts	Pataro et al. (2020)
Grape by-products	E = 3 kV/cm Energy input: 10 kJ/kg Pulse number: 30 Pulse repetition: 2 Hz	Increased (4 times) antioxidant activity of extract compared to traditional extraction High recovery of anthocyanin monoglucosides	Corrales et al. (2008)
	E: about 1 kV/cm Pulse number: 25.2 Pulse duration: 6 μs Pulse repetition:10 Hz	Increased extraction yield of anthocyanins and flavonoids from grape peels, as compared with untreated and US treated samples	Medina-Meza and Barbosa-Cánovas (2015)
Plump peels	E: about 1 kV/cm Pulse number: 25.2 Pulse duration: 6 μs Pulse repetition:10 Hz	Increased extraction yield of flavonoids and total phenols as compared with untreated samples	Medina-Meza and Barbosa-Cánovas (2015)
Rapeseed stems and leaves	E: 0.8–20 kV/cm Energy input: 6.4–160 kJ/kg	Increased extraction of total polyphenols, with the highest purity at 5 kV/cm Increased extraction of proteins when 20 kV/cm	Yu et al. (2015)

(continued)

Table 18.5 (continued)

Product	Process parameters	Results	References
Papaya seed	E: about 13.3 kV/cm Temperature: 20 °C	PEF + supplementary aqueous extraction at 50 °C, pH = 7 for 3 h, had a significant positive effect on protein, TPC and isothiocyanate recovery, as well as led to a higher antioxidant activity	Parniakov et al. (2015)
Cocoa and coffee bean by-products	E = 1.5–4.4 kV/cm, Pulse number: 500–1000 Pulse duration: 5–20 μs	Improved extraction of bioactive compounds from cocoa and coffee by-products by 20% and 21.3%, respectively	Barbosa-Pereira et al. (2018)
Brewers' spent grain	E = 0.5–2.5 kV/cm, Energy input: 0.25–18.75 kJ/kg Pulse repetition: 50–150 Hz	Max. extraction of total free phenolic compounds at 2.5 kV/cm, 50 Hz and 14.5 s Increased extraction of totally free and bound phenolics by 2.7 and 1.7 times as compared with untreated samples	Martín-García et al. (2020)
Microalgae	E: 15, 20, 25 kV/cm Energy input: 100 kJ/kg	Increased extraction of phycocyanin (nine-fold increase in combination to freeze-thawing, comparing to untreated fresh biomass) Higher purity of phycocyanin	Käferböck et al. (2020)
	E = 40 kV/cm Energy input:28, 56, 112 kJ/kg Pulse repetition: 2–6 Hz	Increased phycocyanins yield (up to 76%) Increased proteins yield (up to 102%) Increased antioxidant activity (up to 136%)	Jaeschke et al. (2019)

extraction lines since the industry needs to obtain a dye without the necessity of further processing and purification steps. Moreover, LCA showed that PEF pre-treatment resulted in a more sustainable production of valuable fractions, showing 57–65% lower environmental impacts if compared to the untreated products (Käferböck et al. 2020).

Puértolas and Barba (2016) evaluated the economic cost of PEF treatment necessary to recover valuable compounds from different agri-food wastes (e.g. chicory, grape skin, fennel etc) and it accounted for 0.1–0.5 €/ton and was significantly lower as compared to enzymatic one which accounted for 7.5 €/ton. Similarly, the total cost for apple pulp processing by PEF was around 2.69 €/ton as compared to 8.50 €/ton for enzymatic maceration. Since the majority of the literature examples related to the agri-food waste valorisation was validated only on a laboratory scale, there is a need to perform research at the pilot plants and industrial scale in order to evaluate both the economic and environmental impact of using PEF-assisted processes in comparison to currently available technologies. Moreover, the optimization of PEF treatment and process parameters for each specific application is needed in order to be able to design and optimize the processes before their scaling up (Golberg et al. 2016).

4.2 Ultrasound-Assisted Extraction

There are many literature studies related to the use of ultrasound-assisted extraction (UAE) technology in the extraction of added-value compounds from agri-food waste and by-products, and some of them are reported in Table 18.6.

Various authors found that UAE showed a significantly higher efficiency compared to conventional extraction techniques in terms of reduced extraction time and increased yield of extraction of some bioactive compounds. (Papoutsis et al. 2018) used different models to optimize the UAE of total polyphenols from lemon by-products, while (Ladole et al. 2018) optimised the lycopene extraction from tomato peels. Tomato residue after UAE extraction resulted to be a good source of fatty acids (Ninčević Grassino et al. 2020). The extraction of valuable compounds from grape by-products was extensively studied, showing increased extraction yield of pectins from grape pomace, lipophilic antioxidants from grape seeds and total polyphenols from grape peels, as compared to conventional extraction (Minjares-Fuentes et al. 2014; Medina-Meza and Barbosa-Cánovas 2015; Dimić et al. 2020). Moreover, (Medina-Meza and Barbosa-Cánovas 2015) compared US- and PEF- assisted extractions in plump peels, showing increased recovery of anthocyanins and flavonoids by UAE, as compared with untreated and PEF treated samples. UAE showed also good potential in the recovery of hydrophilic compounds from cocoa by-products (Grillo et al. 2019) and proanthocyanins from brewer's spent grain (Martín-García et al. 2020). Recently, the sustainable extraction of proteins (Vernès et al. 2019) and pigments from microalgae, in particular, red phycoerythrin (Ardiles et al. 2020) and blue Phycocyanin (Minchev et al. 2020), was also reported.

Different studies evaluated the environmental impact of UAE through the LCA studies. (Vauchel et al. 2018) performed a comparative LCA study at different operating conditions for polyphenols extraction from chicory food industry waste. US had a positive effect enabling environmental impacts reduction up to about 25%. However, the more marked positive effect of US application is expected when using different vegetal sources (with less destructured vegetal matrixes). In the optic of industrial implementation, further studies of the economic and environmental impact are necessary to build up the multi-criteria decision tools that would help to select the most suitable technology and associated operating conditions for a given application.

4.3 High Pressure Assisted Extraction

HHP-assisted extraction is a quite recent application of high pressure in the food sector and the literature still lacks the full understanding and validation of this specific application for different by-products. Table 18.7 report some examples of the use of HHP-extraction for different agri-food waste and by-products.

Table 18.6 Effect of US on the agri-food waste and by-products valorisation

Product	Process parameters	Results	References
Lemon by-products	Frequency: 43 kHz Power: 150–250 W Temperature: 45–55 °C Time: 35–45 min	Max. extraction of TPC (18.10 ± 0.24 mg GAE/gdw) with 250 W, 45 min, 50 °C Max. extraction of rutin (3.20 ± 0.12 mg/gdw) with 150 W, 35 min, 48 °C	Papoutsis et al. (2018)
Tomato peels	Frequency: 20 kHz Power: 1–120 W Combined with enzyme co-immobilized AMNPs	Max. extraction of lycopene release at 3% (w/w) enzyme co-immobilized AMNPs, pH 5.0, 50 °C, 10 W and 20 min	Ladole et al. (2018)
	Frequency: 30 kHz Power: 400 W Temperature: 20.8–59.7 °C Time: 5–15 min	Depectinized residues (after HPP treatment) subjected to UAE in 70% ethanol for 15 min contained two times lower values of total phenols than pectinized samples. The residue after UAE is a good source of fatty acids, among which lauric, palmitic and stearic acids are dominant	Ninčević Grassino et al. (2020)
Grape by-products	Frequency: 37 kHz, Power: 140 W	Higher pectin yield from grape pomace (by 20%) compared to traditional extraction Pectin with a higher average molecular weight	Minjares-Fuentes et al. (2014)
	Frequency: 40 kHz, Power: 60 W L^{-1} Temperature: 50 °C Time: 40 min	High extraction yield of lipophilic antioxidants from grape seeds	Dimić et al. (2020)
	Frequency: 24 kHz Power: 400 W Temperature: 50 °C	Increased extraction yield of total polyphenols from grape peels as compared with untreated samples	Medina-Meza and Barbosa-Cánovas (2015)
Plump peels	Frequency: 24 kHz Power: 400 W Temperature: 50 °C	Increased extraction of anthocyanins and flavonoids as compared with untreated of PEF treated samples	Medina-Meza and Barbosa-Cánovas (2015)
Cocoa by-products	Frequency: 19.9 kHz Power: 150 W Temperature: 40 °C Time: 15 min	Good potential in the recovery of hydrophilic mass from cocoa by-products, rich in methylxanthines and polyphenols, and a lipid layer.	Grillo et al. (2019)
Brewers' spent grain	Frequency: 24 kHz Power: 80–400 W Acetone/water: 50%, 75%, and 100% Time: 5–55 min	Max. extraction of proanthocyanidins with 80/20 acetone/water (v/v), 55 min, and 400 W Increased extraction of proanthocyanidins more than twice as compared with non-treated samples.	Martín-García et al. (2020)

(continued)

Table 18.6 (continued)

Product	Process parameters	Results	References
Microalgae	Frequency: 30 kHz Power: 100 W Temperature: 30 °C	Increased extraction yield of phycoerythrin	Ardiles et al. (2020)
	Frequency: 35, 45 kHz Power: 30, 300 W Temperature: 35 °C Time: 1–48 h	Max. Extraction of phycocyanin (2.5 mg/g dw) and purity 0.8 with 35 kHz, 30 W for 3 h	Minchev, Petkova and Milkova-Tomova (2020)
	Frequency: 20 kHz Power: 1000 W Continuous flow rate at 15 mL/h.	Increased recovery of proteins (by 229%)	Vernès et al. (2019)

Table 18.7 Effect of HPP-assisted extraction on agri-food waste and by-products valorisation

Product	Process parameters	Results	References
Lime peel	Pressure: 100–200 MPa Temperature: 50 °C Time: 30 min	Increased enzymatic extraction yield of pectin	Naghshineh et al. (2013)
Tomato waste	Pressure: 100–800 MPa Time: 1–30 min	Higher extraction yields of lycopene (from 2% to 64% increase depending on the solvent used) compared to conventional solvent extraction	Strati et al. (2015)
	Pressure: 300 MPa Temperature: 80 °C Time: 10–45 min	Increased recovery of pectin from tomato peels (by 15% after 45 min), in comparison with the conventional extraction for 180 min	Ninčević Grassino et al. (2020)
Grape by-products	Pressure: 600 MPa Temperature: 70 °C Time: 60 min	Increased (3 times) antioxidant activity of extract compared to traditional extraction Higher recovery of acylated anthocyanin monoglucosides	Corrales et al. (2008)
Olive pomace	Pressure: 200–600 MPa Time: 0–40 min	Increased phenolic concentration of the extract (up to 71.8%) Increased protein extraction (by 88.1% for pressures up to 200 MPa) as compared to the untreated one	Andreou et al. (2020)

HHP has been applied to grape by-products (Corrales et al. 2008), showing a significant increase in total and individual anthocyanin content of the extract (acylated anthocyanin monoglucosides) in comparison to conventional extraction. Moreover, the antioxidant activity of the extract was three-folds higher as compared to the conventionally extracted product. More recently (Strati et al. 2015; Ninčević Grassino et al. 2020) evaluated the effect of HHP on tomato waste, showing improved lycopene extraction from tomato peels and seeds (Strati et al. 2015) and improved pectin extraction from tomato peels (Ninčević Grassino et al. 2020). Improved pectin yield was also observed from lime peel treated by 100–200 MPa (Naghshineh et al. 2013), without polymer chain degradation, allowing scale-up to industrial-scale production of pectin. (Andreou et al. 2020) observed also a higher extraction yield of polyphenols and proteins from olive pomace upon the application of HHP. The available literature confirms that also HHP-assisted extraction allows higher extraction yields with a shorter extraction time and reduces the use of organic solvents, as compared with the conventional extraction. Therefore, the HPP-assisted extraction is a promising technique to improve process efficiency and sustainability, however, for their implementation in the food industry further studies are needed, related to the pilot plant validation and the energetic and economical estimations (Picart-Palmade et al. 2019).

5 Conclusion

This chapter investigated validation and implementation of non-thermal processing (NTP) technologies, such as PEF, US and HHP in food processing chain and agri-food waste valorisation. There as many literatures attesting the validation of these technologies for the specific applications. In fact, all the technologies showed the advantages in the structure modification and mass transfer enhancement, resulting in more sustainable processing (lower energy requirements, shorter processing time, lower solvent usage) if compared with the conventional processing. For example, PEF treatment is already successfully implemented in the potato industry. However, there are also some limits (product quality, investment cost, processing line changes etc.) that need to be overcome before fully implementation of these technologies in the food processing industry.

References

Abbasi S, Farzanmehr H (2009) Optimization of extracting conditions of inulin from Iranian artichoke with/without ultrasound using response surface methodology. J Sci Technol Agric Nat Resour Isfahan Coll Agric Nat Resour Isfahan Univ Technol 13(47(B)):423–436

Abenoza M et al (2013) Effects of pulsed electric field on yield extraction and quality of olive oil. Food Bioprocess Technol 6(6):1367–1373. https://doi.org/10.1007/s11947-012-0817-6

Aganovic K et al (2017) Pilot scale thermal and alternative pasteurization of tomato and watermelon juice: an energy comparison and life cycle assessment. J Clean Prod. https://doi.org/10.1016/j.jclepro.2016.09.015
Alahakoon AU et al (2018) Optimisation of sous vide processing parameters for pulsed electric fields treated beef briskets. Food Bioproc Tech. https://doi.org/10.1007/s11947-018-2155-9
Alahakoon AU et al (2019) Process optimisation of pulsed electric fields pre-treatment to reduce the sous vide processing time of beef briskets. Int J Food Sci Technol. https://doi.org/10.1111/ijfs.14002
Alam MR et al (2018) Effect of pulsed electric field pretreatment on drying kinetics, color, and texture of parsnip and carrot. J Food Sci 83(8):2159–2166. https://doi.org/10.1111/1750-3841.14216
Alarcon-Rojo AD et al (2019) Ultrasound and meat quality: a review. Ultrason Sonochem. https://doi.org/10.1016/j.ultsonch.2018.09.016
Andreou V et al (2020) Application of pulsed electric fields to improve product yield and waste valorization in industrial tomato processing. J Food Eng Elsevier Ltd 270(September 2019):109778. https://doi.org/10.1016/j.jfoodeng.2019.109778
Ardiles P et al (2020) Biochemical composition and phycoerythrin extraction from red microalgae: a comparative study using green extraction technologies. Processes 8(12):1628. https://doi.org/10.3390/pr8121628
Arnal ÁJ et al (2018) Implementation of PEF treatment at real-scale tomatoes processing considering LCA methodology as an innovation strategy in the agri-food sector. Sustainability (Switzerland) 10(4). https://doi.org/10.3390/su10040979
Barba FJ, Parniakov O et al (2015a) Current applications and new opportunities for the use of pulsed electric fields in food science and industry. Food Res Int 77:773–798. https://doi.org/10.1016/j.foodres.2015.09.015
Barba FJ, Terefe NS et al (2015b) New opportunities and perspectives of high pressure treatment to improve health and safety attributes of foods. A review. Food Res Int Elsevier Ltd 77:725–742. https://doi.org/10.1016/j.foodres.2015.05.015
Barbosa-Pereira L, Guglielmetti A, Zeppa G (2018) Pulsed electric field assisted extraction of bioactive compounds from cocoa bean shell and coffee silverskin. Food Bioprocess Technol 11(4):818–835. https://doi.org/10.1007/s11947-017-2045-6
Barekat S, Soltanizadeh N (2018) Effects of ultrasound on microstructure and enzyme penetration in Beef Longissimus lumborum muscle. Food Bioproc Tech. https://doi.org/10.1007/s11947-017-2043-8
Barrett DM (2015) Future innovations in tomato processing. Acta Hortic 1081:49–55. https://doi.org/10.17660/ActaHortic.2015.1081.3
Bekhit AEDA et al (2016) Effect of repeated pulsed electric field treatment on the quality of hot-boned beef loins and topsides. Meat Sci Elsevier Ltd 111:139–146. https://doi.org/10.1016/j.meatsci.2015.09.001
Bhargava N et al (2021) Advances in application of ultrasound in food processing: a review. Ultrason Sonochem. https://doi.org/10.1016/j.ultsonch.2020.105293
Bolumar T et al (2021) High-pressure processing of meat: molecular impacts and industrial applications. Compr Rev Food Sci Food Saf 20(1):332–368. https://doi.org/10.1111/1541-4337.12670
Botero-Uribe M et al (2017) Effect of pulsed electrical fields on the structural properties that affect french fry texture during processing. Trends Food Sci Technol Elsevier Ltd 67:1–11. https://doi.org/10.1016/j.tifs.2017.05.016
Buckow R, Sikes A, Tume R (2013) Effect of high pressure on physicochemical properties of meat. Crit Rev Food Sci Nutr 53(7):770–786. https://doi.org/10.1080/10408398.2011.560296
Cacace F et al (2020) Evaluation of the economic and environmental sustainability of high pressure processing of foods. Innov Food Sci Emerg Technol. https://doi.org/10.1016/j.ifset.2019.102281
Cárcel JA et al (2019) Influence of temperature and ultrasound on drying kinetics and antioxidant properties of red pepper. Dry Technol 37(4):486–493. https://doi.org/10.1080/07373937.2018.1473417

Castagnini JM et al (2020) Pulsed electric fields effect on mechanical and sorption properties of dried apple tissue. Innov Food Sci Emerg Technol 65(July):102442. https://doi.org/10.1016/j.ifset.2020.102442

Celli GB, Brooks MS-L (2017) Impact of extraction and processing conditions on betalains and comparison of properties with anthocyanins — a current review. Food Res Int Elsevier Ltd 100:501–509. https://doi.org/10.1016/j.foodres.2016.08.034

Chang HJ et al (2015) Effects of ultrasound treatment on connective tissue collagen and meat quality of beef semitendinosus muscle. J Food Qual. https://doi.org/10.1111/jfq.12141

Chen Z-G, Guo X-Y, Wu T (2016) A novel dehydration technique for carrot slices implementing ultrasound and vacuum drying methods. Ultrason Sonochem Elsevier BV 30:28–34. https://doi.org/10.1016/j.ultsonch.2015.11.026

Chhikara N et al (2019) Bioactive compounds of beetroot and utilization in food processing industry: a critical review. Food Chem Elsevier Ltd 272:192–200. https://doi.org/10.1016/j.foodchem.2018.08.022

Chian FM et al (2021) Effects of pulsed electric field processing and sous vide cooking on muscle structure and in vitro protein digestibility of beef brisket. Foods 10(3). https://doi.org/10.3390/foods10030512

Corrales M et al (2008) Extraction of anthocyanins from grape by-products assisted by ultrasonics, high hydrostatic pressure or pulsed electric fields: a comparison. Innovative Food Sci Emerg Technol 9(1):85–91. https://doi.org/10.1016/j.ifset.2007.06.002

Dimić I et al (2020) Innovative and conventional valorizations of grape seeds from winery by-products as sustainable source of lipophilic antioxidants. Antioxidants 9(7):568. https://doi.org/10.3390/antiox9070568

El Kantar S et al (2018) Pulsed electric field treatment of citrus fruits: improvement of juice and polyphenols extraction. Innov Food Sci Emerg Technol 46(October 2017):153–161. https://doi.org/10.1016/j.ifset.2017.09.024

European Union (2017) Commission Regulation (EU) 2017/2158: establishing mitigation measures andbenchmark levels for the reduction of the resence of acrylamide in food. Official Journal of the European Union. https://doi.org/http://eur-lex.europa.eu/pri/en/oj/dat/2003/l_285/l_28520031101en00330037.pdf

Fauster T et al (2018) Impact of pulsed electric field (PEF) pretreatment on process performance of industrial French fries production. J Food Eng 235:16–22. https://doi.org/10.1016/j.jfoodeng.2018.04.023

Flores EMM et al (2021) Ultrasound-assisted biomass valorization to industrial interesting products: state-of-the-art, perspectives and challenges. Ultrason Sonochem. https://doi.org/10.1016/j.ultsonch.2020.105455

Gao R et al (2018) A novel two-step ultrasound post-assisted lye peeling regime for tomatoes: reducing pollution while improving product yield and quality. Ultrason Sonochem Elsevier 45(January):267–278. https://doi.org/10.1016/j.ultsonch.2018.03.021

Gengatharan A, Dykes GA, Choo WS (2015) Betalains: natural plant pigments with potential application in functional foods. LWT – Food Sci Technol Academic Press 64(2):645–649. https://doi.org/10.1016/j.lwt.2015.06.052

Genovese J et al (2019) Important factors to consider for acrylamide mitigation in potato crisps using pulsed electric fields. Innov Food Sci Emerg Technol 55. https://doi.org/10.1016/j.ifset.2019.05.008

Golberg A et al (2016) Energy-efficient biomass processing with pulsed electric fields for bioeconomy and sustainable development. Biotechnol Biofuels BioMed Central 9(1):1–22. https://doi.org/10.1186/s13068-016-0508-z

Gómez B et al (2019) Application of pulsed electric fields in meat and fish processing industries: an overview. Food Res Int 123(March):95–105. https://doi.org/10.1016/j.foodres.2019.04.047

Grillo G et al (2019) Cocoa bean shell waste valorisation; extraction from lab to pilot-scale cavitational reactors. Food Res Int. https://doi.org/10.1016/j.foodres.2018.08.057

Jaeschke DP et al (2019) Extraction of valuable compounds from Arthrospira platensis using pulsed electric field treatment. Bioresour Technol 283:207–212. https://doi.org/10.1016/j.biortech.2019.03.035. PMID: 30908985

Käferböck A et al (2020) Sustainable extraction of valuable components from Spirulina assisted by pulsed electric fields technology. Algal Res 48(January). https://doi.org/10.1016/j.algal.2020.101914

Kroehnke J, Musielak G, Boratynska A (2014) Convective drying of potato assisted by ultrasound. PhD Interdiscip J 1:57–65

Ladole MR et al (2018) Synergistic effect of ultrasonication and co-immobilized enzymes on tomato peels for lycopene extraction. Ultrason Sonochem Elsevier 48(May):453–462. https://doi.org/10.1016/j.ultsonch.2018.06.013

Lammerskitten A et al (2019) Impact of pulsed electric fields on physical properties of freeze-dried apple tissue. Innovative Food Sci Emerg Technol 57(July):102211. https://doi.org/10.1016/j.ifset.2019.102211

Lammerskitten A et al (2020) An overview of the potential applications to produce healthy food products based on pulsed electric field treatment. In: Pulsed electric fields to obtain healthier and sustainable food for tomorrow. https://doi.org/10.1016/b978-0-12-816402-0.00002-1

Lingyun W et al (2007) Studies on the extracting technical conditions of inulin from Jerusalem artichoke tubers. J Food Eng 79(3):1087–1093. https://doi.org/10.1016/j.jfoodeng.2006.03.028

Liu T et al (2017) Effect of pulsed electric fields on the structure and frying quality of "kumara" sweet potato tubers. Innov Food Sci Emerg Technol Elsevier Ltd 39:197–208. https://doi.org/10.1016/j.ifset.2016.12.010

Liu C et al (2018) Effects of pulsed electric fields treatment on vacuum drying of potato tissue. LWT 95(May):289–294. https://doi.org/10.1016/j.lwt.2018.04.090

Loginov M et al (2011) Comparison of dead-end ultrafiltration behaviour and filtrate quality of sugar beet juices obtained by conventional and "cold" PEF-assisted diffusion. J Membr Sci Elsevier BV 377(1–2):273–283. https://doi.org/10.1016/j.memsci.2011.05.008

Loginova KV et al (2010) Acceleration of soluble matter extraction from chicory with pulsed electric fields. J Food Eng Elsevier Ltd 96(3):374–379. https://doi.org/10.1016/j.jfoodeng.2009.08.009

Loginova K et al (2011a) Quality and filtration characteristics of sugar beet juice obtained by "cold" extraction assisted by pulsed electric field. J Food Eng 106(2):144–151. https://doi.org/10.1016/j.jfoodeng.2011.04.017

Loginova KV, Lebovka NI, Vorobiev E (2011b) Pulsed electric field assisted aqueous extraction of colorants from red beet. J Food Eng Elsevier Ltd 106(2):127–133. https://doi.org/10.1016/j.jfoodeng.2011.04.019

Loginova K et al (2012) Better lime purification of sugar beet juice obtained by low temperature aqueous extraction assisted by pulsed electric field. LWT – Food Sci Technol Elsevier Ltd 46(1):371–374. https://doi.org/10.1016/j.lwt.2011.10.005

Luengo E, Álvarez I, Raso J (2013) Improving the pressing extraction of polyphenols of orange peel by pulsed electric fields. Innov Food Sci Emerg Technol. https://doi.org/10.1016/j.ifset.2012.10.005

Madhusankha GDMP, Thilakarathna RCN (2021) Meat tenderization mechanism and the impact of plant exogenous proteases: a review. Arab J Chem. https://doi.org/10.1016/j.arabjc.2020.102967

Mannozzi C et al (2020) The influence of different pre-treatments on the quality and nutritional characteristics in dried undersized yellow Kiwifruit. Appl Sci (Switzerland) 10(23):1–13. https://doi.org/10.3390/app10238432

Martín-García B et al (2020) Pulsed electric field (PEF) as pre-treatment to improve the phenolic compounds recovery from brewers' spent grains. Innov Food Sci Emerg Technol Elsevier Ltd 64:102402. https://doi.org/10.1016/j.ifset.2020.102402

Medina-Meza IG, Barbosa-Cánovas GV (2015) Assisted extraction of bioactive compounds from plum and grape peels by ultrasonics and pulsed electric fields. J Food Eng Elsevier Ltd 166:268–275. https://doi.org/10.1016/j.jfoodeng.2015.06.012

Mierzwa D, Kowalski SJ (2016) Ultrasound-assisted osmotic dehydration and convective drying of apples: process kinetics and quality issues. Chem Process Eng 37(3):383–391. https://doi.org/10.1515/cpe-2016-0031
Milani E, Koocheki A, Golimovahhed QA (2011) Extraction of inulin from Burdock root (Arctium lappa) using high intensity ultrasound. Int J Food Sci Technol 46(8):1699–1704. https://doi.org/10.1111/j.1365-2621.2011.02673.x
Minchev I, Petkova N, Milkova-Tomova I (2020) Ultrasound-assisted extraction of chlorophylls and phycocyanin from Spirulina platensis. Biointerface Res Appl Chem 11(2):9296–9304. https://doi.org/10.33263/BRIAC112.92969304
Minjares-Fuentes R et al (2014) Ultrasound-assisted extraction of pectins from grape pomace using citric acid: a response surface methodology approach. Carbohydr Polym. https://doi.org/10.1016/j.carbpol.2014.02.013
Morton JD et al (2017) High pressure processing improves the tenderness and quality of hot-boned beef. Meat Sci. https://doi.org/10.1016/j.meatsci.2017.06.005
Morton JD et al (2018) The physical and biochemical effects of pre-rigor high pressure processing of beef. Meat Sci. https://doi.org/10.1016/j.meatsci.2018.04.021
Mousakhani-Ganjeh A et al (2021) Electro-based technologies in food drying – a comprehensive review. LWT Elsevier Ltd 145(March):111315. https://doi.org/10.1016/j.lwt.2021.111315
Naghshineh M, Olsen K, Georgiou CA (2013) Sustainable production of pectin from lime peel by high hydrostatic pressure treatment. Food Chem Elsevier Ltd 136(2):472–478. https://doi.org/10.1016/j.foodchem.2012.08.036
Neri L et al (2021) Pulsed electric fields (PEF) as hot air drying pre-treatment: effect on quality and functional properties of saffron (Crocus sativus L.). Innov Food Sci Emerg Technol Elsevier Ltd 67(August 2020):102592. https://doi.org/10.1016/j.ifset.2020.102592
Ninčević Grassino A et al (2020) Application of high hydrostatic pressure and ultrasound-assisted extractions as a novel approach for pectin and polyphenols recovery from tomato peel waste. Innovative Food Sci Emerg Technol 64(October 2019):102424. https://doi.org/10.1016/j.ifset.2020.102424
Nowacka M et al (2019) The impact of pulsed electric field on the extraction of bioactive compounds from beetroot. Foods 8(7). https://doi.org/10.3390/foods8070244
Ostermeier R et al (2020) Pulsed electric field as a sustainable tool for the production of healthy snacks. In: Barba FJ, Parniakov O, Wiktor A (eds) Pulsed electric fields to obtain healthier and sustainable food for tomorrow. Academic Press, pp 123–128
Ostermeier R et al (2021) Influence of pulsed electric field (PEF) and ultrasound treatment on the frying behavior and quality of potato chips. Innov Food Sci Emerg Technol 67(November 2020b). https://doi.org/10.1016/j.ifset.2020.102553
Papoutsis K et al (2018) Optimizing a sustainable ultrasound-assisted extraction method for the recovery of polyphenols from lemon by-products: comparison with hot water and organic solvent extractions. Eur Food Res Technol Springer Berlin Heidelberg 244(8):1353–1365. https://doi.org/10.1007/s00217-018-3049-9
Parniakov O et al (2015) New approaches for the effective valorization of papaya seeds: extraction of proteins, phenolic compounds, carbohydrates, and isothiocyanates assisted by pulsed electric energy. Food Res Int Elsevier Ltd 77:711–717. https://doi.org/10.1016/j.foodres.2015.03.031
Pataro G et al (2018) Improved extractability of carotenoids from tomato peels as side benefits of PEF treatment of tomato fruit for more energy-efficient steam-assisted peeling. J Food Eng Elsevier Ltd 233:65–73. https://doi.org/10.1016/j.jfoodeng.2018.03.029
Pataro G et al (2020) Recovery of lycopene from industrially derived tomato processing by-products by pulsed electric fields-assisted extraction. Innov Food Sci Emerg Technol Elsevier 63(April):102369. https://doi.org/10.1016/j.ifset.2020.102369
Pedreschi F et al (2007) Physical properties of pre-treated potato chips. J Food Eng 79(4):1474–1482. https://doi.org/10.1016/j.jfoodeng.2006.04.029
Peña-González EM et al (2017) Quality and sensory profile of ultrasound-treated beef. Ital J Food Sci. https://doi.org/10.14674/1120-1770/ijfs.v604

Peña-Gonzalez E et al (2019) Ultrasound as a potential process to tenderize beef: sensory and technological parameters. Ultrason Sonochem. https://doi.org/10.1016/j.ultsonch.2018.12.045
Pez Jaeschke D et al (2021) Phycocyanin from Spirulina: a review of extraction methods and stability. Food Res Int 143(October 2020). https://doi.org/10.1016/j.foodres.2021.110314
Picart-Palmade L et al (2019) Potentialities and limits of some non-thermal technologies to improve sustainability of food processing. Front Nutr 5(January). https://doi.org/10.3389/fnut.2018.00130
Pourzaki A, Mirzaee H, Hemmati Kakhki A (2013) Using pulsed electric field for improvement of components extraction of saffron (Crocus sativus) stigma and its pomace. J Food Process Preserv 37(5):1008–1013. https://doi.org/10.1111/j.1745-4549.2012.00749.x
Puértolas E, Barba FJ (2016) Electrotechnologies applied to valorization of by-products from food industry: main findings, energy and economic cost of their industrialization. Food Bioprod Process Inst Chem Eng 100:172–184. https://doi.org/10.1016/j.fbp.2016.06.020
Puértolas E, Martínez De Marañón I (2015) Olive oil pilot-production assisted by pulsed electric field: impact on extraction yield, chemical parameters and sensory properties. Food Chem 167:497–502. https://doi.org/10.1016/j.foodchem.2014.07.029
Režek Jambrak A et al (2018) Three pillars of novel nonthermal food technologies: food safety, quality, and environment. J Food Qual 2018. https://doi.org/10.1155/2018/8619707
Ricce C et al (2016) Ultrasound pre-treatment enhances the carrot drying and rehydration. Food Res Int Elsevier Ltd 89:701–708. https://doi.org/10.1016/j.foodres.2016.09.030
Rock C, Yang W, Nooji J et al (2010) Evaluation of Roma Tomato (Solanum lycopersicum) peeling methods: conventional vs. power ultrasound. Proc Fla State Hort Soc 123:241–245
Rock C et al (2012) Conventional and alternative methods for tomato peeling. Food Eng Rev. https://doi.org/10.1007/s12393 011 9047 3
Shinwari KJ, Rao PS (2018) Thermal-assisted high hydrostatic pressure extraction of nutraceuticals from saffron (Crocus sativus): process optimization and cytotoxicity evaluation against cancer cells. Innov Food Sci Emerg Technol Elsevier 48(July):296–303. https://doi.org/10.1016/j.ifset.2018.07.003
Strati IF, Gogou E, Oreopoulou V (2015) Enzyme and high pressure assisted extraction of carotenoids from tomato waste. Food Bioprod Process Inst Chem Eng 94(September):668–674. https://doi.org/10.1016/j.fbp.2014.09.012
Suwandy V et al (2015a) Effect of pulsed electric field on the proteolysis of cold boned beef M: Longissimus lumborum and M: Semimembranosus. Meat Sci. https://doi.org/10.1016/j.meatsci.2014.10.011
Suwandy V et al (2015b) Effect of pulsed electric field treatment on the eating and keeping qualities of cold-boned beef loins: impact of initial pH and fibre orientation. Food Bioproc Tech. https://doi.org/10.1007/s11947-015-1498-8
Suwandy V et al (2015c) Effect of repeated pulsed electric field treatment on the quality of cold-boned beef loins and topsides. Food Bioproc Tech. https://doi.org/10.1007/s11947-015-1485-0
Szadzińska J et al (2017) The effect of high power airborne ultrasound and microwaves on convective drying effectiveness and quality of green pepper. Ultrason Sonochem 34:531–539. https://doi.org/10.1016/j.ultsonch.2016.06.030
Tanabtabzadeh MS, Javanbakht V, Golshirazi AH (2019) Extraction of betacyanin and betaxanthin pigments from red beetroots by chitosan extracted from shrimp wastes. Waste Biomass Valorization Springer Netherlands 10(3):641–653. https://doi.org/10.1007/s12649-017-0086-8
Tekin ZH, Baslar M (2018) The effect of ultrasound-assisted vacuum drying on the drying rate and quality of red peppers. J Therm Anal Calorim Springer Netherlands 132(2):1131–1143. https://doi.org/10.1007/s10973-018-6991-7
Tiwari BK (2015) Ultrasound: a clean, green extraction technology. TrAC Trends Anal Chem 71:100–109. https://doi.org/10.1016/j.trac.2015.04.013
Turantaş F, Kiliç GB, Kiliç B (2015) Ultrasound in the meat industry: general applications and decontamination efficiency. Int J Food Microbiol 198(April):59–69. https://doi.org/10.1016/j.ijfoodmicro.2014.12.026

Vauchel P et al (2018) Comparative LCA of ultrasound-assisted extraction of polyphenols from chicory grounds under different operational conditions. J Clean Prod 196:1116–1123. https://doi.org/10.1016/j.jclepro.2018.06.042

Veneziani G et al (2019) Extra-virgin olive oil extracted using pulsed electric field technology: cultivar impact on oil yield and quality. Front Nutr 6(September):1–8. https://doi.org/10.3389/fnut.2019.00134

Vernès L et al (2019) Application of ultrasound for green extraction of proteins from spirulina. Mechanism, optimization, modeling, and industrial prospects. Ultrason Sonochem. https://doi.org/10.1016/j.ultsonch.2019.02.016

Vorobiev E, Lebovka N (2010) Enhanced extraction from solid foods and biosuspensions by pulsed electrical energy. Food Eng Rev 2(2):95–108. https://doi.org/10.1007/s12393-010-9021-5

Vorobiev E, Lebovka N (2017) Selective extraction of molecules from biomaterials by pulsed electric field treatment. In: Handbook of electroporation. https://doi.org/10.1007/978-3-319-32886-7_163

Warner RD et al (2017) Systematic review of emerging and innovative technologies for meat tenderisation. Meat Sci 132(April):72–89. https://doi.org/10.1016/j.meatsci.2017.04.241

Won YC, Min SC, Lee DU (2015) Accelerated drying and improved color properties of red pepper by pretreatment of pulsed electric fields. Dry Technol 33(8):926–932. https://doi.org/10.1080/07373937.2014.999371

Yilmaz B, Cakmak H, Tavman S (2019) Ultrasonic pretreatment of carrot slices: effects of sonication source on drying kinetics and product quality. An Acad Bras Cienc 91(3). https://doi.org/10.1590/0001-3765201920180447

Yu X et al (2015) A new way for the oil plant biomass valorization: polyphenols and proteins extraction from rapeseed stems and leaves assisted by pulsed electric fields. Ind Crops Prod Elsevier BV 74:309–318. https://doi.org/10.1016/j.indcrop.2015.03.045

Yu Y, Jin TZ, Xiao G (2017) Effects of pulsed electric fields pretreatment and drying method on drying characteristics and nutritive quality of blueberries. J Food Process Preserv 41(6). https://doi.org/10.1111/jfpp.13303

Zhu Z, Mhemdi H (2016) Dead end ultra-filtration of sugar beet juice expressed from cold electrically pre-treated slices: effect of membrane polymer on fouling mechanism and permeate quality. Innovative Food Sci Emerg Technol 36:75–82. https://doi.org/10.1016/j.ifset.2016.05.016

Zhu Z et al (2012) Pilot scale inulin extraction from chicory roots assisted by pulsed electric fields. Int J Food Sci Technol 47(7):1361–1368. https://doi.org/10.1111/j.1365-2621.2012.02981.x

Zhu Z et al (2015) Better damage of chicory tissue by combined electroporation and Ohmic heating for solute extraction. Food Bioprod Process Inst Chem Eng 94(March):248–254. https://doi.org/10.1016/j.fbp.2014.03.003

Zhu Z et al (2016) Recent insights for the green recovery of inulin from plant food materials using non-conventional extraction technologies: a review. Innov Food Sci Emerg Technol Elsevier Ltd 33:1–9. https://doi.org/10.1016/j.ifset.2015.12.023

Part VI
Sustainable Perspective of Nonthermal Technologies

Chapter 19
New Product Development from Marine Sources and Side Streams Valorization Using Nonthermal Processing Technologies

Theofania Tsironi, Ioanna Semenoglou, and Petros Taoukis

1 Nonthermal Processing of Fish and Seafood for Improved Quality and Extended Shelf Life

The development of new food processing and packaging methods or novel combinations of existing technologies is sought by the industry in the pursuit of achieving shelf life extension, producing alternative products, improving food chain management and reducing food waste. In 2014, processed fish and seafood were among the most active new food products in the market. Fish and seafood are very perishable products with limited shelf life. The application of novel processing methods, which would decrease microbial growth rate during storage without affecting their nutritional value and sensory properties, may increase their commercial value. The lack of dissemination of validated laboratory results for the fish and seafood industry is one of the major issues preventing the uptake of nonthermal processing for fresh fish and seafood. Hurdle technology has been proposed as an effective method for preserving fish and seafood products. Based on this preservation approach, the combined application of soft hurdles may inhibit effectively microbial growth, while retaining the sensory properties of food products, compared to the application of a single, more intense preservative factor (e.g. thermal processing) (Tsironi et al. 2020).

T. Tsironi (✉)
Laboratory of Food Process Engineering, Department of Food Science and Human Nutrition, Agricultural University of Athens, Athens, Greece
e-mail: ftsironi@aua.gr

I. Semenoglou · P. Taoukis
Laboratory of Food Chemistry and Technology, School of Chemical Engineering, National Technical University of Athens, Athens, Greece
e-mail: isemen@chemeng.ntua.gr; taoukis@chemeng.ntua.gr

A. Režek Jambrak (ed.), *Nonthermal Processing in Agri-Food-Bio Sciences*, Food Engineering Series, https://doi.org/10.1007/978-3-030-92415-7_19

1.1 Effect of High Pressure (HP) Processing on Fish and Seafood

High pressure (HP) is considered as a high-scored processing method in the Technology Readiness Level (TRL > 7) scale, already applied worldwide in food industries producing meat products, juices, fruit preparations, etc. It is a nonthermal processing method which results in effective inactivation of spoilage bacteria and pathogens at room temperature, while retaining the organoleptic characteristic and nutritional value of fresh foods. For products which cannot be thermally treated, such as highly perishable fresh fish and seafood products, and require very low holding temperatures (0–2 °C) during their very short shelf life, HP treatment could allow the use of higher temperatures (5–7 °C) in the cold chain for an up to three times extended shelf life which will dramatically reduce energy and food waste (Tsironi et al. 2015; Tsironi and Taoukis 2019). Research on the effects of HP on fish tissues has established the preservative effect (microbial inactivation, shelf life extension) and potential drawbacks (effect on colour and texture of fish flesh) of this technology (Ohshima et al. 1993; Murchie et al. 2005). Several studies have evaluated the potential use of HP on different fish species, such as cod (Angsupanich and Ledward 1998), tilapia (Ko et al. 2006), salmon (Gudbjornsdottir et al. 2010), sea bass (Anjos et al. 2019; Chéret et al. 2005; Tsironi et al. 2019a), gilthead seabream (Tsironi et al. 2015) and tuna (Ramirez-Suarez and Morrissey 2006). HP has also been applied to seafoods such as abalone (Briones-Labarca et al. 2012), squid (Paarup et al. 2002), shrimp and clams (Büyükcan et al. 2009; Narwankar et al. 2011). HP has been widely applied in the commercial processing of oysters (Murchie et al. 2005). Previous work has shown that the application of HP in oyster processing offers as additional benefits, HP-induced shucking (He et al. 2002) and maintenance of flavour and nutrients (Cruz-Romero et al. 2004, 2007). In all the reported studies, HP has shown the potential to significantly decrease initial bacterial load, inhibit microbial growth and extend shelf life. However, detrimental changes in appearance, texture and chemicals in fish and seafood flesh due to HP treatment, need to be minimized by systematic investigation of the quality changes and validation of the processing parameters. HP treatment causes a reversible or irreversible structural modification in proteins, leading to protein denaturation, aggregation, or gelatinization. This feature illustrates the potential of HP technology to alter the allergenic potential of food products (Huang et al. 2014). Recently, HP has been proposed as a pre-treatment prior to frozen storage and canning for mackerel for the inhibition of chemical changes during storage (lipid oxidation and fatty acid degradation) (Prego et al. 2021), or in combination with active modified atmosphere packaging for desalted and rehydrated cod for the inactivation of spoilage bacterial (Rode and Rotabakk 2020). Specific main results from the quoted representative studies of HP on fish and seafood is tabulated in Table 19.1.

Table 19.1 Application of nonthermal processing methods for improving quality and extending shelf life of fish and seafood

Fish/ seafood product	NTP methods tested	Main results	References
Cod	High pressure (100–400 MPa, 20 min)	Myosin was denatured at 100–200 MPa and actin at 300 MPa. F or pressures >400 MPa, lipid oxidation was accelerated.	Angsupanich and Ledward (1998)
Squid	High pressure (150–400 MPa, 15 min)	Increased pressure resulted in longer shelf life at 4 °C.	Paarup et al. (2002)
Oyster	High pressure (207–310 MPa, 1–2 min)	HP reduced microbial load and extended shelf life at 4 °C.	He et al. (2002)
Oyster	High pressure (100–800 MPa, 10 min)	HP enhanced shucking and retained quality, especially at higher pressures.	Cruz-Romero et al. (2004)
European sea bass	High pressure (100–500 MPa, 5 min)	HP significantly inhibited microbial growth and extended shelf life.	Chéret et al. (2005)
Tilapia	High pressure (50–300 MPa, 1–12 h)	Extraction of water-soluble proteins was barely affected by HP extraction of salt-soluble proteins decreased 60% after 250 MPa for 1 h.	Ko et al. (2006)
Albacore tuna	High pressure (275 and 310 MPa for 2–6 min)	HP increased shelf life for 22 days at 4 °C and 93 days at −20 °C.	Ramirez-Suarez and Morrissey (2006)
Oyster	High pressure (260 MPa, 3 min)	HP enhanced shucking and retained quality. Significant colour modifications were observed.	Cruz-Romero et al. (2007)
Shrimp	High pressure (200–250 MPa, 10–20 min)	HP resulted in 12 days shelf life extension at 4 °C compared to untreated samples.	Büyükcan et al. (2009)
Clam	High pressure (200–250 MPa, 10–20 min)	HP resulted in 14 days shelf life extension at 4 °C compared to untreated samples.	Büyükcan et al. (2009)
Salmon (cold smoked)	High pressure (400–900 MPa, 10–60 s)	The application of HP for short times was effective to improve the safety of cold smoked salmon.	Gudbjornsdottir et al. (2010)
Clam	High pressure (310 MPa, 3 min)	HP >480 MPa was required to achieve at least 1 log reduction of total microbial population.	Narwankar et al. (2011)
Red abalone	High pressure (500 MPa, 8 min; 550 MPa, 3–5 min)	HP delays quality deterioration and extended shelf life at 4 °C.	Briones-Labarca et al. (2012)
Gilthead seabream	High pressure (600 MPa, 5 min)	Shelf life of the untreated fillets was 11, 7, 4 and 3 days and for the HP-treated fillets 37, 27, 17 and 10 days at 0, 5, 10 and 15 °C, respectively.	Tsironi et al. (2015)

(continued)

Table 19.1 (continued)

Fish/ seafood product	NTP methods tested	Main results	References
European sea bass	High pressure (600 MPa, 5 min)	HP resulted in more than 2 months shelf life of fish fillets at 2 °C.	Tsironi et al. (2019a) and Anjos et al. (2019)
Cod	High pressure (400–600 MPa, 5 min)	The results showed that a shelf life of minimum 49 days can be obtained by HP or by combining HP with MAP or a CO_2 emitter.	Rode and Rotabakk (2020)
Mackerel	High pressure (200–600 MPa, 2 min)	An inhibitory effect on free fatty acids content was observed in canned mackerel previously subjected to frozen storage. This effect increased with HP as a pre-treatment.	Prego et al. (2021)
Gilthead seabream	Osmotic dehydration (maltodextrine, NaCl)	Shelf-life at 5 °C was 4d for untreated fillets and 9, 11 and 13d for fillets treated with 40%, 50% and 60% maltodextrin, respectively.	Tsironi et al. (2009)
Gilthead seabream	Osmotic dehydration (maltodextrine, NaCl, nisin)	OD with the addition of nisin in combination with MAP resulted in 48 days shelf life compared to 10 days for the control samples at 0 °C.	Tsironi and Taoukis (2010)
Gilthead seabream	Osmotic dehydration (maltodextrine, NaCl, carvacrol, glucono-δ-lactone, Citrox)	Shelf life was 7 days for control samples at 5 °C. OD with carvacrol, glucono-d-lactone and Citrox allowed for shelf life extension by 8, 10 and 5 days at 5 °C, respectively.	Tsironi and Taoukis (2012)
Tilapia	Osmotic dehydration (sucrose)	Water loss and solid gain was enhanced by increasing temperature. The temperature effect on mass transfer was limited with increasing the osmotic solution concentration.	Duan et al. (2012)
Gilthead seabream	Osmotic dehydration (maltodextrine, NaCl)	A mathematical model was developed for the calculation of *Pseudomonas* spp. growth and OD fish shelf life as a function of processing conditions and storage temperature.	Tsironi and Taoukis (2014)
Gilthead seabream	Osmotic dehydration (maltodextrine, NaCl, trehalose, glucosamine)	Shelf life was 12, 19, 22 and 22 days at 0 °C for untreated and osmotically pre-treated with maltodextrine, maltodextrine+trehalose and maltodextrine+glucosamine treated fish, respectively, while the respective values at −3 °C were 21, 35, 38 and 38 days.	Tsironi and Taoukis (2017)

(continued)

Table 19.1 (continued)

Fish/ seafood product	NTP methods tested	Main results	References
Chub mackerel	Osmotic dehydration (glycerol and salt)	Storage at refrigeration temperature (7 °C) minimized the color changes of OD fish flesh.	Checmarev et al. (2017)
Tuna	Osmotic dehydration (maltodextrine, NaCl, nisin)	Shelf life was 10 days for untreated and 27 days for OD fish at 5 °C. The addition of nisin increased shelf life to 51 days at 5 °C.	Sofra et al. (2018)
Sea bass	Osmotic dehydration (oligofructose, NaCl, *Rosa damascena* phenolics)	OD extended significantly shelf life of fish in terms of microbial growth, however, it also accelerated lipid oxidation. *Rosa damascena* phenolics delayed lipid oxidation and further extended shelf life.	Giannakourou et al. (2019)
European eel	Osmotic dehydration (glycerol, NaCl, rosemary serum)	Shelf life of fish exhibited a more than ten-fold increase, as compared to control, based on chemical composition and a two to three-fold shelf life improvement, in terms of microbial growth.	Giannakourou et al. (2019)
Sea bass	Osmotic dehydration (glycerol, NaCl) Pulsed electric fields	PEF above 500 pulses resulted in significantly higher diffusion coefficients of water and solids during subsequent osmotic dehydration.	Semenoglou et al. (2020)
Pacific white shrimp	Pulsed electric fields	Shrimp treated with PEF had lower melanosis score than control during 10 days storage at 4 °C.	Shiekh and Benjakul (2020)
Atlantic salmon	Pulsed electric fields	PEF showed the potential of accelerate thawing rate and improve quality of frozen-thawed Atlantic salmon.	Li et al. (2020)
Sea bass	Pulsed electric fields	A strong effect of PEF electroporation on protein oxidation in fish was observed.	Cropotova et al. (2021)
Salmon	Cold plasma (Ar, CO_2)	*Photobacterium phosphoreum* load in cold smoked salmon was reduced up to 3 log CFU/g after 60–120 s of CP treatment.	Chiper et al. (2011)
Salmon	Cold plasma (He, O_2)	The treatment with the mixed gas inhibited *L. monocytogenes* on smoked salmon by 1.0 ± 0.3 log CFU/g.	Lee et al. (2011)
Atlantic herring	Cold plasma (air)	Inhibition of total aerobic mesophilic, total aerobic psychrotrophics, *Pseudomonas*, lactic acid bacteria and *Enterobacteriaceae* growth.	Albertos et al. (2017)
Asian sea bass	Cold plasma (Ar, O_2)	CP extended shelf life at fish for >15 days at 4 °C.	Olatunde et al. (2020)
Ready to eat fish nuggets	Cold plasma (Ar, He)	CP delayed microbial growth in RTE fish samples. *L. innouca* was more sensitive than *S. aureus* to CP.	Hajhoseini et al. (2020)

1.2 Effect of Osmotic Dehydration (OD) on Fish and Seafood

Osmotic dehydration is a nonthermal food processing method used to reduce water activity (a_w) in food products, with the aim to improve stability and extend shelf life. The method consists of an immersion of the product into a concentrated solution (e.g. sugar, salt etc). A driving force for water removal is set up because of a difference in osmotic pressure between the food and its surrounding solution (Collignan et al. 2001; Raoult-Wack 1994; Rastogi et al. 2002). Reduction in water activity of the food matrix may result in inhibition of microbial growth and thus shelf life extension. Furthermore, processing in concentrated aqueous solutions may lead in reduction of microbial load due to slight surface decontamination induced by high solute concentrations at the product/solution interface (Collignan et al. 2001). Several studies in the literature investigate the effect of different osmotic media, such as salt, sucrose and corn starch syrup, on the dehydration rate of fish products (Collignan and Raoult-Wack 1994; Mujaffar and Sankat 2006), as indicated in Table 19.1. Tsironi et al. (2009) evaluated the applicability of a maltodextrine/NaCl osmotic solution for the osmotic dehydration of gilthead seabream fillets. The preservative effect of osmotic dehydration has been also evaluated for chub mackerel (Checmarev et al. 2017) and tilapia (Duan et al. 2012). The synergistic effect of osmotic dehydration and the addition of antimicrobial agents and/or packaging at modified atmospheres or superchilling storage has been extensively studied for gilthead seabream (Tsironi and Taoukis 2010, 2012, 2014, 2017), sea bass (Giannakourou et al. 2019), tuna (Sofra et al. 2018) and eel (Giannakourou et al. 2020). Specific main results from the quoted research is tabulated in Table 19.1.

1.3 Effect of Pulsed Electric Fields (PEF) on Fish and Seafood

Pulsed electric fields (PEF) treatment has been reported as an innovative nonthermal processing method, which involves the application of high-voltage pulses (electric field strength of 0.1–40 kV/cm, and total energy input of 0.5–1000 kJ/kg) to electroconductive foods placed between two electrodes. The resultant electric field induces movements of ions and permeabilization of cell membranes called electroporation and this has been exploited for microbial inactivation and to enhance extraction, mainly, of plant material. It has been studied primarily as a means of microbial inactivation in fish but recently several studies have focused on its application for mass transfer enhancement (Tsironi and Taoukis 2019), as presented in Table 19.1. Some of the advantages of PEF treatment over traditional processing are the very short time (milliseconds to microseconds) of treatment and continuous operability (Johnson et al. 2010). An important recent trend is the introduction of PEF application prior to further processing of fish where mass transfer is required, for example curing, marinating or drying, in order to enhance the process efficacy or significantly reducing treatment time (Blahovec et al. 2017). Shiekh and Benjakul (2020) evaluated the effect of PEF on the inhibition of melanosis in Pacific white shrimp during refrigerated storage. The polyphenol oxidase (PPO) activity in cephalothorax was decreased as specific energy (54–483 kJ/kg) and pulse numbers (200–600) increased. Semenoglou et al. (2020) evaluated the effect of osmotic solution concentration (aquatic solution 40–60% glycerol) and PEF (1.6 kV/cm, up to 1500 pulses/19.7 kJ/kg) on mass transfer during osmotic dehydration of sea bass fillets. Increase of initial concentration of the osmotic solution enhanced water loss, solid gain and water activity reduction and decreased salt content in fish flesh. PEF above 500 pulses resulted in higher diffusion coefficients of water and solids, compared to untreated fish fillets, during subsequent osmotic dehydration. A sigmoidal mathematical model was developed for the calculation of moisture and solute diffusivities, as a function of energy input. The combined effect of pulsed electric field and salting in a brine with 5–10% NaCl on oxidative stability of lipids and proteins and color characteristics of sea bass, was evaluated by Cropotova et al. (2021). PEF has been reported as an effective pre-treatment to decrease the thawing time by 20 min from −2 to 0 °C in frozen Atlantic salmon. In a study by Li et al. (2020), the muscle fiber of PEF-treated sample was preserved better after thawing, resulting in the total loss reduced by 6%. Specific main results from the quoted studies of PEF on fish and seafood is tabulated in Table 19.1.

1.4 *Effect of Cold Plasma (CP) on Fish and Seafood*

Cold plasma processing is an emerging foods preservation technology with high potential (Kim et al. 2016). CP is commonly referred to as the fourth state of matter, where increases in a material's internal energy converts its state from solid to liquid to gas and ultimately to an ionized state of the gas, 'plasma', and exhibits specific properties (Bourke et al. 2018). Recent studies have shown the potential of using CP in order to achieve a limited reduction in the initial microbial load of up to 1 log CFU/g for smoked salmon (Lee et al. 2011) or up to 3 log CFU/g for specific bacteria, such as *Photobacterium phosphoreum* in smoked salmon (Chiper et al. 2011). An overview of studies on application of cold plasma treatment on fish products and main results are presented in Table 19.1. The achieved disinfection by CP treatment is depending on the selected processing conditions. Hajhoseini et al. (2020) reported a significant inhibitory effect of CP on the growth of *Listeria innocua* in ready to eat fish products. In-package CP treatment has been reported as a promising disinfection method for food, with the aim to reduce surface microbial load and thus extend shelf life. The effect of in-bag dielectric barrier discharge cold plasma on the preservation of Asian sea bass slices has been investigated by Olatunde et al. (2020). Shelf life extension by in-package CP treatment has been also reported for Atlantic herring (Albertos et al. 2017). Plasma-activated water (PAW) may be an effective method of CP processing of foods, as an eco-friendly alternative to conventional disinfection methods (Xiang et al. 2020). However, detrimental effects have been observed in several cases, concerning the acceleration of lipid oxidation in CP treated fish flesh during subsequent refrigerated storage (Misra et al. 2019; Park and Ha 2014).

1.5 Case Study: Effect of Nonthermal Processing on Shelf Life of Chilled Whole and Filleted Gilthead Seabream

The effect of nonthermal processing on microbial growth in filleted and whole gutted gilthead seabream (*Sparus aurata*) is illustrated in Fig. 19.1a, b. Total viable count (TVC) growth in gilthead seabream treated with high pressure (HP, 600 MPa, 3 min, 25 °C) or osmotic dehydration (OD, osmotic solution 50% maltodextrine+5% NaCl, 45 min, 15 °C) is compared with the application of modified atmosphere packaging (MAP, 50% CO_2, 10% O_2, 40% N_2) and with untreated and aerobically packed (Control) samples at refrigerated storage (5 °C). The shelf life of fish at 5 °C defined as the time to reach a TVC of10^7 CFU/g correlated to the limit of sensory acceptability (ICMSF 1986; Tsironi et al. 2019b), was calculated as 4, 8, 38 and 12 days for Control, OD, HP and MAP filleted gilthead seabream, respectively. For gutted gilthead seabream, the shelf life at 5 °C was calculated as 7, 14, >40 and 13 days for Control, OD, HP and MAP samples, respectively.

2 Nonthermal Processing for Valorization of Fish and Seafood Side Streams

2.1 Fish and Seafood Side Streams

There is a steadily increasing demand for fish and seafood products with the annual fish consumption from 9.0 kg per capita in 1961, reaching 20.5 kg in 2018. The increase of their consumption due to their high nutritional value in combination with the increasing population led aquaculture to be one of the fastest growing food production sectors with 3.1% average annual growth rate the last decades compared with 2.1% of all other animal protein food sectors. According to FAO (2020b), in 2018 total production of fish and seafood approached 180 million tons of which 46% came from aquaculture and the remaining amount from marine (7%) and inland capture (47%). Aquaculture production consisted of finfish (54.3 million ton), molluscs (17.7 million ton) and crustaceans (9.4 million ton) (FAO 2020a, b).

Consumers demand for high quality, added value and ready-to-eat fish products and the enormous fish production generate significant quantities of nonedible by-products every year which are discarded as waste or underutilized. Fish side streams produced from harvesting, processing and distribution in the supply chain can be up to 35% of the global catches. Fish discards refer to the part of catches which are

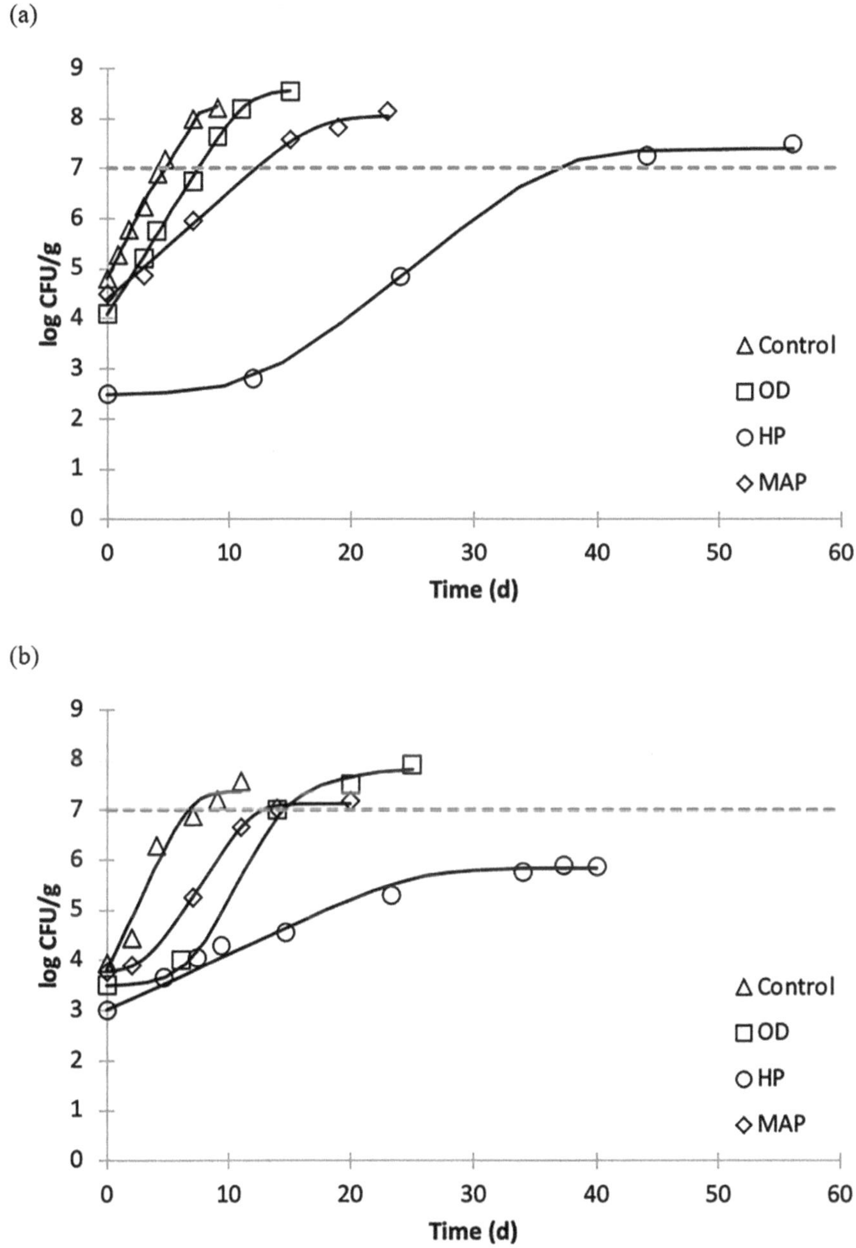

Fig. 19.1 Total viable count in (**a**) filleted and (**b**) gutted gilthead seabream (*Sparus aurata*) during isothermal storage at 5 °C. Control: Untreated and aerobically packed; OD: Osmotically dehydrated using an aquatic osmotic solution with 50% maltodextrine+5% NaCl (45 min, 15 °C) and aerobically packed; HP: in-pack (vacuum) treated with high pressure (600 MPa, 3 min, 25 °C); MAP: untreated and packed in modified atmosphere (50% CO_2, 10% O_2, 40% N_2). The dotted, grey line indicates the limit of acceptability = 10^7 CFU/g for TVC

returned to the sea during fishing due to unmarketable, damaged or undersized species or due to regulations. These wastes have been significantly reduced the last decades mainly because of policies aiming to elimination of these quantities, however they still constitute 9–15% of the total losses (FAO 2011, 2018). In terms of processing, the applied methods after landing depend on the species and the desired final products. The main stages for finfish include fish filleting, evisceration and scaling and lead to large amount of side streams which can represent 20–70% of the initial mass of processed fish. The produced side streams are mainly viscera which represent 12–18% of total fish, heads (9–12%), bones (9–15%) with attached flesh (frames), skin (1–3%), tails, fins, scales (5%), gills and blood and their amount depends on species, season, size and fishing area (FAO 2018, 2020b; Rustad et al. 2011). Moreover, crustaceans and molluscs processing also leads to huge quantities of underutilized side streams such as shells, heads and tails. These by-products represent about 35–70% of the total weight and correspond to 0.5 million ton per year (Hamed et al. 2016; Lopez et al. 2015).

Fish and seafood by-products are susceptible to spoilage due to chemical reactions, microbial growth and enzymatic load and thus their immediate management or valorization is crucial in order to reduce the environmental impact of fish industries side streams. These materials are rich in bioactive compounds such as polyunsaturated (PUFA) and monounsaturated fatty acids (MUFA), proteins, amino acids, enzymes and minerals similar to the edible parts of seafood. As current practice they are thrown away or underutilized basically for the production of animal feed, fish oil and organic fertilizers, while they could optimally be used as high added value ingredients in food, pharmaceutical, nutraceutical and biotechnological applications. The last decades, improved nonthermal and environmentally friendly processes including supercritical fluid extraction (SFE), pulsed electric field and high pressure, have been investigated to optimize the recovery of these bioactive components in the framework of a more sustainable utilization of fish processing side streams (Al Khawli et al. 2019; Bruno et al. 2019; FAO 2018, 2020b).

2.2 *Application of Supercritical Fluid Extraction on Fish and Seafood Side Streams*

SFE is a nonthermal technology that has been introduced in food industry as an alternative to conventional solvent extraction in order to recover heat-sensitive compounds. CO_2 is an appropriate green solvent due to the mild conditions at its critical point and the easy separation from the extract. Supercritical CO_2 extraction presents significant selectivity for non-polar molecules because of the low solvent polarity. In case of polar compounds, the addition of co-solvent such as ethanol has been mentioned to change solvent polarity (Ivanovs and Blumberga 2017). The last decades, SFE research has focused on fatty acids (FA) and other lipid-soluble molecules from fish and seafood side streams (Table 19.2).

Table 19.2 Application of supercritical fluid extraction on fish and seafood side streams

Fish/seafood side streams	Bioactive compounds	SFE conditions	Main SFE results	Reference
Crawfish (*Procambarus clarkii*) shells	Astaxanthin	P: 13.8–31.0 MPa T: 50–70 °C F: 1.0–1.5 L min^{-1} Ethanol: 10% (w/w)	Optimal conditions: P = 31 MPa, T = 60 °C Max yield: 207.6 mg kg^{-1} Co-solvent enhanced extraction	Charest et al. (2001)
Indian mackerel (*Rastrelliger kanagurta*) skin	Oil	P: 20–35 MPa T: 45–75 °C F: 2 ml min^{-1} Cont: 6 h Co-solv: 6 h Sk: soaking for 10 h, SFE for 5 h PS: pressurization for 5 min and SFE for 1 h repeated 10 times	Optimal conditions: P = 35 MPa, T = 75 °C Similar yield and PUFA consistency with Soxhlet (lower on continuous) Sk and PS are most effective to recover AA, DHA, EPA Highest CO_2 consumption in cont-SFE	Sahena et al. (2010a)
Indian mackerel (*Rastrelliger kanagurta*) flesh, skin, viscera and heads	Oil	P: 35 MPa T: 60 °C F: 2 ml min^{-1} t: Cont: 5–8 h Co-solv: 4–6 h Sk: soaking for 10 h, SFE for 3.5–5 h PS: pressurization for 2 h and SFE for 3 h repeated 5–6 times	All techniques had similar yield and PUFA consistency with Soxhlet Sk and PS are highly effective to recover AA, DHA, EPA	Sahena et al. (2010b)
Redspotted shrimp (*Farfantepenaeus paulensis*) heads, tails and shells	Lipids and carotenoids	P: 30 MPa T: 50 °C Ethanol: 10% (w/w)	Co-solvent enhanced lipids and astaxanthin extraction but yields were lower compared to solvent extraction	Sánchez-Camargo et al. (2011a)
Redspotted shrimp (*Farfantepenaeus paulensis*) heads, tails and shells	Lipids and carotenoids	P: 20–40 MPa T: 40–60 °C F: $4.17*10^{-5}$ kg s^{-1}	Optimal conditions: P = 37 MPa, T = 43 °C Max astaxanthin recovery: 39%	Sánchez-Camargo et al. (2011b)
Northern shrimp (*Pandalus borealis* Kreyer) heads, tails and shells	PUFA	15 MPa, 50 °C; 35 MPa, 40 °C F: 3–5 L min^{-1} t: 90 min	Lower yield compared to Soxhlet extraction	Amiguet et al. (2012)

(continued)

Table 19.2 (continued)

Fish/seafood side streams	Bioactive compounds	SFE conditions	Main SFE results	Reference
Yellow cracker (*Larimichthys polyactis*) muscle	Oil	P: 15–25 MPa T: 35–45 °C F: 27.79 g min^{-1} t: 1.5 h	Optimal conditions: P = 25 MPa, T = 35 °C Max yield: 21.36% Higher yield, PUFA and stability (lower PV and AV) compared to Soxhlet with hexane	Lee et al. (2012)
Hake (*Merluccius capensis–Merluccius paradoxus*) off-cuts (skin and muscles)	Oil	P: 25 MPa T: 40 °C	Prevention of lipid oxidation and reduction of impurities. Higher protein concentration on the remaining fish meal.	Rubio-Rodríguez et al. (2012)
Orange roughy (*Hoplostethus atlanticus*) off-cuts (skin and muscles)	Oil	P: 25 MPa T: 40 °C	Similar colour and FA, lower acidity value and impurities, better prevention of lipid oxidation and sensory properties compared to CE, EE and WR	Rubio-Rodríguez et al. (2012)
Salmon (*Salmo salar*) off-cuts (skin and muscles)	Oil	P: 25 MPa T: 40 °C	Similar colour and FA, lower acidity value and impurities, better prevention of lipid oxidation and sensory properties compared to CE, EE and WR	Rubio-Rodríguez et al. (2012)
Jumbo squid (*Dosidicus gigas*) liver	Oil	P: 25 MPa T: 40 °C	SFE prevents lipid oxidation and reduces impurities Higher protein concentration on the remaining fish meal	Rubio-Rodríguez et al. (2012)
Indian mackerel (*Rastrelliger kanagurta*) viscera	Oil	P: 20–35 MPa T: 45–75 °C F: 1–3 ml min^{-1} Cont: 6 h Co-solv: 4 h Sk: soaking for 10 h, SFE for 4 h PS: pressurization for 2 h and SFE for 3 h repeated 5 times	Optimal conditions: P = 35 MPa, T = 60 °C, 2 ml min^{-1} Sk and PS are most effective for oil recovery. Highest CO_2 consumption in Cont.	Sahena et al. (2012)
Redspotted shrimp (*Farfantepenaeus paulensis*) heads, tails and shells	Lipids and carotenoids	P: 30 MPa T: 50 °C F: CO_2: 3 L min^{-1} Ethanol: 5–15% (w/w) t: 100 min	Co-solvent significantly enhanced lipids and astaxanthin extraction (93.8% and 65.2% recovery respectively).	Sánchez-Camargo et al. (2012)

(continued)

Table 19.2 (continued)

Fish/seafood side streams	Bioactive compounds	SFE conditions	Main SFE results	Reference
African Catfish viscera	Oil	P: 10–40 MPa T: 35–80 °C F: 1–3 ml min^{-1} t: 1–4 h	Optimal conditions: P = 40 MPa, T = 57.5 °C, F = 2 ml min^{-1}, t = 2.5 h Max yield: 67% Lower yield compared to Soxhlet with hexane.	Sarker et al. (2012)
Common carp (*Cyprinus carpio* L.) viscera	Oil (PUFA)	P: 20–40 MPa T: 40–60 °C F: 0.194–0.354 kg h^{-1} t: 30–180 min	Optimal conditions: P = 40 MPa, T = 60 °C, F = 0.354 kg h^{-1} Max PUFA yield: 25.24%	Lisichkov et al. (2014)
Tuna (from 3 species) head, skin, viscera	Oil	P: 400 T: 65 °C F: 3 ml min^{-1} 20% Ethanol t: 120 min	Similar yield and FA with Soxhlet n-hexane. Lower FFA and PV.	Sahena et al. (2014)
Salmon (*Salmo salar Linnaeus*) offcuts (heads, trimmings, frames)	Oil	P: 15–35 MPa T: 40–80 °C F: 0.18–0.48 kg h^{-1} t: 120 min d: 125–710 μm	Optimal conditions: P = 35 MPa, T = 60 °C, F = 0.18 kg h^{-1} Max yield: 21.36%	Adeoti and Hawboldt (2015)
Rock Lobsters (*Jasus edwardsii*) livers	PUFA	P: 25–35 MPa T: 50–60 °C F: 0.434 kg h^{-1} t: 240 h	Optimal conditions: P = 35 MPa, T = 50 °C Max oil recovery: 94% Higher PUFA and reduction of heavy metals compared to Soxhlet.	Nguyen et al. (2015)
Atlantic salmon belly part, muscle, frame bone and skin	Oil	P: 25 MPa T: 45 °C F: 27 g min^{-1} t: 180 min	Lower yield compared to hexane extraction and higher than pressed oil. Better colour and viscosity, oil stability (lower PV, AV, FFA) and antioxidant activity.	Haq et al. (2017)
Common carp (*Cyprinus carpio* L.) flesh, viscera and caviar	Oil (MUFA, PUFA)	P: 20–40 MPa T: 40–60 °C t: 15–180 min	Higher yield compared to Soxhlet with petroleum ether and lower yield than methylene chloride and hexane extracts. Highest concentration of MUFA and PUFA.	Kuvendziev et al. (2018)

The yield and the quality of extracts from fish industry by-products mainly depend on the applied pressure (P), temperature (T), solvent flow rate (F) and time (t). Temperature increase under constant pressure results in lower solvent density and higher solutes vapour pressure which have opposite influence on oil solubility. So, the final yield depends on the stronger effect between these two. The complex impact of pressure and temperature has been reported for extracted oil from different sources such as from common carp viscera (Kuvendziev et al. 2018), salmon offcuts (Adeoti and Hawboldt 2015), common carp viscera (Lisichkov et al. 2014), catfish viscera (Sarker et al. 2012), yellow cracker muscle (Lee et al. 2012), Indian mackerel viscera and skin (Sahena et al. 2010a, 2012). Authors have concluded that increasing temperature at lower pressures had negative effect on the yield, while at higher pressures enhanced the extraction. In terms of solvent flow, it was observed that CO_2 flow rate positively affected the recovery of oils (Adeoti and Hawboldt 2015; Lisichkov et al. 2014; Sahena et al. 2012). Moreover, Adeoti and Hawboldt (2015) have concluded that reducing particle size enhanced extraction.

Most of the SFE studies have evaluated the recovery of bioactive compounds using continuous extraction with or without co-solvent, while application of different methods is extremely limited in the literature. Some studies have evaluated the use of co-solvent (co-solv), the soaking (Sk) and pressure swing technique (PS) compared to continuous (cont) on different side streams. Sahena et al. (2010a) have studied oil extraction from Indian mackerel skin using the four methods in a wide range of pressures (20–35 MPa) and temperatures (45–75 °C). Finally, the optimal conditions for each method were at 35 MPa and 75 °C. Co-solvent, soaking and PS method resulted in significant recovery of oil and PUFA, similar to the Soxhlet, while continuous extraction was not sufficient. Moreover, Sahena et al. (2012) have investigated viscera oil produced with the same techniques at the optimal conditions (35 MPa, 60 °C). The maximum oil yield with the different SFE methods was close to the yield with Soxhlet extraction. In addition, the highest CO_2 consumption found in the continuous method and the lowest in soaking and PS. They have also concluded that soaking and PS methods were appropriate for PUFA recovery such as the docosahexaenoic acid (DHA), the eicosapentaenoic acid (EPA) and the arachidonic acid (AA). The same conclusions in terms of yield and PUFA have been reported by Sahena et al. (2010b).

Comparison of SFE with other extraction technologies has been reported in different studies which mainly focused on solvent extraction and specifically on Soxhlet with hexane, petroleum ether or methylene chloride as solvents. Oils obtained by SFE from common carp by-products (Kuvendziev et al. 2018), Atlantic salmon by-products (Haq et al. 2017), African catfish viscera (Sarker et al. 2012) and yellow cracker by-products (Lee et al. 2012) resulted in lower lipid recovery compared to Soxhlet extraction. On the other hand, some studies referred to extracted oil from tuna head, skin and viscera (Sahena et al. 2014) and Indian mackerel side streams (Sahena et al. 2010b) concluded that SFE and Soxhlet yielded similar amounts of extracts. Apart from oil and FA recovery, quality of extracts plays a crucial role in the utilization of SFE for the valorization of fish and seafood industry side streams. In this framework, it has been noted that SFE extracts were characterized by higher antioxidant activity and oil stability compared to convention extraction, with regard to peroxide value (PV), anisidine value (AV), and free fatty acids (FFA) (Haq et al. 2017; Lee et al. 2012; Sahena et al. 2014). Rubio-Rodríguez et al. (2012) have compared SFE of various fish by-products (salmon, orange roughy, hake off-cuts and jumbo squid liver) with cold extraction (CE), wet reduction (WR) and enzymatic extraction (EE). Extraction at 25 MPa and 40 °C resulted in similar FA composition and colour compared to CE, WR and EE. However, SFE preserved lipid oxidation and reduced the pollutants (heavy metals) concentration on the final extract which were presented in the raw material.

In case of seafood side streams, SFE has been used to extract lipids and carotenoids which are present in these materials. Nguyen et al. (2015) have studied PUFA extraction from lobster livers with SFE (25–35 MPa, 50–60 °C) and Soxhlet extraction. SFE resulted in slightly lower lipids recovery but higher PUFA concentration in the final extract. In addition, SFE significantly reduced the presence of heavy metals in the extract. The use of ethanol as co-solvent has been reported in different studies referred to the recovery of astaxanthin with or without the simultaneous lipid recovery. It has been reported that co-solvent enhanced astaxanthin extraction compared to SFE with pure CO_2 (Charest et al. 2001; Sánchez-Camargo et al. 2011a, 2012). Sánchez-Camargo et al. (2011a) have concluded that use of ethanol enhanced astaxanthin extraction while did not affect lipid extraction.

2.3 Application of Pulsed Electric Field on Fish and Seafood Side Streams

PEF is a promising nonthermal processing method for the valorization of fish and seafood by-products through extraction of high added value compounds. The application of high-voltage pulses resulted in pores formation to cells which can enhance extractability of compounds from fish and seafood side streams (He et al. 2017; Semenoglou et al. 2020). Despite the advantages of this technique especially for the extraction of heat-sensitive ingredients such as PUFA, PEF has not been widely used for the valorization of fish by-products as shown in Table 19.3. The effect of PEF on the final extract or product depends on several process parameters such as the pulse shape and number (n), electric field strength (E) and solid material to liquid ratio (s/l) or liquid to solid ratio (l/s).

Several studies have focused on extraction of valuable compounds from fish bones such as chondroitin sulfate (CS), calcium (Ca) and collagen. He et al. (2014) examined the effect of 3 different factors on extraction of chondroitin sulfate. The maximum recovery (6.9 g L^{-1}) was determined by single-factor analysis and the optimal conditions were found at 16.88 kV cm^{-1}, 9 pulses and 1:15 g mL^{-1} material to liquid ratio. Comparing the results at optimal conditions with enzymatic, alkali and ultrasonic (US) extraction, PEF accelerated the extraction and resulted in higher yield 2.02, 1.84 and 1.42 times, respectively and the extracts were characterized by

Table 19.3 Application of pulsed electric field processing on fish and seafood side streams

Fish/seafood side streams	Bioactive compounds	PEF conditions	Main PEF results	Reference
Fish bones	Chondroitin sulphate	E: 5–25 kV cm^{-1} n: 2–12 pulses s/l: 1:5–1:25	Optimal conditions: E = 16.88 kV cm^{-1}, n = 9, s/l = 1:15 Max yield = 6.9 g L^{-1} Higher yield (up to 2 times) and purity compared to EE, US and alkali method.	He et al. (2014)
Fish bones	Chondroitin sulphate, calcium, collagen	E: 20–25 kV cm^{-1} n: 8–10 pulses l/s ratio: 10–12	Optimal conditions: E = 22.79 kV cm^{-1}, n = 11, l/s = 11 Max yield: 19.8, 39.3 and 3.9 mg mL^{-1} for Ca, CS and collagen respectively.	He et al. (2016)
Abalone (*Haliotis discus hannai* Ino) viscera	Protein hydrolysates	E: 5–20 kV cm^{-1} t: 100–800 μs s/l: 1:8–1:14	Optimal conditions: E = 20 kV cm^{-1}, t = 600 μs, s/l = 4:1 Max yield = 39.99% Higher DH, better emulsifying properties and solubility than EE.	Li et al. (2016a)
Sea bass skin	Defatting	E: 16–24 kV cm^{-1} t: 36–108 ms s/l: 1:8	Optimal conditions: E = 24 kV cm^{-1}, t = 72 μs Max defatting: 86.9% Lower MUFA and PUFA in the treated skin. Lower fishy odor/flavor and volatile compounds.	Chotphruethipong et al. (2019)
Bighead carp (*Hypophthalmichthys nobilis*) bones	Collagen	E: 5–25 kV cm^{-1} n: 2–12 pulses s/l: 1:8–1:14	Optimal conditions: E = 20 kV cm^{-1}, n = 8, s/l = 1:10 Max yield = 16.1 mg mL^{-1}	He et al. (2019)
Sea bass and sea bream heads, bones and gills	Peptides and amino acids	E: 1.4 kV cm^{-1} n: 100 pulses s/l: 1:10	Improved antioxidant activity compared to conventional extraction.	Franco et al. (2020)
Pacific white shrimp (*Litopenaeus vannamei*) cephalothorax	Lipids	E: 4–16 kV cm^{-1} n: 120–240 pulses s/l: 1:5	Enhanced yield and decreased PV, TBARs and FFA compared to SE and UAE.	Gulzar and Benjakul (2020)

higher purity. He et al. (2019) evaluated collagen extraction in a wide range of electric field strengths (5–25 kV cm^{-1}), number of pulses (2–12) and solid to solvent ratio (s/l: 1:5–1:25). The highest yield was achieved at 20 kV cm^{-1}, 8 pulses and 1:10 ratio and it was equal to 16.13 mg mL^{-1}. In another study, the simultaneous extraction of CS, collagen and calcium (Ca) from fish bones has been investigated (He et al. 2016). The optimum conditions (22.79 kV cm^{-1}, 9 pulses, 11 liquid to solid ratio) were determined through the Response Surface Methodology (RSM). Under these conditions, CS concentration was 39.3 mg mL^{-1}, Ca was 19.8 mg mL^{-1} and extracted collagen was 3.87 mg mL^{-1}.

Protein enzymatic extraction assisted by PEF from viscera has been developed by Li et al. (2016a) as an alternative for the valorization of fish side streams. They studied several conditions of field strength (up to 20 kV cm^{-1}), treatment time (up to 800 μs) and solid to liquid ratio (3:1–10:1). The highest extraction yield of protein hydrolysates was equal to 39.99% and was achieved after treatment at 20 kV cm^{-1}, for 600 and a solid to liquid ratio of 4:1. This recovery was higher than conventional enzymatic extraction and the extract was characterized by improved solubility, emulsifying properties and higher degree of hydrolysis (DH). Peptides and amino acids extraction with and without PEF treatment has been evaluated by Franco et al. (2020) using sea bass and sea bream side streams consisted of heads, bones and gills. The pre treatment enhanced the antioxidant capacity compared to conventional solvent extraction with water or methanol. In addition, PEF has been proposed as an effective method for defatting of sea bass skin followed by enzymatic removal of lipids (Chotphruethipong et al. 2019). Optimal conditions were defined at 24 kV cm^{-1} and 72 ms in which lipid removal reached 86.93%. The defatted skin was characterized by lower MUFA and PUFA concentration and could further be used for collagen hydrolysis with reduced fishy odor and volatile compounds.

PEF treatment is extremely limited on seafood side streams. Gulzar and Benjakul (2019) evaluated the effect of this processing on the recovery of lipids from Pacific white shrimp cephalothorax. The increase of both number of pulses and field strength up to 12 kV cm^{-1}, increased the yield of oil. At optimal PEF conditions, the yield was higher compared to conventional solvent extraction (SE) and ultrasound-assisted extraction (UAE) and the extracts were less oxidized, in terms of PV, TBARs and FFA and specifically, the higher the electric field strength, the lower the oxidation level due to inactivation of oxidative enzymes.

2.4 *Application of High Pressure on Fish and Seafood Side Streams*

HP treatment on fish and seafood industry has been used to extend their shelf life or to develop new products as mentioned on Sect. 1.1. HP could also effectively be applied as an alternative for valorization of industry side streams as it can affect cellular membranes enhancing mass transfer rates and extraction of high added value biomolecules. Application of HP on fish or seafood by-products has been limited (Bruno et al. 2019; Tsironi et al. 2019a). The yield and quality of final product are affected by different parameters such as the pressure (P), process time (t) and solid to solvent ratio (s/l) or liquid to solid ratio (l/s).

Gómez-Guillén et al. (2005) have examined the effect of HP (up to 400 MPa, 10–20 min) on collagen recovery from Dover sole (*Solea vulgaris*) skin either as pre-treatment or during the extraction. Authors mentioned that HP as pre-treatment increased the recovery compared to conventional extraction but peptides of lower molecular weight (MW ≤100 kDa) were found on the extract. In addition, increasing pressure or time during extraction, decreased the yield and resulted in high MW polymers. However, HP took place in a matter of minutes while conventional treatment required many hours of extraction. In another study, Hemker et al. (2020) have studied HP-assisted enzymatic hydrolysis (with Alcalase) of tilapia (*Oreochromis niloticus*) by-products consisted of heads, tails and fins. They concluded that maximum soluble protein content (5.7 mg mL^{-1}) was achieved at 250 MPa and 35 min and the hydrolysates recovery was higher compared to enzymatic extraction. At mild pressures, HP enhanced enzyme activity while at higher pressures, proteins aggregation negatively affected Alcalase activity. Moreover, HP improved antioxidant activity of extract and solubility of amino acids and peptides of lower MW, while decreased water and oil holding capacity.

HP is also useful for recovering compounds from seafood side streams. Li et al. (2016b) studied the effect of HP (0–600 MPa, 0–20 min, 10–50 mL g^{-1} solvent to solid ratio) on astaxanthin extraction from shrimp waste. The optimum yield was appeared at 200 MPa, 5 min and s/l ratio equal to 20 mL g^{-1}. The recovered astaxanthin by HP was higher and the extract was characterized by higher antioxidant activity compared to solvent extraction. Huang and Tsai (2020) proposed HP as an alternative to extract chitosan from squid pen and specifically, the applied pressure (100–500 MPa) and time (5–15 min) were optimized through RSM. Finally, extraction at 500 MPa and 10 min resulted in highest yield of chitosan (81.9%) and improved antioxidant activity and physicochemical properties in terms of oil and water holding capacity and solubility.

3 Nonthermal Processing of Marine Organisms for Energy Production

The utilization of marine biomass for the development of added value products has received researchers' attention over the last 20 years, although algae have been harvested extensively for the pharmaceutical and food sectors. Plant biomass in general, is the result of the interaction of atmospheric CO_2, H_2O and dissolved salts with the light, with production of phytomass accompanied by atmospheric O_2 production. Referring to marine or freshwater algae, CO_2 and O_2 are dissolved into the water, where the plants live and is absorbed. Seaweed, in addition to algae, can be used as a biomass source of energy (Charlier and Justus 1993). There are no accumulations of living biomass in the marine environment that compare with the forests and grasslands on land. Ocean biology is responsible for the storage of more carbon away from the atmosphere than is the terrestrial biosphere. This is achieved by the sinking of organic matter out of the surface ocean and into the ocean interior before it is returned to dissolved inorganic carbon and dissolved nutrients by bacterial decomposition (Sigman and Hain 2012).

Productivity is affected by wave action and temperature (mainly above 20 °C), sediment deposition and chemical reactions in the surrounding environment. Plant growth may be stimulated mainly by water motion enhance (Charlier and Justus 1993). Process of biorefinery, manipulation of metabolic pathways and advanced genetic engineering of marine organisms have been investigated in order to improve the feasibility of industrial biomass and biofuel production. Manipulation of metabolic pathways may result in stimulation of the growth or redirection of the cellular metabolism towards the production of a specific component. Alternative methods are also being investigated with the aim to efficiently seed several kelp plants (Charlier and Justus 1993; Almarashi et al. 2020). The application of cold plasma has been reported as a potential pre-processing step to enable the random mutagenesis of microalgae and enhancement of a certain metabolite (An et al. 2017; Cao et al. 2017). Almarashi et al. (2020) introduced recently an alternative procedure to enhance biodiesel recovery from the green microalga *Chlorella vulgaris* by the application of low doses of cold atmospheric plasma. Treatment for 60 s resulted in the maximus gross energy output (approximately 80% higher than the respective value at 30 s). PEF technology has been additionally reported as a potentially effective pretreatment method for microalgae cell disruption for biodiesel production. A quick and efficient approach to cell lysing may enhance the process of biofuel production and extraction of components, decreasing the total production costs (Joannes et al. 2015; Kempkes 2016).

4 Conclusions

Nonthermal processing technologies such as OD, PEF, HP, CP and SFE can effectively be applied as alternative methods in fish and seafood industry in terms of inhibiting microbial growth and enzymatic reactions, improved quality and valorization of the produced side streams. The economic impact of nonthermal technologies for fish and seafood is likely to result in a significant increase in shelf life that is also reflected in a reduction in distribution costs (approx. 25%). Shelf life extension of fish and seafood products may open for new distant markets, currently inaccessible to fresh products, providing a significant commercial opportunity to exploit the production capacity of the fish and seafood industry, which currently in Europe exceeds demand (due to competition from low cost import species) often leading to waste and below cost selling. Moreover, nonthermal processing could be effectively used to extract high-quality and high purity bioactive compounds such as unsaturated fatty acids, proteins, amino acids, antioxidants. These methods are environmentally friendly and compared to the conventional technologies can increase the recovery, reduce processing time and the use of toxic solvents. In this framework, it can effectively be applied as an alternative for valorization and a more sustainable utilization of fish and seafood side streams.

References

Adeoti A, Hawboldt K (2015) Experimental and mass transfer modelling of oil extraction from salmon processing waste using SC-CO_2. J Supercrit Fluids 104:160–170

Al Khawli F, Pateiro M, Domínguez R, Lorenzo J, Gullón P, Kousoulak K, Ferrer E, Berrada H, Barba F (2019) Innovative green technologies of intensification for valorization of seafood and their by-products. Mar Drugs 17(12):689. https://doi.org/10.3390/md17120689

Albertos I, Martin-Diana AB, Cullen PJ, Tiwari BK, Shikha Ojha K, Bourke P, Rico D (2017) Shelf-life extension of herring (*Clupea harengus*) using in-package atmospheric plasma technology. Innov Food Sci Emerg Technol 53:85–91. https://doi.org/10.1016/j.ifset.2017.09.010

Almarashi JQM, El-Zohary SE, Ellabban MA, El-Fatah Abomohra A (2020) Enhancement of lipid production and energy recovery from the green microalga *Chlorella vulgaris* by inoculum pretreatment with low-dose cold atmospheric pressure plasma (CAPP). Energy Convers Manag 204:112314

Amiguet VT, Kramp KL, Mao J, McRae C, Goulah A, Kimpe LE, Blais JM, Arnason JT (2012) Supercritical carbon dioxide extraction of polyunsaturated fatty acids from Northern shrimp (*Pandalus borealis* Kreyer) processing by-products. Food Chem 130(4):853–858

An J, Gao F, Ma Q, Xiang Y, Ren D, Lu J (2017) Screening for enhanced astaxanthin accumulation among Spirulina platensis mutants generated by atmospheric and room temperature plasmas. Algal Res 25:464–472

Angsupanich K, Ledward DA (1998) High pressure treatment effect on cod (Gadus morhua) muscle. Food Chem 63(1):39–50

Anjos L, Pinto PIS, Tsironi T, Dimopoulos G, Santos S, Santa C, Manadas B, Canario A, Taoukis P, Power DM (2019) Experimental data from flesh quality assessment and shelf life monitoring of high pressure processed European sea bass (*Dicentrarchus labrax*) fillets. Data Brief 26:104451

Blahovec J, Vorobiev E, Lebovka N (2017) Pulsed electric fields pretreatment for the cooking of foods. Food Eng Rev 9:71–81

Bourke P, Ziuzina D, Boehm D, Cullen PJ, Keener K (2018) The potential of cold plasma for safe and sustainable food production. Trends Biotechnol 36(6):615–626. https://doi.org/10.1016/j.tibtech.2017.11.001

Briones-Labarca V, Perez-Won M, Zamarca M, Aguilera-Radic JM, Tabilo-Munizaga G (2012) Effects of high hydrostatic pressure on microstructure, texture, colour and biochemical changes of red abalone (*Haliotis rufecens*) during cold storage time. Innov Food Sci Emerg Technol 13:42–50

Bruno SF, Ekorong FJAA, Karkal S, Cathrine MSB, Kudre T (2019) Green and innovative techniques for recovery of valuable compounds from seafood by-products and discards: a review. Trends Food Sci Technol 85:10–22. https://doi.org/10.1016/j.tifs.2018.12.004

Büyükcan M, Bozoglou F, Alpas H (2009) Preservation and shelf-life extension of shrimps and clams by high hydrostatic pressure. Int J Food Sci Technol 44(8):1495–1502

Cao S, Zhou X, Jin W, Wang F, Tu R, Han S et al (2017) Improving of lipid productivity of the oleaginous microalgae *Chlorella pyrenoidosa* via atmospheric and room temperature plasma (ARTP). Bioresour Technol 244:1400–1406

Charest DJ, Balaban M, Marshall MR, Cornell JA (2001) Astaxanthin extraction from crawfish shells by supercritical CO_2 with ethanol as cosolvent. J Aquat Food Prod Technol 10(3):81–96
Charlier RH, Justus JR (1993) Chapter 10 Marine biomass energy. Elsevier Oceanogr Ser Elsevier 56:407–432
Checmarev G, Casales MR, Yeannes MI (2017) Kinetics of color change of osmotically dehydrated chub mackerel (*Scomber japonicus*) during storage at different temperatures. Food Sci Technol 37(1):119–123. https://doi.org/10.1590/1678-457X.10516
Chéret R, Chapleau N, Delbarre-Ladrat C, Verrez-Bagnis V, Lamballerie M (2005) Effects of high pressure on texture and microstructure of sea bass (Dicentrarchus labrax L.) fillets. J Food Sci 70(8):477–483
Chiper AS, Chen W, Mejlholm O, Dalgaard P, Stamate E (2011) Atmospheric pressure plasma produced inside a closed package by a dielectric barrier discharge in Ar/CO_2 for bacterial inactivation of biological samples. Plasma Sources Sci Technol 20:025008
Chotphruethipong L, Aluko R, Benjakul S (2019) Effect of pulsed electric field-assisted process in combination with porcine lipase on defatting of seabass skin. J Food Sci 84(7):1799–1805. https://doi.org/10.1111/1750-3841.14687
Collignan A, Bohuon P, Deumier F, Poligne I (2001) Osmotic treatment of fish and meat products. J Food Eng 49:153–162
Cropotova J, Tappi S, Genovese J, Rocculi P, Dalla Rosa M, Rustad T (2021) The combined effect of pulsed electric field treatment and brine salting on changes in the oxidative stability of lipids and proteins and color characteristics of sea bass (*Dicentrarchus labrax*). Heliyon 7(1):e05947
Cruz-Romero M, Smiddy M, Hill C, Kerry JP, Kelly AL (2004) Effects of high pressure treatment on physicochemical characteristics of fresh oysters (*Crassostrea gigas*). Innov Food Sci Emerg Technol 5(2):161–169
Cruz-Romero M, Kelly AL, Kerry JP (2007) Effects of high-pressure and heat treatments on physical and biochemical characteristics of oysters (*Crassostrea gigas*). Innov Food Sci Emerg Technol 8(1):30–38
Collignan A, Raoult-Wack AL (1994) Dewatering and salting of cod by immersion in concentrated sugar/saltsolutions. LWT 27:259–264
Duan D, Wang J, Wu Y, Peng J, Yang Y, Pan Y (2012) Study on characteristic of osmotic dehydration and mass transfer of tilapia fillet in sucrose solution. Adv Mater Res 554–556:1332–1336
FAO (2011) Global food losses and food waste – extent, causes and prevention. Rome
FAO (2018) The state of world fisheries and aquaculture 2018 – meeting the sustainable development goals. Rome. Licence: CC BY-NC-SA 3.0 IGO
FAO (2020a) GLOBEFISH highlights January 2020 ISSUE, with Jan. – Sep. 2019 Statistics – a quarterly update on world seafood markets. Globefish highlights no. 1-2020. Rome. https://doi.org/10.4060/ca7968en
FAO (2020b) The state of world fisheries and aquaculture. Sustainability in action. Rome. https://doi.org/10.4060/ca9229en
Franco D, Munekata P, Agregán R, Bermúde R, López-Pedrouso M, Pateiro M, Lorenzo J (2020) Application of pulsed electric fields for obtaining antioxidant extracts from fish residues. Antioxidants 9(2):90. https://doi.org/10.3390/antiox9020090
Giannakourou M, Tsironi T, Thanou I, Tsagri A-M, Katsavou E, Lougovois V, Kyrana V, Kasapidis G, Sinanoglou VJ (2019) Shelf life extension and improvement of the nutritional value of fish fillets through osmotic treatment based on the sustainable use of the by-products of aromatic herb and flower industry. Foods 8:421
Giannakourou MC, Stavropoulou N, Tsironi T, Lougovois V, Kyrana V, Konteles SJ, Sinanoglou VJ (2020) Application of hurdle technology for the shelf life extension of European eel (*Anguilla anguilla*) fillets. Aquac Fish (Accepted, in press). https://doi.org/10.1016/j.aaf.2020.10.003
Gómez-Guillén MC, Giménez B, Montero P (2005) Extraction of gelatin from fish skins by high pressure treatment. Food Hydrocoll 19(5):923–928
Gudbjornsdottir B, Jonsson A, Hafsteinsson H, Heinz V (2010) Effect of high-pressure processing on *Listeria* spp. and on the textural and microstructural properties of cold smoked salmon. LWT-Food Sci Technol 43(2):366–374

Gulzar S, Benjakul S (2019) Effect of pre-treatments on yield and properties of lipid extracted fromcephalothorax of Pacific white shrimp (*Litopenaeus vannamei*) by ultrasonic assisted process. LWT – Food Sci Tech 100:106–113

Gulzar S, Benjakul S (2020) Impact of pulsed electric field pretreatment on yield and quality of lipid extracted from cephalothorax of Pacific white shrimp (*Litopenaeus vannamei*) by ultrasound-assisted process. Int J Food Sci Technol 55(2):619–630. https://doi.org/10.1111/ijfs.14316

Hajhoseini A, Sharifan A, Housefi HR (2020) Effects of atmospheric cold plasma on microbial growth of Listeria innocua and Staphylococcus aureus in ready-to-eat fish products. Iran J Fish Sci 19(1):262–271

Hamed I, Özogul F, Regenstein JM (2016) Industrial applications of crustacean by-products (chitin, chitosan, and chitooligosaccharides): a review. Trends Food Sci Technol 48:40–50. https://doi.org/10.1016/j.tifs.2015.11.007

Haq M, Ahmed R, Cho Y-J, Chun B-S (2017) Quality properties and bio-potentially of edible oils from Atlantic Salmon by-products extracted by supercritical carbon dioxide and conventional methods. Waste Biomass Valorization 8(6):1953–1967. https://doi.org/10.1007/s12649-016-9710-2

He H, Adams RM, Farkas DF, Morrissey MT (2002) Use of high-pressure processing for oyster shucking and shelf-life extension. J Food Sci 67(2):640–645

He G, Yin Y, Yan X, Yu Q (2014) Optimisation extraction of chondroitin sulfate from fish bone by high intensity pulsed electric fields. Food Chem 164:205–210. https://doi.org/10.1016/j.foodchem.2014.05.032

He G, Yin Y, Yan X, Wang Y (2016) Semi-bionic extraction of effective ingredient from fishbone by high intensity pulsed electric fields. J Food Process Eng 40(2):e12392. https://doi.org/10.1111/jfpe.12392

He G, Yin Y, Yan X, Wang Y (2017) Application of pulsed electric field for treatment of fish and seafood. In: Miklavčič D (ed) Handbook of electroporation. Springer, Cham, pp 2637–2655

He G, Yan X, Wang X, Wang Y (2019) Extraction and structural characterization of collagen from fishbone by high intensity pulsed electric fields. J Food Process Eng 42(6):ee13214. https://doi.org/10.1111/jfpe.13214

Hemker AK, Nguyen LT, Karwe M, Salvi D (2020) Effects of pressure-assisted enzymatic hydrolysis on functional and bioactive properties of tilapia (*Oreochromis niloticus*) by-product protein hydrolysates. LWT Food Sci Technol 122:109003

Huang Y-L, Tsai Y-H (2020) Extraction of chitosan from squid pen waste by high hydrostatic pressure: effects on physicochemical properties and antioxidant activities of chitosan. Int J Biol Macromol 160:677–687. https://doi.org/10.1016/j.ijbiomac.2020.05.252

Huang HW, Hsu CP, Yang BB, Wang CY (2014) Potential utility of high-pressure processing to address the risk of food allergen concerns. Compr Rev Food Sci Food Saf 13(1):78–90. https://doi.org/10.1111/1541-4337.12045

ICMSF (1986) International commission on microbiological specifications for foods, sampling plans for fish and shellfish. In: ICMSF (ed) Microorganisms in foods, sampling for microbiological analysis: principles and scientific applications, 2nd edn. University of Toronto Press, Toronto, pp 181–196

Ivanovs K, Blumberga D (2017) Extraction of fish oil using green extraction methods: a short review. Energy Procedia 128:477–483

Joannes C, Sipaut CS, Dayou J, Yasir SM, Mansa RF (2015) The potential of using pulsed electric field (PEF) technology as the cell disruption method to extract lipid from microalgae for biodiesel production. Int J Renew Res 5(2):598–621

Johnson PE, Vander Planeken I, Balasa A, Husband FA, Grauwet T, Hendrickx M, Knorr D, Mills EN, Mackie AR (2010) High pressure, thermal and pulsed electric-field-induced structural changes in selected food allergens. Mol Nutr Food Res 54:1701–1710. https://doi.org/10.1002/mnfr.201000006

Kempkes MA (2016) Pulsed electric fields for algal extraction and predator control. In: Miklavcic D (ed) Handbook of electroporation. Springer, Cham. https://doi.org/10.1007/978-3-319-26779-1_215-1

Kim HJ, Jayasena DD, Yong HI, Jo C (2016) Chapter 11: Quality of cold plasma treated foods of animal origin. In: Misra NN, Oliver S, Cullen PJ (eds) Cold plasma in food and agriculture: fundamentals and applications. Elsevier, pp 273–291

Ko WC, Jao CL, Hwang JS, Hsu KC (2006) Effect of high-pressure treatment on processing quality of tilapia meat fillets. J Food Eng 77(4):1007–1011

Kuvendziev S, Lisichkov K, Zeković Z, Marinkovski M, Musliu Z (2018) Supercritical fluid extraction of fish oil from common carp (Cyprinus carpio L.) tissues. J Supercrit Fluids 133(1):528–534

Lee HB, Noh YE, Yang HJ, Min SC (2011) Inhibition of foodborne pathogens on polystyrene, sausage casings and smoked salmon using nonthermal plasma treatments. Korean J Food Sci Technol 43:513–517

Lee J-H, Asaduzzaman AKM, Yun J-H, Yun J-H, Chun B-S (2012) Characterization of the yellow croaker *Larimichthys polyactis* muscle oil extracted with supercritical carbon dioxide and an organic solvent. Fish Aquat Sci 15(4):275–281

Li M, Lin J, Chen J, Fang T (2016a) Pulsed electric field-assisted enzymatic extraction of protein from abalone (*Haliotis Discus Hannai* Ino) viscera. J Food Process Eng 39(6):702–710. https://doi.org/10.1111/jfpe.12262

Li J, Sun W, Ramaswamy H, Yu Y, Zhu S, Wang J, Li H (2016b) High pressure extraction of astaxanthin from shrimp waste (Penaeus vannamei Boone): effect on yield and antioxidant activity. J Food Process Eng 40(2):e12353. https://doi.org/10.1111/jfpe.12353

Li J, Shi J, Huang X, Zou X, Li Z, Zhang D, Zhang W, Xu Y (2020) Effects of pulsed electric field on freeze. Innov Food Sci Emerg Technol 65:102454

Lisichkov K, Kuvendziev S, Zeković Z, Marinkovski M (2014) Influence of operating parameters on the supercritical carbon dioxide extraction of bioactive components from common carp (*Cyprinus carpio* L.) viscera. Sep Purif Technol 138:191–197

Lopes C, Antelo L, Franco-Uría A, Alonso A, Pérez-Martín R (2015) Valorisation of fish by-products against waste management treatments – comparison of environmental impacts. Waste Manag 46:103–112. https://doi.org/10.1016/j.wasman.2015.08.017

Misra NM, Yadav B, Roopesh MS, Jo C (2019) Cold plasma for effective fungal and mycotoxin control in foods: mechanisms, inactivation effects and applications. Compr Rev Food Sci Food Saf 18:106–120

Mujaffar S, Sankat CK (2006) The mathematical modelling of the osmotic dehydration of shark fillets at different brine temperatures. Int J Food Sci Technol 41:405–416

Murchie LW, Cruz-Romero M, Kerry JP, Linton M, Patterson MF, Smiddy M, Kelly AL (2005) High pressure processing of shellfish: a review of microbiological and other quality aspects. Innov Food Sci Emerg Technol 6(3):257–270

Narwankar SP, Flimlin GE, Schaffner DW, Tepper BT, Karwe MV (2011) Microbial safety and consumer acceptability of high-pressure processed hard clams (*Mercenaria mercenaria*). J Food Sci 76(6):M375–M380

Nguyen TT, Zhang W, Barber AR, Su P, He S (2015) Significant enrichment of polyunsaturated fatty acids (PUFAs) in the lipids extracted by supercritical CO_2 from the livers of Australian rock lobsters (*Jasus edwardsii*). J Agric Food Chem 63(18):4621–4628

Ohshima T, Ushio H, Koizumi C (1993) High-pressure processing of fish and fish products. Trends Food Sci Technol 4:370–375

Olatunde OO, Benjakul S, Vongkamjan K (2020) Shelf-life of refrigerated Asian sea bass slices treated with cold plasma as affected by gas composition in packaging. Int J Food Microbiol 324:108612

Paarup T, Sanchez JA, Peláez C, Moral A (2002) Sensory, chemical and bacteriological changes in vacuum-packed pressurized squid mantle (*Todaropsis eblanae*) stored at 4°C. Int J Food Microbiol 74:1–12

Park SY, Ha SD (2014) Application of cold oxygen plasma for the reduction of Cladosporium cladosporioides and Penicillium citrinum on the surface of dried filefish (*Stephanolepis cirrhifer*) fillets. Int J Food Sci Technol 50:966–973

Prego R, Fidalgo LG, Saraiva JA, Vázquez M, Aubourg SP (2021) Impact of prior high-pressure processing on lipid damage and volatile amines formation in mackerel muscle subjected to frozen storage and canning. LWT Food Sci Technol 135:109957

Ramirez-Suarez JC, Morrissey MT (2006) Effect of high pressure processing (HP) on shelf life of albacore tuna (*Thunnus alalunga*) minced muscle. Innov Food Sci Emerg Technol 7:19–27

Raoult-Wack AL (1994) Recent advances in the osmotic dehydration of foods. Trends Food Sci Technol 5:255–260

Rastogi NK, Raghavarao KSMS, Niranjan K, Knorr D (2002) Recent developments in osmotic dehydration: methods to enhance mass transfer. Trends Food Sci Technol 13:48–59

Rode TM, Rotabakk BT (2020) Extending shelf life of desalted cod by high pressure processing. Innov Food Sci Emerg Technol 69:102476. https://doi.org/10.1016/j.ifset.2020.102476

Rubio-Rodríguez N, de Diego S, Beltrán S, Jaime I, Sanz MT, Rovira J (2012) Supercritical fluid extraction of fish oil from fish by-products: a comparison with other extraction methods. J Food Eng 109(2):238–248. https://doi.org/10.1016/j.jfoodeng.2011.10.011

Rustad T, Storrø I, Slizyte R (2011) Possibilities for the utilisation of marine by-products. Int J Food Sci Technol 46(10):2001–2014. https://doi.org/10.1111/j.1365-2621.2011.02736.x

Sahena F, Zaidul ISM, Jinap S, Yazid AM, Khatib A, Norulaini NAN (2010a) Extraction of fish oil from the skin of Indian mackerel using supercritical fluids. J Food Eng 99(1):63–69

Sahena F, Zaidul ISM, Jinap S, Yazid AM, Khatib A, Norulaini NAN (2010b) Fatty acid compositions of fish oil extracted from different parts of Indian mackerel (*Rastrelliger kanagurta*) using various techniques of supercritical CO_2 extraction. Food Chem 120(3):879–885. https://doi.org/10.1016/j.foodchem.2009.10.055

Sahena F, Zaidul ISM, Norulaini NAN, Jinap S, Rezaul K, Tariq AR, Omar AKM (2012) Fish oil recovery from viscera of Indian Mackerel (*Rastrelliger kanagurta*) by supercritical fluid: an optimization approach. J Chin Chem Soc 59(11):1421–1429. https://doi.org/10.1002/jccs.201100235

Sahena F, Sarker ZI, Norulaini N, Oliveira A, Yunus K, Chowdury AJ, Akanda J, Omar M (2014) Quality of tuna fish oils extracted from processing the by-products of three species of neritic tuna using supercritical carbon dioxide. J Food Process Preserv 39(4):432–441. https://doi.org/10.1111/jfpp.12248

Sánchez-Camargo AP, Almeida Meireles MÂ, Lopes BLF, Cabral FA (2011a) Proximate composition and extraction of carotenoids and lipids from Brazilian redspotted shrimp waste (*Farfantepenaeus paulensis*). J Food Eng 102(1):87–93

Sánchez-Camargo AP, Martinez-Correa HA, Paviani LC, Cabral FA (2011b) Supercritical CO_2 extraction of lipids and astaxanthin from Brazilian redspotted shrimp waste (*Farfantepenaeus paulensis*). J Supercrit Fluids 56(2):164–173

Sánchez-Camargo AP, Meireles MÂA, Ferreira ALK, Saito E, Cabral FA (2012) Extraction of ω-3 fatty acids and astaxanthin from Brazilian redspotted shrimp waste using supercritical CO_2 + ethanol mixtures. J Supercrit Fluids 61:71–77

Sarker MZI, Selamat J, Habib ASMA, Ferdosh S, Akanda MJH, Jaffri JM (2012) Optimization of supercritical CO_2 extraction of fish oil from viscera of African catfish (*Clarias gariepinus*). Int J Mol Sci 13(9):11312–11322

Semenoglou I, Dimopoulos G, Tsironi T, Taoukis P (2020) Mathematical modelling of the effect of solution concentration and the combined application of pulsed electric fields on mass transfer during osmotic dehydration of sea bass fillets. Food Bioprod Process 121:186–192. https://doi.org/10.1016/j.fbp.2020.02.007

Shiekh KA, Benjakul S (2020) Effect of pulsed electric field treatments on melanosis and quality changes of Pacific white shrimp during refrigerated storage. J Food Process Preserv 44(1):ee14292

Sigman DM, Hain MP (2012) The biological productivity of the ocean. Nat Educ Knowl 3(10):21

Sofra C, Tsironi T, Taoukis PS (2018) Modeling the effect of pre-treatment with nisin enriched osmotic solutionon the shelf life of chilled vacuum packed tuna. J Food Eng 216:125–131
Tsironi T, Taoukis P (2010) Modelling microbial spoilage and quality of gilthead seabream fillets: combined effect of osmotic pretreatment, modified atmosphere packaging and nisin on shelf life. J Food Sci 75(4):M243–M251
Tsironi TN, Taoukis PS (2012) Shelf-life extension of gilthead seabream fillets by osmotic treatment and antimicrobial agents. J Appl Microbiol 112(2):316–328
Tsironi T, Taoukis P (2014) Effect of processing parameters on water activity and shelf life of osmotically dehydrated fish filets. J Food Eng 123:188–192
Tsironi TN, Taoukis PS (2017) Effect of storage temperature and osmotic pre-treatment with alternative solutes on the shelf-life of gilthead seabream (*Sparus aurata*) fillets. Aquac Fish 2:39–47
Tsironi TN, Taoukis PS (2019) Advances in conventional and nonthermal processing of fish for quality improvement and shelf life extension. In: Reference module in food science. Elsevier, pp 1–7. https://doi.org/10.1016/B978-0-08-100596-5.22618-8
Tsironi T, Salapa I, Taoukis PS (2009) Shelf life modelling of osmotically treated chilled gilthead seabream fillets. Innov Food Sci Emerg Technol 10:23–31
Tsironi T, Maltezou I, Tsevdou M, Katsaros G, Taoukis PS (2015) High pressure cold pasteurization of gilthead seabream fillets: selection of process conditions and validation of shelf-life extension. Food Bioprocess Technol 8:681–690. https://doi.org/10.1007/s11947-014-1441-4
Tsironi T, Anjos L, Pinto PIS, Dimopoulos G, Santos S, Santa C, Manadas B, Canario A, Taoukis P, Power D (2019a) High pressure processing of European sea bass (*Dicentrarchus labrax*) fillets and tools for flesh quality and shelf life monitoring. J Food Eng 262:83–91
Tsironi T, Ntzimani A, Gogou E, Tsevdou M, Semenoglou I, Dermesonlouoglou E, Taoukis P (2019b) Modelling the effect of active modified atmosphere packaging on the microbial stability and shelf life of gutted sea bass. Appl Sci 9:5019
Tsironi T, Houhoula D, Taoukis P (2020) Hurdle technology for fish preservation. Aquac Fish 5:65–71
Xiang Q, Fan L, Li Y, Dong S, Li K, Bai Y (2020) A review on recent advances in plasma-activated water for food safety: current applications and future trends. Crit Rev Food Sci Nutr. https://doi.org/10.1080/10408398.2020.1852173

Chapter 20
Efficient Production of Functional and Bioactive Compounds and Foods for Use in Food, Pharma, Cosmetic and Other Industries

Avi Shpigelman and Zoya Okun

1 Introduction

1.1 Natural Bioactive Compounds and Their Sources

Bioactive compounds are described as essential and nonessential compounds (e.g., vitamins or polyphenols) that occur in nature, usually are part of the food chain, and can be shown to promote human health. They are referred to as nutraceuticals that are present as natural constituents in food and provide health benefits beyond the basic nutritional value of the product (Biesalski et al. 2009). Natural bioactive compounds are characterized by a broad diversity of molecular structures and functionalities that provide an arsenal of components that can be utilized for the production of food additives, nutraceuticals, key components in pharma and cosmetic industries, and more. Some of the bioactive compounds can be found in nature at relatively high concentrations, but others can only be found in very limited quantities. Due to the complicated, time demanding, and often unprofitable organic synthesis, as well as the consumers' growing demand for the utilization of natural products having a clean label, massive plant harvesting is needed for obtaining sufficient amounts of the desired bioactive compounds. The challenges in the isolation and purification of these compounds contributed to the development of advanced processing methods and novel technologies based on pressure, mechanical, electrical, and thermal effects, as will be presented and discussed. The final yields of bioactive compounds extracted by various methods and technologies can be controlled and improved by the solvents variability, incubation times, and temperature steps, as well as by combination with enzymatic treatments, novel and green extraction procedures capable of improving the mass transfer of the bio-actives from the matrix,

A. Shpigelman · Z. Okun (✉)
Faculty of Biotechnology and Food Engineering, Technion, Haifa, Israel
e-mail: gzoya@bfe.technion.ac.il

A. Režek Jambrak (ed.), *Nonthermal Processing in Agri-Food-Bio Sciences*,
Food Engineering Series, https://doi.org/10.1007/978-3-030-92415-7_20

the combination of several extraction technologies for a more efficient and satisfactory result is also widely examined. A great variety of bioactive compounds are extracted from agricultural produce, food wastes, and industrial by-products are widely explored. Among them are vitamins, pigments, lignans, polyphenolic compounds, carotenoids, bioactive oligosaccharides, and more. Few recently published examples of studies focusing on extraction of bioactive compounds, their sources, and suggested utilization are presented in Table 20.1.

This chapter will focus on various extraction techniques of bioactive compounds from two diverse families as model components based on aqueous solubility: (1) extraction of polyphenols from grape seeds and skins as a source of hydrophilic bioactive compounds and (2) Astaxanthin, a member of the carotenoid family, as a representative of the lipophilic bioactive compounds.

Global grape production reached 77.8 million tons in 2018, with China, Italy, Spain, and France being the major producers, and more than 70% of the grapes utilized for wine, must, and juice production (International Organisation of Vine and Wine 2019). One of the major by-products in the wine and juice industry are grape seeds, which can provide a rich source for lipids, proteins, carbohydrates, and flavonoids that are well-known polyphenolic antioxidants. Oil, extract and flour are the three products derived from grape seeds and further utilized by various industries. The extract consists of antioxidant compounds, possessing a vast assortment of health-promoting capabilities such as anti-inflammatory, anti-bacterial, and anti-viral effects mostly attributed to the high polyphenolic content of the extract. During the years, more than 8000 polyphenolic compounds have been identified, and a broad range of beneficial activities was reported, highlighting the promising role of these compounds in the prevention of chronic, neurodegenerative, and cardiovascular diseases and for their antioxidant and anti-inflammatory activities (Denaro et al. 2020).

Table 20.1 Types of bioactive compounds from natural sources and their utilization

Source	Bioactive compound	Utilization	Reference
Red wine pomace	Polyphenols	Antioxidants	Croxatto Vega et al. (2021)
Cluster mallow	Natural pigment	Pigment for dye-sensitized solar cells	Golshan et al. (2020)
Mushroom waste	Vitamin D_2, ergosterol	Dietary supplement	Papoutsis et al. (2020)
Aloe	Polysaccharides	Prebiotic	Liu et al. (2021)
Sesame seed	Lignan	Antioxidant	Eom et al. (2021)
Guava's pulp and waste powders	Carotenoids	Lipophilic antioxidants	Da Silva Lima et al. (2020)
Stevia rebaudiana Bertoni	Steviol glycosides	Sweetener	Ahmad et al. (2020)
Crustacean by-products	Astaxanthin, lipids	Antioxidant, pigment	Ahmadkelayeh and Hawboldt (2020)

Astaxanthin is a carotenoid bioactive, part of a large group of lipophilic health-promoting components. This orange-reddish pigment is biosynthesized in various microorganisms. It has a high antioxidant capacity and was reported as an immune response enhancer, contributed to an improvement of skin health and tissue damage, reported to play a role in the treatment of diabetes and cardiovascular diseases, presented anticarcinogenic properties, and much more. The beneficial and promising results were also presented in human clinical trials (Donoso et al. 2021). In nature, astaxanthin is synthesized by microorganisms, both aquatic and non-aquatic, such as bacteria, fungi, yeasts, and microalgae. Its production and extraction from natural sources is being explored due to the consumers' concern regarding synthetic food additives and the demand for clean label and healthier food. The main source of astaxanthin for extraction is the freshwater green microalgae *Haematococcus pluvialis* which is recognized by FDA as a safe astaxanthin source for human consumption.

Astaxanthin contains 40 carbon atoms with two terminal ring systems joined by a chain of conjugated double bonds, responsible for light absorbance in the visible region (~500 nm) and results in characteristic red color (Higuera-Ciapara et al. 2006). Astaxanthin has two chiral centers at positions C-3 and C-3′ (marked red in Fig. 20.1), resulting in a mixture of two enantiomers (3S,3′S & 3R,3′R) and the *meso* form (3R,3′S) as presented in Fig. 20.1. It can be produced naturally by extraction from microorganisms or chemically synthesized. In recent years, the market for natural astaxanthin has been growing substantially, with a purity dependant value varying from 2500 to 7000 USD/kg to about 15,000 USD/kg for human

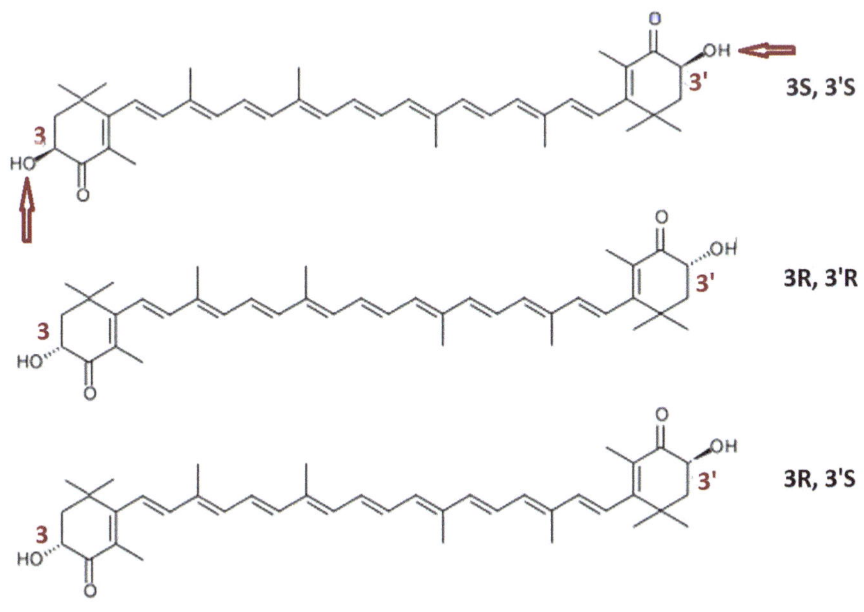

Fig. 20.1 Astaxanthin stereoisomers, esterification sites marked by arrows

applications such as cosmetics, food, and dietary supplements. The synthetic compound is less expensive ~1000$/kg, but is not approved for human consumption and is usually utilized in animal feed (Rodríguez-Sifuentes et al. 2021). The main differences between natural and synthetic astaxanthin are (1) Esterification degree, the natural extract will usually be found in mono or di-ester form, while the synthetic compound will be in its free unesterified form (esterification sites marked by red arrows in Fig. 20.1) and (2) Stereochemical composition, the synthetic form consists of a mixture of isomers in ratio 1:2:1 (3S,3S), (3R,3S), and (3R,3R) respectively, while the composition of natural extract depends on its source. These differences are suggested to affect the health-promoting performance of the synthetic compound.

The goal to extract astaxanthin from natural sources more competitively when compared with the synthetic route is challenging due to the need to overcome the critical points related to biomass concentration, the environmental impact, and product stability, compromising yield and quality of the extracted material.

The reviewed in this chapter technologies may provide an innovative and economical approach for the enhanced production of bioactive compounds for use in food, pharma, cosmetics, and other industries.

2 Industrial Utilization of Bioactive Compounds (Example of Grape Seed Polyphenols and Astaxanthin)

The global market for Carotenoids reached $1.5 billion in 2017 and expected to reach $2.0 billion by 2022, at a compound annual growth rate (CAGR) of 5.7% for the period of 2017–2022, while the astaxanthin market only was valued at 512.8 Million USD in 2016 and projected to grow at a CAGR of 6.73% from 2017, to reach a projected value of 814.1 million USD by 2022 (McWilliams 2018).The driving factors in the astaxanthin market are a high demand due to its antioxidant properties and increasing applications in the aquaculture industry, growing demand for natural food coloring agents with lesser adverse effects as compared to other chemical products, are the key factors estimated to boost the continues market growth. Some of the players in the astaxanthin market are Alga Technologies; Cyanotech Corporation; Fuji Chemicals Industries Co., Ltd; Fenchem; Beijing Gingko Group (BGG) (Astaxanthin Market Size 2020–2027). Astaxanthin derived commercially available products can be found in the form of capsules, powders, soft gels, oil extracts, and creams, few examples of such products are presented in Table 20.2.

As mentioned before, grape seed extract is an industrial derivative of grape seeds. It is rich in antioxidants and oligomeric proanthocyanidin complexes and has been linked to a wide range of potential health benefits. The grape seed extract is available as a dietary supplement in a liquid form, tablets, or capsules. Supplements commonly contain 50–100 mg of the extract. Few studies also suggested the

Table 20.2 Variety of astaxanthin derived products

Manufacturer	Brand name	Form	Astaxanthin content	Application
Algatech	Astapure®	Oleoresin	5%, 10%, 20%	Topical cream, emulsions
Solgar	Natural astaxanthin	Soft gel	5 mg	Antioxidant support,Skin health
Cyanotech	BioAstin®	Capsule	4 mg	Food supplement
BGG	AstaZine®	Powder	2%	Antioxidant
True Botanica	Face Cream – with Astaxanthin	Cream	n.a.	Face moisturizing cream
Astalif	Astaxanthin	Soft gel	12 mg	Antioxidant, food supplement
Sinacon	Red plus⁺	Powder	n.a.	Color enhancer, fish food additive

utilization of grape seed extracts not only as a food additive with health-promoting capabilities, but also as potent dyes and flame retardants in the fabric industry (Guo et al. 2020), and as functional ingredients in meat and fish products (Mainente et al. 2019).

3 Technological Methods for the Production of Bioactive Compounds

Among the common methods used for the extraction are the conventional liquid-liquid or solid-liquid extraction utilizing a broad range of organic solvents by the principle of Soxhlet extraction (Fig. 20.2). Soxhlet is an apparatus utilized for one of the most conventional extraction methods in the case of solid-liquid extractions (also named leaching especially in industrial applications) and usually serves as an accepted standard for comparison with other extraction approaches. In this technique, the plant matrix containing the bioactive compounds is placed into a disposable thimble, positioned in the soxhlet apparatus as presented in Fig. 20.2. The extraction solvent is added to the flask and being refluxed, the vaporized solvent rises in the vapor tube, and cooled by the condenser circulated with cooling water. The condensed solvent flows back into the thimble containing the sample and dissolves the bioactive compounds. Once the siphon is filled, the solvent with the extract is transferred back to the flask. The extraction solvent is continuously cycled through the plant matrix, by boiling and condensation, with the bioactive compounds being extracted and collected in the flask.

Soxhlet method used as a batch process on small scales, and converted into a continuous extraction procedure on medium or large scales. Among the samples for industrial leaching equipment are Bollman, Rotocel and Kennedy extractors (Smith 2011).

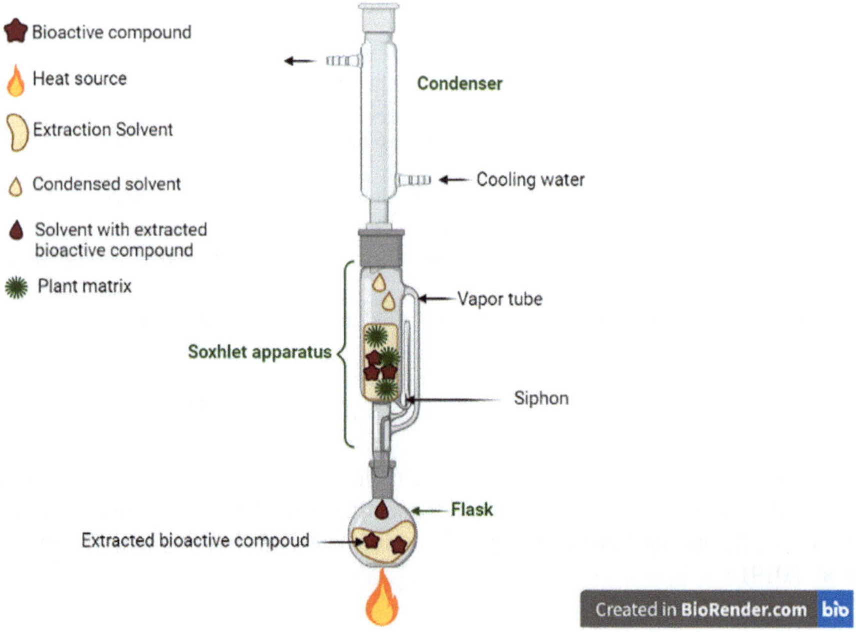

Fig. 20.2 Operation principle of the Soxhlet apparatus

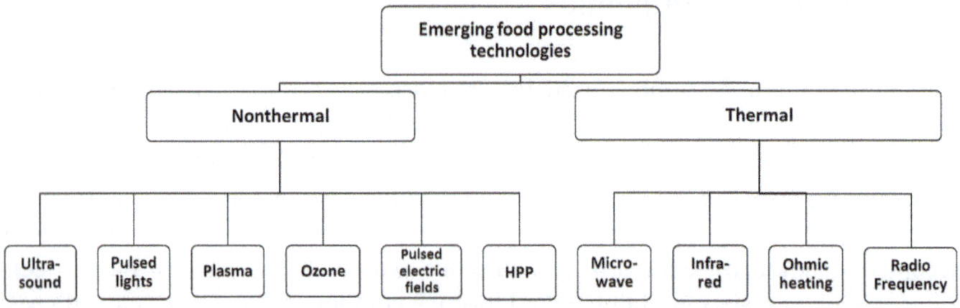

Fig. 20.3 Emerging processing technologies studied for potential to support extraction

Among the drawbacks of Soxhlet is the elevated solvent temperatures used for efficient extraction which makes this method incompatible with thermally sensitive bioactive compounds, long extraction times, utilization of organic solvents accompanied with the need for complete further removal due to their toxicity and their negative environmental impact, as well as high cost of the solvents and their time demanding evaporation.

The emerging processing technologies developed and optimized during the years are classified as thermal and nonthermal methods (presented in Fig. 20.3), in the field of extraction the leading representatives are high pressure, microwave and ultrasound-assisted extractions, pulsed electric field assited extraction and more.

While the nonthermal technologies often include a partial thermal aspect, this aspect is not suggested to the main one reposnible for the beneficial effect.

3.1 Non-thermal Technologies

Due to the increased consumer demand for high-quality foods, new technologies were developed and introduced to the food industry, in many cases for microbial inactivation and shelf life enhancement, and some of these technologies were explored and found useful for improved extraction of bioactive compounds. The advantages of such technologies are often the improvement of matrix disruption, lower processing temperatures, and/or shorter treatment times, which is highly favored for thermolabile compounds. In addition, they can in some cases replace the need for organic solvents. The leading representatives among the nonthermal (or partially thermal) technologies are Pulsed Electric Fields (PEF), Ultrasound (US), and High Hydrostatic Pressure (HHP). In the following section, we will review some of the nonthermal technologies and their most recent utilization for the extraction of astaxanthin and polyphenols.

3.1.1 Pulsed Electric Fields (PEF)

The operation principle of PEF is the application of high voltage pulses to permeate cell membranes or to increase membrane porosity (electroporation). Electroporation is defined as the process of pore formation in cellular membranes due to the utilization of an electric field. This technology involves the operation of high voltage electrical pulses, typically with a field strength from 0.5 to 20 kV/cm to samples/products placed in PEF chamber between two electrodes for a short time (10^{-5}–10^{-2} seconds) to avoid a significant heating effect. Exposure to an external electric field induces the increase of transmembrane potential due to the charges of opposite polarities from both sides of the cell membrane, the electrostatic attraction between the opposite charges results in the formation of pores in the membrane favoring the release of intracellular valuable contents (Fig. 20.4a). Among the parameters affecting the extraction efficacy are electric field intensity, reaction time, number of pulses, pH, and ionic strength of the solvent.

PEF was recently utilized for the extraction of astaxanthin from *H. pluvialis*. Various PEF treatment times, number of pulses, and field strengths were examined to identify optimal conditions. The treatment of 10 pulses 1 kV/cm for 5 ms each at 20 °C resulted in 1.2-fold more efficient extraction compared to classical disruption methods like bead-beating, freezing-thawing, thermal treatment. In this research cells treated by PEF were incubated for 6 hours in a growth medium at room temperature in the absence of light, after the incubation the suspension was centrifuged, and the biomass was resuspended in ethanol that was identified as the optimal

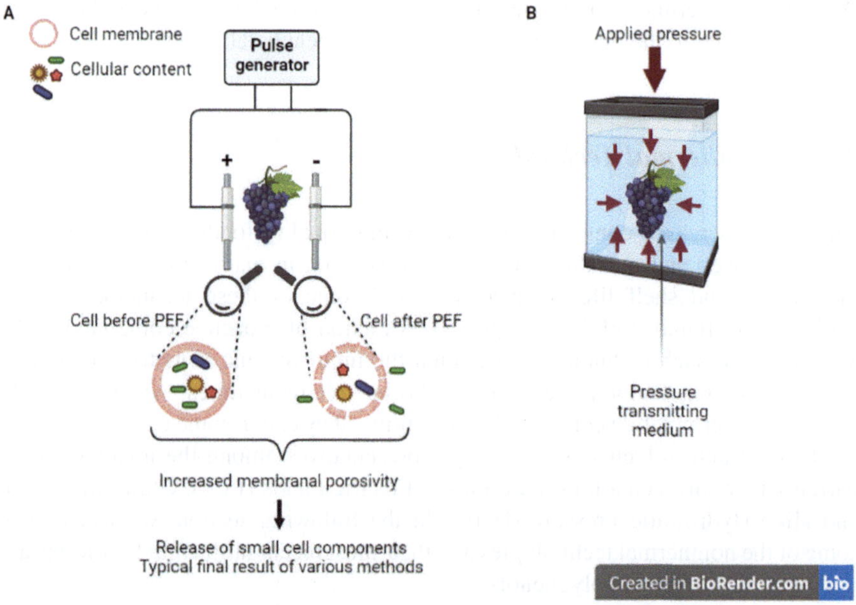

Fig. 20.4 Schematic operation principle of PEF (**a**) and HHP (**b**) assisted extraction

extraction solvent, this resulted in remarkable extraction of 96% carotenoid content compared to 80% using other physical methods (Martínez et al. 2019).

3.1.2 Ultrasound (US)

The operation principle of US is based on the acoustic cavitation phenomenon generated by the sound waves in the range of 200–100 MHz. Ultrasonic cavitation leads to extreme temperatures, yet only locally, pressure differentials, and high shear forces. When cavitation bubbles implode on the surface of plant cells, the perforation of cell walls and cell membranes takes place and cell disruption occurs. The mechanical effects of ultrasound-induced cavitation, intensify the penetration of the solvent into the cell interior and improves the mass transfer between cell and solvent so that the intercellular materials are transferred into the solvent affecting the extraction efficiency. Among the main parameters affecting the extraction efficacy, are matrix factors like particle size and moisture content, process factors like power and reaction time, and solvent factors like polarity and viscosity. The two most popular and researched goals for extraction modes for ultrasound-based instruments are an ultrasonic probe system, where the probe is inserted directly into the sample and an ultrasonic water bath where no direct content between the sample and the probe occurs.

Ultrasound was recently utilized for the extraction of polyphenols from grape seeds. Kinetic studies aiming for an optimized industrial implementation suggested that the optimal model includes parameters such as hydrodynamics, mixing, and mass transfer which control the effectiveness in the physical/chemical processing applications. Such a methodology can promote optimal economic scale-up strategies required for the industrial application of ultrasound assisted extraction.

In extraction trials, aliquots of grape seeds were mixed with several solvents at different extraction conditions. The ultrasonic probe was placed at the center of the extraction vessel and submerged about 5 cm under the mixture surface. The best extraction yield of TPC for all the samples was obtained using UAE with an ethanol-water mixture (57:43 v/v) as the solvent, at 50 °C, 200 W and 26 kHz after 30 min (Natolino and Da Porto 2020).

3.1.3 High Hydrostatic Pressure (HHP) and High Pressure Homogenization (HPH)/(UltraHPH)

HHP is a non-thermal technology increasingly used in the food industry as a cold pasteurization method. Its operation is based on Pascal's principle where an applied change in pressure is transmitted uniformly and immediately in the high-pressure vessel containing the product in a vessel filled with pressure-transmitting medium (Fig. 20.4b). By subjecting the product, usually in the final package, to elevated pressures up to 1000 MPa (industrially up to 600 MPa) for a short holding time (mostly 3–5 min) pathogens and spoilage bacteria can be inactivated, while effects on low-molecular-weight molecules (e.g. sugar, vitamins, pigments, flavor compounds) are minimal. High hydrostatic pressure can damage cellular membranes, disrupting tissues and organelles, thereby favoring the release of bioactive compounds.

Homogenization is a physical process in which a dispersed system, suspension or emulsion, is forced to flow at a high velocity through a narrow passage, a disruption valve, producing a smaller and narrow particle size distribution. High-pressure homogenization (HPH) is one of the emerging technologies being studied and developed for various applications in the food industry. It was suggested as an effective tool for achieving microbial safety and extending the product shelf life of liquid foods in a continuous process while minimizing some negative attributes of thermal processing. The valve geometry, pressure level, inlet temperature, and the number of homogenization cycles are all factors affecting the level of microbial inactivation and the extent of the impact on techno-functionalities of food biopolymers and matrices. Turbulence, high shear, cavitation, and temperature increase induced by HPH treatments enhance emulsion stability, stabilize proteins in solutions, reduce particle size distributions, and increase the accessibility of health-promoting compounds. A major difference between HPH and conventional homogenization is the maximum pressure level reached, and it is dependent on the homogenizer design and characteristics such as gap size, seals, and valve geometry. Ultra HPH reaches pressure levels up to 400 MPa, while HPH is usually defined to reach pressure levels between 50 and 200 MPa.

Recently HPH was utilized for exploring morphological changes and cell disruption of *H. pluvialis* cells for astaxanthin extraction. When pressures of 10,000–30,000 psi (~70–200 MPa) were applied the intact cyst cells were significantly disrupted or fully ruptured, releasing the cytoplasmic components, thereby facilitating the extraction and successful separation of astaxanthin. Number of passages (cycles) of HPH (1–3 times) could significantly improve the cell disruption efficiency. The maximum astaxanthin recovery was estimated to be 1.1% (weight of dry cells) (Praveenkumar et al. 2020).

3.2 *Thermal Technologies*

3.2.1 Microwave (MW)

Microwaves are part of the electromagnetic spectrum in the frequency range 300 MHz–300 GHz (Fig. 20.5c). The operation principle is based on magnetic and electric fields acting on components with the ability to convert the absorbed energy to heat. The induced energy is further transferred via dipole rotation and ionic conduction. In the plant matrix, the moisture evaporates due to the adsorption of electromagnetic energy. Evaporation leads to pressure generation inside the cell resulting in swelling and structural changes improving the porosity of the matrix which favors

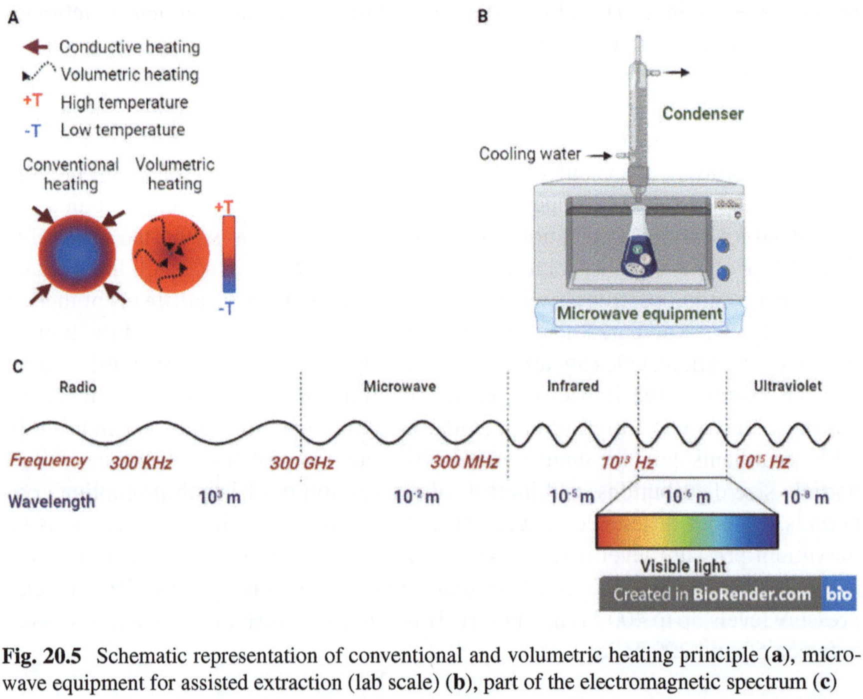

Fig. 20.5 Schematic representation of conventional and volumetric heating principle (**a**), microwave equipment for assisted extraction (lab scale) (**b**), part of the electromagnetic spectrum (**c**)

mass transfer and better extraction of the bioactive compounds. Schematic representation of microwave-assisted extraction procedure is presented in Fig. 20.5b.

Recently a study involving microwave-assisted extraction was published dealing with the combined effects of various techniques and compares it to traditional Soxhlet extraction. The combined enzymatic, microwave andsalting-out extraction method has been developed and evaluated as an effective method to extract polyphenols from grape seeds. The results showed that different extraction methods not only affect the extraction yield of polyphenols but also have significant effects on the composition of monophenols. Ten kinds of monomer phenolic compounds were analyzed by HPLC, only four were identified by Soxhlet extraction *vs* eight different monomeric phenolic compounds with antioxidant activity that were extracted by combined techniques (Jia et al. 2021).

3.2.2 Ohmic Heating (OH)

OH (also termed Joule heating) is based on the application of an alternating electric current for rapid and evenly distributed heat generation. In accordance with Ohm's law, the amount of energy (dissipated heat) is related to both applied voltage and electrical conductivity of the treated matrix. Due to the thermal destruction, the permeability of the cell membrane increases leading to the promotion of the extraction of bioactive materials from the plant matrix.

Ohmic Heating was utilized for the extraction of anthocyanins, from winemaking residues by using the grape skins as natural electrical conductors allowing the internal heat dissipation through OH. Recently, several processing parameters were evaluated: (1) mild temperatures at 40 °C during 20 min; (2) flash heating from 40 to 100 °C in less than 20 s. These treatments were followed by aqueous extraction in water at room temperature. Independently of the temperature applied, OH allowed boosting extraction levels, increasing concentration of total phenolic compounds, soluble solids, and red color intensity of the obtained extracts. OH treatments at high-temperature short-time (HTST), due to the fast internal heating of grape skin structure, resulted in an increase of total concentration of anthocyanins from 756 to 1349 μg/g, with malvidin-3-O-glucoside being the main compound identified and quantified in the aqueous extracts (corresponding to about 60% of the total). These results reveal that OH might be considered an efficient and environmentally friendly technology to improve the extraction of polyphenolic compounds. Due to the volumetric heating effect, the OH technique reduces treatment times and the use of water and lowers energy consumption when compared with conventional thermal processes, furthermore, it can assist in avoiding the use of organic solvents. The authors suggested that the flash-heating extraction process allows permeabilization of tissues without promoting thermal degradation of the bioactive compounds (Pereira et al. 2020).

3.2.3 Infra-red (IR)

The infrared electromagnetic spectrum ranges from 0.75 to 1000 μm. Infrared-assisted extraction, utilizes the IR wavelengths to achieve a required heating effect. During infrared-assisted extraction, the heating leads to cell bursting. Internal heat is generated as a result of molecular collisions that absorb and dissipate energy from the electromagnetic field. The high efficiency of this technology is attributed to the high capacity of IR penetration. Infrared radiation can achieve high efficiency by matching the wavelength to the material absorption characteristics. Infrared has been widely explored for the extraction of bioactive compounds from medicinal herbs with the advantages of shorter time, high efficiency, simple operation, and environmental safety.

IR assisted extraction was utilized for the extraction of catechin, epicatechin, and procyanidin B2 from grape seeds. Three factors were examined for process optimization, (1) solvent, (2) solid/liquid ratio, and (3) illumination time. The chosen infrared-assisted extraction conditions were: 50% methanol solution as extraction solvent, solid/liquid ratio of 1:150 g/mL, and illumination time of 30 min. The extraction efficiency of IR was compared with microwave-assisted extraction (MW), ultrasonic extraction (US), and classical electrical heating (CH) methods, the results are summarized in Table 20.3. The IR method was found as the most effective. This might be explained by the infrared wave possessing a unique heating mechanism resulting in cell bursting which was created during the infrared heating. Cell bursting favors the entry of extracting solvent leading to solubilization of the target molecules. The solution of 50% methanol is a polar extraction solvent that can efficiently absorb infrared wave energy and leads to efficient heating, which eventually results in cell walls rupture. The IR extraction yield of epicatechin was found to be higher than for US (ultrasound), but lower than for MW (microwave). The reason for this phenomenon (the modification of not only total yield but also the ratio between components) may be related to the molecular structure of the extracted

Table 20.3 Comparison of various extraction methods re. polyphenols extraction from grape seeds

Extraction method	Catechin(mg/g)	Epicathechin (mg/g)	Pro. B3 (mg/g)	Examined parameters	Optimal conditions
MW	37	33	10	Solid/liquid ratio, power, working time	1:50, 600 W, 11 min
IR	47.9	30	12.1	Solvent, solid/liquid ratio, illumination time	50% MeOH, 1:150, 30 min
US	36	27	12	Solid/liquid ratio, power, working time	1:50, 90 min, 90 W
CH	30.7	26	7.2		n.a.

Data from Cai et al. (2011)
MW microwave, *IR* infrared, *US* ultrasound, *CH* conventional heating

molecules. The results indicated that IR has a great potential for offering an alternative technique for extraction of bioactive compounds (Cai et al. 2011).

3.2.4 Radio Frequency Heating (RFH)

RFH like IR differs from conventional conduction or convection heating by heat generation within the sample (volumetric heating) (Fig. 20.5a). Radio frequency is the electromagnetic wavelength positioned in the range of 300 kHz to 300 MHz (Fig. 20.5c). Despite the broad frequency range, only three frequencies are legally allowed to be used in the US (13.56 MHz, 27.12 MHz, and 40.68 MHz) to avoid interference with telecommunications. Since the heat in this method is generated within the material, the time required to obtain the required heating is relatively shorter than the time needed by conventional heating methods. The internal heating assists in cellular deformation which allows the enhanced movement of compounds into the extraction solvent. RFH is known as more effective in the case of semi-solids than the conventional heating, because of a more homogeneous heat distribution inside the treated material. RFH efficiency depends on the electromagnetic properties and shape of the sample and electrodes specification.

4 Summary

Natural bioactive compounds are common in numerous plants, algae and microorganisms, but their amounts arc usually limitcd. Thc utilization of wastc and agro-industrial by-products as a rich source of natural bioactives is highly desirable, both in terms of economic and environmental impact and in terms of potential sources for recovery of desired compounds. A growing area of consumers' interest and expectation, when it comes to cleaner labels and natural food components, leads to the need to improve and enhance the extraction methods and technologies to become more environmentally friendly, faster and resulting in a more competitive process than the chemical synthesis. The traditional extraction techniques have gradually switched to novel extraction technologies developed for more efficient and facilitated extraction of bioactive compounds, based on thermal and non-thermal principles which are widely explored for recovery of various bioactive compounds from a vast majority of natural sources and waste streams. In this chapter, thermal and non-thermal technologies were introduced through the examples of polyphenol extraction from grape seeds-the waste of wine and grape juice industry, andastaxanthin extraction, both bioactive compounds known as potential antioxidants with various health-promoting activities and with different aqueous solubility. The innovative extraction technologies present advantages in terms of shorter reaction times, lower temperatures (of extreme importance in the case of thermolabile bioactive compounds), reduced volumes of environmentally non-friendly organic solvents, and higher yields and purity of the desired compound. Different extraction methods not only affect the extraction yield but also have significant effects on the composition of the extract, possibly contributing to diversifying the bioactivity of the extract. Successful membrane permeability, improved diffusion, and enhanced mass transfer are among the main parameters involved in proper extraction. The novel technologies present a range of advantages in the field of bioactive compounds extraction, but significant knowledge gaps still exist, in particular, optimization of technological operation conditions relevant for industrial scales and levels.

References

Ahmad J, Khan I, Blundell R, Azzopardi J, Mahomoodally MF (2020) Stevia rebaudiana Bertoni.: an updated review of its health benefits, industrial applications and safety. Trends Food Sci Technol 100:177–189

Ahmadkelayeh S, Hawboldt K (2020) Extraction of lipids and astaxanthin from crustacean by-products: a review on supercritical CO_2 extraction. Trends Food Sci Technol 103:94–108

Astaxanthin Market Size, Share I Industry Report, 2020–2027. https://www.grandviewresearch.com/industry-analysis/global-astaxanthin-market

Biesalski HK et al (2009) Bioactive compounds: definition and assessment of activity. Nutrition 25:1202–1205

Cai Y, Yu Y, Duan G, Li Y (2011) Study on infrared-assisted extraction coupled with high performance liquid chromatography (HPLC) for determination of catechin, epicatechin, and procyanidin B2 in grape seeds. Food Chem 4:1872–1877

Croxatto Vega G et al (2021) Insights from combining techno-economic and life cycle assessment – a case study of polyphenol extraction from red wine pomace. Resour Conserv Recycl 167:105318

da Silva Lima R, Nunes IL, Block JM (2020) Ultrasound-assisted extraction for the recovery of carotenoids from guava's pulp and waste powders. Plant Foods Hum Nutr 75:63–69

Denaro M et al (2020) Antiviral activity of plants and their isolated bioactive compounds: an update. Phytother Res 34:742–768

Donoso A, González-Durán J, Muñoz AA, González PA, Agurto-Muñoz C (2021) Therapeutic uses of natural astaxanthin: an evidence-based review focused on human clinical trials. Pharmacol Res 166:105479

Eom SJ et al (2021) Development of an ultrasonic system for industrial extraction of unheated sesame oil cake. Food Chem 354:129582

Golshan M, Osfouri S, Azin R, Jalali T (2020) Fabrication of optimized eco-friendly dye-sensitized solar cells by extracting pigments from low-cost native wild plants. J Photochem Photobiol A Chem 388:112191

Guo L, Yang ZY, Tang RC, Yuan HB (2020) Grape seed proanthocyanidins: novel coloring, flame-retardant, and antibacterial agents for silk. ACS Sustain Chem Eng 8(15):5966–5974. https://doi.org/10.1021/acssuschemeng.0c00367

Higuera-Ciapara I, Félix-Valenzuela L, Goycoolea FM, Félix-Valenzuela L (2006) Astaxanthin: a review of its chemistry and applications. Crit Rev Food Sci Nutr 46:185–196

International Organisation of Vine and Wine (2019) 2019 Statistical report on world vitiviniculture. 2019 Statistical report on world vitiviniculture. http://www.oiv.int/public/medias/5029/world-vitiviniculture-situation-2016.pdf

Jia MZ, Fu XQ, Deng L, Li ZL, Dang YY (2021) Phenolic extraction from grape (Vitis vinifera) seed via enzyme and microwave co-assisted salting-out extraction. Food Biosci 40:100919

Liu C et al (2021) Extraction, characterization of aloe polysaccharides and the in-depth analysis of its prebiotic effects on mice gut microbiota. Carbohydr Polym 261:117874

Mainente F, Menin A, Alberton A, Zoccatelli G, Rizzi C (2019) Evaluation of the sensory and physical properties of meat and fish derivatives containing grape pomace powders. Int J Food Sci Technol 54:952–958

Martínez JM et al (2019) Use of pulsed electric field permeabilization to extract astaxanthin from the Nordic microalga Haematococcus pluvialis. Bioresour Technol 289:121694

McWilliams A (2018) Global carotenoids market research report to 2022. The global market for carotenoids. https://www.bccresearch.com/market-research/food-and-beverage/the-global-market-for-carotenoids.html

Natolino A, Da Porto C (2020) Kinetic models for conventional and ultrasound assistant extraction of polyphenols from defatted fresh and distilled grape marc and its main components skins and seeds. Chem Eng Res Des 156:1–12

Papoutsis K et al (2020) Recovery of ergosterol and vitamin D2 from mushroom waste – potential valorization by food and pharmaceutical industries. Trends Food Sci Technol 99:351–366

Pereira RN et al (2020) Using Ohmic Heating effect on grape skins as a pretreatment for anthocyanins extraction. Food Bioprod Process 124:320–328

Praveenkumar R et al (2020) Morphological change and cell disruption of haematococcus pluvialis cyst during high-pressure homogenization for astaxanthin recovery. Appl Sci 10:513

Rodríguez-Sifuentes L, Marszalek JE, Hernández-Carbajal G, Chuck-Hernández C (2021) Importance of downstream processing of natural astaxanthin for pharmaceutical application. Front Chem Eng 2:29

Smith PG (2011) Mass transfer operations. In: Introduction to food process engineering. Springer US, pp 437–465

Chapter 21
Decontamination of Fruit Juices by Combination of High Intensity Pulsed Light and Other Nonthermal Technologies

Taner Baysal and Özge Taştan

Abbreviations

AJ	Apple juice
EO	Essential oil
GRAS	Generally recognized as safe
HHP	High hydrostatic pressure
HIPL	High intensity pulsed light
MIC	Minimum inhibition concentration
MTS	Manothermosonication
NE	Nanoemulsion
NEBI	Non-enzymatic browning index
PEF	Pulsed electric field
PME	Pectin methyl esterase
RIV	Relative inactivation value
TS	Thermosonication
TSS	Total soluble solids
US	Ultrasound
UV	Ultraviolet light

T. Baysal
Ege University, Faculty of Engineering, Department of Food Engineering, İzmir, Turkey

Ö. Taştan (✉)
Yeditepe University, Faculty of Engineering, Department of Food Engineering, İstanbul, Turkey
e-mail: ozge.tastan@yeditepe.edu.tr

A. Režek Jambrak (ed.), *Nonthermal Processing in Agri-Food-Bio Sciences*, Food Engineering Series, https://doi.org/10.1007/978-3-030-92415-7_21

1 Introduction

Fruit juices gain popularity for consumers due to their fresh-like characteristics. Fruit juice consumption has increased worldwide due to the supply of a wide range of micronutrients, phenolics, and antioxidants of juices. However, they are susceptible to microbial, enzymatic, physicochemical deterioration. In addition, main factors like pH, water activity, redox potential, and nutrient content are essential for the spoilage of juices (Santhirasegaram et al. 2016).

Due to the higher acidity value, fruit juices can be spoiled by yeasts, especially *Saccharomyces* spp.; moulds like *Aspergillus* spp., and lactic acid bacteria like *Lactobacillus* and *Leuconostoc* spp. However, *Alicyclobacillus acidoterrestris* is a thermoacidophilic, spore-forming, heat-resistant bacteria that can survive the pasteurized fruit juices (Ferrario et al. 2015; De Cássia Martins 2018).

Thermal resistance studies conducted by *Alicyclobacillus* spores have shown that it is generally able to survive after heat treatment applied to commercial fruit juices.

In the traditional pasteurization of apple juices, a heat treatment reduces *Cryptosporidium parvum* by 5 logs (FDA 2004) or more intense pasteurization to reduce *Alicyclobacillus acidoterrestris* spores by 2 logs (Silva and Gibbs 2001) is required to optimize the process. In addition, it has been reported that heat treatment is required for the control of *Alicyclobacillus* species, which can cause a decrease of 5 times of the D value, that is, 5 logarithmic units at a certain temperature (Baysal and İçier 2012).

To control *Alicyclobacillus* spores in fruit juices could be problematic because their spores survive at pasteurization temperatures of juices, germinate and grow after processing if conditions are appropriate. *A. acidoterrestris* cells are able to grow at a pH between 2.0 and 6.0 at a temperature of 20–55 °C. In addition, *A. acidoterrestris* is known as the most important species and is suggested as the target microorganism for pasteurization in fruit juices and concentrates (Sokołowska et al. 2020).

The presence of *A. acidoterrestris* causes the production of a distinct antiseptic, disinfectant, or medicinal off-odor, resulting in the deterioration of acidic juices with significant economic losses for fruit juice companies. Thus, it is crucial to design control measures to effectively limit bacterial contamination without affecting the quality of juices (Zhao et al. 2021).

After the outbreaks associated with raw juices, FDA introduced a regulation for juice manufacturers to apply a process that achieves a minimum 5 log reduction of the "the most resistant microorganism of public health significance that is likely to occur in the juice". *Salmonella* is generally accepted as the target pathogen in citrus juices, whereas *E. coli* O157:H7 and *Cryptosporidium parvum* are considered a target for AJ (USFDA 2001).

Pasteurization is a critical step for fruit juice production technology as it provides microbial safety of juices. Although thermal pasteurization is the most applied method for the fruit juice industry, it is well-known that thermal pasteurization causes a negative impact on the organoleptic, nutritional, and physicochemical properties of juices. Therefore, the fruit juice industry, in the last decades, has been seeking novel alternative technologies to replace thermal pasteurization and inactivate microorganisms (mainly lactic acid bacteria, yeasts, molds, and *Alicyclobacillus* spores), while avoiding harmful effects on juice quality (Cavender and Kerr 2011; Carrillo et al. 2017; Roobab et al. 2018).

Over the last decades, studies have been increasingly addressed non-thermal technologies due to the potential of inactivating microorganisms and enzymes while maintaining the quality of the juices. Nonthermal technologies such as PEF, HHP, US and continuous or pulsed UV-C, and HIPL can successfully be used, as alone or combined treatment, for decontamination of fruit juices (Caminiti et al. 2012; Ferrario and Guerrero 2017; Roobab et al. 2018).

The growing interest in fresh-like and juices microbiologically safe has promoted the efforts to develop or design a combined use of innovative nonthermal methods. One of the important nonthermal technology is HIPL and the use of this technology in fruit juices has promising results as an alternative to thermal pasteurization. However, such limited literature exists to combine HIPL and other nonthermal technologies for the decontamination of fruit juices. It has still been explored for the efficacy of combined treatments to extend the shelf life, improve the safety of fresh juice, and changes in quality characteristics of juices during storage.

Therefore, this chapter highlights recently published results of the research reported in the combined use of HIPL and the other nonthermal technologies (US, PEF, UV-C, and NE) for microbial decontamination of fruit juices as a hurdle technology.

2 HIPL Technology for Decontamination of Fruit Juices

HIPL is a non-thermal technology for microbial inactivation, approved by FDA (21CFR179.41). HIPL is also called pulsed UV light, pulsed white light, high intensity light, intense pulsed light, or broad-spectrum white light in the scientific literature. HIPL uses high-intensity light emitted by gas discharge lamps that generate radiation in the wavelength range 200–1000 nm, containing the ultraviolet (UV) (200–400 nm), visible (380–780 nm), and near-infrared (NIR) (700–1100 nm) spectral ranges. It comprises a broad spectrum of "white light" with short (100–400 μs) and intense duration pulses. According to the FDA regulation, PL treatments that are used in the food industry must use xenon flash lamps, typically operated at 1–20 pulses per second, with a pulse duration no longer than 2 ms, and the total energy doses of process shall not exceed 12 J/cm^2 (Pataro et al. 2011; Muñoz et al. 2012; De Moraes and Moraru 2018; Preetha et al. 2021; Aaliya et al. 2021).

HIPL inactivates microorganisms mainly due to the absorption of UV light by microbial DNA, but other mechanisms may also contribute to inactivation. The HIPL process also leads to microbial decontamination by photothermal, photochemical, and photophysical effects. The absorption of radiation leads to the antimicrobial effects of UV light by the conjugated carbon-carbon double bonds in proteins, nucleic acids, and structural changes in DNA or RNA. HIPL causes the formation of pyrimidine dimers, which damages the process of cell replication (a photochemical mechanism). Furthermore, due to the result of overheating, membrane disruption was also determined. This is attributed to a difference in UV light absorption between the microorganism and its environment (a photothermal effect). In addition, structural damage in cells, such as the shrinkage of the cytoplasmic membrane, was also indicated (a photophysical effect). However, the main effect of HIPL on microorganisms is because of the photochemical action of the UV-C (200–280 nm), which causes thymine dimerization in the DNA chain preventing cell replication and finally leading to death. The impact of each mechanism in microbial inactivation depends on process parameters such as the pulse fluence, pulse width, frequency, process time, and peak power or food adsorption characteristics and target microorganism (Pataro et al. 2011; Muñoz et al. 2012; Ferrario et al. 2013; Ferrario et al. 2015; Ferrario and Guerrero 2017; De Moraes and Moraru 2018; Aaliya et al. 2021; De Souza Pedrosa et al. 2021).

Nonetheless, some studies have also presented evidence of non-UV-related death mechanisms, such as physical microbial-membrane destruction caused by cell overheating. The relative importance of these inactivation mechanisms likely depends on the fluence, wavelength, and target microorganism, and need more studies to be well understood (De Moraes and Moraru 2018).

In general, factors affecting the inactivation efficacy of HIPL technology for decontamination of fruit juices can be given as:

- Total energy dose (or fluence)
- The composition of the emitted light spectrum,
- The distance of the sample from the light source,
- The thickness, color, opacity, and viscosity of juices
- Flow conditions of the fruit juices
- The inoculum size of target microorganism,
- The geometry of the processing unit
- The number of lamps

Research on fruit juice decontamination by HIPL reported that the microbial inactivation level obtained from clear juices are higher than cloudy juices. This means that turbid substances in juices, which may apply a shadow effect to light penetration in the microbial cells, decrease the process efficiency. Other parameters such as treatment time, the distance of the sample from the light source, the composition of the emitting spectrum, the volume of the sample, number of lamps, and lamp design, mainly affect the sample-light interaction. Fluence decay is another limitation of HIPL treatments for fruit juices. Since fluence decays away from the lamp source because of the light absorption and scattering, it is required to quantify the fluence received within juices to design a HIPL system. Moreover, how light behaves inside the juice must be specified to obtain the successful design of HIPL treatment (Hsu and Moraru 2011; Pataro et al. 2011).

These limitations have recommended that HIPL should be used under a hurdle approach. Combining HIPL with other nonthermal techniques to decontaminate

Table 21.1 Microbial inactivation level of HIPL processed juice

Type of juice	Fluence (J/cm^2)	Log reduction	References
Apple	2–12 J/cm^2, 20 °C, in batch process	Max 4.3 log reduction for *A. acidoterrestris* vegetative cells	Taştan (2019)
Apple	6 J/cm^2, 20 °C, in continuous-flow	2.3 log reduction for *A. acidoterrestris* vegetative cells and 0.5 log reduction for spores	Taştan (2019)
Apple	2.4–71.6 J/cm^2, <12 °C	3 log reduction for *Alicyclobacillus* spores, 4.4 log reduction *S. cerevisiae* cells	Ferrario et al. (2015)
Grape	0.97–29.21 J/cm^2	Max 1.9 log reduction for *Ps. aeruginosa*	Hwang et al. (2015)
Plum	0.97–29.21 J/cm^2	Max 7 log reduction for *Ps. aeruginosa*	Hwang et al. (2015)
Melon	2.4–71.6 J/cm^2, <20 °C	0.3 to 6.9 log reduction of natural occuring microorganisms	Ferrario et al. (2013)
Strawberry	2.4–71.6 J/cm^2, <20 °C	<1 log reduction for *E.coli, L.innocua, Salmonella, S.cerevisiae*	Ferrario et al. (2013)

microorganisms has been successfully investigated in the last few years. If HIPL is combined with other techniques, microbial damage and then inactivation will probably be increased (Ferrario and Guerrero 2016).

Some authors have suggested that HIPL treatment can carry out higher levels of microbial inactivation than continuous UV-C. As can be seen in Table 21.1, HIPL treatment in fruit juices is promising results as an alternative to thermal pasteurization to inactivate pathogenic and spoilage microorganisms.

Sauer and Moraru (2009) reported that turbulence had an apparent effect on inactivation level in juices treated by HIPL, and process effectiveness increased significantly with increasing turbulence. In this study, AJ and cider were inoculated with nonpathogenic *E. coli* ATCC 25922 and pathogenic *E. coli* O157: H7. The inoculated juices were exposed to HIPL (both in static mode and under turbulence), placed in a thin layer (1.3 mm) sample holder, fluences up to 13.1 J/cm^2. For static treatments, 2.66 log reduction was found for *E. coli* ATCC 25922 and 2.52 log for *E. coli* O157:H7 in AJ at static mode. In cider, 2.32 log and 3.22 log reductions were determined for the nonpathogenic and pathogenic strains. Interestingly, 5.76 log reduction in cider and 7.15 log reduction in the juice was obtained by HIPL conducted under turbulence, which satisfies the FDA requirement of a 5 log reduction of *E. coli*. This research shows that the use of HIPL under turbulence is promising for the microbial decontamination of AJ and cider (Sauer and Moraru 2009).

Pataro et al. (2011) used a continuous-flow unit (tubes of 1 mm diameter) to inactivate *E. coli* and *L. innocua* in apple and orange juices. A fluence of 4 J/cm^2 resulted in *E. coli* reductions of 4.00 and 2.90 log CFU/ml and *L. innocua* reductions of 2.98 and 0.93 log CFU/ml in AJ and orange juice.

Preetha et al. (2021) investigated that coconut water, orange and pineapple juice inoculated with *E. coli* (MTCC 433) processed by a continuous-flow HIPL (0.18, 2 and 5.6 W/cm^2) with a flow rate of 100 ml/s. The *E. coli* inactivation level was found as 4.0, 4.5 and 5.33 log cycles for orange, pineapple juice and coconut water, respectively, when treated with the pulsed light doses of 95.2 J/cm^2 (Preetha et al. 2021).

Ferrario et al. (2013) stated that the HIPL inactivation kinetics for *E. coli* ATCC 35218, *L. innocua* ATCC 33090, *S. Enteritidis* MA44 and *S. cerevisiae* KE162 in commercial juices and freshly-squeezed juices. A negative relationship was observed between the absorbance of juices and HIPL effectiveness. HIPL treatment (2.4–71.6 J/cm^2) was not effective in natural strawberry and orange juices. Finally, inactivation of 0.3–6.9 log reduction cycles was obtained after 60 s-HIPL treatment.

As stated by Farag et al. (2021), a lower inactivation level was determined after HIPL exposure to bacterial spores, and spores were found more resistant than vegetative forms to HIPL treatment (Farag et al. 2021).

One of the potential drawbacks of HIPL treatment is that there is limited information on sublethal damage to microbial cells subject to HIPL (Pataro et al. 2011). Besides, another disadvantage is the generation of heat during extended processes, a destructive matter to the quality of the juices that must be balanced by the equipped with a cooling system (Pataro et al. 2011; Ferrario et al. 2014; Bevilacqua et al. 2018).

In order to increase the microbial inactivation level processed by HIPL technology, it should be investigated the use of combined treatments of HIPL and the other nonthermal technologies. Previous studies on the combinations of HIPL with PEF, MTS, and UV-C have already been reported for microbial inactivation of juices (Palgan et al. 2011; Munoz et al. 2011; Munoz et al. 2012).

3 HIPL Combined with a Hurdle Technology

Novel methods to improve safety and minimize quality loss during processing are the most important goal for the food industry. The limitations of different technologies cause the development of hurdle technology, commonly known as a combined treatment. The hurdle technology is defined as combining different technologies or factors (hurdles) to produce minimally processed food with fresh-like properties, safe, nutritious, stable, and economical. The hurdles can be divided into groups as physical, physicochemical, and microbial. A synergistic bactericidal effect exhibited by the hurdle technology contributes to decreasing process time and intensity. In addition, scientists have also used hurdle technologies to preserve and extend the shelf life of various food products (Santhirasegaram et al. 2016; Roobab et al. 2018; Kim et al. 2019; Aaliya et al. 2021).

The antimicrobial effect of nonthermal technologies can be optimized by combining them as hurdles in an overall preservation strategy. The effect of combined technologies can be additive, antagonistic, or synergistic. An individual process can cause injuries to microbial cells, and they may repair during storage. Nevertheless, if cell repairing is prevented by the combined use of technologies that interfere with cellular homeostasis maintenance, the cell cannot grow, and the inactivation level might be higher (Ross et al. 2003; Caminiti et al. 2012; Dixit et al. 2018).

There is a requirement for the combined use of HIPL and other nonthermal processing techniques for the pasteurization of fruit juices to be adapted on a pilot scale. These methods can be improved as an alternative to traditional pasteurization. Additional studies are required to determine the impacts of technologies on microbial stability during storage. The combined use of nonthermal technologies as a hurdle strategy would be a novel direction to preserve the juices that improve the quality and safety in the future (Santhirasegaram et al. 2016).

3.1 *HIPL Combined with US for Decontamination of Fruit Juices*

The US technology has been used for quality improvement and microbial inactivation in the fruit juice industry. This technology acts by inducing cavitation, and then gas bubbles are produced and burst inward, forming the shock waves which cause inactivation through cell membrane disintegration and free radical production. Besides, cavitation generates local changes in temperature and pressure, leading to shear stress in cell walls and then cause cell lysis (Muñoz et al. 2012; Ferrario et al. 2015).

US devices are composed of ultrasonic bath and probe systems. Several factors, such as the source of ultrasound, frequency, and acoustic energy density, effects the inactivation level. Efficiency is also influenced by treatment volume, temperature, viscosity, and gas concentration (Swamy et al. 2018). The mechanism of microbial inactivation by US treatment, generally called sonication. The use of ultrasonic waves as a unique technique can not effectively inactivate all microorganisms; if high levels of ultrasound power are required, the US could negatively change the nutritional and sensory properties of the food. This limitation has suggested that the US could be more efficient when combined with other techniques (Ferrario et al. 2015; Roobab et al. 2018). Advantageous combinations include TS (heat and ultrasound), manosonication (pressure and ultrasound), and MTS (pressure, heat, and ultrasound) (Demirdöven and Baysal 2008; Muñoz et al. 2012).

The US is effective against target microorganisms of fruit juices, and it can satisfy the requirement of a 5 log reduction of pathogens, such as *Escherichia coli*, established by the FDA (Zinoviadou et al. 2015; Santhirasegaram et al. 2016). Furthermore, previous studies have stated the US is effective to reduce the number of foodborne pathogens, such as in orange (500 kHz, 240 W, for 15 min), apple (25 kHz for 60 and 90 min), and carrot juice (20 kHz, 750 W, for 2 min) (Santhirasegaram et al. 2016). Especially, it is difficult to inactivate bacterial spores and several enzymes by US technology; therefore, US used as a single hurdle has some limitations (Ferrario and Guerrero 2017).

US and HIPL can effectively be used to increase the microbial inactivation in fruit juices. In a study conducted by Muñoz et al. (2012), HIPL and TS were applied individually or combined using a continuous-flow system to investigate their effect on *Escherichia coli* inactivation in AJ. Several quality characteristics (pH, TSS, colour, NEBI, and antioxidant activity were also analyzed. The juice was treated by HIPL with fluences of 4.03 J/cm^2 ('low' (L)) and 5.1 J/cm^2 ('high' (H)) and also treated by TS at 40 °C for 2.9 min (L) or 50 °C for 5 min (H). The effect of the resulting four energy levels and sequence (PL + TS and TS + PL) was carried out. As the technologies were applied alone, 2.7 and 4.9 log CFU/mL reductions for TS (H) and PL (H) were obtained. At the same time, most of the combined treatments obtained reductions of 6 log CFU/ml, showing an additive effect for both technologies when combined, regardless of the sequence applied. In addition, both treatments and the sequence in which the technologies were applied affected the colour of the juice (Muñoz et al. 2012).

Ferrario et al. (2015) evaluated the effect of US (10 or 30 min at 20, 30 or 44 °C) and HIPL (2.4 J/cm^2–71.6 J/cm^2; temperature 2, 30, 44 °C) on the inactivation of *A. acidoterrestris* ATCC 49025 spores and *S. cerevisiae* KE162 inoculated in commercial and natural squeezed AJs. The combined treatment causes 3 log cycles of spore reduction in commercial AJ and 2.0 log cycles in natural juice; while for *S. cerevisiae*, 6.4 and 5.8 log cycles of reduction were determined in AJs, both commercial and natural. In natural AJ, the US combined with HIPL at the highest temperature (56 °C) was the most efficient for microbial inactivation. In commercial AJ, the US did not contribute to the inactivation of spores but significantly reduced yeast load. The combinations of US and HIPL showed good microbial stability during storage under refrigerated conditions (Ferrario et al. 2015).

Further research is required for combined treatments of HIPL and US as an alternative method to thermal pasteurization. It should be investigated the effect of combined treatment on decontamination efficiency of various fruit juices, both clear and cloudy, and identifying the mechanisms when used in combination with other technologies as different sequences, and the changes in quality of juices during storage both room temperature and refrigerated conditions.

3.2 HIPL Combined with PEF for Decontamination of Fruit Juices

Processing fruit juices with PEF is one of the nonthermal technologies used for microbial and enzyme inactivation, extending the shelf-life and preserving the nutritional and sensory properties and aroma compounds. PEF treatment of juices uses short pulses (1–100 μs) with high voltage (10–50 kV/cm) to various fruit juices in a continuous-flow. A PEF system is composed of a high-voltage power supply, a pulse generator, a treatment unit, and a switch to discharge energy to electrodes. Moreover, a cooling system can be used to prevent temperature increase during processing (Akdemir Evrendilek 2016; Santhirasegaram et al. 2016).

Several limitations of PEF are reported that it cannot be used effectively on products with lower electrical conductivity and products that contain air bubbles. Also, PEF has limited effects on microbial spores. PEF processing depends on pulse width, electric field strength, flow rate, treatment temperature, and time of exposure. PEF makes the pores in the cell membrane of microorganisms. This electroporation process leads to leakage of cellular contents, then resulting in microbial cell damage. Furthermore, longer pulse width and higher-intensity pulse fields were more efficient in inactivating microbial growth (Santhirasegaram et al. 2016).

HIPL and PEF are innovative non-thermal technologies applied to fruit juices that have antimicrobial effects when applied alone or combined (Caminiti et al. 2011). In a study performed by Caminiti et al. (2011), HIPL combined with PEF were performed to reconstituted AJ in a continuous-flow. PEF field strengths (24 kV/cm or 34 kV/cm) were selected corresponding to "high" (H) and a "low" (L) energy inputs (261.9 and 130.5 J/ml). The juice was exposed to the HILP with fluences of 5.1 J/cm^2 (H) or 4.0 J/cm^2 (L). Finally, PEF (L) followed by either HILP (L or H) suggests a synergistic effect on microbial decontamination. Mainly, the quality features were not affected by the treatments, and sensory analysis showed that the HILP(L)/PEF(L) combination was the most preferable of the selected non-thermal treatments. This study indicated a synergistic effect obtained when combining low PEF intensities with HIPL in the inactivation of *E. coli* in AJ (Caminiti et al. 2011).

3.3 *HIPL Combined with UV-C for Decontamination of Fruit Juices*

UV-C treatment can be used for the inactivation of spoilage and pathogenic microorganisms in various fruit juices. UV-C is effective against pathogens and spoilage microorganisms in fruit juices and apple cider, while maintaining the quality features of juices (Keyser et al. 2008; Hakguder 2009; Pala and Toklucu 2010; Baysal et al. 2013; Ünlütürk and Atılgan 2014; Kaya et al. 2015; Baysal 2018). A single treatment of UV-C (75.04 mJ/cm^2) applied to white grape juice decreased the microbial load was more than 5 log CFU/ml (Hakguder 2009).

Only one study is available in the literature to combine HIPL and UV-C for decontamination of orange-carrot juice blend (Caminiti et al. 2012). The juice was exposed to three selected processes combined with PEF (24 kV/cm, 18 Hz, 93 μs), UV-C (10.6 J/cm^2) or HILP (3.3 J/cm^2), in each case, with MTS technology (400 kPa, 35 °C, 1000 W, 20 kHz). It was reported that no significant changes were obtained in NEBI or antioxidant activity. However, total phenolics were significantly decreased after the treatments. Each combination achieved on average 78% inactivation of PME. Sensory analysis showed no differences in the juice's odour, sweetness, or acidity (Caminiti et al. 2012).

3.4 HIPL Combined with NE for Decontamination of Fruit Juices

EOs are defined as a concentrated hydrophobic liquid containing volatile chemical compounds from plants, mainly terpenes. The antimicrobial activity of EOs is dependent on their characteristics like chemical compounds and the type of microorganism. EOs have been proposed as natural food preservatives due to their strong, wide-spectrum activity against microorganisms. EOs are categorized as GRAS by the FDA (FDA 2016; Dutra et al. 2019; De Souza Pedrosa et al. 2021).

Moreover, EOs can be used for juice stabilization and show strong antimicrobial activity against spoilage and pathogenic microorganisms in juices. However, because of their negative impacts on the sensory properties, some EOs are not recommended to apply fruit juices. Therefore, other preservation methods must be combined with EO to decrease their impact on food flavor (Pandey and Negi 2018).

Nanoencapsulation of EOs depicts an efficient strategy to improve the stability of the bioactives, protecting them from the interactions with the food ingredients, increasing their solubility and bioactivity, and thus enhance the antimicrobial activity (Donsi et al. 2011). For example, NE of the terpenes was added to orange and pear juices inoculated with *L. delbrueckii*. The addition of low concentrations of the NE was able to postpone the microbial growth (1 g/l terpenes) or completely inactivate the microorganisms (5 g/l terpenes) while preserving the organoleptic properties of the fruit juices (Donsi et al. 2011).

Limited literature studies were reported in the literature that combined HIPL and NE for the microbial inactivation. In a study, Tastan et al. (2017) reported that HIPL has a synergistic effect when combined with carvacrol NE for the inactivation of *E. coli* ATCC 26 in a vegetable. The synergistic effect of HIPL and NE can be attributed to the ability of EOs to increase membrane permeability, promoting the HIPL to the cell's genetic material. Further studies are required to assess the effect of combined use on liquid food products such as fruit juices (De Souza Pedrosa et al. 2021).

Tastan (2019) investigated the effect of HIPL combined orange NE on *Alicyclobacillus acidoterrestris* inactivation in clear AJ. 0.25 and 0.9 log cycle reduction was found after adding orange NE at a concentration of MIC and 2*MIC, respectively. In addition, HIPL at 6 J/cm^2 (HIPL6), HIPL6 + MIC, and HIPL6 + 2MIC treatments decreased *A. acidoterrestris* cells as 2.46, 2.75, and 3.45 logs, respectively. Moreover, 3.18, 3.50, and 4.50 logs reductions were obtained by combined treatments as HIPL at 6 J/cm^2 (HIPL12), HIPL12 + MIC ve HIPL12 + 2MIC. This study indicated that combined treatment as HIPL6 + 2MIC and HIPL12 + 2MIC showed synergistic effects on the inactivation of *A. acidoterrestris* vegetative cells in clear AJ.

Combining natural antimicrobials and encapsulated forms with other nonthermal technologies can accomplish more effective antimicrobial activity for food preservation and safety (Pandey and Negi 2018). Therefore, the combination of HIPL and nanoemulsion systems to preserve fruit juices should be further investigated.

3.5 RIV of Combined Non-thermal Technologies

In order to compare the effect of different nonthermal technologies, RIV can be calculated as the ratio of the difference between the inactivation level obtained by combined treatment and that of the non-thermal treatment-1 (NT1) as an individual, and the inactivation level of the non-thermal treatment-2 (NT2) as an individual (Eq. (21.1)) (Severino et al. 2014; Donsì et al. 2015).

$$RIV = \frac{\frac{\log N_0 - \log N_{combined}}{\log N_0} - \frac{\log N_0 - \log N_{NT1}}{\log N_0}}{\frac{\log N_0 - \log N_{NT2}}{\log N_0}} = \frac{\log N_{NT1} - \log N_{combined}}{\log N_0 - \log N_{NT2}} \quad (21.1)$$

In the RIV referred to a selected combined treatment, log N_0 is the initial microbial load (control sample), log N_{comb} is the microbial population resulting from the combined treatment, log N_{NT1} is the microbial population resulting from the non-thermal treatment-1 alone, and log N_{NT2} is the microbial population resulting from the nonthermal treatment-2 alone. Being RIV a normalized inactivation value allows us to compare the results obtained by different treatments.

RIV <1 shows an antagonist effect between the non-thermal treatment-1 and the non-thermal treatment-2, while RIV >1 indicates a synergistic effect between the combined treatments. RIV ≈ 1 defines an additive effect between the tested combined treatments (Severino et al. 2014; Donsì et al. 2015; Tastan et al. 2017).

Combining various nonthermal technologies is a promising way to impart a synergistic effect against spoilage microorganisms and can be used as an alternative way to overcome several limitations of technologies. The combination of various methods as hurdles puts the microorganisms in a hostile environment and disturbs the homeostasis of microorganisms in food temporarily or permanently. Optimization of the hurdle technologies are emerging in the food industry (Aaliya et al. 2021).

4 Conclusion and Future Remarks

There is growing demand of consumers for fresh-like products as traditional thermal processing may have undesirable effects on fruit juices' sensory and nutritional properties. Therefore, scientists currently have investigated the potential combined use of HIPL with other nonthermal technologies that have to increase the efficiency of microbial inactivation to obtain additive or synergistic effects.

From the fruit juice industry perspective, the development of combined nonthermal technologies is essential to produce novel products with improved quality, processed by a continuous-flow HIPL and the other hurdles. Although the combined use of HIPL and other nonthermal technologies is gaining attention from scientists and the food industry, future studies still need to be done to identify possible undesirable side effects, commercial scale-up, and process validation. In addition, legislation of these technologies is a potential obstacle that could delay their commercial uptake. Moreover, future studies should be addressed to assess the potential to form any secondary or toxic by-products after combined treatment as a critical research area. The commercial success of a given novel processing is directly related to summarize scientific knowledge, which contains engineering principles to focus on microbial safety and quality challenges.

Conflict of Interest The authors declare that there is no conflict of interest.

References

Aaliya B, Sunooj KV, Navaf M, Akhila PP, Sudheesh C, Mir SA, Sabu S, Sasidharan A, Hlaing MT, George J (2021) Recent trends in bacterial decontamination of food products by hurdle technology: a synergistic approach using thermal and non-thermal processing techniques. Food Res Int 147:110514. https://doi.org/10.1016/j.foodres.2021.110514

Akdemir Evrendilek G (2016) Change regime of aroma active compounds in response to pulsed electric field treatment time, sour cherry juice apricot and peach nectars, and physical and sensory properties. Innov Food Sci Emerg Technol 33:195–205. https://doi.org/10.1016/j.ifset.2015.11.020

Baysal T, İçier F (2012) Nonthermal technologies on food engineering. Nobel Academic Research Publisher, Ankara, 434p. ISBN: 9786051333304

Baysal AH, Molva C, Unluturk S (2013) UV-C light inactivation and modeling kinetics of Alicyclobacillus acidoterrestris spores in white grape and apple juices. Int J Food Microbiol 166(3):494–498. https://doi.org/10.1016/j.ijfoodmicro.2013.08.015

Baysal AH (2018) Short-wave ultraviolet light inactivation of pathogens in fruit juices, chapter 24. In Rajauria G, Tiwari BK (ed) Fruit juices. Academic Press, p. 463–510, ISBN 9780128022306. https://doi.org/10.1016/B978-0-12-802230-6.00024-2

Bevilacqua A, Petruzzi L, Perricone M, Speranza B, Campaniello D, Sinigaglia M, Corbo MR (2018) Nonthermal technologies for fruit and vegetable juices and beverages: overview and advances. Compr Rev Food Sci Food Saf 17:1–62. https://doi.org/10.1111/1541-4337.12299

Caminiti IM, Palgan I, Noci F, Muñoz A, Whyte P, Cronin DA, Morgan DJ, Lyng JG (2011) The effect of pulsed electric fields (PEF) in combination with high intensity light pulses (HILP) on Escherichia coli inactivation and quality attributes in apple juice. Innov Food Sci Emerg Technol 12:118–123. https://doi.org/10.1016/j.ifset.2011.01.003

Caminiti IM, Noci F, Morgan DJ, Cronin DA, Lyng JG (2012) The effect of pulsed electric fields, ultraviolet light or high intensity light pulses in combination with manothermosonication on selected physico-chemical and sensory attributes of an orange and carrot juice blend. Food Bioprod Process 90:442–448. https://doi.org/10.1016/j.fbp.2011.11.006

Carrillo MG, Ferrario M, Guerrero S (2017) Study of the inactivation of some microorganisms in turbid carrot-orange juice blend processed by ultraviolet light assisted by mild heat treatment. J Food Eng 212:213–225. https://doi.org/10.1016/j.jfoodeng.2017.06.005

Cavender GA, Kerr WL (2011) Inactivation of vegetative cells by continuous high-pressure processing: new insights on the contribution of thermal effects and release device. J Food Sci 76:525–529. https://doi.org/10.1111/j.1750-3841.2011.02325.x

De Cássia Martins Salomão B (2018) Pathogens and spoilage microorganisms in fruit juice: an overview, chapter 16. In Rajauria G, Tiwari BK (ed) Fruit juices. Academic Press, p. 291–308, ISBN 9780128022306. https://doi.org/10.1016/B978-0-12-802230-6.00016-3

De Moraes JO, Moraru CI (2018) High-intensity pulsed light processing. In: Proctor A (ed) Alternatives to conventional food processing, 2nd edn. The Royal Society of Chemistry, Cambridge, pp 355–396. https://doi.org/10.1039/9781849730976-00367

De Souza Pedrosa GT, Pimentel TC, Gavahian M, de Medeiros LL, Pagan R, Magnani M (2021) The combined effect of essential oils and emerging technologies on food safety and quality. LWT Food Sci Technol 147:111593. https://doi.org/10.1016/j.lwt.2021.111593

Demirdöven A, Baysal T (2008) The use of ultrasound and combined technologies in food preservation. Food Rev Intl 25(1):1–11

Dixit Y, O'Sullivan C, Cullen PJ, Tiwari BK (2018) Hurdle technologies for fruit juices, chapter 27. In Rajauria G, Tiwari BK (ed) Fruit juices. Academic Press, p. 539–554, ISBN 9780128022306. https://doi.org/10.1016/B978-0-12-802230-6.00027-8

Donsi F, Annunziata M, Sessa M, Ferrari G (2011) Nanoencapsulation of essential oils to enhance their antimicrobial activity in foods. LWT Food Sci Technol 44:1908–1914. https://doi.org/10.1016/j.lwt.2011.03.003

Donsì F, Marchese E, Maresca P, Pataro G, Vu KD, Salmieri S, Lacroix M, Ferrari G (2015) Green beans preservation by combination of a modified chitosan based coating containing nanoemulsion of mandarin essential oil with high pressure or pulsed light processing. Postharvest Biol Technol 106:21–32. https://doi.org/10.1016/j.postharvbio.2015.02.006

Dutra TV, Castro JC, Menezes JL, Ramos TR, Prado IN, Junior MM, Mikcha JMG, Filho BAA (2019) Bioactivity of oregano (Origanum vulgare) essential oil against Alicyclobacillus spp. Ind Crops Prod 129:345–349. https://doi.org/10.1016/j.indcrop.2018.12.025

Farag MA, Mesak MA, Saied DB, Ezzelarab NM (2021) Uncovering the dormant food hazards, a review of foodborne microbial spores' detection and inactivation methods with emphasis on their application in the food industry. Trends Food Sci Technol 107:252–267. https://doi.org/10.1016/j.tifs.2020.10.037

Ferrario M, Alzamora SM, Guerrero S (2013) Inactivation kinetics of some microorganisms in apple, melon, orange and strawberry juices by high intensity light pulses. J Food Eng 118:302–311. https://doi.org/10.1016/j.jfoodeng.2013.04.007

Ferrario M, Alzamora SM, Guerrero S (2014) Study of the inactivation of Saccharomyces cerevisiae in apple juice by pulse light and ultrasound: assessment of the physiological status by flow cytometry and transmission electron microscopy. International Association for Food Protection annual meeting. J Food Protect Suppl A 77:132–132. (P1-183)

Ferrario M, Alzamora SM, Guerrero S (2015) Study of the inactivation of spoilage microorganisms in apple juice by pulsed light and ultrasound. Food Microbiol 46:635–642. https://doi.org/10.1016/j.fm.2014.06.017

Ferrario M, Guerrero S (2016) Effect of a continuous flow-through pulsed light system combined with ultrasound on microbial survivability, color and sensory shelf life of apple juice. Innov Food Sci Emerg Technol 34:214–224. https://doi.org/10.1016/j.ifset.2016.02.002

Ferrario M, Guerrero S (2017) Impact of a combined processing technology involving ultrasound and pulsed light on structural and physiological changes of Saccharomyces cerevisiae KE 162 in apple juice. Food Microbiol 65:83–94. https://doi.org/10.1016/j.fm.2017.01.012

Food and Drug Administration (FDA) (2004) Guidance for industry: Juice HACCP hazards and controls guidance (1st ed.). http://www.fda.gov/Food/GuidanceRegulation/GuidanceDocumentsRegulatoryInformation/Juice/ucm072557.htm

Food and Drug Administration (FDA) (2016) Substances Generally Recognized as Safe, Part 812, CFR Title 21. https://www.accessdata.fda.gov/scripts/cdrh/cfdocs/cfcfr/cfrsearch.cfm?fr=182.20

Hakguder B (2009) UV disinfection of some of the fruit juices. Master thesis, Izmir Institute of Technology, Graduate school of natural and applied sciences, Food Engineering Department, 95 p

Hsu L, Moraru CI (2011) Quantifying and mapping the spatial distribution of fluence inside a pulsed light treatment chamber and various liquid substrates. J Food Eng 103:84–91. https://doi.org/10.1016/j.jfoodeng.2010.10.002

Hwang H, Cheigh C, Chung M (2015) Relationship between optical properties of beverages and microbial inactivation by intense pulsed light. Innovative Food Science & Emerging Technologies 3191–96. https://doi.org/10.1016/j.ifset.2015.06.009
Kaya Z, Yıldız S, Ünlütürk S (2015) Effect of UV-C irradiation and heat treatment on the shelf life stability of a lemon–melon juice blend: multivariate statistical approach. Innov Food Sci Emerg Technol 29:230–239. https://doi.org/10.1016/j.ifset.2015.03.005
Keyser M, Műller IA, Cilliers FP, Nel W, Gouws PA (2008) Ultraviolet radiation as a non-thermal treatment for the inactivation of microorganisms in fruit juice. Innov Food Sci Emerg Technol 9:348–354. https://doi.org/10.1016/j.ifset.2007.09.002
Kim SS, Park SH, Kim SH, Kang DH (2019) Synergistic effect of ohmic heating and UV-C irradiation for inactivation of Escherichia coli O157:H7, Salmonella Typhimurium and Listeria monocytogenes in buffered peptone water and tomato juice. Food Control 102:69–75. https://doi.org/10.1016/j.foodcont.2019.03.011
Munoz A, Palgan I, Noci F, Morgan D, Cronin D, Whyte P, Lyng J (2011) Combinations of high intensity light pulses and thermosonication for the inactivation of Escherichia coli in orange juice. Food Microbiol 28:1200–1204. https://doi.org/10.1016/j.fm.2011.04.005
Munoz A, Caminiti I, Palgan I, Pataro G, Noci F, Morgan D, Cronin D, Whyte P, Ferrari G, Lyng J (2012) Effects on Escherichia coli inactivation and quality attributes in apple juice treated by combinations of pulsed light and thermosonication. Food Res Int 45:299–305. https://doi.org/10.1016/j.foodres.2011.08.020
Muñoz A, Palgan I, Noci F, Cronin DA, Morgan DJ, Whyte P, Lyng JG (2012) Combinations of selected non-thermal technologies and antimicrobials for microbial inactivation in a buffer system. Food Res Int 47:100–105. https://doi.org/10.1016/j.foodres.2012.02.001
Pala ÇU, Toklucu AK (2010) The application of ultraviolet light on fruit juices. Acad Food J 8(1):17–22. Retrieved from https://dergipark.org.tr/en/pub/akademik-gida/issue/55835/764820
Palgan I, Caminiti IM, Muñoz A, Noci F, Whyte P, Morgan DJ, Cronin DA, Lyng JG (2011) Combined effect of selected non-thermal technologies on Escherichia coli and Pichia fermentans inactivation in an apple and cranberry juice blend and on product shelf life. Int J Food Microbiol 151:1–6. https://doi.org/10.1016/j.ijfoodmicro.2011.07.019
Pandey A, Negi PS (2018) Use of natural preservatives for shelf life extension of fruit juices, chapter 29. In Rajauria G, Tiwari BK (ed) Fruit juices. Academic Press, p. 571–605, ISBN 9780128022306. https://doi.org/10.1016/B978-0-12-802230-6.00029-1
Pataro G, Muñoz A, Palgan I, Noci F, Ferrari G, Lyng JG (2011) Bacterial inactivation in fruit juices using a continuous flow Pulsed Light (PL) system. Food Res Int 44:1642–1648. https://doi.org/10.1016/j.foodres.2011.04.048
Preetha P, Pandiselvam R, Varadharaju N, Kennedy ZJ, Balakrishnan M, Kothakota A (2021) Effect of pulsed light treatment on inactivation kinetics of Escherichia coli (MTCC 433) in fruit juices. Food Control 121:107547. https://doi.org/10.1016/j.foodcont.2020.107547
Roobab U, Aadil RM, Madni GM, El-Din Bekhit A (2018) The impact of nonthermal technologies on the microbiological quality of juices: a review. Compr Rev Food Sci Food Saf 17:437–457. https://doi.org/10.1111/1541-4337.12336
Ross AIV, Griffiths MW, Mittal GS, Deeth HC (2003) Combining nonthermal technologies to control foodborne microorganisms. Int J Food Microbiol 89:125–138. https://doi.org/10.1016/s0168-1605(03)00161-2
Santhirasegaram V, Razali Z, Somasundram C (2016) Safety improvement of fruit juices by novel thermal and nonthermal processing. In: Kotzekidou P (ed) Food hygiene and toxicology in ready-to-eat foods. Elsevier ; Academic, Amsterdam, pp 209–223. https://doi.org/10.1016/B978-0-12-801916-0.00012-1
Sauer A, Moraru C (2009) Inactivation of Escherichia coli ATCC 25922 and Escherichia coli O157:H7 in apple juice and apple cider, using pulsed light treatment. J Food Prot 72:937–944. https://doi.org/10.4315/0362-028x-72.5.937

Severino R, Vu KD, Donsi F, Salmieri S, Ferrari G, Lacroix M (2014) Antimicrobial effects of different combined non-thermal treatments against Listeria monocytogenes in broccoli florets. J Food Eng 124:1–10. https://doi.org/10.1016/j.jfoodeng.2013.09.026

Silva FVM, Gibbs P (2001) Alicyclobacillus acidoterrestris spores in fruit products and design of pasteurization processes. Trends Food Sci Technol 12:68–74. https://doi.org/10.1016/S0924-2244(01)00070-X

Sokołowska B, Połaska M, Dekowska A (2020) Alicyclobacillus-still current issues in the beverage industry. In: Safety issues in beverage production. Woodhead Publishing, Duxford, pp 105–146. https://doi.org/10.1016/B978-0-12-816679-6.00004-8

Swamy GJ, Muthukumarappan K, Asokapandian S (2018) Ultrasound for fruit juice preservation, chapter 23. In Rajauria G, Tiwari BK (ed) Fruit juices. Academic Press, p. 451–462, ISBN 9780128022306. https://doi.org/10.1016/B978-0-12-802230-6.00023-0

Taştan Ö, Pataro G, Donsi F, Ferrari G, Baysal T (2017) Decontamination of fresh-cut cucumber slices by a combination of a modified chitosan coating containing carvacrol nanoemulsions and pulsed light. Int J Food Microbiol 260:75–80. https://doi.org/10.1016/j.ijfoodmicro.2017.08.011

Taştan Ö (2019) Investigation of combined use of high intensity pulsed light, low voltage electrical treatment and natural antimicrobial substances for nonthermal pasteurization of clear apple juice. PhD thesis, Ege University, Graduate school of natural and applied sciences, Food Engineering Department, 182p

US Food and Drug Administration (USFDA) (2001) Federal Register Final Rule – 66 FR 6137, January 19, 2001: Hazard Analysis and Critical Control Point (HACCP); procedures for the safe and sanitary processing and importing of juices. Fed Reg 66:6137–6202

Ünlütürk S, Atılgan MR (2014) UV-C irradiation of freshly squeezed grape juice and modeling inactivation kinetics. J Food Process Eng 37:438–449. https://doi.org/10.1111/jfpe.12099

Zhao N, Zhang J, Qi Y, Xu J, Wie X, Fan M (2021) New insights into thermo-acidophilic properties of Alicyclobacillus acidoterrestris after acid adaptation. Food Microbiol 94:103657. https://doi.org/10.1016/j.fm.2020.103657

Zinoviadou KG, Galanakis CM, Brnčić M, Grimi N, Boussetta N, Mota MJ, Saraiva JA, Patras A, Tiwari B, Barba FJ (2015) Fruit juice sonication: implications on food safety and physico chemical and nutritional properties. Food Res Int 77:743–752

Chapter 22
Food-On-A-Chip: Relevance of Microfluidics in Food Processing

M. S. Giri Nandagopal, Sriram Krishnamurthy, and Thulasiraman Venkatesh

1 Introduction

Microfluidics is comprised of two terms – "micro" meaning small and "fluid" denoting flow. It is the science of studying and applying flow of liquids in the micron (μm) scale. Over the years, this branch of science has found applications in various other fields ranging from printing technologies to biomedical systems. With the advent of new fabrication technologies, higher machining precision and advanced materials, this field is finding applications to a plethora of problems. A typical microfluidic channel has width ranging from a few hundred nanometres to a few micrometres. At these length scales, the flow of fluid is laminar and can be controlled with very high precision. The Reynold's number for such flows are very small and therefore, the inertial forces are significant compared to the viscous forces. The consequences of such a flow is high surface to volume ratio which manifests into surface effects dominating bulk properties and lower reaction times (Convery and Gadegaard 2019). Thus, all the necessary chemical process systems can be shrunk and integrated to a very small scale that can be easily controlled and manipulated. These unique phenomena are an advantage for researchers in varied fields of study like biology, chemistry, engineering etc. The food processing and allied industries in particular have immense scope for the use of this technology for

M. S. Giri Nandagopal
Department of Mechanical Engineering, Indian Institute of Technology, Kharagpur, India

S. Krishnamurthy
Department of Chemical Engineering, Indian Institute of Technology, Madras, India

T. Venkatesh (✉)
Agro Processing and Technology Division, CSIR-National Institute for Interdisciplinary Science and Technology, Thiruvananthapuram, India

Academy of Scientific and Innovative Research (AcSIR), Ghaziabad, India

A. Režek Jambrak (ed.), *Nonthermal Processing in Agri-Food-Bio Sciences*, Food Engineering Series, https://doi.org/10.1007/978-3-030-92415-7_22

diverse applications like micro-emulsion, quality control and detection of harmful contaminants and adulterants. Conventional methods of detection use techniques like Gas Chromatography (GC), High Performance Liquid Chromatography (HPLC), Mass Spectroscopy (MS) etc. which involve the use of hazardous chemicals and need expensive equipment and trained personnel (Guo et al. 2015). The use of microfluidics for detection makes it cheaper, environment friendly and highly portable.

Microfluidics comprises of wide range of interdisciplinary area (Antony et al. 2014a, 2015a), which includes, but not limited to fluid mechanics (Antony et al. 2014a, 2015a), thermodynamics, electrostatics, chemistry (Antony et al. 2015b; Nandagopal et al. 2016; Giri Nandagopal et al. 2014a) to material science (Bayraktar and Pidugu 2006; Stone et al. 2004). "Lab on a chip (LOC)" devices have seen immense growth in the recent years which involves micro flow, microstructures (Chen et al. 2012)and nanomaterials (Abraham et al. 2017; Giri Nandagopal et al. 2014b; Mao and Koser 2006; Zhang and Wang 2013). For effective design of LOC devices, clear understanding of microfluidics physics plays an important role (Antony et al. 2014b; Giri Nandagopal and Selvaraju 2016; Nandagopal et al. 2017). To quantize the performance of LOC devices, an understanding of various dimensionless numbers become necessary. Some of them are:

1. Reynolds number, defined as the ratio between inertial force and the viscous force, is an important dimensionless number in microfluidics. Since the characteristic length of microfluidic devices is a few microms and flow is predominantly laminar, Reynolds number is very small, making viscous forces more dominant.

$$R = \frac{Dv\rho}{\mu}$$

Where, Re = Reynolds Number; D = Diameter of the chamber or pipe carrying the fluid; ρ = Density of the fluid and μ = viscosity of the fluid.

2. Peclet number defines to correlate convective and diffusive transport caused by heat transfer. If Peclet number is lesser than unity, diffusion would be dominant over convection, thereby making the fluid flow and target detection very slow and inefficient (Squires et al. 2008).

$$Pe = \frac{Lu}{\alpha}$$

Where, Pe = Peclet Number; L = Characteristic Length (Hydraulic Diameter for non-circular pipes); u = flow velocity and α = Thermal Diffusivity.

3. Capillary number is extensively used in multiphase flow, which describes the relationship between viscous and interfacial forces. For multiphase flow, since the interfaces between different fluids is a significant factor, surface tension

plays an important role. This interplay between surface forces and the viscous forces is responsible for droplet-based or bubble-based microfluidics, (Giri Nandagopal and Selvaraju 2016; Nandagopal et al. 2017).

$$Ca = \frac{\mu V}{\sigma}$$

Where, Ca = Capillary Number; μ = viscosity of the fluid; V = Characteristic velocity of the fluid and σ = Surface tension at the interface.

4. Weber number is the ratio of inertial to interfacial forces, concerning multiphase flows. Deformation of drops /bubbles will be high when the flow rate is high, which in turn results in high weber number in a microfluidic device (Xu et al. 2010).

$$We = \frac{\rho v^2 l}{\sigma}$$

Where, We = Weber Number; v = Velocity of the Fluid; l = Characteristic Length and σ = Surface tension at the interface.

Conventionally, microfluidic devices are fabricated using silicon, glass, polymers and even paper-based materials. Polydimethylsiloxane (PDMS) is a polymer that is majorly used for fabricating such systems. PDMS has low thermal conductivity, transparency and highly compatible with biological systems (Giri Nandagopal et al. 2017). Recently, paper-based materials have also been explored for their feasibility and portability (Martinez et al. 2010). The use of microfluidics in the food industry faces a unique challenge. The techniques used should be a viable alternative to already used techniques in the industry, safe for use in food and related materials and must be cheaper than the existing systems. Use of modern materials and food safe polymers can lead to excellent alternatives.

This book chapter reviews about the various fabrication methods for microfluidic devices and their application in detection of pathogens, additives, pesticides, herbicides and heavy metals in food.

2 Fabrication Process

2.1 *Micro-Molding*

PDMS is a polymeric material, widely used in the fabrication of microfluidic chip. In this micromolding process. SU-8 photoresist is commonly used as a mold which is further replicated on to a PDMS substrate (Sia and Whitesides 2003). In the conventional process of fabrication, the photoresist is spun-coated on a silicon chip and photoetched. By using different types of photoreceptors and by controlling the speed of spin coating, the thickness of the mold can be adjusted over a range of 10 to 200 microns. Now once the mold is ready, PDMS solution and is typically mixed with a hardener and the air bubbles are removed using a vacuum desiccator. The ratio of hardener to PDMS can be varied to obtain devices of variable hardness and flexibility. Then the PDMS mixture is slowly poured onto the mold to avoid formation fo cavities and bubbles and then heat treated to harden. Once the PDMS is properly hardened, the PDMS can be carefully removed from the mold. This molded PDMS can be further bonded with any substrate made of glass using oxygen plasma treatment. The mold can be used for future fabrication of such devices.

2.2 *Laser Ablation*

In laser ablation technique, desired microflow channels are ablated and micromachined by using a carbon dioxide laser mostly on a polymeric surface with a wavelength of 10.6 μm (van den Driesche et al. 2018). They are applicable in fabricating microchannels on polymeric materials and glasses. Laser ablation has both advantages and disadvantages. The advantages are its relative simplicity, lesser time consumption and one-time ablation is sufficient to complete the entire processing. Disadvantages include formation of uneven surface on the inner wall of microchannel and formation of large number of bubbles which requires further chemical treatment (Wang et al. 2011). This method is suitable only for fabricating the system which has a channel width and depth greater than 80 μm.

2.3 *2D/3D Printing*

2D printing is a conventional printing procedure commonly used in offices, such as laser printer (Garcia-Cordero et al. 2010), inkjet printer (Bsoul et al. 2016), wax printer (Pearce et al. 2016), screen printing (Wee et al. 2015) etc. This technique is usually adopted in paper-based microfluidic chips. Here, the channels in microscale are imprinted in a hydrophilic paper material like paper using a hydrophobic ink. This technique could provide an accuracy between 80 and 400 μm. Moreover, this technique can also be extended for microchannels made of SU-8, PDMS etc. that can be deposited right onto a polymer substrate or glass by screen printing or ink-jet printing (Shangguan et al. 2017). Also, for fabricating channels with electrodes, an appropriate conductive ink can be used to print on the surface of the microfluidic channel. Typically silver nano-particles have been used for this purpose (Tran et al. 2017). There is immense scope for using other advanced and cheaper alternatives for fabricating microchannels with electrical conductivity.

When it comes to 3D printing, two methods are used for the fabrication of microchannels namely stereo-lithography (He et al. 2015) and fused deposition modelling (FDM) (Gaal et al. 2017). The fused deposition modelling 3D printer are relatively low in price making them a suitable option to fabricate low-cost 3D microfluidic chips. In the fused deposition molding technique printing can directly be done on materilas like polylactic acid, polycarbonate, acrylonitrile butadience styrene etc. to make 3D microfluidic chips (Kataoka et al. 2017). An accuracy of 100–500 μm can be attained while using this process.

2.4 Injection Molding

Injection molding is a technique conventionally used for plastic processing. Recently, this technique is adopted for the manufacture of microfluidic based devices. The common injection materials for microfluidic fabrication are polymethylmethacrylate (PDMS), cyclic olefin copolymer and polydimethylsiloxane (Szydzik et al. 2016). However, fabrication of microfluidic based chips using injection molding is comparatively an expensive and time-consuming process. Hence, in order to overcome this Hansen et al. (2010) developed a technique in which an injection mold made of nickel where su-8 photoresist is imprinted on its surface was designed. This mold can be potentially reused for around 300 times making them cost effective. The major advantages of this techniques are good repeatability, fast processing speed, large-scale fabrication etc. while the disadvantage include poor flexibility and high cost.

3 Detection of Foodborne Pathogens

Foodborne pathogens and the illness caused by them is a huge matter of concern for the humans and other living beings (Lin et al. 2014). The major foodborne pathogens that cause foodborne illness outbreaks are *Salmonella Spp., Listeria monocytogenes, Escherichia coli O157:H7, Campylobacter Spp., Clostridium perfringens, and Staphylococcus aureus*. Conventionally these pathogens are detected using two techniques – an ELISA assay or a Polymerase Chain Reaction (PCR) test (Law et al. 2014). However, these methods are conducted in sophisticated laboratory setup with tedious and laborious assay procedures, providing a false negative when the sample delivery is delayed, moreover, the process is time-consuming and expensive too. A majority of these drawbacks can be overcome by using specialized techniques with microfluidics (Prabhu et al. 2020a, 2020b; Ramesh et al. 2021).

A microfluidics based LOC for the detection of *Salmonella Sp.* in food samples was reported by Sun et al. (2015). They reported a platform with on-chip sample preparation by using magnetic beads followed by loop-mediated isothermal amplification (LAMP). The effectiveness of this protocol was demonstrated on eight pork meat samples induced with Salmonella sp. within a time frame of 40 minutes. The demonstrated limit of detection (LOD) was found to be as low as 50 cells per test. This system was further improved by by using nano magnetic beads and quantum dots (QDs). The QDs were used as a fluorescent label. This procedure was able to finish the assay within within 30 min with a LOD of 10^3 CFU/mL.

Clime et al., introduced a microfluidic set-up for the preparation of sample, filtration, and extraction of microorganism (Clime et al. 2015). They demonstrated a novel approach where the debris are removed from the biological sample using hydrodynamic focusing and inertial lateral migration effects. In their study, they could demonstrate that 50% of the debris in ground beef samples were removed by the microfluidic filtration and extraction chip. They observed that up to 70% of the initial microorganisms were retained at the device outlet. Fronczek et al. 2013 developed a microfluidic hand-held device for detection of *Salmonella typhimurium* on poultry packages using polycarbonate (Fronczek et al. 2013) using polycarbonate. In this device, Mie scatter signals in the microfluidic channels were generated from immuno-agglutination between *Salmonella typhimurium* and micron sized carboxylated polystyrene. The device was used to read the conjugated anti-Salmonella. The assay time reported by them for the study was about 10 min and LOD of 10 CFU/mL.

The field of microfluidics have diversified itself into an area called centrifugal microfluidics which is the most powerful platforms in the field of microfluidics (Kwok et al. 2016; Strohmeier et al. 2015). In centrifugal microfluidics various process such as valving, pumping, mixing, separation, and other unit operations can be performed easily (Gorkin et al. 2010). Considering their potential, researchers has explored their potential in developing devices for food borne pathogens. Sayad et al. (2016) demonstrated an automatic detection system for Salmonella by developing a LAMP-based microfluidic lab-on-a-disc. In the study the liquid was dispersed on to the disc by using Centrifugal force. Using the device, they demonstrated the presence of Salmonella in a tomato inducted with Salmonella DNA. The detection of 70 min with a LOD of 5×10^{-3} ng DNA/mL was observed, which is significantly lesser than the conventional method which takes about 3–4 h.

Another interesting microfluidic approach are the paper-based devices. These paper-based approaches have gained wide attention for monitoring food safety as they are simple to fabricate and use, have low-cost, and the possibility of on-site detection of foodborne contaminants. Microfluidic paper based analytical devices (μPADs) developed for detecting pathogens in food mainly rely on enzymatic assay-based optical methods. Such methods predominantly use visual inspection - either with the naked eye or digitally enhanced images and subsequent use of image analysis techniques. In image analysis, the RGB (red-green-blue) values from the image are extracted and a desired algorithm is used to manipulate and arrive at a decision. Jokerst et al. (2017) reported a μPAD for the microspot assay of *Escherichia coli O157:H7, Listeria monocytogenes* and *Salmonella Typhimurium* in meat samples. In this method, a swab of the sample is collected and cultured in appropriate media. The cultured media is further added to a chromogen-impregnated μPAD. An enzyme associated with the pathogen can induce a change in the colour of this μPAD much like a pH paper. The LOD for each of the bacterial assays after image analysis were 104 CFU mL^{-1} for *Salmonella Typhimurium,* 108 CFU mL^{-1} for *Listeria monocytogenes and* 106 CFU mL^{-1} for *E. coli.* The only drawback to this method is that the sample preparation time is approximately 12 hours. But this is significantly lesser than that of the gold standard method for bacterial detection which takes several days.

Zuo et al. (2013) worked on food samples which are enriched with three pathogens namely *L. acidophilus, S. aureus, and S. enterica.* The intention of their study was to demonstrate a homogenous one step assay for the detection of multiplexed pathogens. For that they designed a hybrid biochip made up of PDMS, paper and glass. In the hybrid biochip they integrated a graphene oxide which is functionalized with aptamer, which when interacted with with microbial pathogens causes a change in their fluorescence properties. These change in the fluorescence is caused when the Cy3 dye which is labelled on the aptamer undergo quenching. When they studied this approach on the samples enriched with three pathogens namely L. *acidophilus, S. aureus, and S. enterica,* they could obtain a percent recovery between 92.9 to 107.8%. For a single assay the detection limit for the study obtained was 11.0 CFU mL^{-1} and the reaction time for the single assay is 10 min. While, studies to detect *S. aureus* and *S. enterica* simultaneously using this method showed a detection limit of 800.0 CFU mL^{-1} and 61.0 CFU mL^{-1}.

Jin et al. (2015) worked on the detection of salmonella on a μPAD by adenosine triphosphate (ATP) quantification method. In this method, the paper surface is chemically modified and ATP aptamer with an associated HRP-tagged DNA is attached into it. They performed experiments by culturing the *Salmonella* and then lysed them by boiling. When the sample is added to the modified paper colour change is observed only when *Salmonella* was present. The fundamental concept behind that was, in the presence of ATP associated with *Salmonella* catalytic oxidation of 3-amino-9-ethylcarbazole by HRP/H_2O_2 occurs. The limit of detection reported by them while performing the study was 2×10^7 CFU mL^{-1}. Similarly, Park et al. (2013) performed a study in which the Salmonella samples mixed with Salmonella conjugated particles were added to a μPAD. Then the results are optically observed through a Mie scattering strategy where a smartphone was used to analyze the scatter intensities at the optimized Mie scatter angle. While performing the experiments, they obtained a detection limit for *Salmonella* as 10^2 CFU mL^{-1}.

4 Detection of Additives in Food

Using pigments in food for enhancing its colour for commercial purpose has become a regular routine these days. But, most of these additives used in food are harsh chemicals which can potentially impact serious health conditions to the humans, which may be even teratogenic and carcinogenic (Shaw 2014). Hence, a simple, frugal and accurate detection device for food additives is for great significance. With the advent of microfluidics technology various efforts has been reported in this perspective. Researchers have reported a novel method using μPAD for the separation of pigments from beverages and to detect the amount of pigments present in them (Zhu et al. 2015). In the reported study, a conventional filter paper is taken and it is coated with polyelectrolytes such as polyallylamine hydrochloride and polysodium styrene sulfonate with suitable modifications using silver nanoparticles. Further detection is done using an integrated surface-enhanced Raman spectroscopy (SERS) with the appropriate functionalized paper-based carriers. They performed this study to detect the presence of commonly used additives such as sunset yellow and tartazine in the grape and orange juices respectively. The results revealed that the limit of detection for detection of sunset yellow and tartrazine were 10^{-5} $molL^{-1}$ and 10^{-4} $molL^{-1}$ respectively.

In the meat industry, maintaining the freshness of the meat by protecting them from microbial contamination is a highly important task. For this, there used to be an excess usage of nitrite, as nitrite possess high microbial inhibition properties. However, these nitrite has a nature of reacting with secondary and tertiary amines to produce nitrosamine compounds which can cause severe health hazards which may further lead to death (Feng et al. 2019; Zhang et al. 2014). He et al. (2013) developed a μPAD for the detection of nitrite by integrating Griess-colour nitrite assay with the conventional Whatman filter paper. The basic concept underlying his study was that when nitrite reacts with a Griess reagent in a μPAD, they undergo certain reaction and produces colours of varies intensities based on the nitrite concentration. The studies showed that the method adopted by them could have a dynamic range of 0.156–2.50 mM when quantified using image processing. A similar sort of approach was demonstrated by Jayawardane et al. (2014) for the determination of nitrite and nitrate in food samples and they could demonstrate a limit of detection(LOD) of 1.0 μm and 19 μm for nitrite and nitrate respectively. While, Cardoso et al. (2015) reported a μPADs integrated with colorimetry for the determination of nitrite in ham, sausage and river water. In the device developed by them, they created a container by stamping the appropriate geometry and the sample inlets are connected using channels of microscale dimensions. The LOD measured in their study was 5.6 μM which was in coherence with the spectrophotometry readings. Similar approaches, of integrating Griess method in μPAD for the detection of nitrite in water and food was reported by Lopez-Ruiz et al. (2014), Cardoso et al. (2015) and Jayawardane et al. (2014)

Among the various additives used, food additives such as sucrose, glucose and fructose has become an unavoidable additive widely used to enhance the flavour and colour of food and beverage. Lawrence et al. (2014) demonstrated the detection of glucose in carbonated drinks. They took a fibre-based paper disk and immobilized glucose oxide, while the carbon electrodes were screen printed. The limit of detection achieved while performing the device with the carbonated drinks was about 0.18 mmol/L which was verified using a HPLC and found the results to be consistent.

Similarly, Adkins et al. (2015) developed a μPAD with copper as a working electrode. This setup was studied for the detection of glucose, fructose and sucrose in commercial beverage samples such as Red Bull, Coca-Cola, Orange Powerade, and Vitamin Water. The studies performed to detect the glucose, fructose and sucrose in the sample revealed that the amount of glucose, fructose and sucrose is 270 nM, 340 nM and 430 nM respectively. Colletes et al.(2016) studied to detect the glucose in liquors. They took a paraffin stamped paper substrate and incorporated ionization method for their device design. The detection limit obtained by them using this method was 2.77 mmol.L^{-1}.

5 Detection of Pesticides and Herbicides in Food

The usage of pesticides has been for decades in agricultural sector as it plays a significant role in food quality and production. However, they imply serious health concerns to the people, this further open up a need for an effective detection device (Gilbert-López et al. 2009). Wang et al. (2013) worked on to detect the presence of 2,4 dichlorophenoxyacetic acid in food. For which they fabricated a paper-based multidisc device which is molecular imprinted polymer-grafted (MIP). The MIP approach was proposed as an alternative technique to conventional immunoassays. In immunoassay technique which widely depend on the antibodies. But during production and transportation there is a high chance for denaturation and instability. Hence they took a polymer-grafted device in which tobacco peroxidase (TOP)-labelled 2,4-D was incorporated. Hence during the process an enzyme catalyzed cathodoluminescence emission was achieved from the luminol-TOP-H_2O_2 CL system. The detection limit obtained for the system is 1.0 pM. Similarly, Liu et al. (2014) studied on the detection of DDV in fruits and vegetables by integrating μPAD with cathodoluminescence. The method was successfully studied for food spiked with cucumber, tomato and cabbage and obtained a detection limit of 3.6 ng.mL^{-1} (Liu et al. 2015). This method was successfully demonstrated in cucumber and tomato and they obtained the limit of detection as 0.8 ng.mL^{-1}.

Sicard et al. (2015) worked on to develop μPAD for the detection of organophosphate pesticides such as paraoxon and malathion. The concept adapted by them was the inhibition properties of pesticides by Acetylcholinesterase (AChE), ie, AChE has the potential to hydrolyze indoxyl acetate which is a colourless substrate into a product which is of indigo in colour, when there no presence of pesticides. By recording and processing the result in a smartphone they observed a detection limit of 10 nM. Following this, Nouanthavong et al. (2016) demonstrated a novel method to detect methyl-paraoxon and chlorpyrifos-oxon. They used μPAD coated with nanoceria for the detection in which they integrated enzyme-inhibition assay with AChE and choline oxidase. When, studies were performed to analyse the presence of methyl-paraoxon and chlorpyrifos-oxon with the device, they obtained a LOD of 18 ng.mL^{-1} and 5.3 ng.mL^{-1} for methyl-paraoxon and chlorpyrifos-oxon respectively. The method was adapted for the detection of methyl-paraoxon in dried green mussel and cabbage. They obtained a recovery ~95% for both of them.

Pentachlorophenol (PCP) is an organochlorine pesticide which is widely used in agriculture seeds. It is carcinogenic in nature and has acute toxic effects (Dai and Copley 2004, Fuentes et al. 2013). Sun et al.(2014) developed a method to detect PCP using sensors that work on photo electro chemical (PEC) principle, which uses MIP technique on a μPAD. The electrode paper is coated with a double layer coating of gold nano particles (AuNPs) and a layer of polypyrrole (Ppy)-functionalized

ZnO nanoparticles. Upon irradiation by visible light, the excitation of electrons occurs from highest occupied molecular orbital of Ppy to the lowest unoccupied molecular orbital of ZnO. Due to such excitation of electrons, the molecular orbital levels of Ppy and ZnO matches and the electron finally reaches the gold-paper electrode (Au-PWE) surface, which ultimately leads to an increase in photocurrent and the photo current generation capacity of the electrode is increased. However, such increase in photocurrent is inhibited by the presence of PCP, thereby causing a hindrance in generation of photocurrent, which can be sensed by the LOC upto the level of 4 $pg.mL^{-1}$.

Su et al. (2015) has also developed an another paper based approach for the appliction of herbicide detection that makes use of FL for detection of methyl viologen. The method was based on the interaction of CdTe in presence of the target methyl viologen. CdTe QDs can be impregnated on the paper based devices and its quenching effect is studied against the target. Having a LOD of 0.16 $\mu mol.L^{-1}$ for a CdTe paper based sensors. The darkness of the color on the μPAD is proportional to the methyl viologen concentration.

6 Detection of Heavy Metals

Arsenic, mercury, copper and lead are notable heavy metal contamination of food which has proven to possess human health risk and food safety threat. Moreover, exposure to such heavy metals cause cancer or other associated diseases (Dong et al. 2014) making the development of rapid, reliable detection device for heavy metals highly important. However, the conventional methods available for the detection of these heavy metals are atomic absorption spectrometry (AAS) and inductively coupled plasma mass spectroscopy (ICP-MS). But they have serious limitations that they cannot be used for the in-situ monitoring because the size of equipment, cost of analysis and analysis time is pretty high. These limitations can be overcome by using microfluidic devices. Chaiyo et al. (2015) fabricated a device using wax printing method to detect the presence of copper (Cu^{2+}) ions. In their reported study, they demonstrated the presence of Cu^{2+}in tomato and rice with a LOD of 0.35 ng/mL and linear range between 0.5 to 200 ng/mL. Zheng et al. (2012) concentrated on heavy metal toxicity in marine environment. They developed a microfluidic device provided with a concentration gradient generator for toxicant and a module for the microalgal chemostatics. And they finally made a coorelation between the microalgal motility with that of the toxicity of heavy metals. Gomez-de Pedro et al. (2014) established an automatic, low cost optical based microfluidic system for observing the presence of Hg (II) by employing modified gold nanoparticles. While, Zhang et al. (2015) made a μPAD integrated with single-stranded DNA (ssDNA) and a functionalized graphene oxide sensor for the simultaneous detection of heavy metals such as mercury (II) ions (Hg^{2+}), silver (I) ions (Ag^{+}), and aminoglycoside antibiotic residue in food.

7 Conclusion

Thus, the growth of microfluidics is immense in the area of food processing, especially in the area of food adulteration detection and methods. Increase in demand for processed foods and industrialized food processing sectors resulted in contamination of heavy metals, adulteration, excessive use of preservatives and presence of pesticide residues in agro food produces. Though efforts are being taken to reduce the ill effects, need for identification of any form of adulterants in food produces becomes inevitable. Development of cheap and easy to use diagnostic devices becomes necessary and hence the concept of microfluidics will play a great role development of paper based diagnostic devices in the near future. On the other hand, 3D printing of food is a growing field in food processing with immense potential for marketing and commercialization, where the concepts of microfluidics will play an immense role. Moreover, efforts are also needs to be taken for development of suitable microfluidic devices for development of novel food products and identification of interaction between various molecules present in novel food products.

References

Abraham E, Mukunthan Sulochana GN, Soundarajan B, Narayanasamy S (2017) Experimental investigation on microfluidic reactive extraction of citric acid using Trioctylamine/1-Decanol system in uniform and nonuniform circular microchannels. Ind Eng Chem Res 56:10845–10855

Adkins JA, Henry CS (2015) Electrochemical detection in paper-based analytical devices using microwire electrodes. Anal Chim Acta 891:247–254

Antony R, Giri Nandagopal MS, Sreekumar N, Rangabhashiyam S, Selvaraju N (2014a) Liquid-liquid slug flow in a microchannel reactor and its mass transfer properties - a review. Bull Chem React Eng Catal 9:207 223

Antony R, Nandagopal MSG, Rangabhashiyam S, Selvaraju N (2014b) Probabilistic neural network prediction of liquid- liquid two phase flows in a circular microchannel. J Sci Ind Res (India) 73:525–529

Antony R, Nandagopal G, Rangabhashiyam S, Anu N, Selvaraju N (2015a) Numerical investigation of stratified flow-parallel reaction microchannel system: a deterministic approach. Indian J Chem Technol 22:258–263

Antony R, Nandagopal MSG, Manikrishna C, Selvaraju N (2015b) Experimental comparison on efficiency of alkaline hydrolysis reaction in circular microreactors over conventional batch reactor. J Sci Ind Res 74:390–394

Bayraktar T, Pidugu SB (2006) Characterization of liquid flows in microfluidic systems. Int J Heat Mass Transf 49:815–824

Bsoul A, Pan S, Cretu E, Stoeber B, Walus K (2016) Design, microfabrication, and characterization of a moulded PDMS/SU-8 inkjet dispenser for a lab-on-a-printer platform technology with disposable microfluidic chip. Lab Chip 16:3351–3361

Cardoso TMG, Garcia PT, Coltro WKT (2015) Colorimetric determination of nitrite in clinical, food and environmental samples using microfluidic devices stamped in paper platforms. Anal Methods 7:7311–7317

Chaiyo S, Siangproh W, Apilux A, Chailapakul O (2015) Highly selective and sensitive paper-based colorimetric sensor using thiosulfate catalytic etching of silver nanoplates for trace determination of copper ions. Anal Chim Acta 866:75–83

Chen W, Lam RHW, Fu J (2012) Photolithographic surface micromachining of polydimethylsiloxane (PDMS). Lab Chip 12:391–395

Clime L et al (2015) Microfluidic filtration and extraction of pathogens from food samples by hydrodynamic focusing and inertial lateral migration. Biomed Microdevices 17:17

Colletes TC et al (2016) A new insert sample approach to paper spray mass spectrometry: a paper substrate with paraffin barriers. Analyst 141:1707–1713

Convery N, Gadegaard N (2019) 30 years of microfluidics. Micro Nano Eng 2:76–91

Dai MH, Copley SD (2004) Genome shuffling improves degradation of the anthropogenic pesticide pentachlorophenol by Sphingobium chlorophenolicum ATCC 39723. Appl Environ Microbiol 70:2391–2397

Dong Y, Xu Y, Yong W, Chu X, Wang D (2014) Aptamer and its potential applications for food safety. Crit Rev Food Sci Nutr 54:1548–1561

Feng XZ et al (2019) A facile electrochemical sensor labeled by ferrocenoyl cysteine conjugate for the detection of nitrite in pickle juice. Sensors (Switzerland) 19:268

Fronczek CF, You DJ, Yoon JY (2013) Single-pipetting microfluidic assay device for rapid detection of salmonella from poultry package. Biosens Bioelectron 40:342–349

Fuentes MS et al (2013) Enhanced removal of a pesticides mixture by single cultures and consortia of free and immobilized streptomyces strains. Biomed Res Int 2013:392573

Gaal G et al (2017) Simplified fabrication of integrated microfluidic devices using fused deposition modeling 3D printing. Sensors Actuators B Chem 242:35–40

Garcia-Cordero JL et al (2010) Optically addressable single-use microfluidic valves by laser printer lithography. Lab Chip 10:2680–2687

Gilbert-López B, García-Reyes JF, Molina-Díaz A (2009) Sample treatment and determination of pesticide residues in fatty vegetable matrices: a review. Talanta 79:109–128

Giri Nandagopal MS, Selvaraju N (2016) Prediction of liquid-liquid flow patterns in a Y-junction circular microchannel using advanced neural network techniques. Ind Eng Chem Res 55:11346

Giri Nandagopal MS, Antony R, Rangabhashiyam S, Selvaraju N (2014a) Advance approach on environmental assessment and monitoring. Res J Chem Environ 18:78–90

Giri Nandagopal MS, Antony R, Rangabhashiyam S, Sreekumar N, Selvaraju N (2014b) Overview of microneedle system: a third generation transdermal drug delivery approach. Microsyst Technol 20:1249

Giri Nandagopal MS, Nakkeeran E, Prasanna Venkatesh R, Selvaraju N (2017) Advance microfluidic approach over conventional batch and CTR for improving the efficiency of E-coli cell lysis by CuO nanoparticles. Int J Chem React Eng 15 https://doi.org/10.1515/ijcre-2016-0105

Gómez-De Pedro S et al (2014) Optical microfluidic system based on ionophore modified gold nanoparticles for the continuous monitoring of mercuric ion. Sensors Actuators B Chem 194:19–26

Gorkin R et al (2010) Centrifugal microfluidics for biomedical applications. Lab Chip 10:1758–1773

Guo L, Feng J, Fang Z, Xu J, Lu X (2015) Application of microfluidic 'lab-on-a-chip' for the detection of mycotoxins in foods. Trends Food Sci Technol 46:252–263

Hansen TS, Selmeczi D, Larsen NB (2010) Fast prototyping of injection molded polymer microfluidic chips. J Micromech Microeng 20:015020

He Q, Ma C, Hu X, Chen H (2013) Method for fabrication of paper-based microfluidic devices by alkylsilane self-assembling and UV/O3-patterning. Anal Chem 85:1327–1331

He Y, Wu WB, Fu JZ (2015) Rapid fabrication of paper-based microfluidic analytical devices with desktop stereolithography 3D printer. RSC Adv 5:2694–2701

Jayawardane BM, Wei S, McKelvie ID, Kolev SD (2014) Microfluidic paper-based analytical device for the determination of nitrite and nitrate. Anal Chem 86:7274–7279

Jin SQ, Guo SM, Zuo P, Ye BC (2015) A cost-effective Z-folding controlled liquid handling microfluidic paper analysis device for pathogen detection via ATP quantification. Biosens Bioelectron 63:379–383

Kataoka ÉM et al (2017) Simple, expendable, 3D-printed microfluidic Systems for Sample Preparation of petroleum. Anal Chem 89:3460–3467

Kuek Lawrence CS, Tan SN, Floresca CZ (2014) A 'green' cellulose paper based glucose amperometric biosensor. Sensors Actuators B Chem 193:536–541

Kwok HC et al (2016) Allergy testing and drug screening on an ITO-coated lab-on-a-disc. Micromachines 7:38

Law JWF, Mutalib NSA, Chan KG, Lee LH (2014) Rapid metho ds for the detection of foodborne bacterial pathogens: principles, applications, advantages and limitations. Front Microbiol 5:1–19

Lin F et al (2014) A novel microfluidic chip electrophoresis strategy for simultaneous, label-free, multi-protein detection based on a graphene energy transfer biosensor. Analyst 139:2890–2895

Liu W, Kou J, Xing H, Li B (2014) Paper-based chromatographic chemiluminescence chip for the detection of dichlorvos in vegetables. Biosens Bioelectron 52:76–81

Liu W et al (2015) A molecularly imprinted polymer based a lab-on-paper chemiluminescence device for the detection of dichlorvos. Spectrochim Acta Part A Mol Biomol Spectrosc 141:51–57

Liu J et al (2017) Development and application of paper spray ionization mass spectrometry. Prog Chem 29:659–666

Lopez-Ruiz N et al (2014) Smartphone-based simultaneous pH and nitrite colorimetric determination for paper microfluidic devices. Anal Chem 86:9554–9562

Mao L, Koser H (2006) Towards ferrofluidics for μ-TAS and lab on-a-chip applications. Nanotechnology 17:S34

Martinez AW, Phillips ST, Whitesides GM, Carrilho E (2010) Diagnostics for the developing world: microfluidic paper based analytical devices. Anal Chem 82:3 10

Nandagopal MSG, Antony R, Selvaraju N (2016) Comparative study of liquid-liquid extraction in miniaturized channels over other conventional extraction methods. Microsyst Technol 22:349–356

Nandagopal MSG, Abraham E, Selvaraju N (2017) Advanced neural network prediction and system identification of liquid-liquid flow patterns in circular microchannels with varying angle of confluence. Chem Eng J 309:850–865 https://doi.org/10.1515/ijcre-2016-0105

Nouanthavong S, Nacapricha D, Henry CS, Sameenoi Y (2016) Pesticide analysis using nanoceria-coated paper-based devices as a detection platform. Analyst 141:1837–1846

Park TS, Li W, McCracken KE, Yoon JY (2013) Smartphone quantifies salmonella from paper microfluidics. Lab Chip 13:4832–4840

Pearce JM, Anzalone NC, Heldt CL (2016) Open-source wax RepRap 3-D printer for rapid prototyping paper-based microfluidics. J Lab Autom 21:510–516

Prabhu A et al (2020a) Thread integrated smart-phone imaging facilitates early turning point colorimetric assay for microbes. RSC Adv 10:26853–26861

Prabhu A, Giri Nandagopal MS, Peralam Yegneswaran P, Singhal HR, Mani NK (2020b) Inkjet printing of paraffin on paper allows low-cost point-of-care diagnostics for pathogenic fungi. Cellulose 27:7691–7701

Ramesh H, Prabhu A, S, G. N. M., Dheivasigamani, T. & Kumar, N. (2021) One-dollar microfluidic paper-based analytical devices: Do-It-Yourself approaches. Microchem J 165:106126

Sayad AA et al (2016) A microfluidic lab-on-a-disc integrated loop mediated isothermal amplification for foodborne pathogen detection. Sensors Actuators B Chem 227:600–609

Shangguan JW et al (2017) Microfluidic PDMS on paper (POP) devices. Lab Chip 17:120–127

Shaw IC (2014) Chemical residues, food additives and natural toxicants in food - the cocktail effect. Int J Food Sci Technol 49:2149–2157

Sia SK, Whitesides GM (2003) Microfluidic devices fabricated in poly(dimethylsiloxane) for biological studies. Electrophoresis 24:3563–3576

Sicard C et al (2015) Tools for water quality monitoring and mapping using paper-based sensors and cell phones. Water Res 70:360–369

Squires TM, Messinger RJ, Manalis SR (2008) Making it stick: convection, reaction and diffusion in surface-based biosensors. Nat Biotechnol 26:417–426
Stone HA, Stroock AD, Ajdari A (2004) Engineering flows in small devices: microfluidics toward a lab-on-a-chip. Annu Rev Fluid Mech 36:381–411
Strohmeier O et al (2015) Centrifugal microfluidic platforms: advanced unit operations and applications. Chem Soc Rev 44:6187–6229
Su Y, Ma S, Jiang K, Han X (2015) CdTe-paper-based visual sensor for detecting methyl viologen. Chinese J Chem 33:446–450
Sun G et al (2014) Photoelectrochemical sensor for pentachlorophenol on microfluidic paper-based analytical device based on the molecular imprinting technique. Biosens Bioelectron 56:97–103
Sun Y et al (2015) A lab-on-a-chip system with integrated sample preparation and loop-mediated isothermal amplification for rapid and quantitative detection of salmonella spp. in food samples. Lab Chip 15:1898–1904
Szydzik C et al (2016) Fabrication of complex PDMS microfluidic structures and embedded functional substrates by one-step injection moulding. RSC Adv 6:87988–87994
Tran VT, Wei Y, Liau WJ, Yang H, Du H (2017) Preparing of interdigitated microelectrode arrays for AC electrokinetic devices using inkjet printing of silver nanoparticles ink. Micromachines 8:106
van den Driesche S, Lucklum F, Bunge F, Vellekoop MJ (2018) 3D printing solutions for microfluidic chip-to-world connections. Micromachines 9:1–12
Wang ZK, Zheng HY, Lim RYH, Wang ZF, Lam YC (2011) Improving surface smoothness of laser-fabricated microchannels for microfluidic application. J Micromech Microeng 21:095008
Wang S et al (2013) Molecularly imprinted polymer grafted paper-based multi-disk micro-disk plate for chemiluminescence detection of pesticide. Biosens Bioelectron 50:262–268
Wee WH et al (2015) Fabrication of dielectrophoretic microfluidic chips using a facile screen-printing technique for microparticle trapping. J Micromech Microeng 25:105015
Xu J, Vaillant R, Attinger D (2010) Use of a porous membrane for gas bubble removal in microfluidic channels: physical mechanisms and design criteria. Microfluid Nanofluidics 9:765–772
Zhang M, Wang Z (2013) Nanostructured silver nanowires-graphene hybrids for enhanced electrochemical detection of hydrogen peroxide. Appl Phys Lett 102:213104
Zhang H et al (2014) A sensitive colorimetric method for the determination of nitrite in water supplies, meat and dairy products using ionic liquid-modified methyl red as a colour reagent. Food Chem 151:429–434
Zhang Y, Zuo P, Ye BC (2015) A low-cost and simple paper-based microfluidic device for simultaneous multiplex determination of different types of chemical contaminants in food. Biosens Bioelectron 68:14–19
Zheng G, Wang Y, Qin J (2012) Microalgal motility measurement microfluidic chip for toxicity assessment of heavy metals. Anal Bioanal Chem 404:3061–3069
Zhu Y, Zhang L, Yang L (2015) Designing of the functional paper-based surface-enhanced Raman spectroscopy substrates for colorants detection. Mater Res Bull 63:199–204
Zuo P, Li X, Dominguez DC, Ye BC (2013) A PDMS/paper/glass hybrid microfluidic biochip integrated with aptamer-functionalized graphene oxide nano-biosensors for one-step multiplexed pathogen detection. Lab Chip 13:3921–3928

Part VII
Food Waste Management and Sustainable Parameters Analysis

Chapter 23
Analysis and Comparison of Environmental Impacts of Nonthermal Food Technologies

Ilija Djekic and Igor Tomasević

1 Introduction

Strong evidence that food processing technologies have been an integral part of mankind date since the period of ancient Egypt and have largely contributed to the progressions of Greek, Chaldean, and Roman civilizations. Anthropologists and evolutionary biologists argue that a single thermal food processing technology (cooking) is responsible for the evolution of *Homo sapiens* and permitted humans to sustain big brains (Wrangham 2010). Without a doubt, traditional and contemporary food processing technologies are still shaping the world we live in today in many ways. However, Western civilization and its consumers are now demanding minimally processed foods and substitution of heat (at least partially) in food processing (Ortega-Rivas 2012), making nonthermal food processing technologies (NTFPT) popular in food science and food industry of the twenty-first century. Common denominator for all NTFPT is the objective to retain nutritional and sensory quality of food, while achieving microbial and enzymatic inactivation (food preservation) as a standalone technique or together with other (nonthermal) "hurdle" processing technologies.

Extensive commercialization of NTFPTs is still anticipated for several reasons (lower effectiveness of microbial inactivation, higher cost of exploitation and bigger complexity of operation) including the legislative one. Namely, food regulations are lacking a joint classification of these processing technologies (Režek Jambrak et al. 2018) averting its adoption by the food industries.

When we tackle the main advantages we can specify microbial inactivity in many of the mentioned technologies as "number #1" food safety benefit (Režek Jambrak et al. 2021). Also, these technologies prevent negative effects of heat on

I. Djekic (✉) · I. Tomasević
University of Belgrade, Faculty of Agriculture, Belgrade, Serbia
e-mail: idjekic@agrif.bg.ac.rs; tbigor@agrif.bg.ac.rs

A. Režek Jambrak (ed.), *Nonthermal Processing in Agri-Food-Bio Sciences*, Food Engineering Series, https://doi.org/10.1007/978-3-030-92415-7_23

nutritional properties of foods improving its health component. It is of note that, some have certain technological issues in maintaining sensorial attributes of food (Režek Jambrak et al. 2018).

However, environmental dimension is rather only touched since comparison of environmental impacts of NTFPTs is difficult due to differences in the scale (lab – scale vs industrial application) and type of food processed (Hospido et al. 2010). Several studies compared NTFPTs and conventional technologies (Pardo and Zufía 2012) or Valsasina et al. (2017) with estimated proof of environmental potentials when using NTFPTs. On the contrary, Aganovic et al. (2017) studied the energy balance of pulsed electric fields and high pressure processing technologies in comparison to conventional thermal processing and both NTFTPs presented higher energy consumption expressed per liter of treated juice compared to conventional. In general, environmental analysis of NTFTPs highlights two main advantages: lower energy / resource consumption and shorter treatment time.

Based on above mentioned, can we conclude that NTFPTs are environmentally friendly? Do we have enough proof? How can we measure (and compare) their environmental footprint? This chapter gives a critical perspective of environmental impacts of NTFPTs.

2 Nonthermal Food Processing Technologies

Based on their mode of action and source of energy used (Fig. 23.1), NTFPTs can be divided into: (a) technologies based on mechanical action (hydrodynamic effects / hydrodynamic cavitation, ultrasound and irradiation); (b) electro-magnetic fields-based technologies (pulsed electric fields, pulsed ultraviolet light, cold plasma, radiofrequency, oscillating magnetic fields, electrohydrodynamic processing and electron beam processing) and (c) pressure-based technologies (high-pressure processing, ultrahigh-pressure homogenization, supercritical fluid drying / extraction). Among these, some specific technologies have been broadly investigated in recent years and are foreseen as most promising methods for competition or complementation of traditional food preservation technologies.

Pulsed ultraviolet light (PL) has a strong potential for the commercial application because it provides reductions of up to 7 log cycles for vegetative microorganisms on smooth, non-porous surfaces such as those of food contact (Rajkovic et al. 2010) and food packaging materials (Gómez-López 2016). Even though PL is less effective on surfaces such as those of most food materials, it can still provide a microbial reductions above 2 log cycles, successfully preventing food cross-contamination without affecting quality attributes by lipid and protein oxidation (Rajkovic et al. 2017). In addition to the ability of reducing the microbial load in wide range of food products (Barba et al. 2015) with low electrical conductivity and no air bubbles in order to avoid the dielectric breakdown (Chauhan and Unni 2015) (different liquid and semi-solid food products such as fruit or vegetable juices), pulsed electric fields (PEFs) treatment can also be a good strategy to change food textural properties including the improvement of tenderization and aging of meat (Gómez et al. 2019).

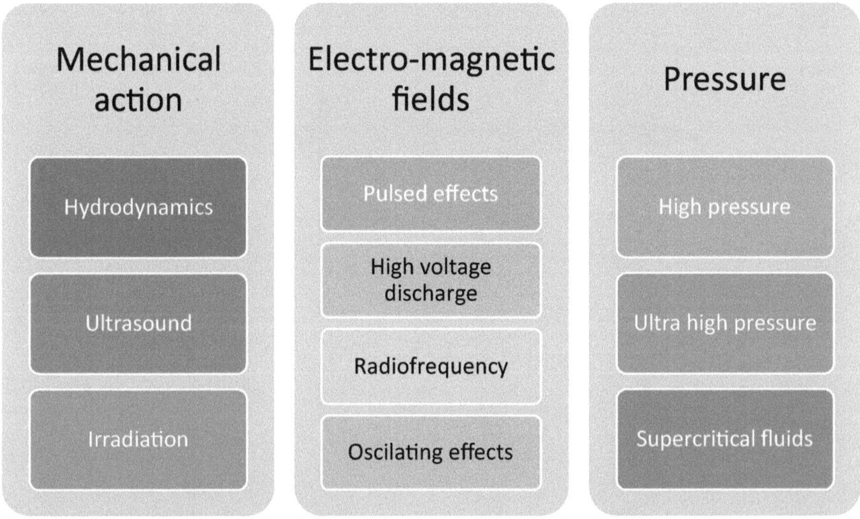

Fig. 23.1 Categorization of different NTFPTs

High-pressure processing (HPP) was first used in 1899 in the United States (Tewari et al. 1999). However, in those days, the equipment was not very dependable and research was discontinued. It was only a century later, and in the 1990's, that a modern HPP equipment was developed and with its successfully application for the processing of fruits and vegetables (Denoya et al. 2020), juices (Yildiz et al. 2021) and meat (Warner and Ha 2019) we would like to argue that this is the most widely adopted NTFPT by the food industry of today.

3 Environmental Impact

To analyze environmental impact of a NTFPT, the easiest way is to deploy a mass-energy balance model a of a lab-scale equipment using this known environmental tool (Režek Jambrak et al. 2018). Figure 23.2 depicts a simplified Sankey diagram, where the inputs are: (i) food samples that are treated, (ii) consumption of natural resources for operating the technology (water / various types of energy) and (iii) depending on the type of food and technology different other inputs such as chemicals for cleaning the equipment, or different fluids (Nitrogen or Argon for high voltage electrical discharge, or CO_2 for supercritical fluids). Depending on the technology, two types of outputs can be recognized: "intended" and "unintended". Intended outputs are treated samples (with all its positive and negative physical, chemical and microbial characteristics) while all other outputs may be categorized as unintended, comprising of potential emission of greenhouse gasses, wastewater discharge occurring during sanitation of the equipment, disposal of different types of waste, etc.

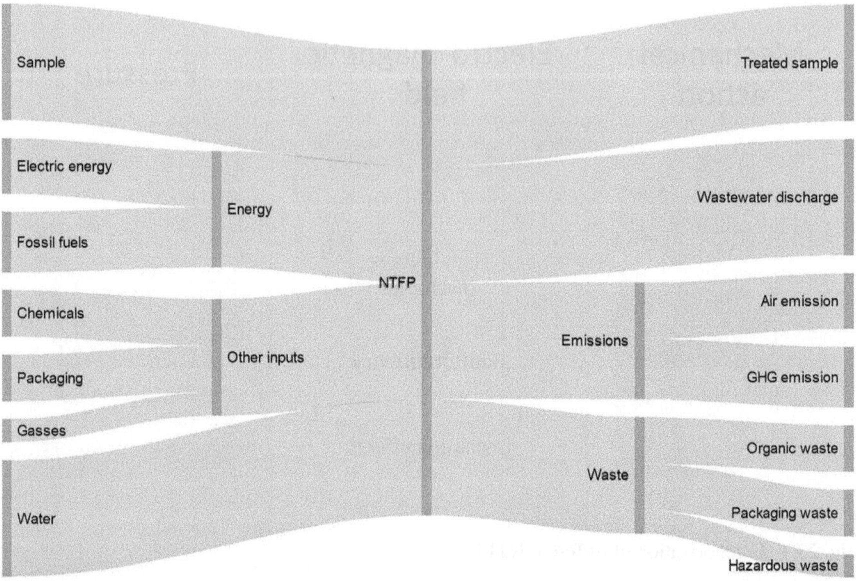

Fig. 23.2 Sankey diagram depicting mass-energy balance of a lab-scale NTFPT

As outlined in ISO 14001, when determining environmental aspects of a process (in our case NTFTP), the following should be considered: emissions to air, releases to water and land, use of raw materials, natural resources and energy, emitted energy, waste disposal (ISO 2015). This is the first step in classifying all environmental impacts. Table 23.1 gives an overview of aspects associated with different NTFTPs based on various literature sources explaining the mechanism of NTFTPs (Srivastav et al. 2020; Djekic et al. 2018b; Režek Jambrak et al. 2010).

First common denominator of all NTFTPs is use of electric energy for running the equipment, clearly associated with resource depletion. Food samples used for treatment, originate from nature as raw materials, either as plant or animal origin. Also, depending on type of raw material, certain impacts may be associated with biodiversity loss if production of raw materials affects the ecosystem. After treatment, the samples are either consumed (for sensory analysis) or further tested (using different physical/chemical/microbial methods) and treated as (organic) waste.

When it comes to specific NTFTPs, the ones using hydrodynamic effects and cavitation use different liquids depending on their physicochemical properties such as surface tension, viscosity and vapor pressure of liquid (Gogate and Pandit 2001; Gogate 2011). Usage of such liquids is categorized in "other" since several liquids with different properties are in use. Similar situation is with high voltage electrical discharge extraction because different gases are used for generation of plasma, such as argon and nitrogen, as presented in work of Nutrizio et al. (2020). It is known that both gases are considered as environmentally friendly, however, their use during extraction may cause some impacts.

Table 23.1 Environmental aspects associated with different NTFTPs

Activity	Aspect	Impact	RE	WA	AI	LA	BI	CL	WS	OT
All NTFPTs	Use of (electric) energy	Resource depletion	☑							
All NTFPTs	Use of food samples as raw materials	Resource depletion	☑							
All NTFPTs	Use of food samples as raw materials	Ecosystem degradation					☑			
All NTFPTs	Generation of organic waste	Waste disposal							☑	
Hydrodynamic NTFTP	Use of various fluids	Depends on the liquid								☑
Cold plasma NTFTP	Use of various fluids	Depends on the gas								☑
Irradiation NTFPT	Irradiation of food	Ionizing radiation								☑
Supercritical fluid NTFTP	Use of various fluids	Depends on the equipment								☑
	Use of various adsorbents	Depends on the equipment								☑

Legend: *RE* resource depletion, *WA* wastewater discharge, *AI* air emission, *LA* land contamination, *BI* biodiversity loss, *CL* climate change effects, *WS* waste disposal, *OT* other

Use of radiation in treating food is achieved by applying gamma rays, electron beams or X-rays (Morehouse and Komolprasert 2004). Effects on food (and surrounding environment) depend on the type of the radiation and its energy level raising two issues of concern – radiation of food and radiation of people operating the equipment. Although use of ionizing radiation in food industry is covered with various food regulations, environmental impacts still exist.

When using supercritical fluid technology, the most often used fluid is carbon dioxide, due to its positive characteristic such as low price, it is considered as a non-toxic and non-flammable gas with high density and low viscosity, and generally recognized as safe (Smigic et al. 2019). Application of this technology for drying, at mild temperatures and higher pressures also require usage of water adsorbents such as zeolites or ionic liquid sorbents for removing water from the products (Djekic et al. 2018b).

In general, we can conclude that all NTFTPs have two common impacts – resource depletion (use of electric energy and natural raw materials) and ecosystem degradation (depending on type of food). However, specificity of technology directly influences "other" impacts – ionized radiation and use of different fluids (liquids / gasses). Since all NTFTPs after use need to be cleaned and sanitized as one of important activities, Table 23.2 presents environmental aspects associated with hygiene of equipment.

Table 23.2 Environmental aspects associated with hygiene of NTFPT equipment

Activity	Aspect	Impact	RE	WA	AI	LA	BI	CL	WS	OT
Cleaning and sanitation	Use of water	Resource depletion	☑							
	Use of chemicals	Disposal of hazardous packaging							☑	
	Water discharge	Water contamination		☑						

Legend: *RE* resource depletion, *WA* wastewater discharge, *AI* air emission, *LA* land contamination, *BI* biodiversity loss, *CL* climate change effects, *WS* waste disposal, *OT* other

4 Life-Cycle Assessment

It is known that life cycle assessment (LCA) is a tool that consists of compiling and evaluating inputs, outputs and recognizing potential environmental impacts of a product system where typical subsystems are: (i) cradle-to-gate studies; gate-to-gate studies and gate-to-disposal studies (ISO 2006). In case of NTFPTs, the best approach is to consider it as a "gate-to-gate study" since most of equipment are at Universities located in laboratories and/or R&D units. As outlined in ISO 14040, stages of a LCA comprise of defining goal and scope, inventory analysis, impact assessment and interpretation of the results (Fig. 23.3).

It is obvious that scope and boundaries are set at pilot / lab scale. Also, functional unit (FU) is set as one (1) treatment of selected food product. FU is used as a unit for expressing results and a basis for benchmarking (Djekic et al. 2018a). Inventory analysis should specify all inputs and output from the mass-balance analysis of NTFPTs. When it comes to impact assessment, the main question is "which environmental impacts should be calculated?" Emissions of greenhouse gasses? Ozone depletion potential? Human toxicity? Water and energy consumption? All of these should be calculated per functional unit (in our case 1 treatment of selected food product).

Although life-cycle assessment is a scientifically validated methodology, still, there are certain challenges that need to be resolved (Djekic et al. 2019). In the 'scope and goal stage' (Table 23.3), three potential problems are 'inadequate system boundaries', 'inappropriate data collection method' and 'low level of data quality'. First issue is regarding upscaling lab/scale NTFPT to food processing level (for benchmarking with conventional food processing technologies) and further upscaling to analyze the entire food supply chain. Second issue of concern is how data are collected (by exact measuring or by estimations). Finally, depending on the measurement methods, are primary sources or secondary data used? Since this can also bring uncertainty.

Within 'Inventory analysis' the main question is 'have all material and energy flows been identified'? This is directly connected with mass-energy balance and classification of all inputs and outputs. 'Impact assessment' once more highlights the need of choosing appropriate environmental impacts and providing answer to where should we focus our LCA?

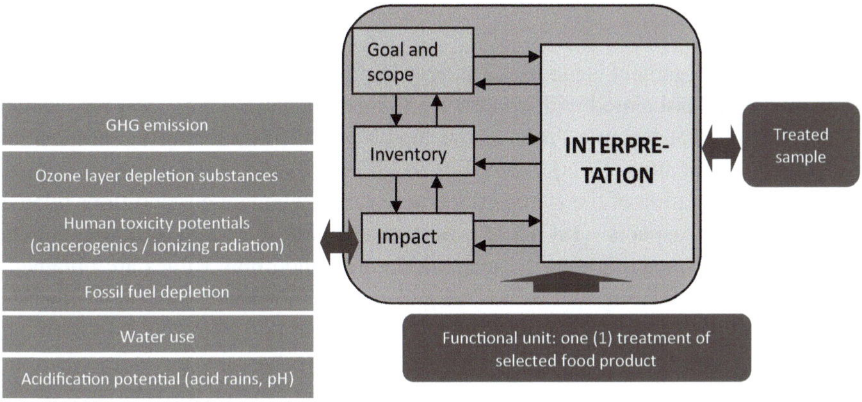

Fig. 23.3 Partial life-cycle assessment of NTFPTs

Table 23.3 Challenges associate with performing LCA of a NTFPT

LCA Stage	Issue of concern	Challenge
Scope and goal	Inadequate system boundaries	Can we upscale the model to the entire food chain?
Scope and goal	Inappropriate data collection method	Inadequate collection methods (measurement, estimation, combination, re-calculation, ...)
Scope and goal	Low level of data quality	Lab scale approximations of primary/ secondary data
Inventory analysis	Material and energy flows	Have we included all material/energy flows?
Impact assessment	Lack of an accepted official list of environmental impacts of NTFPT	Subjective choice of environmental impacts
Interpretation of results	Uncertainty	Quality of data, subjective choice system boundaries, functional units, environmental impacts
Interpretation of results	High sensitivity	What input data (lab scale / entire chain) mostly influence LCA?

Finally, in the interpretation stage, ‘uncertainty’ may occur depending on the quality of data / functional units and choice of environmental impacts. This is a ‘highly sensitive’ activity since some data may greatly influence the results (both at lab scale and at entire food chain level).

5 Environmental Indicators

Since impact assessment is identified as very important, its quantification and evaluation is directly associated with approach and methods used (Carvalho et al. 2014). One approach that became "rule of the thumb" is by introducing environmental performance indicators (EPIs). According to ISO 14001 they are defined as "a measurable representation of the status of operations, management or conditions related to environmental aspects" (ISO 2015). According to Đekić I. (2009) they should be (i) measurable – meaning that there should be a method how values are obtained; (ii) objective – indicating that values are gathered from trusted sources; (iii) verifiable – enabling verification of the results; (iv) repeatable – allowing measurement of indicators under all operating conditions and (v) technically feasible – demonstrating all necessary measuring device in place. Depending on the approach, three levels of EPIs may be generated (Đekić and Tomašević 2017).

6 First and Second Level Indicators

First level are considered as basic indicators (Djekic and Tomasevic 2020) and Table 23.4 gives an overview based on mass-energy balance (Fig. 23.2).

Second level indicators are generated from at least two first level indictors, usually expressed in functional units. Such an approach shows the relationship between use of NTFPTs and the environment, describing environmental impacts. Focus of determining this type of indicators is associated with NTFPTs and calculation of consumptions and discharges per FUs such as energy / water consumption per one treated sample, wastewater discharge or chemical usage. Table 23.5 depicts most common second level indicators originating in using NTFPTs.

Energy is used for machines and equipment associated with NTFPTs and for controlling temperature regimes (heating / refrigerating of samples) (IPPC 2006). If needed, energy should be deployed using "top-down" approach by clarifying

Table 23.4 First level EPIs (associated with mass-energy balance)

Indicator	Unit
Use of samples	[kg / units]
Consumption of electric energy	[MJ / kWh]
Consumption of fossil fuels	[L / kg]
Consumption of water	[L]
Consumption of chemicals	[L / kg]
Wastewater discharge	[L]
Waste disposal	[kg]
Usage of packaging	[units / kg]

Table 23.5 Overview of the most common second level environmental indicators

Second level indicator	Formula [unit]
Consumption of water per FU	$\frac{Consumption\ of\ water[L]}{FU[1\,treatment]}$
Consumption of energy per FU	$\frac{Consumption\ of\ energy[MJ]}{FU[1\,treatment]}$
Consumption of fossil fuels per FU	$\frac{Consumption\ of\ fuels[L]}{FU[1\,treatment]}$
Consumption of chemicals per FU	$\frac{Consumption\ of\ chemicals[L/kg]}{FU[1\,treatment]}$
Discharge of wastewater per FU	$\frac{Discharge\ of\ wastewater[L]}{FU[1\,treatment]}$
Discharge of waste per FU	$\frac{Discharge\ of\ waste[kg]}{FU[1\,treatment]}$

Legend: *FU* functional unit

consumption of all types of energy (electric, thermal etc.), as well as types and quantities of fossil fuels (Djekic and Tomasevic 2018). Water can be used in three possible ways: for sample preparation, as part of processing and for cleaning and sanitizing the NTFPTs.

Waste management can be cascades by quantities and different types of waste (organic / inorganic, hazardous / non-hazardous, etc.) or depending on material (plastic, metal, wooden, paper, food-waste, cardboard). Finally, it should be treated by applying waste management hierarchy (EC 2008). Wastewater is a result of cleaning and sanitation as an important prerequisite program in any type of food production (CAC 2003). If needed wastewater quality index can be used as an indicator that depends on legislation / methodology applied and parameters measured (Djekic and Tomasevic 2020).

7 Third Level Indicators – Path to Footprints

Third level of EPIs are mainly linked with different types of environmental footprints (Đekić and Tomašević 2017). As defined by Čuček et al. (2015), footprints are tools for calculating level of pollution prevention and/or environmental improvements. Herva et al. (2011) suggests three footprint family members – ecological, water and carbon footprints. These footprints are developed by environmental scientists and are generic regardless of the type of companies, processes or products (Djekic et al. 2018a).

The first one is comprised of various natural, social, cultural and economic parameters and is not directly linked with NTFPTs (especially when analyzed at lab-scale level). The water footprint builds on the concept of virtual water and refers to total quantity used during the treatment. It should be further deployed to 'blue' (linked with surface and groundwater), 'green' (associated with rainwater stored within the soil), and 'gray' (volume of freshwater) footprints (Čuček et al. 2015; Mekonnen and Hoekstra 2010).

Carbon footprint is an indicator that calculates all greenhouse gas (GHG) emissions. It is expressed as CO_2 equivalent (CO_2e) since carbon dioxide is recognized as the largest single contributor to climate change (Herva et al. 2011). GHG comprise of the following: carbon dioxide, methane, nitrous oxide, hydro-chlorofluorocarbons, hydro-fluorocarbons, and ozone in the lower atmosphere (WMO 2017).

Acidification potential (expressed in SO_2 equivalents) computes the potential of acidifying pollutants (SO_2, NO_x, HCl, NH_3, HF) to form H^+ ions and damage plants, animals, and the eco-system, mainly associated with Ammonia occurring in animal production released from manure in farms and during manure handling. Eutrophication potential is expressed in PO_3^{4-} equivalents and links aquatic plant growth attributable of nutrients (nitrogen and/or phosphorus) left by over-fertilization of water and soil. Ozone depletion potential (calculated in CFC-11 or R11 equivalents) estimates the potential for reducing the protective stratospheric ozone layer from use of ozone-depleting substances such as freons (used as refrigerants in the cold chain), chlorofluorocarbons, carbon tetrachloride, and methyl chloroform (Čuček et al. 2015; Djekic and Tomasevic 2018). Table 23.6 shows main formula needed for calculating these footprints.

Table 23.6 Environmental footprints

Footprint	Formula
Global warming potential	$GWP = \sum_{i}^{n} GWP_i \, x m_i \left[\text{kgCO}_{2\text{e}}\right]$
Acidification potential	$AP = \sum_{i}^{n} AP_i \, x m_i \left[\text{kgSO}_{2\text{e}}\right]$
Eutrophication potential	$EP = \sum_{i}^{n} EP_i \, x m_i \left[\text{kgPO}_{4\text{e}}\right]$
Ozone depletion potential	$ODP = \sum_{i}^{n} ODP_i \, x m_i \left[\text{kgR11}_{\text{e}}\right]$

Legend: m_i mass of emitted gas (kg), GWP_i global warming potential of the emitted gas, AP_i acidification potential of the emitted substance, EP_i eutrophication potential of the emitted substance, ODP_i ozone depletion potential of the emitted gas

8 Are NTFPTs Sustainable?

First, let us remind ourselves what is sustainable development? As defined by the World Commission on Environment and Development, back in 1987, sustainable development is defined as "development that meets the needs of the present without compromising the ability of future generations to meet their own needs" (Hariem Brundtland 1985). It consists of three main pillars named 'environmental pillar', 'economic pillar', and 'social pillar', whilst priority should not be given to any of the three pillars but rather a balance of economic, social, and environmental needs and goals (Gast et al. 2017). Sustainable development has advanced through two stages – Agenda 21 (UN 1992) and Agenda 2030 (UN 2016) with its 17 sustainable development goals (UN 2019). The environmental pillar often gets the most attention and is focused on improving the environmental performance either through pollution prevention or by decreasing environmental impacts, as mentioned and explained in detail above.

The economic pillar of sustainability is targeting businesses in terms of profitability but without compromising other two pillars, covering also legal compliance and proper governance. In general, within this pillar one must consider two types of financial assets. First, there are initial investments in new non-thermal food processing technology. How do we calculate this investment and what should be included depending on the technology readiness level from proof of concept to industrial application? What should be included when calculating return of investment with regard to evaluating the efficiency of an investment? Since NTFPTs bring novelties in terms of both technologies and food, it also causes additional initial costs in promoting such novelties. The working assets can be divided as directs costs, indirect costs and social costs (White et al. 1992). Direct costs cover all operational (or production) costs joint with the annual depreciating costs. The indirect costs relate to the environmental dimension related to negative effects (such as pollution), as well as health damage costs caused by the new technologies (for example irradiation) and how are irradiation prevention costs calculated. Social costs cover various liability costs associated with environmental impacts such as payment for transportation / treatment of hazardous waste if insufficiently prevented (Krozer 2017).

Finally, the social pillar is the most "indefinite" meaning that such a sustainable business should have support and approval all stakeholders where the business should also be a responsible community member, both locally and globally. Some authors agree that „although as equally important as economic or environmental sustainability – social sustainability still lacks broad recognition "(Spangenberg and Omann 2006). Since this pillar enables creativity, in terms of non-thermal food processing technology, we can analyze it in two ways. From a technological point of view analyzing its effect on employees in terms of creating new jobs and health and safety issues associated with the use of new technology. On the other side we can promote the treated food product and the benefits it brings to the community regarding safety of the product, its nutritional values and sensorial attributes.

When speaking about NTFPTs technologies, it must be clarified that there is no clear definition what is considered as a "sustainable non-thermal food processing technology". As discussed by Mulder et al. (2017), all new technologies are considered as an "open design challenge" because many different aspects of sustainability should be taken into account. Here authors present some open questions regarding sustainability of NTFPTs seeking for the answers:

- Is its sustainability correlated whether technology is 'new' or 'improved design' of a well-known food processing technology?
- Does the technology readiness level (proof of concept, lab-scale or industrial application) affect its sustainability?
- What criteria are considered when analyzing pros and cons of the NTFPTs technology, compared to conventional already in use within the food sector?
- Are there any legal constraints (covering environmental or food safety dimension)?
- Are cost assessments associated with initial costs for setting a new technology or does it cover working assets during the entire life-cycle of exploitation?
- What are the social challenges regarding employees, i.e. does the new technology create new jobs or shuts down old jobs?

Basically, to understand NTFPTs one should look on the new technology "out of the box" differentiating sustainable processes and sustainability of the product. There is a need for development of a clear tool that will enable initial validation of NTFPT's sustainability and later verification of its sustainability during exploitation.

9 Conclusion

Progress in NTFPT innovations is driven by many benefits obtained from their current use, both in food industry and academia. Revelation of the environmental dimension and overall, of its sustainability, is gaining attention and needs to be supported. However, none of emerging NTFPTs that have appeared in the food technology arena in the recent decades, have demonstrated the potential to displace thermal processing counterparts completely. Heat and fire, that shaped and defined humans through centuries, still remain an integral part of our food processing for the foreseeable future.

References

Aganovic K, Smetana S, Grauwet T, Toepfl S, Mathys A, Van Loey A, Heinz V (2017) Pilot scale thermal and alternative pasteurization of tomato and watermelon juice: an energy comparison and life cycle assessment. J Clean Prod 141:514–525

Barba FJ, Parniakov O, Pereira SA, Wiktor A, Grimi N, Boussetta N, Saraiva JA, Raso J, Martin-Belloso O, Witrowa-Rajchert D (2015) Current applications and new opportunities for the use of pulsed electric fields in food science and industry. Food Res Int 77:773–798

CAC (2003) CAC/RCP 1-1969, Rev.4- 2003 Recommended International code of Practice – General principles of food hygiene Codex Alimentarius Commission,

Carvalho A, Mimoso AF, Mendes AN, Matos HA (2014) From a literature review to a framework for environmental process impact assessment index. J Clean Prod 64:36–62. doi:http://dx.doi.org/https://doi.org/10.1016/j.jclepro.2013.08.010

Chauhan OP, Unni LE (2015) 9 – Pulsed electric field (PEF) processing of foods and its combination with electron beam processing. In: Pillai SD, Shayanfar S (eds) Electron beam pasteurization and complementary food processing technologies. Woodhead Publishing, pp 157–184. https://doi.org/10.1533/9781782421085.2.157

Čuček L, Klemeš JJ, Kravanja Z (2015) Overview of environmental footprints. In: Assessing and measuring environmental impact and sustainability. Elsevier, pp 131–193

Đekić I (2009) Upravljanje zaštitom životne sredine u proizvodnji hrane (Environmental management in food production) (trans: Production EMiF), 1st edn. Poljoprivredni fakultet Univerziteta u Beogradu, Beograd, Srbija

Đekić I, Tomašević I (2017) Environmental footprints in the meat chain. IOP Conf Ser Earth Environ Sci 85(1):012015

Denoya GI, Xia Q, Xie F (2020) Chapter 12 – HPP of fruit and vegetable products: impact on quality and applications. In: Barba FJ, Tonello-Samson C, Puértolas E, Lavilla M (eds) Present and future of high pressure processing. Elsevier, Amsterdam, pp 273–293. https://doi.org/10.1016/B978-0-12-816405-1.00012-1

Djekic I, Tomasevic I (2018) Chapter 3 – environmental indicators in the meat chain. In: Muthu SS (ed) Quantification of sustainability indicators in the food sector. Environmental footprints and eco-design of products and processes, 1st edn. Springer Singapore, Singapore, pp 55–82. https://doi.org/10.1007/978-981-13-2408-6

Djekic I, Tomasevic I (2020) Role of potable water in food processing. In: Leal Filho W, Azul AM, Brandli L, Lange Salvia A, Wall T (eds) Clean water and sanitation. Springer International Publishing, Cham, pp 1–10. https://doi.org/10.1007/978-3-319-70061-8_136-1

Djekic I, Sanjuán N, Clemente G, Jambrak AR, Djukić-Vuković A, Brodnjak UV, Pop E, Thomopoulos R, Tonda A (2018a) Review on environmental models in the food chain – current status and future perspectives. J Clean Prod 176:1012–1025. https://doi.org/10.1016/j.jclepro.2017.11.241

Djekic I, Tomic N, Smigic N, Udovicki B, Hofland G, Rajkovic A (2018b) Hygienic design of a unit for supercritical fluid drying – case study. Br Food J 120(9):2155–2165. https://doi.org/10.1108/BFJ-01-2018-0052

Djekic I, Pojić M, Tonda A, Putnik P, Bursać Kovačević D, Režek-Jambrak A, Tomasevic I (2019) Scientific challenges in performing life-cycle assessment in the food supply chain. Foods 8(8):301

EC (2008) Directive 2008/98/EC of the European Parliament and of the Council of 19 November 2008 on waste and repealing certain Directives L 312. European Co\mmission, Brussels, Belgium

Gast J, Gundolf K, Cesinger B (2017) Doing business in a green way: a systematic review of the ecological sustainability entrepreneurship literature and future research directions. J Clean Prod 147:44–56. https://doi.org/10.1016/j.jclepro.2017.01.065

Gogate PR (2011) Hydrodynamic cavitation for food and water processing. Food Bioprocess Technol 4(6):996–1011. https://doi.org/10.1007/s11947-010-0418-1
Gogate PR, Pandit AB (2001) Hydrodynamic cavitation reactors: a state of the art review. Rev Chem Eng 17(1):1–85. https://doi.org/10.1515/REVCE.2001.17.1.1
Gómez B, Munekata PES, Gavahian M, Barba FJ, Martí-Quijal FJ, Bolumar T, Campagnol PCB, Tomasevic I, Lorenzo JM (2019) Application of pulsed electric fields in meat and fish processing industries: an overview. Food Res Int 123:95–105. https://doi.org/10.1016/j.foodres.2019.04.047
Gómez-López VM (2016) Pulsed light and packaging. In: Reference module in food science. Elsevier, Amsterdam. https://doi.org/10.1016/B978-0-08-100596-5.03279-0
Hariem Brundtland G (1985) World Commission on environment and development. Environ Policy Law 14(1):26–30. https://doi.org/10.1016/S0378-777X(85)80040-8
Herva M, Franco A, Carrasco EF, Roca E (2011) Review of corporate environmental indicators. J Clean Prod 19(15):1687–1699. https://doi.org/10.1016/j.jclepro.2011.05.019
Hospido A, Davis J, Berlin J, Sonesson U (2010) A review of methodological issues affecting LCA of novel food products. Int J Life Cycle Assess 15(1):44–52
IPPC (2006) Integrated pollution prevention and control. Reference document on best available techniques in the food, drink and milk industries. European Commission, Seville
ISO (2006) ISO 14040:2006 environmental management — life cycle assessment — principles and framework. International Organization for Standardization, Geneva
ISO (2015) ISO 14001:2015 environmental management systems – requirements with guidance for use. International Organization for Standardization, Geneva
Krozer Y (2017) Economics of sustainable technologies: private and public costs and benefits. Encyclopedia of Sustainable Technologies, Elsevier, pp 11–21
Mekonnen MM, Hoekstra AY (2010) The green, blue and grey water footprint of farm animals and animal products, vol 48. Value of water research report 48. UNESCO-IHE Institute for Water Education Elft, the Netherlands
Morehouse KM, Komolprasert V (2004) Irradiation of food and packaging: an overview. American Chemical Society, Washington, D.C.
Mulder K, Ferrer D, Van Lente H (2017) What is sustainable technology?: perceptions, paradoxes and possibilities. Routledge
Nutrizio M, Pataro G, Carullo D, Carpentieri S, Mazza L, Ferrari G, Chemat F, Banović M, Režek Jambrak A (2020) High voltage electrical discharges as an alternative extraction process of phenolic and volatile compounds from Wild Thyme (Thymus serpyllum L.): In Silico and Experimental Approaches for Solubility Assessment. Molecules 25(18):4131
Ortega-Rivas E (2012) Classification of food processing and preservation operations. In: Non-thermal food engineering operations. Springer US, Boston, pp 3–10. https://doi.org/10.1007/978-1-4614-2038-5_1
Pardo G, Zufía J (2012) Life cycle assessment of food-preservation technologies. J Clean Prod 28:198–207. https://doi.org/10.1016/j.jclepro.2011.10.016
Rajkovic A, Tomasevic I, Smigic N, Uyttendaele M, Radovanovic R, Devlieghere F (2010) Pulsed UV light as an intervention strategy against Listeria monocytogenes and Escherichia coli O157:H7 on the surface of a meat slicing knife. J Food Eng 100(3):446–451. https://doi.org/10.1016/j.jfoodeng.2010.04.029
Rajkovic A, Tomasevic I, De Meulenaer B, Devlieghere F (2017) The effect of pulsed UV light on Escherichia coli O157:H7, Listeria monocytogenes, Salmonella Typhimurium, Staphylococcus aureus and staphylococcal enterotoxin A on sliced fermented salami and its chemical quality. Food Control 73:829–837. https://doi.org/10.1016/j.foodcont.2016.09.029
Režek Jambrak A, Lelas V, Herceg Z, Badanjak M, Werner Z (2010) Application of high power ultrasound in drying of fruits and vegetables. Kemija u industriji 59(4):169–177
Režek Jambrak A, Vukušić T, Donsi F, Paniwnyk L, Djekic I (2018) Three pillars of novel non-thermal food technologies: food safety, quality, and environment. J Food Qual 2018:8619707. https://doi.org/10.1155/2018/8619707

Režek Jambrak A, Nutrizio M, Djekić I, Pleslić S, Chemat F (2021) Internet of nonthermal food processing technologies (IoNTP): food industry 4.0 and sustainability. Appl Sci 11(2):686

Smigic N, Djekic I, Tomic N, Udovicki B, Rajkovic A (2019) The potential of foods treated with supercritical carbon dioxide (sc-CO_2) as novel foods. Br Food J 121(3):815–834. https://doi.org/10.1108/BFJ-03-2018-0168

Spangenberg JH, Omann I (2006) Assessing social sustainability: social sustainability and its multicriteria assessment in a sustainability scenario for Germany. Int J Innov Sustain Dev 1(4):318–348

Srivastav PP, Verma DK, Patel AR, Al-Hilphy AR (2020) Emerging thermal and nonthermal technologies in food processing. Apple Academic Press & CRC Press. ISBN 978-1-77188-831-8. eISBN 978-1-42929-733-5. Palm Bay, Florida, USA

Tewari G, Jayas D, Holley R (1999) High pressure porcessing of foods: an overview. Sci Aliments 19:619–661

UN (1992) Agenda 21. United Nations conference on environment & development Rio de Janerio, Brazil, 3 to 14 June 1992. United Nations, Rio de Janeiro, Brazil

UN (2016) Transforming our world: The 2030 agenda for sustainable development. United Nations, New York, USA

UN (2019) Sustainable development goals. About the sustainable development goals. United Nations, New York, USA

Valsasina L, Pizzol M, Smetana S, Georget E, Mathys A, Heinz V (2017) Life cycle assessment of emerging technologies: the case of milk ultra-high pressure homogenisation. J Clean Prod 142:2209–2217

Warner RD, Ha M (2019) Utilising high pressure processing for meat tenderisation. In: Reference module in food science. Elsevier, Amsterdam. https://doi.org/10.1016/B978-0-08-100596-5.21348-6

White A, Becker M, Goldstein J (1992) Total cost assessment: accelerating industrial pollution prevention through innovative project financial analysis. US EPA, Office of Pollution Prevention and Toxics, Washington, DC. Product Design to Reduce Restricted Substances

WMO (2017) Global atmosphere watch program – greenhouse gas research. World Meteorological Organization, Atmospheric Environment Research Division, Research Department, Geneva

Wrangham RW (2010) Catching fire : how cooking made us human. Basic Books, New York

Yildiz S, Pokhrel PR, Unluturk S, Barbosa-Cánovas GV (2021) Shelf life extension of strawberry juice by equivalent ultrasound, high pressure, and pulsed electric fields processes. Food Res Int 140:110040. https://doi.org/10.1016/j.foodres.2020.110040

Chapter 24
Emerging Non-thermal Processing of Food Waste and by-Products for Sustainable Food Systems - Selected Cases

Aleksandra P. Djukić-Vuković, Dragana D. Mladenović, Jelena D. Pejin, and Ljiljana V. Mojović

1 Introduction

One in every nine people in the world suffered from hunger before COVID-19 crisis. With COVID-19 crisis, additional 132 million people may go hungry during 2020 and 2021 which is almost 20% increase (FAO 2020). Hunger is again rising in the world, after decades of decline in undernourishment. At the same time, around one third of all produced food is wasted (Morone et al. 2019) and decreasing food waste is among 17 UN Sustainable Development Goals. The food industry should provide enough food for world growing population, but it is clear that rational use of food and raw materials is the missing link in tackling both problems - hunger and food waste. This becomes even more challenging in COVID-19 crisis with rise of unemployment and food prices and general increase in poverty. Limitations in transport and logistics affect safety of transported food and the amount of food being discarded. Disruptions along the food supply chain may induce food losses, but also, some studies show decrease in food wasting on consumer level, since income often correlates with food wasting (Pappalardo et al. 2020; Roe et al. 2021). It can be expected that pandemic will decrease investments in prevention and reduction of food waste and temporarily put it off the agenda of governments mitigating primarily health and financial risks related to COVID-19. A fragile balance between food consumption and food waste or loss is severely disturbed in any crisis, being

A. P. Djukić-Vuković (✉) · L. V. Mojović
Faculty of Technology and Metallurgy, University of Belgrade, Belgrade, Serbia
e-mail: adjukic@tmf.bg.ac.rs

D. D. Mladenović
Innovation center, Faculty of Technology and Metallurgy, University of Belgrade, Belgrade, Serbia

J. D. Pejin
Faculty of Technology, University of Novi Sad, Novi Sad, Serbia

A. Režek Jambrak (ed.), *Nonthermal Processing in Agri-Food-Bio Sciences*, Food Engineering Series, https://doi.org/10.1007/978-3-030-92415-7_24

caused by weather, war or pandemic. Present crises could be used to faster rethink the food supply chain, including smart and innovative solutions in food waste and by-products processing in order to return the resources into the human food chain.

Estimates of generated food waste are ranging from 194 to 389 kg/person/year globally and from 158 to 290 kg/person/year in EU (Corrado and Sala 2018). Food waste and by-products are generated in different stages throughout the food supply chain and the estimates strongly depend on how food waste, food loss and by-products are defined and used in literature. Comprehensive reviews of food waste hierarchy could be found in literature (Papargyropoulou et al. 2014), but few definitions are most often used. The first is that food waste is the edible fraction of food not eventually eaten by humans (FAO 2011; Gustavsson et al. 2013). The second definition is that food waste additionally considers edible material that is intentionally fed to animals or is a by-product of food processing diverted away from the human food chain, including by-products diverted into feed production (Stuart 2009; Papargyropoulou et al. 2014). The latest definition by Food and Agriculture Organisation of the United Nations (FAO) refer to food loss as the decrease in the quantity or quality of food resulting from decisions and actions by food suppliers in the chain, excluding retailers, food service providers and consumers (FAO 2019). Food waste refers to the decrease in the quantity or quality of food resulting from decisions and actions by retailers, food service providers and consumers (FAO 2019). Estimates of generated food waste are often based only on the collected food waste and on waste treated in waste management facilities, although other methodologies have been applied (Corrado et al. 2019) which also causes variations in reported data.

The aim of this chapter is twofold: to analyze the roles that non-thermal processing had until now in food waste valorization and to highlight their potential for new applications based on the results from other fields where they are successfully applied. Bibliometric coupling and mapping were used for analysis. Non-thermal processing technologies can be combined and this is advantageous characteristic for cascade and integrated waste processing. Selected cases of biorefinery solutions for processing of wastes from breweries and other fermentation facilities are presented. The roles of non-thermal technologies in these processes were emphasized.

2 Methodology

The VOSviewer software tool was used to construct a network of keywords based on co-occurrence data aiming to address the questions such as what are the main research topics connected with non-thermal processing and how do they relate to each other. This tool applies a distance-based approach for visualizing Bibliometric co-occurring networks, meaning that the nodes representing the keywords in a network are positioned in such a way that the distance between two nodes approximately indicates their relatedness (van Eck and Waltman 2010, 2014). This way, the clusters of related keywords can be easily identified and linked to specific research topics that are connected with non-thermal processing technologies.

Data for visualization were collected from the ISI Web of Science database. Co-occurrence networks were constructed based on keywords of selected publications. Before constructing the network, the data cleaning was performed by merging typographical variations of the same keyword. The search in ISI Web of science for networks presented in Fig. 24.1 covered original research articles and review articles reporting on food waste and the most prominent non-thermal technologies (TOPIC: "food waste$" or "fruit waste$" or "fruit residues$" or "vegetable waste$" or "vegetable residue$" or "food by-product$" or "agro-industrial residues$" or "industrial by-product$" AND TOPIC: "pulsed electric field" or "cold plasma" or "high pressure process" or "high hydrostatic pressure" or "ultrasound" or "ultrasonication") in the title, abstract, or keywords for articles or reviews. Keywords with a threshold of 5 occurrences were included in this network. In total, 252 documents published until 19th of February, 2021, were selected for data visualization.

The search related to network presented in Fig. 24.2 included „food waste" and related terms (TOPIC: "food waste$" or "fruit waste$" or "fruit residues$" or "vegetable waste$" or "vegetable residue$" or "food by-product$" or "agro-industrial residues$" or "industrial by-product$") in the title, abstract, or keywords for articles or reviews published in 2020, giving a total number of 2526 publications. Only author keywords were used for construction of network with a threshold of 9 occurrences.

The search related to network presented in Fig. 24.3 was performed using term "non-thermal processing" and variations (TOPIC: "non-thermal process*" or "nonthermal process*" or "non-thermal technolog*" or "nonthermal technolog*") in the title, abstract, or keywords for articles or reviews published until 19th of February 2021, giving a total number of 1334 publications. All keywords (author and indexed keywords) were used for network construction, with a threshold of 15 occurrences.

3 Processing of Food Waste and Food by-Products

The strategies to decrease and manage wasting of food are mostly rooted in segmentation into avoidable and unavoidable food waste (Morone et al. 2019) or food loss. In general, minimisation of avoidable food waste would be preferred, but the complexity of the food supply chain and differences in food industry and waste management between undeveloped, developing, and developed countries determine scales and routes for minimisation, treatment or value recovery from food waste (Lopez Barrera and Hertel 2020).

The main intentions of food processing companies are to decrease amount of generated food waste or to divert all side streams into valuable products (Hamelin et al. 2019). But capability of food processing companies to adopt innovations for waste reduction and valorization of by-products is generally low, even in developed

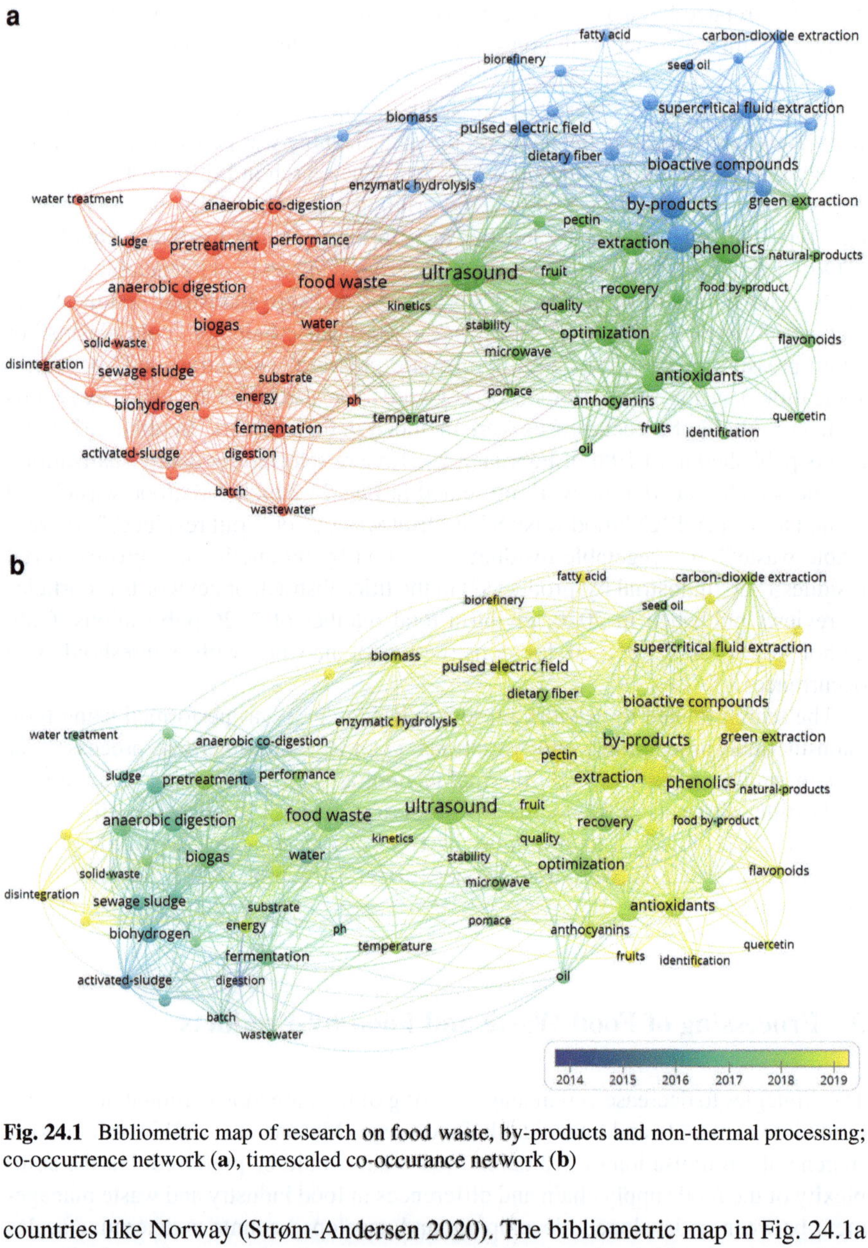

Fig. 24.1 Bibliometric map of research on food waste, by-products and non-thermal processing; co-occurrence network (**a**), timescaled co-occurance network (**b**)

countries like Norway (Strøm-Andersen 2020). The bibliometric map in Fig. 24.1a presents the main strategies for management and processing of food waste and by-products reported in scientific literature, while the trends and progress are presented in Fig. 24.1b.

In practice, food waste is most often subjected to: composting, anaerobic digestion, incineration, landfilling etc. (Fig. 24.1a). Sustainability of these options depends on many reasons (Bernstad and Jansen 2012; Slorach et al. 2019), but they proved the feasibility to be implemented and scaled up. However, the recovery of bioactive compounds and their recycling back by above technologies into the food chain is very low, so the alternatives are being tested. The alternatives are to extract bioactive compounds from food waste and by-products and to purify them before being returned into the food chain. The options for valorization of food wastes and by-products can be further expanded by means of biotechnology. The components from food waste and by-products can be converted by microorganisms and enzymes and the terms such as *biorefineries, fermentation, enzymatic hydrolysis* start to emerge in the networks related to food wastes since 2015 (Fig. 24.1b). Three clusters were identified under the applied conditions (Fig. 24.1a). More conventional strategies for food waste treatment such as anaerobic digestion, biogas and biohydrogen production were mostly published until 2015. During the last 5 years, ultrasound becomes the first non-thermal technology studied in a context of food waste, with ultrasound-assisted extraction playing the major role in the treatment of both food waste and by-products (Fig. 24.1b). New clusters, green cluster-mostly related to the extractions of antioxidants (*flavonoids*, *polyphenols*, *anthocyanins* etc.) and blue cluster -mostly related to technologies used for extractions from by-products accrued (Fig. 24.1a). As the most significant technologies- *ultrasound*, *pulsed electric field,* and *high pressure processing* are emerging.

The new value-added chemicals which can be obtained by means of extraction or bioprocessing can be streamed back to food and feed industries or into applications in medicine, pharmacy, cosmetics, packaging. The recovery of resources from waste and by-products contributes to rational utilization of raw materials in *biorefinery* processes, *sustainability*, *life cycle assessment* and *circular economy*, which are evidently becoming more significant as it is visible in the most recent bibliometric map of publications in 2020, presented in Fig. 24.2. Some specific aspects, like those related to COVID-19 crises and its links with food safety and security are also recognized in literature (Fig. 24.2).

The roles of non-thermal technologies in future waste and by-products valorization depend on their potential and scalability. Non-thermal processing was first implemented in food industry and has many advantages explained in detail in other chapters of this book. The pattern established for applications in food processing can be translated to some extent into the food waste processing. Bibliometric analysis of search using "non-thermal processing" reveals also other emerging technologies like *UV irradiation*, *pulsed light* and *cold plasma*, as presented in Fig. 24.3. All identified non-thermal technologies in this network are strongly and closely linked to *microbial inactivation* and related terms (e.g. *food preservation*, *food safety*, *sterilization*, *decontamination*, *pasteurization*). Ultrasound is also linked to extractions of *antioxidants* and other *bioactive compounds* in food processing, similarly to *ultrasound-assisted extractions* from *food waste* and *by-products* as identified in Figs. 24.1 and 24.2. The new fields of research oriented towards *physicochemical* or *functional properties* of treated food, *extraction* of specific ingredients or mechanical characteristics of food are emerging in Fig. 24.3, and could be also expected to expand in food waste treatment. Among all identified technologies presented in Fig. 24.3, *cold plasma* is the most recently introduced in food processing. Cold

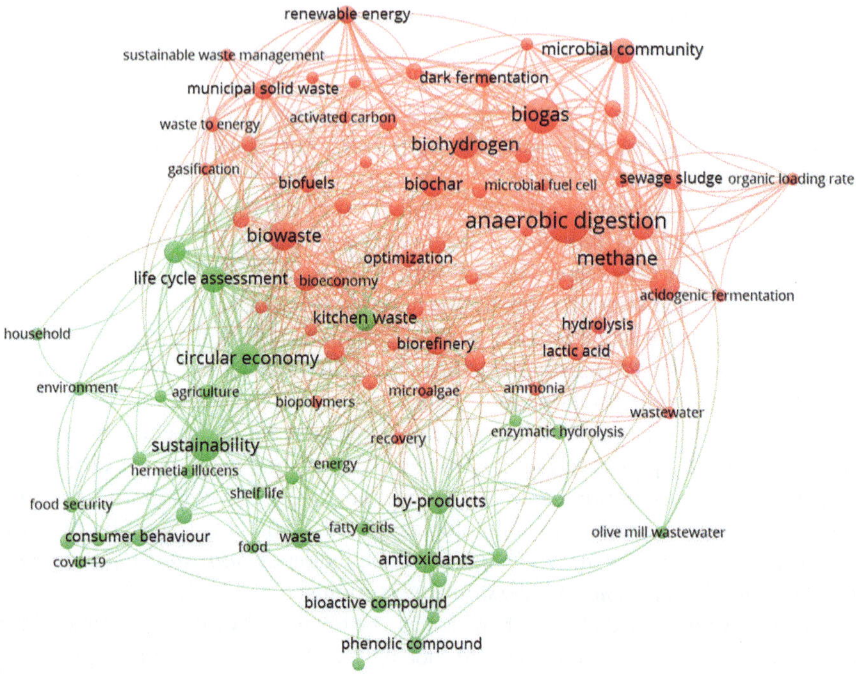

Fig. 24.2 Bibliometric map of food waste and by-products related publications, published in 2020 – co-occurrence network

plasma is not present on the maps related to food waste (Fig. 24.1), but detailed search within publications database reveals few publications related to cold plasma in last 2 years. It is expected that the cold plasma will be more studied in future for resource recovery, including waste valorization.

4 Why Food Waste Is Not more Often Converted into High-Value Products?

Agricultural residues or by-products are generally more suitable than food waste for biorefineries where the stability in supply and the quality consistency are highly required. They are produced in larger batches, often contain less water and therefore can be stored for longer time. Segregated food waste from households or restaurants varies more in chemical composition and indigenous microbiota, limiting options for valorization (Girotto et al. 2015). Proteins, sugars, fibers and antioxidants are among the most valuable compounds in food wastes and extraction, separation and purification to some extent are necessary. Selection and segregation of food waste and by-products on time determines the potential of these substrates to be valorized.

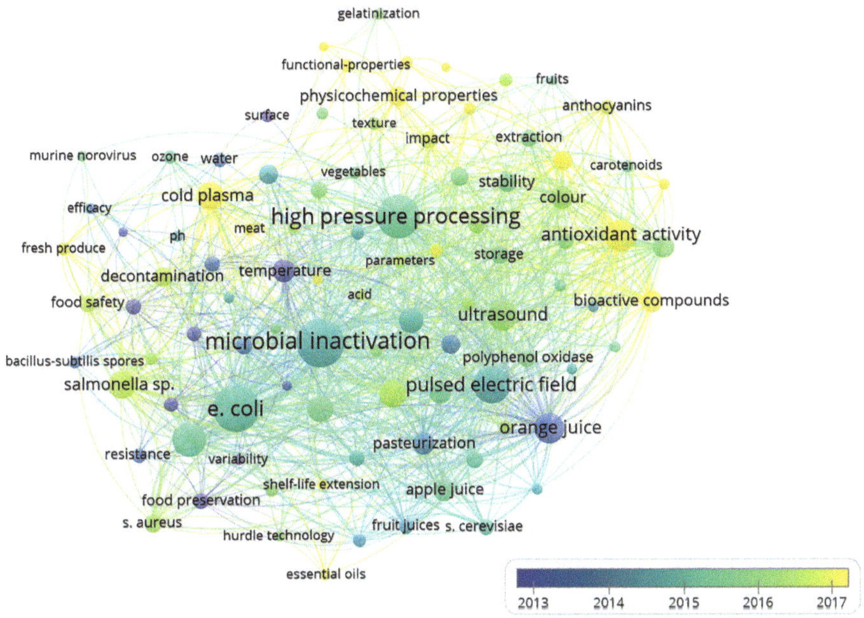

Fig. 24.3 Bibliometric map of non-thermal processing related publications- timescaled co-occurrence network

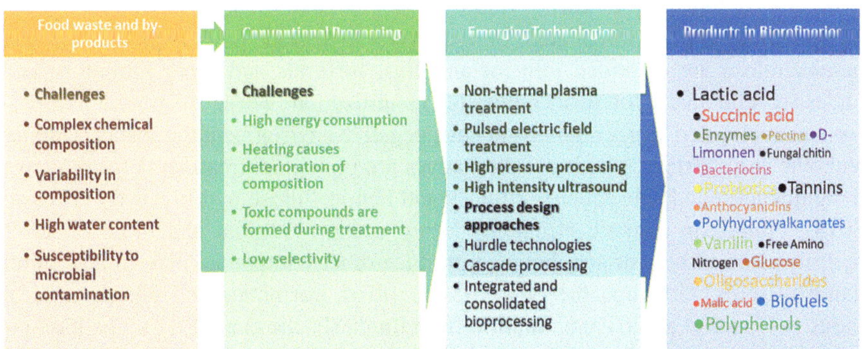

Fig. 24.4 Possibilities for application of emerging processing technologies in valorization of food waste and by-products

For example, for efficient conversion of wastes into lactic acid (LA) by LA bacteria, it is necessary to provide an adequate C/N ratio, to avoid growth inhibitors, to prevent contamination of media or adequately treat these complex substrates while enabling desired chemical composition (Fig. 24.4).

4.1 Complex Chemical Composition and Variations in Food Waste and by-Products

A complex chemical composition of residues, including high water content in some cases, highly influence the feasibility of waste and by-products valorization. Primary, secondary, and tertiary residues are generated at different stages of the food supply chain. The primary residues are solid agricultural remains, mainly of lignocellulosic composition and are usually left in the field after harvest. The secondary residues are discarded during food processing stage and they can be both liquid and solid. These are rich in easily assimilative sugars, oligosaccharides, proteins, antioxidants etc. and usually contain more water than primary residues. The tertiary residues remain after the product consumption in households or restaurants (e.g., food leftovers). These are in general very susceptible to contamination, produced in smaller batches and very difficult to collect on time and treat properly. This segmentation is based on the origin of generated wastes and residues, but it also correlates with the chemical composition of the residues. Some emerging processing technologies are more advantageous over others for the valorization of primary, secondary or tertiary residues and the most promising applications will be emphasized.

The main limitation for utilization of lignocellulose rich primary residues as sources of sugar in biorefinery is a high content of lignin and recalcitrant structure of lignocellulose in general. To be used as feedstock in biorefinery processes, they require pretreatment that can break down the lignin protective barrier and facilitate hydrolysis of the complex carbohydrates to fermentable sugars. Conventional treatments include application of acids or hydroxides to degrade lignin, but cellulose and hemicellulose are degraded into melanoidins, furfurals and other inhibitory and undesired chemicals for further biorefinery utilization. These treatments are performed at elevated temperatures and have negative environmental impacts (Djukić-Vuković et al. 2019). Promising alternatives are biological treatments for selective delignification and subsequent cellulose and hemicellulose hydrolysis (Jović et al. 2018). Fungi can degrade efficiently lignin and provide not just more effective hydrolysis of carbohydrates, but also produce other compounds like fungal chitin nanostructures with tuneable properties for food, cosmetics and pharmaceutical industry (Jones et al. 2019) or vanillin from lignin (Fache et al. 2016). The biological treatments have a huge potential in valorisation of primary residues, however, rather slow growth of fungi is the main limiting factor for their faster adoption (Troiano et al. 2020). On industrial scale, the production of biomass of edible fungi on food wastes has been already established, where metabolic engineering could provide economically feasible approaches for novel products from fungi, firstly enzymes but also some others (Mahboubi et al. 2017; Jones et al. 2019; Troiano et al. 2020).

Physical or physico-chemical treatments are also possible for primary residues. High voltage electric discharge has shown promising results. It provides sufficient disruption of lignocellulose and improves access for enzymes during hydrolysis

(Puértolas and Barba 2016). The costs of currently used enzymatic treatments are high, up to 8.5 € per ton of substrate (Zhang et al. 2011). Any strategy to decrease the amount of required enzyme and energy is valuable. Cold plasma treatment of lignocellulose substrates has shown some selectivity towards lignin (Zanini et al. 2005; Baltazar-Y-Jimenez and Bismarck 2007; Baltazar-y-Jimenez et al. 2008) and efficiency in depolymerisation of cellulose (Benoit et al. 2012; Jérôme et al. 2016). A comprehensive assessment of novel, potentially inhibitory compounds produced during these treatments in complex waste materials was not performed so far, to the best of our knowledge. These can act inhibitory on the growth of microorganisms employed in subsequent biorefinery process and this presents a gap in the current knowledge.

Secondary residues have higher water content besides the fluctuating composition of carbohydrates, proteins, and other essential nutrients. Valuable data on composition of available by-products generated from soybean, sugar beet, cereal, and cassava processing have been previously reviewed and summarized (Elmekawy et al. 2013; Zhang et al. 2016; Loman and Ju 2016; Tomaszewska et al. 2018). The dry matter content is often below 20% in these substrates and transportation from producing to treatment facility is costly with a small time window due to susceptibility to contamination. Drying as an alternative is energy consuming and expensive processing step. Substrate sterilization and application of so called closed fermentations are also energy consuming, however, wherever possible, the open fermentations are considered as better option (Qureshi et al. 2017; Qiu et al. 2017; Wang et al. 2019; Djukic-Vukovic et al. 2019).

The application of non-thermal technologies with intention to decrease growth or load of undesired microorganisms in substrate is particularly attractive. Moreover, this was the primary field of application for numerous non thermal technologies in food science and technology, starting few decades ago (Fig. 24.3). Emerging technologies like pulsed electric field (PEF), non-thermal or cold plasma treatments (NTP) or high pressure processing (HPP) were mostly studied for microbial inactivation (Fig. 24.3). This corresponds with keywords *microbial inactivation, food safety, preservation, decontamination* etc. This approach was validated for liquid foods like juices (Fig. 24.3; *apple juice, orange juice),* primarily using PEF and HPP. All these technologies were studied as alternatives to conventional *thermal processing* for prevention of *foodborne pathogens* (Fig. 24.3). PEF, HPP or ultrasound treatments can enable reduction of several logN units of various microbial species which can be present as undesired in liquid substrates (Wouters et al. 2001; Mahnič-Kalamiza et al. 2014; Raso et al. 2016). The bibliometric analysis suggests expansion of these technologies into the food waste or liquid by-products processing (Fig. 24.1). Storage time can be increased and the range of possibilities for valorisation of secondary residues can be expanded. They can be used to decrease microbial load before open fermentations or other bioprocessing technologies. However, for a complete inactivation of microorganisms in media, these technologies are most often combined as *hurdle technology* in food processing and it is recognized in literature (Fig. 24.1). The hurdle approach enables sufficient safety and quality of the substrate obtained from residues while preventing chemical deterioration by intensive heating (Khan et al. 2017). In that case, the overall energy consumption should be lower and the treatment time should be shorter.

The tertiary residues are the remains after the final product consumption (e.g., food leftovers). Previous studies have reported that mixed food and bakery waste in general consists of 29–60% starch, 5–14% proteins, and 10–40% lipids proving food waste as a promising feedstock for the production of lactic (Kwan et al. 2016), succinic acid (Leung et al. 2012; Zhang et al. 2013) and other platform chemicals (Trivedi et al. 2020). Fungal hydrolysis of food waste in submerged fermentation showed exceptional opportunities to reduce the dry weight up to 90% and to recycle glucose, free amino nitrogen, and phosphate by their assimilation in microbial biomass (Pleissner et al. 2014). This can be further used for high-value chemicals, energy as well as for food and feed production (Pleissner et al. 2014). Homogenization of food waste and release of the content from cells is also important and energy consuming, although mechanical grounding is currently the most commonly used (Puértolas and Barba 2016). High intensity ultrasound or high pressure homogenization can successfully improve the extraction and homogenization of food waste (Putnik et al. 2017) (Fig. 24.1) mostly by significantly altering the composition of liquid fraction of treated waste (Jiang et al. 2014). An increase in concentrations of reducing sugars and proteins in the liquid fraction was higher in comparison to lipid concentration, which is a favorable effect for many biorefineries (Jiang et al. 2014).

The techno-economic analysis of potential routes for the valorization of a specific type of food waste generated in particular locations is of utmost importance (Cristóbal et al. 2018). This is recognized, and novel rigorous studies are performed to support best decisions in the selection of processes on wastes and by-products (Puértolas and Barba 2016; Aganovic et al. 2017). For techno-economic assessment, data on available amounts of particular waste are needed and they strongly depend on the applied strategies for collection and separation of waste. The collection and separation of waste are the most challenging steps of waste management in general, and it also stands for food waste. A high awareness related to separation of food waste in households, at the consumer level, can significantly decrease the costs and improve recovery of resources buried in food waste. The dedication to this aspect of waste management and non-compliance to related policies are certainly the limiting factors in valorization of food waste in some countries.

4.2 *Cascade and Integrated Processing of Food Waste and By-Products in Biorefinery*

The cascade processing enables extraction or conversion of desired compounds from complex media in fractions. Desired ingredients are separated, which significantly simplifies downstream purification. Antioxidants, mostly reported as polyphenols, tannins or anthocyanidins, important in human diet, are present in processing waste of plant origin (fruits, vegetables, herbs). In order to valorise them on the market, it is crucial to provide sufficiently purified standardised extracts. Technologies which have shown selectivity are attractive for such extractions. Antioxidants, most often polyphenols were extracted from fruit peels or pomace, in subsequent steps (Wijngaard et al. 2012; Galanakis 2012; Kammerer et al. 2014; Herrero et al. 2015; Barba et al. 2016; Pradal et al. 2016; Putnik et al. 2017) or simultaneously (Galanakis et al. 2016).

The PEF proved successful since it provides an increase in permeability of cell membranes and leaking of intracellular content. Optimization of process conditions can enable selectivity. It is mainly performed in water based media, but was also applied in mixtures with ethanol with lower conductivity (Shorstkii et al. 2017). The utilization of green solvents and minimal heating during PEF processing preserves the chemical structure and activity of thermo labile antioxidants in extracts (Barba et al. 2016; Putnik et al. 2017). After the extraction of water-soluble active compounds by PEF, residues can be subjected to additional treatments for effective downprocessing.

Therefore, physical and biochemical processes are most often combined. For example, in the first step, low-temperature hydrothermal processing of waste pomegranate peel enables simultaneous recovery of pectin and phenolic compounds. It is followed by enzymatic hydrolysis and yeast fermentation for bioethanol production. As a result, the high methoxyl food quality pectin, punicalagin rich aqueous phenolic extract, and bioethanol were produced. Two of three obtained products were suitable to be recycled into the food chain (Talekar et al. 2018).

Martinez et al. (2016) developed the multi-purpose cascading biorefinery process for the valorization of red grape pomace and the production of polyphenols, biopolymer, and biomethane. In the first step, supercritical CO_2 extraction was applied for the recovery of polyphenols. The remaining dephenolised grape pomace was further anaerobically digested under batch low pH conditions. Obtained volatile fatty acid-rich liquid fraction was employed as the substrate for producing polyhydroxyalkanoates, while the solid leftover underwent further methanogenic process dedicated to the production of a methane-rich biogas (Martinez et al. 2016). Other examples could be found in literature (Goula and Lazarides 2015), but wastes from citrus fruits stand out as very good candidates for cascade processing (Pfaltzgraff et al. 2013).

For making biorefinery more sustainable, complementary technologies and substrates should be combined. This decreases equipment and transportation costs and opens new markets. The production of chemicals should be coupled with biofuel production and/or extraction of functional compounds, contributing significantly to the overall process profitability and productivity (Schieber et al. 2001; Koutinas et al. 2014). A coupled production should be based upon existing production facilities like biodiesel, sugar or ethanol facilities. The combined production enables the widening of the product spectrum of existing factories to chemical and other markets increasing their competitiveness. The fermentative production of LA or succinic acid, for example, opens a broad range of new products such as polymer materials, solvents, food-feed additives, etc., what paves the way for closing the loop and making truly circular processes (Lyko et al. 2009).

Very comprehensive overview of integrated biorefinery by means of fungal producers of various chemicals confirmed that consolidated bioprocessing where the pretreatments were simultaneously applied with fermentation as one processing step, were very successful in the case of solid state residues (Troiano et al. 2020). Microbial conversion of lignin into vanillin as aroma compound in food industry is a good example of a route for integrated processing and recovery of valuable food ingredients from residues (Banerjee and Chattopadhyay 2019).

For assessing the feasibility of scale-up with a combined processing, Dimou et al. (2016) performed techno-economic analysis of a wine lees refining process for the production of ethanol, an antioxidant-rich extract, calcium tartrate, and a solid fraction rich in yeast cells. The final products of the proposed biorefinery could be intended for use in various industrial sectors, including food, feed, chemical, and cosmetic industries. This study revealed the lowest selling price of the antioxidant-rich extract that should be achieved in order to develop a cost-effective wine lees biorefinery. The extraction yield and the purity of the antioxidant extracts strongly affect their market price determining the cost-competitiveness of the integrated process (Dimou et al. 2016). Every new step introduced into the process significantly affects its profitability. Availability of resources - quantity and quality of food waste and by-products as substrates in the process and the quality and safety of obtained products are the most important data to be taken into account when considering novel process design.

5 Example of brewer's Spent Grain as a Substrate in Biorefineries

Brewer's spent grain (BSG) is the most abundant brewing by-product, which corresponds to around 85% of the total by-products generated in breweries (Mussatto et al. 2013). Per hectoliter of beer, approximately 15–20 kg of BSG is produced, which results in annual worldwide production of over 30 million tonnes (Niemi et al. 2013). Despite being generated in huge amounts, its use is still limited and predominantly streamed into animal feed (Bohnsack et al. 2011). BSG has a favorable chemical composition and efforts have been put towards efficient extraction, purification, and reuse in the food industry and other industries.

Hemicellulose, cellulose and starch as fractions of BSG can be converted into xylose, arabinose and glucose as building-blocks and platform chemicals. A near-critical CO_2 was previously shown as a promising technology to assist enzymes in the hydrolysis of BSG polysaccharides into fermentable sugars (Luft et al. 2018). Similarly, a new pretreatment method based on non-thermal plasma technology was developed to improve the enzymatic hydrolysis of BSG, and obtained hydrolysates were subsequently used for bioethanol production (Ravindran et al. 2019). These technologies were mainly implemented to increase surface area of BSG or to increase its porosity and make cellulose and hemicelullose fibers more accessible to enzymes. Additionally, LA fermentation of sugars derived from BSG using homofermentative *Lactobacillus* species was also studied as an additional step in bioprocessing (Pejin et al., 2015). This way produced LA may find a wide range of applications in the food and beverage sector. Similarly, hemicellulosic hydrolysate derived from BSG has been used for the production of bacteriocins, efficient antimicrobials in the fields of health, agriculture, and food preservation (Paz et al. 2018).

In addition to polysaccharides, BSG is an important source of phenolic compounds originated from the barley grain husk. Major phenolic acids in BSG are ferulic, p-coumaric, sinapic and caffeic acids, which have been linked with many health-promoting effects such as antioxidant, anticarcinogenic, and anti-inflammatory activities (Ikram et al. 2017). Several studies applied PEF to facilitate the extraction of these bioactive compounds from BSG. With optimized PEF parameters such as electric field strength, frequency and treatment time, concentrations of total free and bound phenolic compounds were 2.7 and 1.7 fold higher, respectively, compared to the control BSG samples without PEF, indicating an improvement in the phenolic recovery with the use of PEF as a pretreatment strategy (Martín-García et al. 2020). PEF assisted extraction of light and dark BSG enhanced levels of phenolics, amino acids and proteins in extracts and their antioxidative, antimicrobial and immunomodulatory activities (Kumari et al. 2019). These studies on BSG open up new avenues for a broad spectrum of value-added products intended for food/feed market. And these can be produced within the beer facilities. BSG derived products could be applied in other markets. The production of volatile fatty acids (Ribau Teixeira et al. 2020), biobutanol (López-Linares et al. 2019) or biohydrogen (Zhang and Zang 2016) are examples of other biorefinery processes based on BSG.

6 Example of Combined LA and Probiotic Biomass Production on by-Products from Agri-Food Industry

LA biorefineries gained interest of scientific community during the last decade because of high demand for LA in polymer market. LA is an important chemical with wide range of applications and its polymers are thermostable, biocompatible and suitable for medical and pharmaceutical applications (Nampoothiri et al. 2010; Murariu and Dubois 2016; Djukić-Vuković et al. 2019). LA concentrations and productivities obtained in LA fermentations on various food wastes and agri-food industry by-products are given in Fig. 24.5. A combination of complementary substrates such as distillery stillage from bioethanol production from waste potato (rich in nitrogen source) and sugar beet molasses (rich in carbohydrates) proved advantageous in LA fermentations by *Lactobacillus paracasei* NRRL B-4564 (Mladenović et al. 2016). Sugar beet molasses is a by-product of the sugar industry that is produced in huge amounts. For instance, processing of 1 t of sugar beet generates 140 kg of sugar, 58 kg of dried beet pulp, 40 kg of molasses, and 15 kg of primary beet residues (Capper et al. 2013). Stillage is a liquid waste product from bioethanol

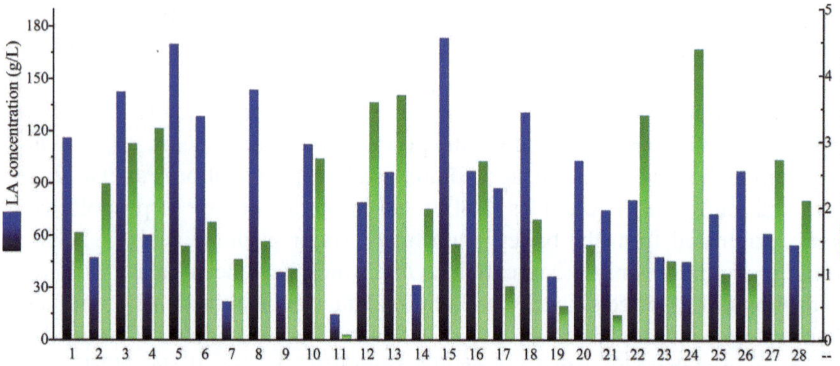

Fig. 24.5 LA concentrations and productivities in fermentations on different agro-food residues reported in literature. Meaning of symbols on x-axis: (1) Sugarcane juice (Dey et al. 2012), (2) Cashew apple juice (Silveira et al. 2012), (3) Sweet sorghum juice (Liu et al. 2013), (4) Date juice (Nancib et al. 2009), (5) Sugar beet molasses (Mladenović et al. 2018a), (6) Sugar cane molasses (Thakur et al. 2017), (7) Carob pod waste (Bahry et al. 2019), (8) Dairy waste (Bernardo et al. 2016), (9) Cassava wastewater (Coelho et al. 2010), (10) Cassava bagasse (Chen et al. 2020), (11) Potato peel waste (Liang et al. 2014), (12) Broken rice (Nakano et al. 2012), (13) Orange waste (de la Torre et al. 2020), (14) Brewer's spent grain and malt rootlets (Radosavljević et al. 2019), (15) Tobacco waste (Zheng et al. 2016), (16) Waste *Curcuma longa* biomass (Nguyen et al. 2013), (17) Wheat bran (Li et al. 2017), (18) Wheat straw (Zhang et al. 2014), (19) Rice straw (Yadav et al. 2021), (20) Corncob (Kong et al. 2019), (21) Corncob molasses (Wang et al. 2010), (22) Empty fruit bunch of oil palm trees (Ye et al. 2014), (23) Coffee pulp (Pleissner et al. 2016), (24) Residue from coffee production (Neu et al. 2016), (25) Sugarcane bagasse (Unrean 2018), (26) Corn stover (Qiu et al. 2017), (27) Organic fraction of municipal solid waste (López-Gómez et al. 2020), (28) Food waste (Pleissner et al. 2017)

production, which is generated in an amount of up to 20 L per liter of bioethanol. A common strategy for stillage valorization is its drying and using in animal nutrition. However, taking into account a high water content, this strategy is not economically feasible (Mojović et al. 2012; Djukić-Vuković et al. 2019). NTP pretreatment of stillage for LA fermentation was proved successful in reduction of initial microbial load. It enabled open LA fermentation with unaffected stereoselectivity of produced L-LA of above 95% (Djukic-Vukovic et al. 2019).

After successful attempts to combine low-cost and abundant agro-food residues, a significant improvement of the process efficiency with combined solid and liquid residues was made by immobilized cell systems. This way primary residues where combined with secondary. Sunflower seed hull, brewer's spent grains, and sugar beet pulp acted as carriers for *L. paracasei* immobilization. The comparison of the stability and efficiency proved sugar beet pulp immobilized system superior to the other two, achieving LA productivity of 1.48 g/Lh, with maximal LA concentration of 80.1 g/L (Mladenović et al. 2018b). Another study showed an efficient immobilization of *Lactobacillus rhamnosus* ATCC 7469 onto zeolite type X or zeolites modified by Mg^{2+}, resulting in up to three-fold higher LA productivity compared to free cells system (Djukić-Vuković et al. 2013, 2016).

Many LA bacteria strains are also traditionally used as probiotic food/feed additives, which opens the possibility to valorise them as added-value products contributing to the overall process sustainability. Food and feed production are interlinked and food/feed competition should be always taken into account when analysing options for biorefineries (van Hal et al. 2019). The majority of LA bacteria are classified as Generally Recognized As Safe by United States Food and Drug Administration and accepted by the European Food Safety Authority, suggesting that even LA bacteria which have not yet been proven as probiotics can be considered as safe. The strains *L. rhamnosus* ATCC 7469 and *L. paracasei* NRRL B-4564 have already shown favourable probiotic characteristics (Djukić-Vuković et al. 2015; Mladenović et al. 2019). Besides, according to the chemical analysis, the side streams of the processes applying immobilized cell systems based on lignocellulosic residues can be valorised as a probiotic enriched source of nutrients in compounded ruminant feeds (Mladenović et al. 2019).

7 Examples of LA Production on Food Wastes

Another example of complementary substrates used for LA production is a combination of bakery waste and lucerne green juice as an alternative nitrogen source (Alexandri et al. 2020). Batch fermentation of bakery waste hydrolysate combined with lucerne green juice with *Bacillus coagulans* resulted in a final LA concentration of 62.2 g/L, with a productivity of 2.59 g/Lh and a yield of 0.57 g LA/g bakery waste. The same work reported fivefold higher LA productivity in a continuous fermentation system coupled with cell retention membranes (Alexandri et al. 2020), proving the advantage of continuous over batch mode. High productivity obtained in continuous mode using low-cost substrate paves the way for further process optimization and improvements to ensure sustainable industrial implementation.

A very prospective one-step process for L (+)-LA production from mixed restaurant food waste was proposed by (Pleissner et al. 2017). In simultaneous saccharification and fermentation *Streptococcus* sp. efficiently degraded the food waste and produced LA at a maximum rate of 2.16 g/Lh and yield of 0.81 g/g. This study revealed a linear relationship between LA concentration and the solid-to-liquid ratio of food waste, as well as a possibility to perform the process under non-sterile conditions without considerable production of other organic acids, such as acetic acid (Pleissner et al. 2017).

Further improvement of the LA production process from food waste was studied through sequential production of LA and biogas. Solids generated in LA production using either simultaneous saccharification and fermentation or separate enzymatic hydrolysis and fermentation could be valorized in a composting process, making the LA production economically more feasible and at the same time solving the issue of fermentative residues (Demichelis et al. 2017).

The high optical purity of produced LA is a prerequisite for most commercial applications. Isomer L (+) is preferable for the food and pharmaceutical industry since it is the only LA isomer produced in the human body. The final optical purity of LA produced from biowaste, especially in the open fermentation process, is usually affected by naturally occurring LA bacteria which can proliferate and produce a racemic mixture of LA, thus limiting commercial applications of the product. To solve this issue, a pretreatment with monopolar electrodialysis membranes was proposed for the removal of D (−) LA initially present in biowaste hydrolysate. Subsequent fermentation of purified hydrolysate resulted in enantiomeric purity of over 98% L (+)-LA (López-Gómez et al. 2020), encouraging future studies on LA production from biowaste, including downstream and purification methods, as well as an assessment of the economic feasibility of the whole process.

8 Conclusions

Sustainability of the food chain strongly correlates with responsible utilization of available resources. Estimate is that the third of all resources which enter food supply chain is wasted and this could be even enhanced by pandemic, climate changes or wars. Therefore, it is essential to find a way to recover as much as possible, and return it back into the food chain, while simultaneously preventing food wasting in general.

Non-thermal technologies and bioprocessing are identified as very promising approaches since the complexity of food wastes and by-products impose many limitations for conventionally used technologies for resource recovery. The bibliometric mapping based on co-occurrence networks revealed the main trends, obstacles and limitations in the converting food waste and by–products into higher value products. Non-thermal technologies were first introduced in food science for microbial inactivation and extraction. A similar pattern can be expected now when they are being applied in food waste valorization. Bioprocessing of brewers' spent grains and LA based biorefinery confirms a potential of emerging technologies in resource recovery. This concept, together with the proper design of cascade processing and process integrations specific for various substrates, will result in more effective conversion into higher value products suitable to be recycled into the food chain. Understanding of the current potentials and limitations of emerging technologies in processing of waste provides a starting point for creating more sustainable and resilient food systems in future.

Acknowledgments This work was supported by the Ministry of Education, Science and Technological Development of the Republic of Serbia (Contract No. 451-03-68/2020-14/200135, Contract No. 451-03-68/2020-14/200287, Project of Technological cooperation Serbia-China #I-1).

References

Aganovic K, Smetana S, Grauwet T et al (2017) Pilot scale thermal and alternative pasteurization of tomato and watermelon juice: an energy comparison and life cycle assessment. J Clean Prod 141:514–525. https://doi.org/10.1016/j.jclepro.2016.09.015

Alexandri M, Blanco-Catalá J, Schneider R et al (2020) High L(+)-lactic acid productivity in continuous fermentations using bakery waste and lucerne green juice as renewable substrates. Bioresour Technol 316:123949. https://doi.org/10.1016/j.biortech.2020.123949

Bahry H, Abdalla R, Pons A et al (2019) Optimization of lactic acid production using immobilized Lactobacillus Rhamnosus and carob pod waste from the Lebanese food industry. J Biotechnol 306:81–88. https://doi.org/10.1016/j.jbiotec.2019.09.017

Baltazar-Y-Jimenez A, Bismarck A (2007) Surface modification of lignocellulosic fibres in atmospheric air pressure plasma. Green Chem 9:1057–1066. https://doi.org/10.1039/b618398k

Baltazar-y-Jimenez A, Bistritz M, Schulz E, Bismarck A (2008) Atmospheric air pressure plasma treatment of lignocellulosic fibres: impact on mechanical properties and adhesion to cellulose acetate butyrate. Compos Sci Technol 68:215–227. https://doi.org/10.1016/j.compscitech.2007.04.028

Banerjee G, Chattopadhyay P (2019) Vanillin biotechnology: the perspectives and future. J Sci Food Agric 99:499–506

Barba FJ, Zhu Z, Koubaa M et al (2016) Green alternative methods for the extraction of antioxidant bioactive compounds from winery wastes and by-products: a review. Trends Food Sci Technol 49:96–109

Benoit M, Rodrigues A, De Oliveira VK et al (2012) Combination of ball-milling and non-thermal atmospheric plasma as physical treatments for the saccharification of microcrystalline cellulose. Green Chem 14:2212–2215. https://doi.org/10.1039/c2gc35710k

Bernardo MP, Coelho LF, Sass DC, Contiero J (2016) L-(+)-Lactic acid production by Lactobacillus rhamnosus B103 from dairy industry waste. Braz J Microbiol 47:640–646. https://doi.org/10.1016/j.bjm.2015.12.001

Bernstad A, Jansen J (2012) Review of comparative LCAs of food waste management systems–current status and potential improvements. Waste Manag 32:2439–2455

Bohnsack C, Ternes W, Büsing A, Drotleff AM (2011) Tocotrienol levels in sieving fraction extracts of brewer's spent grain. Eur Food Res Technol 232:563–573. https://doi.org/10.1007/s00217-010-1419-z

Capper JL, Berger L, Brashears MM, Jensen H (2013) Animal feed vs. human food: challenges and opportunities in sustaining animal agriculture toward 2050. Counc Agric Sci Technol 53:1–16

Chen H, Chen B, Su Z et al (2020) Efficient lactic acid production from cassava bagasse by mixed culture of Bacillus coagulans and lactobacillus rhamnosus using stepwise pH controlled simultaneous saccharification and co-fermentation. Ind Crop Prod 146:112175. https://doi.org/10.1016/j.indcrop.2020.112175

Coelho LF, de Lima CJB, Bernardo MP et al (2010) Improvement of L(+)-lactic acid production from cassava wastewater by lactobacillus rhamnosus B 103. J Sci Food Agric 90:1944–1950. https://doi.org/10.1002/jsfa.4039

Corrado S, Sala S (2018) Food waste accounting along global and European food supply chains: state of the art and outlook. Waste Manag 79:120–131

Corrado S, Caldeira C, Eriksson M et al (2019) Food waste accounting methodologies: challenges, opportunities, and further advancements. Glob Food Sec 20:93–100

Cristóbal J, Caldeira C, Corrado S, Sala S (2018) Techno-economic and profitability analysis of food waste biorefineries at European level. Bioresour Technol 259:244–252. https://doi.org/10.1016/j.biortech.2018.03.016

de la Torre I, Ladero M, Santos VE (2020) D-lactic acid production from orange waste enzymatic hydrolysates with L. delbrueckii cells in growing and resting state. Ind Crops Prod 146:112176. https://doi.org/10.1016/j.indcrop.2020.112176

Demichelis F, Pleissner D, Fiore S et al (2017) Investigation of food waste valorization through sequential lactic acid fermentative production and anaerobic digestion of fermentation residues. Bioresour Technol 241:508–516. https://doi.org/10.1016/j.biortech.2017.05.174

Dey P, Sikder J, Roy S, Pal P (2012) Fermentative lactic acid production from a renewable carbon source under response surface optimized conditions without alkali addition: a membrane-based green approach. Clean Techn Environ Policy 14:827–835. https://doi.org/10.1007/s10098-011-0448-z

Dimou C, Vlysidis A, Kopsahelis N et al (2016) Techno-economic evaluation of wine lees refining for the production of value-added products. Biochem Eng J 116:157–165. https://doi.org/10.1016/j.bej.2016.09.004

Djukić-Vuković AP, Mojović LV, Jokić BM et al (2013) Lactic acid production on liquid distillery stillage by lactobacillus rhamnosus immobilized onto zeolite. Bioresour Technol 135:454–458. https://doi.org/10.1016/j.biortech.2012.10.066

Djukić-Vuković AP, Mojović LV, Semenčenko VV et al (2015) Effective valorisation of distillery stillage by integrated production of lactic acid and high quality feed. Food Res Int 73:75–80. https://doi.org/10.1016/j.foodres.2014.07.048

Djukić-Vuković AP, Jokić BM, Kocić-Tanackov SD et al (2016) Mg-modified zeolite as a carrier for lactobacillus rhamnosus in L(+) lactic acid production on distillery wastewater. J Taiwan Inst Chem Eng 59:262–266. https://doi.org/10.1016/j.jtice.2015.07.035

Djukic-Vukovic A, Lazovic S, Mladenovic D et al (2019) Non-thermal plasma and ultrasound-assisted open lactic acid fermentation of distillery stillage. Environ Sci Pollut Res Int 26:35543–35554. https://doi.org/10.1007/s11356-019-04894-9

Djukić-Vuković A, Mladenović D, Ivanović J et al (2019) Towards sustainability of lactic acid and poly-lactic acid polymers production. Renew Sust Energ Rev 108:238–252. https://doi.org/10.1016/J.RSER.2019.03.050

Elmekawy A, Diels L, De Wever H, Pant D (2013) Valorization of cereal based biorefinery byproducts: reality and expectations. Environ Sci Technol 47:9014–9027

Fache M, Boutevin B, Caillol S (2016) Vanillin production from lignin and its use as a renewable chemical. ACS Sustain Chem Eng 4:35–46. https://doi.org/10.1021/acssuschemeng.5b01344

FAO (2011) Global food losses and food waste – extent, causes and prevention. https://www.google.com/search?q=Food+and+Agriculture+Organization+of+the+United+Nations+(FAO)%2C+2011.+Global+Food+Losses+and+Food+Waste+–+Extent%2C+Causes+and+Prevention&rlz=1C1GCEA_enRS871RS871&oq=Food+and+Agriculture+Organization+of+the+United+Nations+. Accessed 29 Mar 2020

FAO (2019) The state of food and agriculture 2019 (SOFA): moving forward on food loss and waste reduction. In: Food and Agriculture Organization of the United Nations. FAO, Washington, D.C

FAO (2020) COVID-19 pandemic – impact on food and agriculture | FAO | Food and Agriculture Organization of the United Nations.. http://www.fao.org/2019-ncov/q-and-a/impact-on-food-and-agriculture/en/. Accessed 15 Feb 2021

Galanakis CM (2012) Recovery of high added-value components from food wastes: conventional, emerging technologies and commercialized applications. Trends Food Sci Technol 26:68–87

Galanakis CM, Castro-Muñoz R, Cassano A, Conidi C (2016) Recovery of high-added-value compounds from food waste by membrane technology. Membrane Technologies for Biorefining. Elsevier Inc., In, pp 189–215

Girotto F, Alibardi L, Cossu R (2015) Food waste generation and industrial uses: a review. Waste Manag 45:32–41. https://doi.org/10.1016/j.wasman.2015.06.008

Goula AM, Lazarides HN (2015) Integrated processes can turn industrial food waste into valuable food by-products and/or ingredients: the cases of olive mill and pomegranate wastes. J Food Eng 167:45–50. https://doi.org/10.1016/j.jfoodeng.2015.01.003

Gustavsson J, Cederberg C, Sonesson U, Emanuelsson A (2013) The methodology of the FAO study: "Global Food Losses and Food Waste – extent, causes and prevention"; -FAO, 2011 By SIK -The Swedish Institute for Food and Biotechnology

Hamelin L, Thomsen M, Alexa Teigiserova D, Hamelin L (2019) Review of high-value food waste and food residues biorefineries with focus on unavoidable wastes from processing

Extraction and reuse of phosphorous in wastewater View project Macroalgae biorefining into high value products View project Review of high-value food waste and food residues biorefineries with focus on unavoidable wastes from processing. Elsevier https://doi.org/10.1016/j.resconrec.2019.05.003

Herrero M, del Sánchez-Camargo AP, Cifuentes A, Ibáñez E (2015) Plants, seaweeds, microalgae and food by-products as natural sources of functional ingredients obtained using pressurized liquid extraction and supercritical fluid extraction. TrAC Trends Anal Chem 71:26–38

Ikram S, Huang LY, Zhang H et al (2017) Composition and nutrient value proposition of brewers spent grain. J Food Sci 82:2232–2242

Jérôme F, Chatel G, De Oliveira VK (2016) Depolymerization of cellulose to processable glucans by non-thermal technologies. Green Chem 18:3903–3913

Jiang J, Gong C, Wang J et al (2014) Effects of ultrasound pre-treatment on the amount of dissolved organic matter extracted from food waste. Bioresour Technol 155:266–271. https://doi.org/10.1016/j.biortech.2013.12.064

Jones MP, Lawrie AC, Huynh TT et al (2019) Agricultural by-product suitability for the production of chitinous composites and nanofibers utilising Trametes versicolor and Polyporus brumalis mycelial growth. Process Biochem 80:95–102. https://doi.org/10.1016/j.procbio.2019.01.018

Jović J, Buntić A, Radovanović N et al (2018) Lignin-degrading abilities of novel autochthonous fungal isolates Trametes hirsuta F13 and stereum gausapatum F28. Food Technol Biotechnol 56:354–365. https://doi.org/10.17113/ftb.56.03.18.5348

Kammerer DR, Kammerer J, Valet R, Carle R (2014) Recovery of polyphenols from the by-products of plant food processing and application as valuable food ingredients. Food Res Int 65:2–12. https://doi.org/10.1016/j.foodres.2014.06.012

Khan I, Tango CN, Miskeen S et al (2017) Hurdle technology: a novel approach for enhanced food quality and safety – a review. Food Control 73:1426–1444

Kong X, Zhang B, Hua Y et al (2019) Efficient L-lactic acid production from corncob residue using metabolically engineered thermo-tolerant yeast. Bioresour Technol 273:220–230. https://doi.org/10.1016/j.biortech.2018.11.018

Koutinas A, Vlysidis A, Pleissner D et al (2014) Valorization of industrial waste and by-product streams via fermentation for the production of chemicals and biopolymers. Chem Soc Rev 43:2587–2627. https://doi.org/10.1039/c3cs60293a

Kumari B, Tiwari BK, Walsh D et al (2019) Impact of pulsed electric field pre-treatment on nutritional and polyphenolic contents and bioactivities of light and dark brewer's spent grains. Innov Food Sci Emerg Technol 54:200–210. https://doi.org/10.1016/j.ifset.2019.04.012

Kwan TH, Hu Y, Lin CSK (2016) Valorisation of food waste via fungal hydrolysis and lactic acid fermentation with lactobacillus casei Shirota. Bioresour Technol 217:129–136. https://doi.org/10.1016/j.biortech.2016.01.134

Leung CCJ, Cheung ASY, Zhang AYZ et al (2012) Utilisation of waste bread for fermentative succinic acid production. Biochem Eng J 65:10–15. https://doi.org/10.1016/j.bej.2012.03.010

Li J, Sun J, Wu B, He B (2017) Combined utilization of nutrients and sugar derived from wheat bran for D-Lactate fermentation by Sporolactobacillus inulinus YBS1-5. Bioresour Technol 229:33–38. https://doi.org/10.1016/j.biortech.2016.12.101

Liang S, McDonald AG, Coats ER (2014) Lactic acid production with undefined mixed culture fermentation of potato peel waste. Waste Manag 34:2022–2027. https://doi.org/10.1016/j.wasman.2014.07.009

Liu Q, Wang S, Zhi JF et al (2013) Efficient production of lactic acid from sweet sorghum juice by a newly isolated lactobacillus salivarius CGMCC 7.75. Indian J Microbiol 53:332–336. https://doi.org/10.1007/s12088-013-0377-0

Loman A, Ju LK (2016) Soybean carbohydrate as fermentation feedstock for production of biofuels and value-added chemicals. Process Biochem 51:1046–1057

Lopez Barrera E, Hertel T (2020) Global food waste across the income spectrum: implications for food prices, production and resource use. Food Policy 98:101874. https://doi.org/10.1016/j.foodpol.2020.101874

López-Gómez JP, Alexandri M, Schneider R et al (2020) Organic fraction of municipal solid waste for the production of L-lactic acid with high optical purity. J Clean Prod 247:119165. https://doi.org/10.1016/j.jclepro.2019.119165
López-Linares JC, García-Cubero MT, Lucas S et al (2019) Microwave assisted hydrothermal as greener pretreatment of brewer's spent grains for biobutanol production. Chem Eng J 368:1045–1055. https://doi.org/10.1016/j.cej.2019.03.032
Luft L, Confortin TC, Todero I et al (2018) Transformation of residual starch from brewer's spent grain into fermentable sugars using supercritical technology. J Supercrit Fluids 140:85–90. https://doi.org/10.1016/j.supflu.2018.06.006
Lyko H, Deerberg G, Weidner E (2009) Coupled production in biorefineries-combined use of biomass as a source of energy, fuels and materials. J Biotechnol 142:78–86. https://doi.org/10.1016/j.jbiotec.2009.03.016
Mahboubi A, Ferreira JA, Taherzadeh MJ, Lennartsson PR (2017) Value-added products from dairy waste using edible fungi. Waste Manag 59:518–525. https://doi.org/10.1016/j.wasman.2016.11.017
Mahnič-Kalamiza S, Vorobiev E, Miklavčič D (2014) Electroporation in food processing and biorefinery. J Membr Biol 247:1279–1304. https://doi.org/10.1007/s00232-014-9737-x
Martinez GA, Rebecchi S, Decorti D et al (2016) Towards multi-purpose biorefinery platforms for the valorisation of red grape pomace: production of polyphenols, volatile fatty acids, polyhydroxyalkanoates and biogas. Green Chem 18:261–270. https://doi.org/10.1039/c5gc01558h
Martín-García B, Tylewicz U, Verardo V et al (2020) Pulsed electric field (PEF) as pre-treatment to improve the phenolic compounds recovery from brewers' spent grains. Innov Food Sci Emerg Technol 64:102402. https://doi.org/10.1016/j.ifset.2020.102402
Mladenović DD, Djukić-Vuković AP, Kocić-Tanackov SD et al (2016) Lactic acid production on a combined distillery stillage and sugar beet molasses substrate. J Chem Technol Biotechnol 91:2474–2479. https://doi.org/10.1002/jctb.4838
Mladenović D, Pejin J, Kocić-Tanackov S et al (2018a) Enhanced lactic acid production by adaptive evolution of lactobacillus paracasei on agro-industrial substrate. Appl Biochem Biotechnol. https://doi.org/10.1007/s12010-018-2852-x
Mladenović D, Pejin J, Kocić-Tanackov S et al (2018b) Lactic acid production on molasses enriched potato stillage by Lactobacillus paracasei immobilized onto agro-industrial waste supports. Ind Crop Prod 124:142–148. https://doi.org/10.1016/j.indcrop.2018.07.081
Mladenović D, Djukić-Vuković A, Stanković M et al (2019) Bioprocessing of agro-industrial residues into lactic acid and probiotic enriched livestock feed. J Sci Food Agric. https://doi.org/10.1002/jsfa.9759
Mojović L, Pejin D, Rakin M et al (2012) How to improve the economy of bioethanol production in Serbia. Renew Sust Energ Rev 16:6040–6047. https://doi.org/10.1016/j.rser.2012.07.001
Morone P, Koutinas A, Gathergood N et al (2019) Food waste: challenges and opportunities for enhancing the emerging bio-economy. J Clean Prod 221:10–16
Murariu M, Dubois P (2016) PLA composites: from production to properties. Adv Drug Deliv Rev 107:17–46. https://doi.org/10.1016/j.addr.2016.04.003
Mussatto SI, Moncada J, Roberto IC, Cardona CA (2013) Techno-economic analysis for brewer's spent grains use on a biorefinery concept: the Brazilian case. Bioresour Technol 148:302–310. https://doi.org/10.1016/j.biortech.2013.08.046
Nakano S, Ugwu CU, Tokiwa Y (2012) Efficient production of D-(−)-lactic acid from broken rice by lactobacillus delbrueckii using ca(OH)2 as a neutralizing agent. Bioresour Technol 104:791–794. https://doi.org/10.1016/j.biortech.2011.10.017
Nampoothiri KM, Nair NR, John RP (2010) An overview of the recent developments in polylactide (PLA) research.pdf. Bioresour Technol 101:8493–8501
Nancib A, Nancib N, Boudrant J (2009) Production of lactic acid from date juice extract with free cells of single and mixed cultures of lactobacillus casei and Lactococcus lactis. World J Microbiol Biotechnol 25:1423–1429. https://doi.org/10.1007/s11274-009-0029-z

Neu AK, Pleissner D, Mehlmann K et al (2016) Fermentative utilization of coffee mucilage using Bacillus coagulans and investigation of down-stream processing of fermentation broth for optically pure l(+)-lactic acid production. Bioresour Technol 211:398–405. https://doi.org/10.1016/j.biortech.2016.03.122

Nguyen CM, Kim JS, Nguyen TN et al (2013) Production of l- and d-lactic acid from waste Curcuma longa biomass through simultaneous saccharification and cofermentation. Bioresour Technol 146:35–43. https://doi.org/10.1016/j.biortech.2013.07.035

Niemi P, Martins D, Buchert J, Faulds CB (2013) Pre-hydrolysis with carbohydrases facilitates the release of protein from brewer's spent grain. Bioresour Technol 136:529–534. https://doi.org/10.1016/j.biortech.2013.03.076

Papargyropoulou E, Lozano R, Steinberger JK et al (2014) The food waste hierarchy as a framework for the management of food surplus and food waste. J Clean Prod 76:106–115. https://doi.org/10.1016/j.jclepro.2014.04.020

Pappalardo G, Cerroni S, Nayga RM, Yang W (2020) Impact of Covid-19 on household food waste: the case of Italy. Front Nutr 7:585090. https://doi.org/10.3389/fnut.2020.585090

Paz A, da Silva SS, Vallejo M et al (2018) Using brewer's spent grain to formulate culture media for the production of bacteriocins using Patagonian strains. LWT 96:166–174. https://doi.org/10.1016/j.lwt.2018.05.027

Pejin J, Radosavljević M, Mojović L et al (2015) The influence of calcium-carbonate and yeast extract addition on lactic acid fermentation of brewer's spent grain hydrolysate. Food Res Int 73:31–37. https://doi.org/10.1016/j.foodres.2014.12.023

Pfaltzgraff LA, De Bruyn M, Cooper EC et al (2013) Food waste biomass: a resource for high-value chemicals. Green Chem 15:307–314

Pleissner D, Kwan TH, Lin CSK (2014) Fungal hydrolysis in submerged fermentation for food waste treatment and fermentation feedstock preparation. Bioresour Technol 158:48–54. https://doi.org/10.1016/j.biortech.2014.01.139

Pleissner D, Neu AK, Mehlmann K et al (2016) Fermentative lactic acid production from coffee pulp hydrolysate using Bacillus coagulans at laboratory and pilot scales. Bioresour Technol 218:167–173. https://doi.org/10.1016/j.biortech.2016.06.078

Pleissner D, Demichelis F, Mariano S et al (2017) Direct production of lactic acid based on simultaneous saccharification and fermentation of mixed restaurant food waste. J Clean Prod 143:615–623. https://doi.org/10.1016/j.jclepro.2016.12.065

Pradal D, Vauchel P, Decossin S et al (2016) Kinetics of ultrasound-assisted extraction of antioxidant polyphenols from food by-products: extraction and energy consumption optimization. Ultrason Sonochem 32:137–146. https://doi.org/10.1016/j.ultsonch.2016.03.001

Puértolas E, Barba FJ (2016) Electrotechnologies applied to valorization of by-products from food industry: Main findings, energy and economic cost of their industrialization. Food Bioprod Process 100:172–184

Putnik P, Bursać Kovačević D, Režek Jambrak A et al (2017) Innovative "green" and novel strategies for the extraction of bioactive added value compounds from citrus wastes—a review. Molecules 22:680. https://doi.org/10.3390/molecules22050680

Qiu Z, Gao Q, Bao J (2017) Constructing xylose-assimilating pathways in Pediococcus acidilactici for high titer D-lactic acid fermentation from corn stover feedstock. Bioresour Technol 245:1369–1376. https://doi.org/10.1016/j.biortech.2017.05.128

Qureshi AS, Khushk I, Naqvi SR et al (2017) Fruit waste to energy through open fermentation. Energy Procedia, In, pp 904–909

Radosavljević M, Pejin J, Pribić M et al (2019) Utilization of brewing and malting by-products as carrier and raw materials in l-(+)-lactic acid production and feed application. Appl Microbiol Biotechnol 103:3001–3013. https://doi.org/10.1007/s00253-019-09683-5

Raso J, Frey W, Ferrari G et al (2016) Recommendations guidelines on the key information to be reported in studies of application of PEF technology in food and biotechnological processes. Innov Food Sci Emerg Technol 37:312–321. https://doi.org/10.1016/j.ifset.2016.08.003

Ravindran R, Sarangapani C, Jaiswal S et al (2019) Improving enzymatic hydrolysis of brewer spent grain with nonthermal plasma. Bioresour Technol 282:520–524. https://doi.org/10.1016/j.biortech.2019.03.071
Ribau Teixeira M, Guarda EC, Freitas EB et al (2020) Valorization of raw brewers' spent grain through the production of volatile fatty acids. New Biotechnol 57:4–10. https://doi.org/10.1016/j.nbt.2020.01.007
Roe BE, Bender K, Qi D (2021) The impact of COVID-19 on consumer food waste. Appl Econ Perspect Policy 43:401–411. https://doi.org/10.1002/aepp.13079
Schieber A, Stintzing FC, Carle R (2001) By-products of plant food processing as a source of functional compounds - recent developments. Trends Food Sci Technol 12:401–413. https://doi.org/10.1016/S0924-2244(02)00012-2
Shorstkii I, Mirshekarloo MS, Koshevoi E (2017) Application of pulsed electric field for oil extraction from sunflower seeds: electrical parameter effects on oil yield. J Food Process Eng 40:7. https://doi.org/10.1111/jfpe.12281
Silveira MS, Fontes CPML, Guilherme AA et al (2012) Cashew apple juice as substrate for lactic acid production. Food Bioprocess Technol 5:947–953. https://doi.org/10.1007/s11947-010-0382-9
Slorach PC, Jeswani HK, Cuéllar-Franca R, Azapagic A (2019) Environmental and economic implications of recovering resources from food waste in a circular economy. Sci Total Environ 693:133516. https://doi.org/10.1016/j.scitotenv.2019.07.322
Strøm-Andersen N (2020) Innovation and by-product valorization: a comparative analysis of the absorptive capacity of food processing firms. J Clean Prod 253:119943. https://doi.org/10.1016/j.jclepro.2019.119943
Stuart T (2009) Waste : uncovering the global food scandal. W.W. Norton & Co
Talekar S, Patti AF, Vijayraghavan R, Arora A (2018) An integrated green biorefinery approach towards simultaneous recovery of pectin and polyphenols coupled with bioethanol production from waste pomegranate peels. Bioresour Technol 266:322–334. https://doi.org/10.1016/j.biortech.2018.06.072
Thakur A, Panesar PS, Saini MS (2017) L(+)-Lactic acid production by immobilized lactobacillus casei using low cost agro-industrial waste as carbon and nitrogen sources. Waste Biomass Valoriz:1–11. https://doi.org/10.1007/s12649-017-0129-1
Tomaszewska J, Bieliński D, Binczarski M et al (2018) Products of sugar beet processing as raw materials for chemicals and biodegradable polymers. RSC Adv 8:3161–3177. https://doi.org/10.1039/c7ra12782k
Trivedi J, Bhonsle AK, Atray N (2020) Processing food waste for the production of platform chemicals. In: Refining Biomass Residues for Sustainable Energy and Bioproducts. Elsevier Science & Technology, San Diego, pp 427–448
Troiano D, Orsat V, Dumont MJ (2020) Status of filamentous fungi in integrated biorefineries. Renew Sustain Energy Rev 117:109472
Unrean P (2018) Optimized feeding schemes of simultaneous saccharification and fermentation process for high lactic acid titer from sugarcane bagasse. Ind Crop Prod 111:660–666. https://doi.org/10.1016/j.indcrop.2017.11.043
van Eck NJ, Waltman L (2010) Software survey: VOSviewer, a computer program for bibliometric mapping. Scientometrics 84:523–538. https://doi.org/10.1007/s11192-009-0146-3
van Eck NJ, Waltman L (2014) Visualizing bibliometric networks. In: Measuring Scholarly Impact. Springer International Publishing, Cham, pp 285–320
van Hal O, Weijenberg AAA, de Boer IJM, van Zanten HHE (2019) Accounting for feed-food competition in environmental impact assessment: towards a resource efficient food-system. J Clean Prod 240:118241. https://doi.org/10.1016/j.jclepro.2019.118241
Wang L, Zhao B, Liu B et al (2010) Efficient production of l-lactic acid from corncob molasses, a waste by-product in xylitol production, by a newly isolated xylose utilizing Bacillus sp. strain. Bioresour Technol 101:7908–7915. https://doi.org/10.1016/j.biortech.2010.05.031

Wang Y, Cao W, Luo J et al (2019) One step open fermentation for lactic acid production from inedible starchy biomass by thermophilic Bacillus coagulans IPE22. Bioresour Technol 272:398–406. https://doi.org/10.1016/j.biortech.2018.10.043

Wijngaard H, Hossain MB, Rai DK, Brunton N (2012) Techniques to extract bioactive compounds from food by-products of plant origin. Food Res Int 46:505–513. https://doi.org/10.1016/j.foodres.2011.09.027

Wouters P, Alvarez I, Raso J (2001) Critical factors determining inactivation kinetics by pulsed electric field food processing. Trends Food Sci Technol 12:112–121

Yadav N, Nain L, Khare SK (2021) One-pot production of lactic acid from rice straw pretreated with ionic liquid. Bioresour Technol 323:124563. https://doi.org/10.1016/j.biortech.2020.124563

Ye L, Bin HMS, Li Z, Wu JC (2014) Simultaneous detoxification, saccharification and co-fermentation of oil palm empty fruit bunch hydrolysate for l-lactic acid production by Bacillus coagulans JI12. Biochem Eng J 83:16–21. https://doi.org/10.1016/j.bej.2013.12.005

Zanini S, Riccardi C, Canevali C et al (2005) Modifications of lignocellulosic fibers by Ar plasma treatments in comparison with biological treatments. Surf Coat Technol 200:556–560. https://doi.org/10.1016/j.surfcoat.2005.01.090

Zhang J, Zang L (2016) Enhancement of biohydrogen production from brewers' spent grain by calcined-red mud pretreatment. Bioresour Technol 209:73–79. https://doi.org/10.1016/j.biortech.2016.02.110

Zhang HQ, Barbosa-Cánovas G V., Balasubramaniam VM, et al (2011) Nonthermal processing technologies for food, Wiley-Blackwell, Ames, USA, ISBN: 9780813816685

Zhang AYZ, Sun Z, Leung CCJ et al (2013) Valorisation of bakery waste for succinic acid production. Green Chem 15:690–695. https://doi.org/10.1039/c2gc36518a

Zhang Y, Chen X, Qi B et al (2014) Improving lactic acid productivity from wheat straw hydro lysates by membrane integrated repeated batch fermentation under non-sterilized conditions. Bioresour Technol 163:160–166. https://doi.org/10.1016/j.biortech.2014.04.038

Zhang M, Xie L, Yin Z et al (2016) Biorefinery approach for cassava-based industrial wastes: current status and opportunities. Bioresour Technol 215:50–62

Zheng Y, Wang Y, Zhang J, Pan J (2016) Using tobacco waste extract in pre-culture medium to improve xylose utilization for l-lactic acid production from cellulosic waste by Rhizopus oryzae. Bioresour Technol 218:344–350. https://doi.org/10.1016/j.biortech.2016.06.071

Chapter 25
Strategies for Commercializing Scientific Results and Combining Separate Processes Into Complex Technologies

Umm E. Ruman, Muhammad Zubair, Mujahid Farid, Sajjad Hussain Summra, Muhammad Faizan Nazar, and Muhammad Asif Hanif

1 Introduction

1.1 What Is Scientific Data?

Scientific data is scholarly, public access journal for the elucidation of scientifically valuable datasets and the research that proceed the sharing and reuse of scientific data. It is the information that has been gathered by using particular methods for studying and probing some discrete purpose. Experiments are performed in the lab under certain conditions and the results are recorded in the form of scientific data. Scientific data is the outcome of scientific research, which is performed by involving scientific methods. It refers to the information which is established on research performed by the scientists and then published in scholarly reviewed journal. The tremendous data has been created using scientific equipment and computer reproductions. The new scientific instruments have wonderful accuracy and precision so the data quality as well as data quantity is rapidly improving. The data magnitude is getting twice every year. The data stored previously, derived from various researches is the foundation for the development of new science (Gray et al. 2005). Scientific data basically publishes Data Descriptors, which is a new type of publication that

U. E. Ruman · M. Zubair (✉) · S. H. Summra
Sustainable AgriFoodtech Innovation & Research (SAFIR), Arras, France
e-mail: muhammad.zubair@uog.edu.pk

M. Farid
Department of Environmental Sciences, University of Gujrat, Gujrat, Pakistan

M. F. Nazar
Department of Chemistry, University of Education, Multan Campus, Multan, Pakistan

M. A. Hanif
Department of Chemistry, University of Agriculture, Faisalabad, Pakistan

A. Režek Jambrak (ed.), *Nonthermal Processing in Agri-Food-Bio Sciences*, Food Engineering Series, https://doi.org/10.1007/978-3-030-92415-7_25

provides the detailed explanations of research datasets and also illustrate those methods which are used to collect the data. The main purpose of data descriptors is helping others to reuse data, instead of testing hypothesis or introducing new interpretations. Wide range of natural sciences can make submissions to scientific data, including scientific data from life, biomedical and environmental science. The submissions involve small or big data from new experiments or from the assemblage of existing data. The complete reporting of scientific data is based on the exactly tracked data from the origin of scientific research (Usama Fayyad 1996).

1.2 Innovation and Novelty in Scientific Data

Innovation is the process with the help of which new products, new processes or methods can be created. It is the application of better solutions to fulfil the new requirements or already existing needs. Novelty in scientific data refers to the elements that are new in research, it also includes new methods or new observation which result in the discovery of new knowledge. In order to find the novelty in the areas of research, researcher should undergo a thorough literature review to find what has been studied what are the gaps which are needed to be clarified (Cohen 2017). There are many topics in literature that are studied but still there is confliction. Researching on this can lead to the novelty in scientific data. So novelty is simply an additional change in the existing products to differentiate them from other market products in this era of competition. One common example that how novelty in scientific data leads to commercialization is of *Toothpaste*. In the past toothpaste manufacturers have one or two types of toothpastes, but due to competition in market, the manufacturers reviewed the scientific data and introduce some novel features in toothpaste such as fluoride, tarter control, whitening and stripes etc. But this improvement was temporary, because these changes didn't change the foundation of dental health industries, these only improved the products for a while. The manufacturers confused the novelty with innovation (Elg 2014). Innovation is long search process before and after the new product or methodology has been introduced. To focus on the competitiveness for long duration at national or regional level is the strategy of innovation. Innovation in scientific data changes the way, the industries execute. Thus, innovation is the process which involves constant interaction with the present and future customers, with suppliers, with competitors and with the academic researchers (Elg 2014).

One example that how innovation in scientific data leads to commercialization, Michael Krauser developed a sensor that can measure glucose level. This innovation has been commercialized in the form of instrument called Free Style Liber reader (A freestyle libre is a flash glucose monitoring system measures your sugar (glucose) levels continuously throughout the day. It can help you and your diabetes team see: if your sugar levels are going up or down. How your sugar levels change over time) or it is in the form of Free Style Liber Link app installed in the smart phones. Across the bluetooth enabled sensor. This sensor enables the users to set alarms when their glucose level is too low or high, these alarms are extremely helpful and users can maintain their glucose level by eating or by injecting insulin. This innovation is a great solution for diabetics, because they no more need finger sticks for checking glucose level (Staff 2019).

1.3 Validity of Scientific Data and Its Approval for Commercialization

Validity in scientific data is mainly concerned with the the accuracy and sincerity of scientific findings. A lot of scientific data that has been collected during the scientific research is not valid. The scientific research firstly undergo testing to prove it for industrial products. Hence, validity is very important for commercialization (Brink 1993).

1.3.1 Validity of Scientific Data

Validity refers to the evidence presented to support or refute the meaning or interpretation assigned to results of scientific research. It is often thought that the results of scientific research are often valid because the study is scientific, but unluckily that does not always happen. Researchers who perform scientific studies are stimulated by external factors such as to get desired results, to receive funding and to get published their findings. As a result number of scientific studies are unauthentic (Deterdin 2019). Dr. Ioannidis who is one of the [1]meta-researchers in medical field reported that scientific studies that get published are those with noticeable outcomes, which when studied under diligent conditions disintegrate under the burden of contradictory data. Dr. Ioannidis has spent his career in exposing the validity of scientific studies. He found that many biomedical researchers conclude the results of their research, which in turn are used by many doctors, when they prescribe antibiotics or other medications are often invalid and imprecise.(Freedman 2010). In scientific data or scientific research we came across two types of validity i.e. internal validity or external validity. Internal validity refers to the design of the scientific method or procedure we are adopting to get scientific data is standard or making logical sense, while the external validity is about whether the conclusion we have drawn from the experiment is real explanation for our procedure in this world and is there any other alternative explanation to that result. Most scientists and researchers remain successful at improving the problems related to internal validity, but external validity is somewhat difficult to achieve (2016).

1.3.2 Approval of Scientific Data for Commercialization

Commercialization of scientific data is not different from that of commercializing anything, but to put scientific data into something practical has been found somewhat difficult. The reason is that to bring something in the form of product in market is harder than to design a product. There are some simple rules that should must be followed to bring scientific research into commercialization (Fletcher and Bourne 2012)

- **There is no single path to commercialization**
  - There are many ways to bring your scientific research into commercialization including licensing, incubation, royalties and in-house development. There are many routes to go to market from laboratory bench, so commercialization seems like a business process. Most routes are fundamentally mechanistic, some work and some don't.
- **Make decision how much of your research you want to give**
  - You have the opportunity that you can give your scientific research completely and nothing to do with the upcoming commercialization or on the other end you can completely involve in the company that is commercializing your research.
- **Be realistic, Make the difference between Basic research and the Development of research**
  - There is huge difference between basic research and the development of research. Most commonly this development is done by the organization which is commercializing the scientific research and this development is considered the midpoint between the academic research and its commercialization. This is one of the important steps needed for the approval of scientific research.
- **Consider the difference between 'the Need' and 'the Want'**
  - The academic researcher should know that the product that will result from the commercialization of his scientific research will be expressed by people as a want or need. People often prefer the wants but they buy needs.
- **Make your research outcomes clear and comprehensible**
  - The people who will claim to fund your research for commercialization will be businessmen not the scientists so you should make an elevator statement from your research outcomes that will set the base why one should purchase your research.
- **Ultimate Peer Review is the key for the approval of scientific data**
  - Peer review of the research publications estimates the novelty, accuracy of scientific methods, reproducibility and its value to the society. The example of Henri Poincare´ is beneficial to quote there, the first version of his book on ' The Three Body Problem' contained a serious mistake that was caught and removed during peer review.

1.4 How Entrepreneur Work to Commercialize Scientific Data

According to Giordan, an entrepreneur should possess the properties i.e. a person should be trained and skilled and should have high vocabulary. But whether the person can do the range of work that is required to be a successful entrepreneur is a different story. Albers explains, entrepreneurs need to ask a core set of questions about what type of scientific research they want to commercialize: Is there a market need? If so, does the technology provide the solution to that need? Does any other researcher have a better solution? And finally, how much money is required for bringing the technology to market and how much will they earn from the commercialization of that research? (Morrissey 2012). Universities are commonly known as the source of technological innovation and play an important role in transforming the university invented technology to the commercial product in the market. Now entrepreneurs start up companies and make profit based on these innovations. The intelligence, skills and ability of an academic entrepreneurs to recognize technological opportunity, to utilize commercial opportunity and to develop technological advanced product are primary characteristics that highlight their role in technology entrepreneurship. There are various [1]challenges that entrepreneur has to face while bringing the research into market, so the attitude of entrepreneur should be creative, innovative, risk taking and should be self -confident (Abd Rahim et al. 2015). Entrepreneur has to work through four different stages to commercialize an innovation in research (Yimamu 2018) (Fig. 25.1):

- **To pick up an innovation**

  It may include one's own innovation or the entrepreneurs look for the innovation or for the market opportunity.
- **Triggering Event**
  - It involves different aspects such as motivation to take a decision to start business, business plans, locating different resources and risk assessment.
- **Implementation**
  - It includes setting up the business strategies and then executing business, providing resources and managing progress.
- **Growth**
  - It is the time to maximize profit, the stage to have rewards, and the stage to continuously grow your venture with involving other opportunities.

1.4.1 Patent and Commercialization

A patent is a form of intellectual property that gives its owner the legal right to exclude others from making, using or selling an invention for a particular time period. One main reason that why inventors are awarded patents is that they commercialize the invention and encourage R and D investments. All inventions are not

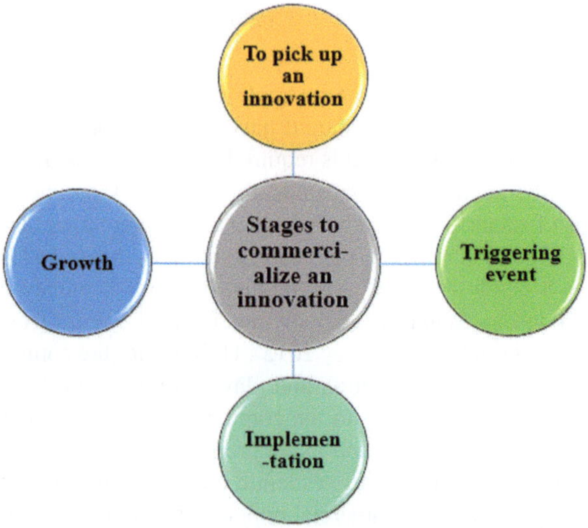

Fig. 25.1 Different stages for the commercialization of an innovation

able to get patent, because it is difficult to find these new ideas, products and developments among the companies where all patents are registered and if the inventions are found it is difficult to find out whether these are sufficient improvements qualifying as inventions (Svensson 2012). Commercialization of patent means that the owner of the patent has either sold the patent, licensed the patent or has introduced the new product based on the same patent. Commercialization simply means that patent has provided some income to the owner. Each patent has at least one owner and one applying firm. The inventors and the applying firm can be the owner of patent. Sometimes the inventors are only as employee in the firm that owns the patent (Svensson 2012).

World Wide Recognized Patents
The patent protection treaty (PCT) helps the applicants in finding out the worldwide patent protection for their inventions. This treaty helps the patent offices in making decisions for granting patents. It encourages the public approach to the wealth of technical information concerning to those inventions. In order to get worldwide patent, international patent application is filled under the PCT hence applicants can seek protection of their innovation in number of countries. For world wide recognized patents the following criteria must be satisfied: novelty, inventive step, industrial applicability. It means that invention should not have gone public before the date of application, our product or process should be an inventive solution and it should be possiblethat we can apply our new invention to industries (Sichelman 2009) (Fig. 25.2).

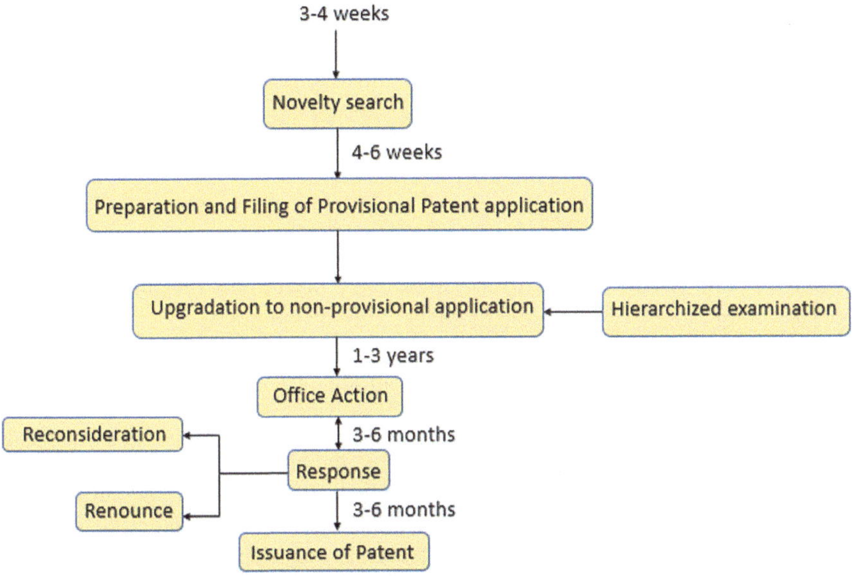

Fig. 25.2 Schematic diagram for complete patent filing

1.5 Benefits of Commercialization to Future Research and Problems in Commercializing Future Research

1.5.1 Commercialization Benefits

Commercialization of the academic research is the most important way through which universities and research organizations meet their [1]key performance indicator and they are adding value to the industry. There are numerous benefits that commercialization can award us: (Importance of Commercialization-Benefits for You, Your Research and Your Research Community, 2019)

- Commercialization provides the way how your research relates to the wider industry and when we are promoting and commercializing research we are also promoting our research tools.
- When we are commercializing our reagents, it is a guarantee for the safety of our reagents in future. Our research tools are located in multiple locations so that we can restock our research tools, if something happens to stock.
- Commercialization supports our future research and our institution or research organization. The rewards or output generated from the research commercialization is shared with the researcher and the institution. This ensures you not to lose any revenue that has been produced from your research commercialization and provides you funding to support your current research.
- Furthermore, investigating the Intellectual property background of your research and research tools makes you able to manage your commercialization process,

saves time and also ensures you are protecting both yourself and your institute (Ogbogu 2015).

1.5.2 Problems in Commercialization Novel Scientific Development

The innovators face the following problems while commercializing their innovations (Al Natsheh et al. 2015):

- The attempts made by the inventors to make perfect product may cause the entrance of any other competitive product into the market.
- The funding may exhaust in the pre-sales activities.
- The distribution and supply chains of the product can take time and experience to establish.

Review of literature has demonstrated that there are number of challenges that commercialization has to be faced. Parker and Mainelli (2001) discovered certain mistakes made during the commercialization of technology and scientific research which are: (1) using top-down market analysis, (2) inappropriate testing of technology and (3) assigning of unskilled supervisor for the commercialization process (Parker and Mainelli 2001).

Rosa and Rose (2007) find out that two main challenges in commercializing novel researches are the financial problems that results due to insufficient funding and human resource problems in the form of lack of skillful persons that are being hired to sell and promote the innovating products.(Rosa and Rose 2007).

Tahvanainen and Nikulainen (2011) reported that a lack of time and interest, negative attitude in the research environment, financial risks, bureaucratic disturbance, lack of business and commercialization knowledge and issues faced during the ownership of rights are the main problems that are confronting commercialization (Tahvanainen and Nikulainen 2011).

Problems related to the improvement of ability of science to commercialize the results of basic and applied research are the following (Komkov and Bondareva 2007):

- To monitor the financing of innovations with a high commercial potential.
- To stimulate government to provide support for innovative projects.
- To improve innovation legislation.
- To use the advanced international techniques of commercializing new technologies including both the diversification of research and production.

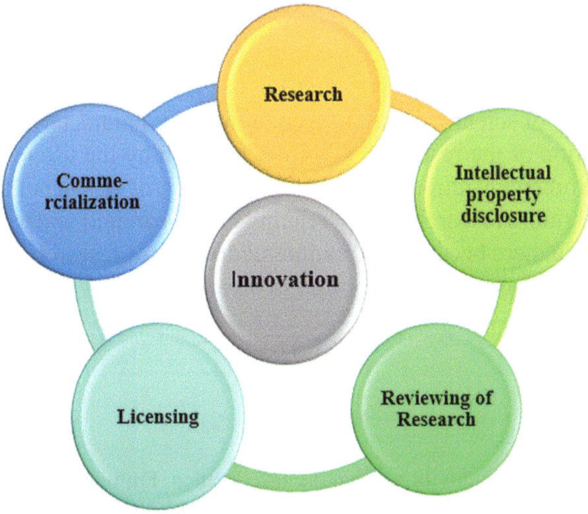

Fig 25.3 Simpler process leads to a complex process

1.6 Multiple Processes Involved in Technology-Based Product Innovation

The model of innovation that has been promoted in the form of complex process is usually stimulated by the interactions that are being developed with new product development(NPD) (Slotnick and Sobel 2002). Innovation involves simple processes which combine together to give rise to a complicated process. These simple processes are research, intellectual property disclosure, reviewing of research, Licensing and commercialization (Kalle Lyytinen 2016) (Fig. 25.3).

2 Copyrights

Copyright is the dedicated legal right to produce, reproduce, publish or perform original literary or drama work, performances and sound recordings. This type of protection is available to both published as well as unpublished work (Vaidhyanathan 2003). Copyrights protect the creators create their creations but it they do not systems or methods of operations. For example in a computer program the copyright covers the program's lines of code but does not protect its use and function. (Reichman 1989).

3 Summary

Novelty and innovation in scientific data leads to the commercialization of scientific data. The validity of scientific research is an important parameter that results in the commercialization of the results of scientific research by the use of different techniques. The commercialization of scientific research provides benefit by supporting our current and future research and by relating our research to wider industry. Some problems are also there because the commercialization of research product requires time and hence increases the competition in the market.

References

(2016) How to Improve Validity of a Scientific Investigation. From https://study.com/academy/lesson/how-to-improve-validity-of-a-scientific-investigation.html

Abd Rahim N, Mohamed ZB, Amrin A (2015) Commercialization of emerging technology: the role of academic entrepreneur. Procedia Soc Behav Sci 169:53–60

Al Natsheh A, Gbadegeshin SA, Rimpiläinen A, Imamovic-Tokalic I, Zambrano A (2015) Identifying the challenges in commercializing high technology: a case study of quantum key distribution technology. Technol Innov Manag Rev 5(1)

Brink HI (1993) Validity and reliability in qualitative research. Curationis 16(2):35–38

Cohen BA (2017) Point of view: how should novelty be valued in science? elife 6:e28699

Deterdin DG (2019) Validity threats in quantitative data collection with games: a narrative survey. HCI Res Games 50:302–329

Elg L (2014) Innovations and new technology (VINNOVA analysis 2015.5)

Fletcher AC, Bourne PE (2012) Ten simple rules to commercialize scientific research. PLoS Comput Biol 8(9):e1002712

Freedman DH (2010) Lies, damned lies, and medical science. Atlant 306(4):76–84

Gray J, Liu DT, Nieto-Santisteban M, Szalay A, DeWitt DJ, Heber G (2005) Scientific data management in the coming decade. ACM SIGMOD Rec 34(4):34–41

Kalle Lyytinen YY (2016) Digital product innovation within four classes of innovation networks. Info Syatems:47–75

Komkov N, Bondareva N (2007) Problems in the commercialization of scientific research and ways of addressing them. Stud Russ Econ Dev 18(1):2–18

Morrissey SR (2012) Scientists who are driven to start their own business and take the risks a start-up entails must think carefully about their technology, personal goals. In: What researchers who want to be entrepreneurs need to know, vol 90

Ogbogu TC (2015) The commercialization of university-based research: balancing risks and benefits. BMC Med Ethics:1–7

Parker K, Mainelli M (2001) Great mistakes in technology commercialization. Strateg Chang 10(7):383–390

Reichman JH (1989) Computer programs as applied scientific know-how: implications of copyright protection for commercialized university research. Vand L Rev 42:639

Rosa J, Rose A (2007) Report on Interviews on the Commercialization of Innovation, Science, Innovation and Electronic Information Division, Statistics Canada

Slotnick SA, Sobel MJ (2002) New product innovation with multiple features and technology constraints. In: Scholarship and professional work-business, pp 1268–1284

Sichelman T (2009) Commercializing patents. Stan L Rev 62:341

Staff TS (2019) 2019 Top 10 Innovations. The Scientist, Exploring life, Inspiring innovation

Svensson R (2012) Commercialization, renewal, and quality of patents. Econ Innov New Technol 21(2):175–201

Tahvanainen A-J, Nikulainen T (2011) Commercialization at Finnish universities: researchers' perspectives on the motives and challenges of turning science into business, ETLA Discussion Papers

Usama Fayyad DH (1996) Mining scientific data. Commun ACM 39

Vaidhyanathan S (2003) Copyrights and copywrongs: the rise of intellectual property and how it threatens creativity. New York University Press

Yimamu N (2018) Entrepreneurship and entrepreneurial motivation

Chapter 26
Sustainable Processing Through Efficient Use of Energy and Minimizing Waste Production

Anubhav Pratap-Singh, Shaba Noore, Ronit Mandal, and Anika Singh

1 Introduction

In ancient history, culture and religion, energy has been designated as one of the most important element for the existence of life on planet Earth. Processing of food started long ago by the ancient civilization from discovery of heat from fire about 1.5 million years ago. Animals are hunted and roasted from approximately 1.8 million years; wine/beer was invented between 7000 and 5000 BC while bread was invented almost 30k years ago. Olive oil and palm oil was introduced in the middle of 5400 and 3000 BC (Du Pisani 2006). Several other food products were familiarized in early BC era including pickles, chocolate, bacon, sugar, noodles, etc., (Fellows 2004). Since that era processing of food product begins and it was further classified into two fractions (a) primary (drying, milling, grinding), (b) secondary processing (where food primarily processed is formulated further into the processed products with it prolonged shelf life). Presently, there are plethora of food companies that works on the processing of food products to meet the demand of the consumers. Moreover, these processing companies not only make the food available for the consumer with its enhanced nutrition level, but it also generates source of income and employment (Kim 2013).

A. Pratap-Singh (✉) · R. Mandal
Food Nutrition & Health Program, Faculty of Land & Food Systems,
The University of British Columbia, Vancouver, BC, Canada
e-mail: anubhav.singh@ubc.ca

S. Noore
School of Biosystems and Food Engineering, University College Dublin, Dublin, Ireland

A. Singh
Natural Health and Food Products Research Group, Centre for Applied Research & Innovation (CARI), British Columbia Institute of Technology, Burnaby, BC, Canada

A. Režek Jambrak (ed.), *Nonthermal Processing in Agri-Food-Bio Sciences*,
Food Engineering Series, https://doi.org/10.1007/978-3-030-92415-7_26

Table 26.1 Conventional unit operation in food processing

Preliminary operations	Cleaning, sorting, grading, peeling, dehusking, shelling, hulling, coating, scarifying, rendering, butchering, evisceration, filleting, coring, pitting, trimming etc.,
Conversion	
Physical	Size reduction: Cutting, slicing, dicing, grating, milling, grading, emulsification, homogenization. Size enlargement: Agglomeration, granulation, coagulation, flocculation, pelleting Reshaping: sheeting, rounding, flaking, puffing
Chemical	Fermentation, enzymatic conversions, chemical reactions such as acidification (pickling), hydrolysis, sulphating, caramelization, depolymerization etc.
Physico-chemical	Cooking, baking, roasting, frying, extrusion, smoking
Preservation Techniques	Elevated temperature processes: Blanching, pasteurization, sterilization, drying. Ambient temperature processes: Salting, fumigation, irradiation, high pressure application Low temperature process: chilling, freezing, freeze-concentration, freeze drying. Packaging: canning/bottling (aseptic/non-aseptic/modified atmosphere), waxing
Separation Techniques	Crystallization, precipitation, filtration, centrifugation, evaporation, distillation, membrane processes (ultra/microfiltration, reverse/forward osmosis, per-evaporation) osmotic dehydration

At present, food processing industry consist of multi-disciplinary activities involving the application of chemistry, biochemistry, biophysics, nutrition, microbiology and several branches of engineering. A detailed list of unit operations in food processing has being listed in Table 26.1. These operation are further divided into four different categories viz., (a) Preliminary operations, (b) Conversion, (c) Preservation Techniques and (d) Separation Techniques. Presently, these technologies are widely utilized in every food se**ctor.**

However, these food processing techniques, which maintain balance of energy in humans, necessitates the continuous flow of energy supply, for instance to deliver 1 J of food energy approximately 10 J of energy is consumed from natural resources. Moreover, the vivid growth in the human population and its ever increasing nutritional demand has enhanced the consumption of total energy input by 40% over the last few decades (Food 2019).

The given line graph (Fig. 26.1) illustrates the source of energy consumed by the food industries is divided in several sectors including gas-diesel oil, natural gas, fuel, electricity, motor gasoline, liquefied petroleum gas and coal over the past ten years. It can be seen that maximum amount of energy is being consumed by gas-diesel while consumption of natural gas is limited to 50K (TJoules) from the year 2009 to 2019. Other sources of energy were utilized moderately and the gradually increase can be seen over the years and hence the energy intake by the agro-food industries are not sustainable in the long run and about 30% of world's total energy if being consumed by the food sector and generate about 20% of world's

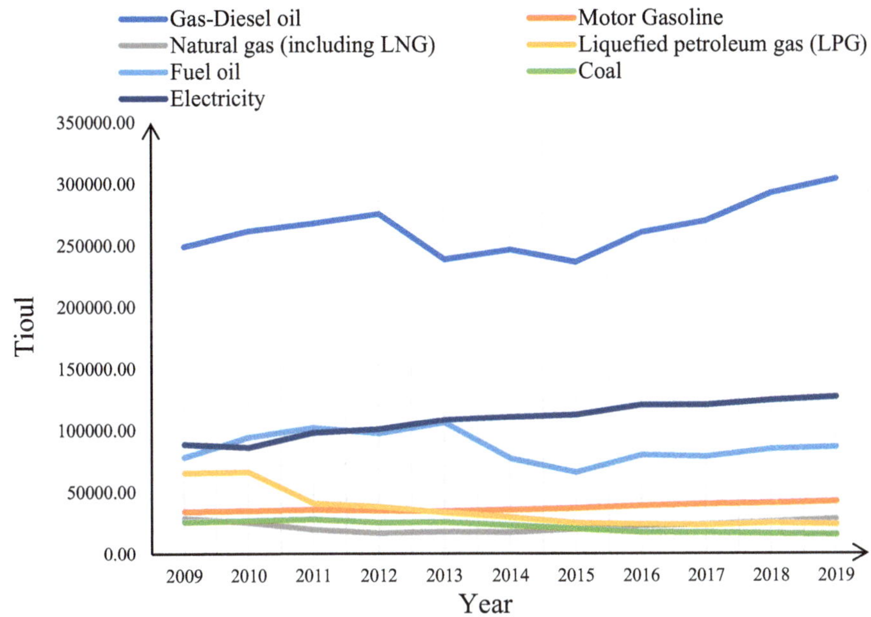

Fig. 26.1 Energy supply in food industries from 2009 to 2019. (Source: energy consumption in food processing FAO statistics, 2019)

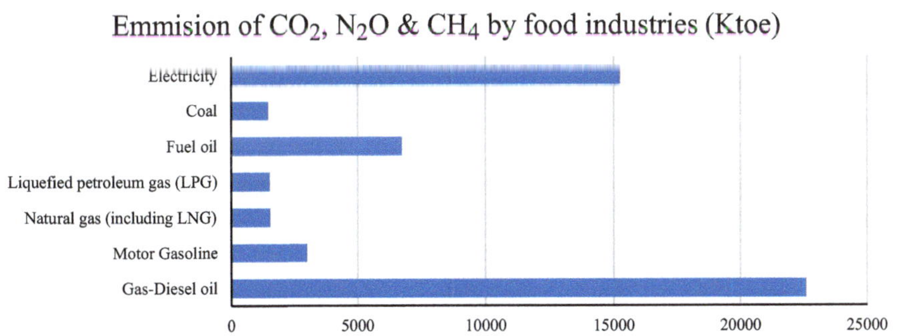

	Gas-Diesel oil	Motor Gasoline	Natural gas (including LNG)	Liquefied petroleum gas (LPG)	Fuel oil	Coal	Electricity
CO_2	22590.8866	2990.6351	1551.4993	1529.0868	6733.9684	1471.1547	15234.1424
N_20	8.5189	0.0827	0.0028	0.003	0.0545	0.0233	0.2027
CH_4	2.7936	3.8785	0.1383	0.2472	1.815	4.6654	38.0361

Fig. 26.2 Emmision of CO_2, N_2O and CH_4 by food industries by using different energy sources for the year 2019 (energy consumption in food processing FAO statistics)

greenhouse gas emission. Additionally, due to the excessive utilization of energy, several environmental problems have been reported including enhanced emission of carbon dioxide, methane and nitrous oxide gases, dust, black carbon in the atmosphere and combustion processes waste. Energy consumption in food processing FAO (Food 2019) statistics (Fig. 26.2) clarifies about the emission of CO_2, N_2O and

CH_4 by the food industries and it can be clearly seen that CH_2 emission is maximum by gas-diesel oil, followed by electricity, fuel oil and motor gasoline. Natural gas and coal reflected in less emission of CO_2 and hence proves to be better source of energy compared with other sources and can be used as sustainable option in the coming future. Not much emission was seen in case N_2O and CH_4, however natural gas reflected in minimum release (0.0028 Ktoe; 0.1383 Ktoe for N_2O and CH_4 respectively) which further promotes utilization of natural gas.

Moreover, countries with high GDP (Gross Domestic Products) consume more energy for processing and transportation as compared to low-GDP territories, generally in cooking. One fifth of energy demand is from farm and fishery sector but its excruciation to elaborate that it emits two-third of greenhouse gases which is huge global challenge to irradiate (2019). In order to maintain an equilibrium balance between the environmental condition and the energy utilized to meet the demand of the nutritious food products, it is important to introduce efficient technologies which will not only be helpful in meeting the public food demand and reducing energy input but also it will be efficient enough to minimize the cost of production in minimum processing time.

Several novel technologies such as thermodynamic cycles (absorption, heat pump/pipes, refrigeration cycles), non-thermal food processes (irradiation, high pressure processing, pulse electric field) and unique heating strategies (microwave, infrared and ohmic heating) are perfect substitute for the conventions techniques and can be introduced in food industries (Sun and Wang 2001; Ozyurt et al. 2004; Kuzgunkaya and Hepbasli 2007). These emergent strategies not only reduce energy consumption but it also helps in cost reduction of the developed product. Additionally, these techniques are time efficient with no waste generation, thereby enhancing the sustainability of food processing industries. This chapter focuses on presenting basic evidence and references, to illustrate the consumption of energy in several food processing industries. It also provides details about the sustainable novel technologies practiced to reduce the consumption of energy and as food by-product waste (Toepfl et al. 2006; Brown et al. 2008; Yang et al. 2010; Nguyen et al. 2013).

2 Energy Consumption in Food Processing

2.1 Energy Consumption in Food Industry

Processing of food initially started with thermal treatment on natural resources to develop a processed product. It works on the principle of high and low temperature application for processing as well as preservation of the products. William Cullen back in 1750 invented a cold press system which in present date has been modified to energy efficient system known as freezing and refrigeration. In the same way, Nicholas Appert in early 1800s developed a food preservation method to extent the shelf life of the food products to be delivered to French army and presently that

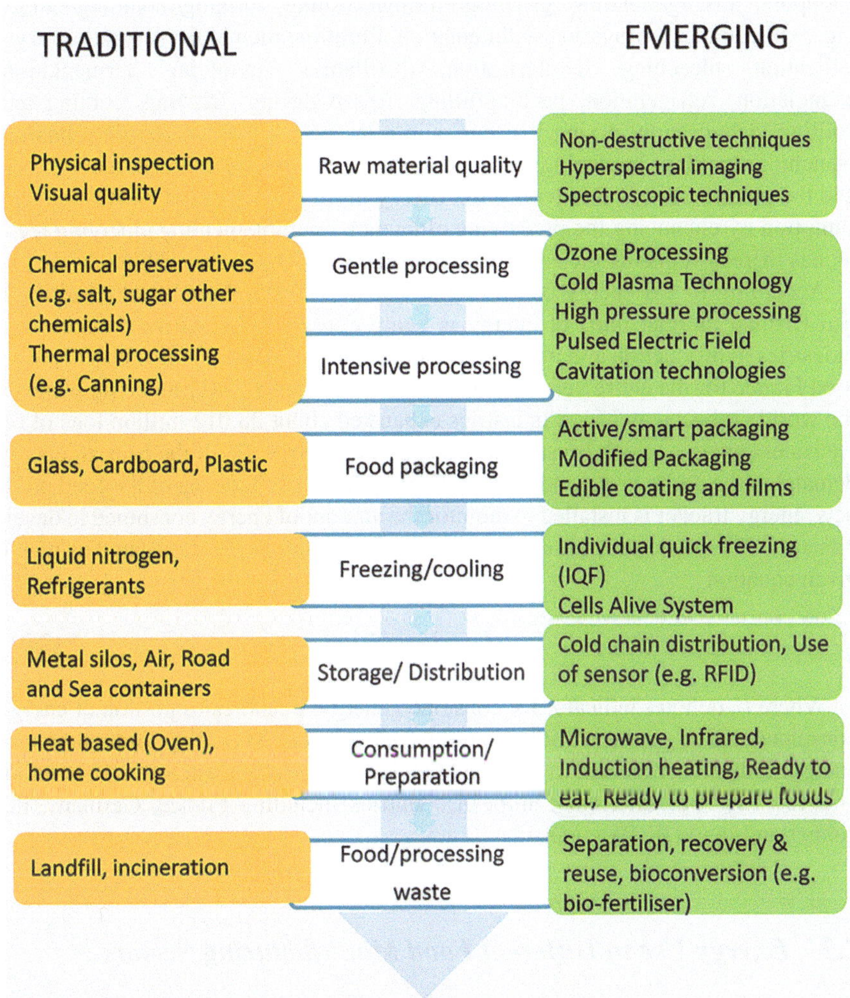

Fig. 26.3 Traditional and emerging technologies and approaches used along the food chain. (Adopted with permission Knorr et al.(2020))

method has been modified to thermal-processing equipment. Figure 26.3 recapitulates some of the conventional and emerging strategies that are practiced in food processing industries. The prime focus to find a suitable alternative to thermal and chemical preservation was to reduce the consumption of energy and waste generated (Brown and Pariser 1975).

In food processing industry there are ample number of processes including material handling/storage, sorting/screening, peeling, washing/thawing, cutting/slicing/

chopping, mixing/blending, grinding/milling/crushing, forming/moulding/extruding, extraction, centrifugation sedimentation, filtration, membrane separation, crystallization, bleaching, deodorization, distillation, dissolving, fermentation, coagulation, refrigeration, packing/filling, freeze drying, freezing, cooling and chilling, dehydration, drying, pasteurization, tempering, frying, roasting, baking, blanching, melting, hardening, carbonation etc., which require huge energy input and the possible solution to reduce the energy input is to optimize the energy consumption by enhancing the processing efficiency by implementing emerging techniques thereby reducing the emission and effluents (Tiwari et al. 2013).

According to US energy consumption statistics 2018, about 101.1 quadrillion Btu (British thermal unit) of energy is being consumed by the food processing industries in production and transportation of food which is equal to the total power used France for the entire year and thus US has the largest food producing sector in the world. Tobacco and food industries exhausted about 28.012 million tons of oil equivalent energy, which is estimated to approximately 9.8% of the over-all energy demand in the entire production sector. During the manufacturing of the food products, energy tracker is installed to monitor the amount of energy consumed in development of the particular product. The energy consumed can be calculated using the given equation

$$I'_e = U_e / M_p \tag{26.1}$$

Where I'_e denotes indicator for consumed energy, U_e indicates portion of energy consumed and M_p is the portion of products developed. According to the literature energy indicator reflected a tremendous increase in the energy consumption (14–48%) in four different European countries including France, Germany, the Netherlands and UK (Wang 2014)

2.2 *Energy Use in Different Food Manufacturing Sectors*

There are six major food processing sectors which includes (a) grain and oil milling, (b) sugar and confectionary processing, (c) meat and fish processing, (d) fruits and vegetable processing, (e) bakery and (f) dairy processing industries. The major portion of enhanced energy consumption is from meat processing industries as it consumes excess energy to process the frozen and cut meat products. While in case on rice processing industry, energy consumption is reported to be minimum (0.43 MJ/Kg) (Arendt and Zannini 2013). Further to develop breakfast cereals it was reported to consume 66 MJ/kg of energy which includes grinding, milling, wetting, baking and drying (Aguilera et al. 2011). Other food industries such as sugar and confectionary industries consumed almost 6.90 MJ/kg of energy where 65% of thermal energy was used by evaporative crystallizer and other 25% or the total electricity was consumed for melting and centrifugal drying.

At present sustainable use of energy is carried out by eradicating juice purification step and replacing evaporative crystallization method by cool crystallization

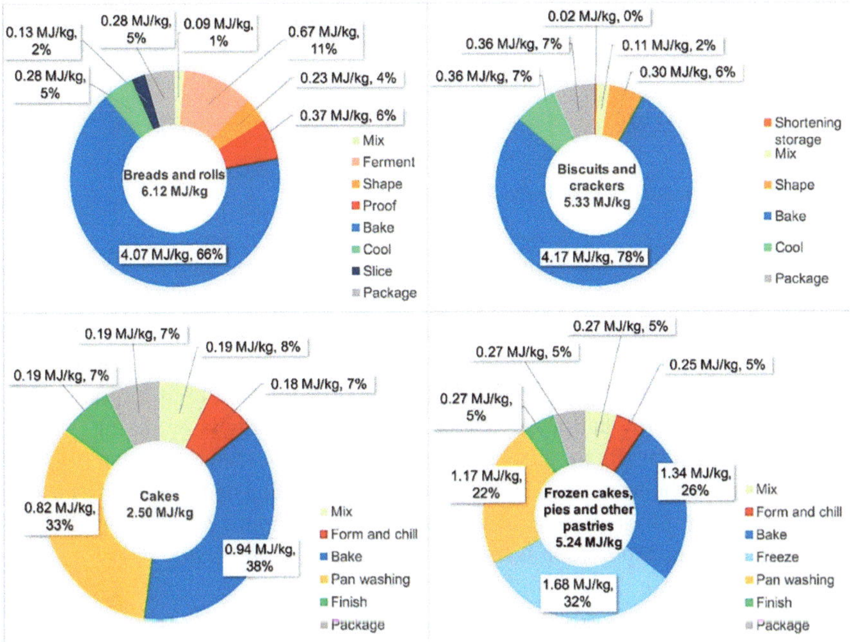

Fig. 26.4 Energy consumed to produce bakery products. (Adopted with permission from Ladha-Sabur et al. (2019))

using concentrated fresh juice. Also, energy is regenerated by using by-products of sugar as a biomass to generate biofuel. Fruits and vegetable, as they are rich in water content, excessive amount of energy is generally consumed to process them into final products. However, in case of cheese and dairy products energy consumed in the range of 10.30–13.85 MJ/kg while Bakery products energy consumption was reported to be limited to 5.21 MJ/kg (Singh 2013). Figure 26.4 educates about the several sections of bakery processing along with its energy consumption.

2.3 *Energy Use for Production of Different Food Products*

Consumption of energy totally depends on the type of food being processed, for instance to produce one ton of milk powder the amount of energy consumed is reported to be approximately 9385 and 9870 MJ while to prepare one ton pasta energy consumption is limited to 2 MJ. Further to produce one ton of wheat starch 2960 MJ of electricity us consumed whereas to produce one ton of beet pulp the energy consumption is limited to 5 MJ.

In development of different products several unit operations are taken into practice including freezing, drying, blanching etc., now in order to freeze one kilogram of product the amount of energy used is 1 MJ but to dry the same product the amount of energy used is consumed six time more to remove one liter of water from the product (Wang 2008).

2.4 Energy Sources for Food Processing

The major source of energy at present that are extensively utilized by the manufacturing industries includes gas diesel oil, fuel oil, motor oil, natural gas, liquefied petroleum gas, electricity, coal. Although natural gas is known to be the largest source of energy still consumption of renewable source of energy tends to grow at a rapid pace as food industries are utilizing there agro-food waste as biofuel for the development of energy and about 9% of the electric energy is produces by using onsite power system. In food manufacturing industries, steam is the major unit which consumes the one third of the total energy consumed by the other unit operations. Motor and refrigeration energy consumption is about 12 to 16% while heating, lighting and transportation is limited to 8% (Okos et al. 1998) (Fig. 26.5).

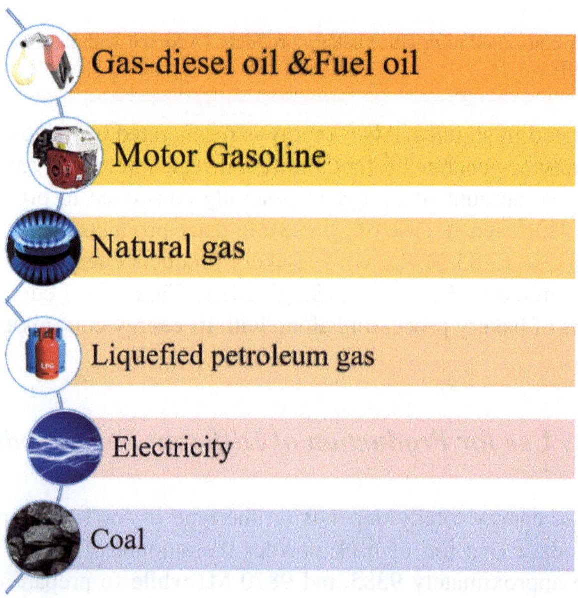

Fig. 26.5 Different sources of energy

3 Energy Conservation for the Utilities in Food Processing

3.1 Energy Savings in Steam Supply

Majority of food manufacturing industries comprises of huge demand for steam and use boilers to generate steam that liberates heat energy for the processing (pasteurization, sterilization, and dehydration) of food products. It consumes almost one-third of the total energy generated and approximately 57% of the total fuel used by the food industries (Einstein et al. 2001). Some porting of energy is lost during adverse conditions such as steam leakage, bad surface insulation etc., Interestingly, by practicing steam technology to generate heat energy, almost 20% or energy is being saved and also emission of CO_2 has been reduced which proved to be a sustainable solution that can be enhanced in the coming future to bring down the level of energy consumption in food producing sector (Walker et al. 2013).

Furthermore, this technology is majorly sub divided into two separate groups (a) operation optimization and (b) waste heat recovery. In case of operation optimization it comprises of several operational conditions (boiler size, boiler pressure and temperature, optimal excess air and blowdown stalk) that needs to be optimized in order to enhance the efficiency of the boiler to approximately 15%.

If the conditions are optimized and equipment is efficiently used, about 78% of stored heat energy can be recuperated which can save 1.3% fuel preserved for boiler use (Einstein et al. 2001).

3.2 Energy Savings in Compressed Air Supply

Compressed air is another extremely interesting processing technique used in food processing basically for wrapping machine, colour sorter, blow drying for removal of water from cans before labelling, pneumatic controls and lid machines. To develop compressed air, it is considered to be extremely expensive but efficient and proper use of air compressor can save 20—50% of energy (Mull 2001). Moreover, energy can be saved by using high speed and efficient motors, low inlet air temperature, waste heat recovering unit, low air pressure, no air leakage and utilization of local air delivery strategies.

3.3 Energy Savings in Power Supply

Two main components of food industries which drive huge amount of electricity are motor drives and refrigerators and account to be approximately 48–25% of entire electricity (Okos et al. 1998). Moreover, in meat industries this energy utilization percent reflects a tremendous rise to 90% during the production time and reaches to

100% at the time is non-production (Ramirez et al. 2006). Motor and pumps play a vital role in food processing industries as pumps are responsible for the movement of the fluid or semi-fluid from one unit to different units in production section and motors are required to provide energy to pump. Amount of money spent on electricity can be saved by monitoring the demand, enhancing the power factor, and decreasing the utilization of electricity as much as possible. Since motor consumes about 5–30% of power therefore to conserve energy it is necessary to find sustainable method. Motor works on the principle which required two types of power (a) resistance power and (b) inductance power and it is designed to work at rated loads. Hence the most actual way to conserve energy is to match the essential loads along with the rated motor load. Additionally, when buying a new motor, parameters which need to be taken care of in order to keep the energy consumption as low as possible is (a) high power factor, and (b) adjustable speed motor.

3.4 Energy Savings in Heat Exchanger

In food processing sector, numerous unit operation including freezing, heating, thawing, thermal sterilization, refrigeration, drying/evaporation are practiced in order to transfer the amount of heat between heating object and food product. This phenomena of exchange of heat takes place in heat exchangers which also plays a vital role in waste heat recovery. Different energy saving strategies such as optimized heat exchange network/ design and improved heat transfer is being utilized to enhance the effectiveness of heat exchangers (Zimparov 2002). Literature indicates that about 30% of energy consumed in meat and slaughterhouse can be preserved after proper configuration of heat exchangers (Smith 2000; Fritzson and Berntsson 2006).

3.5 Energy Savings by Recovering Waste Heat

Any food processing treatment which is above ambient temperature, requires energy. Leftover energy including boiler fuel gas, boiler blowdown water, ovens, air compressor water, drier exhaust gas, and cooker vapours are some of the major source of waste heat and in US about 50% of the consumed energy was liberates as waste heat into the environments (Mull 2001). About 40% of the energy can be saved by implementing waste heat recovery as it not only reduces the energy expense but also it will diminish the environmental pollution caused by thermal air liberated from industries and factories. Nevertheless, the value of recovered waste relies on heat transfer equipment used and the quality and quantity of the recovered heat waste.

4 Energy Conservation in Energy-Intensive Unit Operations of Food Processes

4.1 Energy Savings in Thermal Food Processing

Thermal food processing involves high-power consuming unit operations including sterilization, chilling, freezing, pasteurization, evaporation and drying which are implemented to preserve and extent the shelf life of the food products. In this processing sector energy can be conserved in three different protocols (a) enhancement of energy effectiveness in current unit, (b) exchange of energy unit with novel techniques, (c) consumption of renewable energy sources specifically derived from agro-food waste.

4.2 Energy Savings in Concentration, Dehydration, and Drying

In order to prevent spoilage and extent the shelf life of the food products, food undergoes certain treatments including dehydration, concentration dying thereby changing its physico-chemical properties which includes moisture content, weight and volume of the final products and hence prevents it from spoilage. Several other techniques of drying includes microwave drying, vacuum drying and freeze drying and the energy required to practice these techniques in food industry is approximately 2.5–2.7 MJ/kg, while energy required for conventional drying methods is reported to be 4–6 MJ/kg (Midilli and Kucuk 2003; Dincer and Sahin 2004; Akpinar 2004; Akpinar et al. 2005, 2006). However, membrane technology has the potential to bring down the total energy consumption during the evaporating /drying operation and thus reducing the overall energy consumption (Kumar et al. 1999; Onsekizoglu et al. 2010; Cassano et al. 2011).

4.3 Energy Savings in Refrigeration and Freezing

In food processing industries there is a dense utilization of refrigeration unit. According to statistical analysis about 15% of total world's energy is consumed by refrigeration. In United States around 25% of the electricity is consumed only for refrigeration and process cooling (Okos et al. 1998). Moreover, dairy and meat processing sectors are the major users of refrigeration and hence the consumption of energy by the refrigeration can be reduced by improving insulation with enhanced refrigeration cycles powered with waste heat.

5 Utilizations of Energy Efficiency Technologies in Food Processing

As the thermal processing operations are energy intensive, newer ways of conservation of energy application of novel thermodynamic cycles, non-thermal processes are being used. Some of the recent cutting-edge and energy-efficient technologies that are being used in the food industry are discussed in this section.

5.1 Application of Novel Thermodynamic Cycles

5.1.1 Heat Pump

Heat pump is a thermodynamic device that basically heats up a space by transferring energy from cooler body to a warmer body. It utilizes the application of external power to deliver heat from colder body to warmer body by using the refrigeration cycle (Fig. 26.6).

A working substance (a liquid refrigerant) vaporizes at temperatures below boiling point by application of low pressure which is then compressed to higher pressure. The compressed vapour is condensed during which it releases heat to the surroundings. The energy efficiency of heat pumps is typically termed as "coefficient of performance" or COP, which is the ratio of the heat supplied to the expressed as:

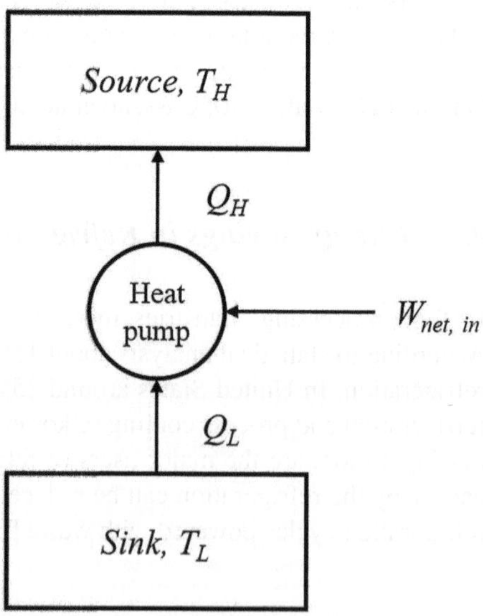

Fig. 26.6 Heat pump thermodynamic cycle

$$COP_{Heat\ pump} = Q_H / W_{net,in} = Q_H / (Q_H - Q_L) \tag{26.2}$$

The heat pumps have been long used for space heating applications and desalinations (Goh et al. 2011). Heating applications of heat pump in food processing has been mainly carried out to assist in thermal processes (Ozyurt et al. 2004). The researchers designed a milk pasteurization for cheese production unit whereby the incoming cold raw milk was heated by the heat pump condenser, while hot pasteurized milk was cooled down by the evaporation unit. For the system operating between pasteurization temperature of 72 °C and 32 °C, the COP was 2.3–3.1. They showed that the heat pump assisted pasteurization system reduced the energy consumption by 66% compared to the thermal pasteurization techniques.

Drying is one of the energy intensive unit operation in food processing (Pratap Singh et al. 2020). In the recent decades, heat pumps have been studied for their application in food dehydration systems. In a typical food dehydration system, te heat pump is incorporated as an augment to recover heat and dehumidify the hot air used in drying food materials. Warm and moist air form the drying unit is cooled and dehumidified in the heat recovery unit and evaporator, respectively while the condenser unit heats up the dry air for drying process (Fig. 26.3) (BudžakI et al. 2019) (Fig. 26.7).

The condenser latent heat supplied the sensible heat to the air that is input to the drying unit. Some of the advantages of the heat pump assisted dryers include higher

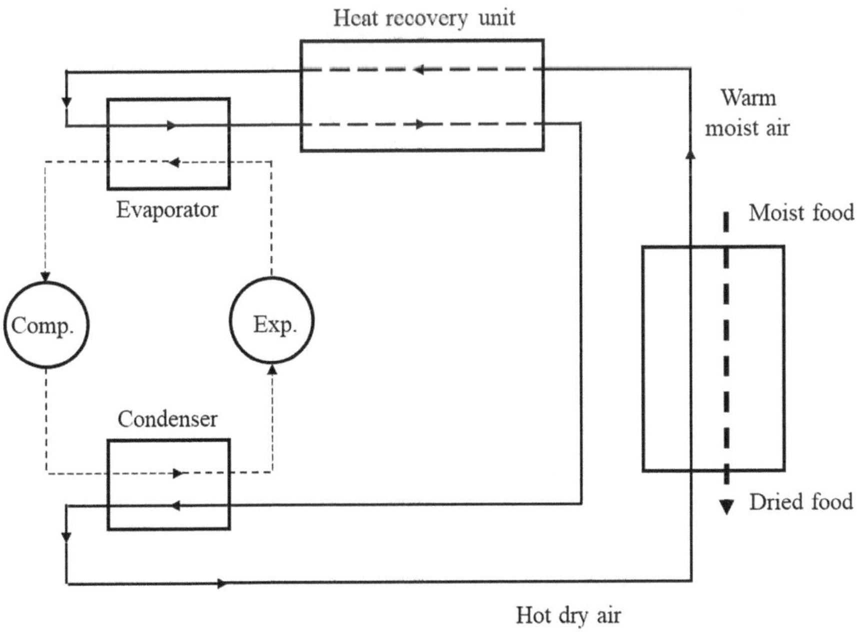

Fig. 26.7 Heat pump assisted hot air drying cycle

energy efficiency, efficient temperature, and humidity control of the air, suitable for drying of sensitive agricultural commodities. The drying system approaches efficiency of 100% because of total recirculation as compared to hot-auir drying alone (35–40%) (Perera and Rahman 1997). However, drying should be carried out at temperatures <60 °C and relative humidities ≥30%. Perera and Rehman (Perera and Rahman 1997) formulated the efficiency value for heat pump assisted dryer in terms of the COP of the dehumidifier as a function of specific moisture extraction rate (SMER, kg/kW-h) and latent heat of vaporization (h_{fg}, kJ/kg)

$$COP = 1 + SMER * h_{fg} \tag{26.3}$$

In a study, Aktaş et al. (2017) dried grated carrot using heat pump assisted air dryer where they obtained an energy efficiency and exergy efficiency of ~50%. Also, due to lower temperature of drying, the process could be beneficial for retaining the overall quality of the products. For instance, there was an observed potential for drying paddy in terms of energy saving and maintaining quality characteristics (Jinjiang and Yaosen 2010). Low temperature (30–37 °C) heat pump drying of mint leaves led to retention of ascorbic acid (Venkatachalam et al. 2020) Some of the applications of the heat pump assisted drying have been summarized in Table 26.2.

5.1.2 Novel Refrigeration Cycles

Several newer refrigeration cycles based on absorption and adsorption (liquid-liquid or liquid-solid) have emerged in the recent decades. Typical absorption-desorption (Fig. 26.8) works on the basis that partial pressure of refrigerant vapour changes according to the temperature and concentration of a refrigerant and sorbent solution. H_2O/NH_3 and LiBr -H_2O mainly used as absorbent solutions. The low pressure

Table 26.2 Food commodities dried using heat pump assisted drying in the recent years

Product	Geometry	Final moisture content (%)	Drying temperature (°C)	Humidity (%)	Velocity (m/s)	Time (min)	References
Potato	Chips: 2*3 mm		35, 45, 55	10–30	1–2		Zlatanović et al. (2017)
Banana	Cylinders: 6*7 mm						
Tomato slices	Diameter: 40 mm		35–45		1		Coşkun et al. (2017)
Coffee beans		11% (wet basis)	50	70	2	4320	Dong et al. (2017)
Shiitake mushrooms			35-62		2	1020	Liu et al. (2018)
					3	1200	

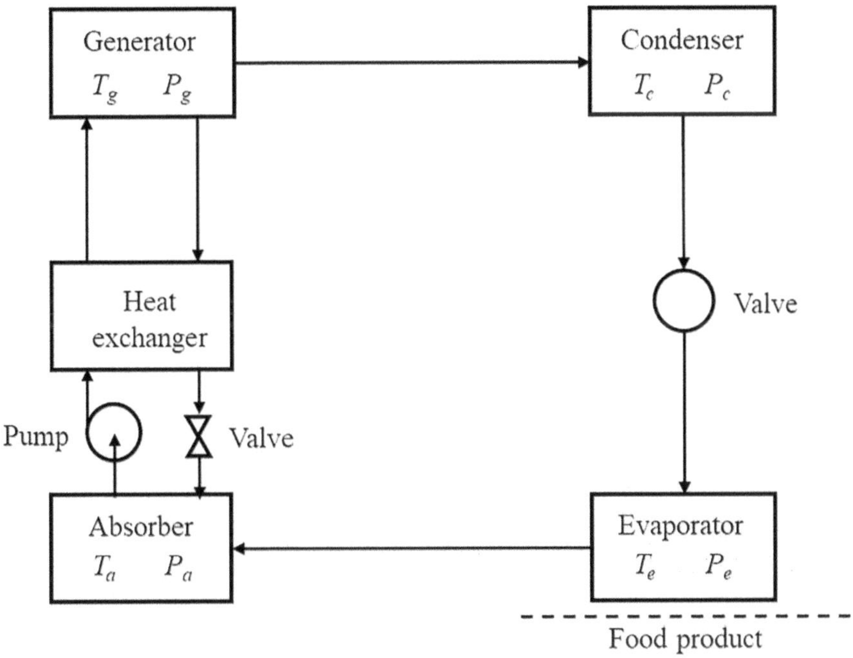

Fig. 26.8 Vapor absorption refrigeration cycle

refrigerant vapor like H_2O from the evaporator (where the product is chilled) is absorbed by LiBr to form a refrigerant-sorbent system, which is pumped to the generator at higher pressure. The generator heats up the refrigerant to remove the sorbent which is throttled back to absorber by a throttling valve at low pressure. The refrigerant moves to the condenser to be condensed into liquid form (Sun and Wang 2001). The COP of the refrigeration system is less than that of a vapor compression cycle system.

Adsorption type refrigeration system (Fig. 26.9) consists typically of an evaporator, condenser and an adsorber. Adsorbent system replaces the compressor system of the mechanical vapor refrigeration cycle and the absorption system of the vapor absorption cycle system. Some of the used adsorbents are hydrides of metal, activated carbon, silica gel, zeolite etc. with the adsorption system involving the lkes of zeolite-water or activated carbon-methanol. During a cycle, the refrigerant from the evaporator is adsorbed by the adsorbent bed, which cools the product. Desorption is carried out by heating the sorbent bed and the refrigerant vapor is condensed in the condenser. There exist a slow heat transfer in the system leading to longer cycle times and low COP compared to compression systems (Sun and Wang 2001).

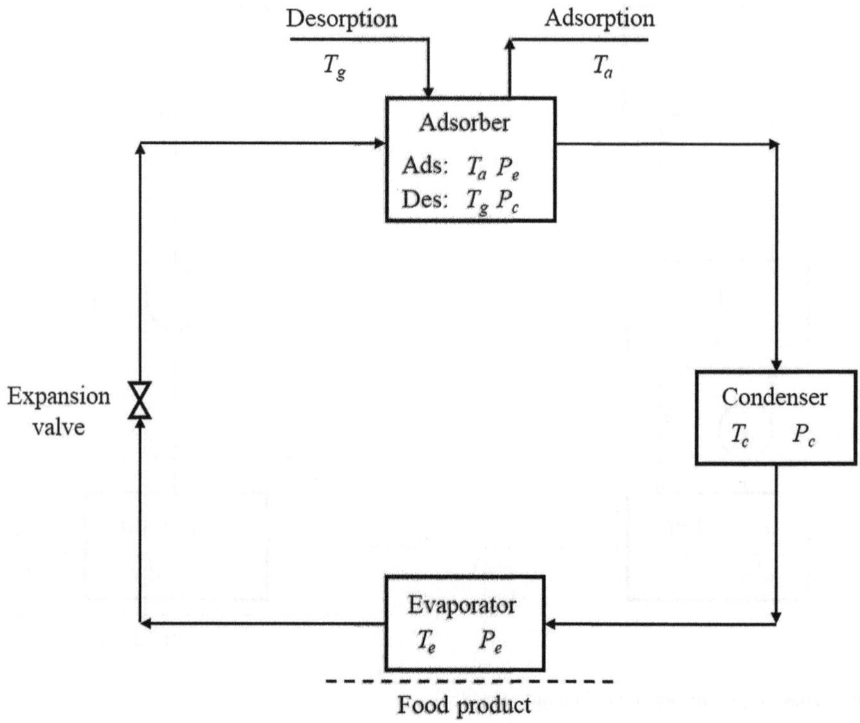

Fig. 26.9 Vapor adsorption refrigeration cycle

5.1.3 Heat Pipes

Heat pipes involves the use of a long metallic tube with a wick and working fluid like water inside the tube. There exists a finite temperature difference between the either ends of the tube. One end serves as the evaporator whereby the liquid evaporates and moves to the other end as vapor (Fig. 26.10). The vapor are condensed on the other side by removal of the latent heat. The condensed liquid then travels back into the evaporator end by capillary action of the wick or by gravity. Since the operation involves simultaneous heat and mass transfer with release of the latent heat by phase change, the heat transfer is very quick (Chaudhry et al. 2012). Some of the used working liquids are water, heptane, ammonia, acetone, pentane etc. (Chaudhry et al. 2012). Studied technologies for heat pipe include tubular heat pipe, variable conductance heat pipe, thermal diode, pulsating heat pipe, loop heat pipe, micro heat pipe, sorption heat pipe (BudžakI et al. 2019).

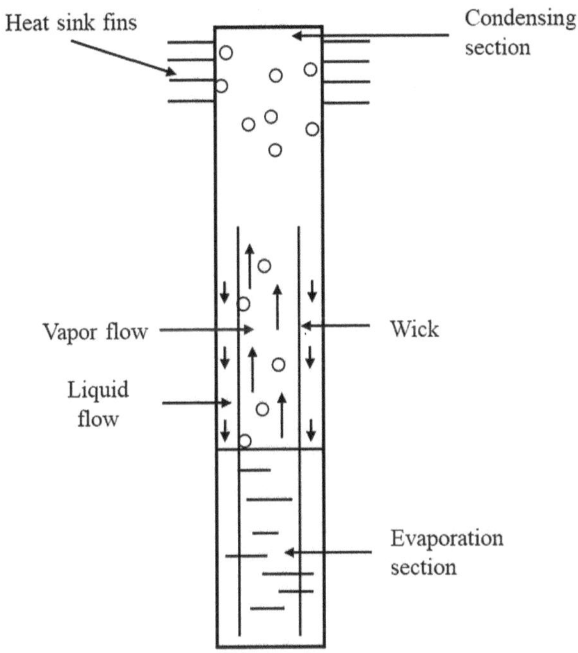

Fig. 26.10 Heat pipe operation

5.2 *Application of Non-thermal Food Processes*

Many thermal processes like pasteurization, sterilization are energy intensive processes. They are characterized by slow heating processes. Non-thermal processes, which include food irradiation, pulsed electric fields processing, high pressure processing etc. do not require heating and are quick and less energy intensive.

5.2.1 Food Irradiation

Food irradiation is the process of treatment of food products by exposing them with high intensity ionizing electromagnetic radiations including the X-rays, electron beams, gamma rays. They have very high frequency with short wavelengths ranging from 100 picometers to 100 nanometers. Owing to high energy, they tend to knock electrons from the molecules they are targeted with. Their germicidal nature arises from their capability to interact with the DNA of the microbial cell and thereby inactivating them. Irradiation with high energy X-rays and gamma rays do not produce much heat (up to 3 °C rise in temperature) (Loaharanu 1996). Also, it consumes much less energy than conventional thermal processing. Energy imparted to the food during irradiation is quantified in terms of KiloGray or kGy. Food and agricultural products are usually treated with less than 10 kGy (equivalent to 10 kJ/kg).

5.2.2 Pulsed Electric Fields

Pulsed electric fields is the application of an external electric field as short pulses (1–100 μs) to the sample for inactivation of microorganisms. The fundamental mechanism of microbial inactivation is based on electroporation. When an external electric field is applied to a cell, an electric potential develops across the cell membrane. When the electrical potential exceeds the critical value, structural changes in the cell membrane take place. Such structural changes in the membrane lead to formation of pores in the membrane, thus the term electroporation (Wiktor et al. 2020). Due to electroporation, the cellular materials start leaking out and cell dies. Pulsed electric field has been used for the cold pasteurization of liquid foods, imparting a fresh like taste and higher nutritional value than conventional processes. Pulsed electric field for the treatment of foods at around 30 °C demands a specific energy input of 100 kJ/kg, which is considered to be higher than the conventional pasteurization (Wang 2014). In a study, Aganovic et al. (Aganovic et al. 2017) observed that tomato and water melon juice treated by pulsed electric fields required 0.2 kWh/L of juice compared to thermal processing (0.04 kWh/L). Additionally, it has been observed that processing at elevated temperatures lower the energy conception considerably (40 kJ/kg compared to 100 kJ/kg) (Heinz et al. 2003). However, it has been shown to assist in drying processes of food commodities compared to conventional drying. One drawback of the technology is its high capital cost even for low capacity units.

5.2.3 High-Pressure Processing

High pressure processing is the application of pressure for the treatment of foods. At high pressure, the microbial cells are inactivated due to permeabilization of their cell membranes, which leaks out the cellular organelles. Membrane consists of proteins and lipids. When pressure is applied on the cells, the protein chains get altered in configuration, leading to permeabilization. During pressurization of liquids, the compression work tends to increase the temperature of the liquid by adiabatic compression. For examples, in case of compression of water, 3 °C rise in temperature occurs every 100 °C. For 600 MPa pressurization, 18 °C temperature rises will occur. Energy input for high pressure processing at 600 MPa would be 122- kJ/kg.

5.2.4 Membrane Processing

The principle of membrane processing is to separate liquid or gas food into two streams by applying pressure through a semipermeable membrane. The stream with materials passing through the membrane is called permeate. At the same time, the stream with molecules that are cast off by the membranes is retentate. Depending on the properties of the membrane and the design of the system, this technique enables the separation or concentration of specific molecules based on the molecular size, diffusion, shape, and charge (Rajendran et al. 2021).

In decades, the high energy consumption of many traditional separation methods has become a great concern in food film operations. As a perfect alternative for food separation, membrane processing can save 30–50% of the potential energy compare to evaporation and distillation. Nikmaram and Rosentrater's study announced that the evaporation with mechanical vapour-recompression consumed 50 kJ/kg for water removal, while membrane filtration only uses around 14–36 kJ/kg. About 90% of those saving is because no heating is involved for phase change in membrane processing (Nikmaram and Rosentrater 2019). No heat involvement in membrane processing also improves the retention rate of molecules that are sensitive to heat, such as protein, vitamins, and phytochemicals. The quality and nutritional value of the product from the membrane processing are thus improved. Furthermore, microfiltration membrane processing can also enhance electricity production in a system with a salinity environment. This feature improves the energy efficiency in a way recovering the energy from the processing system (Nikmaram and Rosentrater 2019).

5.2.5 Supercritical Fluid Processing

Super critical fluids are used for extraction of certain components from a product. Supercritical fluids for example supercritical CO_2 have a low critical temperature and pressure (31.1 °C and 7.3 MPa). At such state, the fluids possess excellent solvent properties. Due to low critical temperature and pressure, the process is far better than the conventional solvent extraction processes. This is advantageous in terms of the applied low operational cost of the process.

5.3 Application of Novel Heating Methods

5.3.1 Microwave and Radio Frequency Heating

Microwaves are electromagnetic radiation that travels in straight lines through the air, glass, paper, and plastic but are reflected by metals. Radio frequency (RF) refers to the radio waves that are at the low-energy end of the electromagnetic spectrum. Because of the larger penetration depth of the RF rays, RF heating is more suitable to be used for large volume products. When those electromagnetic radiations are absorbed by food, heat is generated through the intermolecular fiction from the water vibration (Vadivambal and Jayas 2010). Besides, both microwave and radio frequency heating perform dielectric heating, in which heat is produced throughout the dielectric material (Sun et al. 2016). Since the heat is directly transferred to the food component without heating up the whole oven, microwave and radio frequency heating can be a more energy-efficient processing method comparing to the conventional heating method (Sun et al. 2016).

To understand the energy efficiency of microwave heating, several studies had demonstrated comparison experiments to distinguish the energy usage difference between microwave and conventional heating. In terms of drying, microwave heating presented approximately 75% faster processing time than the conventional oven, electric oven, hot air drying and vacuum oven (Li et al. 2019). Menon et al. also reported that the application of microwave drying saves nearly 50% of the energy consumption compared to hot air drying (Menon et al. 2020). Further, the improvement of the drying processing efficiency of microwave heating was discovered in various food products and wave frequency compared to conventional drying (Li et al. 2019). Those evidence has proven that microwave heating can be widely used to decrease energy consumption in food industries. The combinational use of microwave heating with other heating methods is another strategy to reduce energy usage in food companies. Using convective drying with microwave drying had significantly reduced 70% of the overall energy consumption in contrast with hot air drying. It is noticeable that hybrid drying applications even had a greater energy reduction rate than microwave drying alone (Menon et al. 2020).

Regarding to RF heating, Jiang et al. observed an average of 76.1% heating efficiency for pasteurization and disinfestation on low moisture food (Jiang et al. 2020). Besides, the general conversion rate between electricity to thermal energy in an RF heating system is nearly 60% higher than the traditional heating system. The equipment used in RF heating is also inexpensive compare to the conventional method. That characteristic ensures RF heating perform with less charge in energy and operation (Jiang et al. 2020).

Apart from drying processing, microwave heating can also be used for multiple purposes. Microwave-assisted extraction (MAE) is a popular novel application that provides benefits such as short processing time, reduction of solvent usage, and extraction yield improvement to the food companies. In addition, microwaves can cause a reorientation effect and different heating rates between different molecules. This phenomenon shows the potential of microwaves to perform separation, which includes microwave-assisted pervaporation and microwave-assisted reactive distillation (Li et al. 2019). With respect to radio frequency, blanching and thawing are the major application for egg, vegetable, and fruit processing. Radio frequency provides high uniform heating with limited negative effects on food quality and nutritional values. However, the telecommunication present in the food industry may use frequencies that can interfere with the RF used in food processing. It is necessary to design proper shielding to prevent interference through processing in RF applications (Altemimi et al. 2019).

5.3.2 Ohmic Heating

Ohmic heating generates heat by directly passing an electric current through an electrically conducting food product. The heating energy transfer internally within food particles indicates the heating efficiency will not be affected by the solid-liquid phase system of the food product (Wang et al. 2021). Ohmic heating has a great advantage in energy saving due to its exceptionally short processing time (few seconds to several minutes) and close to 100% energy transfer efficiency (Chizoba Ekezie et al. 2017). In contrast with the traditional heating method, ohmic heating can reduce 82–97% of the energy consumption and decrease the heating time by 90–95% (Stojceska et al. 2019). Even comparing the energy efficiency of ohmic heating with other novel heating methods like microwaves, the energy conversion rate of ohmic heating can be at least 35% higher than microwave heating (Chizoba Ekezie et al. 2017). Using ohmic heating as a pre-treatment method for drying can also improve the energy efficiency of the process as well as the food structure of the final product. According to Cokgezme et al., the combinational use of ohmic heating with vacuum drying can improve the energy efficiency by up to 27.88% comparing to vacuum drying alone. Generally, ohmic heating can decrease the processing cost of a company by 3.5–5 times (Çokgezme et al. 2017).

Excepting the energy efficiency benefits from ohmic heating, the transferred energy dissipates directly without undermining any internal solid particles in the food, which enhances the quality, nutrition value and value of the final products (Chizoba Ekezie et al. 2017). Among the novel heating methods, ohmic heating is the most suitable option for bread fermentation. As the gas cells developed during fermentation, the conductivity of the bread dough decreases linearly, which allows the company to demonstrate a simple numeric model to control the temperature through ohmic heating (Wang et al. 2021). From the environmental aspect, ohmic heating reduces about 3–5 times of the greenhouse gas emission than traditional drying methods (Çokgezme et al. 2017). Those characteristics make ohmic heating a sustainable, economically beneficial, high product quality technique for various type of food industries.

5.3.3 Infrared Radiation Heating

Infrared radiation (IR) belongs to the electromagnetic wave that is located between visible light and microwave electromagnetic radiation, having a wavelength of 0.78–1000 μm (Lee 2020). In recent years, IR heating application on food production has bought people's attention due to its convenience, versatility and rapid heating characteristics. Heat transfer of IR rays occurs by radiative heating at the surface and conductive heating on the inside of the food (Vaidyanathan and Krishnamurthy 2020). Currently, IR heating applications at industrial and pilot levels include baking, drying, thawing, pasteurization, sterilization, blanching, and peeling (Lee 2020).

Comparing to the conventional heating method, IR heating provides distinct advantages of efficient heat transfer to food products by reducing process time and energy costs (Lee 2020). Neglect the possible energy absorption of the air, infrared radiation can transfer heating energy directly to the food substance with no energy loss to the surrounding medium. Hence, IR heating can provide uniform, energy-efficient heating into the sample (Vaidyanathan and Krishnamurthy 2020). As summarized in Table 26.3, IR heating has improved the processing time, productivity, and energy efficiency when comparedwith convection heating. The use of an

Table 26.3 The comparison of energy consumption between IR heating and conventional heating (Lee 2020)

Attribute	Infrared heating process	Conventional heating process	Description
Billet heating time (mins)	20	240	The processing time using IR heating[1] is reduced by 12 times
Throughput (production rate) (lbs/hr)	350	80	IR heating[1] provides a greater production rate
Energy used for Al-billet heating (BTU/lb)	500	1500	The IR heating[1] uses 3 times less energy than the conventional heating[2]
Overall efficiency of the system (%)	~30	~10	IR heating[1] process is 3 times more efficient than the conventional heating[2] process

[1]Infrared heating process is performed by continuous-belt hybrid rapid infrared furnace
[2]Conventional heating process is performed by conventional convective gas-fired Furnace

infrared furnace has increased the system's overall efficiency by three times compared to the convective gas-fired furnace (Lee 2020). IR heating application on roasting also reduced 33% and 66% of processing time compared to sand and drum roasting, respectively (Bagheri 2020). However, designing a proper IR heating system is the key to improve the efficiency of food processing. A single electrical IR lamp emits IR energy in all directions, and the minority of them can get to the surface of the food. Insufficient heating also occurred in using gas-fired IR heat since the IR energy cannot emit to the backside. Hence, focusing the IR rays emitted onto the target surface will be necessary for establishing an effective heating process. As an example, the use of elliptical and parabolic reflectors with IR lamps can improve the emitting rate of IR rays to the desired surface. An IR system designed by Lee's team uses reflectors to concentrate IR energy to a limited surface, which saves more than 90% of the energy compare to the conventional heating method (Lee 2020).

One of the limitations for IR application is the low penetration ability of the infrared radiation, which creates difficulty in heating up the center of the food. Combining IR application with additional heating techniques can help to exceed the limitations.(Bagheri 2020). The combination of hot-air conventional roasting with infrared heating shows higher roasting efficiency and lower roasting time compare to using these methods alone (Bagheri 2020). A synergistic effect has distinguished between the Infrared radiation energy and hot air, which the hybrid application can reduce 31% of the energy usage compared to the conventional hot air method (Bagheri 2020).

On the other hand, the application of IR heat also demonstrates the advantage of uniform heating, reducing quality loss and the absence of solute migration. The devices used for IR heating are commonly inexpensive and smaller compared to other commercial heating methods, which can save more working space for the firm. Since the application of infrared heating does not require any medium, it brings convenience for the industry to clean up the working environment. Those advantages make IR heating a great practice for industry food processing (Lee 2020).

References

Aganovic K, Smetana S, Grauwet T, Toepfl S, Mathys A, Van Loey A, Heinz V (2017) Pilot scale thermal and alternative pasteurization of tomato and watermelon juice: an energy comparison and life cycle assessment. J Clean Prod 141:514–525

Aguilera J, Simpson R, Welti-Chanes J, Aguirre D, Barbosa-Cánovas G (2011) Food engineering interfaces. Springer, New York

Akpinar EK (2004) Energy and exergy analyses of drying of red pepper slices in a convective type dryer. Int Comm Heat and Mass Transfer 31:1165–1176

Akpinar EK, Midilli A, Bicer Y (2005) Energy and exergy of potato drying process via cyclone type dryer. Energy Convers Manag 46:2530–2552

Akpinar EK, Midilli A, Bicer Y (2006) The first and second law analyses of thermodynamic of pumpkin drying process. J Food Eng 72:320–331

Aktaş M, Khanlari A, Amini A, Şevik S (2017) Performance analysis of heat pump and infrared–heat pump drying of grated carrot using energy-exergy methodology. Energy Convers Manag 132:327–338

Altemimi A, Aziz SN, Al-Hilphy ARS, Lakhssassi N, Watson DG, Ibrahim SA (2019) Critical review of radio-frequency (RF) heating applications in food processing. Food Qual Safety 3(2):81–91. https://doi.org/10.1093/fqsafe/fyz002

Arendt E, Zannini E (2013) Cereal grains for the food and beverage industries. Woodhead Pub, Oxford, UK

Bagheri H (2020) Application of infrared heating for roasting nuts. J Food Qual 2020:1–10. https://doi.org/10.1155/2020/8813047

Brown NL, Pariser E (1975) Food science in developing countries. Science 188:589–593. https://doi.org/10.1126/science.188.4188.589

Brown ZK, Fryer PJ, Norton IT, Bakalis S, Bridson RH (2008) Drying of foods using supercritical carbon dioxide—investigations with carrot. Innovative Food Emerg Technol 9:280–289

BudžakI S, Leko J, Jovanović K, Viszmeg J, Koški I (2019) Air source heat pump assisted drying for food applications: a mini review. Croatian J Food Sci Technol 11(1):122–130

Cassano A, Conidi C, Drioli E (2011) Clarification and concentration of pomegranate juice (Punica granatum L.) using membrane process. J Food Eng 107:366–373

Chaudhry HN, Hughes BR, Ghani SA (2012) A review of heat pipe systems for heat recovery and renewable energy applications. Renew Sust Energ Rev 16(4):2249–2259

Chizoba Ekezie F, Sun D, Han Z, Cheng J (2017) Microwave-assisted food processing technologies for enhancing product quality and process efficiency: a review of recent developments. Trends Food Sci Technol 67:58–69. https://doi.org/10.1016/j.tifs.2017.05.014

Çokgezme Ö, Sabanci S, Çevik M, Yildiz H, Icier F (2017) Performance analyses for evaporation of pomegranate juice in ohmic heating assisted vacuum system. J Food Eng 207. https://doi.org/10.1016/j.jfoodeng.2017.03.015

Coşkun S, Doymaz I, Tunçkal C, Erdoğan S (2017) Investigation of drying kinetics of tomato slices dried by using a closed loop heat pump dryer. Heat Mass Transf 53(6):1863–1871

Dincer I, Sahin AZ (2004) A new model for thermodynamic analysis of a drying process. Int J Heat Mass Transf 47:645–652

Dong W, Hu R, Chu Z, Zhao J, Tan L (2017) Effect of different drying techniques on bioactive components, fatty acid composition, and volatile profile of robusta coffee beans. Food Chem 234:121–130

Du Pisani JA (2006) Sustainable development–historical roots of the concept. Environ Sci 3:83–96. https://doi.org/10.1080/15693430600688831
Einstein D, Worrell E, Khrushch M (2001) Steam systems in industry: energy use and energy efficiency improvement potentials. Lawrence Berkeley National Laboratory. Paper LBNL-49081. online: http://repositories.cdlib.org/lbnl/LBNL-49081
Fellows P (2004) Processed foods for improved livelihoods. FAO, Rome
https://www.theconsciouschallenge.org/ecologicalfootprintbibleoverview/food-and-energy; Food, May, 2019
Fritzson A, Berntsson T (2006) Efficient energy use in a slaughter and meat processing plant—opportunities for process integration. J Food Eng 76:594–604
Goh LJ, Othman MY, Mat S, Ruslan H, Sopian K (2011) Review of heat pump systems for drying application. Renew Sust Energ Rev 15(9):4788–4796
Heinz V, Toepfl S, Knorr D (2003) Impact of temperature on lethality and energy efficiency of apple juice pasteurization by pulsed electric fields treatment. Innovative Food Sci Emerg Technol 4(2):167–175
http://www.fao.org/faostat/en/#data/GN/metadata, 2019
Jiang H, Gu Y, Gou M, Xia T, Wang S (2020) Radio frequency pasteurization and disinfestation techniques applied on low-moisture foods. Crit Rev Food Sci Nutr 60(9):1417–1430. https://doi.org/10.1080/10408398.2019.1573415
Jinjiang Z, Yaosen W (2010) Experimental study on drying high moisture paddy by heat pump dryer with heat recovery. Int J Food Eng 6
Kim E (2013) The amazing multimillion-year history of processed food. Sci Am 309:50–55. https://doi.org/10.1038/scientificamerican0913-50
Knorr D, Augustin MA, Tiwari B (2020) Advancing the role of food processing for improved integration in sustainable food chains. Front Nutr 7:1–8. https://doi.org/10.3389/fnut.2020.00034
Kumar A, Croteau S, Kutowy O (1999) Use of membranes for energy efficient concentration of dilute steams. Appl Energy 64:107–115
Kuzgunkaya EH, Hepbasli A (2007) Exergetic performance assessment of a ground-source heat pump drying system. Int J Energy Res 31:760–777
Ladha-Sabur A, Bakalis S, Fryer PJ, Lopez-Quiroga E (2019) Mapping energy consumption in food manufacturing. Trends Food Sci Technol 86:270–280. https://doi.org/10.1016/j.tifs.2019.02.034
Lee E (2020) A review on applications of infrared heating for food processing in comparison to other industries. https://doi.org/10.1016/B978-0-08-100596-5.22670-X
Li H, Zhao Z, Xiouras C, Stefanidis GD, Li X, Gao X (2019) Fundamentals and applications of microwave heating to chemicals separation processes. Renew Sust Energ Rev 114:109316. https://doi.org/10.1016/j.rser.2019.109316
Liu Y, Zhao KZ, Jiu M, Zhang Y (2018) A heat pump system for Lentinula edodes drying and its drying property. Therm Sci 22(4):1759–1764
Loaharanu P (1996) Irradiation as a cold pasteurization process of food. Vet Parasitol 64:71–82
Menon A, Stojceska V, Tassou SA (2020) A systematic review on the recent advances of the energy efficiency improvements in non-conventional food drying technologies. Trends Food Sci Technol 100:67–76. https://doi.org/10.1016/j.tifs.2020.03.014
Midilli A, Kucuk H (2003) Energy and exergy analyses of solar drying process of pistachio. Energy 28:539–556
Mull TE (2001) Practical guide to energy management for facilities engineers and plant managers. ASME Press, New York
Nguyen LT, Choi W, Lee SH, June S (2013) Exploring the heating patterns of multiphase foods in a continuous flow, simultaneous microwave and ohmic combination heater. J Food Eng 116:65–71
Nikmaram N, Rosentrater KA (2019) Overview of some recent advances in improving water and energy efficiencies in food processing factories. Front Nutr 6. https://doi.org/10.3389/fnut.2019.00020

Okos M, Rao N, Drecher S, Rode M, & Kozak J (1998) Energy usage in the food industry. American Council for an Energy-Efficient Economy. Online: http://www.aceee.org/pubs/ie981.htm
Onsekizoglu P, Bahceci KS, Acar MJ (2010) Clarification and the concentration of apple juice using membrane processes: a comparative quality assessment. J Membr Sci 352:160–165
Ozyurt O, Comakli O, Yilmaz M, Karsli S (2004) Heat pump use in milk pasteurization: an energy analysis. Int J Energy Res 28:833–846
Perera CO, Rahman MS (1997) Heat pump dehumidifier drying of food. Trends Food Sci Technol 8(3):75–79
Pratap Singh A, Mandal R, Shojaei M, Singh A, Kowalczewski PŁ, Ligaj M, Pawlicz J, Jarzębski M (2020) Novel drying methods for sustainable upcycling of brewers' spent grains as a plant protein source. Sustainability 12(9):3660
Rajendran SRCK, Mason B, Doucette AA (2021) Review of membrane separation models and technologies: processing complex food-based biomolecular fractions. Food Bioprocess Technol 14:415–428. https://doi.org/10.1007/s11947-020-02559-x
Ramirez CA, Blok K, Neelis M, Patel M (2006) Adding apples and oranges: the monitoring of energy efficiency in the Dutch food industry. Energy Policy 34:1720–1735
Singh B (2013) Biofuel crop sustainability. Wiley 9781118635643, Ames
Smith R (2000) State of the art in process integration. Appl Therm Eng 20:1337–1345
Stojceska V, Atuonwu J, Tassou SA (2019) Ohmic and conventional drying of citrus products: energy efficiency, greenhouse gas emissions and nutritional properties. Energy Procedia 161:165–173. https://doi.org/10.1016/j.egypro.2019.02.076
Sun DW, Wang LJ (2001) Novel refrigeration cycles, chapter 1. In: Sun DW (ed) Advances in food refrigeration. Leatherhead Publishing, UK, pp 1–69
Sun J, Wang W, Yue Q (2016) Review on microwave matter interaction fundamentals and efficient microwave-associated heating strategies. Materials 9:231. https://doi.org/10.3390/ma9040231
Tiwari BK, Norton T, Holden NM (2013) Sustainable food processing. Wiley, Chichester
Toepfl S, Mathys A, Heinz V, Knorr D (2006) Review: potential of high hydrostatic pressure and pulsed electric fields for energy efficiency and environmentally friendly food processing. Food Rev Intl 22:405–423
Vadivambal R, Jayas DS (2010) Non-uniform temperature distribution during microwave heating of food materials—a review. Food Bioprocess Technol 3:161–171. https://doi.org/10.1007/s11947-008-0136-0
Vaidyanathan JS, Krishnamurthy K (2020) Infrared heating for decontamination. Reference module in food science. Elsevier. https://doi.org/10.1016/B978-0-08-100596-5.22348-2
Venkatachalam SK, Thottipalayam Vellingri A, Selvaraj V (2020) Low-temperature drying characteristics of mint leaves in a continuous-dehumidified air drying system. J Food Process Eng 43(4):e13384
Walker ME, Lv Z, Masanet E (2013) Industrial steam systems and the energy-water nexus. Environ Sci Technol 47:13060–13067
Wang LJ (2008) Energy efficiency and management in food processing facilities. Taylor and Francis, Boca Raton
Wang L (2014) Energy efficiency technologies for sustainable food processing. Energ Effic 7:791–810. https://doi.org/10.1007/s12053-014-9256-8
Wang Z, Huang J, Ma S, Wang X, Sun B, Wang F, Li L, Bao Q (2021) Novel heating technologies to improve fermentation efficiency and quality in wheat products: a short review. Grain Oil Sci Technol 4(2):81–87. https://doi.org/10.1016/j.gaost.2021.01.001
Wiktor A, Singh AP, Parniakov O, Mykhailyk V, Mandal R, Witrowa-Rajchert D (2020) PEF as an alternative tool to prevent thermolabile compound degradation during dehydration processes. In: Pulsed electric fields to obtain healthier and sustainable food for tomorrow. Academic, pp 155–202
Yang J, Bingol G, Pan Z, Brandl MT, McHugh TH, Wang H (2010) Infrared heating for dry-roasting and pasteurization of almonds. J Food Eng 101:273–280

Zimparov V (2002) Energy conservation through heat transfer enhancement techniques. Int J Energy Res 26:675–696

Zlatanović I, Komatina M, Antonijević D (2017) Experimental investigation of the efficiency of heat pump drying system with full air recirculation. J Food Process Eng 40(2):e12386

Chapter 27
Food Safety and Security (HACCP and HAZOP) for Consumers and Workers (Nonthermal Technologies and Their Use)

Muhammad Kamran Khan, Muhammad Imran, Muhammad Haseeb Ahmad, Rabia Shabir Ahmad, Farid Chemat, and Aliza Zulifqar

1 Introduction

In past few years, research shown more interest in applying non thermal technologies as compared to thermal technologies, these technologies fulfil the required function as well as preserve the functional properties of food (Van Impe et al. 2018). Innovative non-thermal technologies include pulsed electric field, ultrasound, high pressure processing and gamma rays etc. These technologies mainly used in extraction, processing, preservation and modification; these technologies overcome all negative impact of thermal processing that traditionally used (Daniela Pingret and Catherine Renard 2012). Non-thermal technologies, which have a Green chemistry principle, extensively studied as innovative techniques. When comparing the performance of the conventional heating with non-thermal techniques, two of the advantages should be considered, high rate of energy transfer and reduced nutrient losses. An alternative option to conventional thermal processing is ultrasound. The proliferation of ultrasound in any liquid product engenders bubbles which are mainly due to changes in pressure. In result of these pressure changes, induced bubbles disintegrate and cause a sudden temperature and pressure increases. This sudden powerful and localized pressure and energy convey pasteurization effect to product without heavy momentous change in temperature (Jiménez-Sánchez et al. 2017). Among

M. K. Khan (✉) · M. Imran · M. H. Ahmad · R. S. Ahmad · A. Zulifqar
Department of Food Science, Faculty of Life Sciences, Government College University, Faisalabad, Pakistan
e-mail: mk.khan@gcuf.edu.pk

F. Chemat
Université d'Avignon et des Pays de Vaucluse, UMR 408 INRA-UAPV, Sécurité et Qualité des Produits d'Origine Végétale, GREEN (Groupe de Recherche en Eco-Extraction des produits Naturels), Avignon, France

A. Režek Jambrak (ed.), *Nonthermal Processing in Agri-Food-Bio Sciences*, Food Engineering Series, https://doi.org/10.1007/978-3-030-92415-7_27

other innovative non-thermal technologies, PEF is also efficient technology which is being used nowadays for the safe beverage production (Rawson et al. 2011). HPP is another non thermal technology which is used for both solids and liquid products (mostly in batch system). This technology uses different ranges of pressure on products which are pre packed before this step, fruits and vegetables mostly preserved by applying this technique (Betoret et al. 2015). Gamma rays, being non-thermal technique, directly damage the DNA of microorganisms, thus preserving food from microbial contamination (Eissa et al. 2014) (Table 27.1).

2 Non-Thermal Technologies and Their Industrialization

It is clearly evident the role of technology in society from the past. With time being new development in industry continues towards the urge of producing high performance technology with low cost. Such development replaces the older one and become a necessity. In these developments, safety for material handler, environmental concerns of surrounding areas should be balanced too along with performance and quality (Chen et al. 2015). The scale and direction of further technological development should also be considered, considering that these process phases are critically dependent on energy transfer within reactions to providing the maximum level of process performance and final-product consistency.

In general, non-thermal methods use relative permeability measurements by a network analyzer, which is an indicator of the heat produced in a sample (Pereira et al. 2008). To scale up these methods from to laboratory to industry processes, quantitative classification done successfully for a) Electric field profile, magnetic field and radio waves in the device and (b) changes in the resulted product. In collaboration with Cooperative National ND de Chime in Montpellier, Serem has set up a unique platform in Europe for industrial scale experimentation. The main parameters of a pilot-scale installation were maximum induction of ultrasound power and high achievable US power density. Process intensity is actively adapted based on non-thermal technology to achieve better positions on an industrial scale (Motarjemi and Mortimore 2005). High response rates and advanced selectivity, along with the opportunity of repetitive procedures, show the useful implementation of these permitting strategies. A vital role of scaling up is continuous flow processing. The main benefit of continuous processing is the convenience of amplifying responses through the working of a couple of systems in parallel or correlated strategies, in order that product-level volumes can be easily achieved (Hill 1998).

Table 27.1 Applications of non-thermal technologies in food processing

Technology	Applications	Advantages	Products	Reference
Ultrasound	Preservation	Improve the shelf life Minimal negative effects on fruit juice Lower processing cost	Pomegranate	Pala et al. (2015)
	Processing	Cultivate the probiotic strain *L. casei* which was then able to ferment sonicated pineapple juice without any nutrient supplementation Improve shelf life	Pineapple	Costa et al. (2013)
Pulsed electric field	Preservation	Log reduction of *Z. bailii* - Retention of ascorbic acid and antioxidant capacity	Amla	Bansal et al. (2015)
		Microbial stability for 31 days Conservation of nutritional values and fresh-like characteristics	Fruit juice	Morales-de La Peña et al. (2010)
		The maximum inactivation of native microflora Retention of carotene content and overall acceptability	Mango	Kumar et al. (2015)
High pressure processing	Processing	Minimal negative effects on nutritional quality of fruit juice Improve the shelf life Lower processing cost	Apple juice	(Sokołowska 2013)
	Blending	Product was similar to unprocessed controls and appeared to retain fresh like characteristics Minimal effects on nutritional quality	Red fruit-based smoothies	Hurtado et al. (2017)
	Production	Microbial reduction of *E. coli, Salmonella typhimurium* and *L. monocytogenes* Significant changes in the glucose, fructose and sucrose activity after any of the treatments	Coconut	Lukas (2013)
Gamma rays	Packaging	Reduction in microbial count Improve shelf life, elevated color Off flavor development	Red pepper powder	Jung et al. (2015)
	Preserving	Prolong shelf life by decreasing microbial count No significant effect on color and flavor Increased antioxidant contents No significant nutritional loss	Grape juice blends	Carvalho Mesquita et al. (2020)

3 How HACCP and HAZOP Become Important for New Technologies Adopted at Industrial Scale?

The quality and safety protocols of food products were accomplished by utilizing the idea of quality control in any food sector. By keeping this concept, hazard analysis critical control point as well as hazard operability strategies are utilized as executive's safety and quality management system. In any activity, nothing is a higher priority than the wellbeing of personal, particularly safety of workers along with quality of finished product. In a running industry different safety and quality parameters are already being applied but whenever a new technology is adopted there are different limitations and unknown hazards which may not be safe for both the operators and end product. For such technologies hazard analysis critical control point (HACCP) becomes mandatory to access the hazards, critical limits and corrective action to be taken. Presently hazard operability technique as a reciprocal framework, is being endeavored to be utilized in the food industry with the integration of HACCP investigation. Thus, this aggregate conceptualizing exertion can make a careful survey of the process. Three fundamental approaches are focused in food industry in any operations. First of all, to ensure the safety measures that will viably keep food from contaminations and other perils, secondly to understand a successful utilization of man power and materials in order to achieve the highly economical production mode and at the last, it's very important to get an appropriate food quality and safety to raise hygiene managing level of practitioners. Above stated approaches are principle preferences of HACCP framework which demonstrate the need of industry to actualize HACCP system for any novel technologies (Kushwah and Kumar 2017). Nowadays the food business isn't just liable for delivering safe food, a job which has consistently been perceived, yet additionally for showing in a straightforward way how food handling has been arranged such as to measure the hazards and ensure the product safety. This is done by improving the HACCP studies and plantings as a feature of the food handling confirmation system (Motarjemi and Mortimore 2005). As customers have few assumptions for the food supply, including that it should be nutritious, healthy, unadulterated and safe. Lately customers have set expanded accentuation on sanitation and expect that food ought not to add to persistent illness like cancer and heart diseases. No industry would need to see customers becoming sick from utilization of the food it produces while simultaneously it loses its standing and the trust of consumer (Knowlton 1981). Guaranteeing food safety implies a consistent fight against microbial and chemical risks. To combat this effectively HACCP system gets compulsory for food industries.

As HACCP framework counts all the risks regarding safety of the product and process, the HAZOP framework connects significance to the security and safety of workers along with equipment. Contrary to conventional safety contemplations, there are some other safety concerns novel to non-thermal methods, for example, direct impacts of ultrasound waves, execution attributes, and processed materials and impact of rays. Consequently, it is hard to build up a single standard for recently received advanced implemented techniques. Effective use of men force and their safety are major concerns for any operating industry. In any industry there are a number of reasons which can lead towards need of HAZOP implementation for innovative/newly adopted technologies; workers safety and safety of equipment's and items are the most important one. Safety of operating procedures of a process steps, identified constructive nonconformities in apparatus and system, overall will improve the safety of already existed plant and will help to decide that where a plant should be installed. To check and assure all these points HAZOP application becomes complementary for industries (Boonthum et al. 2014). To meet above protection assessments any enterprise calls for a systematic evaluation of protection, hazards, and associated risks within different operating units (storage room, processing hall, etc.), within the chemical process industry or in different words, calculation of risk to the general public health, plant and surroundings due to risky operation/mishap in a chemical system plant. HAZOP is one such study, wherein potential risks to the plant, to the personnel, and to the plant environment are recognized, assessed, and various chance factors are computed primarily based on the chance of prevalence and the effects of such dangers (Khan and Abbasi 1997). HAZOP mainly identify all momentous accident results and major reasons behind these accidents including the which can be occur and the majority of accident causes, including all the cases which are likely to occur i.e. have a full-size danger contribution (Taylor 2017).

4 Hazard Analysis Critical Control Point HACCP

The hazard analysis critical control point (HACCP) system is a procedure that identifies and perceives the dangers and risks connected with the processing, storage and delivery of products, and implements the best controls focused on the elimination of the risks at specific point in processing (Kushwah and Kumar 2017). HACCP is main element of modern food safety management system as layout, application, control and management. HACCP system is very critical and important for the production of safe food products. HACCP (abbreviation of Hazard Analysis Critical Control Point) was developed by the Natick Army Laboratory of the National Aeronautics and Space Administration of USA with association of non-public effective food company Pillsbury for starting a food producing management system that's ensure the food safety of astronauts (Motarjemi et al. 1996). The idea was recognized and acclaimed between professionals in food safety on its application in 1960. From 1960s, HACCP had become the gold standard in valuing food safety from farm to fork.

HACCP food safety system considered as a cost-effective system for food safety. It is a safety tool and it must be implemented as a safety tool in complete production line (Knowlton 1981). Food technologist take the food in terms of hazards and risks, "Hazard" known as any biological, chemical or physical agent which cause an adverse health effect in human. Hazard Analysis and Critical Control Point (HACCP), a food safety tool, is mainly known as a powerful way to reduce the risks in the food industry. HACCP stimulates an efficient, protective approach to improve food safety regarding biological, chemical, and physical hazards tangled in food production line (Boonthum et al. 2014). The key and novel function of HACCP is to control product quality over the critical steps in the production line than normally inspection of finished products. Producers got a scientific methodology for finding and avoiding potential hazards, and compromised product quality (Khan and Abbasi 1997). Literature of safety and management showed that HACCP is a food safety tool which enables the safety of food products and significant protection for both customers and enterprises.

4.1 Principles

The HACCP is a food safety system which basically comprises of seven principles. A working framework developed by using these seven principles along with quality and safety characteristics of system (Taylor 2017). This combination of food safety approach and systematic quality approaches known as an important aspect of HACCP implementation in any industry by food scientists.

These are the universal principles of HACCP

- Draw a process flow diagram and conduct a hazard analysis
- Pointing the critical control points in process flow diagram (CCP's)
- Defining critical limits for identified critical control points
- Development of a system for monitor/control of the CCP which were identified by establishing principle 2 of HACCP
- Establish a plan to take a corrective action when the previous step indicates that any of the CCP's is not under control.
- Development of a system that verify that implemented HACCP system is working competently.
- Establishment of a documentation plan which keeps all the procedures, and keep the record of deviation which occur and handled according to principles (Dunjó et al. 2010)

4.2 HACCP and Non-thermal Processing

4.2.1 Quality Assurance

One of the pillars in food manufacturing is quality assurance during production, monitoring which involves a number of analyses of product (Casola et al. 2019). Before the implementation of non-thermal technologies in food sector, some of the analysis was performed with destructive thermal technologies, which were also time taking and long testing procedures. Non-thermal technologies are helpful for such analysis, for instance, ultrasound with low-intensity has been used to assess the quality of watermelons, avocados and mangoes by estimating ultrasonic restrictions such as speed and attenuation relative to the physical properties of the medium. The quality of red meat, white meat, cod, pork, milk, alcohol, sugar solutions and ghee/oil is assessed using ultrasonic parameters, applying the principle that how the wave structure (speed and attenuation) relays the physical structure of the medium (Cuihua 2014). Cheese texture, vegetables (cooked) texture, and ripening level of fruit all have been accessed using ultrasonic waves (Sicaire et al. 2015). High pressure processing (HPP) inactivates the microbes by applying the high pressure on the cell membranes which results in rupturing of cell. The Pulsed Electric Field (PEF) make negligible changes in food properties while ensuring optimal safety. Pulsed light is used in the treatment of various (transparent) fluids. Both high pressure processing and pulsed electric field treatments kill vegetative bacteria, mainly by mechanical destruction of cell structure (Kushwah and Kumar 2017). Food quality is an important component that compels the development of innovative technologies to lessen thermal decomposition, remove foodborne pathogens from food products and provide maximum nutrient products. Non-thermal techniques can be used, but important control procedures of non-thermal processing are required (Motarjemi et al. 1996). The HACCP program needs to be set up for use in food engineering, in which critical control points of food processing can be identified so that potential risks can be controlled in the production of a safe quality product.

4.2.2 Process Control

Wherever HACCP is implemented it covers all the going on process in food processing plant from transportation of raw material to packaging of end product as well its storage or placement in warehouse. As HACCP is systematic and following approach, so for proper working flow charts are necessary which contain all steps that have to be followed in any specified process. This flow chart must be confirmed by the HACCP team members. In non-thermal novel techniques different equipment's used which basically modulate and preserve the foods which are of liquid and semiliquid consistency (Arvanitoyannis and Stratakos 2010). In non-thermal processing modification or change in characteristics like emulsification, foaming etc. and preservation, needs a heat exchanger and a treatment area which varies according to technology being used for example it may be use of high pressure, ultrasonic waves, gamma rays or pulsed electric field. Hazards which are mainly focused are of three types i.e., physical, chemical or biological. HACCP team selects these hazards according to process and equipment. In pulsed electric field treatment, electric field strength, pulse waveform, energy input, temperature, food composition and volume to be processed are considered as critical control points. In ultrasound treatment of food, intensity of ultrasonic waves, time of contact, type and

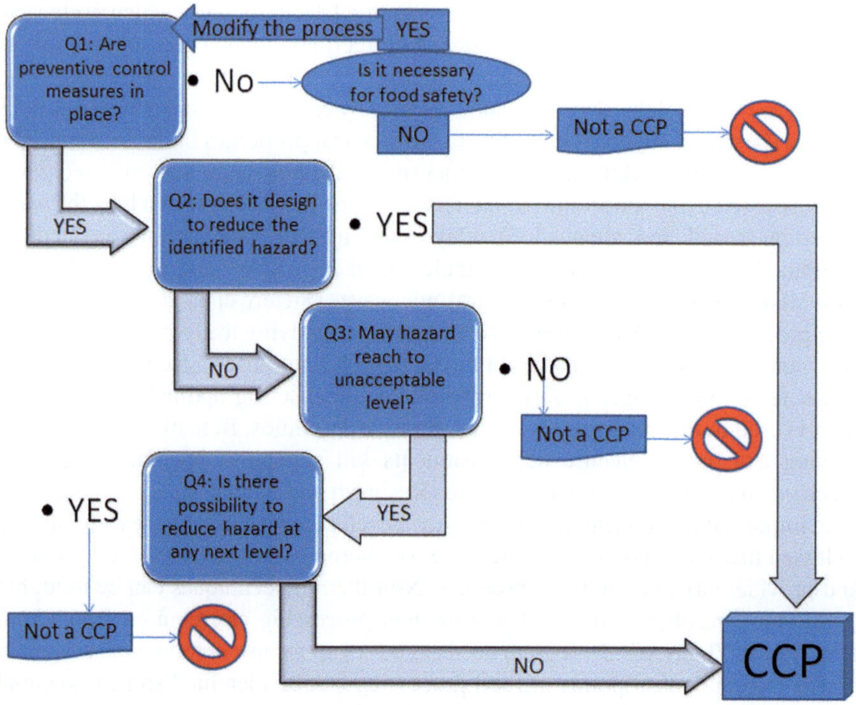

Fig. 27.1 CCP decision tree for HACCP system

nature of microorganism, composition and volume of food to be processed are considered as critical control points. In processing by gamma rays, critical control points are assumed to be the dose of rays, exposure time, type of product, type and nature of microorganism Therefore, the HACCP team needs knowledge of the potential microbial, chemical, physical and biological hazards associated with the processes under evaluation. After identifying current hazards and control measures, a step or process is defined so that food safety risk can be avoided, eliminated, or reduced to an acceptable level (Barbosa-Cánovas et al. 1999). This stage or control process is an important control point (CCP). CCPs are selected using the CCP Decision Tree (Fig. 27.1). This decision tree is designed to assist the HACCP team in determining the critical control point against a control point that can handle specific, logical questions. This assumed decision tree is not perfect and main purpose is to get attention of HACCP team members, to identify the critical control points and what action should be taken to control the hazard (Table 27.2).

Table 27.2 Critical control points according to decision tree

	Questions from the CCP decision tree				CCP or not a CCP	
Processing steps	Question 1	Question 2	Question 3	Question 4		Hazards associated
Reception of raw material	Yes	Yes			CCP	Microbes, physical pollutants, chemicals
Storage	Yes	No	Yes	Yes	Not a CCP	
Preparation	No	No			Not a CCP	
Mixing	Yes	No	Yes	Yes	Not a CCP	Adulterants, excess of food additives
Pulsed electric field treatment	Yes	Yes			CCP	Biological, physical and chemical
Ultrasound treatment	Yes	Yes			CCP	Biological, physical and chemical
Irradiation treatment	Yes	Yes			CCP	Biological, physical and chemical
Drying	Yes	No	Yes	Yes	Not a CCP	
Aseptic packaging	Yes	Yes			CCP	Pathogenic microbes
Finished product storage	Yes	Yes			CCP	Pathogenic microbes

4.2.3 HACCP and Preservation Through No Thermal Processing

In food preservation main aim is to kill or inactivate unwanted microorganisms. There are many treatments used for this purpose like sterilization and pasteurization but this process may contribute to quality of food by affecting its characteristics (Soriano et al. 2002). Non-thermal food preservation includes the innovative technologies which basically target the initial population of undesirable microorganisms to a level that is considered to be safe, without effecting or minimal effect on food quality characteristics.

In HACCP approach the key attention is given to avoid contamination and deterioration of products by keeping a check on coming raw material, processing ingredients, packing materials and subsequent processing in the food processing plant, cross-contamination and degradation of products, their subsequent inspection, acceptance and storage until they are used, and inspections or complete check on semi processed and fully processed products from the time when they were produced (Notermans and Mead 1996; Vanne et al. 1996). Nevertheless, with some emerging technologies, microorganisms become more resistant over time to the action of specific factors such as pressure, electricity, or sound waves. In these cases, some reports show a positive effect on some microorganisms and enzymes when combined with other preservation techniques (Piyasena et al. 2003). However, with some new processing technologies, microbes become more resistant over time to the action of specific factors such as electricity or high intensity sound waves. In such cases these techniques can be used in combination to produces the desired results (Rodríguez et al. 2003). High pressure will have a lethal effect on undesirable microbes when applied in combination with ultrasound treatment. If ultrasound used as single process, its effects are too gentle towards microbial population. To increase the effectiveness of processing or preservation, ultrasound can be combined with pressure or electricity too and same with the other non-thermal food preservation technologies (Soria and Villamiel 2010).

5 Hazard Analysis and Operability

Hazard analysis and operability (HAZOP) is a systematic methodology used to analyze the process to identify ways in which parameters or specific conditions of the process or equipment can deviate from their intended specification or standards and thus cause hazards or problems with operability (Single et al. 2019). It is the most commonly used Process Hazard Analysis (PHA) method accepted by regulators all over the world today. The main objective of the HAZOP is to identify potential process, plant, equipment, or operating system hazards. These hazards may include those that are relevant to a particular localized area of a process, equipment, or operating system, as well as those that may affect or influence upstream and downstream plant processes (Baybutt 2015). HAZOP also identifies possible operational problems and the causes of operational disturbances and production abnormalities within the process, equipment and operating systems. This method identifies the deviation from the intended specific conditions by using guide words such as No, More, Less, As well as, Reverse, and Other Than to the process parameters such as pressure, temperature, flow, reactions, etc. on the specific portion of the process at which specific parameters are being investigated for deviation called Nodes (Hyatt 2018). A significant advantage of HAZOP is that the resulting expertise, gathered in a standardized and comprehensive way to recognizing possible hazards and operability issues, provides the investigation team with more details when assessing suitable remedial steps and thereby processing safety barriers or safeguards (Taylor 2017).

5.1 Principles

Basic underlying principles are as follow:

- **Definition and preparation –** Scope and objectives of HAZOP analysis are defined comprehensively and teams with different expertise are selected for assigning specific responsibilities. Plan for HAZOP study, collect data, determine the style of recording and arrange meetings (Hyatt 2018).

- **Design representation -** A descriptive model adequately describing the system, equipment or process under analysis, its parts and elements, and identifying their process parameters. The representation may be of the physical design or logical design that gives the function of each part and element qualitatively and quantitatively (Siddiqui et al. 2014).
- **Examination –** Divide the system into parts or sections (Nodes) and then select the nodes and define node intent (desired process behavior) and process parameters. Identify deviation by using guide word on process parameter, identify and record the cause and consequences of this deviation and then estimate its severity, like hood and risk (Rossing et al. 2010). After this, identify and establish protection and corrective actions to mitigate this deviation and set up monitoring procedures to check the efficiency of corrective actions.
- **Record keeping and documentation**- record the examination process and document the final output report of the analysis and follow up on actions required.

5.2 HAZOP and the Potential Hazard of Using Non-thermal Equipment

5.2.1 Ultrasonication

Ultrasonication is a non-thermal processing technique that creates alternating low- and high-pressure waves, causing the development and vigorous collapse of micro-bubbles called cavitation. Recently, much attention has been centered on ultrasound application in food processing, preservation and active compound extraction. Many research studies indicate technological benefits from the integration of conventional techniques with ultrasound. Ultrasonication generally requires processes that can reduce processing time and increase rates, improve quality and safety, (Ravikumar et al. 2017).

Direct effect on operators Accidental operator's contact exposure to ultrasonic waves is the key threat that operators can face during the processing operation. Much industrial application of ultrasound often incidentally produces and transmits sound waves at high-pressure level in the sonic and ultrasonic range in the air can cause a hazard to be arising from the airborne ultrasound reception of ears (Smagowska and Pawlaczyk-Łuszczyńska 2013). Operator contact exposure occurs due to the direct intimate interaction between the transducer and the operator tissues when there is no air gap between them. An air gap, however, can decrease the transmitted ultrasonic energy. Accidental immersion of the operator's body parts i.e., hands in an ultrasonically excited water bath may cause tissue damage and approximately 65 percent of the energy transmitted to the bone from radiations (Sicaire et al. 2015). Ultrasound equipment running at a low frequency will readily cause tissue damage through contact exposure. The ultrasonic wave emitting from the humidifier can cause severe pain in the finger within a second due to overheating of bones. However, when the ultrasound equipment operating at high frequencies between 20 and 40 kHz can cause pain in the operator's hands on direct exposure to ultrasonication (Shankar et al. 2011). Ultrasound equipment is an electrical system that provides operators with a possible threat. Electricity is an invisible energy that, during electrical installation, makes it much more dangerous because it can readily cause electric shocks and burns, and even pose a possible fire hazard (Jafari et al. 2012).

The indirect effect on operators Airborne ultrasound waves primarily cause a seriously detrimental effect on the nervous system and cause harm to the operator's ear. Several other health consequences caused by ultrasonic waves such as fatigue, nausea, headache, tinnitus, hearing loss, and disruption of neuromuscular synchronization have been reported (Kühler et al. 2019). Direct exposure to both ultrasonic waves and audible acoustic energy can also cause hearing loss, nausea, headache, fatigue and tinnitus in females working near the area of an ultrasonic cleaner. The same symptoms have been experienced by the operator working with laboratory equipment's operating at high-frequency acoustic/ultrasonic waves (Ahmadi et al. 2012). The use of acoustical energy of 165 dB can cause the local heating and burning sensation in the gaps between the fingers and excruciating heating and burning sensation on the hand palm after several second of exposure, primarily due to sound waves absorption and heating (Chen et al. 2006).

5.2.2 Pulsed Electric Field Technology (PEF)

PEF is a non-thermal processing technique operating at moderate temperature by exposing the food material to short pulses at high electric field intensities (Toepfl et al. 2014). The food industry has strong prospects of using PEF processing because promising results have been obtained on the nutritional improvement, product safety and inactivation of unwanted microorganisms. Besides, problems with nutritional losses in PEF treated foods seem to be minimal. In conjunction with other non-thermal processing techniques or mild heating techniques, the use of this technology will increase microbial destruction (Syed et al. 2017).

Direct effect on operator The operator can face a serious threat of accidental contact exposure to the pulsed electric field during the processing operation. Direct exposure to an electric field can damage the skin tissue, redness and burning sensations (Pakhomova et al. 2012). Cataract development has been investigated due to exposure of the eye to high-intensity electric field radiation (Liu et al. 2017). The electric field in the processing area is more dangerous because it can cause electrical shocks, burns, muscular spasms, headaches, convulsions, nervous disorders, cardiac arrests and even pose a possible fire hazard in the processing area.

The indirect effect on operator The electric field can affect the cardiovascular system by slightly increasing and decreasing the heartbeat rate (3–5 beats/minutes).

Nevertheless, no obvious acute or long-term cardiovascular-related risks have been proven at levels below a certain threshold level. When a certain threshold level of the field strength has been passed, it can affect the brain nerve tissues, change in response time for complex reasoning tasks and change in time perception has been occurred in worker or operator in the pulsed electric field processing area. It has also been suggested that electric fields might decrease the level of certain hormones such as melatonin in operators that ultimately induce the symptoms of cancer (Lerchl et al. 2015).

5.2.3 Irradiation

Food irradiation is a non-thermal food preservation process. It is a treatment of food exposition on an amount of energy in the form of rays. Irradiation processing results in improved microbiological safety reduced storage losses, and prolong shelf life depending on radiation dose (Ehlermann 2016). Irradiation treatment is a physical process in which controlled doses of the energy of X-rays, gamma rays or electron beams are delivered to food commodities, bulk or pre-packaged. Irradiation affects living organisms directly and indirectly, such as microorganisms, insects and parasites, which results in its preservative effect. The following ionizing radiation has been approved for food treatment: cobalt-60, gamma radiation with a maximum energy of 1.17 and 1.33 MeV, accelerated electrons beam with a maximum energy of 10 MeV, cesium-137 with an energy of 0.66 MeV, and X-rays with a maximum energy of 5 MeV, respectively (Roberts 2014).

Direct effect on operator Being exposed to a lot of radiation can cause skin burn, skin redness and radiation burns. The direct exposure may also lead to headaches and diarrhea in operators and workers dealing with irradiation processing operations. Radiation exposure beyond a certain threshold level can impair the functioning of certain tissues and organs of the body of the operator (Mostafavi et al. 2012). During Irradiation processing, electrical shock can occur that is a physical effect and a violent electrical force reaction entering the body. The main electrical injury indicating tissue damage occurs after the electric current is encountered. Electricity can also cause serious burns to the body. The reason is concealed from the electrical resistance of the body in the power dissipation. The symptoms of shock can be cardiac arrest, tissue and organ burns, muscle spasms, extreme nervous system effects, and any unexpected consequences (Mettler 2012). Other disorders, depending on which organs the current passed through, can occur within weeks or months after the shack.

The indirect effect on operator Extremely high radiation exposure can cause acute health problems such as radiation sickness, headache, hearing loss, and skin redness and burns, just like being exposed, like being exposed to an atomic explosion (Dörr and Meineke 2011). Long-term exposure may cause health problems such as cardiovascular disease and cancer. However, the low ambient radiation

Table 27.3 Operability hazards associated with non-thermal technologies

Non-thermal processing techniques	HAZOP and process parameters where hazards can occur	Hazards
Ultrasonication	Sound energy density, intensities and frequencies, temperature, ultrasound power, size and shape of ultrasound reactor, solvent, external pressure and presence of dissolved gases, properties of the matrix	Electric shock, airborne ultrasound waves, increasing in process temperature, high ultrasound power, rusting in the chamber
Pulsed electric field (PEF)	Electric field intensity and field strength, wave type, pulsed numbers, frequency and width, processing time and temperature, solvent	High electric field intensity, process temperature, changes in wave type and shape,
Irradiation	Irradiation source, dose, time, temperature, radiation wavelength, frequency, intensity dose distribution, irradiation chamber and detector	Overdosage beyond a threshold level, detector malfunctioning, radiation sources

exposure level does not cause immediate health consequences but is a small contributor to the overall risk of cancer (Singh and Seed 2018). Radiation exposure over a short period can cause symptoms like nausea and vomiting within hours and often result in death over the following days and weeks which is called acute radiation syndrome. Radiation exposure can damage operators' tissues and organs depending on the dose of radiation received or absorbed, which is generally expressed in an international unit called gray (Gy). The potential damage from an absorbed dose also depends on the radiation type and the sensitivity of different tissues and/or organs. More than 0.75 Gy exposure can induce the symptoms of acute radiation syndrome. However, exposure to the low radiation level does not cause immediate health effects but can lead to a small increase in cancer risks over a lifetime (Kamiya et al. 2015). These health consequences are more severe at higher doses rates and longer exposure (Morgan and Sowa 2015) (Table 27.3).

6 Nonthermal Technologies for Safe Use

6.1 Ultrasonication

Limited occupancy and direct or indirect contact exposure to high-power ultrasound operating areas must be always prohibited. The operating staff and industrial safety inspectors should be informed about the potential adverse effects of high-power ultrasound waves and the appropriate protective actions to be taken. Only operators trained for the use of ultrasound equipment or persons under strict control should be permitted to enter the facility when the equipment is running. Safety labels must be displayed on the high-power ultrasound systems that could cause a person to receive ultrasound contact exposure and the equipment should not be touched when it is working or a certain part of it (Badri et al. 2012). The low-power ultrasound levels, in general, cause little risk of harm from accidental contact exposure. However, excessive contact exposure should be avoided as biological impact data are still inconclusive. It is important to put warning signs on all high-power ultrasound products.

There should also be a statement accompanying each alert sign that shows safety and precautionary measure to be taken when the equipment is operating. Safety labels should be displayed on all ultrasonic cleaning tanks to alert operators not to touch or immerse hands or other parts of the body in the tank when the cleaning process is running (Houston et al. 2011). The safety protocols for staff protection must ensure that the ambient sound pressure levels should not exceed the recommended maximum allowable level for the operator. This is done first by measuring the acoustic energy and where possible, by lowering exposure levels with certain engineering controls, by installing sound-absorbing materials and containment baffles on or in its path. If it is not possible to provide engineering controls, then ear protection should be used that reduces the amount of ultrasound in their ears (Chen et al. 2013).

6.2 *Pulsed Electric Field Technology*

During PEF treatment, voltage strength, which is in the kilovolt range, is the main concern for the operators working in the PEF facility. A high voltage power supply is required to charge the capacitor and the discharge switch releases the stored electric energy in the form of the electric field through the product (Nimunkar and Webster 2008). The power supply, capacitor and treatment chamber must be confined in the restricted access area with interlocking gates. If opened when the power supply is on, the gates will shut the pulser off. There should be emergency switches with easy access in case of a process failure and discharges locks must be given to discharge the products from the facility before the repair or inspection of the equipment's take place (Knoerzer et al. 2012). All chamber connections must be separated and the material carrying pipes to or from the treatment chamber must be attached to the ground in case of high-voltage leakage through any fluid in contact with the treatment chamber.

Detailed standard operating procedures for operating and maintaining the equipment should be in place and training must be conducted to educate and taught these procedures to the operator involved in the PEF. Proper warning signs for safety measures and risks must be in place in the manufacturing areas. The cleaning solutions i.e., detergents and sanitizers must be chosen that comply with the regulations of the FDA and UDSA or those of similar organizations in other countries. Operators must use safety equipment such as face masks, caps, apron, gloves and goggles when employing and eliminating the cleaning solutions. There must be a complete method to specify when, where, which and how to use cleaning and sanitizing solutions. Proper record-keeping is required to avoid contamination of the goods with detergents and cleaning solutions. There should be a complete layout of the facility/plant including details on the location of equipment and emergency exits. The layout must reflect modifications to the configuration of the facility (Vega-Mercado et al. 2004).

6.3 Irradiation Processing

The basic criteria for designing safe radiation processing equipment must ensure that access to the radiation room is denied during exposure to the irradiation source and ensure that no person is exposed to the radiators when inside the irradiation processing area. Irradiator operators work closely with a certain irradiator and are usually responsible for their safe operation day after day (Ic and Cetinkaya 2020). The irradiator should only be approved for use by trained irradiator operators. Education and training should be given to operators together with their responsibilities and duties for operating irradiating equipment. The personnel who routinely enter i.e., irradiation equipment operator and maintenance staff, should be subjected to dose monitoring. This personnel should wear whole-body monitoring gadgets such as dosimeter, film badges, optically stimulated luminescent dosimeter(Hine and Brownell 2013).

The International Atomic Energy Agency (IAEA) has regulated the safety standers that can establish safety requirements to ensure the health safety of operators and environmental safety from the harmful effects of high levels of ionizing radiations. Fundamental principles and procedures are established to monitor the exposure of humans to radiations and the release of radioactive materials to the atmosphere. These principles and procedures reduce the risk of events that may lead to radiation and minimize the effects of events if they occur. The protective measure should be implemented to the activities and processing area facilities that can cause radiation threats which include the use of radiation and its radioactive sources, transport of radioactive materials, and its waste management, etc. The safety assessment of the plant and its equipment should be conducted regularly for identifying and assessing the plant's predictive response to postulated equipment failure or malfunctions, a common cause of equipment failure and expected human errors and any other external events that could lead to an accident. Combinations of certain malfunctions, failures, errors and incidents should be included in this HAZOP analysis (Abdul-Halim and Davey 2016).The operating organization can consult a professional expert for regulating the radiation safety measure and for solving issues relating to radiation safety maintenance of engineering features and other facilities, health and safety optimization, dosimetry and radiation control, abnormally high exposure and over-exposure investigation, safety evaluation and contingency arrangements, improvements to a facility, equipment or facility (Parlato et al. 2014).

Radiation rooms are equipped with monitors that can detect high levels of radiation. High radiation levels occur in case of an accident such as when the electron beam is expected to be switched off or when the radioactive source is expected to be in the fully shielded position. Some fixed detectors are also designed to detect a high level of radiation when the product leaves the radiation room, a potential indicator of the dislodging of radioactive material from the radiation room. The radiation dose rates may be temporarily higher after the equipment has been shut down due to the dark current and the presence of activation products in the shielded, accelerator and ancillary parts of equipment. Radiation level survey of irradiation processing area should be carried out before personnel are permitted to enter.

References

Abdul-Halim N, Davey KR (2016) Impact of suspended solids on Fr 13 failure of UV irradiation for inactivation of Escherichia coli in potable water production with turbulent flow in an annular reactor. ChEnS 143:55–62

Ahmadi F, McLoughlin IV, Chauhan S et al (2012) Bio-effects and safety of low-intensity, low-frequency ultrasonic exposure. PBMB 108:119–138

Arvanitoyannis I, Stratakos A (2010) Applications of irradiation on meat and meat products. Irradiation of Food Commodities:171–221

Badri A, Nadeau S, Gbodossou A (2012) Proposal of a risk-factor-based analytical approach for integrating occupational health and safety into project risk evaluation. Accid Anal Prev 48:223–234

Bansal V, Sharma A, Ghanshyam C et al (2015) Influence of pulsed electric field and heat treatment on Emblica officinalis juice inoculated with Zygosaccharomyces bailii. Food Bioprod Process 95:146–154

Barbosa-Cánovas GV, Pothakamury UR, Gongora-Nieto MM et al (1999) Preservation of foods with pulsed electric fields. Elsevier

Baybutt P (2015) A critique of the hazard and operability (HAZOP) study. J Loss Prev Process Indust 33:52–58

Betoret E, Betoret N, Rocculi P et al (2015) Strategies to improve food functionality: structure–property relationships on high pressures homogenization, vacuum impregnation and drying technologies. Trends Food Sci Technol 46:1–12

Boonthum N, Mulalee U, Srinophakun T (2014) A systematic formulation for HAZOP analysis based on structural model. Reliab Eng Syst Saf 121:152–163

Carvalho Mesquita T, Evangelista Vasconcelos Schiassi MC, Maria Teixeira Lago A et al (2020) Grape juice blends treated with gamma irradiation evaluated during storage. RaPC 168:108570

Casola V, De Benedictis A, Rak M et al (2019) Toward the automation of threat modeling and risk assessment in IoT systems. Internet of Things 7:100056

Chen C, Chang B-S, Lund RA et al (2013) Process control for a sonication cleaning tank. Google Patents,

Chen H, Li X, Wan M (2006) The inception of cavitation bubble clouds induced by high-intensity focused ultrasound. Ultra 44:e427–e429

Chen J, Zheng X, Dong J et al (2015) Optimization of effective high hydrostatic pressure treatment of Bacillus subtilis in Hami melon juice. LWT-Food Sci Technol 60:1168–1173

Costa MGM, Fonteles TV, de Jesus ALT et al (2013) Sonicated pineapple juice as substrate for L. casei cultivation for probiotic beverage development: process optimisation and product stability. Food Chem 139:261–266

Cuihua Q (2014) Establish central kitchen under HACCP control in food and beverage industry to ensure food safety and hygiene. In: SHS Web of Conferences. EDP Sciences, p 03005

Daniela Pingret A-SF-T, Le Bourvellec-Samour C, Catherine Renard FC (2012) Lab and pilot-scale ultrasound-assisted water extraction of polyphenols from apple pomace. J Food Eng 111:73–81

Dörr H, Meineke V (2011) Acute radiation syndrome caused by accidental radiation exposure-therapeutic principles. BMC Med 9:1–6

Dunjó J, Fthenakis V, Vílchez JA et al (2010) Hazard and operability (HAZOP) analysis. A literature review. J Hazard Mater 173:19–32

Ehlermann DA (2016) The early history of food irradiation. RaPC 129:10–12

Eissa HA, Shaheen MS, Botros HW (2014) Impact of γ-irradiation on aroma flavour, bio-active constituents and quality attributes of water melon juice. J Plant Pathol Microbiol 5:227

Hill A (1998) Microwave ovens

Hine GJ, Brownell GL (2013) Radiation dosimetry. Elsevier

Houston LE, Allsworth J, Macones GA (2011) Ultrasound is safe... Right? Resident and maternal-fetal medicine fellow knowledge regarding obstetric ultrasound safety. J Ultrasound Med 30:21–27

Hurtado A, Guàrdia MD, Picouet P et al (2017) Stabilization of red fruit-based smoothies by high-pressure processing. Part A. Effects on microbial growth, enzyme activity, antioxidant capacity and physical stability. J Sci Food Agric 97:770–776

Hyatt N (2018) Guidelines for process hazards analysis (PHA, HAZOP), hazards identification, and risk analysis. CRC press

Ic E, Cetinkaya N (2020) Food safety and irradiation related sanitary and phytosanitary approaches-Chinese perspective. RaPC:109324

Jafari D, Cody K, Panebianco N et al (2012) Small retained foreign bodies: what is the limit of detection using current emergency ultrasound equipment? Crit Ultrasound J 1. Springer:1–13

Jiménez-Sánchez C, Lozano-Sánchez J, Segura-Carretero A et al (2017) Alternatives to conventional thermal treatments in fruit-juice processing. Part 2: effect on composition, phytochemical content, and physicochemical, rheological, and organoleptic properties of fruit juices. Crit Rev Food Sci Nutr 57:637–652

Jung K, Song B-S, Kim MJ et al (2015) Effect of X-ray, gamma ray, and electron beam irradiation on the hygienic and physicochemical qualities of red pepper powder. LWT Food Sci Technol 63:846–851

Kamiya K, Ozasa K, Akiba S et al (2015) Long-term effects of radiation exposure on health. Lancet 386:469–478

Khan FI, Abbasi S (1997) OptHAZOP—an effective and optimum approach for HAZOP study. J Loss Prev Process Indust 10:191–204

Knoerzer K, Baumann P, Buckow R (2012) An iterative modelling approach for improving the performance of a pulsed electric field (PEF) treatment chamber. Comput Chem Eng 37:48–63

Knowlton RE (1981) An introduction to hazard and operability studies: the guide word approach. Chemetics International Limited

Kühler R, Weichenberger M, Bauer M et al (2019) Does airborne ultrasound lead to activation of the auditory cortex? Biomed Eng/Biomedizinische Technik 64:481–493

Kumar R, Bawa A, Kathiravan T et al (2015) Optimization of pulsed electric field parameters for mango nectar processing using response surface methodology. Int Food Res J 22:1353

Kushwah A, Kumar R (2017) HACCP–its need and practices. Acta Chemica Malaysia 1:01–05

Lerchl A, Klose M, Grote K et al (2015) Tumor promotion by exposure to radiofrequency electromagnetic fields below exposure limits for humans. BBRC 459:585–590

Liu J, Yan XL, Zheng XL et al (2017) Electric field exposure promotes epithelial-mesenchymal transition in human lens epithelial cells via integrin β1-FAK signaling. Mol Med Rep 16:4008–4014

Lukas AR (2013) Use of high pressure processing to reduce foodborne pathogens in coconut water. Virginia Tech

Mettler FA (2012) Medical effects and risks of exposure to ionising radiation. JRP 32:N9

Morales-de La Peña M, Salvia-Trujillo L, Rojas-Graü M et al (2010) Impact of high intensity pulsed electric field on antioxidant properties and quality parameters of a fruit juice–soymilk beverage in chilled storage. LWT-Food Sci Technol 43:872–881

Morgan WF, Sowa MB (2015) Non-targeted effects induced by ionizing radiation: mechanisms and potential impact on radiation induced health effects. Cancer Lett 356:17–21

Mostafavi HA, Mirmajlessi SM, Fathollahi H (2012) The potential of food irradiation: benefits and limitations. Trends Vital Food Control Eng 5:43–68

Motarjemi Y, Käferstein F, Moy G et al (1996) Importance of HACCP for public health and development the role of the World Health Organization. Food Control 7:77–85

Motarjemi Y, Mortimore S (2005) Industry's need and expectations to meet food safety, 5th international meeting: Noordwijk food safety and HACCP forum 9–10 December 2002. Food Control 16:523–529

Nimunkar AJ, Webster JG (2008) Safety of pulsed electric devices. PhyM 30:101

Notermans S, Mead G (1996) Incorporation of elements of quantitative risk analysis in the HACCP system. Int J Food Microbiol 30:157–173

Pala CU, Zorba NND, Ozcan G (2015) Microbial Inactivation and physicochemical properties of ultrasound processed pomegranate juice. J Food Protec 78(3):531–539. https://doi.org/10.4315/0362-028X.JFP-14-290

Pakhomova ON, Khorokhorina VA, Bowman AM et al (2012) Oxidative effects of nanosecond pulsed electric field exposure in cells and cell-free media. Arch Biochem Biophys 527:55–64

Parlato A, Giacomarra M, Galati A et al (2014) ISO 14470: 2011 and EU legislative background on food irradiation technology: the Italian attitude. Trends Food Sci Technol 38:60–74

Pereira R, Martins RC, Vicente A (2008) Goat milk free fatty acid characterization during conventional and ohmic heating pasteurization. J Dairy Sci 91:2925–2937

Piyasena P, Mohareb E, McKellar R (2003) Inactivation of microbes using ultrasound: a review. Int J Food Microbiol 87:207–216

Ravikumar M, Suthar H, Desai C et al (2017) Ultrasonication: an advanced technology for food preservation. Int J Pure App Biosci 5:363–371

Rawson A, Patras A, Tiwari B et al (2011) Effect of thermal and non thermal processing technologies on the bioactive content of exotic fruits and their products: review of recent advances. Food Res Int 44:1875–1887

Roberts PB (2014) Food irradiation is safe: half a century of studies. RaPC 105:78–82

Rodríguez JJ, Barbosa-Cánovas GV, Gutiérrez-López GF et al (2003) An update on some key alternative food processing technologies: microwave, pulsed electric field, high hydrostatic pressure, irradiation and ultrasound. Food Sci Food Biotechnol 26:279–312

Rossing NL, Lind M, Jensen N et al (2010) A functional HAZOP methodology. Comput Chem Eng 34:244–253

Shankar H, Pagel PS, Warner DS (2011) Potential adverse ultrasound-related biological effects: a critical review. J Am Soc Anesthesiol 115:1109–1124

Sicaire A-G, Fine F, Vian M et al (2015) HACCP and HAZOP in ultrasound food processing. In: Handbook of Ultrasonics and Sonochemistry, pp 1–19

Siddiqui D, Nandan A, Sharma M et al (2014) Risk management techniques HAZOP and HAZID study. Int J Occup Health Saf, Fire Environ Allied Sci 1:5–8

Singh V, Seed T (2018) An update on sargramostim for treatment of acute radiation syndrome. Drugs Today (Barc) 54:679–693

Single JI, Schmidt J, Denecke J (2019) State of research on the automation of HAZOP studies. J Loss Prev Process Indust 62:103952

Smagowska B, Pawlaczyk-Łuszczyńska M (2013) Effects of ultrasonic noise on the human body—a bibliographic review. Int J Occup Saf Ergonomics 19:195–202

Sokołowska BSAS, Fonberg-Broczek M, Niezgoda J, Chotkiewicz M, Dekowska A, Rzoska SJ (2013) Factors influencing the inactivation of Alicyclobacillus acidoterrestris spores exposed to high hydrostatic pressure in apple juice. High Pressure Res 33:73–82

Soria AC, Villamiel M (2010) Effect of ultrasound on the technological properties and bioactivity of food: a review. Trends Food Sci Technol 21:323–331

Soriano JM, Rico H, Moltó J et al (2002) Effect of introduction of HACCP on the microbiological quality of some restaurant meals. Food Control 13:253–261

Syed QA, Ishaq A, Rahman U et al (2017) Pulsed electric field technology in food preservation: a review. J Nutr Health Food Eng 6:168–172

Taylor J (2017) Automated HAZOP revisited. Process Saf Environ Prot 111:635–651

Toepfl S, Siemer C, Saldaña-Navarro G et al (2014) Overview of pulsed electric fields processing for food. In: Emerging technologies for food processing. Elsevier, pp 93–114

Van Impe J, Smet C, Tiwari B et al (2018) State of the art of nonthermal and thermal processing for inactivation of micro-organisms. J Appl Microbiol 125:16–35

Vanne L, Karwoski M, Karppinen S et al (1996) HACCP-based food quality control and rapid detection methods for microorganisms. Food Control 7:263–276

Vega-Mercado H, Gongora-Nieto MM, Barbosa-Canovas GV et al (2004) Pulsed electric fields in food preservation. Food Sci Technol-New York-Marcel Dekker 167:783

Part VIII
Success Stories of Industrial Implementation of Nonthermal Technologies

Chapter 28
Innovative Success Stories on Commercial Non-thermal Technologies - Interviews of Major Food Industries Working in This Area

Yuthana Phimolsiripol, Kasemsak Uthaichana, Noppol Leksawasdi, and Choncharoen Sawangrat

1 Cold Plassma Technology for Startup Company

High quality and safe foodstuffs are one of the fundamental achievements by this technology. The quality of foodstuffs has gained considerable attention with expectation of uncompromising hygiene level. The Food and Agriculture Organization (FAO) estimated that nearly 25% of food is contaminated by either undesirable chemicals, such as pesticides or mycotoxins, or undesirable microorganisms, like fungi and bacteria (Batos et al. 2017). These unwholesome substances and microbes are transferred to food due to pre-harvest or post-harvest contamination. A number of contaminated microorganisms and chemicals in food and feed can be mitigated by keeping appropriate measures during the primary agricultural production as well as further post-harvest and processing (Phan et al. 2017a). Non-thermal cold plasma technology has been applied in several foodstuffs such as mango in terms of pesticides mitigation (Phan et al. 2017b, 2018) and microorganism contamination removal (Phan et al. 2019). In addition, the modification of rice using cold plasma technology was also investigated by Noppakun et al. (2021). The success story of startup company for cold plasma technology for pesticide residue decontamination was initiated by one PhD dissertation from Chiang Mai University, Chiang Mai, Thailand. The flagship product of the company is a plasma activated water generator in the household scale as called "REMO" as presented in Fig. 28.1.

The developed technology was the startegic partnership of the Science and Technology Park (STeP) and Thai-Korean Research Collaboration Center (TKRCC),

Y. Phimolsiripol (✉) · N. Leksawasdi
Faculty of Agro-Industry, Chiang Mai University, Chiang Mai, Thailand
e-mail: yuthana.p@cmu.ac.th

K. Uthaichana · C. Sawangrat
Faculty of Engineering, Chiang Mai University, Chiang Mai, Thailand

A. Režek Jambrak (ed.), *Nonthermal Processing in Agri-Food-Bio Sciences*, Food Engineering Series, https://doi.org/10.1007/978-3-030-92415-7_28

Fig. 28.1 House-hold machine of plasma activated water generator (REMO) produced by Thai Startup company

Chiang Mai University. The PhD candidate started the project by obtaining the cold plasma treatment from TKRCC and applied the treatment for pesticide residue decontamination to various fruits and vegetables. The key point of pesticide decontamination is reactive oxygen species (ROS) production by cold plasma while combining the massive amount of microbubble for improving the machine efficiency. The budget for commercial version of the prototype was about US$ 50,000 funded by TED Fund Ministry of Higher Education, Science, Research and Innovation, Thailand.

Before startup launching, the founders participated in a business model competition called "Research to Market" (R2M) supported by STeP in 2014. The development of commercial purpose machinery was soon followed and resulted in the first prize of startup Thailand league in 2017. The additional prize was also received from the Government Savings Bank in the same year. In this stage, the developed machine is still required to be registered in accordance to Thai Industrial Standards with expert company from Italy. Although the machine is not relevant to the FDA approvement, the process of future application in food industry is still required for novel processing.

2 Radio Frequency (RF) in Rice Industry

Radio frequency (RF) is a rate of oscillation in the range of 3 kHz to 300 GHz, which corresponds to the frequency of radio waves. RF heating is the process in which a high-frequency alternating electric field and electromagnetic radiation rotates molecular dipole of dielectric material generating thermal energy. RF dielectric heating at intermediate frequencies, due to its greater penetration over microwave heating, shows greater promise than microwave systems due to the relatively short processing time. Uniform heating is also observed with extensive elimination of parasites and pests in certain harvested crops (Vearasilp et al. 2015).

This story started from the research group from Faculty of Agriculture, Chiang Mai University, Chiang Mai, Thailand for more than 15 years ago. The fruitful collaboration between Professor Dr. Suchada Vearasilp and Professor in Gottingen University, Germany by Thai-German Research Platform was supported by Asia Development Bank Fund which resulted in the laboratory scale RF machine. The insect decontamination experiments were then carried out in Thailand as a representative of suitable tropical countries. The seed treatment, insect elimination strategy, decontamination of mold, and drying process were investigated thoroughly. Subsequently, Vearasilp et al. (2015) developed a prototype of RF heating system at 27 MHz which was further developed to control rice weevil contamination in milled rice. The insects at all four life stages, especially the eggs and adult insects were completely eradicated at 50–55 °C with exposure time for at least 2 min. In term of rice physical quality, the color of treated rice at 50 °C for 0 and 3 min indicated no significant difference ($p > 0.05$). This prototype is thus highly effective in completely eliminate insects in rice.

As the conventional chemical strategy of applying phosphene to rice was no longer desirable due to awareness of consumers towards possible chemical residue in rice. The development of RF technology as alternative clean and green method with chemical-free process in rice treatment industry has become very promising. Once the small-scale technology for 5 kg capacity production (batch type) was proven, the continued success has gone further by the support of STeP and the Science Park Promotion Agency (SPA) Thailand to construct the RF pilot plant with the capacity of 1,000 kg/h per module or 8,000 kg/day capacity production as shown in Fig. 28.2. This could perfectly match the demands by several co-operatives in many provinces. The module can also be stacked to meet higher demands by large-scale rice milling production.

After the success in feasibility study of 1000 kg/h production per module (1 RF source), the RF technology was adopted by a rice machinery company namely “Yon Phol Dee” whose owner is also a PhD candidate in Faculty of Agriculture, Chiang Mai University. The main business of Yon Phol Dee is production of rice and seed milling machines. Yon Phol Dee has developed the BIO-Q for commercialization (Fig. 28.3) with approval from Ministry of Industry for Thailand Industrial Standards. The Innovative Product from National Innovation Agency (NIA) also recognized BIO-Q and induct the unit as innovative product. This is another success story case of non-thermal processing in industrial commercialization.

Fig. 28.2 Pilot plant RF technology for controlling rice weevil in milled rice

Fig. 28.3 Commercial BIO-Q RF technology machine controlling rice weevil in milled rice

3 High-Pressure Processing (HPP) in Tuna Industry

Another interview of success story is tuna industry in which a representative from the Thai Union (TU) Group PCL has participated. TU Group PCL is a world's seafood leader for more than 40 years with the aims of pioneering the sustainable high quality, healthy, tasty, and innovative seafood products to customers across the world. Today, TU is regarded as the world's largest producer of shelf-stable tuna products with annual sales exceeding US$ 4.03 billion and a global workforce of over 49,000 personnels. The company's global brand portfolio also includes market-leading international brands such as Chicken of the Sea, John West, Petit Navire, Parmentier, Mareblu, King Oscar, and Rügen Fisch and Thai-leading brands SEALECT, Fisho, Qfresh, Monori, Bellotta and Marvo.

The beginning of TU success story arises from the global vision for innovation and establishment of the Global Innovation Unit. TU allocated large budget to construct the pilot plant with interests of adopting novel non-thermal food processing technique such as HPP. This technology applies high hydrostatic pressure through water with multiple advantages for foods and beverages processing. The pilot plant is equipped by leading-edge technology to innovate sustainable and value-added products. Global Innovation Incubator (Gii) was in charge of this seafood pilot plant and Gii extension during 2014–2015. The results of US$ 6.3 million investment boost in this project by TU has helped facilitating vision of the Royal Thai Goverment to become a hub for food research and innovation. The TU's novel food product - **"Premium Yellowfin Tuna Slices"**, a ready-to-eat, pre-sliced tuna product made from whole tuna loins, was launched in 2018 as presented in Fig. 28.4. This product has been awarded as the Best New Foodservice Product Category of 2018 Seafood Excellence Award in at Boston's Seafood Expo, USA.

TU's Yellowfin tuna slices are the world's first pre-sliced, pre-seasoned tuna made from whole yellowfin tuna loins. This product is developed to provide healthier and convenient alternative to traditional luncheon meats which is suitable for sandwiches, salads, charcuterie and cold-cut platters. HPP technology is used to modify the texture of normal tuna to novel texturized tuna slices. As mentioned by Scott Solar, TU director of food service, nutritionists have encouraged the public to add more seafood into their diet, but food service has offered few new alternatives to the conventional beef, pork, chicken and turkey cuisine. The challenge is to modify the classic deli experience by making sliced tuna a ready-to-eat option with added health benefits to sandwiches and salads. TU developed Yellowfin Tuna Slices at Gii where the company utilized innovative and patent-pending technology to deliver a fresh and novel way of processing and consuming tuna.

The story started by the idea of develop new tuna product by Gii which seeking the novel non-thermal processing such as HPP as alternative to conventional thermal processing to obtain the new product. The main reason for developing the new product is also influenced by high amount of water consumption system and high temperature. Since 2015, Gii started sending tuna sample to Commonwealth Scientific and Industrial Research Organisation (CSIRO), Australia for evaluation on possibility of HPP implementation. Laboratory scale processing indicated that the processing should combine high pressure and sterilization. The first trial of

Fig. 28.4 TU's Yellowfin tuna slices. (Source: Thai Union (2021))

product development resulted in overcooked product with rubberlike texture that could be sliced in similar manner as ham. Gii team concluded that the texture was very different from common tuna and the HPP should be optimized for better texture. The optimization of temperature and pressure conditions were soon followed. The research and development team also sought collaboration with other research agents aroud the world.

The second step was to establish linkage with HPP machine company such as Hiperbaric, Spain, and other research centers such as CSIRO, Australia and Campden BRI, UK. During that time, industrial scale HPP was relatively new and there was no industrial player of this technology in Thailand. The optimization of HPP process for tuna slice was carried out in Spain for more than a year before achieving the desirable product. The next obstacle is find the way to sell this novel product with FDA approval. TU has a strategic partner with the objective of exporting and marketing this product in USA which requires FDA agreement. The US-FDA recommended that the product must pass food safety evaluation, especially *Clostridium botulinum* and *Listeria monocytogenase,* by the challenge study. TU contacted Campden BRI to perform this safety test in parallel with TU laboratory. The test for *C. Botulinum* was done first and subsequently followed by *L. monocytogenase* which was examined in the USA. After food safety has been proven, the permission for full scale-up production was granted. The construction was carried out for 6 months before initial full-scale production. The shelf life of this product was about 14–21 days in refrigerator. The approval from US-FDA was beneficial for TU leading to the successful application from Thai FDA. This is a success story of creating diversification in tuna product by non-thermal processing such as HPP by

TU with the total budget of US$ 4 million. The lesson learned is useful for TU for future development of innovative products.

In summary, cold plasma technology in fruits and vegetables industry, radio frequency in rice industry and high-pressure processing in tuna industry are innovative success stories on commercial non-thermal technologies. The continual dedicated collaboration and strong partnership between academic and industrial sectors as well as visions of the company all play important factors towards these success case studies.

Acknowledgements Authors would like to sincerely thank all participants of the interview from REMO startup company, Yon Phol Dee Company and Thai Union Group PCL. All gratitudes also go to STeP team support for all relevant information.

Conflict of Interest Authors declare that they have no conflict of interest.

References

Batos P, Kriz P, Havelka Z, Bohata A, Olsan P, Špatenka P, Čurn V, Dienstbier M (2017) Plasma Technology in Food Industry: mini-review. Kvasny Prum 63(3):134–138

Noppakun M, Naruenartwongsakul S, Seesuriyachan S, Phimolsiripol Y, Boonyawan D, Intipunya P (2021) Effects of plasma treatment on cooking and physical qualities of pigmented Thai rice. Chiang Mai Univ J Nat Sci 20(1):e2021004

Phan KKT, Phan HT, Brennan CS, Phimolsiripol Y (2017a) Non-thermal plasma for pesticides and microbial elimination on fruits and vegetables: an overview. Int J Food Sci Technol 52:2127–2137

Phan KKT, Phan HT, Uthaichana K, Phimolsiripol Y (2017b) Effect of non-thermal plasma on physicochemical properties of Nam Dok Mai mango. Int J Adv Sci Eng Inf Technol 7:263–268

Phan KKT, Phan HT, Boonyawan D, Inthipunya P, Brennan CS, Regenstein JM, Phimolsiripol Y (2018) Non-thermal plasma for elimination of pesticides residues in mango. Innov Food Sci Emerg Technol 48:164–171

Phan KKT, Phan HT, Brennan CS, Regenstein JM, Jantansakulwong K, Boonyawan D, Phimolsiripol Y (2019) Gliding arc discharge non-thermal plasma for retardation of mango anthracnose. LWT- Food Sci Technol 105:142–148

Thai Union (2021) Thai Union to bring award-winning Yellowfin tuna slices to seafood expo global in Brussels. [Internet] Retrived 2 April 2021, from: https://www.thaiunion.com/en/newsroom/press-release/799/thai-union-to-bring-award-winning-yellowfin-tuna-slices-to-seafood-expo-global-in-brussels

Vearasilp S, Thanapornpoonpong S, Krittigamas N, Suriyong S, Akaranuchat P, von Hörsten D (2015) Vertical operating prototype development supported radio frequency heating system in controlling rice weevil in milled rice. Agric Agric Sci Proc 5:184–192

Index

A. Režek Jambrak (ed.), *Nonthermal Processing in Agri-Food-Bio Sciences*, Food Engineering Series, https://doi.org/10.1007/978-3-030-92415-7

Q

R

MIX
Papier aus verantwortungsvollen Quellen
Paper from responsible sources
FSC® C105338

If you have any concerns about our products,
you can contact us on
ProductSafety@springernature.com

In case Publisher is established outside the EU,
the EU authorized representative is:
Springer Nature Customer Service Center GmbH
Europaplatz 3, 69115 Heidelberg, Germany

Printed by Libri Plureos GmbH
in Hamburg, Germany